T0300942

NEW WINDOWS ON MASSIVE STARS, ASTEROSEISMOLOGY, INTERFEROMETRY AND SPECTROPOLARIMETRY

IAU SYMPOSIUM No. 307

COVER ILLUSTRATION:

Zoom on the symposium poster

IAU SYMPOSIUM PROCEEDINGS SERIES

Chief Editor

THIERRY MONTMERLE, IAU General Secretary
Institut d'Astrophysique de Paris,
98bis, Bd Arago, 75014 Paris, France
montmerle@iap.fr

Editor

PIERO BENVENUTI, IAU Assistant General Secretary
University of Padua, Dept of Physics and Astronomy,
Vicolo dell'Osservatorio, 3, 35122 Padova, Italy
piero.benvenuti@unipd.it

INTERNATIONAL ASTRONOMICAL UNION

UNION ASTRONOMIQUE INTERNATIONALE

NEW WINDOWS ON MASSIVE STARS, ASTEROSEISMOLOGY, INTERFEROMETRY AND SPECTROPOLARIMETRY

PROCEEDINGS OF THE 307th SYMPOSIUM OF THE INTERNATIONAL ASTRONOMICAL UNION HELD IN GENEVA, SWITZERLAND JUNE 23–27, 2014

Edited by

GEORGES MEYNET
Geneva Observatory, University of Geneva, CH-1290 Versoix, Switzerland

CYRIL GEORGY
Astrophysics group, EPSAM, Keele University, Lennard-Jones Labs, Keele, ST5 5BG, UK

JOSÉ GROH
Geneva Observatory, University of Geneva, CH-1290 Versoix, Switzerland

and

PHILIPPE STEE
Observatoire de la Côte d'Azur - CNRS - UNSA and Boulevard de l'Observatoire, CS 34229, F06304 Nice Cedex 4, France

CAMBRIDGE
UNIVERSITY PRESS

University Printing House, Cambridge CB2 8BS, United Kingdom

One Liberty Plaza, 20th Floor, New York, NY 10006, USA

477 Williamstown Road, Port Melbourne, VIC 3207, Australia

314-321, 3rd Floor, Plot 3, Splendor Forum, Jasola District Centre, New Delhi - 110025, India

103 Penang Road, #05-06/07, Visioncrest Commercial, Singapore 238467

Cambridge University Press is part of the University of Cambridge.

It furthers the University's mission by disseminating knowledge in the pursuit of education, learning and research at the highest international levels of excellence.

www.cambridge.org
Information on this title: www.cambridge.org/9781107078581

First published 2014

A catalogue record for this publication is available from the British Library

ISBN 978-1-107-07858-1 Hardback

Table of Contents

Preface xiii

The Organizing Committee xv

Conference photograph xvi

Participants xvii

Address by the Local Organizing Committee xix

The advanced phases of massive stars and the explosive yields 1
    A. Chieffi & M. Limongi

Physics of rotation: problems and challenges 9
    A. Maeder & G. Meynet

The Physics of Convection in Massive Stars 20
    C. A. Meakin

Physics of Mass Loss in Massive Stars 25
    J. Puls, J. O. Sundqvist & N. Markova

A binary progenitor for the Type Ib Supernova iPTF13bvn 37
    M. C. Bersten

Winds of metal-poor OB stars: Updates from HST-COS UV spectroscopy 41
    M. García, A. Herrero, F. Najarro, D. J. Lennon & M. A. Urbaneja

Combining observational techniques to constrain convection in evolved massive star
    models 47
    C. Georgy, H. Saio & G. Meynet

Massive stars near the Eddington-limit, pulsations & mass-loss 52
    G. Gräfener

Discovery of a Thorne-Żytkow object candidate in the Small Magellanic Cloud . 57
    E. M. Levesque, P. Massey, A. N. Żytkow & N. Morrell

A New Class of Wolf-Rayet Stars: WN3/O3s 64
    P. Massey, K. F. Neugent, N. Morrell & D. J. Hillier

New prescriptions of turbulent transport from local numerical simulations 70
    V. Prat, F. Lignières & G. Lesur

Rotational velocities of single and binary O-type stars in the Tarantula Nebula . 76
    O. H. Ramírez-Agudelo, H. Sana, A. de Koter, S. Simón-Díaz,
    S. E. de Mink, F. Tramper, P. L. Dufton, C. J. Evans, G. Gräfener,
    A. Herrero, N. Langer, D. J. Lennon, J. Maíz Apellániz, N. Markova,
    F. Najarro, J. Puls, W. D. Taylor & J. S. Vink,

Stellar Yields of Rotating First Stars: Yields of Weak Supernovae and Abundances of Carbon-enhanced Hyper Metal Poor Stars 82
K. Takahashi, H. Umeda & T. Yoshida

The Gaia-ESO Survey and Massive Stars 88
R. Blomme, Y. Frémat, E. Gosset, A. Herrero, A. Lobel, J. Maíz Apellániz, T. Morel, I. Negueruela, T. Semaan, S. Simón-Díaz & D. Volpi

Non-LTE Abundances in OB stars: Preliminary Results for 5 Stars in the Outer Galactic Disk 90
G. A. Bragança, T. Lanz, S. Daflon, K. Cunha, C. D. Garmany, J. W. Glaspey, M. Borges Fernandes, M. S. Oey, T. Bensby & I. Hubeny

Luminous Infrared Sources in the Local Group: Identifying the Missing Links in Massive Star Evolution 92
N. Britavskiy, A. Z. Bonanos & A. Mehner

Chemical abundances of fast-rotating OB stars............................ 94
C. Cazorla, T. Morel, Y. Nazé & G. Rauw

Massive star archeology in globular clusters 96
W. Chantereau, C. Charbonnel & G. Meynet

Linking 1D Stellar Evolution to 3D Hydrodynamic Simulations 98
A. Cristini, R. Hirschi, C. Georgy, C. Meakin, D. Arnett & M. Viallet

First Results of the Analysis of the Wolf-Rayet Star WR6 100
A. C. Gormaz-Matamala, A. Hervé, A. Chené, M. Curé & R. Mennickent.

Evolution of the rotational properties and nitrogen surface abundances of B-Type stellar populations 102
A. Granada, G. Meynet, S. Ekström, C. Georgy & L. Haemmerlé

Delta-slow solution to explain B supergiant stars' winds 104
M. Haucke, I. Araya, C. Arcos, M. Curé, L. Cidale, S. Kanaan, R. Venero & M. Kraus

Massive OB stars at varying Z 106
A. Herrero, M. Garcia, S. Simón-Díaz, I. Camacho, C. Sabín-Sanjulián & N. Castro

Massive stars: flare activity due to infalls of comet-like bodies 108
S. Ibadov & F. S. Ibodov

Study of environment and photosphere of 51 Oph........................ 111
N. Jamialahmadi, Ph. Berio, B. Lopez, A. Meilland & Ph. Stee

Line profile variability in spectra of hot massive stars 113
A. Kholtygin, N. Sudnik & V. Dushin

Discrete absorption components in the massive LBV Binary MWC 314......... 115
A. Lobel, C. Martayan, M. Corcoran, J. H. Groh & Y. Frémat

The mass discrepancy problem in O stars of solar metallicity. Does it still exist? 117
N. Markova & J. Puls

Investigation of the brightest stars in the Cyg OB2 association 119
    O. Maryeva & S. Parfenov

OHANA: Eta Carinae's Variability in the Near-IR........................ 121
    A. Mehner, W.-J. de Wit, T. Rivinius & the Paranal VLTI group

Markov Chain Monte-Carlo Models of Starburst Clusters 123
    J. Melnick

A spectroscopic and photometric study of the interacting binary and double period
    variable HD 170582.. 125
    R. E. Mennickent, G. Djurašević, M. Cabezas, A. Cséki, J. Rosales,
    E. Niemczura, I. Araya & M. Curé

The Close Binary Frequency of Wolf-Rayet Stars as a Function of Metallicity in
    M31 and M33 127
    K. F. Neugent & P. Massey

Fundamental parameters of B type stars................................ 129
    M.-F. Nieva

A Search for Hot Subdwarf Companions to Rapidly-Rotating Early B Stars.... 131
    G. J. Peters, D. R. Gies, L. Wang & E. D. Grundstrom

An empirical pipeline for determining the viscosity parameter for Be star disks . 133
    L. R. Rímulo, A. C. Carciofi, T. Rivinius & X. Haubois

Westerlund 1 is a Galactic Treasure Chest: The Wolf-Rayet Stars 135
    C. K. Rosslowe & P. A. Crowther

Herschel/PACS: Constraining clumping in the intermediate wind region of OB stars 137
    M. M. Rubio-Díez, F. Najarro, J. O. Sundqvist, A. Traficante, J. Puls,
    L. Calzoletti, A. Herrero, D. Figer & J. Martin-Pintado

NGC 3293 revisited by the Gaia-ESO Survey 140
    T. Semaan, T. Morel, E. Gosset, J. Zorec, Y. Frémat, R. Blomme &
    A. Lobel

Revisiting the Hunter diagram with the Geneva Stellar Evolution Code 142
    R. Simoniello, G. Meynet, S. Ekström, C. Georgy & A. Granada

The properties of single WO stars 144
    F. Tramper, S. M. Straal, G. Gräfener, L. Kaper, A. de Koter, N. Langer,
    H. Sana & J. S. Vink

Spectral analysis of LBV stars in M31: AF And and Var 15................. 146
    A. F. Valeev, O. Sholukhova & S. Fabrika

Variable C – "a typical" LBV in M 33? 148
    K. Weis, R. M. Humphreys, B. Burggraf & D. J. Bomans

Variational approach for rotating-stellar evolution in Lagrange scheme 150
    N. Yasutake & S. Yamada

Wolf-Rayet stars from Very Massive Stars.............................. 152
    N. Yusof

Massive Star Asteroseismology in Action............................   154
   C. Aerts

Asteroseismology of red giants to constrain angular momentum transport......   165
   P. Eggenberger

Photometric Variability of OB-type stars as a New Window on Massive Stars ..   171
   M. Kourniotis, A. Z. Bonanos, I. Soszyński, R. Poleski, G. Krikelis & the
   OGLE team

Behaviour of Pulsations in Hydrodynamic Models of Massive Stars...........   176
   C. C. Lovekin & J. A. Guzik

Asteroseismic Diagnostics for Semi-Convection in B Stars in the Era of K2   182
   E. Moravveji

Are the stars of a new class of variability detected in NGC 3766 fast rotating SPB
   stars?...   188
   S. J. A. J. Salmon, J. Montalbán, D. R. Reese, M.-A. Dupret &
   P. Eggenberger

Asteroseismology of OB stars with hundreds of single snapshot spectra (and a few
   time-series of selected targets)....................................   194
   S. Simón-Díaz

Probing high-mass stellar evolutionary models with binary stars   200
   A. Tkachenko

Rotation and the Cepheid Mass Discrepancy   206
   R. I. Anderson, S. Ekström, C. Georgy, G. Meynet, N. Mowlavi & L. Eyer

Tidal interactions in rotating multiple stars and their impact on their evolution   208
   P. Auclair-Desrotour, S. Mathis & C. Le Poncin-Lafitte

Constraints on stellar evolution from white dwarf asteroseismology   211
   A. Bischoff-Kim

Radiative Levitation in Massive Stars: A self-consistent approach   213
   D. D'souza & A. Weiss

Leaky-wave-induced disks around Be stars: a pulsational analysis on their formation   215
   M. Godart, H. Shibahashi & M.-A. Dupret

Time Resolved Photometric and Spectroscopic Analysis of Chemically Peculiar
   Stars   218
   S. Joshi, G. C. Joshi, Y. C. Joshi & R. Aggrawal

Stochastic excitation of gravity waves in rapidly rotating massive stars........   220
   S. Mathis & C. Neiner

An attempt of seismic modelling of β Cephei stars in NGC 6910   222
   D. Moździerski, Z. Kołaczkowski & E. Zahajkiewicz

Pulsation Period Change & Classical Cepheids: Probing the Details of Stellar Evolution...   224
   H. R. Neilson, A. C. Bisol, E. Guinan & S. Engle

Pulsations of massive stars beyond TAMS: effects of mass loss, diffusion, overshoot-
    ing   226
    J. Ostrowski & J. Daszyńska-Daszkiewicz

Deep Photospheric Emission Lines as Probes for Pulsational Waves.   228
    Th. Rivinius, M. Shultz & G. A. Wade

Stability boundaries for massive stars in the sHR diagram   230
    H. Saio, C. Georgy & G. Meynet

Asteroseismology of the SPB star HD 21071   232
    W. Szewczuk & J. Daszyńska-Daszkiewicz

Spectral Effects of Pulsations in Blue Supergiants   235
    S. Tomić, M. Kraus & M. E. Oksala

Is λ Cep a pulsating star?   237
    J. M. Uuh-Sonda, P. Eenens & G. Rauw

Seismic analysis of the massive β Cephei star 15 Canis Majoris   239
    P. Walczak & G. Handler

An interferometric journey around massive stars   241
    A. Meilland & P. Stee

Basics of Optical Interferometry: A Gentle Introduction   252
    G. T. van Belle

The photosphere and circumstellar environment of the Be star Achernar   261
    D. M. Faes, A. Domiciano de Souza, A. C. Carciofi & P. Bendjoya

Zooming into Eta Carinae with interferometry   267
    J. H. Groh

Evidences for a large hot spot on the disk of Betelgeuse (α Ori).   273
    M. Montargès, P. Kervella, G. Perrin, A. Chiavassa & J. B. Le Bouquin

On the atmospheric structure and fundamental parameters of red supergiants . .   280
    *M. Wittkowski, B. Arroyo-Torres, J. M. Marcaide, F. J. Abellan,
    A. Chiavassa, B. Freytag, M. Scholz, P. R. Wood & P. H. Hauschildt*

Amplitude Modulation of Cepheid Radial Velocity Curves as a Systematic Source
    of Uncertainty for Baade-Wesselink Distances.   286
    R. I. Anderson

The impact of the rotation on the surface brightness of early-type stars   288
    *M. Challouf, N. Nardetto, A. Domiciano de Souza, D. Mourard, H. Aroui,
    P. Stee & A. Meilland*

The circumstellar environment of the B[e] star GG Car: an interferometric modeling   291
    *A. Domiciano de Souza, M. Borges Fernandes, A. C. Carciofi & O.
    Chesneau*

Angular Diameters of O- and B-type Stars   293
    K. Gordon, D. Gies & G. Schaefer

Binarity of the LBV HR Car   295

Th. Rivinius, H. M. J. Boffin, W. J. de Wit, A. Mehner, Ch. Martayan,
S. Guieu & J.-B. Le Bouquin

AMBER/VLTI Snapshot Survey on Circumstellar Environments   297
Th. Rivinius, W. J. de Wit, Z. Demers, A. Quirrenbach & the VLTI Science
Operations Team

Recent highlights of spectropolarimetry applied to the magnetometry of massive
stars   301
J. H. Grunhut

Basics of spectropolarimetry   311
J. D. Landstreet

Magnetic Field - Stellar Winds Interaction   321
A. ud-Doula

The BinaMIcS project: understanding the origin of magnetic fields in massive stars
through close binary systems   330
E. Alecian, C. Neiner, G. A. Wade, S. Mathis, D. Bohlender, D. Cébron,
C. Folsom, J. Grunhut, J.-B. Le Bouquin, V. Petit, H. Sana, A. Tkachenko,
A. ud-Doula & the BinaMIcS collaboration

Revealing the Mass Loss Structures of Four Key Massive Binaries Using Optical
Spectropolarimetry   336
J. R. Lomax

The B Fields in OB Stars (BOB) Survey   342
T. Morel, N. Castro, L. Fossati, S. Hubrig, N. Langer, N. Przybilla,
M. Schöller, T. Carroll, I. Ilyin, A. Irrgang, L. Oskinova,
F. R. N. Schneider, S. Simon Díaz, M. Briquet, J. F. González,
N. Kharchenko, M.-F. Nieva, R.-D. Scholz, A. de Koter, W.-R. Hamann,
A. Herrero, J. Maíz Apellániz, H. Sana, R. Arlt, R. Barbá, P. Dufton,
A. Kholtygin, G. Mathys, A. Piskunov, A. Reisenegger, H. Spruit, &
S.-C. Yoon

Unraveling the variability of σ Ori E   348
M. E. Oksala, O. Kochukhov, J. Krtička, M. Prvák & Z. Mikulášek

Constraining general massive-star physics by exploring the unique properties of
magnetic O-stars: Rotation, macroturbulence & sub-surface convection . . .   353
J. O. Sundqvist

Linear line spectropolarimetry as a new window to measure 2D and 3D wind ge-
ometries   359
J. S. Vink

Discovery of Secular Evolution of the Atmospheric Abundances of Ap Stars   365
J. D. Bailey, J. D. Landstreet & S. Bagnulo

The magnetic field of ζ Ori A   367
A. Blazère, C. Neiner, J-C. Bouret, A. Tkachenko & the MiMeS
collaboration

Spectropolarimetric study of selected cool supergiants   369
V. Butkovskaya, S. Plachinda & D. Baklanova

Beam me up, Spotty: Toward a new understanding of the physics of massive star
  photospheres 371
  A. David-Uraz, G. Wade & S. Owocki

Impact of rotation on the geometrical configurations of fossil magnetic fields . . . 373
  C. Emeriau & S. Mathis

A Simple Mean-Field Diagnostic from Stokes V Spectra 375
  K. G. Gayley & S. P. Owocki

Linear Polarization and the Dynamics of Circumstellar Disks of Classical Be Stars 377
  R. J. Halonen & C. E. Jones

Multiple, short-lived "stellar prominences" on O stars: the supergiant λ Cephei. 379
  H. F. Henrichs & N. Sudnik

Project VeSElkA : Preliminary results for CP stars recently observed with ES-
  PaDOnS 381
  V. Khalack & F. LeBlanc

Abundance analysis of HD 22920 spectra 383
  V. Khalack & P. Poitras

Fundamental properties of single O stars in the MiMeS survey 385
  F. Martins, A. Hervé, J.-C. Bouret, W. L. F. Marcolino, G. A. Wade,
  C. Neiner, E. Alecian & the MiMeS collaboration

Spectropolarimetry and modeling of WR156 387
  O. Maryeva

The UVMag space project: UV and visible spectropolarimetry of massive stars 389
  C. Neiner & the UVMag consortium

Magnetic main sequence stars as progenitors of blue supergiants 391
  I. Petermann, N. Castro & N. Langer

Magnetic CP stars in Orion OB1 association 393
  I. I. Romanyuk & E. A. Semenko

Stellar magnetic fields from four Stokes parameter observations 395
  N. Rusomarov, O. Kochukhov & N. Piskunov

Plasma Leakage from the Centrifugal Magnetospheres of Magnetic B-Type Stars 397
  M. Shultz, G. Wade, T. Rivinius, J. Grunhut, V. Petit & the MiMeS
  Collaboration

ξ^1 CMa: An Extremely Slowly Rotating Magnetic B0.7 IV Star 399
  M. Shultz, G. Wade, T. Rivinius, W. Marcolino, H. Henrichs, J. Grunhut &
  the MiMeS Collaboration

Magnetic fields and internal mixing of main sequence B stars 401
  G. A. Wade, C. P. Folsom, J. Grunhut, J. D. Landstreet & V. Petit

Links between surface magnetic fields, abundances, and surface rotation in clusters
  and in the field 404
  N. Przybilla

Massive Star Astrophysics with the new Magellanic Cloud photometric survey
    MCSF 414
    D. J. Bomans, A. Becker & K. Weis

Asteroseismology and spectropolarimetry: opening new windows on the internal
    dynamics of massive stars 420
    S. Mathis & C. Neiner

The Massive Star Population at the Center of the Milky Way 426
    F. Najarro, D. de la Fuente, T. R. Geballe, D. F. Figer & D. J. Hillier

Accretion Signatures on Massive Young Stellar Objects..................... 431
    F. Navarete, A. Damineli, C. L. Barbosa & R. D. Blum

The X-ray properties of magnetic massive stars 437
    Y. Nazé, V. Petit, M. Rinbrand, D. Cohen, S. Owocki, A. ud-Doula &
    G. Wade

Combining seismology and spectropolarimetry of hot stars 443
    C. Neiner, M. Briquet, S. Mathis & P. Degroote

X-rays From Centrifugal Magnetospheres in Massive Stars 449
    C. Bard & R. Townsend

Abundance study of two magnetic B-type stars in the Orion Nebula Cluster ... 451
    T. Morel

Circumstellar Environments of MYSOs Revealed by IFU Spectroscopy 453
    F. Navarete, A. Damineli, C. L. Barbosa & R. D. Blum

An X-ray surprise in a magnetic pulsator 455
    Y. Nazé

New insights on Be shell stars from modelling their Hα emission profiles........ 457
    J. Silaj, C. E. Jones, T. A. A. Sigut & C. Tycner

3D and Some Other Things Missing from the Theory of Massive Star Evolution   459
    W. D. Arnett

Asteroseismology of Massive Stars : Some Words of Caution................. 470
    A. Noels, M. Godart, S. J. A. J. Salmon, M. Gabriel, J. Montalbán & A.
    Miglio

Interferometry of massive stars: the next step........................... 480
    Ph. Stee, A. Meilland & O. L. Creevey

Spectropolarimetry of massive stars: Requirements and potential from today to
    2030.. 490
    G. A. Wade

Observing programs, what are the priorities? 499
    G. Meynet & H. Henrichs

Stellar Models: What is the future direction? 504
    A. ud-Doula

Author index 505

Preface

A Universe without massive stars would be very different from the one we can observe. Indeed these stars are important drivers for the photometric and chemical evolution of galaxies, they are the sources of important elements for the building of living bodies, they feed with their strong winds and supernova explosion the interstellar medium with momentum and kinetic energy having thus an impact on the star formation rate. They are the progenitors of core collapse events and of the most energetic stellar explosions in the Cosmos, the Gamma Ray Bursts. They give birth to compact objects as neutron stars and black holes.

From what precedes, one can figure out that knowing the evolution of massive stars is not only important for stellar physics, but also for probing the evolution of galaxies and their star formation history along the whole cosmic history. This subject also connects in a particularly strong and direct way the observations of nearby objects with that of far distant galaxies.

In this context, the possibilities that became operative only recently to probe the interior of stars, to constrain the size of their convective cores and the way they rotate in their interiors through asteroseismology, to determine the strength and topology of their surface magnetic fields through spectropolarimetry and to measure their shape and the distribution of their circumstellar environments through interferometry opened new paths for investigating their properties. Associated with other more classical methods, as photometry and/or spectroscopy, these technics will change our understanding of massive star evolution and may show that beside the initial mass and the metallicity, the evolution of massive stars, either single or in close binary systems, may also depend on their axial rotation and on their surface magnetic field.

Although for some of these technics, the application to massive stars is still in its infancy, we thought that it was time to convey astronomers from these different areas in order 1) to investigate how these technics can guide us towards new and innovative solutions to the most topical questions regarding the evolution of massive stars, 2) to allow the participants of different disciplines, and hopefully the readers of these proceedings, to grasp the essential of these observing methods and 3) to stimulate new ideas for using synergies between different observational technics.

At the end of 2012, a letter of intent, followed by a detailed description of the project was sent to IAU. We were very pleased to receive in May 2013 the announcement that our project of Symposium was one of the 9 selected symposia among the 17 proposals. Began then the work of organizing the sessions and selecting the invited reviewers with the SOC.

The present conference conveyed 138 astronomers from 28 countries. The scientific programs consisted in 6 sessions (Challenges in massive star evolution, Asteroseismology, Interferometry, Spectropolarimetry, Synergies between different techniques, and Towards a synthetic view), with 17 review talks, 33 contributed talks, 2 general discussions, and 2 poster sessions. Each session dedicated to one observational technics began with two review talks providing first a simple but rigorous presentation of the principles on which the observations are based and second a discussion of the main results obtained so far. We hope that the readers of these proceedings will take benefit from all these presentations as much as the participants of the symposium. We had also the pleasure to organize an outreach conference given by Coralie Neiner entitled *Le magnétisme stellaire : son rôle sur la vie des étoiles et la nôtre*, which attracted about 100 people.

The present conference belongs to the family of the following recent meetings: the IAU Symposium 272 entitled *Active OB stars: structure, evolution, mass loss, and critical limits*, held in Paris, in July 2010, the IAU Symposium 302 dedicated to *Magnetic fields throughout stellar evolution*, held in August 2013, at Biarritz, France, the conference *Magnetic Fields in the Universe IV: From Laboratory and Stars to the Primordial Structures*, held in February 2013, at Playa del Carmen in Mexico and the conference *Massive Stars: From Alpha to Omega*, held in June 2013, at Rhodes in Greece.

It is a great pleasure to acknowledge the financial support of our sponsors listed on page *xv* of these Proceedings and the active and efficient support of the members of the LOC, in particular Chantal Taçoy and Sylvia Ekström (Department of Astronomy of the Geneva University).

We dedicate these proceeding to our dear friends and colleagues Olivier Chesneau (left picture) and Stan Stefl (right picture), who passed away this year.

Georges Meynet and Phillipe Stee, co-chairs SOC,
Geneva, August 31, 2014

THE ORGANIZING COMMITTEE

Scientific

D. Arnett (USA)
R. Hirschi (UK)
M. Limongi (Italy)
P. Massey (USA)
C. Neiner (France)
S. Owocki (USA)
H. Saio (Japan)
R. Townsend (USA)

L. Cidale (Argentina)
E. Levesque (USA)
A. Maeder (Switzerland)
G. Meynet (co-chair, Switzerland)
A. Noels (Belgium)
T. Rivinius (Chile)
P. Stee (co-chair, France)
G. Wade (Canada)

Local

P. Eggenberger
C. Georgy
J. Groh
G. Meynet
G. Simond

S. Ekström (co-chair)
A. Granada
L. Haemmerlé
G. Priviterra
C. Tacoy (co-chair)

Acknowledgements

The symposium is sponsored and supported by the IAU Division G (Stars and Stellar Physics), by the IAU Commissions No. 25 (Astronomical Photometry and Polarimetry), No. 35 (Stellar Constitution), No. 36 (Theory of Stellar Atmospheres), the IAU Working Group on Active B Stars and the IAU Working Group on Massive Stars.

The Local Organizing Committee operated under the auspices of the Geneva Observatory, University of Geneva.

Funded by the
International Astronomical Union,
Swiss National Science Foundation,
Commission administrative, Geneva University,
Geneva Observatory, University of Geneva,
Faculty of Science, Geneva University,
Rectorate of Geneva University,
Swiss Society of Astronomy and Astrophysics,
Société de Physique et d'Histoire Naturelle, Geneva.

Participants

Conny **Aerts**, Institute of Astronomy, University of Leuven, Belgium — conny@ster.kuleuven.be
Evelyne **Alecian**, IPAG - Observatoire de Grenoble, France — evelyne.alecian@obs.ujf-grenoble.fr
Richard I. **Anderson**, Department of Astronomy, Geneva University, Switzerland — richard.anderson@unige.ch
Rainer **Arlt**, Leibniz Institute for Astrophysics Potsdam, Germany — rarlt@aip.de
W. David **Arnett**, Steward Observatory, University of Arizona, USA — wdarnett@gmail.com
Dietrich **Baade**, European Southern Observatory, Garching, Germany — dbaade@eso.org
Jeffrey **Bailey**, Max Planck Institut für extraterrestrische Physik, Germany — jeffbailey@mpe.mpg.de
Fabio **Barblan**, Department of Astronomy, Geneva University, Switzerland — fabio.barblan@unige.ch
Christopher **Bard**, University of Wisconsin-Madison, USA — bard@astro.wisc.edu
Melina **Bersten**, Kavli IPMU, Japan — melina.bersten@ipmu.jp
Aurore **Blazere**, LESIA, France — aurore.blazere@obspm.fr
Ronny **Blomme**, Royal Observatory of Belgium, Brussels, Belgium — Ronny.Blomme@oma.be
Dominik **Bomans**, Astronomical Institute, Ruhr-University Bochum, Germany — bomans@astro.rub.de
Gustavo **Bragança**, Observatório Nacional/Observatoire de la Côte d'Azur, Nice, France — ga.braganca@gmail.com
Nikolay **Britavskiy**, IAASARS, National Observatory of Athens, Greece — britvavskiy@gmail.com
Varvara **Butkovskaya**, Crimean Astrophysical Observatory, Taras Shevchenko National University, Kyiv, Ukraine — itbiz@mail.ru
Matteo **Cantiello**, KITP, USA — matteo@kitp.ucsb.edu
Alex **Carciofi**, Instituto de Astronomia, Geofísica e Ciências Atmosféricas, Brazil — carciofi@usp.br
Constantin **Cazorla**, Institut d'astrophysique, géophysique et océanographie, University of Liège, Belgium — cazorla@astro.ulg.ac.be
Mounir **Challouf**, Observatoire de la Côte d'Azur, Nice, France — mounir.challouf@oca.eu
William **Chantereau**, Department of astronomy, Geneva University, Switzerland — william.chantereau@unige.ch
Corinne **Charbonnel**, Department of astronomy, Geneva University, Switzerland, and CNRS, France — Corinne.Charbonnel@unige.ch
Alessandro **Chieffi**, INAF - Istituto di Astrofisica e Planetologia Spaziale, Italy — alessandro.chieffi@inaf.it
Andrea **Cristini**, Keele University, UK — a.j.cristini@keele.ac.uk
Alexandre **David-Uraz**, Queen's University/Royal Military College, Canada — adavid-uraz@astro.queensu.ca
Alex **de Koter**, Anton Pannekoek Institute for Astronomy, University of Amsterdam, Netherlands — A.deKoter@uva.nl
Jacqueline **den Hartogh**, Keele University, UK — j.den.hartogh@keele.ac.uk
Armando **Domiciano de Souza**, Observatoire de la Côte d'Azur, Nice, France — armando.domiciano@oca.eu
Durand **D'souza**, Max Planck Institute for Astrophysics, Germany — durand@mpa-garching.mpg.de
Philippe **Eenens**, University of Guanajuato, Mexico — eenens@gmail.com
Patrick **Eggenberger**, Department of astronomy, Geneva University, Switzerland — patrick.eggenberger@unige.ch
Sylvia **Ekström**, Department of astronomy, Geneva University, Switzerland — sylvia.ekstrom@unige.ch
Daniel **Faes**, IAG-USP, Brazil — moser@usp.br
Gaston **Folatelli**, Kavli IPMU, Japan — gaston.folatelli@ipmu.jp
Luca **Fossati**, Argelander-Institut für Astronomie, Germany — lfossati@uni-bonn.de
Miriam **García**, Center for Astrobiology, Spain — mgg@cab.inta-csic.es
Ken **Gayley**, University of Iowa, USA — kenneth-gayley@uiowa.edu
Marcus **Gellert**, Leibniz Institute for Astrophysics, Potsdam, Germany — mgellert@aip.de
Cyril **Georgy**, Keele University, UK — c.georgy@keele.ac.uk
Melanie **Godart**, Tokyo University, Japan — melanie.godart@gmail.com
Katie **Gordon**, Georgia State University, USA — kgordon@chara.gsu.edu
Alex **Gormaz-Matamala**, University of Concepción, Chile — agormaz@astro-udec.cl
Götz **Gräfener**, Armagh Observatory, UK — ggr@arm.ac.uk
Anahí **Granada**, Department of astronomy, Geneva University, Switzerland — anahi.granada@unige.ch
Nicolas **Grevesse**, Centre Spatial de Liège, Université de Liège, Belgium — nicolas.grevesse@ulg.ac.be
José **Groh**, Department of astronomy, Geneva University, Switzerland — jose.groh@unige.ch
Jason **Grunhut**, European Southern Observatory, Garching, Germany — jgrunhut@eso.org
Lionel **Haemmerlé**, Department of astronomy, Geneva University, Switzerland — lionel.haemmerle@unige.ch
Robbie **Halonen**, University of Western Ontario, Canada — rhalone@uwo.ca
Maximiliano **Haucke**, Facultad de Ciencias Astronómicas y Geofísicas, UNLP, Argentina — mhaucke@fcaglp.unlp.edu.ar
Huib **Henrichs**, University of Amsterdam, Netherlands — h.f.henrichs@uva.nl
Artemio **Herrero**, Instituto de Astrofísica de Canarias, Spain — ahd@iac.es
Raphael **Hirschi**, Keele University, UK — r.hirschi@keele.ac.uk
Subhon **Ibadov**, Institute of Astrophysics, Tajik Academy of Sciences, Tajikistan — ibadovsu@yandex.ru
Narges **Jamialahmadi**, Observatoire de la Côte d'Azur, Nice, France — jami@oca.eu
Santosh **Joshi**, ARIES, Nainital, India — santosh@aries.res.in
Viktor **Khalack**, Université de Moncton, Canada — khalakv@umoncton.ca
Agnes **Kim**, Penn State Worthington Scranton, USA — axk55@psu.edu
Oleg **Kochukhov**, Uppsala University, Sweden — oleg.kochukhov@physics.uu.se
Michalis **Kourniotis**, IAASARS, National Observatory of Athens, Greece — mkourniotis@astro.noa.gr
Karolina **Kubiak**, Institut für Astrophysik, Austria — karolina.kubiak@gmail.com
John **Landstreet**, Department of Physics & Astronomy, University of Western Ontario, Canada — jlandstr@uwo.ca
Norbert **Langer**, Argelander-Institut für Astronomie, Universität Bonn, Germany — nlanger@astro.uni-bonn.de
Emily **Levesque**, University of Colorado, Boulder, USA — Emily.Levesque@colorado.edu
Jean-Christophe **Leyder**, European Space Agency (ESA), Spain — jc.leyder@esa.int
Sergey **Lisakov**, Observatoire de la Côte d'Azur, Nice, France — lisakov57@gmail.com
Alex **Lobel**, Royal Observatory of Belgium, Belgium — alobel@sdf.lonestar.org
Jamie **Lomax**, University of Oklahoma, USA — Jamie.R.Lomax@ou.edu
Catherine **Lovekin**, Mount Allison University, Canada — clovekin@mta.ca
André **Maeder**, Department of astronomy, Geneva University, Switzerland — andre.maeder@unige.ch
Nevena **Markova**, Institute of astronomy with National Astronomical Observatory, Bulgaria — nmarkova@astro.bas.bg
Fabrice **Martins**, LUPM, CNRS & Montpellier University, France — fabrice.martins@univ-montp2.fr
Olga **Maryeva**, Special Astrophysical Observatory of the Russian Academy of Sciences, Russian Federation — olga.maryeva@gmail.com
Philip **Massey**, Lowell Observatory, USA — phil.massey@lowell.edu
Stéphane **Mathis**, AIM Paris-Saclay, CEA/DSM/IRFU/SAp, France — stephane.mathis@cea.fr
Casey **Meakin**, University of Arizona, USA — casey.meakin@gmail.com
Redouane **Mecheri**, Centre de Recherche en Astronomie, Astrophysique et Géophysique (CRAAG), Algeria — r.mecheri@craag.dz
Andrea **Mehner**, European Southern Observatory, Chile — amehner@eso.org
Anthony **Meilland**, Observatoire de la Côte d'Azur, Nice, France — ame@oca.eu

Jorge **Melnick**, European Southern Observatory, Chile
jmelnick@eso.org

Ronald **Mennickent**, Universidad de Concepción, Chile
rmennick@udec.cl

Georges **Meynet**, Department of astronomy, Geneva University, Switzerland
georges.meynet@unige.ch

Andrea **Miglio**, School of Physics and Astronomy, University of Birmingham, UK
a.miglio@bham.ac.uk

Miguel **Montargès**, LESIA - Observatoire de Paris, France
miguel.montarges@obspm.fr

Ehsan **Moravveji**, Institute of Astronomy, KU Leuven, Belgium
ehsan.moravveji@ster.kuleuven.be

Thierry **Morel**, Institut d'Astrophysique et de Géophysique, Liège, Belgium
morel@astro.ulg.ac.be

Dawid **Moździerski**, Astronomical Institute, University of Wrocław, Poland
mozdzierski@astro.uni.wroc.pl

Francisco **Najarro**, Centro de Astrobiología, Spain
najarro@cab.inta-csic.es

Felipe **Navarete**, University of Sao Paulo, Brazil
navarete@usp.br

Yaël **Nazé**, Institut d'astrophysique, géophysique et océanographie, University of Liège, Belgium
naze@astro.ulg.ac.be

Hilding **Neilson**, East Tennessee State University, USA
neilsonh@etsu.edu

Coralie **Neiner**, LESIA, Paris-Meudon Observatory, France
coralie.neiner@obspm.fr

Kathryn **Neugent**, Lowell Observatory, USA
kathrynneugent@gmail.com

Maria-Fernanda **Nieva**, Institute for Astro- and Particle Physics, University of Innsbruck, Austria
nieva@sternwarte.uni-erlangen.de

Tetiana **Nikolaiuk**, National academy of fine arts and architecture, Ukraine
nikolaiukt@gmail.com

Arlette **Noels**, Institut d'astrophysique et de géophysique - Université de Liège, Belgium
Arlette.Noels@ulg.ac.be

Mary **Oksala**, Astronomical Institute, ASČR, Czech Republic
meo@udel.edu

Jakub **Ostrowski**, Astronomical Institute, University of Wrocław, Poland
ostrowski@astro.uni.wroc.pl

Ilka **Petermann**, Argelander Institut für Astronomie, Germany
ilka@astro.uni-bonn.de

Geraldine **Peters**, University of Southern California, USA
gjpeters@mucen.usc.edu

Vincent **Prat**, Max-Planck-Institut für Astrophysik, Germany
vprat@mpa-garching.mpg.de

Giovanni **Privitera**, Department of astronomy, Geneva University, Switzerland
giovanni.privitera@unige.ch

Norbert **Przybilla**, Institute for Astro- and Particle Physics, University of Innsbruck, Austria
norbert.przybilla@uibk.ac.at

Joachim **Puls**, University Observatory Munich, Germany
uh101aw@usm.uni-muenchen.de

Oscar Hernan **Ramírez Agudelo**, Anton Pannekoek Institute, Netherlands
o.h.ramirezagudelo@uva.nl

Jared **Rice**, University of Nevada Las Vegas, USA
jrice@physics.unlv.edu

Thomas **Rivinius**, European Southern Observatory, Chile
triviniu@eso.org

Iosif **Romanyuk**, Special Astrophysical Observatory of Russian Academy of Sciences, Russian Federation
roman@sao.ru

Christopher **Rosslowe**, Department of Physics and Astronomy, University of Sheffield, UK
chris.rosslowe@sheffield.ac.uk

Maria del Mar **Rubio Díez**, Centro de Astrobiología (CSIC-INTA), Spain
mmrd@cab.inta-csic.es

Naum **Rusomarov**, Uppsala University, Sweden
naum.rusomarov@physics.uu.se

Sophie **Saesen**, Department of astronomy, Geneva University, Switzerland
sophie.saesen@unige.ch

Hideyuki **Saio**, Astronomical Institute, Graduate School of Science, Tohoku University, Japan
saio@astr.tohoku.ac.jp

Sébastien **Salmon**, Université de Liège, Belgium
sebastien.salmon@doct.ulg.ac.be

Thierry **Semaan**, Institut d'Astrophysique et de Géophysique, Université de Liège, Belgium
thierry.semaan@ulg.ac.be

Tomer **Shacham**, Hebrew University of Jerusalem, Israel
tomer.shacham@phys.huji.ac.il

Olga **Sholukhova**, Special Astrophysical Observatory, Russian Federation
olgasao@mail.ru

Matthew **Shultz**, European Southern Observatory, Chile / Queen's University, Canada
mshultz@eso.org

Jessie **Silaj**, University of Western Ontario, Canada
jsilaj@uwo.ca

Sergio **Simón-Díaz**, Instituto de Astrofísica de Canarias, Spain
ssimon@iac.es

Rosaria **Simoniello**, CEA - Service d'Astrophysique (SAP), France
rosaria.simoniello@cea.fr

Philippe **Stee**, Observatoire de la Côte d'Azur, Nice - CNRS, France
Philippe.Stee@oca.eu

Natallia **Sudnik**, Saint Petersburg State University, Russian Federation
snata.astro@gmail.com

Jon **Sundqvist**, University of Munich, Germany
mail@jonsundqvist.com

Wojciech **Szewczuk**, Astronomical Institute, Wrocław University, Poland
szewczuk@astro.uni.wroc.pl

Koh **Takahashi**, University of Tokyo, Japan
ktakahashi@astron.s.u-tokyo.ac.jp

Andrew **Tkachenko**, Instituut voor Sterrenkunde, KU Leuven, Belgium
andrew@ster.kuleuven.be

Sanja **Tomić**, Astronomical Institute AVČR, Czech Republic
sanja@sunstel.asu.cas.cz

Frank **Tramper**, Astronomical Institute Anton Pannekoek, University of Amsterdam, Netherlands
F.Tramper@uva.nl

Asif **ud-Doula**, Penn State Worthington Scranton, USA
auu4@psu.edu

Jorge **Uuh-Sonda**, Department of Astronomy, University of Guanajuato, Mexico
juuh@astro.ugto.mx

Gerard **van Belle**, Lowell Observatory, USA
gerard@lowell.edu

Jorick **Vink**, Armagh Observatory, UK
jsv@arm.ac.uk

Gregg **Wade**, RMC, Canada
wade-g@rmc.ca

Przemek **Walczak**, Nicolaus Copernicus Astronomical Center, Poland
pwalczak@camk.edu.pl

Kerstin **Weis**, Astronomical Institute, Germany
kweis@astro.rub.de

Markus **Wittkowski**, European Southern Observatory, Germany
mwittkow@eso.org

Nobutoshi **Yasutake**, Chiba Institute of Technology, Japan
nobutoshi.yasutake@p.chibakoudai.jp

Norhasliza **Yusof**, University of Malaya, Malaysia
norhaslizay@um.edu.my

Address by the Local Organizing Committee

Dear colleagues,

This is a great pleasure to welcome you all here in Geneva and more precisely in Geneva University for this IAU Symposium number 307. Marcel Proust, the famous french author has written: *The true exploration does not consist in discovering new landscapes, but in having new eyes.* This sentence underlines the fact that the capacity to see what is around us with new methods or from a different viewpoint, opens the way to discoveries.

The discovery of the principles of spectroscopy, in the mid of the nineteen century by Bunsen and Kirchoff, well illustrates this sentence. The stars had not changed but the way to look at them was new. Spectroscopy allowed to unveil the nature of stars and their surface composition. This discovery happened about 14 years after Auguste Comte a prominent French philosopher, made one of the worst intellectual predictions regarding the limits of astrophysics. He wrote, about the observations of stars: *All investigations which are not ultimately reducible to simple visual observations are ... necessarily denied to us. While we can conceive of the possibility of determining their motions, we shall never be able by any means to study their chemical composition.* Today, we use spectroscopy to measure chemical abundances, temperatures, velocities, rotations, ionization states, magnetic fields, pressure, turbulence, density, and many other properties of distant planets, stars, and galaxies. Spectroscopy is the richest source of information about the universe. So we have to be cautious in front of unbalanced statements that close for ever a field. We should never forget that new windows can open unexpectedly.

This conference is dedicated to three observational technics that provide new views on stars. The technics which is the nearest from spectroscopy is **spectropolarimetry**. Measuring the polarization of the radiation field allows us to obtain complementary information about astrophysical objects that may remain hidden to the ordinary intensity spectrum. The polarized spectrum, in contrast to intensity, enables us to determine vector quantities, e.g. the magnetic field vector.

Astronomical interferometers can produce higher-resolution astronomical images than any other type of telescope. At radio wavelengths, image resolutions of a few micro-arcseconds have been obtained, and image resolutions of a fractional milliarcsecond have been achieved at visible and infrared wavelengths. This allows to study the size, the shape and the circumstellar environment of stars, close enough for allowing this technic to be applied.

Asteroseismology provides the tool to find the internal structure of stars. The pulsation frequencies give the information about the density profile of the region where the waves originate and travel. Asteroseismology helps to constrain other characteristics of stars such as mass and radius.

We are here 138 astronomers from 29 countries for understanding a little better how nature is working and how these new windows provide new and complementary guidelines that can sharpen our knowledge. As organizers, we hope that through the talks and discussions, each of us will go back home with new ideas and new eyes to observe the stars.

Georges Meynet, for the LOC
Geneva, 23 June 2014

New windows on massive stars: asteroseismology, interferometry, and spectropolarimetry
Proceedings IAU Symposium No. 307, 2014
G. Meynet, C. Georgy, J. H. Groh & Ph. Stee, eds.

© International Astronomical Union 2015
doi:10.1017/S174392131400619X

The advanced phases of massive stars and the explosive yields

Alessandro Chieffi[1]† and Marco Limongi[2,3,4]

[1] Istituto di Astrofisica e Planetologia Spaziale - INAF
email: alessandro.chieffi@inaf.it

[2] Osservatorio Astronomico di Roma - INAF
email: marco.limongi@oa-roma.inaf.it

[3] Kavli Institute for the Physics and Mathematics of the Universe, Todai Institutes for Advanced Study, The University of Tokyo, Kashiwa 277-8583, Japan

[4] European Southern Observatory, Karl-Schwarzschild-Str. 2, 85748 Garching bei Munchen, Germany

Abstract. I will briefly review the dependence of the explosive yields on the initial mass, metallicity and initial rotational velocity.

Keywords. Stars: evolution, Stars: interiors, Stars: rotation, Stars: Supernovae, Stars: abundances

1. Introduction

Massive stars play an active and fundamental role in both the physical and chemical evolution of the Galaxies. Among the others, they strongly contribute to the progressive enrichment of the interstellar gas in elements with $Z >= 2$. In particular they dominate the production of the intermediate elements O to Ca, contribute to the synthesis of C, of the Fe-peak nuclei (Sc to Zn) and of the S-weak component (Ga to Zr). Viceversa these stars are not supposed to be primary producers of N, F and the main S component. In some specific cases, like at $Z = 0$, a consistent amount of N may be produced as a consequence of the penetration of the He convective shell in the H rich mantle.

The amount of matter synthesized by each star depends on its initial mass, chemical composition and rotational velocity. In the following we will briefly present our latest grid of models and associated explosive yields. The code will be presented in Sec. 2 while the new yields will be briefly discussed in Sec. 3.

2. The FRANEC code

All present models have been computed with the latest version of the FRANEC evolutionary code whose latest release has been described in Chieffi & Limongi (2013) and references therein. This version of the code includes the effects of rotation and takes into account two instabilities: meridional circulation and shear. The adopted nuclear network extends from H to Bi and follows explicitly the temporal evolution of 335 nuclear species. We added in these computations an additional mass loss when the luminosity exceeds the Eddington one: in particular we assumed that all the mass zones where $L/L_{\rm edd} \geqslant 1$ are instantaneously lost from the star. The adopted solar chemical composition is the

† Present address: IAPS-via fosso del cavaliere, 100 - 00133 Roma, Italy

Table 1. Elements assumed to be overabundant at metallicities lower than solar. The overabundances where determined according to Cayrel *et al.* (2004) and Spite *et al.* (2005)

overabundance		
[C/Fe]	=	0.18
[O/Fe]	=	0.47
[Mg/Fe]	=	0.27
[Si/Fe]	=	0.37
[S/Fe]	=	0.35
[Ar/Fe]	=	0.35
[Ca/Fe]	=	0.33
[Ti/Fe]	=	0.23

Asplund *et al.* (2009) one. At lower initial Fe abundances, i.e. $[Fe/H] = -1, -2$ and -3, some elements are assumed to be overabundant with respect to the solar value, see Table 1. The global metallicities therefore are $Z = 1.345 \cdot 10^{-2}$, $3.236 \cdot 10^{-3}$, $3.236 \cdot 10^{-4}$ and $3.236 \cdot 10^{-5}$ The corresponding adopted initial He abundances are $Y = 0.265$ ($[Fe/H]=0$), $Y = 0.25$ ($[Fe/H]=-1$) and $Y = 0.24$ for the two lowest ones.

3. The grid of models and the yields

We computed a grid of models extending in mass between 13 and $80 \, M_\odot$ (13, 15, 20, 25, 30, 40, 60 and 80), in metallicity between $[Fe/H]=0$ and $[Fe/H]=-3$ (0, -1, -2 and -3) and for three initial equatorial rotational velocities $v = 0$, 150 and 300 km/s. All models where followed from the Hayashi track up to the moment of the core collapse. Figs. 1 and 2 show the logarithm of the net yields in solar masses of the elements included in the network as a function of the initial mass for solar metallicity models. Figs. 3 and 4 show the same quantities for $[Fe/H] = -2$. In all Figures, the black lines refer to non rotating models while the red one to models initially rotating at 300 km/s.

The first thing worth noting in Figs. 1 to 4 is that on average the elemental yields tend to increase with the initial mass at all metallicities. This is true for both the elements produced in the hydrostatic and the explosive burnings. The basic reason is that the convective mixing control both the amount of matter processed by a given burning and the final mass-radius relation. Since the sizes of the convective cores in central H and He burnings and the C convective shells usually scale directly with the initial mass (at least until the mass loss is not so efficient to reduce significantly the He core mass), the amount of mass processed by the H, He and C burnings increases with the initial mass and therefore the products of these burnings as well. In addition to this, since the size of the H convective core determines also the mass of the He core, which in turn controls the final compactness of the star at the moment of the core bounce, also the final mass-radius relation is largely controlled by the extension of the H convective core. Since the amount of mass processed by the explosive burnings scales directly with the final mass-radius relation, it follows that also the yields of the elements produced in the explosive burnings basically increase with the initial mass of a star. It goes without saying that, given the pivotal role of mixing in the synthesis of many elements, any modification of the border of the unstable areas or the details of the mixing could significantly affect the final yields of many elements. By the way it is clear that we are not considering here the main s-component which is basically synthesized in a radiative environment within the ^{13}C pocket in stars of 1 to $3 \, M_\odot$ (Straniero *et al.* 1995) as well as the nuclei produced by the r-processes.

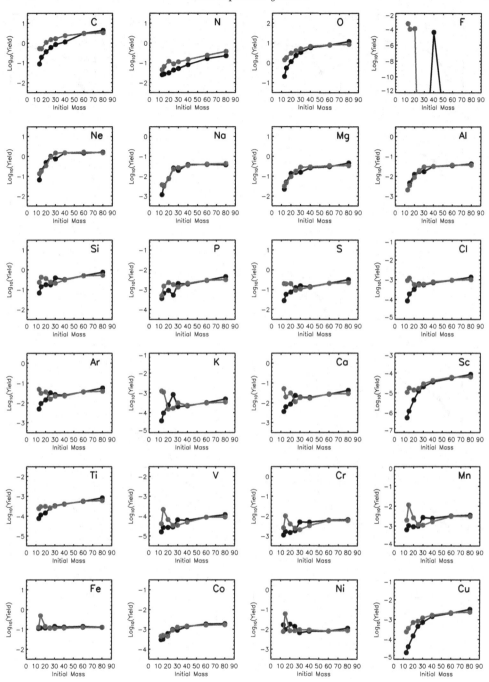

Figure 1. See text.

Since the main effect of rotation is that of favoring an additional mixing that sums to that of the classical thermal instabilities, we can expect that in general the effect of rotation will be that of increasing the yields of the elements. Before proceeding further, it is however important to stress that the amount of mixing induced by the rotational instabilities obviously scales directly with the adopted initial rotational velocity: the

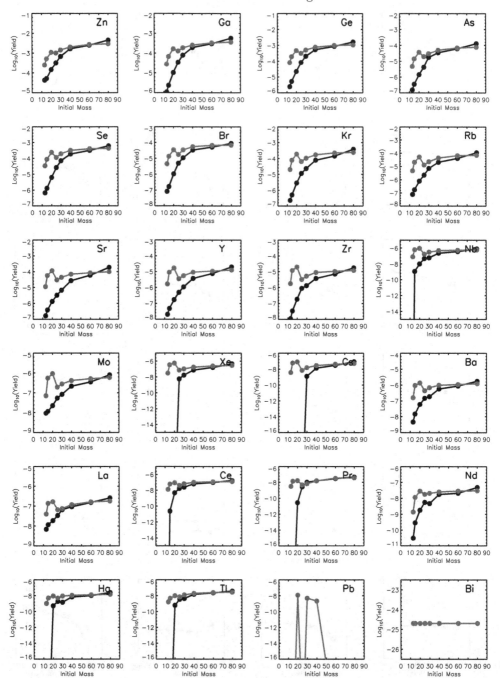

Figure 2. See text.

larger the initial velocity the stronger the effect of the mixing. For sufficiently large initial velocities, the stars can be even forced to be fully mixed and follow an homogeneous evolution (Brott *et al.* 2011) qualitatively similar to that of very low mass stars ($0.5\,M_\odot$ or less) that are fully convective in central H burning. For the specific initial rotational velocities chosen here, the influence of rotation on the yields is not very strong, the

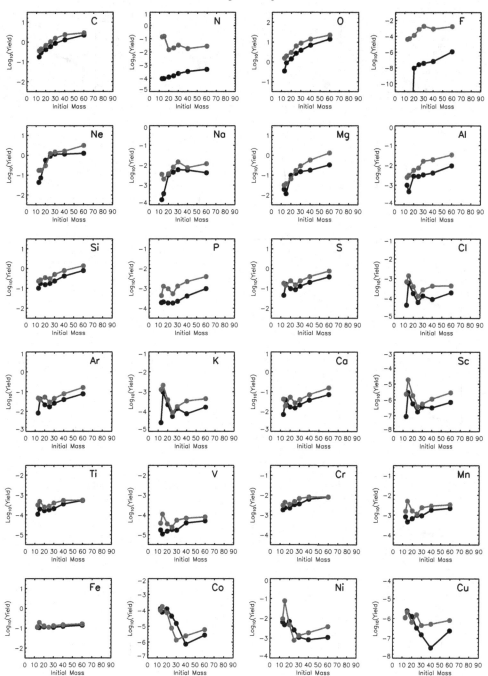

Figure 3. See text.

largest differences occurring for F and the s-process elements in the sense that rotating models tend to increase significantly the yields of these elements. As we turn to lower metallicities, this effect becomes much stronger, so that at [Fe/H] = −2, for example, F and all the s-process elements are largely overproduced with respect to their respective non rotating model. Also N is largely overproduced by rotating models at sub solar Fe

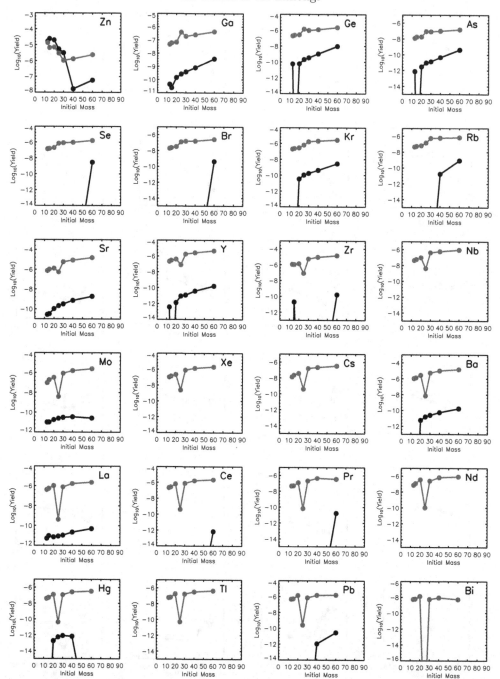

Figure 4. See text.

abundances. We will discuss in detail all such differences in a forthcoming paper. Here, as an example, we will focus only on Fluorine.

In massive stars F production occurs in the He convective shell that forms after the central He burning in the H exhausted core, above the outer edge of the He convective core. It is produced by the sequence $^{14}N(\alpha,\gamma)^{18}F(\beta+)^{18}O(p,\alpha)^{15}N(\alpha,\gamma)^{19}F$ (Goriely *et al.* 1989; Lugaro *et al.* 2004). The protons necessary to activate this sequence are produced by the ^{14}N(n,p)^{14}C nuclear reaction that activates also an additional source of ^{18}O through the ^{14}C(α,γ)^{18}O. The neutron flux may be produced in principle by either the ^{13}C(α,n)^{16}O and/or the ^{22}Ne(α,n)^{25}Mg. The nuclear sequence sketched above implies that the key, basic fuel for the F production is ^{14}N because it may be either the source of the protons and/or of the neutrons through the ^{22}Ne channel. However, in standard non rotating models ^{14}N and ^{22}Ne cannot be present simultaneously because, as it is well known, the second one is produced by the burning of the first one. The ^{13}C channel, on the other hand, cannot be efficient because of the low ^{13}C equilibrium abundance left by the CNO cycle in the He core. The only minor production may occur when the He convective shell, after having converted all the ^{14}N in ^{22}Ne, advances in mass engulfing fresh ^{14}N from the radiative region above it. It goes without saying that this minor production is in any case possible only if the temperature in the He convective shell reaches at least 300 MK. This explains why F is basically destroyed or at most slightly overproduced in massive stars at all (non zero) metallicities. This scenario changes drastically in presence of rotation because of the continuous slow mixing of matter between the He convective core and the active H burning shell. In particular, freshly produced ^{12}C is continuously brought from the convective core up to the tail of the H burning shell where it is converted in ^{13}C and ^{14}N by the CNO cycle and then spread out within the He core. Vice versa the fresh ^{14}N brought back in the He convective core is quickly converted in ^{22}Ne and spread out again outside the convective core. The net result is that at the end of the central He burning the abundances of ^{13}C , ^{14}N and ^{22}Ne are all largely enhanced in the He core with respect to the non rotating models. In stars less massive than $25 M_\odot$ it is the higher concentration of ^{13}C responsible for the large and systematic F production. As the initial mass increases the temperature at the base of the convective shell increases as well so that the ^{22}Ne neutron source becomes progressively more important.

All what has been described above does not take into account the fundamental role played by mass loss in the synthesis of F. While the F net yield increases with the initial mass of the star at [Fe/H] $= -2$ (see Fig. 3) in both the rotating and non rotating models, at solar metallicity F is destroyed in the rotating models (for masses larger that $20\,M_\odot$) as well as in the non rotating ones (see Fig. 1). The reason is that, while at [Fe/H] $= -2$ mass loss does not fully remove the H mantle before the end of the central He burning, at solar metallicity mass loss is so efficient that it removes the whole H rich mantle (and part of the He core as well) well before the central He exhaustion, strongly inhibiting the interplay between the central He burning and the H shell that is necessary to raise the abundances of the key elements necessary to F production. By the way, the peak corresponding to the $40\,M_\odot$ non rotating star visible in Fig. 1 is an artifact due to the fact that this is the only non rotating model that does not destroy F, i.e. for which the net yield is not negative but just slightly positive.

All other features and properties of these models are under analysis and will be presented and discussed in detail as soon as possibile.

Acknowledgements

A.C. warmly thanks the organizers for the excellent hospitality and for financial support. We acknowledge also the PRIN MIUR 2010-2011 project *The Chemical and dynamical Evolution of the Milky Way and Local Group Galaxies*, prot. 2010LY5N2T.

References

Angulo, C., Arnould, M., Rayet, M., *et al.* 1999, *Nuclear Physics A* 656, 3

New windows on massive stars: asteroseismology, interferometry, and spectropolarimetry
Proceedings IAU Symposium No. 307, 2014
G. Meynet, C. Georgy, J. H. Groh & Ph. Stee, eds.

© International Astronomical Union 2015
doi:10.1017/S1743921314006206

Physics of rotation: problems and challenges

Andre Maeder and Georges Meynet

Geneva Observatory, University of Geneva
email: `andre.maeder@unige.ch`

Abstract. We examine some debated points in current discussions about rotating stars: the shape, the gravity darkening, the critical velocities, the mass loss rates, the hydrodynamical instabilities, the internal mixing and N–enrichments. The study of rotational mixing requires high quality data and careful analysis. From recent studies where such conditions are fulfilled, rotational mixing is well confirmed. Magnetic coupling with stellar winds may produce an apparent contradiction, i.e. stars with a low rotation and a high N–enrichment. We point out that it rather confirms the large role of shears in differentially rotating stars for the transport processes. New models of interacting binaries also show how shears and mixing may be enhanced in close binaries which are either spun up or down by tidal interactions.

Keywords. Stellar Physics, Stellar rotation, Stellar evolution

---

1. Introduction

Interferometry, asteroseismology and spectropolarimetry have brought new evidences about the high impact of stellar rotation on the stellar structure and evolution. We can say that rotation influences all observational properties as it has an impact on all model outputs, whether for single or binary stars. The question is whether models and observations are in agreement. We concentrate in this review on current problems and new challenging questions regarding the effects of axial rotation on stellar structure and evolution. For basic developments on the effects of rotation on stellar structure, evolution and nucleosynthesis, the reader may see for example Maeder & Meynet (2012).

2. Shape, gravity darkening, critical velocities and mass loss

2.1. *Shape of rotating stars*

The classical Roche model assumes that the gravitational potential is only due to a central mass concentration. At critical rotation, i.e. when the outwards centrifugal force just compensates central gravity, the Roche model leads to an extreme ratio of the equatorial radius to the polar radius $R_e/R_p = 1.5$. (Structure models predict that the polar radius generally decreases by a few percents for extreme rotation, but this does not affect the critical ratio R_e/R_p).

The VLTI observations of fast rotating stars have led to many discussions, particularly in the case of the Be star Achernar. Domiciano de Souza *et al.* (2003) first found a value of 1.56 for the ratio of the equatorial to the polar radius of Achernar, which was a problem for the Roche model. Kervella & Domiciano de Souza (2006) have studied the oblateness of Achernar and shown that the observations are influenced by the presence of a circumstellar envelope along the polar axis, in addition to the rotational flattening of the photosphere. Carciofi *et al.* (2008) pointed out that the controversial observations may be better interpreted with the account of gravity darkening with in addition a small equatorial disk making the transition between the photosphere and the circumstellar environment. Delaa *et al.* (2013) also demonstrated in the case of α Cep the importance of a good determination of the position angle of the rotation axis, in addition to the other mentioned effects.

On the theoretical side, Zahn *et al.* (2010) went a step beyond the Roche model by accounting for the quadrupolar moment of the mass distribution for a star of $7\,M_\odot$ corresponding to Achernar. They showed that at critical velocity, the ratio R_e/R_p exceeds the standard value of 1.50. In the case of uniform rotation, they found that the extreme ratios R_e/R_p ranges from 1.526 to 1.516 as the star evolves on the Main Sequence (MS). In the case of shellular rotation (with angular velocity Ω constant on level surface and increasing with depth), the values range from 1.560 to 1.535 over the MS phase. Thus, accurate interferometric observations of stars at critical rotation might potentially provide internal constraints on their internal rotation.

2.2. *Gravity darkening*

The von Zeipel theorem (von Zeipel 1924) states that the flux $\vec{F}(\Omega,\vartheta)$ at given angular velocity Ω and colatitude ϑ on a uniformly rotating varies like the effective gravity $\vec{g}_{\text{eff}}(\Omega,\vartheta)$, which is the sum of the Newtonian gravity and of the centrifugal acceleration. The von Zeipel theorem in the case of shellular rotation leads to (Maeder 1999),

$$\vec{F}(\Omega,\vartheta) = -\frac{L}{4\,\pi\,G\,M^\star}\,\vec{g}_{\text{eff}}(\Omega,\vartheta)[1+\zeta(\Omega,\vartheta)] \quad \text{with} \quad M^\star = M\left(1-\frac{\Omega^2}{2\,\pi\,G\,\overline{\varrho}_M}\right). \quad (2.1)$$

L is the luminosity and $\overline{\varrho}_M$ the average density over the mass M. The reduced mass $M^\star$ was generally forgotten in previous studies, despite the fact that it should also be there in case of uniform rotation. The term $\zeta(\Omega,\vartheta)$ is only present in differentially rotating stars, it brings correcting terms depending on the Ω–gradient, the opacities and the gradient of μ (which also depends on ionization). Without the term ζ and the mass reduction, Eq. (2.1) implies that T_{eff} behaves likes g_{eff}^β, with $\beta = 0.25$ in the classical case. Claret (2012) has also found significant deviations from the classical case, which depend on the optical depth, on T_{eff} and on the adopted atmosphere model.

Several authors have attempted to determine the parameter β from interferometric observations, for recent references see Zhao *et al.* (2009); Che *et al.* (2011); Delaa *et al.* (2013). Che *et al.* support a value $\beta = 0.19$ for stars with $T_{\text{eff}} > 7500$ K. Below, convective envelopes tend to appear and imply low β-value as shown long ago by Lucy (1967).

We emphasize that gravity darkening affects all photometric and spectroscopic observations. A rotating star may be seen as a composite star made of thousands local atmosphere models with different local values of g_{eff} and T_{eff}. Gravity darkening says how these parameters are distributed over the stellar surface. In addition, all these local models are seen with different limb–darkening effects.

2.3. *Critical velocities*

The classical expression of the critical or break–up velocity of a rotating star is

$$v_{\text{crit},1} = \left(\frac{GM}{R_{e,\text{crit}}}\right)^{\frac{1}{2}} = \left(\frac{2}{3}\frac{GM}{R_{p,\text{crit}}}\right)^{\frac{1}{2}}. \quad (2.2)$$

In massive stars, the high radiation pressure may add its outwards force to the centrifugal force and modify the expression of the critical velocity. One often finds in literature the following expression with the Eddington factor Γ,

$$v_{\text{crit}} = \sqrt{\left(\frac{GM}{R_{e,\text{crit}}}\right)(1-\Gamma)}, \quad (2.3)$$

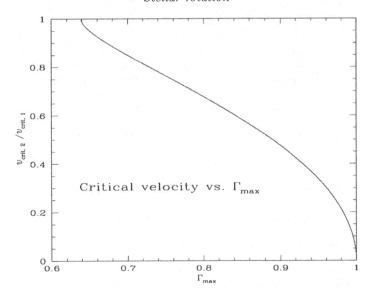

Figure 1. The ratio $(v_{\mathrm{crit},2}/v_{\mathrm{crit},1})$ of the second root of Eq. (2.5) to the first root as a function of $\Gamma_{\max}$ (the maximum over the surface). Eq. (2.6) describes this relation to better than 1%.

which is erroneously supposed to account for the reduction of the critical velocity by radiation pressure. Indeed, this expression ignores the von Zeipel theorem, which decreases the effect of radiation pressure at equator.

To obtain a correct expression, one has first to express the Eddington factor in a rotating star, accounting that both the actual radiation flux and the limiting maximum flux are influenced by the von Zeipel effect. This leads to (Maeder 2009)

$$\Gamma(\Omega, \vartheta) = \frac{\kappa(\Omega, \vartheta)\, L}{4\pi\, c\, G\, M \left(1 - \frac{\Omega^2}{2\pi G \bar{\varrho}_M}\right)}\,. \tag{2.4}$$

This is the local Eddington factor, which varies as a function of rotation and colatitude ϑ. Now, the correct expression of the critical velocity is defined by

$$\vec{g}_{\mathrm{eff}}\,[1 - \Gamma_\Omega(\vartheta)] = \vec{0}. \tag{2.5}$$

This is a second degree equation, which thus has 2 roots. For each couple (Ω, Γ), one has to consider the smallest of the two roots. For $\Gamma < 0.639$, the smallest root is just given by Eq. (2.2). Due to the von Zeipel theorem, the radiative flux at equator decreases in the same way as the effective gravity as rotation increases. The second root has a relatively complex expression (Maeder 2009), which is illustrated in Fig. 1.

We see that for an Eddington factor tending towards 1, the critical velocity tends to zero, because even for a zero velocity the surface is unbound. This reduction of the critical velocity comes from the reduction of the effective mass in Eq. (2.1). The exact expression of $v_{\mathrm{crit},2}$, given by Eq. (4.42) in (Maeder 2009), is a bit difficult to handle. A very good fit of the line in Fig. 1 can be made by a combination of a linear and an exponential relation,

$$\frac{v_{\mathrm{crit},2}}{v_{\mathrm{crit},1}} = 1.00 - 1.93\,(\Gamma - 0.63) - 0.286\,\Gamma^{40}. \tag{2.6}$$

For $\Gamma > 0.639$, it gives a fit with an accuracy better than 0.01.

PHYSICAL EFFECTS	CRITERION
- Effects of the thermal gradients	SCHWARZSCHILD
- Thermohaline mixing	ULRICH, KIPPENHAHN
- Semiconvective diffusion	LEDOUX
- Semiconvection + rotation	SOLBERG - HOILAND
- Shear mixing by differential rotation	RICHARDSON
- Distribution of angular momentum	RAYLEIGH-TAYLOR
- Radiative losses	DUDIS
- Heat transport by hzt. turbulence - Element diffusion by hzt. turbulence	ZAHN

Figure 2. The hydrodynamical effects accounted for in the new criterion.

2.4. *Mass loss*

How does rotation affect mass loss? We are now facing two opposite answers to this important question. From the application of the force multipliers, an increase of the mass loss rates $\dot{M}$ with Ω is predicted (Maeder 2009),

$$\frac{\dot{M}(\Omega)}{\dot{M}(0)} = \left[1 - \frac{4}{9}\left(\frac{v}{v_{\mathrm{crit},1}}\right)^2\right]^{\frac{7}{8}-\frac{1}{\alpha}}. \tag{2.7}$$

The force multipliers take values $\alpha = 0.64, 0.64, 0.59, 0.565$ for $T_{\mathrm{eff}} = 50'000, 40'000, 30'000, 20'000\,\mathrm{K}$ (Pauldrach *et al.* 1986). The term $\frac{4}{9}\left(\frac{v}{v_{\mathrm{crit},1}}\right)^2$ comes from $\frac{\Omega^2}{2\pi G \bar{\varrho}_M}$ in M^*.

Oppositely, a decrease of the mass loss with rotation is predicted by Müller & Vink (2014), who apply a parameterized description of the line acceleration that only depend on radius at a given latitude. We note that they use the incorrect expression (2.3) of the critical velocity and do not account for the mass reduction M^*. Whether, this explains the difference in their conclusions is uncertain. In this context, we note that there are enough data on rotational velocities and mass loss rates to allow an observational test.

3. A new global approach about mixing

The instabilities producing chemical mixing and transport of angular momentum, such as those listed in Fig. 2, are generally considered separately and the global diffusion diffusion coefficient is taken as the sum of the various particular coefficients. This is incorrect and ignores the interactions between the various instabilities, which can amplify or damp each other.

A new stability condition may be written, accounting for the various instabilities mentioned (Maeder *et al.* 2013). It is a quadratic equation of the form $Ax^2 + Bx + C > 0$,

$$\left[N_{\mathrm{ad}}^2 + N_\mu^2 + N_{\Omega-\delta v}^2\right] x^2 +$$
$$\left[N_{\mathrm{ad}}^2 D_{\mathrm{h}} + N_\mu^2(K + D_{\mathrm{h}}) + N_{\Omega-\delta v}^2(K + 2D_{\mathrm{h}})\right] x +$$
$$N_{\Omega-\delta v}^2(D_{\mathrm{h}}K + D_{\mathrm{h}}^2) > 0, \tag{3.1}$$

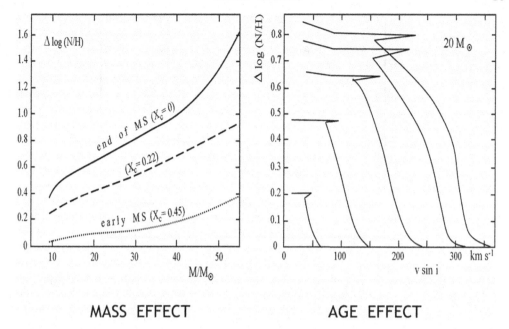

MASS EFFECT   AGE EFFECT

Figure 3. Left: The relative variations of log (N/H) for average rotational velocities as a function of the initial stellar masses of model at $Z = 0.02$ at 3 different evolutionary stages on the MS labeled by X_c the central H–content. Right: The relative variations of log (N/H) as function of the rotational velocities along the MS phase, starting from different initial velocities, in the case of a model of $20\,M_\odot$ with $Z = 0.02$ (Maeder *et al.* 2009).

$$\text{with} \quad N^2_{\Omega - \delta v} = \frac{1}{\varpi^3} \frac{\mathrm{d}\left(\Omega^2\,\varpi^4\right)}{\mathrm{d}\varpi} \sin\vartheta - \mathcal{R}i_c \left(\frac{\mathrm{d}v}{\mathrm{d}r}\right)^2. \tag{3.2}$$

Expression of $N^2_{\Omega-\delta v}$ is a modified form of the Rayleigh oscillation frequency, accounting for both the angular momentum distribution and the excess of energy in the shear. The quantity ϖ is the distance to the rotation axis, K the thermal diffusivity and D_h the diffusion coefficient of horizontal turbulence. Various forms of D_h are existing (Maeder 2009). Eqn. (3.1) is the general equation, which should be considered in a rotating star to account for the various effects mentioned above. *The global diffusion coefficient D_{tot} is just given by $D_{tot} = 2x$ and not by the sum of the specific coefficients. The solution of an equation of the second degree is not the sum of some peculiar solutions !*

We notice several interesting consequences. - All usual criteria need to account for rotation. - The horizontal turbulence has an effect on various transport processes. - Interestingly enough, a stable distribution of angular momentum according to the Rayleigh–Taylor may reduce the effect of a shear. - Conversely, the simultaneous account of shears may also enhance the Rayleigh–Taylor instability. Globally, the interactions of the various effects may either damp or enhance the instabilities.

4. Rotation and N-enhancements: on some confusions

4.1. *The physical effects influencing chemical mixing*

The relations between rotational velocities and N–enrichments have recently been the object of many confusing statements. It is thus necessary to clarify the matter. Let us first say a few words about the physical effects which enter into the game. The models

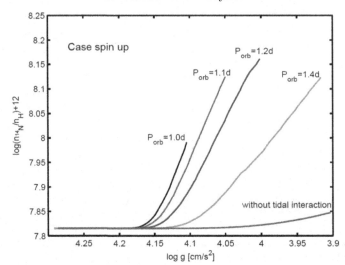

Figure 4. Variation of the N content as a function of log g over the MS phase for the primary of a binary system of a $15 + 10\,M_\odot$ with different initial orbital periods P_{orb} which spin-up during the MS phase starting from an initial rotation velocity equal to 20% of the critical value. The mixing is mainly due to tidally induced shear mixing (Song *et al.* 2013).

of rotating intermediate and massive stars made since 15 years clearly show the role of several effects, confirmed by more recent works (Ekström *et al.* 2012; Georgy *et al.* 2013), as well as by observations (Bouret *et al.* 2013). In particular, *the N-enrichments resulting from internal mixing are a multivariate function of several parameters: first of the rotation velocity, but also of the mass, age, binarity, metallicity and magnetic fields.* Let us note that some models of rotating stars consider the transport (particularly of angular momentum) by meridional circulation as a diffusion, while it is an advection. The two processes are physically not the same: a diffusion transports something from a region where there is a lot to a place where there is very little. An advection (or circulation) can transport matter in any directions, even from where there is a little to a place where there is a lot. Thus, treating meridional circulation as a diffusion, as often done, can even give the wrong sign for the transport !

Besides the rotational velocities which directly influence mixing, there are several other parameters which play a role.

– 1. Mass: Fig. 3 (left) shows the variations of N–enrichment with stellar mass for an average rotation velocity. Depending on the masses considered, the enrichments may vary by more than an order of magnitude.

–2. The age effect is illustrated on the right: we see that for a given mass and for the same observed velocity, the enrichments may also differ by more than an order of magnitude depending on the evolutionary stage considered (Maeder *et al.* 2009).

–3. The tidal interaction in binaries, even before mass transfer, may also drastically enhance the N–enrichments (Song *et al.* 2013). Fig. 4 illustrates a case of tidal interaction in binaries, where the primary is spun–up from a low initial rotation velocity by tidal interaction. The N–enrichments are considerable. The tides accelerate the external layers, this increases the shear throughout the star and thus the N–enrichments at the surface. The opposite case may also occur (Song *et al.* 2013): the tides may produce a spin–down of initially fast rotators, the resulting shears also contribute to N–enrichments. Works in progress show that due to tidal mixing some stars may even become chemically homogeneous, thus they evolve contracting and never reach a stage of mass transfer.

–4. Metallicity Z influences mixing: low Z stars being more compact experience stronger shears, which favor mixing and N–enrichments (Maeder & Meynet 2001). The effect applies down to very low metallicities, as shown by current studies on the origin of the primary nitrogen. In addition, there are clear indications that at lower metallicities, stars have on the average faster rotational velocities than at solar composition (Maeder *et al.* 1999; Martayan *et al.* 2007). This makes the rotational mixing even a more important effect in lower metallicity populations and earlier galactic evolutionary stages.

–5. The relation between magnetic field and N–enrichments is complex both theoretically and observationally. Models with the Tayler–Spruit dynamo predict significant N–enrichments, however the existence of this dynamo in radiative zones of differentially rotating stars is still very debated (Zahn *et al.* 2007). The observations indicate that some magnetic stars are N–enriched and some are not (cf. presentations by Neiner, by Henrichs, by Wade at this meeting). It is clear that magnetic fields transport the angular momentum and favor the internal coupling of the rotation, but as far as the chemical mixing is concerned the situation remains uncertain. Things are more clear concerning the effects of magnetic fields at the stellar surface, which are observed in about 10% of the massive stars. These fields produce a mechanical coupling of the stellar surface with the stellar winds. This removes angular momentum from the star (ud-Doula & Owocki 2002; Ud-Doula *et al.* 2008). The prescriptions of these authors were used in models by Meynet *et al.* (2011), who showed that the magnetic coupling can produce an efficient braking. This braking at the surface considerably increases the internal shear and thus makes a strong mixing. This leads to massive stars with low rotation velocities and at the same time strong enrichments. Thus, the surface effects of magnetic fields may produce stars which just seem to contradict the current expectations about a simple relation between N-enrichments and $v \sin i$. However, it rather confirms the large role of shears in differentially rotating stars for the transport processes.

As a conclusion, for a careful observational study of the rotational mixing, it is indispensable to strongly limit the range of the other intervening parameters, mass, age, binarity, metallicity and magnetic filed.

4.2. *Some current observations on N-enrichments*

The VLT–Flames survey of massive stars in the Magellanic Clouds lead the authors to challenge the existence of rotational mixing (Hunter *et al.* 2007, 2009; Brott *et al.* 2011b). They found two groups of stars, which do not follow a relation between $v \sin i$ and N/H ratios. Group 1 (25 % of the stars) with little enrichments and high $v \sin i$ and Group 2 (16 %) with high enrichments and low $v \sin i$. The sample was rediscussed by Maeder *et al.* (2009), who showed that the sample by Hunter *et al.* is in fact a mixture of masses between 10 and 30 $M_\odot$ and of evolutionary stages, with stars close to the Main–Sequence and stars out of it. When a limited range of masses and ages is analyzed, the correlation improves considerably (Maeder *et al.* 2009), as far as the accuracy of the data permit it. Group 1 essentially vanishes, while Group 2 may be explained by the effect of magnetic braking mentioned above, it also contains some evolved stars.

Regarding possible tests about the accuracy of the chemical abundances, the properties of the N/C vs. N/O diagram have recently been discussed by Maeder *et al.* (2014). They show that up to an enrichment of N/O by a factor of 4, the relation between N/C and N/O is essentially model independent, being determined only by nuclear physics. Up to the mentioned enrichment, the scatter around the mean relation, if anyone, is essentially observational or coming from the analysis process. Such a test beautifully confirms the high quality of the data by Przybilla *et al.* (2010), while the quality of the abundance data from Hunter *et al.* (2007, 2009) appears much lower.

A detailed analysis of the chemical abundances of O–stars in the SMC was made by Bouret *et al.* (2013). They found that 65% of the observed stars well fit the predicted relations of N–enhancement versus $v \sin i$, 26% are marginal and 9% do not fit. This fraction of outliers is much lower than that found by Hunter *et al.* (2009) and Brott *et al.* (2011a). In particular, Bouret *et al.* confirm the existence of Group 2 with slowly rotating N–rich stars. As we have stressed, this group may result fom magnetic braking. They confirm that faster rotators have higher N/C ratios, that more massive stars have higher N/C than less massive stars. The same is true for the N/O ratios.

Finally, we must also mention a recent study of 68 bright stars of all over the sky with high–precision spectrocopic and asteroseismic measurements by Aerts *et al.* (2014). These authors conclude "We deduce that neither the projected rotational velocity nor the rotational frequency has predictive power for the measured nitrogen abundance". An examination of the data used in this analysis shows stars with spectral types from A0 to O5, i.e. with masses between about 3 and 30 $M_\odot$. Fig. 3 left indicates that this range of masses is much too large to allow reasonable comparisons. In addition, the sample contain stars of all luminosity classes: dwarfs, giants and supergiants, which should also lead to some difficulties in the comparisons according to Fig. 3 right. These stars are all over the sky, spanning a range of galactocentric distances of several kpc, thus with significant metallicity differences. Therefore, the conclusions by Aerts *et al.* regarding the the effects of rotation on nitrogen abundances are a little premature.

5. Conclusions: the next great challenge

We hope that the above lines help to clarify some points in current discussions about rotating stars, concerning their shape, gravity darkening, critical velocities, mass loss, instabilities, mixing and N–enrichments. These points must not prevent us to see another big problem we are now facing concerning rotation.

Recent asteroseismic observations indicate too slow rotation of stellar cores in red giants compared to model predictions. The precise determinations from KEPLER of the rotational frequency splittings of mixed modes provide information on the internal rotation of red giants. Beck *et al.* (2012) showed that in the case of the red giant KIC 8366239 the core rotates about 10 times faster than the surface. Deheuvels *et al.* (2012, 2014) analyzed the rotation profiles of six red giants and found that the cores rotate between 3 and 20 times faster than the envelope. They also found that the rotation contrast between core and envelope increases during the subgiant branch. The crucial point is that the observed differences of rotation between cores and envelopes are much smaller than predicted by the evolutionary models of rotating stars. Analyses with stellar modeling by several groups (Eggenberger *et al.* 2012; Marques *et al.* 2013; Ceillier *et al.* 2013; Tayar & Pinsonneault 2013) clearly demonstrate that some generally unaccounted physical process is at work producing an important internal viscosity.

In this context, we must also remember previous results showing that the evolutionary models leading to pulsars always rotate much too fast compared to the observed rotation rates of pulsars (Heger *et al.* 2004; Hirschi *et al.* 2005). This is even the case when the magnetic field of the Tayler–Spruit dynamo (Spruit 2002), which could act in rotating radiative regions, is accounted for. A similar problem appears for white dwarfs, which generally show rotation velocities much lower than predicted by standard models (Berger *et al.* 2005). Thus, there is an ensemble of observations pointing in favor of some additional internal coupling in star evolution.

These new windows provided by recent technologies are a gift for our better understanding of stellar physics. To be fully enjoyed, the gift must receive great care in reduction and analysis.

References

Aerts, C., Molenberghs, G., Kenward, M. G., & Neiner, C. 2014, *ApJ* 781, 88

Beck, P. G., Montalban, J., Kallinger, T., *et al.* 2012, *Nature* 481, 55

Berger, L., Koester, D., Napiwotzki, R., Reid, I. N., & Zuckerman, B. 2005, *A&A* 444, 565

Bouret, J.-C., Lanz, T., Martins, F., *et al.* 2013, *A&A* 555, A1

Brott, I., de Mink, S. E., Cantiello, M., *et al.* 2011a, *A&A* 530, A115

Brott, I., Evans, C. J., Hunter, I., *et al.* 2011b, *A&A* 530, A116

Carciofi, A. C., Domiciano de Souza, A., Magalhães, A. M., Bjorkman, J. E., & Vakili, F. 2008, *ApJ (Letters)* 676, L41

Ceillier, T., Eggenberger, P., García, R. A., & Mathis, S. 2013, *A&A* 555, A54

Che, X., Monnier, J. D., Zhao, M., *et al.* 2011, *ApJ* 732, 68

Claret, A. 2012, *A&A* 538, A3

Deheuvels, S., Doğan, G., Goupil, M. J., *et al.* 2014, *A&A* 564, A27

Deheuvels, S., García, R. A., Chaplin, W. J., *et al.* 2012, *ApJ* 756, 19

Delaa, O., Zorec, J., Domiciano de Souza, A., *et al.* 2013, *A&A* 555, A100

Domiciano de Souza, A., Kervella, P., Jankov, S., *et al.* 2003, *A&A* 407, L47

Eggenberger, P., Montalbán, J., & Miglio, A. 2012, *A&A* 544, L4

Ekström, S., Georgy, C., Eggenberger, P., *et al.* 2012, *A&A* 537, A146

Georgy, C., Ekström, S., Eggenberger, P., *et al.* 2013, *A&A* 558, A103

Heger, A., Woosley, S. E., Langer, N., & Spruit, H. C. 2004, in A. Maeder & P. Eenens (eds.), *Stellar Rotation*, Vol. 215 of *IAU Symposium*, p. 591

Hirschi, R., Meynet, G., & Maeder, A. 2005, *A&A* 443, 581

Hunter, I., Brott, I., Langer, N., *et al.* 2009, *A&A* 496, 841 (H+09)

Hunter, I., Dufton, P. L., Smartt, S. J., *et al.* 2007, *A&A* 466, 277

Kervella, P. & Domiciano de Souza, A. 2006, *A&A* 453, 1059

Lucy, L. B. 1967, *ZfA* 65, 89

Maeder, A. 1999, *A&A* 347, 185

Maeder, A. 2009, *Physics, Formation and Evolution of Rotating Stars*

Maeder, A., Grebel, E. K., & Mermilliod, J.-C. 1999, *A&A* 346, 459

Maeder, A. & Meynet, G. 2001, *A&A* 373, 555

Maeder, A. & Meynet, G. 2012, *Reviews of Modern Physics* 84, 25

Maeder, A., Meynet, G., Ekström, S., & Georgy, C. 2009, *Communications in Asteroseismology* 158, 72

Maeder, A., Meynet, G., Lagarde, N., & Charbonnel, C. 2013, *A&A* 553, A1

Maeder, A., Przybilla, N., Nieva, M.-F., *et al.* 2014, *A&A* 565, A39

Marques, J. P., Goupil, M. J., Lebreton, Y., *et al.* 2013, *A&A* 549, A74

Martayan, C., Frémat, Y., Hubert, A.-M., *et al.* 2007, *A&A* 462, 683

Meynet, G., Eggenberger, P., & Maeder, A. 2011, *A&A* 525, L11

Müller, P. E. & Vink, J. S. 2014, *A&A* 564, A57

Pauldrach, A., Puls, J., & Kudritzki, R. P. 1986, *A&A* 164, 86

Przybilla, N., Firnstein, M., Nieva, M. F., Meynet, G., & Maeder, A. 2010, *A&A* 517, A38

Song, H. F., Maeder, A., Meynet, G., *et al.* 2013, *A&A* 556, A100

Spruit, H. C. 2002, *A&A* 381, 923

Tayar, J. & Pinsonneault, M. H. 2013, *ApJ (Letters)* 775, L1

ud-Doula, A. & Owocki, S. P. 2002, *ApJ* 576, 413

Ud-Doula, A., Owocki, S. P., & Townsend, R. H. D. 2008, *MNRAS* 385, 97

von Zeipel, H. 1924, *MNRAS* 84, 665

Zahn, J.-P., Brun, A. S., & Mathis, S. 2007, *A&A* 474, 145

Zahn, J.-P., Ranc, C., & Morel, P. 2010, *A&A* 517, A7

Zhao, M., Monnier, J. D., Pedretti, E., *et al.* 2009, *ApJ* 701, 209

Discussion

WADE: In our poster we describe precise magnetic field measurements of the sample of 20 early B stars for which Nieva and Przybilla obtained high-quality nitrogen abundances. Of the 5 stars we identify as N-rich, 3 are magnetic. But 2 are not, with upper limits below 10 G. To make the situation more confusing, Thierry Morel has reported N abundances for 2 other early B/late O magnetic stars (NGC 2244-201, HD 57682) which seem to be normal. So there seem to be a great diversity of objects!

MAEDER: Your point is an important one. Some years ago, the N/H excesses were considered as a signature of magnetic fields. Such results would have been in agreement with our evolution models based on the Tayler-Spruit dynamo. However, evidences seem to accumulate in favor of a fossil origin of magnetic fields (cf. Mathis). In this context, an explanation is more difficult since the magnetic field lines are frozen in the medium (if the field is large) and this does not seem to favor the transport.

BAADE: To supplement Gregg Wade's comment: quite some years ago, Johann Kolb and I tried to measure N abundances of broad-line Be stars *relative to Bn stars*, which seem to rotate about as rapidly as Be stars but not suffer outbursts leading to the formation of circumstellar disks. Moreover, Be stars are the only group of stars investigated by MIMES and found to be 100% free of detectable magnetic fields. Yet, we found Be stars more nitrogen rich than Bn stars. Unfortunately, this project remained unfinished since Johann left astronomy. As well, Thomas Rivinius told me that he and Maria-Fernanda Nieva measured N in some narrow-lined Be stars. So far, the stars studied have solar N abundances. Therefore, one of the two preliminary results is in error or the N abundances depend on stellar latitude.

AERTS: How would the coupling of a magnetic field impact on the internal rotation profile for a B star with a radiative envelope? (see our latest $\Omega(r)$ profile deduced from seismology of a B star with gravity waves.)

MAEDER: The models we have at present are based on the Tayler-Spruit dynamo. They indicate that for the fields created by this dynamo, the internal profiles of $\Omega(r)$ are very close to solid body rotation. We have not yet models for weak fossil fields, but below some critical field value, it is likely that some significant differential rotation may exist.

ARLT: It should be noted that magnetic instabilities deliver two different diffusion coefficients for angular-momentum transport and mixing. The slow-down and nitrogen peculiarities may therefore develop on different time-scales.

MAEDER: The equations of transport of the angular momentum and of chemical transport are different and the transport coefficients are not necessarily the same. However, this depends on the magnetic instabilities and horizontal turbulence.

NOELS: If most massive stars undergo this "needed" strong braking already in the early stages of the main sequence, what are the consequences on the nitrogen enrichments?

MAEDER: The external braking by magnetic fields enhances differential rotation and thus the shears, in turn these drive mixing and surface N-enrichment. On the contrary, internal magnetic coupling generally reduces the shears and thus the mixing. The question is what is the balance between the internal and external effects. There are not yet relevant models. Thus, at the present time, it is difficult to answer your most interesting question.

André Maeder and Arlette Noels

Casey Meakin

New windows on massive stars: asteroseismology, interferometry, and spectropolarimetry
Proceedings IAU Symposium No. 307, 2014
G. Meynet, C. Georgy, J. H. Groh & Ph. Stee, eds.
© International Astronomical Union 2015
doi:10.1017/S1743921314006218

The Physics of Convection in Massive Stars

Casey A. Meakin[1,2,3]

[1]New Mexico Consortium, Los Alamos, NM 87544 USA

[2]Los Alamos National Laboratory, Los Alamos, NM 87545 USA

[3]University of Arizona, Tucson, AZ 85721 USA
email: casey.meakin@gmail.com

Abstract. I summarize current state-of-the-art methods for treating the difficult problem of turbulent convection in stellar interiors and I discuss a powerful approach for analysis that allows one to leverage the most from 3D stellar models.

Keywords. convection, turbulence, stars: evolution, supernovae

1. Introduction

The shortcomings of turbulent mixing models for stars, including the mixing length theory of convection and various types of overshoot mixing prescriptions, are now widely acknowledged. Of particular concern are the errors related to accurately predicting the amount of mixing that takes place *across* convective boundaries which result in a wide range of modeling errors, including predicted nucleosynthetic yields, supernova progenitor structures, and stellar lifetime estimates, to name just a few. The problem of turbulent mixing is not confined to massive stars but impacts stars of nearly every mass. Therefore, one of the most pressing frontier questions in modern stellar structure and evolution is, *how can we develop a better model of stellar turbulence?* In this short article I briefly recount a promising line of inquiry into this problem that incorporates simulation modeling and classical turbulence analysis.

2. RANS: A Powerful Tool for Investigating Stellar Turbulence

As with any frontier science, defining terms and posing useful questions is an essential first step. And while research on the compressible, stratified, multi-species, reactive turbulence relevant to stellar interiors remains parked at the scientific frontier, much work has been done on simpler, related problems out of which has grown a powerful set of techniques for discussing the most important issues.

Of special importance is the method of Reynolds decomposition and the development of a set of statistical mean-field equations known as Reynolds-Averaged Navier-Stokes equations or RANS equations. The basic principle behind Reynold decomposition is to write a fluid property as the sum of its time-averaged and instantaneous-fluctuation components $\varphi(x,t) = \overline{\varphi}(x) + \varphi'(x,t)$, where x and t are the space and temporal coordinates, respectively; the overbar denotes time (and sometimes also space) averaging; and the prime indicates the time dependent, fluctuating component of the field. RANS equations are then developed by time or space averaging the conservation laws after having rewritten the fluid properties in terms of Reynolds decomposed fields; and a complementary set of moment equations are developed by multiplying the conservation equations by products of fluctuating components prior to applying the averaging operator. This procedure is the workhorse for turbulence research, and remains one of the primary windows into the nature of turbulent flow and has been elaborated upon in standard texts written over the last several decades (e.g. Landau & Lifshitz 1959; Hinze 1959; Tennekes & Lumley 1972; Frisch 1995; Pope 2000; Chassaing 2002).

The RANS equations developed in this way both define a set of turbulence quantities precisely (such as turbulent kinetic energy, turbulent stress, and energy flux), and include a set of time dependent equations describing the evolution of these quantities and their relationships. But as excellent a result as this is, Mother Nature does not give up her secrets so easily. The nonlinear terms in the conservation laws ultimately result in an infinite set of coupled equations wherein the evolution for each statistical moment depends on moments of a higher order, and we are faced with a *closure problem* akin to the BBGKY hierarchy in statistical physics. The research problem is to find a way to truncate the set of equations by modeling higher order terms as a function of lower order terms.

3. Confronting the Closure Problem

While the RANS equations offer a precise way to discuss and analyze turbulent systems, we are ultimately left with an incomplete description whenever we truncate our system of equations to a finite, and therefore usable, set of statistical moments. Therefore, in order to develop a useful model we must find a way to *close* the system. In physical terms, this means that we need to develop a closed form description of how fluctuations correlate with each other.

The mixing length theory (MLT) is in fact a closure model: it prescribes, albeit in a simplistic fashion, how velocity and temperature fluctuations are correlated. MLT, then, allows us to close the mean field equations of stellar structure and evolution and find solutions. More sophisticated models, such as those developed by Kuhfuss (1986) and Canuto (1992), have also been produced over the years and draw on a more detailed picture of how turbulence behaves. However, in all cases, it remains unclear which model is best suited to treating stellar turbulence.

Without strong justification for one model over the next, it makes sense to keep on using the most user-friendly available, which remains MLT. Today, however, we are starting to witness a transformation in our ability to study turbulent flow that is being driven by computer simulations of ever finer granularity and that are now classifiable as representing truly turbulent flow.

4. Leveraging 3D Simulation

The beauty of the RANS approach to stellar turbulence is that it provides a well defined set of turbulence quantities and evolution equations that can be systemically investigated from a physical perspective. Combining this analytic tool with simulation data (or numerical experiments) of stellar convection is starting to provide justification for one type of turbulence model over another. Modern stellar simulation data (e.g. Meakin 2006; Arnett & Meakin 2011) is of sufficient quality to begin testing the basic building-block assumptions that go into turbulence models, like MLT, thus providing a rational for choosing one over the next.

In Mocák *et al.* (2014), we have provided a "nuts and bolts" look at how this approach can be pursued. First, we have developed a set of RANS equations which are of particular interest for stellar evolution: those that are derived from the conservation laws of momentum, energy, and mass (including a multi-species compositional description) and are cast against a fixed (or slowly time varying) spherical background base state. Second, we have calculated a large collection of statistical mean fields from our diverse collection of 3D stellar simulation data, which includes a supernova progenitor model, as well as lower mass models including a helium-shell flash and a reg giant convective envelope.

As discussed in some detail in Meakin & Arnett (2006, 2007a,b,c), Meakin (2006, 2008), Arnett *et al.* (2009), and Meakin *et al.* (2011), and then more explicitly in terms of RANS by Murphy & Meakin (2011) and Viallet *et al.* (2013), this approach provides a path towards developing physical intuition on what might otherwise appear to be very elusive characteristics of turbulence, including the origin of composition and energy fluctuations, the mechanisms of transport, and the geometric structures underlying these flows. While the equations presented in Mocák *et al.* (2014) may appear similar to those recently published by Canuto (2011a,b,c,d,e), in other places we have adopted a different approach in our work. In particular, we present our equation set completely unmodeled and unclosed and without approximation. Our intention is to provide exact evolution equations together with what is in essence *raw data* to be used to check approximations used in models such as those presented in Canuto's and similar works. *We have chosen this approach because we believe that the community's knowledge about turbulent mixing is still too limited to develop adequate and robust models.* There is still much work to be done!

5. Data Availability on the Web

In an attempt to facilitate developments in this challenging field of study, we are striving to continue to make our simulation data available for download and use by as broad an audience as possible. At present the subset of data discussed in Viallet *et al.* (2013) is available at http://www.stellarmodels.org and includes as many as 113 mean fields for study. In addition, we have supplied a Python library for reading and plotting data together with a short sample script to get started working with our data.

References

Arnett, D., Meakin, C., & Young, P. A. 2009, *ApJ* 690, 1715

Arnett, W. D. & Meakin, C. 2011, *ApJ* 733, 78

Canuto, V. M. 1992, *ApJ* 392, 218

Canuto, V. M. 2011a, *A&A* 528, A76

Canuto, V. M. 2011b, *A&A* 528, A77

Canuto, V. M. 2011c, *A&A* 528, A78

Canuto, V. M. 2011d, *A&A* 528, A79

Canuto, V. M. 2011e, *A&A* 528, A80

Chassaing, P. 2002, *Variable Density Fluid Turbulence*, Springer Science and Business Media

Frisch, U. 1995, *Turbulence: The Legacy of A. N. Kolmogorov*, Cambridge University Press

Hinze, J. O. 1959, *Turbulence*, McGraw-Hill

Kuhfuss, R. 1986, *A&A* 160, 116

Landau, L. D. & Lifshitz, E. M. 1959, *Fluid Mechanics*, Elesvier

Meakin, C. & Arnett, D. 2007a, in F. Kupka, I. Roxburgh, & K. L. Chan (eds.), *IAU Symposium*, Vol. 239 of *IAU Symposium*, pp 296–297

Meakin, C. A. 2006, *Ph.D. thesis*, The University of Arizona, Arizona, USA

Meakin, C. A. 2008, in L. Deng & K. L. Chan (eds.), *IAU Symposium*, Vol. 252 of *IAU Symposium*, pp 439–449

Meakin, C. A. & Arnett, D. 2006, *ApJ (Letters)* 637, L53

Meakin, C. A. & Arnett, D. 2007b, *ApJ* 665, 690

Meakin, C. A. & Arnett, D. 2007c, *ApJ* 667, 448

Meakin, C. A., Sukhbold, T., & Arnett, W. D. 2011, *Ap&SS* 336, 123

Mocák, M., Meakin, C., Viallet, M., & Arnett, D. 2014, *ArXiv e-prints*
Murphy, J. W. & Meakin, C. 2011, *ApJ* 742, 74
Pope, S. B. 2000, *Turbulent Flows*, Cambridge University Press
Tennekes, H. & Lumley, J. L. 1972, *First Course in Turbulence*, MIT Press
Viallet, M., Meakin, C., Arnett, D., & Mocák, M. 2013, *ApJ* 769, 1

Discussion

AERTS: What would be the effect of internal gravity waves caused by a convective core on the internal rotation profile of a star, e.g. starting from a rigid $\Omega(r)$ profile: what would $\Omega(r)$ look like after a nuclear timescale?

MEAKIN: The mean field framework that I have described is well suited to study questions such as this, but there is still a lot of work to do to characterize the relevant fluid dynamics. It has become increasingly clear over the past several decades that there is not a "fundamental" theory of turbulence and mixing but that instead each instability and process needs to be paid special attention. Our 3d simulation work to date has only included one case with a differential rotation profile and since it is representative of an advanced shell burning stage of evolution (oxygen burning) it is not relevant to main sequence evolution. Therefore, my own work does not speak directly to the question you ask. But I would add that the a definitive answer to your question will depend on both unspecified parameters, such as the absolute value of the initial rotational velocity profile, the mass of the star, it's composition, etc., as well as a variety of highly uncertain fluid dynamical processes such as mixing driven by double-diffusive instabilities and core-boundary interactions (e.g., "overshoot").

PULS: To make things even more complex: what about the radiation field (also for outer convective zones)?

MEAKIN: The character of radiative effects are very different between deep interior convection and surface convection. A diffusive treatment is sufficient to treat interiors while surface convection requires a more complete description of the radiation field, as you point out. A number of research groups have been simulating surface convection in 3D with great success. Our complimentary efforts have to date been focused instead on deep interior convection with an emphasis on the late burning stages in massive stars (post helium burning). Under these conditions, not only can radiation be treated very accurately as a diffusive process but the overall importance of the radiation field is significantly reduced due to the increased role played by neutrino emission which effectively takes over as the dominant "transport" process during these phases.

YASUTAKE: What is the role of rotation in 3D simulations? Is it important or not? If important, we need 3D to 2D applications for stellar evolution instead of 3D to 2D simulations. Is it correct?

MEAKIN: From the point of view of 3D simulation, rotation is simply treated as an azimuthal velocity field. Whether it is important or not depends on the strength of the rotation and its gradients as is usually the case. As you correctly point out, a self-consistent treatment of the mean velocity field for a star that possesses dynamically significant angular momentum requires at least a 2D description. In some cases, 3D might be required.

YASUTAKE: Can you calculate the front of combustion in your 3D HD simulation? You showed the effects of resolution. I am afraid you did not check the convergence of scale. In my opinion we need AMR technique for hydrodynamic simulation including nuclear burning front, right?

MEAKIN: The nuclear burning that occurs during the late burning stages in a massive star is not characterised by a thin flame front like the deflagrations that takes place in Type Ia supernovae explosions, so resolving the "front" is not a challenge. Nevertheless, the question of numerical resolution is still an important one for stellar convection simulations and this remains an active area of study. Our resolution studies, together with some basic fluid dynamical arguments, give us confidence that we are resolving the important processes within the convective layers that we are simulating (see, e.g., Section 4.7 of Viallet *et al.* 2014). The convective boundaries, on the other hand, have been more difficult to understand. Whether or not we have numerically converged mixing rates at these boundaries remains an open question and we are indeed beginning to study this part of the problem with AMR.

NOELS: You showed us how important it was to go from simple MLT to more realistic 3D models of convection. Would it be possible to come back from there towards an "adjusted" MLT with "adjusted" convection criteria? This would be very convenient to deal with stability analysis.

MEAKIN: It is important to be able to construct simplified models for whatever hydro-dynamic processes we discover in our 3D simulations. Whether or not it is possible to incorporate all of the important physics as modifications to the current generation of MLT-based 1D models remains an open question. It is possible that an implementation based on a Reynolds-Averaged Navier-Stokes (RANS) approach is required and I would argue that there are already signs that we are starting to see the limitations of the current generation of codes, even those with "embellished" MLT style models. It is only through hard work and experience that we will be able to answer this question with certainty so we need to keep pushing ahead.

New windows on massive stars: asteroseismology, interferometry, and
spectropolarimetry
Proceedings IAU Symposium No. 307, 2014
G. Meynet, C. Georgy, J. H. Groh & Ph. Stee, eds.

© International Astronomical Union 2015
doi:10.1017/S174392131400622X

Physics of Mass Loss in Massive Stars

Joachim Puls[1], Jon O. Sundqvist[1] and Nevena Markova[2]

[1]Universitätssternwarte, Scheinerstr. 1, D-81679 München; Germany
email: uh101aw@usm.uni-muenchen.de

[2]Institute of Astronomy with NAO, BAS, PO Box 136, 4700 Smolyan, Bulgaria

Abstract. We review potential mass-loss mechanisms in the various evolutionary stages of massive stars, from the well-known line-driven winds of O-stars and BA-supergiants to the less-understood winds of Red Supergiants. We discuss optically thick winds from Wolf-Rayet stars and Very Massive Stars, and the hypothesis of porosity-moderated, continuum-driven mass loss from stars formally exceeding the Eddington limit, which might explain the giant outbursts from Luminous Blue Variables. We finish this review with a glance on the impact of rapid rotation, magnetic fields and small-scale inhomogeneities in line-driven winds.

Keywords. stars: early-type, stars: mass loss, stars: winds, outflows

1. Introduction

Stellar winds from massive stars are fundamentally important in providing energy and momentum input into the interstellar medium, in creating wind-blown bubbles and circumstellar shells, and in triggering star formation. The presence and amount of mass loss decisively controls the evolution and fate of massive stars† (e.g., the type of final Supernova explosion), by modifying evolutionary timescales, surface abundances, and stellar luminosities. Moreover, mass loss also affects the atmospheric structure, and only by a proper modeling of stellar winds it is possible to derive accurate stellar parameters by means of quantitative spectroscopy. In the following, we review the *status quo* of our knowledge about the physics of these winds.

2. Basic considerations

Since massive stars have a high luminosity, they are basically able to generate a large radiative acceleration in their atmospheres.

Global energy budget – The photon tiring limit. Thus, a first question regards the maximum mass-loss rate that can be radiatively accelerated. By equating the mechanical luminosity in the wind at infinity with the available photospheric luminosity L_* (in this case, all photons have used up their energy/momentum and $L(\infty) = 0$), one obtains

$$\dot{M}_{\max} = \frac{2L_*}{v_\infty^2 + v_{\rm esc}^2} = \frac{\dot{M}_{\rm tir}}{1 + (v_\infty/v_{\rm esc})^2} \approx \dot{M}_{\rm tir} \ldots \frac{\dot{M}_{\rm tir}}{10}, \qquad (2.1)$$

where v_∞ is the terminal wind speed (typically on the order of one to three times the photospheric escape velocity, $v_{\rm esc} = \sqrt{2GM/R}$), and $\dot{M}_{\rm tir}$ the 'photon tiring mass-loss rate' (Owocki & Gayley 1997), i.e., the maximum mass-loss rate when the wind just

---

† A change of only a factor of two in the mass-loss rates can have a dramatic effect (Meynet *et al.* 1994).

escapes the gravitational potential, with $v_\infty \to 0$. In convenient units,

$$\dot{M}_{\rm tir} = \frac{2L_*}{v_{\rm esc}^2} = 0.032 \frac{M_\odot}{\rm yr} \frac{L_*}{10^6 L_\odot} \frac{R/R_\odot}{M/M_\odot}. \tag{2.2}$$

This maximum mass-loss rate is much larger than the mass-loss rates of winds from OB-stars/A-supergiants/WR-stars/Red Supergiants(RSGs) (by a factor of 10^3 and larger), whereas it is on the order of the mass-loss rates estimated for the giant eruptions from Luminous Blue Variables (LBVs, see Sect. 6).

Global momentum budget – optically thick/thin winds. The dominating terms governing the equation of motion of a massive star wind are the inward directed gravitational pull and the outward directed radiative acceleration, $g_{\rm rad}$†. In spherical symmetry, the latter can be written as

$$g_{\rm rad}(r) = \int_0^\infty d\nu \frac{\kappa_\nu(r)F_\nu(r)}{c} = \kappa_{\rm F}(r)\frac{L_*}{4\pi r^2 c}, \tag{2.3}$$

with frequential flux F_ν, mass absorption coefficient κ_ν, and flux-weighted mass absorption coefficient $\kappa_{\rm F}$. By integrating the equation of motion over $dm = 4\pi r^2 \rho dr$ between the sonic point, $r_{\rm s}$, and infinity (and neglecting pressure terms), one can express the wind-momentum rate, $\dot{M}v_\infty$, in terms of the total momentum rate of the radiation field, L_*/c, and flux mean optical depth of the wind, $\tau_{\rm F}$,

$$\eta = \frac{\dot{M}v_\infty}{L_*/c} \approx \tau_{\rm F}(r_{\rm s}) - \frac{\tau_{\rm e}(r_{\rm s})}{\Gamma_{\rm e}} \tag{2.4}$$

(cf. Abbott 1980). The second term on the rhs is a typically small correction for overcoming the gravitational potential, consisting of the electron-scattering optical depth $\tau_{\rm e}$, and the conventional Eddington-Gamma, $\Gamma_{\rm e} \propto L/M$, evaluated for electron-scattering. Note that this relation is only valid for $\dot{M} \ll \dot{M}_{\rm tir}$.

Because of its definition, η is called the wind performance number. For optically thin winds, defined by $\tau_{\rm F}(r_{\rm s}) < 1$, the performance number is lower than unity, $\eta < 1$, which is typical for OB-stars and A-supergiants, whilst for optically thick winds (e.g., WR-winds) with $\tau_{\rm F}(r_{\rm s}) > 1$ also $\eta > 1$. In a line-driven wind (see Sect. 3), η becomes roughly unity when each photon in the wind is scattered once. $\eta > 1$ then indicates that most photons have been scattered more than once (multi-line scattering).

3. (Optically thin) Line-driven winds

The winds from OB-stars and A-supergiants, with typical mass-loss rates $\dot{M} \approx 10^{-7}\ldots10^{-5}\ M_\odot/{\rm yr}$, and terminal velocities, v_∞, ranging from 200 … 3,500 km s^{-1}, are thought to be accelerated by radiative line-driving.

Photospheric light is scattered/absorbed in spectral lines, and momentum is transferred to the absorbing ions, predominantly into the radial direction. Note that there is no momentum loss or gain during the (re-)emission process, at least in a spherically symmetric configuration, since this process is fore-aft symmetric. Most of the momentum-transfer is accomplished via metallic resonance lines, and this momentum is then further transferred from the accelerated metal-ions (with a low mass-fraction) to the wind bulk plasma, H and He, via Coulomb collisions (e.g., Springmann & Pauldrach 1992).

Since the complete process requires a large number of photons (i.e., a high luminosity), such winds occur in the hottest stars, like O-type stars of all luminosity classes, but

† as long as pressure terms can be neglected, i.e., when $v(r) \gg v_{\rm sound}$ for the largest part of the wind.

also in cooler BA-supergiants, because of their larger radii. Efficient line-driving further requires a large number of spectral lines close to the flux-maximum and a high interaction probability (i.e., a significant line optical depth). Since most spectral lines originate from various metals, a strong dependence of $\dot{M}$ on metallicity is thus to be expected, and such *line-driven* winds should play a minor role (if at all) in the early Universe (but see Sect. 6).

The theory of line-driven winds has been pioneered by Lucy & Solomon (1970) and particularly by Castor *et al.* (1975, 'CAK'), with essential improvements regarding a quantitative description and application provided by Friend & Abbott (1986) and Pauldrach *et al.* (1986). Line-driven winds have been reviewed by Kudritzki & Puls (2000) and more recently by Puls *et al.* (2008).

In the following, we will briefly consider some relevant aspects, mostly in terms of the 'standard model', assuming a steady-state, spherically symmetric, and homogeneous outflow (i.e., neglecting rotation, magnetic fields, and density inhomogeneities, considered later in Sect. 7).

Calculating the radiative acceleration by means of the Sobolev approximation (Sobolev 1960), assuming well-separated lines (justified for most optically thin winds, e.g., Puls 1987), and a distribution of line-strengths following a power-law with exponent $\alpha - 2$ (for details, see Puls *et al.* 2000), the *total* radiative line acceleration from all participating lines can be expressed by

$$g_{\rm rad}(\text{all lines}) \propto \left(\frac{{\rm d}v/{\rm d}r}{\rho}\right)^{\alpha}. \tag{3.1}$$

Particularly because of the dependence on ρ, this leads to a self-regulation of the mass-loss rate, and an analytic solution of the equation of motion is possible. In compact notation, and neglecting the effects of an ionization stratification† (typically weak for O stars), one finds the following scaling laws

$$\dot{M} \approx \frac{L_*}{c^2}\frac{\alpha}{1-\alpha}\left(\frac{\bar{Q}\Gamma_{\rm e}}{1-\Gamma_{\rm e}}\right)^{1/\alpha-1}\frac{1}{(1+\alpha)^{1/\alpha}} \tag{3.2}$$

(Owocki 2004, and references therein),

$$v(r) = v_\infty\left(1 - \frac{R_*}{r}\right)^{\beta} \qquad v_\infty \approx 2.25\frac{\alpha}{1-\alpha}v'_{\rm esc} \tag{3.3}$$

(Kudritzki *et al.* 1989), where $v'_{\rm esc} = v_{\rm esc}\sqrt{1-\Gamma_{\rm e}}$ is the effective escape velocity corrected for Thomson acceleration. For typical O-stars, one has the Gayley's (1995) dimensionless line-strength parameter $\bar{Q} \approx 2000$, $\alpha \approx 0.6$, $\beta \approx 0.8$, and $\dot{M}$ is on the order of $10^{-6}\,M_\odot{\rm yr}^{-1} \ll \dot{M}_{\rm tir}$. Note that $\bar{Q}$ scales with metallicity, $\bar{Q} \propto Z/Z_\odot$, such that $\dot{M} \propto (Z/Z_\odot)^{0.7}$ for the above α.

For quantitative results, the most frequently used theoretical mass-loss rates are based on the wind models by Vink *et al.* (2000, 2001), calculated by means of approximate non-local thermodynamical equilibrium (NLTE) occupation numbers and a Monte Carlo transport (i.e., without invoking any line statistics). From interpolating the mass-loss rates derived in this way for a large model grid, the provided 'mass-loss recipe' becomes $\dot{M} = \dot{M}(L_*, M, T_{\rm eff}, v_\infty/v_{\rm esc}, Z)$, with a similar metallicity dependence as above. For alternative models and calculation methods, see, e.g., Krtička & Kubát (2000), Pauldrach *et al.* (2001), and Kudritzki (2002).

† corresponding to a 'force-multiplier' parameter $\delta = 0$, see Abbott (1982)

The wind-momentum luminosity relation (WLR). By using the scaling relations for $\dot{M}$ and v_∞ (Eqs. 3.2, 3.3), and approximating $\alpha \approx 2/3$, one obtains the so-called wind-momentum luminosity relation (WLR; Kudritzki *et al.* 1995),

$$\log D_{\mathrm{mom}} = \log\left(\dot{M}v_\infty \left(\frac{R}{R_\odot}\right)^{\frac{1}{2}}\right) \approx \frac{1}{\alpha}\log\left(\frac{L_*}{L_\odot}\right) + \mathrm{offset}(Z, \text{spectral type}), \qquad (3.4)$$

which relates the *modified* wind-momentum rate D_{mom} with only the stellar luminosity. The mass-dependence (due to Γ_e) becomes negligible as long as α is close to $2/3$. The offset in Eq. 3.4 depends on metallicity and spectral type, mostly because the effective number of driving lines and thus $\bar{Q}$ depend on these quantities (e.g., Puls *et al.* 2000), via different opacities and contributing ions.

Though derived from simplified scaling relations, the WLR concept has also been confirmed by numerical model calculations, e.g., those from Vink *et al.* (2000, their Fig. 9).

An impressive *observational* confirmation of this concept has been provided by Mokiem *et al.* (2007), compiling observed stellar and wind parameters from Galactic, LMC and SMC O-stars, and analyzing the corresponding WLRs. Accounting for wind-inhomogeneities (see Sect. 7.3) in an approximate way, they derive $\dot{M} \propto (Z/Z_\odot)^{0.72\pm0.15}$, in very good agreement with theoretical predictions.

4. (Optically thick) Winds from WR- and Very Massive Stars†

From early on, the mass-loss rates of Wolf-Rayet stars posed a serious problem for theoretical explanations, since they are considerably larger (by a factor of ten and more) compared to mass-loss rates from O-stars of similar luminosity. Though Lucy & Abbott (1993) showed that line-overlap effects, coupled with a significantly stratified ionization balance, can help a lot to increase the mass-loss, it was Gräfener & Hamann (2005, 2006, 2007, 2008) who were the first to calculate consistent WR-wind models with the observed large mass-loss rates in parallel with high terminal velocities (2000 - 3000 $\mathrm{km\,s^{-1}}$). They showed that a high Eddington-Γ is necessary to provide a low effective gravity and to enable a deep-seated sonic point at high temperatures. Then, a high mass-loss rate leading to an *optically thick* wind can be initiated either by the 'hot' Fe-opacity bump (around 160 kK, for the case of WCs and WNEs) or the cooler one (around 40 to 70 kK, for the case of WNLs)‡. The high initiated mass-loss rates can then be further accelerated by efficient multi-line scattering in a stratified ionization balance (see above), at least if the outer wind is significantly clumped.

Alternative wind models for Very Massive Stars in the range of 40 $M_\odot < M < 300\ M_\odot$ (i.e., including models which should display WR spectral characteristics) have been constructed by Vink *et al.* (2011) (but see also Pauldrach *et al.* 2012), who argue that for $\Gamma_e > 0.7$ these line-driven winds become optically thick already at the sonic point, which enables a high $\dot{M}$ with a steeper dependence on Γ_e than for optically thin winds¶. Recently, Bestenlehner *et al.* (2014) investigated the mass-loss properties of a sample of 62 O, Of, Of/WN, and WNh stars within the Tarantula nebula, observed within the VLT FLAMES Tarantula Survey (Evans *et al.* 2011) and other campaigns. Indeed, they found a change in the slope of $\mathrm{d}\log\dot{M}/\mathrm{d}\log\Gamma_e$ towards higher values. However, this change

† see also Gräfener, this Volume
‡ The importance of these opacity bumps had already been pointed out by Nugis & Lamers (2002).
¶ The actual origin of this behavior is still unclear, but the authors of this review speculate about a higher efficiency of multi-line effects.

occurs already at $\Gamma_e = 0.25$, i.e., (much) earlier than predicted by Vink *et al.* (2011), and more consistent with the models by Gräfener & Hamann (2008). Moreover, at least for Of and Of/WN stars there is still the possibility that the conventional CAK theory (which already includes a tight dependence on Γ_e, cf. Eq. 3.2) remains applicable, though with a lower α (0.53 instead of 0.63) than for typical O-stars. Thus, a number of issues still need to be worked out before these optically thick winds are fully understood.

5. Winds from Red Supergiants

Typical mass-loss rates from RSGs† range from 10^{-5} to 10^{-4} $M_\odot \mathrm{yr}^{-1}$, with terminal velocities on the order of 20 to 30 km s^{-1} ($\approx v_{esc}/3$). Note that the atmospheres of RSGs consist of giant convective cells, with diameters scaling with the vertical pressure scale height (Stein & Nordlund 1998; Nordlund *et al.* 2009). Though the physics of RSG winds is still unknown, a similarity to the dust-driven winds from (carbon-rich = C-type) AGB-stars is often hypothesized‡. In these stars, stellar (p-mode) pulsations or large scale convective motions lead to the formation of outward propagating shock waves, that lift the gas above the stellar surface, intermittently creating dense, cool layers where dust may form. Similar to line-driven winds, these dust grains are radiatively accelerated, and drag the gas via collisions (for a review, see Höfner 2009).

Josselin & Plez (2007) alluded to some problems of this scenario when applied to RSG-winds. Namely, RSGs have only irregular, small-amplitude variations, which makes lifting the gas difficult in the first place, and moreover the dust seems to form much further out than in AGB-stars. On the other hand, they also pointed out that turbulent pressure related to convection helps in lowering the *effective* gravity¶ and thus the effective escape velocity, $g_{eff} \approx g/(1 + \mu v_{turb}^2/(2k_B T))$, with μ the mean molecular weight, and suggested that radiative acceleration provided by molecular lines might help lift the material to radii where dust can form, though without any quantitative estimate. To conclude, further investigations and simulations to explain RSG-winds are urgently needed.

6. Continuum-driven winds

As a prelude to the following scenario, let us check in how far a hot stellar wind can be also driven by pure continuum processes, with major opacities due to bound-free absorption and Thomson scattering.

The simple picture. Since these opacities (per volume) scale mostly with linear density, the corresponding mass absorption coefficients (frequential and flux-weighted, see Eq. 2.3) do not display any explicit density dependence. Consequently, the *total* Eddington-Gamma,

$$\Gamma_{tot}(r) = \frac{g_{rad}(r)}{g_{grav}(r)} = \frac{\kappa_F(r)L}{4\pi c G M} \rightarrow \Gamma_{cont}(r) \tag{6.1}$$

is density-independent as well (contrasted, e.g., to the case of line-driving), and it seems that basically *any* $\dot{M}$ might be accelerated∥ as long as $\Gamma_{cont}(r)$ increases through the sonic point, with $\Gamma_{cont}(r_s) = 1$, and remains beyond unity above.

---

† for structure and stellar parameters, see Wittkowski, this Volume

‡ According to Höfner (2008), dust-driving might be also possible in oxygen-rich (M-type) AGB-atmospheres, if prevailing conditions allow forsterite grains (Fe-free olivine-type, $Mg_2 SiO_4$) to grow to sizes in the micro-meter range.

¶ As a side note, we might ask whether the well-known mass-discrepancy for O-type dwarfs might be related to the neglect of a potentially large turbulent pressure in present atmospheric models (see Markova & Puls, this Volume).

∥ since the equation of motion does no longer depend on ρ

As pointed out by Owocki & Gayley (1997), however, photon tiring (Sect. 2) decreases the available luminosity, $L(r) < L_*$, and thus the mass-loss rate is still restricted by $\dot{M} \leqslant \dot{M}_{\text{tir}}$. Moreover, the complete process requires a substantial fine-tuning to reach and *maintain* $\Gamma_{\text{cont}} \geqslant 1$ in (super-) sonic regions, and such a 'simple' continuum-driving is rather difficult to realize.

Super-Eddington winds moderated by porosity. Whilst, during 'quiet' phases, LBVs lose mass most likely via ordinary line-driving (cf. Sect. 3), they are also subject to one or more phases of much stronger mass loss. E.g., the giant eruption of η Car with a cumulative loss of $\sim 10~M_\odot$ between 1840 and 1860 (Smith *et al.* 2003) corresponds to $\dot{M} \approx 0.1$-$0.5~M_\odot \text{yr}^{-1}$, which is a factor of 1000 larger than that expected from a line-driven wind at that luminosity. Such strong mass loss has been frequently attributed to a star approaching or even exceeding the Eddington limit.

Building upon pioneering work by Shaviv (1998, 2000, 2001b), Owocki *et al.* (2004) developed a theory of "porosity-moderated" continuum driving in such stars, where the dominating acceleration is still due to continuum-driving, mostly due to electron scattering, i.e., $\Gamma_{\text{tot}} \rightarrow \Gamma_{\text{cont}} \approx \Gamma_{\text{e}} > 1$.

For stars near or (formally) above the Eddington limit, non-radial instabilities will inevitably arise and make their atmospheres inhomogeneous (clumpy), see, e.g., Shaviv (2001a). As noted by Shaviv (1998), the *porosity* of such a structured medium can reduce the radiation acceleration significantly (photons 'avoid' regions of enhanced density), by lowering the *effective* opacity in deeper layers, $\Gamma_{\text{cont}}^{\text{eff}} < 1$ for $r < r_{\text{s}}$, thus enabling a quasi-hydrostatic photosphere, but allowing for a transition to a supersonic outflow when the over-dense regions become optically thin due to expansion, $\Gamma_{\text{cont}}^{\text{eff}} \rightarrow \Gamma_{\text{cont}} > 1$ for $r > r_{\text{s}}$.

The effective opacity in a porous medium consisting of an ensemble of clumps can be derived from rather simple arguments (Owocki *et al.* 2004), but here it is sufficient to note that for *optically thick clumps* and ρ-dependent opacities

$$\kappa_{\text{F}}^{\text{eff}}(r) = \frac{1}{h\langle\rho\rangle(r)} \ll \kappa_{\text{F}}(r), \qquad (6.2)$$

the *effective* opacity (here: the effective mass absorption coefficient) becomes grey and much smaller than the original one. In this equation, $\langle\rho\rangle(r)$ is the *mean* density of the medium, and h the so-called porosity length, which is the photon's mean free path for a medium consisting of optically thick clumps. Thus, $\kappa_{\text{F}}^{\text{eff}}(r)$ and consequently $\Gamma_{\text{cont}}^{\text{eff}}(r)$ have a specific density dependence around the sonic point ($\propto \langle\rho\rangle^{-1}$), and there is a corresponding, well-defined $\dot{M}$ which can be initiated and accelerated.

If one now considers clumps with a range of optical depths, distributed according to an exponentially truncated power-law with index α_{p} (Owocki *et al.* 2004), one obtains for sound speed a and pressure scale height H,

$$\dot{M}(\alpha_{\text{p}} = 2) = \left(1 - \frac{1}{\Gamma_{\text{cont}}}\right) \frac{H}{h} \frac{L_*}{ac} = \frac{\dot{M}(\alpha_{\text{p}} = 1/2)}{4\Gamma_{\text{cont}}}, \qquad (6.3)$$

where the mass-loss rate for a canonical $\alpha_{\text{p}} = 2$ model (obtained from a clump-ensemble that follows Markovian statistics, Sundqvist *et al.* 2012a; Owocki 2014) saturates for very high Γ_{cont}, but where the alternative $\alpha_{\text{p}} = 1/2$ model can give an even higher mass loss for such cases, approaching the tiring limit under certain circumstances, and being on order the mass loss implied by the ejecta of η Car for an assumed $h \approx H$ (Owocki *et al.* 2004; Owocki 2014). Detailed simulations are needed here to further constrain the clump-distribution function and the porosity length in these models.

Nonetheless, together with quite fast outflow speeds, $v_\infty \approx \mathcal{O}(v_{\text{esc}})$, and a velocity law corresponding to $\beta = 1$, the derived wind structure in such a porosity-moderated wind model may actually explain the observational constraints of giant outbursts in η Car and other LBVs. Moreover, the porosity model retains the essential scalings with gravity and radiative flux (the von Zeipel theorem, cf. Sect. 7.1) that would give a rapidly rotating, gravity-darkened star an enhanced polar mass loss and flow speed, similar to the bipolar Homunculus nebula. Note that continuum driving (if mostly due to Thomson scattering) does *not* require the presence of metals in the stellar atmosphere. *Thus, it is well-suited as a driving agent in the winds of low-metallicity and First Stars, and may play a crucial role in their evolution.*

7. Additional physics in line-driven winds

In this last section we now return to line-driven winds, and discuss specific conditions and effects which might influence their appearance.

7.1. *Rapid rotation*

When stars rotate rapidly, their photospheres become oblate (because of the centrifugal forces, see Collins 1963; Collins & Harrington 1966), the effective temperature decreases from pole towards equator ('gravity darkening', von Zeipel 1924; Maeder 1999), and the winds from typical O-stars are predicted to become *prolate* (because of the larger illuminating polar fluxes), with a fast and dense polar outflow, and a slow and thinner equatorial one (Cranmer & Owocki 1995)†.

Whilst the basic effects of stellar oblateness and gravity darkening have been confirmed by means of interferometry (Domiciano de Souza *et al.* 2003; Monnier *et al.* 2007, see also van Belle, Meilland, Faes, this Volume), the predictions on the wind-structure of rapidly rotating stars have *not* been verified by observations so far (Puls *et al.* 2011 and references therein): first, only few stars in phases with extreme rotation are known (but they exist, e.g., Dufton *et al.* 2011), and second, the tools to analyze the atmospheres and winds (multi-D models!) of such stars are rare. Even though Vink *et al.* (2009) (see also Vink, this Volume) reported, analyzing data obtained by means of linear H_α spectro-polarimetry, that most winds from rapidly rotating O-stars are spherically symmetric (actually, they looked for disks), the asymmetry predicted for the winds of this specific sample should be rather low anyway.

Besides the polar-angle dependence of $\dot{M}$ induced by rotation, also the global mass-loss rate becomes modified; a significant increase, however, is only found for rapid rotation *and* a large Γ_e, with a *formal* divergence of $\dot{M}$ – which at least needs to be corrected for photon tiring effects – at the so-called $\Omega\Gamma$-limit. For details, see Maeder & Meynet (2000).

Finally, for near-critically rotating stars, mass loss might also occur via decretion disks (Krtička *et al.* 2011). The corresponding $\dot{M}$ from such decretion disks can be significantly *less* than the spherical, wind-like mass loss (aka 'mechanical winds') previously assumed in evolutionary calculations.

7.2. *Magnetic fields*

Recent spectropolarimetric surveys, mostly performed by the international collaboration Magnetism in Massive Stars (MiMeS; e.g., Wade *et al.* 2012), and work done by S. Hubrig and collaborators (e.g., Hubrig *et al.* 2013; see also Grunhut, Morrell, this Volume) have

---

† All these effects become significant if the rotational speed exceeds roughly 70% of the critical one. Note also that cooler winds might retain an oblate structure, if the ionization balance decreases strongly from pole to equator, and the effective number of driving line increases in parallel.

revealed that roughly 10% of all massive stars have a large-scale, organized magnetic field in their outer stellar layers, on the order of a couple of hundred to several thousand Gauss. The origin of these fields is still unknown, though most evidence points to quite stable fossil fields formed sometimes during early phases of stellar formation (Alecian *et al.* 2013). The interaction of these fields with a line-driven stellar wind has been investigated by ud-Doula, Owocki and co-workers in a series of publications (summarized in ud-Doula 2013, see also ud-Doula, this Volume). In the following, we concentrate on *slowly* rotating magnetic O-stars (spectroscopically classified as O f?p, Walborn 1972), which give rise to so-called 'dynamic magnetospheres' (Sundqvist *et al.* 2012b).

The most important quantity to estimate the influence of a magnetic field on the wind is the ratio of magnetic to wind energy,

$$\eta = \frac{B^2/8\pi}{\rho v^2/2} = \frac{B_*^2 R_*^2}{\dot{M} v_\infty^2} f(r) := \eta_* f(r), \tag{7.1}$$

with η_* the so-called confinement parameter. E.g., for a typical O-supergiant, a B-field of 300 Gauss is required to reach $\eta_* = 1$ (for $\eta_* < 1$ the wind is not much disturbed). In the case of an η_* significantly greater than unity, to a good approximation the corresponding Alvén radius (which is the maximum radius for closed loops, and determines whether the wind is confined in such loops), can be expressed by $R_A \approx R_* \eta_*^{1/4}$.

An instructive example for a strongly confined wind can be found in, e.g., Sundqvist *et al.* (2012b), who performed hydro-dynamical simulations and Hα radiative transfer calculations for the prototypical Of?p star HD 191612, with a B-field of $\approx 2,500$ Gauss corresponding to $\eta_* = 50$. In this model, the field loops are closed near the equatorial plane ($R_A \approx 2.7 R_*$), and the confined wind is accelerated and channeled upwards from foot-points of opposite polarity. The flows collide near the loop tops, forming strong shocks with hard X-ray emission (Gagné *et al.* 2005). The shocked, very dense material then cools and becomes accelerated *inwards* by the gravitational pull, emitting strongly in, e.g., the optical Hα line. Whilst there are complex infall patterns along the loop lines, the field lines in polar regions are still open and the polar wind remains almost undisturbed. The *infalling* material of dynamical magnetospheres reduces the global mass-loss rates, for large η_* by a factor ~ 5 compared to non-magnetic winds (ud-Doula *et al.* 2008, analytic scaling relations available). Note also that the non-spherical structures require well-suited diagnostic methods.

7.3. *Inhomogeneous winds – a few comments*

We finish this review with few comments about the presence and impact of small-scale wind inhomogeneities (for a detailed discussion and references, also regarding large-scale inhomogeneities, see, e.g., Puls *et al.* 2008).

Over the last two decades, a multitude of direct and indirect indications has been accumulated that hot star winds are inhomogeneous on small scales, i.e., consist of over-dense (compared to the mean-density) clumps and an inter-clump material which is frequently assumed to be void (but see Šurlan *et al.* 2013; Sundqvist *et al.* 2014).

The most likely origin is the line-driven (or line-deshadowing) instability, which, for short-wavelength perturbations, can be summarized by $\delta g_{\rm rad}^{\rm lines} \propto \delta v$, giving rise to strong, outward propagating reverse shocks emitting in the X-ray regime, and a wind-structure consisting of fast and thin material (inter-clump matter), and dense, spatially narrow clumps moving roughly at the speed of smooth-wind models.

In dependence of the considered absorption process and wavelength, clumps can be optically thin ('micro-clumping') or optically thick ('macro-clumping'/ porosity, see Sect. 6). Moreover, line processes are prone to porosity in velocity space. To account for the effects of wind-inhomogeneities, a simplified treatment is typically employed (both within radiative transfer and when calculating the occupation numbers), based on a one- or two-component description, with parameterized clumping properties such as volume-filling factors, over-densities, etc.

The major impact of these inhomogeneities regards the various mass-loss diagnostics. If clumps are optically thin for ρ^2-dependent opacities (e.g., H_α and IR diagnostics in O-stars), the actual rates turn out to be lower than derived from smooth-wind models. If clumps are optically thick and/or velocity porosity needs to be accounted for (e.g., UV-resonance lines), the final rates are larger than derived from models assuming optically thin clumps alone. On the other hand, mass-loss diagnostics based on bound-free (bf) absorption (by the cool wind) of X-ray line emission from the above wind-embedded shocks is particularly robust, since it remains uncontaminated by inhomogeneities in typical O-star winds: first, there is no direct effect from micro-clumping, because the involved bf-opacities (per volume) scale with ρ, and second, porosity effects are negligible or low (e.g., Cohen *et al.* 2010, 2013; Leutenegger *et al.* 2013; Hervé *et al.* 2013).

Comparing now 'observed' O-star mass-loss rates with theoretical ones (from Vink *et al.* 2000) used in stellar evolution, there is the following *status quo*†. These theoretical mass-loss rates are (i) a factor of 2-3 *lower* than those from standard H_α diagnostics assuming a smooth wind; (ii) roughly consistent with radio mass-loss rates assuming a smooth wind; (iii) a factor of 2-3 *larger* than recent diagnostics of Galactic O-stars accounting adequately for wind inhomogeneities (Najarro *et al.* 2011: mostly IR-lines; Cohen *et al.* 2014: absorbed X-ray line emission; Sundqvist *et al.* 2011, Šurlan *et al.* 2013, Sundqvist *et al.* 2014: UV-lines including velocity porosity+optical lines).

Of course, further investigations and larger samples are certainly required to prove this discrepancy, but particularly the X-ray results are a strong argument. Indeed, there are various possibilities for a potential overestimate of theoretical mass-loss rates, summarized by Sundqvist (2013). A rather promising explanation relates to (so far neglected) effects from velocity porosity when calculating the line force, which can lead to reduced theoretical mass-loss rates if already present in the lower wind (Sundqvist *et al.* 2014, see also Muijres *et al.* 2011).

In summary, there is still much to do, and the physics of massive star winds remains a fascinating topic!

Acknowledgements

JOS and JP gratefully acknowledge support by the German DFG, under grant PU117/8-1. A travel grant by the University of Geneva is gratefully acknowledged by NM and JP.

References

Abbott, D. C. 1980, *ApJ* 242, 1183
Abbott, D. C. 1982, *ApJ* 259, 282
Alecian, E., Wade, G. A., Catala, C., *et al.* 2013, *MNRAS* 429, 1001
Bestenlehner, J. M., Gräfener, G., Vink, J. S., *et al.* 2014, *ArXiv:1407.1837*
Castor, J. I., Abbott, D. C., & Klein, R. I. 1975, *ApJ* 195, 157
Cohen, D., Sundqvist, J., & Leutenegger, M. 2013, in *Massive Stars: From alpha to Omega*
Cohen, D. H., Leutenegger, M. A., Wollman, E. E., *et al.* 2010, *MNRAS* 405, 2391
Cohen, D. H., Wollman, E. E., Leutenegger, M. A., *et al.* 2014, *MNRAS* 439, 908

† to, e.g., clarify corresponding statements in the recent review by Smith (2014) that are somewhat simplified in this respect.

Collins, II, G. W. 1963, *ApJ* 138, 1134

Collins, II, G. W. & Harrington, J. P. 1966, *ApJ* 146, 152

Cranmer, S. R. & Owocki, S. P. 1995, *ApJ* 440, 308

Domiciano de Souza, A., Kervella, P., Jankov, S., *et al.* 2003, *A&A* 407, L47

Dufton, P. L., Dunstall, P. R., Evans, C. J., *et al.* 2011, *ApJ (Letters)* 743, L22

Evans, C. J., Taylor, W. D., Hénault-Brunet, V., *et al.* 2011, *A&A* 530, A108

Friend, D. B. & Abbott, D. C. 1986, *ApJ* 311, 701

Gagné, M., Oksala, M. E., Cohen, D. H., *et al.* 2005, *ApJ* 628, 986

Gayley, K. G. 1995, *ApJ* 454, 410

Gräfener, G. & Hamann, W.-R. 2005, *A&A* 432, 633

Gräfener, G. & Hamann, W.-R. 2006, in H. J. G. L. M. Lamers, N. Langer, T. Nugis, & K. Annuk (eds.), *Stellar Evolution at Low Metallicity: Mass Loss, Explosions, Cosmology*, Vol. 353 of *Astronomical Society of the Pacific Conference Series*, p. 171

Gräfener, G. & Hamann, W.-R. 2007, in N. St.-Louis & A. F. J. Moffat (eds.), *Massive Stars in Interactive Binaries*, Vol. 367 of *Astronomical Society of the Pacific Conference Series*, p. 131

Gräfener, G. & Hamann, W.-R. 2008, *A&A* 482, 945

Hervé, A., Rauw, G., & Nazé, Y. 2013, *A&A* 551, A83

Höfner, S. 2008, *A&A* 491, L1

Höfner, S. 2009, in T. Henning, E. Grün, & J. Steinacker (eds.), *Cosmic Dust - Near and Far*, Vol. 414 of *Astronomical Society of the Pacific Conference Series*, p. 3

Hubrig, S., Schöller, M., Ilyin, I., *et al.* 2013, *A&A* 551, A33

Josselin, E. & Plez, B. 2007, *A&A* 469, 671

Krtička, J. & Kubát, J. 2000, *A&A* 359, 983

Krtička, J., Owocki, S. P., & Meynet, G. 2011, *A&A* 527, A84

Kudritzki, R.-P. 2002, *ApJ* 577, 389

Kudritzki, R.-P., Lennon, D. J., & Puls, J. 1995, in J. R. Walsh & I. J. Danziger (eds.), *Science with the VLT*, p. 246

Kudritzki, R.-P., Pauldrach, A., Puls, J., & Abbott, D. C. 1989, *A&A* 219, 205

Kudritzki, R.-P. & Puls, J. 2000, *ARA&A* 38, 613

Leutenegger, M. A., Cohen, D. H., Sundqvist, J. O., & Owocki, S. P. 2013, *ApJ* 770, 80

Lucy, L. B. & Abbott, D. C. 1993, *ApJ* 405, 738

Lucy, L. B. & Solomon, P. M. 1970, *ApJ* 159, 879

Maeder, A. 1999, *A&A* 347, 185

Maeder, A. & Meynet, G. 2000, *ARA&A* 38, 143

Meynet, G., Maeder, A., Schaller, G., Schaerer, D., & Charbonnel, C. 1994, *A & AS* 103, 97

Mokiem, M. R., de Koter, A., Vink, J. S., *et al.* 2007, *A&A* 473, 603

Monnier, J. D., Zhao, M., Pedretti, E., *et al.* 2007, *Science* 317, 342

Muijres, L. E., de Koter, A., Vink, J. S., *et al.* 2011, *A&A* 526, A32

Najarro, F., Hanson, M. M., & Puls, J. 2011, *A&A* 535, A32

Nordlund, Å., Stein, R. F., & Asplund, M. 2009, *Living Reviews in Solar Physics* 6, 2

Nugis, T. & Lamers, H. J. G. L. M. 2002, *A&A* 389, 162

Owocki, S. 2004, in M. Heydari-Malayeri, P. Stee, & J.-P. Zahn (eds.), *EAS Publications Series*, Vol. 13 of *EAS Publications Series*, pp 163–250

Owocki, S. P. 2014, *ArXiv:1403.6745*

Owocki, S. P. & Gayley, K. G. 1997, in A. Nota & H. Lamers (eds.), *Luminous Blue Variables: Massive Stars in Transition*, Vol. 120 of *Astronomical Society of the Pacific Conference Series*, p. 121

Owocki, S. P., Gayley, K. G., & Shaviv, N. J. 2004, *ApJ* 616, 525

Pauldrach, A., Puls, J., & Kudritzki, R. P. 1986, *A&A* 164, 86

Pauldrach, A. W. A., Hoffmann, T. L., & Lennon, M. 2001, *A&A* 375, 161

Pauldrach, A. W. A., Vanbeveren, D., & Hoffmann, T. L. 2012, *A&A* 538, A75

Puls, J. 1987, *A&A* 184, 227

Puls, J., Springmann, U., & Lennon, M. 2000, *A & AS* 141, 23

Puls, J., Sundqvist, J. O., & Rivero González, J. G. 2011, in C. Neiner, G. Wade, G. Meynet, & G. Peters (eds.), *IAU Symposium*, Vol. 272 of *IAU Symposium*, pp 554–565

Puls, J., Vink, J. S., & Najarro, F. 2008, *A&A Rev.* 16, 209

Shaviv, N. J. 1998, *ApJ (Letters)* 494, L193

Shaviv, N. J. 2000, *ApJ (Letters)* 532, L137

Shaviv, N. J. 2001a, *ApJ* 549, 1093

Shaviv, N. J. 2001b, *MNRAS* 326, 126

Smith, N. 2014, *ArXiv:1402.1237*

Smith, N., Gehrz, R. D., Hinz, P. M., *et al.* 2003, *AJ* 125, 1458

Sobolev, V. V. 1960, *Moving envelopes of stars*, Cambridge: Harvard University Press, 1960

Springmann, U. W. E. & Pauldrach, A. W. A. 1992, *A&A* 262, 515

Stein, R. F. & Nordlund, Å. 1998, *ApJ* 499, 914

Sundqvist, J. O. 2013, in *Massive Stars: From alpha to Omega*

Sundqvist, J. O., Owocki, S. P., Cohen, D. H., Leutenegger, M. A., & Townsend, R. H. D. 2012a, *MNRAS* 420, 1553

Sundqvist, J. O., Puls, J., Feldmeier, A., & Owocki, S. P. 2011, *A&A* 528, A64

Sundqvist, J. O., Puls, J., & Owocki, S. P. 2014, *ArXiv:1405.7800*

Sundqvist, J. O., ud-Doula, A., Owocki, S. P., *et al.* 2012b, *MNRAS* 423, L21

ud-Doula, A. 2013, in J.-P. Rozelot & C. . Neiner (eds.), *Lecture Notes in Physics, Berlin Springer Verlag*, Vol. 857 of *Lecture Notes in Physics, Berlin Springer Verlag*, p. 207

ud-Doula, A., Owocki, S. P., & Townsend, R. H. D. 2008, *MNRAS* 385, 97

Šurlan, B., Hamann, W.-R., Aret, A., *et al.* 2013, *A&A* 559, A130

Vink, J. S., Davies, B., Harries, T. J., Oudmaijer, R. D., & Walborn, N. R. 2009, *A&A* 505, 743

Vink, J. S., de Koter, A., & Lamers, H. J. G. L. M. 2000, *A&A* 362, 295

Vink, J. S., de Koter, A., & Lamers, H. J. G. L. M. 2001, *A&A* 369, 574

Vink, J. S., Muijres, L. E., Anthonisse, B., *et al.* 2011, *A&A* 531, A132

von Zeipel, H. 1924, *MNRAS* 84, 665

Wade, G. A., Grunhut, J. H., & MiMeS Collaboration 2012, in A. C. Carciofi & T. Rivinius (eds.), *Circumstellar Dynamics at High Resolution*, Vol. 464 of *Astronomical Society of the Pacific Conference Series*, p. 405

Walborn, N. R. 1972, *AJ* 77, 312

Discussion

DE KOTER: We find that the line-driven winds of O stars at metallicities below that of the SMC do not seem to obey the theory of line driving: the mass-loss rates are higher (Tramper *et al.* 2011). What are your ideas on an explanation of this peculiar behavior?

PULS: Actually, this is not completely clear. As will be shown in the next talk by M. García, there might be a bias on $\dot{M}$ due to variations in the ratio of v_∞/v_{esc}, and in this particular case the iron abundance (driving agent) might be higher than implied by the oxygen abundance.

DE KOTER: The winds of RSG are very difficult to understand, so currently we focus on understanding AGB winds. Though ideas have been put forward to explain the O-rich outflows, in my view there are still fundamental problems. These winds can only be driven through scattering on large ($0.3\,\mu$m) grains and it is not clear at present how to grow such large grains in the warm molecular layer.

PULS: Completely agreed.

KHALAK: Can you explain the reasons for the excitation of atoms that increases line opacity and causes optically thick wind having $\Gamma_e > 0.7$ in WR stars?

PULS: 1. Lines become more easily optically thick because of the higher density. Line overlap effects are particularly effective for optically thick lines. 2. Due to the higher wind density, the ionization/excitation couples closer to the local electron temperature. Thus there is an ionization stratification (which again allows for efficient line overlap), and the occupation numbers of the excited levels increases as well (closer to Boltzmann).

WEIS: You showed how the giant eruption could be explained. Can that model also explain the bipolar structure?

PULS: Indeed, the radiative acceleration due to porosity-moderated continuum driving has a similar dependence on polar angle as in line-driven winds, for rapidly rotating stars/winds. Thus, also here a prolate structure is expected.

NOELS: Would it be possible to have a table with simple formulas of the mass-loss rates across the upper part of HR diagram, with the uncertainties (even large!)? This could appear "as of today" in this proceeding.

PULS: I will try my best, but only scaling relations can be provided in some cases (e.g., excluding RSGs). The absolute numbers of mass-loss rates are heavily debated, e.g. due to the impact of wind inhomogeneities. *Added after review has been finished:* Actually, such a table could not be provided, given the limited space and the state of our knowledge. Anyhow, all relevant references have been provided, but there is still strong disagreement on the uncertainties. And since factors of two *are* important, quoting disputed values with large error bars is not meaningful.

Joachim Puls

New windows on massive stars: asteroseismology, interferometry, and spectropolarimetry
Proceedings IAU Symposium No. 307, 2014
G. Meynet, C. Georgy, J. H. Groh & Ph. Stee, eds.

© International Astronomical Union 2015
doi:10.1017/S1743921314006231

A binary progenitor for the Type Ib Supernova iPTF13bvn

Melina C. Bersten

Kavli Institute for the Physics and Mathematics of the Universe (WPI), The University of Tokyo, 5-1-5 Kashiwanoha, Kashiwa, Chiba 277-8583, Japan

Abstract. The recent detection in archival HST images of an object at the the location of supernova (SN) iPTF13bvn may represent the first direct evidence of the progenitor of a Type Ib SN. The object's photometry was found to be compatible with a Wolf-Rayet pre-SN star mass of $\approx 11\,M_\odot$. However, based on hydrodynamical models we show that the progenitor had a pre-SN mass of $\approx 3.5\,M_\odot$ and that it could not be larger than $\approx 8\,M_\odot$. We propose an interacting binary system as the SN progenitor and perform evolutionary calculations that are able to self-consistently explain the light-curve shape, the absence of hydrogen, and the pre-SN photometry. Our models also predict that the remaining companion is a luminous O-type star of significantly lower flux in the optical than the pre-SN object. A future detection of such star may be possible and would provide the first robust progenitor identification for a Type-Ib SN.

Keywords. stars: evolution, hydrodynamics, supernovae: general, supernovae: individual (iPTF13bvn)

1. Introduction

Determining the nature of the progenitor of core-collapse supernovae (SNe) is a crucial problem in astrophysics. For hydrogen-deficient SN (Types Ib and Ic) this is particularly controversial, and single massive stars and close binary systems are the most appealing alternatives. The search for progenitor stars in deep pre-explosion images is a powerful, direct approach, although it so far yielded no firm detection for Type Ib or Ic SN (Yoon *et al.* 2012; Groh *et al.* 2013b). The recent identification of the Type Ib SN iPTF13bvn with a luminous, blue object may represent the first such case (Cao *et al.* 2013) and it led to the suggestion that the progenitor, assuming it to be single, was a Wolf-Rayet star with an initial mass of $31 - 35\,M_\odot$ (Groh *et al.* 2013a).

Hydrodynamical modeling of SN observations is an alternative method to infer progenitor properties. This methodology is particularly powerful when combined with stellar evolution calculations. A recent example of the predictability of this technique can be seen in our analysis of the Type IIb SN 2011dh (Bersten *et al.* 2012; Benvenuto *et al.* 2013), which allowed us to provide a self-consistent explanation of the progenitor nature that was later confirmed (Van Dyk *et al.* 2013; Ergon *et al.* 2014). Here we use the same approach to address the problem of the progenitor of iPTF13bvn.

2. Hydrodynamical modeling

A set of explosion models was calculated using our one-dimensional, local thermodynamical equilibrium, radiation hydrodynamics code (Bersten *et al.* 2011). Helium stars of different masses were adopted as initial structures. Specifically, we tested models with $3.3\,M_\odot$ (HE3.3), $4\,M_\odot$ (HE4), $5\,M_\odot$ (HE5), and $8\,M_\odot$ (HE8), which correspond to main-sequence masses of 12, 15, 18, and $25\,M_\odot$, respectively (Nomoto & Hashimoto 1988).

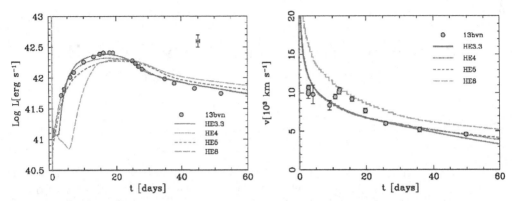

Figure 1. Hydrodynamical modeling of iPTF13bvn. Bolometric light curve (*left panel*) and photospheric velocity evolution (*right panel*) are compared with observations (dots). Models with different masses are shown with different line types and colors. HE3.3 and HE4 give a good representation of the observations but a slightly more massive object, HE5, provides a worse comparison. A model with $8 M_\odot$ (HE8) is clearly not acceptable. The error bars at the top of the figure indicate the nearly constant uncertainty in luminosity and the adopted uncertainty in the explosion time.

Figure 1 shows the results for these models. While HE3.3 and HE4 can reproduce reasonably well the observations (light curve and velocities), a slightly more massive model, HE5, already shows a worse agreement. We adopted as preferred model the one with intermediate parameters between HE3.3 and HE4, i.e. a low-mass He star of $\approx 3.5 M_\odot$, with an ejected mass of $M_{\rm ej} \approx 2.3 M_\odot$, an explosion energy of $E = 7 \times 10^{50}$ erg, and a ^{56}Ni yield of $M_{\rm Ni} \approx 0.1 M_\odot$.

The low progenitor mass suggested by our modeling is in clear contradiction with the range of masses allowed for Wolf-Rayet stars (Langer 2012). Specifically, in Figure 1 we also show the case of a He star with $8 M_\odot$. From the figure it is clear that this model is not able to reproduce the SN observations even considering all the uncertainties related with the model hypotheses and with the observations. Therefore, based on the hydrodynamical modeling we can firmly rule out models with He core mass $\gtrsim 8 M_\odot$ as progenitors of iPTF13bvn.

3. Binary progenitor

The mass we derived from hydrodynamical modeling is difficult to reconcile with the idea of a single progenitor for the Type Ib SN iPTF13bvn. The question is if there are binary configurations capable of simultaneously reproducing the SN properties and the pre-explosion photometry. To address this question we performed binary evolution calculations with mass transfer using a code developed by Benvenuto & De Vito (2003). Figure 2 shows the evolutionary tracks in the Hertzsprung-Russell (H-R) diagram for a system composed by a donor (primary) star of $20 M_\odot$ and an accretor (secondary) star of $19 M_\odot$ on a circular orbit with initial period of 4.1 days, assuming conservative mass transfer (parametrized by the accretion efficiency, $\beta = 1$).

At the time of explosion the primary is a H-free star with a mass of $3.74 M_\odot$ and a radius of $32.3 R_\odot$, in concordance with our hydrodynamical estimations. The companion star reaches a mass of $33.7 M_\odot$, with luminosity and effective temperature [$\log (L_2^{\rm f}/L_\odot) = 5.36$ and $\log (T_{\rm eff}/K) = 4.64$] comparable to a zero-age main sequence (ZAMS) star of

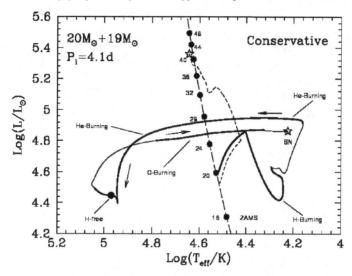

Figure 2. Evolutionary tracks of the binary components of the progenitor of iPTF13bvn for a proposed system with initial masses of $20\,M_\odot$ and $19\,M_\odot$ and an initial orbital period of 4.1 days. The solid line indicates the track of the primary (donor) star (arrows show the evolutionary progress). The short-dashed line shows the evolution of the secondary (accretor) star. Fully conservative accretion ($\beta = 1$) is assumed. The star symbols show the location of both components at the moment of explosion of the primary star. Thick portions of the primary's track indicate the phases of nuclear burning at the stellar core. The long-dashed line shows the locus of the ZAMS, with dots showing different stellar masses (labels in units of $M_\odot$).

$\approx 42\,M_\odot$. Figure 3 shows the composed spectral energy distribution (SED) of the system at the moment of the explosion, compared with the pre-explosion *Hubble Space Telescope* (*HST*) observations of the SN site. A black body spectrum was assumed for the primary star, and an atmosphere model from Kurucz (1993) for the secondary. The synthetic photometry of the progenitor system in the three existing bands is in agreement with the observations within the uncertainties, with differences of less than 0.1 mag. We note that the primary star dominates the flux in the optical regime, so as a result of the SN explosion we predict that the flux in the observed bands will decrease significantly when the SN fades.

The solution to the progenitor system presented here is not unique but it serves to demonstrate the feasibility of the binary progenitor scenario. Based on the pre-explosion photometry and the SN observations, it is possible to analyze the range of allowed binary systems with the aim of predicting the nature of the remaining companion star. We found that the remaining star should necessarily be close to the ZAMS, with a range of luminosities of $4.6 \lesssim \log\left(L_2^{\mathrm{f}}/L_\odot\right) \lesssim 5.6$ (Bersten *et al.* 2014). This means that the companion star may be detected in the future with deep *HST* imaging in the ultraviolet–blue range. The detection of the companion would produce the first robust identification of a hydrogen-deficient SN progenitor as a binary system.

Note

This work is based on an article recently accepted for publication (Bersten *et al.* 2014).

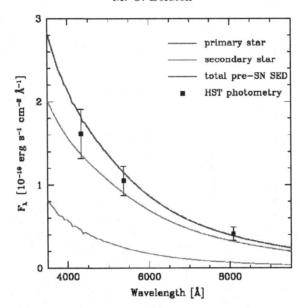

Figure 3. Predicted spectrum of the binary progenitor (solid black line) compared with (*HST*) pre-SN photometry (black squares). The binary spectrum is the sum of a primary star approximated by a black body (red line) and a secondary star represented by an atmosphere model of Kurucz (1993) (blue line). The HST photometry was adopted from Cao *et al.* (2013) and converted to specific fluxes at the approximate effective wavelength of the F435W, F555W, and F814W bands.

References

Benvenuto, O. G., Bersten, M. C., & Nomoto, K. 2013, *ApJ* 762, 74
Benvenuto, O. G. & De Vito, M. A. 2003, *MNRAS* 342, 50
Bersten, M. C., Benvenuto, O., & Hamuy, M. 2011, *ApJ* 729, 61
Bersten, M. C., Benvenuto, O. G., Folatelli, G., *et al.* 2014, *AJ* in press (ArXiv:1403.7288)
Bersten, M. C., Benvenuto, O. G., Nomoto, K., *et al.* 2012, *ApJ* 757, 31
Cao, Y., Kasliwal, M. M., Arcavi, I., *et al.* 2013, *ApJ (Letters)* 775, L7
Ergon, M., Sollerman, J., Fraser, M., *et al.* 2014, *A&A* 562, A17
Groh, J. H., Georgy, C., & Ekström, S. 2013a, *A&A* 558, L1
Groh, J. H., Meynet, G., Georgy, C., & Ekström, S. 2013b, *A&A* 558, A131
Kurucz, R. L. 1993, *Physica Scripta Volume T* 47, 110
Langer, N. 2012, *ARA&A* 50, 107
Nomoto, K. & Hashimoto, M. 1988, *Phys. Rep.* 163, 13
Van Dyk, S. D., Zheng, W., Clubb, K. I., *et al.* 2013, *ApJ (Letters)* 772, L32
Yoon, S.-C., Gräfener, G., Vink, J. S., Kozyreva, A., & Izzard, R. G. 2012, *A&A* 544, L11

Discussion

GROH: You showed very convincingly that the SN lightcurve modeling of iPTF13bvn requires a low mass progenitor, likely from a binary system. Would you be able to user other observables (such as the oxygen mass and radio lightcurve) to further constrain the binary properties such as the mass ratio and initial orbital period?

BERSTEN: The oxygen mass only provides information about the progenitor mass. I think that the only observable that can provide more information about the binary parameters is the detection of the binary companion. We plan to observe the field with HST in the future to do this.

New windows on massive stars: asteroseismology, interferometry, and
spectropolarimetry
Proceedings IAU Symposium No. 307, 2014
G. Meynet, C. Georgy, J. H. Groh & Ph. Stee, eds.

© International Astronomical Union 2015
doi:10.1017/S1743921314006243

Winds of metal-poor OB stars: Updates from HST-COS UV spectroscopy

M. García[1], A. Herrero[2,3], F. Najarro[1], D. J. Lennon[4] and M. A. Urbaneja[5,6]

[1]Centro de Astrobiología, CSIC-INTA
email: mgg@cab.inta-csic.es

[2]Instituto de Astrofísica de Canarias

[3]Departamento de Astrofísica, Universidad de La Laguna

[4]European Space Astronomy Centre

[5]Institute for Astro- and Particle Physics, University of Innsbruck

[6]Institute for Astronomy, University of Hawai'i

Abstract. In the race to break the SMC frontier and reach metallicity conditions closer to the First Stars the information from UV spectroscopy is usually overlooked. New HST-COS observations of OB stars in the metal-poor galaxy IC1613, with oxygen content ~1/10 solar, have proved the important role of UV spectroscopy to characterize blue massive stars and their winds. The terminal velocities (v_∞) and abundances derived from the dataset have shed new light on the problem of metal-poor massive stars with strong winds. Furthermore, our results question the v_∞-v_{esc} and v_∞-Z scaling relations whose use in optical-only studies may introduce large uncertainties in the derived mass loss rates and wind-momenta. Finally, our results indicate that the detailed abundance pattern of each star may have a non-negligible impact on its wind properties, and scaling these as a function of one single metallicity parameter is probably too coarse an approximation. Considering, for instance, that the [α/Fe] ratio evolves with the star formation history of each galaxy, we may be in need of updating all our wind recipes.

Keywords. Galaxies: individual: IC 1613 – Stars: early-type – Stars: massive – Stars: Population III – Stars: winds, outflows – Ultraviolet: stars

1. Introduction

Massive stars impact the dynamics and the energetics of their host galaxy both locally and at galactic-scale. They are also the claimed progenitors of two of the most energetic events of the Universe, long γ-ray bursts (GRBs) and some kinds of supernova (SN). Therefore a number of Astrophysics disciplines require understanding how massive stars live and die, and estimates of their multi-facet feedback.

The radiation-driven winds (RDWs) experienced by blue massive stars not only contribute to mechanical feedback but are also one of the principal agents of their evolution. By removing mass and angular momentum from the star, RDWs alter the physical conditions at the stellar core, the nuclear reaction rates, the duration/sequence of evolutionary stages, the ionizing fluxes, the yields and, eventually, the supernova engine.

Because the Universe is more metal-poor as we go back in time (e.g. Prochaska *et al.* 2003) and because the First Stars were likely very massive and roughly metal-free, the massive star community has taken an earnest interest in metal-poor massive stars. Even though the theoretical framework for as low as $Z \leqslant 10^{-4} Z_\odot$ already existed (e.g. Kudritzki 2002; Schaerer 2002) the Small Magellanic Cloud (SMC), with roughly $1/5 Z_\odot$, stood as the reference for the low metallicity regime until recently. The theoretical predictions for low-Z winds and evolution were contrasted against observations of SMC massive stars, and their spectra were used to simulate metal-poor populations.

The advent of the 8-10m class telescopes, and in particular the instrumentation at the Very Large Telescopes (VLT) and the Gran Telescopio Canarias (GTC), have granted access to resolved massive stars in farther galaxies, hence opening the way to more metal-poor environments (Garcia *et al.* 2011). In particular, the spotlight is currently on the dwarf irregular galaxies IC1613, NGC3109 and WLM, where the oxygen abundance of HII regions scales to a global metallicity of $Z \sim 1/7\, Z_{\odot}$.

The problem of low-Z O-type stars with strong winds:
Because RDWs are powered by the scattering of photons by metallic lines they were expected negligible in galaxies with poorer metal-content than the SMC. However, recent findings suggest otherwise and have puzzled the community over the past 5 years.

Our VLT-VIMOS program on IC1613 revealed a luminous blue variable star with strong P Cygni profiles (Herrero *et al.* 2010), and an Of star whose wind lines suggest a stronger wind momentum than predicted by theory or a rare case of slow wind acceleration (Herrero *et al.* 2012, hereafter H12). Independently, Tramper *et al.* (2011, hereafter T11) studied a sample of O-stars in IC1613, NGC3109 and WLM with VLT-XSHOOTER, and found that their winds were also stronger than the prediction.

The results of these works have large error bars which add to yet unexplored wind inhomogeneity effects. However, the fact that almost all studied stars exceeded the theoretical prediction (see Fig. 1-left), and the large potential impact on the evolution and feedback of low-Z massive stars, granted renewed interest from the community.

2. Analysis of ultraviolet (UV) spectroscopy

H12 and T11 did not directly obtain the mass loss rate ($\dot{M}$) but the parameter $Q = \dot{M}/(v_{\infty} \cdot R_{\star})^{1.5}$ (Kudritzki & Puls 2000). Lacking a value for the terminal velocity (v_{∞}), which can only be measured from UV lines, optical studies usually estimate v_{∞} from calibrations with the escape velocity which suffer from large scatter ($v_{\infty}/v_{esc} = 2.65$, Kudritzki & Puls 2000), and then scale it with Z using the empirical relation $v_{\infty} \propto Z^{0.13}$ (Leitherer *et al.* 1992). This procedure introduces large errors into $\dot{M}$ that propagate to the WLR, the relation between the modified wind momentum ($D_{mom} = \dot{M} \cdot v_{\infty} \cdot (R_{\star}/R_{\odot})^{1/2}$) and stellar luminosity, and our main tool to evaluate the wind strength.

The confirmation of the *strong wind problem* required UV spectroscopy to directly measure terminal velocities and improve mass loss rates and D_{mom}. We obtained 23 orbits of Hubble Space Telescope (HST) to observe a sample of IC1613 OB stars (Tab. 1) with the Cosmic Origins Spectrograph (COS). The data cover with good signal to noise ratio the $\sim$1150-1800Å spectral range, with a resolution of R $\sim$ 2600. These are the first good quality UV spectra of OB stars in an oxygen-poorer galaxy than the SMC.

The terminal velocities were determined with the Sobolev plus Exact Integration (SEI) method (Lamers *et al.* 1987), using J. Puls's implementation of Haser (1995)'s code. More details on the observations and their analysis are provided in Garcia *et al.* (2014).

Table 1. Program stars and derived terminal velocities. ID numbers from Garcia *et al.* (2009).

		69217	62024	65426	67559	63932	69336	62390	60449
Sp. Type		O3-4Vf	O6.5IIIf	O7.5III-V((f))	O8.5III((f))	O9II	B0Ia	B0.5Ia	B1.5Ia
v_{∞}[km s^{-1}]		2200^{+150}_{-100}	1250^{+150}_{-200}	1500^{+250}_{-250}	1500^{+300}_{-200}	1000^{+500}_{-400}	1300^{+100}_{-100}	1075^{+75}_{-75}	875^{+75}_{-75}

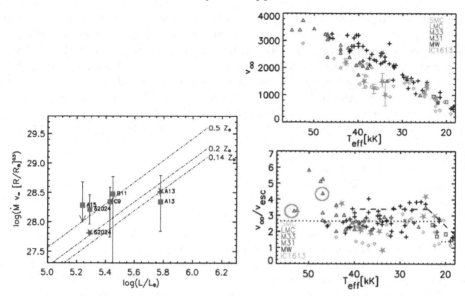

Figure 1. Left: The wind-momentum luminosity relation of IC1613 stars. Squares mark T11 and H12 results, and stars mark our updated values using UV terminal velocities. The dashed–dotted lines are Vink *et al.* (2001)'s predictions for the WLR at $0.14\,Z_\odot$, $0.2\,Z_\odot$ and $0.5\,Z_\odot$ (representing IC1613, SMC and LMC). **Right:** Terminal velocities (up) and the v_∞/v_{esc} ratio (down) of LG O- and early-B stars as a function of effective temperature (rhombus: SMC; triangles: LMC; squares: M33; crosses: M31; plus-signs: MW; stars: IC1613). The dotted lines mark the $v_\infty/v_{esc} = 2.65$ ratio, and the dashed lines the relations found by Crowther *et al.* (2006) and Markova & Puls (2008) for Galactic B-supergiants. The photospheric parameters of the two encircled targets were derived by Rivero-González *et al.* with the latest version of FASTWIND.

2.1. *Results*

The WLR revisited:

We recalculated the wind momentum of two stars included in the T11 and H12 samples with our derived terminal velocities. Their updated D_{mom} better matches the prediction for IC1613 metallicity (Fig. 1-left), illustrating the important role of v_∞ to assess the WLR. We note that final D_{mom} values must await a consistent analysis that constraints photospheric and wind parameters, abundances and clumping, planned as future work.

The v_∞ vs Z dependence:

We compiled all the terminal velocities of OB stars in Local Group -LG- galaxies derived from UV diagnostics, and their photospheric parameters derived by modern works (see Garcia *et al.* 2014, for references). Fig. 1-right shows a clear trend of increasing terminal velocity with increasing effective temperature (T_{eff}). However, we detect no clear dependence of v_∞ with host galaxy. IC1613, SMC and LMC O-stars are found in the same locus, whereas they clearly depart from Milky Way (MW) stars. This contradicts the prediction of a simple $v_\infty \propto Z^{0.13}$ relation: if MW 40000 K O-dwarfs have $v_\infty \sim 3000\,\mathrm{km\,s^{-1}}$, we would expect that LMC/SMC/IC1613 $Z_\odot$ stars have $v_\infty \sim 2740/2430/2220\,\mathrm{km\,s^{-1}}$. No segregation with metallicity is detected in the B-supergiant regime.

The v_∞/v_{esc} scaling relation:

The v_∞/v_{esc} vs T_{eff} plot shows no separation with metallicity, not even with MW stars. O- and B-supergiants cluster around the canonical $v_\infty/v_{esc} = 2.65$ value with some scatter, but the sample of LG dwarf stars exhibits large departures.

v_{esc} is subject to several sources of error, some inherited from gravity ($\log g$ requires accurate T_{eff}'s and a good nebular subtraction) and some, specially in the MW, from distance. Even though we only used photospheric parameters derived with the state-of-the-art codes FASTWIND and CMFGEN, the methodology of different research groups may also introduce some scatter. We note here two stars analyzed by the same team with the latest version of FASTWIND (Rivero González *et al.* 2012a,b) that properly treats the high-temperature regime. These targets, marked within green circles in Fig. 1-right, have v_∞/v_{esc}=3.3 and 4.4 hinting real departures from the wide-spread used 2.65 value.

The iron content of IC1613:

The morphological comparison of IC1613, SMC and LMC stars suggests that the iron content of the IC1613 sample stars is similar to, or even slightly higher than the SMC's. Moreover, the photospheric models that best reproduced the observed UV pseudo-continuum, dominated by iron lines, had SMC metallicity.

While pending confirmation from a full quantitative spectral analysis, this finding agrees with the $0.2\,\mathrm{Fe}_\odot$ abundance measured in three IC1613 red supergiants by Tautvaišienė *et al.* (2007). With the oxygen abundances well established at $\sim 0.12\,\mathrm{O}_\odot$ from HII regions or $\sim 0.16\mathrm{O}_\odot$ from B-supergiants (Bresolin *et al.* 2007), IC1613's abundance ratio of α-element to iron may be sub-solar ([α/Fe] $= -0.1$ dex). Similar chemical mixtures have been found in other LG dwarf irregulars (Tautvaišienė *et al.* 2007; Hosek al. 2014). They are indicative of a tranquil recent star formation history with no major or violent episodes, and without a dominant population of massive stars at late times.

3. Discussion

The revision of the v_∞-v_{esc} dependence with the available data from the LG shows that this empirical relation suffers from such a large scatter that its use may be impractical. The scatter can be partly explained by the systematic uncertainties of the involved stellar properties, however we argue that part may be actually expected in the framework of RDW theory (see Garcia *et al.* 2014, for an extended discussion).

The terminal velocity is determined by the radiative acceleration at the outer wind layers, which is dominated by the strong resonance lines of a small number of light elements that still keep their ionization stage (Vink *et al.* 1999; Puls *et al.* 2000). Apparently subtle differences of [T_{eff},$\log g$] between stars can lead to different local conditions and ionization equilibria in the outer wind that may alter v_∞. In this context v_∞/v_{esc} is not necessarily monotonic, and this translates into scatter in the v_∞/v_{esc} vs T_{eff} plot. Departures from solar chemical mixtures, unaccounted for when constructing the scaling relations of v_∞, may also add to the scatter.

The effect of non-solar chemical mixtures on RDWs has been scarcely studied in the literature. Besides terminal velocities, mass loss rates may also be affected. At solar-like metallicity, iron, with many optically thick lines in the wind, is the main driver of mass loss. At poorer metallicities the iron lines become optically thin while the strong resonance lines of CNO remain optically thick and may drive the wind (Vink *et al.* 2001; Krtička & Kubát 2014). Krtička & Kubát (2014) establish the separation of these two regimes at $Z \lesssim 0.1\,Z_\odot$. In other words: at the metallicity of IC1613, NGC3109 and WLM $\dot{M}$ and v_∞ (hence $\mathrm{D}_{\mathrm{mom}}$) depend not only on the stellar luminosity and global metallicity, but also on detailed abundances.

These points suggest that wind properties should be studied on a star to star basis. Some groups have already taken on this approach and, instead of recipes, provide full wind simulations at several points of the [T_{eff},$\log g$] parameter space of O-stars (e.g.

Muijres *et al.* 2012). If recipes and scaling relations are still to be used we should at least evolve into a two-parameter view of metal content, accounting separately for light elements and iron, especially at the $Z \lesssim 0.1\,Z_\odot$ regime. A view where α-elements and iron are considered separately, better represents the chemical evolution of galaxies. The effects of CNO processing should also be studied in detail.

4. Summary and conclusions

We present the first UV study of low-Z OB-stars beyond the SMC. Our results set an urgent reminder to three points:

• UV spectroscopy holds key information on the winds of OB stars that, if neglected, may lead to erroneous results.

• The v_∞/v_{esc} scaling relation suffers from large scatter. Terminal velocities obtained from this relation may be wrong by a factor of 2. Our results also call into question the dependence of v_∞ with one single global metallicity parameter.

• Oxygen is not a good proxy for metallicity, as the detailed abundance pattern of a galactic region depends on the star formation history of the galaxy.

These three points, long-known but overlooked because of feasibility issues, introduce large uncertainties in the WLR. The problem of low-Z OB stars with strong winds, established from optical spectroscopy only, might serve as illustration. We will provide further evidence on this topic with a joint optical+UV analysis of our IC1613 sample.

Acknowledgments: Funded by Spanish MINECO under grants ESP2013-47809-C3-1-R, FIS2012-39162-C06-01, AYA2012-39364-C02-01, SEV 2011-0187-01, 20105Y1221, AYA2010-21697-C05-01 and by Gobierno de Canarias (PID2010119).

References

Bresolin, F., Urbaneja, M. A., Gieren, W., *et al.* 2007, *ApJ*, 671, 2028
Crowther, P. A., Lennon, D. J., & Walborn, N. R. 2006, *A&A*, 446, 279
Garcia, M., Herrero, A., Vicente, B., *et al.* 2009, *A&A*, 502, 1015 [GHV09]
Garcia, M., Herrero, A., & Najarro, F. 2011, *Ap&SS*, 335, 91
Garcia, M., Herrero, A., Najarro, F., Lennon, D. J., & Urbaneja, M. A. 2014, *ApJ*, 788, 64
Haser, S. M. 1995, *Ph.D. Thesis, Universitäts-Sternwarte der Ludwig-Maximillian Universität*
Herrero, A., Garcia, M., Uytterhoeven, K., *et al.* 2010, *A&A*, 513, A70
Herrero, A., Garcia, M., Puls, J., *et al.* 2012, *A&A*, 543, A85 [H12]
Hosek, M. W., Jr., Kudritzki, R.-P., Bresolin, F., *et al.* 2014, *ApJ*, 785, 151
Krtička, J. & Kubát, J. 2014, *A&A*, 567, A63
Kudritzki, R. P. 2002, *ApJ*, 577, 389
Kudritzki, R.-P. & Puls, J. 2000, *ARA&A*, 38, 613
Lamers, H. J. G. L. M., Cerruti-Sola, M., & Perinotto, M. 1987, *ApJ*, 314, 726
Leitherer, C., Robert, C., & Drissen, L. 1992, *ApJ*, 401, 596
Markova, N. & Puls, J. 2008, *A&A*, 478, 823
Muijres, L. E., Vink, J. S., de Koter, A., Müller, P. E., & Langer, N. 2012, *A&A*, 537, A37
Prochaska, J. X., Gawiser, E., Wolfe, A. M., *et al.* 2003, *ApJ (Letters)*, 595, L9
Puls, J., Springmann, U., & Lennon, M. 2000, *A&ASS*, 141, 23
Rivero González, J. G., Puls, J., Najarro, F., & Brott, I. 2012a, *A&A*, 537, A79
Rivero González, J. G., Puls, J., Massey, P., & Najarro, F. 2012b, *A&A*, 543, A95
Schaerer, D. 2002, *A&A*, 382, 28
Tautvaišienė, G., Geisler, D., Wallerstein, G., *et al.* 2007, *AJ*, 134, 2318
Tramper, F., *et al.* 2011, *ApJ (Letters)*, 741, L8 [T11]
Vink, J. S., de Koter, A., & Lamers, H. J. G. L. M. 1999, *A&A*, 350, 181
Vink, J. S., *et al.* 2001, *A&A*, 369, 574

Discussion

NIEVA: Is there a way you can recognize spectroscopic binaries in your star sample?

GARCÍA: For some stars we have multiple spectra and we can check for radial velocity variations. But mostly, the resolution is too low to recognize binaries. We do know that one of the sample stars is an eclipsing binary, found in a previous photometric survey by Alceste Bonanos.

Miriam García

Cyril Georgy

New windows on massive stars: asteroseismology, interferometry, and spectropolarimetry
Proceedings IAU Symposium No. 307, 2014
G. Meynet, C. Georgy, J. H. Groh & Ph. Stee, eds.

© International Astronomical Union 2015
doi:10.1017/S1743921314006255

Combining observational techniques to constrain convection in evolved massive star models

C. Georgy[1], H. Saio[2] and G. Meynet[3]

[1] Astrophysics group, Lennard-Jones Laboratories, EPSAM, Keele University, Staffordshire
ST5 5BG, UK
email: c.georgy@keele.ac.uk

[2] Astronomical Institute, Graduate School of Science, Tohoku University, Sendai 980-8578,
Japan

[3] Geneva Observatory, University of Geneva, Maillettes 51, CH-1290 Versoix, Switzerland

Abstract. Recent stellar evolution computations indicate that massive stars in the range $\sim 20 - 30\,M_\odot$ are located in the blue supergiant (BSG) region of the Hertzsprung-Russell diagram at two different stages of their life: immediately after the main sequence (MS, group 1) and during a blueward evolution after the red supergiant phase (group 2). From the observation of the pulsationnal properties of a subgroup of variable BSGs (α Cyg variables), one can deduce that these stars belongs to group 2. It is however difficult to simultaneously fit the observed surface abundances and gravity for these stars, and this allows to constrain the physical processes of chemical species transport in massive stars. We will show here that the surface abundances are extremely sensitive to the physics of convection, particularly the location of the intermediate convective shell that appears at the ignition of the hydrogen shell burning after the MS. Our results show that the use of the Ledoux criterion to determine the convective regions in the stellar models leads to a better fit of the surface abundances for α Cyg variables than the Schwarzschild one.

Keywords. stars: abundances–stars: early-type–stars: evolution–stars: mass loss–stars: oscillations.

---

1. Introduction

The post main-sequence (MS) evolution of massive stars is still poorly understood, and the prediction of the simulations with different stellar evolution codes lead to very different results (Martins & Palacios 2013), particularly due to the different way the transport mechanisms (convection, rotational mixing) are implemented. To improve our knowledge of massive stars evolution, it is thus of prime importance to find some observational tests that allow to discriminate between the various existing prescriptions for the internal transport mechanisms.

Some arguments seem to indicate that the mass-loss rates used in the stellar evolution codes during the red supergiant (RSG) phase (often, the rates by de Jager *et al.* 1988) could be underestimated (see the discussions in Georgy 2012; Georgy *et al.* 2012, as well as Vanbeveren *et al.* 1998). Recently, the Geneva group has released a new set of stellar models, including such an increased mass-loss rates during the RSG phase (Ekström *et al.* 2012; Georgy *et al.* 2013). These models show that for stars in the mass range $\sim 20 - 30\,M_\odot$, the evolution after the MS is the following: a first crossing of the Hertzsprung-Russell diagram (HRD) up to the RSG branch, and then, due to the strong mass loss during the RSG phase, a second crossing occurs, the stars ending their life in

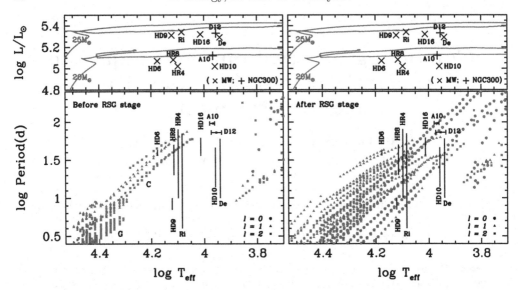

Figure 1. Each side shows the tracks in the HRD *(top panel)* and the pulsation periods of the excited modes *(bottom panel)* of a rotating $20\,M_\odot$ (red) and $25\,M_\odot$ (blue) model at solar metallicity. *Left panel:* Periods computed during the first crossing of the HRD (group 1). *Right panel:* Periods computed during the second crossing of the HRD (group 2). Observational values are also indicated (Firnstein & Przybilla 2012; Moravveji *et al.* 2012; Leitherer & Wolf 1984; van Leeuwen *et al.* 1998; Fraser *et al.* 2010; Kaltcheva & Scorcio 2010; Kaufer *et al.* 1996, 1997; Sterken 1977; Sterken *et al.* 1999; Schiller & Przybilla 2008; Richardson *et al.* 2011; Markova & Puls 2008; Percy *et al.* 2008; Kudritzki *et al.* 2008; Bresolin *et al.* 2004). Figure adapted from Saio *et al.* (2013).

the blue side of the HRD. There is thus a double population of blue supergiant stars (BSG): the first one consists in stars immediately after the MS that are in their first crossing (group 1, see Georgy *et al.* 2014), and the second one consists in stars that are post-RSG stars, that are going from the red side to the blue side of the HRD (group 2).

2. Are α Cyg variables group 2 stars?

The evolution in the HRD of post-MS stars in the range $20 - 30\,M_\odot$ computed in Ekström *et al.* (2012) and Saio *et al.* (2013) occurs at roughly constant luminosity. However, a major difference between the phase where the star is in group 1 and the phase in group 2 is the current mass. Indeed, the star encounters a very strong mass loss during the RSG phase, and is thus considerably less massive once it reaches again the BSG region. For example, a rotating model with an initial mass of $25\,M_\odot$ has a mass of $23.48\,M_\odot$ when it reaches for the first time $\log(T_{\text{eff}}) = 4$ (group 1), and a mass of only $12.68\,M_\odot$ when it has the same T_{eff} during the second crossing (group 2). This makes the luminosity-to-mass ratio bigger for group 2 stars compared to group 1, and considerably changes the pulsationnal properties of these stars.

Figure 1 shows the period of the different excited modes for a group 1 model (left) and group 2 (model). These results indicate that in order to reproduce the observed period of α Cyg stars, it is unavoidable to lose a lot of mass to increase the L/M ratio. This seems to indicate that these stars are post-MS stars and belong to group 2 rather than group 1.

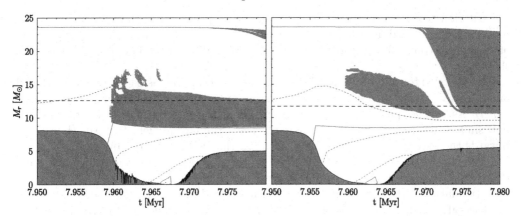

Figure 2. Kippenhahn plot for a $20\,M_\odot$ model computed with the Schwarzschild criterion for convection *(left panel)* and Ledoux criterion *(right panel)*. Only the very end of the MS and the beginning of central He-burning is shown. Grey areas indicate the position of the convective zones. The red line shows the position of the stellar surface. Blue (green) solid line indicate the maximimum of the energy generation rate due to H-burning (He-burning) and the dotted line when the energy generation rate is $100\,\mathrm{erg\,s^{-1}\,g^{-1}}$. The black dashed line shows the layer that will be uncovered by mass loss when the star reaches $\log(T_{\mathrm{eff}}) = 4$ during the second crossing. Adapted from Georgy *et al.* (2014).

3. Surface abundances of group 2 stars

For at least two α Cyg variables (Rigel and Deneb), some measurements of the surface abundances are available. In the following, we will focus on the N/C ratio. The observed ratio are the following (Przybilla *et al.* 2010): N/C = 2.0 (Rigel) and 3.4 (Deneb). These relatively small values indicate that the surface abundances of these stars are partially processed by the CNO cycle (the solar N/C ratio is ~ 0.3), but they are far from the values at the CNO equilibrium, ~ 60).

The "standard" models shown above were computed using the Schwarzschild criterion for convection (see Ekström *et al.* 2012, for the details of the physical processes implemented in these models). When the star becomes for the second time a BSG, and thus presents pulsation periods that are compatible with the observations of α Cyg variables, the surface N/C ratio is 58, much more than the observed values. This is explained by looking at the Kippenhahn plot shown in Fig. 2 *(left panel)*. At the very end of the MS, when H-burning migrates from the centre to a shell, an important intermediate convective zone appears just on top of the previous convective core, bringing towards more external layers (up to a lagrangian coordinate $M_r \sim 15\,M_\odot$) material strongly processed by CNO cycle. Due to the strong mass loss the star will encounter during the RSG phase, these layers will be uncovered when the star enter in the group 2 region of the HRD, explaining the high N/C ratio.

As our model have to go through a previous RSG phase to explain the pulsationnal properties of α Cyg variables, the surface abundances of these stars tell us that something is missing in the treatment of internal mixing we used to compute these models.

4. Model with the Ledoux criterion for convection

As explained previously, the reason of the strong enrichment of the surface in our "standard" models is due to the position of the intermediate convective shell that appears at the ignition of the H-burning shell. A possible solution to this problem is thus to find a way to change the position of this convective zone. This can be achieved by changing the criterion that determines if a zone is convective or not. The Schwarzschild criterion, used in the "standard models", does not account for the stabilising effect of the presence of a chemical composition gradient. On the other hand, the Ledoux criterion account for this effect.

We have tried to compute a model with the Ledoux criterion instead of the Schwarzschild one (Georgy *et al.* 2014). The corresponding Kippenhahn plot is shown in Fig. 2 *(right panel)*. As a consequence of the Ledoux criterion, the intermediate convective shell is now shifted towards the surface compared to the "standard model" *(left panel)*, bringing towards the external layers of the star material that is less affected by the CNO cycle. The zone in the star that will be uncovered by the mass loss during the RSG phase has a C/N ratio of ~ 7, much closer to the observations than previously. In the same time, the pulsationnal properties of group 2 stars are still in good agreement with the observations.

Although the situation is improved, it is still not perfect. The surface gravity of Rigel and Deneb was also determined: $\log(g) = 1.75 \pm 0.1$ (Rigel), and $\log(g) = 1.20 \pm 0.1$ (Deneb, Przybilla *et al.* 2010). At effective temperatures corresponding to these stars, our models have $\log(g) = 1.56$ and $\log(g) = 0.76$ respectively. This quite big discrepancy is not solved to date, and further investigations are needed.

5. Conclusions

The main result of this work is to show how important is a good treatment of the convection in stellar evolution codes. It appeared that changing the criterion applied to determine if a zone is convective or not can drastically change the evolution of the surface properties of a stellar model. In that framework, the current development of observational technics such as asteroseismology combined with other methods (abundances measurements, ...) can bring new constraints for the stellar modeling.

On the theoretical/numerical side, some efforts have also to be made in order to improve our knowledge of convection, and improve how it is implemented in the classical 1d codes. The development of multi-d hydro-simulations of convection (see e.g. Meakin & Arnett 2007; Viallet *et al.* 2013) is an obvious step in that direction, and will probably lead to decisive changes in the way convection is treated in the numerical codes.

References

Bresolin, F., Pietrzyński, G., Gieren, W., *et al.* 2004, *ApJ* 600, 182
de Jager, C., Nieuwenhuijzen, H., & van der Hucht, K. A. 1988, *A & AS* 72, 259
Ekström, S., Georgy, C., Eggenberger, P., *et al.* 2012, *A&A* 537, A146
Firnstein, M. & Przybilla, N. 2012, *A&A* 543, A80
Fraser, M., Dufton, P. L., Hunter, I., & Ryans, R. S. I. 2010, *MNRAS* 404, 1306
Georgy, C. 2012, *A&A* 538, L8
Georgy, C., Ekström, S., Eggenberger, P., *et al.* 2013, *A&A* 558, A103
Georgy, C., Ekström, S., Meynet, G., *et al.* 2012, *A&A* 542, A29
Georgy, C., Saio, H., & Meynet, G. 2014, *MNRAS* 439, L6
Kaltcheva, N. & Scorcio, M. 2010, *A&A* 514, A59
Kaufer, A., Stahl, O., Wolf, B., *et al.* 1997, *A&A* 320, 273
Kaufer, A., Stahl, O., Wolf, B., *et al.* 1996, *A&A* 305, 887
Kudritzki, R.-P., Urbaneja, M. A., Bresolin, F., *et al.* 2008, *ApJ* 681, 269
Leitherer, C. & Wolf, B. 1984, *A&A* 132, 151
Markova, N. & Puls, J. 2008, *A&A* 478, 823
Martins, F. & Palacios, A. 2013, *A&A* 560, A16
Meakin, C. A. & Arnett, D. 2007, *ApJ* 667, 448

Moravveji, E., Guinan, E. F., Shultz, M., Williamson, M. H., & Moya, A. 2012, *ApJ* 747, 108

Percy, J. R., Palaniappan, R., Seneviratne, R., Adelman, S. J., & Markova, N. 2008, *PASP* 120, 311

Przybilla, N., Firnstein, M., Nieva, M. F., Meynet, G., & Maeder, A. 2010, *A&A* 517, A38+

Richardson, N. D., Morrison, N. D., Kryukova, E. E., & Adelman, S. J. 2011, *AJ* 141, 17

Saio, H., Georgy, C., & Meynet, G. 2013, *MNRAS* 433, 1246

Schiller, F. & Przybilla, N. 2008, *A&A* 479, 849

Sterken, C. 1977, *A&A* 57, 361

Sterken, C., Arentoft, T., Duerbeck, H. W., & Brogt, E. 1999, *A&A* 349, 532

van Leeuwen, F., van Genderen, A. M., & Zegelaar, I. 1998, *A & AS* 128, 117

Vanbeveren, D., De Donder, E., van Bever, J., van Rensbergen, W., & De Loore, C. 1998, *New. Astron.* 3, 443

Viallet, M., Meakin, C., Arnett, D., & Mocák, M. 2013, *ApJ* 769, 1

Discussion

NOELS: Comment: In models computed with the Schwarzschild criterion, the layers in the μ-gradient region become convective while they should be semi-convective. This would lower the efficiency of the mixing and would lower the ratios N/C and N/O.

NOELS: What would happen to the ratio if no intermediate convective zone at all is present? It would still be possible to affect the N/C and N/O ratios if mass loss in the RSG phase is large enough to uncover the layers where CNO cycle has affected the CNO abundances already in the early phases of the main sequence.

GEORGY: It has to be checked. As the mass loss uncover quite deep layers, it probably shows some material affected at least partially by CNO cycle. This could be easily checked from the structure at the end of the MS, before the intermediate convective zone develops.

MORAVVEJI: Where does all lost hydrogen rich envelope go during the RSG phase? Do we have observations for that?

GEORGY: Some indirect observations of interactions with a dense circumstellar shell exist. For example, type IIn supernovae, or the presence of ring nebulae around Wolf-Rayet stars.

PRZIBILLA: How do the lifetimes of blue supergiants on the first crossing of the HRD to the red and those of the second crossing towards the blue compare?

GEORGY: It is extremely dependent on the stellar model (e.g. rotation or not) and can change from 10%-90%. So far, it seems to be difficult to assess firm statistics on that point.

New windows on massive stars: asteroseismology, interferometry, and spectropolarimetry
Proceedings IAU Symposium No. 307, 2014
G. Meynet, C. Georgy, J. H. Groh & Ph. Stee, eds.

© International Astronomical Union 2015
doi:10.1017/S1743921314006267

Massive stars near the Eddington-limit, pulsations & mass-loss

G. Gräfener

Armagh Observatory, College Hill, Armagh, BT61 9DG, United Kingdom

Abstract. Very massive stars (in excess of $\sim 100\,M_\odot$) and massive stars in pre-SN phases at the end of their evolution are continuously approaching the Eddington limit. According to our theoretical predictions their high Eddington factors lead to a peculiar sub-photospheric structure and enhanced mass-loss. Their proximity to the Eddington limit is thus likely the reason why these objects appear as LBVs and WR stars. Here we discuss how our predictions relate to the characteristics of strange-mode pulsations, and how rotating massive stars at low metallicities can produce spectroscopic signatures that have recently been observed in a sample of star-forming galaxies at redshifts $z \approx 2 - 4$.

1. Introduction

In this contribution we discuss two processes that are predicted to occur when massive stars are approaching the Eddington limit. A radial inflation of the outer stellar envelope, and strong, Wolf-Rayet (WR) type mass-loss. Both processes are caused by the so-called Fe-opacity peak, i.e. an enhancement of the flux-weighted mean opacity that occurs in a density and temperature regime where the Fe-group elements have very complex electron configurations. We highlight the relation of these processes to stellar pulsations, and their relevance for stellar populations at low metallicity.

2. Envelope inflation near the Eddington limit

Figure 1 shows the Hertzsprung-Russell (HR) diagram of Galactic WR stars of the WN subtype from Hamann *et al.* (2006). The diagram shows two populations of WN stars. A group of extremely luminous H-rich WN stars (WNh stars) near the ZAMS, and a group of less luminous H-poor and H-free WN stars between the ZAMS and the He-ZAMS. Both groups of stars show strong WR-type mass loss. WNh stars are believed to be extremely massive stars ($M \gtrsim 100\,M_\odot$) that are still in the core H-burning phase. They constitute the group of stars with the highest masses known, i.e., the top of the main-sequence (cf., Gräfener & Hamann 2008; Crowther *et al.* 2010; Gräfener *et al.* 2011; Bestenlehner *et al.* 2014).

Here we discuss the group of hot, H-poor WN stars which consists predominantly of core He-burning stars in a post-RSG or post-LBV phase (e.g., Gräfener *et al.* 2012b). Hamann *et al.* (2006) discussed that the majority of such objects would be expected to lie near the He-ZAMS, i.e., at much higher temperatures than observed. Sander *et al.* (2012) have shown that the same problem exists for Galactic WC stars. This "WR radius problem" has often been attributed to the radially extended photospheres of WR stars due to their optically thick stellar winds. However, as the WR wind extension is included in the models used for the spectral analyses above, this argument is most likely wrong and the radius problem persists. The discrepancy between the observed and theoretically predicted WR radii reaches in some cases factors up to 10!

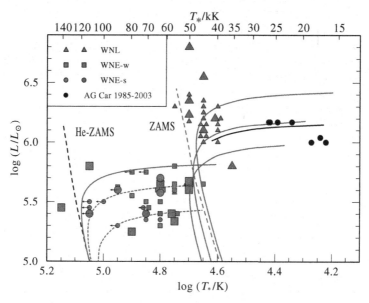

Figure 1. Hertzsprung-Russell diagram of the Galactic WN stars, and the LBV AG Car. Red/blue symbols indicate observed HR diagram positions of WN stars from Hamann *et al.* (2006) (blue: with hydrogen ($X > 0.05$); red: hydrogen-free ($X < 0.05$)). Black symbols indicate the HR diagram positions of AG Car throughout its S Dor Cycle from 1985–2003, according to Groh *et al.* (2009). Large symbols refer to stars with known distances from cluster/association membership. The symbol shapes indicate the spectral subtype (see inlet). Arrows indicate lower limits of T_* for stars with strong mass loss. The observations are compared to stellar structure models from Gräfener *et al.* (2012a) (blue with hydrogen, red hydrogen-free), and for AG Car ($X = 0.36$, black line). The dashed red lines indicate models for which clumping factors of 4 and 16 have been assumed in the sub-photospheric layers.

To investigate this effect Gräfener *et al.* (2012a) performed stellar structure computations for chemically homogeneous stars with and without hydrogen. In agreement with previous works (Ishii *et al.* 1999; Petrovic *et al.* 2006), their models show a substantial envelope inflation at high luminosities (red solid line in Fig. 1). However, to reproduce the observed WR radii it is necessary to invoke the inflation effect already at lower luminosities. Gräfener *et al.* managed to bring the observed WR radii in agreement with the observations by assuming that the sub-photospheric layers of WR stars are inhomogeneous, or clumped (red dashed lines in Fig. 1). Due to the higher density within clumps, clumping leads to an increased mean opacity. The adopted clumping factors are comparable with those observed in the winds of WR stars (cf. Hamann & Koesterke 1998), hinting at a possible sub-photospheric origin of WR wind-clumping (cf. also Cantiello *et al.* 2009).

The underlying reason for the envelope inflation lies in the topology of the Fe-opacity peak near $\sim 160\,\mathrm{kK}$. As the star has to avoid a super-Eddington situation in its sub-photospheric layers, the opacity $\chi(\rho, T)$ near the Fe-peak needs to be lowered. The only way to do this for given T, is to go to low densities ρ. Close to the Eddington limit, the topology of the opacity peak thus leads to the formation of a low-density stellar envelope near $160\,\mathrm{kK}$. Due to the relation between temperature and optical depth in the stellar interior (approximately with $T^4 \propto \tau$), and $\tau \propto \rho\,\Delta R$, a low density automatically implies a large radial extension ΔR. Above the low-density zone the density increases again, so that a cavity is formed (cf. Fig. 2). Gräfener *et al.* (2012a) could describe this effect in an analytical approach which revealed the existence of a stability limit. In this approach the

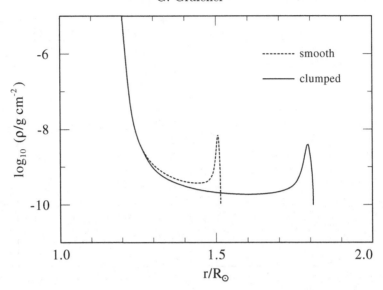

Figure 2. Density structure of inflated stellar envelopes with and without clumping from Gräfener *et al.* (2012a).

radial extension of the inflated layers increases with increasing stellar mass, approaching infinity at the stability limit. For higher masses no hydrostatic solutions exist.

Due to the low densities within the inflated layers the gas pressure falls far below the radiation pressure (note that the radiation pressure is a given quantity for the given Fe-peak temperature). In linear stability analyses the resulting radiation-dominated cavities have been found to be subject to strange-mode pulsations which are believed to be oscillation modes that are trapped within the cavities (e.g. Saio *et al.* 1998). The main characteristic of these pulsations, that the pulsation period increases with increasing mass, is a direct consequence of the increasing envelope extension for increasing mass.

In non-linear simulations Glatzel & Kaltschmidt (2002); Glatzel (2008, 2009) showed that the inflated layers may be subject to strange-mode instabilities. If these instabilities are realized in nature, they will introduce a highly inhomogeneous density structure in the inflated layers. In the same way as in the simplified clumping approach of Gräfener *et al.* (2012a), these inhomogeneities will lead to an enhanced mean opacity and envelope inflation. This has been illustrated in Fig. 3 of Glatzel (2009), where they compare the radial extension of a smooth hydrostatic model with one of their dynamic models. In Fig. 2 we show a very similar case where we compare a smooth and a clumped envelope model from Gräfener *et al.* (2012a). The resulting envelope structure is very similar to Fig. 3 in Glatzel (2009) in the sense that the clumped model shows a lower minimum density and a larger envelope extension.

The inflation effect is expected to be highly dependent on metallicity (Petrovic *et al.* 2006), in notable agreement with the fact that late WR subtypes are tendentially found in high-metallicity environments, such as the Galactic center, while early subtypes dominate in low-metallicity environments such as the LMC.

Another important aspect is the potential role that the envelope inflation may play for more H-rich stars near the Eddington limit. In Fig. 1 we show a comparison of our stellar structure models with the LBV AG Car during its S-Dor cycle (Groh *et al.* 2009). According to our computations, this star is expected to be subject to a substantial radius inflation, and is located near the discussed stability limit. This may potentially explain the nature of the erratic radius variations that are typical for LBVs.

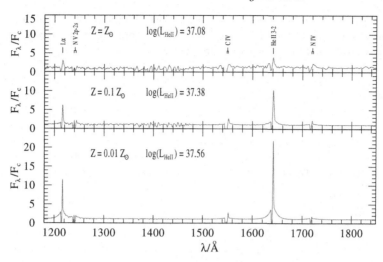

Figure 3. Synthetic UV spectra from hydrodynamically self-consistent wind models for WNh stars at different metallicities Z from Gräfener & Hamann (2008).

We note that the inflation effect is only expected to occur if convective energy transport is inefficient. Moreover, Petrovic *et al.* (2006) discussed that high wind mass-loss rates can destroy the effect. We also expect that the porosity of clumped material within the envelope reduces the effect. Furthermore, the inflation effect does not seem to be observed for the group of very massive WNh stars in the Galaxy (indicated by blue triangles in Fig. 1), and the LMC (cf. Bestenlehner *et al.* 2014). It thus seems that further dedicated studies are necessary to understand in which regime this important effect is realised in nature. Modelling and analysis of the pulsational properties of objects near the Eddington limit may be the key to resolve this question.

3. Enhanced mass loss near the Eddington limit

In addition to the peculiarities in their radii, WR stars and LBVs both show enhanced mass loss through stellar winds. According to our work, both phenomena are most likely related to their proximity to the Eddington limit. E.g., for very massive stars Gräfener & Hamann (2008); Vink *et al.* (2011) predicted a strong mass-loss dependence on the classical Eddington factor $\Gamma_e = \chi_e L/(4\pi c\,GM)$, where χ_e denotes the electron scattering opacity and the other variables have their usual meaning. This dependence provides a natural explanation why these stars display WR-type spectra, although they are still burning hydrogen in their cores. The Γ_e-dependence has been qualitatively confirmed by empirical studies of samples of very massive stars by Gräfener *et al.* (2011); Bestenlehner *et al.* (2014).

In the following we wish to highlight the importance of this effect for stellar populations at low metallicity (Z). In this regime high *effective* values of Γ_e are expected because rotating massive stars tend to approach critical rotation more easily (e.g. Meynet & Maeder 2002). Gräfener & Hamann (2008) predicted that stars of this type can form strong WR-type winds with low terminal wind speeds. In Fig. 3 we show synthetic spectra from Gräfener & Hamann (2008) that demonstrate that stars of this kind can form strong and narrow He II 1640 emission lines at low Z. Similar narrow He II emissions have recently been detected in a sample of star-forming galaxies with redshifts between 2 and 4.6 by Cassata *et al.* (2013). They interpreted them as nebular emission excited by a population of very hot Pop III stars formed in pockets of pristine gas at these redshifts. Gräfener *et al.* (in prep) discuss that stellar emission from a population very massive WNh stars at low Z may provide an alternative explanation.

References

Bestenlehner, J. M., Gräfener, G., Vink, J. S., *et al.* 2014, *A&A*, submitted

Cantiello, M., Langer, N., Brott, I., *et al.* 2009, *A&A* 499, 279

Cassata, P., Le Fèvre, O., Charlot, S., *et al.* 2013, *A&A* 556, A68

Crowther, P. A., Schnurr, O., Hirschi, R., *et al.* 2010, *MNRAS* 408, 731

Glatzel, W. 2008, in A. Werner & T. Rauch (ed.), *Hydrogen-deficient stars*, Vol. 391 of *Astronomical Society of the Pacific Conference Series*, p. 307, San Francisco: Astronomical Society of the Pacific

Glatzel, W. 2009, *Communications in Asteroseismology* 158, 252

Glatzel, W. & Kaltschmidt, H. O. 2002, *MNRAS* 337, 743

Gräfener, G. & Hamann, W.-R. 2008, *A&A* 482, 945

Gräfener, G., Owocki, S. P., & Vink, J. S. 2012a, *A&A* 538, A40

Gräfener, G., Vink, J. S., de Koter, A., & Langer, N. 2011, *A&A* 535, A56

Gräfener, G., Vink, J. S., Harries, T. J., & Langer, N. 2012b, *A&A* 547, A83

Groh, J. H., Hillier, D. J., Damineli, A., *et al.* 2009, *ApJ* 698, 1698

Hamann, W.-R., Gräfener, G., & Liermann, A. 2006, *A&A* 457, 1015

Hamann, W.-R. & Koesterke, L. 1998, *A&A* 335, 1003

Ishii, M., Ueno, M., & Kato, M. 1999, *PASJ* 51, 417

Meynet, G. & Maeder, A. 2002, *A&A* 390, 561

Petrovic, J., Pols, O., & Langer, N. 2006, *A&A* 450, 219

Saio, H., Baker, N. H., & Gautschy, A. 1998, *MNRAS* 294, 622

Sander, A., Hamann, W.-R., & Todt, H. 2012, *A&A* 540, A144

Vink, J. S., Muijres, L. E., Anthonisse, B., *et al.* 2011, *A&A* 531, A132

Discussion

MORAVVEJI: The position of the Fe-bump is at $\log(T/K) \sim 5.2$. Is there a possibility – from theoretical calculations – that the bump is situated closer to the surface or towards the core?

GRÄFENER: No, the bound-bound opacity calculations seem robust and there seems not much freedom from theory to change the position of the iron bump.

DE KOTER: You discussed that the clumping properties that you require in the envelope near the peak iron bump (at 150000 K) are similar to the clumping properties in the outflows (the wind). Are there reasons to expect that these clumping properties should be the same?

GRÄFENER: In a recent work (Gräfener & Vink 2013) we discussed that the location of the sonic point in WR winds follows a relation that requires that the sub-surface layers of most WR stars are clumped and inflated.

HIRSCHI: Where would the wind start in stars with inflated envelopes?

GRÄFENER: The models presented here are static. In the dynamic case, the location of the sonic point (and thus the mass-loss rate) will likely affect the envelope inflation.

New windows on massive stars: asteroseismology, interferometry, and spectropolarimetry
Proceedings IAU Symposium No. 307, 2014
G. Meynet, C. Georgy, J. H. Groh & Ph. Stee, eds.

© International Astronomical Union 2015
doi:10.1017/S1743921314006279

Discovery of a Thorne-Żytkow object candidate in the Small Magellanic Cloud

Emily M. Levesque[1], Philip Massey[2], Anna N. Żytkow[3] and Nidia Morrell[4]

[1] Center for Astrophysics & Space Astronomy, University of Colorado UCB 389, Boulder, CO 80309, USA; Hubble Fellow
email: Emily.Levesque@colorado.edu

[2] Lowell Observatory, 1400 W. Mars Hill Road, Flagstaff, AZ 86001, USA

[3] Institute of Astronomy, University of Cambridge, Madingley Road, Cambridge CB3 0HA, UK

[4] Las Campanas Observatory, Carnegie Observatories, Casilla 601, La Serena, Chile

Abstract. Thorne-Żytkow objects (TŻOs) are a theoretical class of star in which a compact neutron star is surrounded by a large, diffuse envelope. Supergiant TŻOs are predicted to be almost identical in appearance to red supergiants (RSGs), with their very red colors and cool temperatures placing them at the Hayashi limit on the H-R diagram. The only features that can be used at present to distinguish TŻOs from the general RSG population are the unusually strong heavy-element and lithium lines present in their spectra. These elements are the unique products of the stars fully convective envelope linking the photosphere with the extraordinarily hot burning region in the vicinity of the neutron star core. We have recently discovered a TŻO candidate in the Small Magellanic Cloud. It is the first star to display the distinctive chemical profile of anomalous element enhancements thought to be characteristic of TŻOs; however, up-to-date models and additional observable predictions (including potential asteroseismological signatures) are required to solidify this discovery. The definitive detection of a TŻO would provide the first direct evidence for a completely new model of stellar interiors, a theoretically predicted fate for massive binary systems, and never-before-seen nucleosynthesis processes that would offer a new channel for heavy-element and lithium production in our universe.

Keywords. stars: peculiar, supergiants, variables

1. Introduction

Thorne-Żytkow objects (TŻOs) are a class of star originally proposed by Thorne & Żytkow (1975, 1977), comprised of a neutron star core surrounded by a large and diffuse envelope. They are expected to form as the product of a dual-massive-star binary, with the neutron star forming when the more massive star explodes as a supernova. During subsequent evolution of the system, the expanding envelope of the lower-mass companion may lead to a common envelope state and the spiral-in of the neutron star into the companion's core (Taam *et al.* 1978). Alternately, Leonard *et al.* (1994) propose that a TŻO may be produced when a newly-formed neutron star receives a supernova 'kick' velocity in the direction of its red supergiant (RSG) companion and becomes embedded.

TŻOs represent a completely new theoretical class of stellar object, offering a novel model for stellar interiors and an example of unique nucleosynthetic processes. However, there has never been a positive observational identification of a TŻO. They are predicted to be virtually indistinguishable from luminous M-type RSGs, lying at the Hayashi limit (Hayashi & Hoshi 1961) and showing signs of excess mass loss. To date, the only predicted observational signature of a TŻO is an unusual set of atmospheric chemical abundances, produced by the 'interrupted rapid-proton' process or *irp*-process that is uniquely possible in a stellar interior combining a neutron star core and a completely convective envelope (e.g. Zimmerman 1979; Biehle 1991; Cannon 1993; see also Wallace & Woosley 1981). Several specific *irp*-process elements should be observable in a TŻO atmosphere, including lines of Rb I and Mo I (Biehle 1994). ^{7}Li should also be over-abundant in TŻOs due to the ^{7}Be-transport mechanism, which is similarly dependent on the star's internal structure (e.g. Cameron 1955; Podsiadlowski *et al.* 1995). While previous searches have been made for TŻOs, none have observed a star with a chemical profile attributable to a TŻO interior (Vanture *et al.* 1999; Kuchner *et al.* 2002).

Here we present our observations of the TŻO candidate HV 2112 in the Small Magellanic Cloud. This star exhibits the distinctive set of anomalous element enhancements thought to be produced by processes unique to TŻOs. We discuss our sample, observations, and analyses (Section 2), discuss the properties of HV 2112 (Section 3), and consider the implications and potential for future work on TŻOs (Section 4).

2. Observations, Reduction, and Analysis

We constructed our sample of 22 SMC RSGs from objects observed in our past effective temperature (T_{eff}) studies (Levesque *et al.* 2006) and additional stars with 2MASS photometry and colors consistent with RSGs ($K < 8.9$, $J - K > 1$). Closely-orbiting massive binaries are more common at low metallicities (Linden *et al.* 2010), making the predicted binary progenitor scenarios for TŻOs more likely in metal-poor host environments such as the SMC ($\log(\mathrm{O/H}) + 12 = 8.0$, van den Bergh 2000).

Our observations were carried out using the Magellan Inamori Kyocera Echelle (MIKE; Bernstein et al. 2003) and Magellan Echellete (MagE; Marshall *et al.* 2008) spectrographs on the Magellan 6.5-meter Clay telescope at Las Campanas Observatory on 13-15 Sep 2011. We obtained high-resolution spectra with MIKE (using the $0'.7 \times 5'$ slit with 2×2 binning, 'slow' readout, and the standard grating settings to give $R \sim 42,000$) for the purpose of measuring line ratios, as well as lower-resolution MagE spectra (using the $1''$ with 1×1 binning and 'slow' readout) for the purpose of producing flux-calibrated spectrophotometry that could be used to determine contemporaneous physical properties for our stars. Our observations were generally performed with seeing $\lesssim 1''$ and airmasses $\lesssim 1.5$; our MagE spectra were taken at the parallactic angle, and spectrophotometric standards were observed for flux calibration. The MIKE and MagE data were reduced using standard IRAF echelle routines and the `mtools` package (see Massey *et al.* 2012).

To search for signs of anomalous TŻO-like element enhancements in our spectra, we compared the equivalent widths of Li, Rb, and Mo absorption features - all elements expected to be enhanced in TŻOs - to those of nearby spectral features where no significant enhancements were previously predicted, such as K, Ca, Fe, and Ni (Cannon 1993; Biehle 1994; Podsiadlowski *et al.* 1995; however, see also Tout *et al.* 2014). However, line blanketing effects in RSG spectra completely obscure the continuum, rendering traditional measurements of absorption impossible. We therefore adopt the 'pseudo-equivalent width" method detailed in Kuchner *et al.* (2002), and determined pseudo-equivalent width measurements for our lines of interest with systematic errors of $\lesssim 5\%$.

We calculated the the Rb I $\lambda 7800.23$/Ni I $\lambda 7797.58$ (Rb/Ni) and Rb I $\lambda 7800.23$/Fe I $\lambda 7802.47$ (Rb/Fe) line ratios to probe the relative Rb enhancement in our stars; the Li I $\lambda 6707.97$/K I $\lambda 7698.97$ (Li/K) and Li I $\lambda 6707.97$/Ca I $\lambda 6572.78$ (Li/Ca) ratios for the relative Li enhancement; and the Mo I $\lambda 5570.40$/Fe I $\lambda 5569.62$ (Mo/Fe) ratio for the relative Mo enhancement. To ensure that our ratios were truly probing abundances anomalies in TŻO products rather than variations in our comparison features, we also

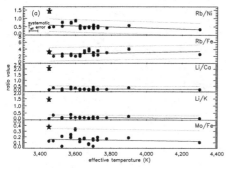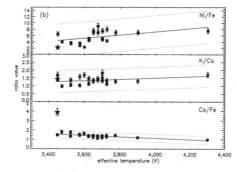

Figure 1. T_{eff} vs. measured line ratios for our SMC RSGs (circles) and HV 2112 (stars). The ratios include features predicted to be enhanced in TŻOs (a) as well as our 'control' features (b). A dark grey line shows the best linear fit for each line ratio as a function of T_{eff}, while light grey lines mark the 3σ deviations from the fit that encompass 99.7 per cent of a normally-distributed sample. Error bars for each point illustrate the systematic errors of $\lesssim 5\%$.

calculated Ni I $\lambda7797.58$/Fe I $\lambda7802.47$ (Ni/Fe), K I $\lambda7698.97$/Ca I $\lambda6572.78$ (K/Ca), and Ca I $\lambda6572.78$/Fe I $\lambda65569.62$ (Ca/Fe) ratios for our stars.

We compared the line ratios for our stars relative to their measured T_{eff}. We then determined the best linear fit for each set of ratios as a function of T_{eff}, and calculated the 3σ variations from these fits that should encompass 99.7% of a normally-distributed sample. Following this method, we considered a ratio to be statistically anomalous if it fell outside the 3σ range. For a more detailed discussion of our observations, reductions, and analysis techniques please see Levesque *et al.* (2014).

3. The TŻO Candidate HV 2112

We found that one star, HV 2112, has anomalously high values of Rb/Ni, Li/K, Li/Ca, and Mo/Fe. The ratios measured for our full sample, including HV 2112, are shown in Figure 1. These ratios present clear evidence of Rb, Li, *and* Mo enhancement in the star's atmosphere, a combination thought to be possible as a result of the exotic stellar interiors of TŻOs. A comparison of the HV 2112 spectrum to a 'typical' SMC RSG is shown in Figure 2 (left); the HV 2112 spectrum exhibits notably stronger TŻO features. We also measure an atypically high Ca/K ratio; recent work suggests that Ca may be produced in the final stages of TŻO formation (Tout *et al.* 2014).

Observations of the Ca II triplet in the HV 2112 allowed us to measure a radial velocity (RV) of $\sim 157\,\text{km}\,\text{s}^{-1}$, consistent with this star being a member of the SMC rather than a foreground dwarf (Neugent *et al.* 2010). From fitting a MagE spectrum of HV 2112 with a MARCS stellar atmosphere model (Gustafsson *et al.* 2008; see Figure 2, right), we determined a T_{eff} of 3450 K and a spectral type of M3 I, along with a V magnitude of 13.7 ± 0.1. Correcting for the SMC distance modulus of 18.9 (van den Bergh 2000) and the T_{eff}-dependent BC in V (Levesque *et al.* 2006) we determined an $M_{\text{bol}} = -7.82 \pm 0.2$; this corresponds to a initial mass of $M \sim 15\,M_{\odot}$ (Maeder & Meynet 2001). We also observed an excess amount of visual extinction in the direction of HV 2112 ($A_V \sim 0.4$ as compared to the average $A_V = 0.24$ for SMC OB stars assuming $R_V = 3.1$; Cardelli *et al.* 1989; Massey *et al.* 1995) and a slight flux excess ($\lesssim 10\,\text{ergs}\,\text{s}^{-1}\,\text{cm}^{-2}\,\text{Å}^{-1}$) in the near-UV. Both of these features are thought to be signatures of excess circumstellar dust associated with strong mass loss in RSGs (Levesque *et al.* 2005; Massey *et al.* 2005). Finally, past work identified HV 2112 as photometrically and spectroscopically variable (Payne-Gaposchkin & Gaposchkin 1966; Wood *et al.* 1983; Reid & Mould 1990; Smith

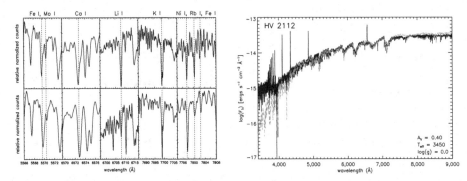

Figure 2. Left: spectral features of HV 2112 (a-e) and [M2002] SMC 005092 (f-j) used in our analyses. [M2002] SMC 005092 is a 'typical' SMC RSG with $T_{\rm eff} = 3475\,$K, comparable to the 3450 K $T_{\rm eff}$ we measure for HV 2112. Spectra are from our MIKE observations and have been corrected to rest-frame. Wavelengths of key absorption features are marked as vertical dashed lines. **Right**: MagE spectral energy distribution of HV 2112 (solid black line) and the best-fit MARCS stellar atmosphere model (dashed gray). The strong Balmer emission in the HV 2112 spectrum is clearly visible, along with the near-UV excess and near-IR deficiency typical of circumstellar dust effects (Levesque *et al.* 2005; Massey *et al.* 2005).

et al. 1995), properties predicted for TŻOs (Thorne & Żytkow 1977; van Paradijs *et al.* 1995).

Rb and Mo have been individually observed in stars and attributed to the *s*-process; however, there are no observed or predicted examples of the *s*-process producing both in a single star along with additional Li enhancement. Simultaneous Rb and Li enhancement is not observed in cool *s*-process stars (e.g. García-Hernández *et al.* 2013), and the combination of Rb, Mo, and Li has never been observed in any other cool massive star. Futhermore, while Rb, Mo, and Li can all individually be produced in super-asymptotic giant branch (SAGB) stars as well as TŻOs (through a combination of the *s*-process and hot bottom burning), an SAGB star cannot synthesize their own Ca and would not produce the Ca enhancement seen in HV 2112 (Tout *et al.* 2014). Finally, Ba is a common *s*-process product but we see no signs of Ba enhancement in the HV 2112 spectrum (see Vanture *et al.* 1999). Combined, our observed element enhancements and physical properties of HV 2112 strongly support its classification as a TŻO candidate.

However, HV 2112 does exhibit several unusual spectral features not previously predicted to be associated with TŻOs. The spectrum exhibit strong hydrogen Balmer emission features than extend from Hα out to H18 (see Figure 2, right) and display a Balmer *increment* (with Hδ as the strongest feature followed by Hγ, Hβ, and Hα) similar to that seen in M-type Mira variables. In Miras, this inverted Balmer emission is thought to be produced by non-LTE radiative transfer effects in the hydrogen lines from formation in a shocked atmosphere (Castelaz *et al.* 2000), with shocks generated by pulsation and propagating through the atmosphere. While the origin of these Balmer emission features in HV 2112 remains unclear, it is likely that the photometric and spectroscopic variability associated with HV 2112 could produce similar pulsationally-driven shocks.

In addition, while the Rb/Ni ratio in HV 2112 is far higher than the average ratio measured for SMC RSGs, the Rb/Fe ratio is quite typical. The Ni/Fe ratio in HV 2112 is also typical of the RSG sample, precluding the possible explanation of a Fe overabundance. The Mo, Li, and Rb features do not, at a glance, exhibit the extremely increased strengths one might expect based on the predicted enhancements for TŻOs (Cannon 1993; Biehle 1994; Podsiadlowski *et al.* 1995); however, relatively weak enhancements could be indicative of an early or short-lived TŻO phase (Thorne & Żytkow 1975, 1977;

Podsiadlowski *et al.* 1995). Convection models for massive star envelopes (particularly at low temperatures) have also advanced significantly in recent years, suggesting that the predicted enhancements for TŻOs could potentially be altered. It is clear that new models of TŻOs are needed; future theoretical work will likely provide updated predictions of element enhancements as well as additional identifying observables that can be used to test whether HV 2112 is indeed a bona fide TŻO.

EML is supported by Hubble Fellowship grant HST-HF-51324.01-A from STScI, which is operated by AURA under NASA contract NAS5-26555. PM acknowledges support from NSF grant AST-1008020. ANŻ thanks the Mitchell Family Foundation for support and Texas A&M University and Cook's Branch Nature Conservancy for their hospitality.

References

Bernstein, R., Shectman, S. A., Gunnels, S. M., Mochnacki, S., & Athey, A. E. 2003, in M. Iye & A. F. M. Moorwood (eds.), *Instrument Design and Performance for Optical/Infrared Ground-based Telescopes*, Vol. 4841 of *Society of Photo-Optical Instrumentation Engineers (SPIE) Conference Series*, pp 1694–1704

Biehle, G. T. 1991, *ApJ* 380, 167

Biehle, G. T. 1994, *ApJ* 420, 364

Cameron, A. G. W. 1955, *ApJ* 121, 144

Cannon, R. C. 1993, *MNRAS* 263, 817

Cardelli, J. A., Clayton, G. C., & Mathis, J. S. 1989, *ApJ* 345, 245

Castelaz, M. W., Luttermoser, D. G., Caton, D. B., & Piontek, R. A. 2000, *AJ* 120, 2627

García-Hernández, D. A., Zamora, O., Yagüe, A., *et al.* 2013, *A&A* 555, L3

Gustafsson, B., Edvardsson, B., Eriksson, K., *et al.* 2008, *A&A* 486, 951

Hayashi, C. & Hoshi, R. 1961, *PASJ* 13, 442

Kuchner, M. J., Vakil, D., Smith, V. V., *et al.* 2002, in M. M. Shara (ed.), *Stellar Collisions, Mergers and their Consequences*, Vol. 263 of *Astronomical Society of the Pacific Conference Series*, p. 131

Leonard, P. J. T., Hills, J. G., & Dewey, R. J. 1994, *ApJ (Letters)* 423, L19

Levesque, E. M., Massey, P., Olsen, K. A. G., *et al.* 2005, *ApJ* 628, 973

Levesque, E. M., Massey, P., Olsen, K. A. G., *et al.* 2006, *ApJ* 645, 1102

Levesque, E. M., Massey, P., Żytkow, A. N., & Morrell, N. 2014, *MNRAS* 443, L94

Linden, T., Kalogera, V., Sepinsky, J. F., *et al.* 2010, *ApJ* 725, 1984

Maeder, A. & Meynet, G. 2001, *A&A* 373, 555

Marshall, J. L., Burles, S., Thompson, I. B., *et al.* 2008, in *Society of Photo-Optical Instrumentation Engineers (SPIE) Conference Series*, Vol. 7014 of *Society of Photo-Optical Instrumentation Engineers (SPIE) Conference Series*

Massey, P., Lang, C. C., Degioia-Eastwood, K., & Garmany, C. D. 1995, *ApJ* 438, 188

Massey, P., Morrell, N. I., Neugent, K. F., *et al.* 2012, *ApJ* 748, 96

Massey, P., Plez, B., Levesque, E. M., *et al.* 2005, *ApJ* 634, 1286

Neugent, K. F., Massey, P., Skiff, B., *et al.* 2010, *ApJ* 719, 1784

Payne-Gaposchkin, C. & Gaposchkin, S. 1966, *Smithsonian Contributions to Astrophysics* 9, 1

Podsiadlowski, P., Cannon, R. C., & Rees, M. J. 1995, *MNRAS* 274, 485

Reid, N. & Mould, J. 1990, *ApJ* 360, 490

Smith, V. V., Plez, B., Lambert, D. L., & Lubowich, D. A. 1995, *ApJ* 441, 735

Taam, R. E., Bodenheimer, P., & Ostriker, J. P. 1978, *ApJ* 222, 269

Thorne, K. S. & Żytkow, A. N. 1975, *ApJ (Letters)* 199, L19

Thorne, K. S. & Żytkow, A. N. 1977, *ApJ* 212, 832

Tout, C. A., Zytkow, A. N., Church, R. P., & Lau, H. H. B. 2014, *ArXiv e-prints*

van den Bergh, S. 2000, *The Galaxies of the Local Group*, Cambridge

van Paradijs, J., Spruit, H. C., van Langevelde, H. J., & Waters, L. B. F. M. 1995, *A&A* 303, L25
Vanture, A. D., Zucker, D., & Wallerstein, G. 1999, *ApJ* 514, 932
Wallace, R. K. & Woosley, S. E. 1981, *ApJS* 45, 389
Wood, P. R., Bessell, M. S., & Fox, M. W. 1983, *ApJ* 272, 99
Zimmerman, M. L. 1979, *Ph.D. thesis*, MASSACHUSETTS INSTITUTE OF TECHNOLOGY.

Discussion

GREVESSE: Why do you not use spectral synthesis around your lines rather than equivalent widths measured with a "pseudo-continuum"?

LEVESQUE: High-resolution modeling of RSG atmospheres is subject to a number of physical complexities, including treatments of atmospheric geometry, optical depth variations, mass loss effects, and host galaxy abundance variations. To avoid potential biases introduced by the assumptions inherent to atmospheric modeling, we instead simply calculated equivalent width ratios for our spectral lines of interest and compared the resulting ratios across our entire sample to identify outliers. This allowed us to do a direct and consistent comparison that is independent of any modeling assumptions.

VAN BELLE: Why doesn't the formation of a TŻO result in explosive detonation of the system? This seems like the ingredients of a cataclysmic variable.

LEVESQUE: This is definitely a good question, and one that we're hoping to explore further in future work with formation models! For now, the Taam *et al.* (1978) and Leonard *et al.* (1994) papers detailing the two most commonly-cited massive binary formation mechanisms are probably the best resources for answer this. An interesting consideration along these lines is whether there would be some kind of non-terminal eruptive event or other observable signature during the actual formation process of a TŻO, and this will also hopefully be addressed by future models.

MEYNET: Do we have any information about the rotation rate of the TŻO candidate?

LEVESQUE: We unfortunately don't have estimates of this from our observations, but future work on binary merger models for producing TŻOs would likely be one means of predicting what kind of surface rotation rate we would expect to see.

GROH: Have you had a chance to look at the circumstellar environment of this object, e.g. in the mid-IR? Also, are you able to quantify $\dot{M}$ at this point?

LEVESQUE: We haven't yet examined the circumstellar environment of HV 2112 in the mir-IR, and we can't quantify $\dot{M}$ with our current data. Imaging of the circumstellar environment would certainly be extremely interesting both for studying the mass loss properties and for seeking out any potential remnants of a recent supernova that could have produced the core neutron star.

KHALACK: Have you taken into account the NLTE effects that can contribute to the formation of line profiles that belong to Li, Ru, Mo, Y, etc.?

LEVESQUE: This isn't something we have examined directly; our ratio determinations are independent of any modeling assumptions, so NLTE effects are not a primary concern for this work. However, these effects would of course be important in future modeling of TŻO atmospheres!

KHALACK: How many lines have you used to determine an average ratio of equivalent widths (for two elements)?

LEVESQUE: We do not determine an "average ratio" of equivalent widths for two elements; instead, we simply measure the ratio of equivalent widths for two spectral features, one from each element that we are interested in, and compare the individual ratios from each star in our sample to identify any anomalous ratios. This is discussed in more detail in the above proceeding

ANDERSON: Given the spectroscopic variability you mentioned, have you considered photometric time-series from the EROS, OGLE, and MACHO surveys? Doing so could help find additional differences to RSGs, semi-regular variables, and Miras.

LEVESQUE: This star has been included in a number of different optical and IR surveys and has long been identified as a photometric variable; we haven't yet examined this data in detail but agree that it would be an interesting approach for potentially identifying other similar TŻO candidates in the future and better understanding HV 2112 in particular.

Emily Levesque

Alex de Koter asking a question

New windows on massive stars: asteroseismology, interferometry, and
spectropolarimetry
Proceedings IAU Symposium No. 307, 2014
G. Meynet, C. Georgy, J. H. Groh & Ph. Stee, eds.

© International Astronomical Union 2015
doi:10.1017/S1743921314006280

A New Class of Wolf-Rayet Stars: WN3/O3s

Philip Massey[1], Kathryn F. Neugent[1], Nidia Morrell[2] and D. John Hillier[3]

[1]Lowell Observatory
email: phil.massey@lowell.edu

[2]Las Campanas Observatory, Carnegie Observatories

[3]Department of Physics and Astronomy & Pittsburgh Particle Physics, Astrophysics, and Cosmology Center, University of Pittsburgh

Abstract. Our new survey for Wolf-Rayet stars in the Magellanic Clouds is only 15% complete but has already found 9 new Wolf-Rayet (WR) stars in the Large Magellanic Cloud (LMC). This suggests that the total WR population in the LMC may be underestimated by 10-40%. Eight of the nine are of the WN subtype, demonstrating that the "observed" WC to WN ratio is too large, and is biased towards WC stars. The ninth is another rare WO star, the second we have found in the LMC in the past two years. Five (and possibly six) of the 8 WNs are of a new class of WRs, which pose a significant challenge to our understanding. Naively we would classify these stars as "WN3+O3V," but there are several reasons why such a pairing is unlikely, not the least of which is that the absolute visual magnitudes of these stars are *faint*, with $M_V \sim -2.3$ to -3.1. We have performed a preliminary analysis with the atmospheric code CMFGEN, and we find that (despite the faint visual magnitudes) the bolometric luminosities of these stars are normal for early-type WNs. Our fitting suggests that these stars are evolved, with significantly enriched N and He. Their effective temperatures are also normal for early-type WNs. What is unusual about these stars is that they have a surprisingly small mass-loss rate compared to other early-type WNs. How these stars got to be the way they are (single star evolution? binary evolution?) remains an open question. For now, we are designating this class as WN3/O3, in analogy to the late-type WN "slash" stars.

Keywords. surveys, stars: Wolf-Rayet, stars: evolution, stars: early-type, galaxies: individual (LMC)

1. A Modern Survey for Wolf-Rayet Stars in the Magellanic Clouds

One of the long-standing tests of massive star evolutionary models is how well they do in predicting the observed ratios of various types of evolved massive stars as a function of metallicity (see, e.g., Maeder *et al.* 1980; Massey & Johnson 1998; Meynet & Maeder 2005; Meynet *et al.* 2007; Eldridge *et al.* 2008; Neugent *et al.* 2012a). Of these tests, possibly the most robust is that of comparing the relative number of WC- and WN-type Wolf-Rayet (WR) stars. Observationally, this quantity has long been known to be low in the Small Magellanic Cloud (SMC), higher in the Large Magellanic Cloud (LMC), and higher still in the Milky Way, in accord with the progression of metallicity. Vanbeveren & Conti (1980) were the first to attribute this to the effects of metallicity on the mass-loss rates of main-sequence massive stars. However, as shown in many previous studies (e.g., Massey & Conti 1983; Armandroff & Massey 1985; Massey & Johnson 1998; Neugent & Massey 2011; Neugent *et al.* 2012a), the "observed" ratios of WCs to WNs can readily be biased towards stars of WC type, as the strongest lines in WCs are significantly ($\sim 10\times$) stronger than the strongest lines in WNs (Massey & Johnson 1998). We have therefore begun new surveys for WRs in the star-forming galaxies of the Local Group (i.e., M33,

Neugent & Massey 2011, and M31, Neugent *et al.* 2012a) to allow more meaningful comparisons with the new generation of evolutionary models now becoming available (e.g., Eldridge *et al.* 2008; Ekström *et al.* 2012; Georgy *et al.* 2013).

As part of this process, our attention was drawn to the fact that the Geneva rotating models at LMC metallicity predict a WC to WN ratio of 0.09 (Neugent *et al.* 2012a) while the observed ratio is 0.23 according to the Breysacher *et al.* (1999) catalog (BAT99), as updated by Neugent *et al.* (2012b). Is this a problem with the single-star Geneva models, or could the problem be observational? For many years, our knowledge of the WR population of the Magellanic Clouds has been considered essentially complete. We realized that in the 15 years since the BAT99 catalog was published, there were seven new WRs found in the LMC, 6 of them of were of WN type. The other one was a rare WO star, found by ourselves (Neugent *et al.* 2012b). The WO star has very strong lines, and was found as part of a spectroscopic study of the stellar content of Lucke-Hodge 41, the home of S Doradus and many other massive stars. This discovery served as pretty much the last straw: we felt it behooved us to conduct a modern search for Wolf-Rayet stars in the Magellanic Clouds (Massey *et al.* 2014).

It was clear from the onset *how* we needed to do this: we would use the same sort of interference filters we had used so successfully to survey M33 and M31 for WRs (Neugent & Massey 2011; Neugent *et al.* 2012a). We even knew *where* we wanted to do this: the Swope 1-m telescope had both a good image scale and a large field-of-view (FOV). Nevertheless, it would take about 800 fields and several hundred hours of observing to completely cover both Clouds. Fortunately they have just replaced the camera with one with an even larger FOV and much reduced overhead. Nevertheless this is a long term, multi-year project. In our first year (with 6 excellent nights) we were able to survey about ∼15% of each Cloud. Image-subtraction techniques then allowed us to identify stars which were brighter in a C III $\lambda4650$ or a He II $\lambda4686$ filter relative to neighboring continuum. We had concentrated on fields where WR stars were already known (following the philosophy attributed apocryphally to "Slick" Willie Sutton) and readily recovered all of the previously known WRs, except in the most crowded regions of the R136 cluster. We also re-discovered many previously known planetary nebulae and Of-type stars, both of which would have He II $\lambda4686$ in emission. And, much to our relief, we also found a number of previously unknown WR candidates. Follow-up spectroscopy with Magellan allowed us to confirm 9 new WRs in the LMC in our first observing season. This increases the number of known WRs in the LMC by about 6%, suggesting that the total WR population of the LMC has been underestimated by 10-40%. While impressive, the greater significance is not in the quantity but the quality of what we found.

2. What We Found

Of the 9 newly found WRs in the LMC, 8 of these are WN stars. The other is another WO star. The WO is located only $9''$ from the one we found two years ago by accident in Lucke-Hodge 41 (Neugent *et al.* 2012b). Of the 8 WNs, 2 are normal WN3+mid-to-late O-type binaries; we were able to obtain two epochs of radial velocities that show the absorption and emission moving in anti-phase, as one would expect.

Five others (and possibly the sixth) are, however, quite peculiar. Naively we would classify their spectra as WN3+O3V. However, such a pairing would be unlikely for several reasons, the most unarguable one being that the absolute visual magnitudes of these six stars are quite faint, with $M_V \sim -2.3$ to -3.1. This is much fainter than that of a typical O3 V star, with $M_V \sim -5.4$ (Conti 1988). Furthermore, we had two radial velocity epochs for three of these which failed to show variations. We will therefore refer to these

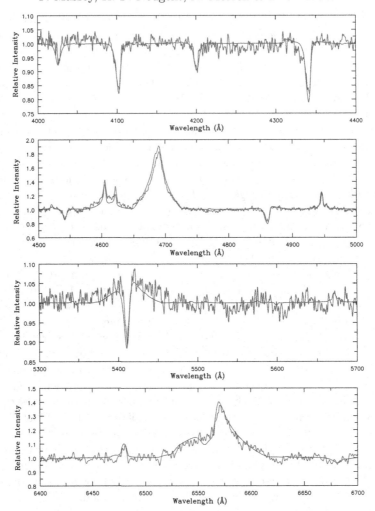

Figure 1. Our best CMFGEN fit to the optical spectrum of LMC170-2 From Massey *et al.* (2014) and used by permission.

stars as WN3/O3 WRs, with the slash reminiscent of the "transition" Ofpe/WN9 stars (Bohannan & Walborn 1989; Bohannan 1990).

Could both the absorption and emission be coming from a single object? To explore this, we attempted to fit our optical spectrum of one of these stars (LMC170-2) with CMFGEN (Hillier & Miller 1998), a stellar atmosphere code designed for hot stars with stellar winds where the usual assumptions of plane-parallel geometry and local thermo-dynamic equilibrium no longer hold. We found that we could obtain a very good fit with a single set of parameters as given in Table 1. Portions of the fit are shown in Fig. 1. A slightly different model also produced an adequate fit, and we give these parameters in Table 1 as well, to demonstrate the current uncertainties.

In Fig. 2 we show how these parameters compare to that of the LMC WN stars recently analyzed by Hainich *et al.* (2014). What we find is that the effective temperature and luminosity are normal for early-type WN stars (Fig. 2, left), but that the mass-loss rates are significantly lower (Fig. 2, right). In many ways this makes sense, as it suggests that the winds are weak enough that we see something like a normal stellar photosphere.

Table 1. Parameters of our CMFGEN fits.

Parameter	Best Fit	Pretty Good Fit
Teff (K)	100,000	80,000
$L/L_\odot$	4×10^5	2.0×10^5
$\dot{M}^1$ ($M_\odot$/yr)	1.2×10^{-6}	7.6×10^{-7}
He/H (by #)	1.0	0.5
N	10.0× solar	5.0× solar
C, O	0.05× solar	0.05× solar

Notes:
[1] Assumes a clumping filling factor of 0.1, wind terminal velocity of v_∞=2400 km/sec, and beta-type velocity law with β=0.8.

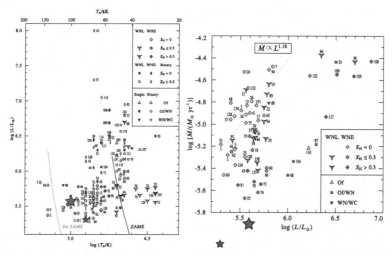

Figure 2. How the parameters of our fit of the WN3/O3 star LMC170-2 compare to those of other LMC WNs analyzed by Hainich *et al.* (2014). Our best fit is shown by a large star, while our "pretty good" fit is shown by a smaller star. Adapted from Figs. 6 and 7 of Hainich *et al.* (2014) and used by permission.

Of course, this now raises significant new questions: Why are the mass-loss rates low, and how did these stars evolve? Are binary models needed to produce such objects, or are these the hitherto unrecognized products of single-star evolution? If the former, then where is the spectroscopic signature of the companion?

We note that at this point the physical parameters are not well constrained (compare the two models in Table 1): the optical spectra lack multiple ionization stages of the same species, impacting our ability to constrain the effective temperature. We just learned that we were successful in obtaining time on *HST* to observe these objects in the UV; these spectra will provide data on additional ionization states (such as N IV λ1718) as well as provide key diagnostics of the stellar wind, such as the C IV λ1550 resonance line. We also plan to obtain higher S/N optical data with Magellan, which should also help. We will continue to monitor these stars for radial velocity variations. We will be carrying out our survey for WRs in the rest of the Magellanic Clouds, and this will show us to see how many more of these WN3/O3s are out there.

And, of course, we do not know what other surprises await! In addition to these new WRs our study has also revealed two O8f?p stars, further examples of this rare class of magnetically-braked oblique rotators (Walborn *et al.* 2010). We have ten further nights scheduled on the Swope in the next Magellanic Cloud observing season (late 2014) with time for follow-up spectroscopy. So, stayed tuned!

This work has been supported by the National Science Foundation under AST-1008020 and by Lowell Observatory's research support fund, thanks to generous donations by Mr. Michael Beckage and Mr. Donald Trantow. D.J.H. acknowledges support from STScI theory grant HST-AR-12640.01. We appreciate the fine support we always receive at Las Campanas Observatory, where these observations were carried out. We are also grateful to our hosts here in Geneva for providing an opportunity to discuss this work in such a nice setting!

References

Armandroff, T. E. & Massey, P. 1985, *ApJ* 291, 685

Bohannan, B. 1990, in C. D. Garmany (ed.), *Properties of Hot Luminous Stars*, Vol. 7 of *Astronomical Society of the Pacific Conference Series*, pp 39–43

Bohannan, B. & Walborn, N. R. 1989, *PASP* 101, 520

Breysacher, J., Azzopardi, M., & Testor, G. 1999, *A & AS* 137, 117

Conti, P. S. 1988, *NASA Special Publication* 497, 119

Ekström, S., Georgy, C., Eggenberger, P., *et al.* 2012, *A&A* 537, A146

Eldridge, J. J., Izzard, R. G., & Tout, C. A. 2008, *MNRAS* 384, 1109

Georgy, C., Ekström, S., Eggenberger, P., *et al.* 2013, *A&A* 558, A103

Hainich, R., Rühling, U., Todt, H., *et al.* 2014, *A&A* 565, A27

Hillier, D. J. & Miller, D. L. 1998, *ApJ* 496, 407

Maeder, A., Lequeux, J., & Azzopardi, M. 1980, *A&A* 90, L17

Massey, P. & Conti, P. S. 1983, *ApJ* 273, 576

Massey, P. & Johnson, O. 1998, *ApJ* 505, 793

Massey, P., Neugent, K. F., Morrell, N., & Hillier, D. J. 2014, *ApJ* 788, 83

Meynet, G., Eggenberger, P., & Maeder, A. 2007, in A. Vazdekis & R. Peletier (eds.), *Stellar Populations as Building Blocks of Galaxies*, Vol. 241 of *IAU Symposium*, pp 13–22

Meynet, G. & Maeder, A. 2005, *A&A* 429, 581

Neugent, K. F. & Massey, P. 2011, *ApJ* 733, 123

Neugent, K. F., Massey, P., & Georgy, C. 2012a, *ApJ* 759, 11

Neugent, K. F., Massey, P., & Morrell, N. 2012b, *AJ* 144, 162

Vanbeveren, D. & Conti, P. S. 1980, *A&A* 88, 230

Walborn, N. R., Sota, A., Maíz Apellániz, J., *et al.* 2010, *ApJ (Letters)* 711, L143

Discussion

SUNDQVIST: It seems to me that another interpretation of your two very interesting objects would be that they are just slightly evolved, still hydrogen-burning O-star with unusual high $T_{\rm eff}$. Is this something you have considered?

MASSEY: It was our first thought – that maybe these were not evolved objects. However, our modeling has shown that N is strongly enhanced, and C and O are way down. The He/H ratio is about 1 by number. I can't say what's going on in the core. But we have never seen an O star like this, and naively to me the N V and He II emission line strengths are similar to those of WN3 stars.

NAJARRO: Phil, were you able to get a handle on $\log(g)$ (masses) for the objects?

MASSEY: We adopted a $\log(g)$ of 5.0. That would lead to a mass of $15 M_\odot$, but the uncertainties are of order factors of 2 or 3. John felt he could rule out a $\log(g)$ of 5.5

for a temperature of 100000 K, but it isn't well constrained. We hope with better optical data (higher S/N) and UV data with *HST* we can get more solid physical parameters.

Philip Massey

Victor Khalack asking a question

New windows on massive stars: asteroseismology, interferometry, and spectropolarimetry
Proceedings IAU Symposium No. 307, 2014
G. Meynet, C. Georgy, J. H. Groh & Ph. Stee, eds.

© International Astronomical Union 2015
doi:10.1017/S1743921314006292

New prescriptions of turbulent transport from local numerical simulations

V. Prat[1,2,3], F. Lignières[2,3] and G. Lesur[4,5]

[1]MPA, Karl-Schwarzschild-Str. 1, 85748, Garching b. München, Germany
email: vprat@mpa-garching.mpg.de

[2]Université de Toulouse; UPS-OMP; IRAP; Toulouse, France

[3]CNRS; IRAP; 14, avenue Édouard Belin, F-31400 Toulouse, France

[4]Univ. Grenoble Alpes, IPAG, 38000, Grenoble, France

[5]CNRS, IPAG, 38000, Grenoble, France

Abstract. Massive stars often experience fast rotation, which is known to induce turbulent mixing with a strong impact on the evolution of these stars. Local direct numerical simulations of turbulent transport in stellar radiative zones are a promising way to constrain phenomenological transport models currently used in many stellar evolution codes. We present here the results of such simulations of stably-stratified sheared turbulence taking notably into account the effects of thermal diffusion and chemical stratification. We also discuss the impact of theses results on stellar evolution theory.

Keywords. Hydrodynamics, turbulence, stars: evolution, stars: rotation

1. Introduction

Transport of chemical elements and transport of angular momentum both have a strong impact on stellar structure and evolution and can be done either by microscopic processes, which are relatively well known, or by macroscopic motions, which are still poorly understood. In particular, as massive stars often experience fast rotation, rotationally induced motions may play a crucial role. These motions are usually decomposed into large-scale axisymmetric motions, such as differential rotation and meridional circulation, and small-scale turbulent motions generated by various instabilities. Whereas the former can be either directly solved in 2D stellar evolution codes or modelled in 1D codes thanks to additional assumptions (e.g. the shellular model of Zahn 1992), the latter are fundamentally three-dimensional and thus need to be modelled in both 1D and 2D stellar evolution codes. The main obstacle to building such models is that these turbulent motions have very small length and time scales as compared to evolutionary ones.

In many current stellar evolution codes, transport of chemical elements due to shear instability is taken into account thanks to a set of diffusion coefficients initially derived by Zahn (1992) from phenomenological arguments. One of these coefficient, the radial transport coefficient generated by radial differential rotation, is given in Zahn's model by

$$D_{\rm t} = \frac{\kappa}{3} \frac{{\rm Ri}_{\rm c}}{N^2} \left(r \sin\theta \frac{{\rm d}\Omega}{{\rm d}r} \right) = \frac{\kappa}{3} \frac{{\rm Ri}_{\rm c}}{{\rm Ri}}, \tag{1.1}$$

where κ is the thermal diffusivity of the fluid, very large in stellar interiors; ${\rm Ri} = (N/S)^2$ the Richardson number, which compares stratification, characterised by the Brunt-Väisälä frequency N, and shear, characterised by the shear rate $S = r\sin\theta {\rm d}\Omega/{\rm d}r$; ${\rm Ri}_{\rm c}$ its critical value; r, θ the spherical coordinates and Ω the angular velocity of the star.

The usual way to constrain models of turbulent transport is to compare measurements of surface chemical abundances to predictions of stellar evolution models. Recently, it has become possible to constrain the internal rotation profile of red (sub-)giant stars using asteroseismology (e.g. Deheuvels *et al.* 2012). Here we use another approach: we perform local direct numerical simulations of steady homogeneous stably stratified sheared turbulence to test existing models and propose new prescriptions for stellar evolution codes.

Our governing equations are presented in Sect. 2. In Sect. 3 we summarize the results we have obtained concerning the turbulent chemical diffusity and in Sect. 4 we present new results about turbulent thermal diffusivity and turbulent viscosity. Finally, we conclude on our results and prospects in Sect. 5.

2. Governing equations

Our mean flow configuration consists of a uniform vertical velocity shear and stable, uniform vertical temperature and concentration gradients. We solve the dimensionless form of the Boussinesq equations

$$\vec{\nabla} \cdot \vec{v} = 0, \tag{2.1}$$

$$\frac{\partial \vec{v}}{\partial t} + (\vec{v} \cdot \vec{\nabla})\vec{v} = -\vec{\nabla}p + (\mathrm{Ri}\theta + \mathrm{Ri}_\mu c')\vec{e}_z + \frac{1}{\mathrm{Re}}\Delta\vec{v}, \tag{2.2}$$

$$\frac{\partial \theta}{\partial t} + \vec{v} \cdot \vec{\nabla}\theta + v_z = \frac{1}{\mathrm{Pe}}\Delta\theta, \tag{2.3}$$

$$\frac{\partial c'}{\partial t} + \vec{v} \cdot \vec{\nabla}c' + v_z = \frac{1}{\mathrm{Pe_c}}\Delta c', \tag{2.4}$$

where $\vec{v}$ and p denotes velocity and pressure, θ and c' temperature and concentration fluctuations around the mean profiles, Ri_μ the chemical equivalent of the Richardson number, $\mathrm{Re} = UL/\nu$ the Reynolds number characterising the viscosity ν and $\mathrm{Pe_c} = UL/D_\mathrm{m}$ the chemical Péclet number characterising the molecular diffusivity D_m. The regime of very high thermal diffusivities is explored using the so-called small-Péclet-number approximation (SPNA, see Lignières 1999), in which Eqs. (2.2) and (2.3) are replaced by

$$\frac{\partial \vec{v}}{\partial t} + (\vec{v} \cdot \vec{\nabla})\vec{v} = -\vec{\nabla}p + (\mathrm{Ri}\,\mathrm{Pe}\psi + \mathrm{Ri}_\mu c')\vec{e}_z + \frac{1}{\mathrm{Re}}\Delta\vec{v}, \tag{2.5}$$

$$v_z = \Delta\psi, \tag{2.6}$$

noting $\psi = \theta/\mathrm{Pe}$.

3. Turbulent chemical diffusivity

Here we are interested in the turbulent chemical diffusivity D_t. In the chemically neutral case ($\mathrm{Ri}_\mu = 0$), we studied its dependence on thermal diffusivity (see Sect. 3.1). Then, we investigated the effect of chemical stratification in the SPNA, as presented in Sect. 3.2.

3.1. *Effect of thermal diffusion*

We performed a series of simulations with different Péclet number in the Boussinesq approximation and one simulation in the SPNA. In the small-Péclet-number regime, we showed (Prat & Lignières 2013) that the quantity $D_\mathrm{t}/(\kappa\mathrm{Ri}^{-1})$ tends to a constant value

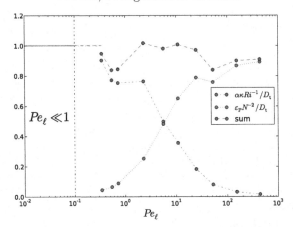

Figure 1. Ratios between models and effective diffusion coefficient as a function of the turbulent Péclet number $\mathrm{Pe}_\ell = u\ell/\kappa$, where u and ℓ are turbulent velocity and length scales (Prat & Lignières 2014)

as the Péclet number decreases, that is sensibly the same as in the SPNA. We can thus write

$$D_\mathrm{t} = \alpha\kappa\mathrm{Ri}^{-1}, \qquad (3.1)$$

with a constant α. This is in agreement with Zahn's model (1.1). Our estimate of the proportionality constant, around 5.6×10^{-2}, is of the same order of magnitude as what Zahn (1992) proposed, but 30% smaller.

In the large-Péclet-number regime (see Prat & Lignières 2014), the generalisation of Zahn's model proposed by Maeder (1995), intended to be also valid in this regime, turned to be incompatible with our simulations. In the opposite, we found a good agreement with the model of Lindborg & Brethouwer (2008), proposed in the geophysical literature, in which the diffusion coefficient is given by

$$D_\mathrm{t} = \frac{\varepsilon_\mathrm{P}}{N^2}, \qquad (3.2)$$

where ε_P is the dissipation rate of turbulent potential energy. As illustrated in Fig. 1, the total diffusion coefficient estimated in our simulations can be seen as the sum of Zahn's model and Lindborg's one. Moreover, according to Osborn & Cox (1972), the diffusion coefficient (3.2) is equal to the turbulent thermal diffusivity κ_t. One can thus write

$$D_\mathrm{t} = \kappa_\mathrm{t} + \alpha\frac{\kappa}{\mathrm{Ri}}, \qquad (3.3)$$

where a prescription for κ_t is still to be given (see Sect. 4).

3.2. Effect of chemical stratification

Our SPNA simulations performed with different values of the chemical Richardson number Ri_μ (also presented in Prat & Lignières 2014) show that the quantity $D_\mathrm{t}/(\kappa\mathrm{Ri}^{-1})$ is well represented by an affine function of Ri_μ, as represented in Fig. 2. As a consequence, one can write

$$\frac{D_\mathrm{t}}{\kappa\mathrm{Ri}^{-1}} = \beta(\mathrm{Ri}_\mathrm{c} - \mathrm{Ri}_\mu), \qquad (3.4)$$

with $\beta = 0.45$ and $\mathrm{Ri}_\mathrm{c} = 0.12$. The turbulent diffusion coefficient is then given by

$$D_\mathrm{t} = \beta\kappa\frac{\mathrm{Ri}_\mathrm{c} - \mathrm{Ri}_\mu}{\mathrm{Ri}}, \qquad (3.5)$$

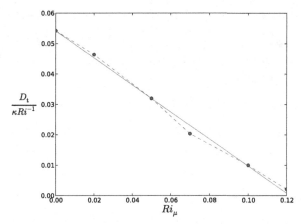

Figure 2. $D_t/(\kappa\mathrm{Ri}^{-1})$ as a function of the chemical Richardson number. Dots correspond to simulations and the solid line to the linear regression (Prat & Lignières 2014).

which is of the same form as what Maeder & Meynet (1996) derived in the small-Péclet-number regime assuming that the stability criterion in presence of a μ-gradient is

$$\mathrm{Ri}\,\mathrm{Pe}_\ell + \mathrm{Ri}_\mu > \mathrm{Ri}_c. \qquad (3.6)$$

According to this criterion, chemical stratification is able to completely inhibit transport in regions of strong stable chemical gradients, such as around convective cores. In contrast, the model of Maeder (1997), which is incompatible with our simulations, allows mixing in such regions and has a better fit with observations (Meynet *et al.* 2013), thus suggesting the existence of an additional transport process or physical ingredient.

4. Turbulent thermal diffusivity and turbulent viscosity

This section is dedicated to new results concerning other diffusion coefficient, namely the turbulent thermal diffusivity κ_t and the turbulent viscosity ν_t. As regards κ_t, recent simulation results displayed in Fig. 3 suggest that at constant Reynolds number, turbulent thermal diffusivity scales as $\mathrm{Ri}^{-4/3}$. This preliminary result has to be verified, and in particular, the dependence on the Reynolds number is currently investigated.

Concerning ν_t, it is important to note that in many evolution codes, it is simply taken to be equal to D_t. In our simulations, we observe between them the following relation:

$$\nu_t \simeq \frac{3}{4}D_t, \qquad (4.1)$$

which is valid for all Re, Pe, Ri, and Ri_μ.

5. Conclusion

Our simulations enabled us to test existing models of turbulent transport in stellar radiative zones. In particular, we recovered Zahn's model and the model of Maeder & Meynet (1996) (in presence of μ-gradients) in the small-Péclet-number regime and the model of Lindborg & Brethouwer (2008) in the large-Péclet-number regime. In addition, we are now about to be able to give new prescriptions for turbulent transport of heat and angular momentum. Again, the fact that recovered models have not the best fit with observations suggests that another significant source of mixing may exist, or that an essential physical ingredient is missing.

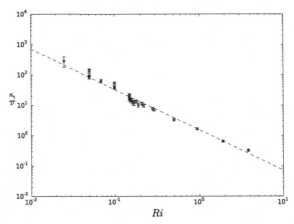

Figure 3. κ_t/ν as a function of the Richardson number. Dots correspond to simulations at $Re = 10^4$ and Pe from $2 \cdot 10^4$ (brown) down to 15 (blue) and the dashed line to the power law $\propto Ri^{-4/3}$.

To go further, one may think of studying the effect of a very efficient horizontal diffusion induced by horizontal differential rotation. As predicted by Talon & Zahn (1997), this would weaken the effect of chemical stratification and thus enhance the transport. Other physical ingredients should be considered, such as the Coriolis force and the magnetic field. By altering shear instability or triggering new instabilities, they might increase the amount of mixing in stellar interiors.

Three-dimensional direct numerical simulations (DNS) can also be used to study other mixing processes, such as thermohaline convection or convective boundary mixing. Moreover, prescriptions from DNS could be used as sub-grid models to improve large-eddy-simulations, that may be useful for non-local processes. Finally, all models coming from simulations must be included in stellar evolution codes to be compared to observations.

References

Deheuvels, S., García, R. A., Chaplin, W. J., et al. 2012, *ApJ* 756, 19

Lignières, F. 1999, *A&A* 348, 933

Lindborg, E. & Brethouwer, G. 2008, *J. Fluid Mech.* 614, 303

Maeder, A. 1995, *A&A* 299, 84

Maeder, A. 1997, *A&A* 321, 134

Maeder, A. & Meynet, G. 1996, *A&A* 313, 140

Meynet, G., Ekstrom, S., Maeder, A., *et al.* 2013, in *Lecture Notes in Physics*, Vol. 865, p. 3

Michaud, G. & Zahn, J.-P. 1998, *Theor. Comput. Fluid Dyn.* 11, 183

Osborn, T. R. & Cox, C. S. 1972, *Geophys. Astrophys. Fluid Dynam.* 3, 321

Prat, V. & Lignières, F. 2013, *A&A* 551, L3

Prat, V. & Lignières, F. 2014, *A&A* 566, A110

Talon, S. & Zahn, J.-P. 1997, *A&A* 317, 749

Zahn, J.-P. 1992, *A&A* 265, 115

Discussion

ARNETT: You said that your DNS calculations were "not turbulent". How would you defend their use for stellar dimensions?

PRAT: I said that when the effect of stratification is strong enough, we are not able to reach a turbulent statistical steady state because turbulence is decaying exponentially. The results I showed are based on turbulent simulations. About the validity of these results to stellar dimensions, I would like to emphasise the fact that the turbulent Reynolds number (i.e. based on turbulent length and velocity scales) is of the order of 100, as expected in stably stratified zones (see e.g. Michaud & Zahn 1998).

KHALAK: Could you comment why chemical diffusion will diminish chemical stratifications caused by thermal diffusion?

PRAT: What I told has not been correctly understood. Chemical stratification is not caused by thermal diffusion, and it is turbulent horizontal chemical diffusion that can weaken the effect of chemical stratification by smoothing vertical chemical fluctuations and hence decreasing the amplitude of the buoyancy force.

Vincent Prat

Oscar Hernan Ramìrez Agudelo

New windows on massive stars: asteroseismology, interferometry, and spectropolarimetry
Proceedings IAU Symposium No. 307, 2014
G. Meynet, C. Georgy, J. H. Groh & Ph. Stee, eds.

© International Astronomical Union 2015
doi:10.1017/S1743921314006309

Rotational velocities of single and binary O-type stars in the Tarantula Nebula

O. H. Ramírez-Agudelo[1], H. Sana[2], A. de Koter[1,3], S. Simón-Díaz[4,5], S. E. de Mink[6,7], F. Tramper[1], P. L. Dufton[8], C. J. Evans[9], G. Gräfener[10], A. Herrero[4,5], N. Langer[11], D. J. Lennon[12], J. Maíz Apellániz[13], N. Markova[14], F. Najarro[15], J. Puls[16], W.D. Taylor[9] and J.S. Vink[10],

[1] Astronomical Institute Anton Pannekoek, University of Amsterdam, The Netherlands
email: o.h.ramirezagudelo@uva.nl

[2] ESA/Space Telescope Science Institute 3700 San Martin Drive, Baltimore, MD21218, USA

[3] Instituut voor Sterrenkunde, Universiteit Leuven, Celestijnenlaan 200D, 3001, Leuven, Belgium

[4] Instituto de Astrofísica de Canarias, C/ Vía Láctea s/n, E-38200 La Laguna, Tenerife, Spain

[5] Departamento de Astrofísica, Universidad de La Laguna, Avda. Astrofísico Francisco Sánchez s/n, E-38071 La Laguna, Tenerife, Spain

[6] Observatories of the Carnegie Institution for Science, 813 Santa Barbara St, Pasadena, CA 91101, USA

[7] Cahill Center for Astrophysics, California Institute of Technology, Pasadena, CA 91125, USA

[8] Astrophysics Research Centre, School of Mathematics and Physics, Queen's University of Belfast, Belfast BT7 1NN, UK

[9] UK Astronomy Technology Centre, Royal Observatory Edinburgh, Blackford Hill, Edinburgh, EH9 3HJ, UK

[10] Armagh Observatory, College Hill, Armagh, BT61 9DG, Northern Ireland, UK

[11] Argelander-Institut für Astronomie, Universität Bonn, Auf dem Hügel 71, 53121 Bonn, Germany

[12] European Space Agency, European Space Astronomy Centre, Camino Bajo del Castillo s/n, Urbanizacin Villafranca del Castillo, 28691 Villanueva de la Caada, Madrid, Spain

[13] Instituto de Astrofísica de Andalucía-CSIC, Glorieta de la Astronomía s/n, E-18008 Granada, Spain

[14] Institute of Astronomy with NAO, Bulgarian Academy of Science, PO Box 136, 4700 Smoljan, Bulgaria

[15] Centro de Astrobiología (CSIC-INTA), Ctra. de Torrejón a Ajalvir km-4, E-28850 Torrejón de Ardoz, Madrid, Spain

[16] Universitätssternwarte, Scheinerstrasse 1, 81679 München, Germany

Abstract. Rotation is a key parameter in the evolution of massive stars, affecting their evolution, chemical yields, ionizing photon budget, and final fate. We determined the projected rotational velocity, $v_e \sin i$, of $\sim$330 O-type objects, i.e. $\sim$210 spectroscopic single stars and $\sim$110 primaries in binary systems, in the Tarantula nebula or 30 Doradus (30 Dor) region. The observations were taken using VLT/FLAMES and constitute the largest homogeneous dataset of multi-epoch spectroscopy of O-type stars currently available. The most distinctive feature of the $v_e \sin i$ distributions of the presumed-single stars and primaries in 30 Dor is a low-velocity peak at around $100 \, \mathrm{km \, s^{-1}}$. Stellar winds are not expected to have spun-down the bulk of the stars significantly since their arrival on the main sequence and therefore the peak in the single star sample is likely to represent the outcome of the formation process. Whereas the spin distribution of presumed-single stars shows a well developed tail of stars rotating more rapidly than $300 \, \mathrm{km \, s^{-1}}$, the sample of primaries does not feature such a high-velocity tail. The tail of the presumed-single star distribution is attributed for the most part – and could potentially be completely due – to spun-up binary products that appear as single stars or that have merged. This would be consistent with the lack of such post-interaction products in the binary sample, that is expected to be dominated by pre-interaction systems. The peak in this distribution is broader and is shifted toward somewhat higher spin rates compared to the distribution of presumed-single stars. Systems displaying large radial velocity variations, typical for short period systems, appear mostly responsible for these differences.

Keywords. stars: rotation, stars: binaries, galaxies: Magellanic Clouds

---

1. Introduction

Rotation impacts the evolution of massive stars, affecting their evolution, chemical yields, budget of ionizing photons and final fate as supernovae and long gamma-ray bursts (e.g. Langer 2012). For massive stars the *initial* distribution of spin rates is especially interesting because so little is known about how these stars form (e.g., Zinnecker & Yorke 2007). Potentially, it can tell us more about the formation process.

The 30 Dor starburst region in the Large Magellanic Cloud contains the richest sample of massive stars in the Local Group and is the best possible laboratory to investigate aspects of the formation and evolution of massive stars, and to establish statistically meaningful distributions of their physical properties. Here we present an analysis of the $v_e \sin i$ properties of the O-type objects, i.e. spectroscopic single stars and binary systems observed in the context of the VLT-FLAMES Tarantula Survey (VFTS; Evans *et al.* 2011; de Koter *et al.* 2011). VFTS is a multi-epoch spectroscopic campaign targeting over 300 O-type objects – singles and binaries. The spectral classification and radial velocity (RV) measurements, relevant for the study at hand, are presented in Walborn *et al.* (2014) and Sana *et al.* (2013). Here we will report on the projected rotational properties, $v_e \sin i$, of the presumed-single O-type stars (Ramírez-Agudelo *et al.* 2013; Ramirez-Agudelo & VFTS Consortium 2013) and primary stars, i.e. the brightest component of binary systems composed of at least one O-type star (Ramírez-Agudelo *et al.* in prep.).

2. Sample and Method

The VFTS project and the data have been described in Evans *et al.* (2011). In short the total Medusa sample contains ∼330 O-type objects. Sana *et al.* (2013), from multi-epoch radial velocity (RV) measurements have identified ∼210 O-type stars that show no significant RV variations (ΔRV) and are presumably single (ΔRV $\leqslant$ 20 km s^{-1}) and ∼110 objects (the rest of the sample) with ΔRV > 20 km s^{-1} that are considered spectroscopic binaries.

For the single sample we use Fourier transform (Gray 1976; Simón-Díaz & Herrero 2007) and line profile fitting methods (Simón-Díaz & Herrero 2014) to measure projected rotational velocities. A discussion of the methods used and the accuracy that can be achieved can be found in Ramírez-Agudelo *et al.* (2013). To obtain the projected rotational velocities of the binary sample we calibrate full width at half maximum (FWHM) measurements versus $v_e \sin i$ for specific spectral lines, using the FWHM measurements of the O-type stars from Sana *et al.* (2013) and the $v_e \sin i$ measurements of the single O-type star presented in Ramírez-Agudelo *et al.* (2013). Details of the method and its accuracy are found in Ramírez-Agudelo *et al.* in prep..

O. H. Ramírez-Agudelo *et al.*

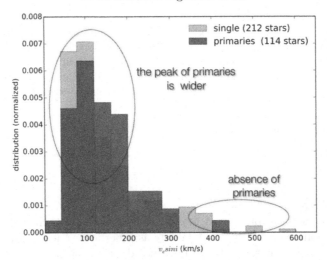

Figure 1. Distribution of the projected rotational velocities, $v_e \sin i$, of the O-type presumably single and primary samples.

3. Results

Figure 1 shows the $v_e \sin i$ distributions of presumed-single stars and primaries. Qualitatively, both distributions display a peak at around $\sim 100\,\mathrm{km\,s^{-1}}$ though we also note some differences. First, the main peak of the primary distribution is wider than that of the presumed-single sample. Second, at $v_e \sin i > 300\ \mathrm{km\,s^{-1}}$ there is a deficiency of rapidly rotating primaries with respect to the presumed-single stars. While for the presumed-single sample 22 stars out of 212 exhibit projected rotational velocities larger than $300\ \mathrm{km\,s^{-1}}$ (corresponding to $10\pm2\,\%$ of that sample), in the primaries we only have 3 stars out of 114 ($3\pm1\,\%$). In Section 4 we argue that the high-velocity tail in the presumed-single star distribution is compatible with post-interaction binary evolution.

The size of the primary sample allows us to explore $v_e \sin i$ distributions for subpopulations. By selecting a radial velocity amplitude limit ($\Delta \mathrm{RV_{limit}}$) of $200\,\mathrm{km\,s^{-1}}$ we divide the primary sample into two subsamples, i.e. stars that display $\Delta \mathrm{RV} \leqslant \Delta \mathrm{RV_{limit}}$ (henceforth low-$\Delta \mathrm{RV}$) and stars with $\Delta \mathrm{RV} > \Delta \mathrm{RV_{limit}}$ (high-$\Delta \mathrm{RV}$). The latter are systems where we expect tidal synchronization to become important during the main sequence phase; given our sample properties they roughly correspond to binaries with an orbital period of less than $10\,\mathrm{days}$ (see Ramírez-Agudelo *et al.* in prep.). Most of the systems in the low-$\Delta \mathrm{RV}$ subsample are wider systems that will not suffer from tides with the potential exception of those that are seen at a low inclination angle. Figure 2 plots the $v_e \sin i$ distributions of the low-$\Delta \mathrm{RV}$ (85 stars) and high-$\Delta \mathrm{RV}$ (29 stars) subsamples. At $v_e \sin i \leqslant 200\ \mathrm{km\,s^{-1}}$ the distribution of high-$\Delta \mathrm{RV}$ sources, compared to the low-$\Delta \mathrm{RV}$ sources, appears shifted by one bin to higher rotational velocities. The weighted mean of the samples (119 and $190\,\mathrm{km\,s^{-1}}$ for the low- and high-$\Delta \mathrm{RV}$ samples respectively) are significantly different from one another, confirming an average faster rotation for the high-$\Delta \mathrm{RV}$ (short periods) systems. These differences may be related to tidal effects, an hypothesis that we will explore further in the next subsection.

3.1. *Tidal interaction*

To investigate the effect of tidal locking in our sample we make use of a diagram showing the relation between projected rotational velocity and amplitude of the RV variations

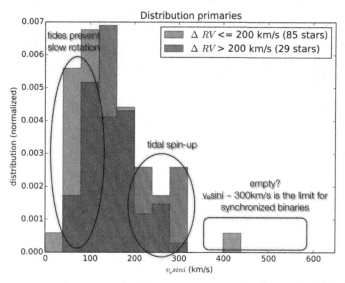

Figure 2. Distribution of the projected rotational velocities, $v_e \sin i$, of the O-type primaries with ΔRV $\leqslant \Delta$RV$_{\mathrm{limit}}$ (low-ΔRV) and ΔRV $> \Delta$RV$_{\mathrm{limit}}$ (high-ΔRV).

(see Fig. 3). The timescale of tidal synchronization is a function of the primary mass M_1, the mass ratio of the primary and secondary $q = M_2/M_1$, and the orbital period P_{orb}. For our sample we lack information about the latter and therefore we use the radial velocity amplitude ΔRV as a proxy of the period. Figure 3 gives the relation between $v_e \sin i$ and ΔRV for the sample of primaries. The green dashed region shows the range of parameter space for which tidal synchronization of a $20\,M_\odot$ primary – a typical mass for our sample – occurs within the main sequence life for q ranging from 0.25 (left-side boundary) to 1 (right-side boundary) and P_{orb} ranging from 10 d (bottom boundary) to 0.25 d (upper boundary), or sooner when the primary fills its Roche lobe prior to reaching the end of the main sequence (de Mink *et al.* 2009, 2013). The thick green line is for $q = 0.5$. The top axis of the figure displays the orbital period $P_{\mathrm{orb}}/\sin i$ associated with the radial velocity amplitude given on the horizontal axis, assuming ΔRV is twice the actual semi-amplitude of the radial velocity curve. Similarly, the $P_{\mathrm{rot}}/\sin i$ displayed on the right vertical axis corresponds to the projected rotational velocity given on the left.

Tides are less effective for smaller mass ratios and longer periods, therefore the timescale of tidal synchronization is longest for systems in the lower left corner of the green zone. In the upper right corner synchronization proceeds the fastest. For instance, for a 0.5 d system with $q = 0.75$ synchronization by turbulent viscosity and radiative dissipation occurs within about 1 percent of the main sequence lifetime (de Mink *et al.* 2009). The gray region shows the M_1 dependence by also displaying the zone as defined here for a $60\,M_\odot$ primary. It is stretched out to both higher $v_e \sin i$ ($\propto R$), and ΔRV ($\propto M_1^{1/3}$), where R is the stellar radius. The green zone that according to the above argument may be expected to contain systems that are synchronized or tend toward synchronization is populated by 40 primaries. The 29 sources that constitute the pink (ΔRV $> 200\,\mathrm{km\,s}^{-1}$) distribution in Fig. 2 are almost exclusively found in the green region, more specifically in the part of that region for which synchronization by tides is the most relevant.

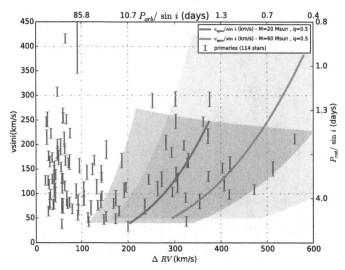

Figure 3. Projected rotational velocity vs. radial velocity amplitude for the primary sample (114 stars). The green and gray regions show where a $20\,M_\odot$ and $60\,M_\odot$ primary is expected to become synchronized with its companion before the primary leaves the main sequence, computed for mass ratios q ranging between 0.25 and 1 and periods $P_{\rm orb}$ from 0.25 to 10 days. The thick green and gray lines are for $q = 0.5$. The top axis displays the orbital period $P_{\rm orb}/\sin i$ associated with the radial velocity amplitude given on the horizontal axis for a $20\,M_\odot$ primary. Similarly, $P_{\rm rot}/\sin i$ displayed on the right vertical axis corresponds to the projected rotational velocity given on the left.

4. Implications

The most distinctive feature of the $v_e \sin i$ distribution of the O-type presumed-single stars and primaries in 30 Dor is its low-velocity peak at around $100\,{\rm km\,s^{-1}}$. For the bulk of the samples, mass loss in a stellar wind and/or envelope expansion is not efficient enough to significantly spin down these stars. Therefore the peak is likely to be the outcome of the formation process (see discussion in Ramírez-Agudelo *et al.* 2013 and in prep.). The presence of a high-velocity tail in the presumed-single sample, and the absence of such a tail in the primary sample, is compatible with predictions of binary interaction. Rapid rotators result from spin-up through mass transfer and mergers (de Mink *et al.* 2013), that mostly appear as, or have become, single objects (de Mink *et al.* 2014). Such a nature of the high-velocity tail has important implications for the evolutionary origin of the progenitors of long gamma-ray bursts, reducing the likelihood that long-GRBs may also result from single stars that are born spinning rapidly. Finally, if post-interaction systems have been removed from the binary sample for the above outlined reason it is dominated by pre-interaction systems. The short period systems among these may suffer from tidal synchronization effects that may qualitatively explain the differences in the low-velocity peak structure of presumed-single stars and of primaries.

References

de Koter, A., Sana, H., Evans, C. J., *et al.* 2011, *Journal of Physics Conference Series* 328(1), 012022
de Mink, S. E., Cantiello, M., Langer, N., *et al.* 2009, *A&A* 497, 243
de Mink, S. E., Langer, N., Izzard, R. G., Sana, H., & de Koter, A. 2013, *ApJ* 764, 166
de Mink, S. E., Sana, H., Langer, N., Izzard, R. G., & Schneider, F. R. N. 2014, *ApJ* 782, 7
Evans, C. J., Taylor, W. D., Hénault-Brunet, V., *et al.* 2011, *A&A* 530, A108

Gray, D. 1976, *The Observation and Analysis of Stellar Photospheres*, Cambridge University Press, third edition edition

Langer, N. 2012, *ARA&A* 50, 107

Ramírez-Agudelo, O. H., Simón-Díaz, S., Sana, H., *et al.* 2013, *A&A* 560, A29

Ramirez-Agudelo, O. H. & VFTS Consortium 2013, in *Massive Stars: From alpha to Omega*

Sana, H., de Koter, A., de Mink, S. E., *et al.* 2013, *A&A* 550, A107

Simón-Díaz, S. & Herrero, A. 2007, *A&A* 468, 1063

Simón-Díaz, S. & Herrero, A. 2014, *A&A* 562, A135

Townsend, R. H. D., Owocki, S. P., & Howarth, I. D. 2004, *MNRAS* 350, 189

Walborn, N. R., Sana, H., Simón-Díaz, S., *et al.* 2014, *A&A* 564, A40

Zinnecker, H. & Yorke, H. W. 2007, *ARA&A* 45, 481

Discussion

MASSEY: Conti & Ebbets (1977) showed that the dwarf O stars had both a low and a high $v_e \sin i$ peak but that the O supergiants have only a low peak. How can post interacting O stars (merged or not) be luminosity class V with modest M_v? Would not you expect them to have high luminosities?

RAMÍREZ-AGUDELO: In Ramírez-Agudelo *et al.* (2013) subdivide the contribution of dwarfs, giants and supergiants to the $v_e \sin i$ distribution. We find that the tail is dominated by the dwarfs and that supergiants are almost missing. The reason is that the break-up velocity of supergiants is considerably lower, about $300\,\mathrm{km\,s^{-1}}$, than that of dwarfs ($\sim 600\,\mathrm{km\,s^{-1}}$), simply because they are bigger. The secondary O V stars that receive the mass may remain dwarfs, the interaction itself does not turn them into supergiants. Later they may evolve towards supergiants – but then their $v_e \sin i$ will decrease because of envelope expansion.

MAEDER: You have shown that you are calibrating your $v_e \sin i$ on the basis of the change of the line profiles. However, as shown long ago by Collins, the equivalent width of the lines may also change since the stellar flux is coming from regions with a variety of g_{eff} and T_{eff}. If this is not accounted for, a large underestimate of the $v_e \sin i$ may result. Could you please comment on what you are doing.

RAMÍREZ-AGUDELO: The gravity darkening effect that you refer to has been studied for B stars by Townsend *et al.* (2004) and becomes important for stars that spin faster than about 95% of critical. In our sample of presumably-single O stars only six sources out of the ~ 210 spin above 60% of critical, and one source out of 114 binaries. Because of the very limited effect on the sample as a whole we did not correct for the darkening effect.

New windows on massive stars: asteroseismology, interferometry, and spectropolarimetry
Proceedings IAU Symposium No. 307, 2014
G. Meynet, C. Georgy, J. H. Groh & Ph. Stee, eds.

© International Astronomical Union 2015
doi:10.1017/S1743921314006310

Stellar Yields of Rotating First Stars: Yields of Weak Supernovae and Abundances of Carbon-enhanced Hyper Metal Poor Stars

Koh Takahashi[1], Hideyuki Umeda[1] and Takashi Yoshida[2]

[1]Department of Astronomy, The University of Tokyo
email: ktakahashi@astron.s.u-tokyo.ac.jp

[2]Yukawa Institute for Theoretical Physics, Kyoto University

Abstract. The three most iron-poor stars known until now are also known to have peculiar enhancements of intermediate mass elements. Under the assumption that these iron-deficient stars reveal the nucleosynthesis result of Pop III stars, we show that a weak supernova model successfully reproduces the observed abundance patterns. Moreover, we show that the initial parameters of the progenitor, such as the initial masses and the rotational property, can be constrained by the model, since the stellar yields result from the nucleosynthesis in the outer region of the star, which is significantly affected by the initial parameters. The initial parameter of Pop III stars is of prime importance for the theoretical study of the early universe. Future observation will increase the number of such carbon enhanced iron-deficient stars, and the same analysis on the stars may give valuable information for the Pop III stars that existed in our universe.

Keywords. nuclear reactions, nucleosynthesis, abundances – stars: abundances – stars: Population III – stars: rotation

1. Introduction

Stars that are firstly formed in the universe are called *first stars* or *Population III (Pop III) stars*. Their evolutionary characteristic is important and interesting topic in astrophysics, since high energy photons emitted from the surface initiate the cosmic reionization, and moreover, they are the first nuclear reactor in the universe (for review, Bromm & Yoshida 2011). Recent development of star formation theory allows us to investigate what is the realistic star formation site of first stars. Hirano *et al.* (2014) showed that the initial mass range of Pop III stars will be wide from $\sim$10 $M_\odot$ to $\sim$ 1000 $M_\odot$. Also, Stacy *et al.* (2013) found that a Pop III star can have a fast rotation speed at its birth.

How can we test the very interesting expectations on Pop III stars, and what are the mass range and the rotational property of Pop III stars that really existed in the universe? A good candidate for the investigation is to use abundance patterns preserved in metal-poor stars. Theoretically, metal-poor stars should be born in a metal-poor, unevolved gas cloud that is expected to show nuclear synthesis features of the metal supplying mother star(s). Therefore, the abundance patterns can be used to constrain theoretical models for nucleosynthesis in the progenitor stars, and the method is called *abundance profiling*. Such a theoretical study benefits from the large number of observed metal-poor stars. So far, three iron-poor stars that have [Fe/H]† < -5 are known, they are HE 0107-5240 of [Fe/H] = -5.3 (Christlieb *et al.* 2002), HE 1327-2326 of [Fe/H] = -5.7 (Frebel *et al.*

† $[A/B] = \log_{10}(N_A/N_B) - \log_{10}(N_A/N_B)_\odot$, N_A is a number abundance of a species A.

2005), and recently found SMSS 0313-6708 of [Fe/H] < -7 (Keller *et al.* 2014). Not only the low iron abundance, but also the enhancement of intermediate mass elements, such as carbon, nitrogen, oxygen, sodium, and magnesium, is the characteristic feature of the abundances.

Until now, several works investigated yields from the Pop III stars and the first supernovae (e.g., Umeda & Nomoto 2005). However, no abundance features has been known that can be used to constrain the initial parameters of Pop III progenitors. Firstly, because of the degeneracy in the explosive yields, the initial mass range of the progenitor has not yet been constrained (Umeda & Nomoto 2005). Secondly, since no rotating Pop III calculations have been conducted for abundance profiling so far, abundance features showing rotational contribution has not yet obtained. Thus, the aim of this work is to find stellar yields by doing abundance profiling, that are useful to constrain the initial masses and the rotation of the first stars.

2. Method

2.1. *Stellar Evolution Calculation*

Evolution of massive zero metallicity stars are calculated by the latest version of the stellar evolution code used in Takahashi *et al.* (2014). The primordial abundance by Steigman (2007) is used for the initial composition. A wide mass range from 12 M$_\odot$ to 140 M$_\odot$ is taken in order to cover a likely mass range for all core collapse supernova. Also, stellar rotation is included in the calculation, which affects the yields by effective internal mixing due to several rotational instabilities.

2.2. *Assumptions on Supernova Explosion*

In this work, stellar matter is assumed to be ejected by supernova explosion at the end of the stellar life. First assumption for the ejection is that the only outer part of the star that is weakly bound by gravity is ejected by the explosion. Second assumption is that the explosion is too weak to modify the composition of outer distributed matter by the shock heating. We call a supernova explosion that is responsible for these assumptions as a *weak supernova*. With these assumptions, stellar yields M_i can be calculated by a simple integration using the parameter for the inner boundary M_{in},

$$M_i(M_{in}) = \int_{M_{in}}^{M_{\text{surface}}} X_i(M)dM, \tag{2.1}$$

where X_i is a mass fraction distribution of i element. In order to show the depth of the ejection, the parameter $f_{in} = M_{in}/M_{\text{CO}}$ is also used in the analysis, here M_{CO} is a mass of the CO core.

2.3. *Abundance Calculation*

Using the integrated mass M_i as the stellar yield, the chemical composition of a second generation star can be calculated as

$$X_{i,\text{2nd}} = \frac{M_i/M_{\text{SN}} + X_{i,\text{ISM}} M_{\text{ISM}}/M_{\text{SN}}}{1 + M_{\text{ISM}}/M_{\text{SN}}} \tag{2.2}$$

where $M_{\text{SN}} = \sum_i M_i$ and M_{ISM} are total masses of the ejecta and the ISM, and $D = M_{\text{ISM}}/M_{\text{SN}}$ is a dilution factor. Dilution factors vary with models, and typically become several hundreds for best fitted models for HMP stars. Comparison between calculated abundances and observation is done by using [i/j] as the indicator. The solar abundance by Asplund *et al.* (2009) is used in the comparison.

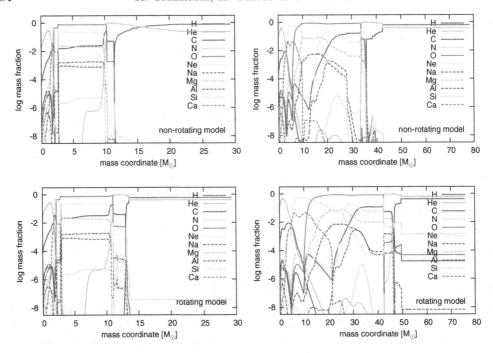

Figure 1. Abundance distributions of rotating and non-rotating, 30 and 80 $M_\odot$ models.

3. Stellar Yields

Fig 1 shows calculated abundance distribution in models of rotating and non-rotating, 30 and 80 $M_\odot$ stars. Important differences are appeared in outer regions of stars, especially in their helium layers.

Carbon and Oxygen All stellar models yield carbon and oxygen in their helium layers. Since abundant helium remains during the evolution in this region, resulting C/O ratio becomes large and always exceeds unity. Thus, carbon and oxygen production with a large [C/O] can be regarded as a general signature of nucleosynthesis in the outer region of a star.

Magnesium and Silicon In the figure, magnesium and Silicon are produced in 80 $M_\odot$ models, and generally, these alpha elements are produced only in massive models. Production and non-production of these alpha elements are, thus, useful to constrain the initial mass of the progenitor.

Sodium and Aluminum Both rotating 30 and 80$M_\odot$ models yield sodium and aluminum in their helium layers, while non-rotators do not. This is due to the rotationally induced mixing in the rotating models. First, carbon and oxygen synthesized by core helium burning are transported into hydrogen burning shell by the rotationally induced mixing. This results in nitrogen production by CNO cycle, and nitrogen distributes in the helium layer. Then alpha capture reactions on nitrogen take place, producing ^{22}Ne, a famous neutron source for massive stellar evolution. Finally neutron capture reactions on ^{22}Ne and ^{26}Mg take place, producing ^{23}Na and ^{27}Al. Sodium and aluminum enhancement are used to constrain the stellar rotation of the progenitor.

Calcium Only very massive models yield calcium, and the mechanisms are different between rotating and non-rotating progenitors. For rotating models, efficient alpha capture reactions are responsible for the production, and the process is similar to the production of magnesium and silicon. For non-rotating models, efficient proton capture

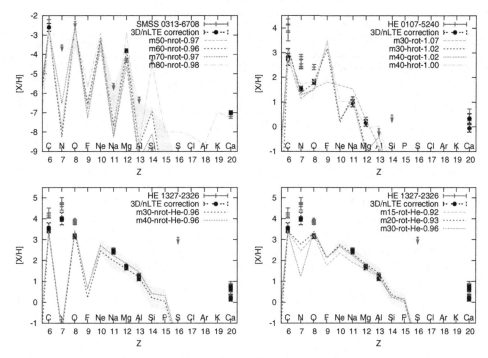

Figure 2. Results of abundance profiling.

reactions take place at the base of the hydrogen burning shell and account for the calcium production.

4. Abundance Profiling

SMSS 0313-6708 SMSS 0313-6708, recently found by Keller *et al.* (2014), is the most iron-deficient star known until now. Iron has not been detected in the star, and very small upper limit of [Fe/H] < −7 has been reported. Interesting feature of the composition of the star is the enhancement of carbon ([C/H] ∼ −2.6) and magnesium ([Mg/H] ∼ −3.9). In addition, upper limits of [Na/Mg] < −1.2 and [Al/Mg] < −1.9 are valuable for the abundance profiling.

First, massive stars of >100 $M_\odot$ stars overproduce magnesium at helium layers. These stars fail to explain the small [Mg/C] and thus rejected as the progenitor star. On the other hand, less massive models of <40 $M_\odot$ produce sodium and aluminum besides magnesium, and do not fit with the observation. This is because, magnesium should be ejected from inner carbon burning region in the less massive stars, however, carbon burning simultaneously synthesizes sodium and aluminum, which are not detected in the star. Similarly, the upper limits of sodium and aluminum are used to constrain rotating models. Due to the rotationally induced mixing, all rotating progenitors produce sodium and aluminum in helium layers. Because of the overproduction, no rotating models can explain the observation.

Therefore, a massive, but not too massive, non-rotating star can be a progenitor of SMSS 0313-6708. The non-rotating 60 $M_\odot$ gives the best fitting to the abundance pattern from carbon to silicon. Calcium has been detected in the star with a small value of [Ca/H] = −7.0. If the future observation reveal the enhancement of calcium as well as

Table 1. Summary of the abundance profiling.

Object	[Fe/H]	M_{ini}	f_{in}	Rotation	Dilution Factor
SMSS 0313-6708	< -7.1	50-80	0.96 ± 0.04 (60 M$_\odot$)	non-rotating	$1.78 \times 10^3 - 6.09 \times 10^2$
			0.98 ± 0.04 (80 M$_\odot$)	non-rotating	$1.62 \times 10^3 - 1.91 \times 10^2$
HE 0107-5240	-5.3	30-40	1.07 ± 0.06 (30 M$_\odot$)	rotating	$7.84 \times 10^2 - 2.23 \times 10^2$
HE 1327-2326	-5.7	20-40	0.96 ± 0.01 (40 M$_\odot$)	non-rotating	$5.00 \times 10^2 - 4.32 \times 10^2$
		15-30	0.93 ± 0.01 (20 M$_\odot$)	rotating	$7.92 \times 10^2 - 7.35 \times 10^2$

other intermediate mass elements, then the non-rotating 80 M$_\odot$ will be a good candidate for the progenitor, since the model produces calcium by fast proton capture reactions.

HE 1327-2326 and HE 0107-5240 The other two iron-deficient stars, HE 1327-2326 ([Fe/H]=-5.7, reported by Frebel et al. (2005)) and HE 0107-5240 ([Fe/H]=-5.3, reported by Christlieb et al. (2002)), are also compared with the theoretical yields. Plotted data are obtained from Aoki et al. (2006); Frebel et al. (2006, 2008); Bonifacio et al. (2012) for HE1327-2326, and from Christlieb et al. (2004); Bessell et al. (2004); Bessell & Christlieb (2005), and for 3D collection from Collet et al. (2006).

Abundance pattern from sodium to aluminum is useful to constraint the progenitor of HE 1327-2326. The pattern, and the relative abundance with carbon (e.g., [Mg/C] = -2.7) can be well represented by ejecting tiny fraction of material in the carbon burning region of $\sim$15-40 M$_\odot$ models. Thus, rotating 15-30M$_\odot$ and non-rotating 20-40 M$_\odot$ models give good fits for HE 1327-2326. HE 0107-5240 has a highly enhanced carbon abundance relative to oxygen, [O/C] = -1.4. Such a pattern can only be explained by a mass ejection from a very outer region of the progenitor star. In addition, sodium abundance is much higher than magnesium abundance in the star, [Na/Mg] = 0.8, and stellar rotation can be responsible for the enhancement. In our models, the rotating 30 M$_\odot$ star with f_{in} = 1.01-1.13 gives the best fit to the star.

References

Aoki, W., Frebel, A., Christlieb, N., et al. 2006, ApJ 639, 897

Asplund, M., Grevesse, N., Sauval, A. J., & Scott, P. 2009, ARA&A 47, 481

Bessell, M. S. & Christlieb, N. 2005, in V. Hill, P. Francois, & F. Primas (eds.), *From Lithium to Uranium: Elemental Tracers of Early Cosmic Evolution*, Vol. 228 of *IAU Symposium*, pp 237–238

Bessell, M. S., Christlieb, N., & Gustafsson, B. 2004, ApJ (Letters) 612, L61

Bonifacio, P., Caffau, E., Venn, K. A., & Lambert, D. L. 2012, A&A 544, A102

Bromm, V. & Yoshida, N. 2011, ARA&A 49, 373

Christlieb, N., Bessell, M. S., Beers, T. C., et al. 2002, Nature 419, 904

Christlieb, N., Gustafsson, B., Korn, A. J., et al. 2004, ApJ 603, 708

Collet, R., Asplund, M., & Trampedach, R. 2006, ApJ (Letters) 644, L121

Frebel, A., Aoki, W., Christlieb, N., et al. 2005, Nature 434, 871

Frebel, A., Christlieb, N., Norris, J. E., Aoki, W., & Asplund, M. 2006, ApJ (Letters) 638, L17

Frebel, A., Collet, R., Eriksson, K., Christlieb, N., & Aoki, W. 2008, ApJ 684, 588

Hirano, S., Hosokawa, T., Yoshida, N., et al. 2014, ApJ 781, 60

Keller, S. C., Bessell, M. S., Frebel, A., et al. 2014, Nature 506, 463

Stacy, A., Greif, T. H., Klessen, R. S., Bromm, V., & Loeb, A. 2013, MNRAS 431, 1470

Steigman, G. 2007, Annual Review of Nuclear and Particle Science 57, 463

Takahashi, K., Umeda, H., & Yoshida, T. 2014, ArXiv e-prints

Takahashi, K., Yoshida, T., & Umeda, H. 2013, ApJ 771, 28

Umeda, H. & Nomoto, K. 2005, ApJ 619, 427

Discussion

LANGER: Would your $50 - 80\,M_\odot$ stars not form black holes rather than supernovae?

TAKAHASHI: Still proper predictions of fates of massive stars are difficult. We just assume in this work that the outer matter of the first stars is ejected somehow. The mechanism may be a weakly energetic supernova, a jet-like explosion, or a failed supernova.

LANGER: How complete is your parameters space, e.g., what about binaries or mixing during the supernova explosion?

TAKAHASHI: Currently, binarity is too complicated to implement in our work, though it could have important effects on the results.

MORAVVEJI: $Z = 0$ stars must have masses around $0.8\,M_\odot$, not $80\,M_\odot$, otherwise they wouldn't have stayed until the present days, so that we observe them. Then, how can you compare yields of $80\,M_\odot$ models with a $\sim 0.8\,M_\odot$ star?

TAKAHASHI: These stars are not essentially first generation, but from next generations. So, they already have some enrichments from previous generations.

MORAVVEJI: To model second generation of metal-free stars, do you start from the predicted yields of the first generation stars?

TAKAHASHI: No, we do not do that here.

MEYNET: Do you have predictions about the $^{12}C/^{13}C$ ratio? This is a very constraining quantity for comparisons with observations.

TAKAHASHI: I will check the data. My expectation is that because the weak supernova yields have moderate amount of ^{12}C from the layers in addition to ^{13}C from CNO regions, $^{13}C/^{12}C$ will show a slight enhancement.

Koh Takahashi

New windows on massive stars: asteroseismology, interferometry, and spectropolarimetry
Proceedings IAU Symposium No. 307, 2014
G. Meynet, C. Georgy, J. H. Groh & Ph. Stee, eds.

© International Astronomical Union 2015
doi:10.1017/S1743921314006322

The Gaia-ESO Survey and Massive Stars

Ronny Blomme[1], Yves Frémat[1], Eric Gosset[2], Artemio Herrero[3,4], Alex Lobel[1], Jesús Maíz Apellániz[5], Thierry Morel[2], Ignacio Negueruela[6], Thierry Semaan[2], Sergio Simón-Díaz[3,4] and Delia Volpi[1]

[1]Royal Observatory of Belgium
email: Ronny.Blomme@oma.be

[2]Université de Liège, Belgium

[3]Instituto de Astrofísica de Canarias, Tenerife, Spain

[4]Universidad de La Laguna, La Laguna, Spain

[5]Instituto de Astrofísica de Andalucía-CSIC, Granada, Spain

[6]Universidad de Alicante, Spain

Abstract. As part of the Gaia-ESO Survey (GES), a number of clusters will be observed that were chosen specifically for their massive-star content. We report on the procedures we followed to determine the stellar parameters from the massive-star spectra of this survey. We intercompare the results from the different techniques used by the nodes of our group to determine these parameters and discuss some of the problems encountered. We present preliminary results for NGC 6705, NGC 3293, and Trumpler 14. We study microturbulence in A-type stars, we use the repeat observation to investigate binarity, and we determine cluster membership from the radial velocity information. The large number of massive-star spectra obtained by the Gaia-ESO Survey will allow us to critically test stellar evolution modelling.

Keywords. surveys, stars: abundances, stars: atmospheres, binaries: spectroscopic, stars: early-type, stars: statistics, open clusters and associations: Trumpler 14, NGC 3293, NGC 6705

1. Introduction

The Gaia-ESO Survey† (GES) groups over 400 Co-Investigators in a project led by Gerry Gilmore and Sofia Randich. It is an ambitious project to study the formation and evolution of the Milky Way and its stellar populations. The FLAMES instrument on VLT-UT2 is being used during 300 nights (spread over 5 years) to collect about 100 000 Giraffe spectra and 10 000 UVES spectra. As part of the survey, a number of clusters are observed that were chosen specifically for their massive-star content. As members of GES Workgroup 13 (WG13), we are responsible for the spectrum analysis of the O-, B- and A-type stars, as well as the analysis of stars in older clusters that have been observed with the hot-star Giraffe gratings.

WG13 is further split up into a number of groups (called nodes, see Table 1) that analyze the spectra with their specific techniques and codes. Overlap between the nodes is encouraged. Once each group has determined the stellar parameters, the results are combined into recommended parameters. For the recommended parameters, preference is given to the ROB, Liège or IAC results, as they are based on more specific atmosphere models and spectra. When these values are not available (which occurs in the majority of the cases!), the ROBGrid values are used. Using these recommended parameters, each

† http://www.gaia-eso.eu/

Table 1. Five nodes contribute to the analysis of the O-, B- and A-type stars.

Node	Details	Spectral type
ROB	Refined grid of Kurucz ATLAS9 models; LTE spectrum synthesis; compare equivalent widths and detailed shapes of selected lines	A stars
Liège	Kurucz or Tlusty models; NLTE spectrum synthesis; compare spectral line shapes	B stars
IAC	FASTWIND models; χ^2 fitting to spectral line shapes	O stars
ROBGrid	Model grids from the literature; χ^2 fitting of full spectral range	all stars (hot+cool)
IAA	Spectral classification of O-type stars	O stars

node then determines the abundances. Currently, we determine iron abundances in A-stars, He, Mg and Si abundances in B-stars, and He abundances in O-type stars.

2. Preliminary results

Comparison nodes. The results of the different nodes are compared to one another, to judge the uncertainty in the parameters. The effective temperatures are generally in good agreement between ROBGrid and the other nodes, except for ROB. The agreement in gravity between ROBGrid and the other nodes is not very good.

Trumpler 14. For all clusters, we analyse plots of the so-called "spectroscopic" HR diagram (Langern & Kudritzki 2014). Specifically for Trumpler 14, the isochrones show a spread in ages, suggesting star-formation during several Myr. A number of ROBGrid determinations fall below the ZAMS because they have incorrect log g values.

Cluster membership NGC 3293. Starting from histograms of the radial velocities derived from the GES spectra, we select those stars with the most frequently occurring radial velocities, and map them back on to their sky coordinates. These presumed members cover a large range in distances from the cluster centre. This suggests that NGC 3293 is larger than assumed so far.

Microturbulence in A-type stars. The ROB node also determined the microturbulence for A-type stars in NGC 3293 and NGC 6705. The highest values for microturbulence are found around 8000 K.

Binarity in NGC 3293. A repeat observation made about 1 month later allows us to look for binarity in these clusters. We measure the radial velocity difference with a cross-correlation technique and test its significance using Monte-Carlo simulations. Specifically for NGC 3293, about 5 % of the stars show clear signs of binarity. To derive the true binary fraction, this number needs to be corrected for the fact that we have only two epochs available. This correction factor is however poorly defined with only two epochs.

3. Conclusions

The Gaia-ESO Survey data will:
- provide critical tests for hot-star evolution,
- provide masses, radial velocities, binary fraction information for cluster studies,
- give specific information about A-, B- and O-type stars,
- contribute to the determination of Galactic abundance gradients.

However, to achieve these goals, the remaining discrepancies among the various node results will need to be clarified.

References

Langer, N. & Kudritzki, R. P. 2014, *A&A* 564, A52

New windows on massive stars: asteroseismology, interferometry, and spectropolarimetry
Proceedings IAU Symposium No. 307, 2014
G. Meynet, C. Georgy, J. H. Groh & Ph. Stee, eds.

© International Astronomical Union 2015
doi:10.1017/S1743921314006334

Non-LTE Abundances in OB stars: Preliminary Results for 5 Stars in the Outer Galactic Disk

Bragança, G. A.[1,2], Lanz, T.[2], Daflon, S.[1], Cunha, K.[1,3], Garmany, C. D.[4], Glaspey, J. W.[4], Borges Fernandes, M.[1], Oey, M. S.[5], Bensby, T.[6] and Hubeny, I.[3]

[1] Observatório Nacional, Rio de Janeiro, Brazil
email: ga.braganca@gmail.com

[2] Observatoire de la Côte d'Azur, Nice, France

[3] Steward Observatory, Tucson, AZ, U.S.A.

[4] National Optical Astronomy Observatory, Tucson, AZ, U.S.A

[5] University of Michigan, Ann Arbor, MI, U.S.A.

[6] Lund Observatory, Lund, Sweden

Abstract. The aim of this study is to analyse and determine elemental abundances for a large sample of distant B stars in the outer Galactic disk in order to constrain the chemical distribution of the Galactic disk and models of chemical evolution of the Galaxy. Here, we present preliminary results on a few stars along with the adopted methodology based on securing simultaneous O and Si ionization equilibria with consistent NLTE model atmospheres.

Keywords. stars: early-type, stars: fundamental parameters, stars: abundances, Galaxy: disk, Galaxy: abundances, Galaxy: evolution

---

The chemical distribution of B stars in the outer Galactic disk is presently poorly probed, based on only a few abundance results for distant B stars (e.g., Daflon *et al.* 2004). In order to enlarge the number of studied stars and to better represent the chemical distribution of the outer Galactic disk, we obtained high-resolution echelle spectra for a sample of 136 OB stars located towards the Galactic anti-center using the MIKE spectrograph on the 6.5m Magellan Clay telescope. A subsample of 50 sharp-lined B stars has been selected for the abundance analysis. High resolution, high signal-to-noise spectra of 3 well studied main-sequence B stars (HD 61068, HD 63922 and HD 74575) were added to the sample in order to test our adopted methodology.

We use an iterative method to obtain simultaneously the stellar parameters (effective temperature ($T_{\rm eff}$), surface gravity ($\log g$), microturbulence, and abundances of Si and O) based on non-LTE synthesis of H, He, Si, and O profiles. The synthetic spectra are computed using SYNSPEC (Hubeny & Lanz 2011), which interpolates in a grid of non-LTE model atmospheres computed with TLUSTY (Hubeny & Lanz 1995) and detailed atomic models of O (69, 219 and 41 levels for OI, II and III, respectively) and Si (70, 122 and 53 levels for SiII, III and IV, respectively). The adopted method, based on Hunter *et al.* (2007), consists of the following steps:

(*a*) Initial values for the stellar parameters are set from the stellar spectral type.

(*b*) Ionization balance of SiII/III/IV and/or OI/II/III provides $T_{\rm eff}$ and abundances of Si and O.

(*c*) $\log g$ is obtained from fits of the pressure broadened wings of the Balmer lines Hα, Hβ and Hγ.

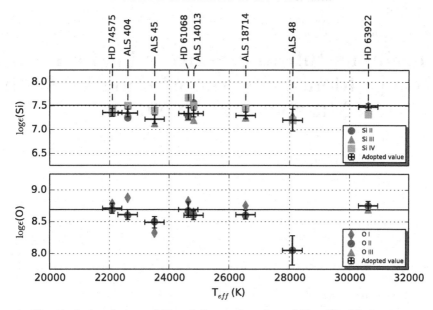

Figure 1. Chemical abundances of Si and O as a function of T_{eff}. The black crosses represent the adopted abundances for Si and O with the respective dispersion. The colored symbols show the abundance obtained for each species (see legend) and the solid lines represent the solar value obtained by Asplund *et al.* (2009).

(*d*) Microturbulence is defined by requiring that the SiIII line abundances are independent of the line strength (equivalent width).

(*e*) We check for convergence of the basic stellar parameters: T_{eff}, $\log g$ and microturbulence. If not, the previous steps are repeated.

(*f*) If converged, we fit the Si and O lines to obtain the adopted abundance values. The parameters T_{eff}, $\log g$, microturbulence and Si and O abundances (in the $\log \epsilon_X = \log(N_X/N_H) + 12$ notation) are obtained with uncertainties of 325 K, 0.07 dex, 3 km/s, 0.08 dex and 0.11 dex, respectively.

Our preliminary abundance results and effective temperatures for 5 ALS sample stars and 3 reference HD stars are shown in Fig. 1. The derived abundances show no trends with effective temperature, indicating the non-LTE calculations are free of systematics. This iterative method will be applied to the whole sub-sample of 50 sharp-lined, distant B stars in order to probe the chemical distribution of the outer Galactic disk.

Acknowledgements

We acknowledge financial support of National Science Foundation (NSF, AST-0448900), Coordenação de Aperfeiçoamento de Pessoal de Nível Superior (CAPES) and the International Astronomomic Union (IAU).

References

Asplund, M., Grevesse, N., Sauval, A. J., & Scott, P. 2009, *ARA&A* 47, 481

Daflon, S., Cunha, K., & Butler, K. 2004, *ApJ* 606, 514

Hubeny, I. & Lanz, T. 1995, *ApJ* 439, 875

Hubeny, I. & Lanz, T. 2011, *Synspec: General Spectrum Synthesis Program*, Astrophysics Source Code Library

Hunter, I., Dufton, P. L., Smartt, S. J., *et al.* 2007, *A&A* 466, 277

New windows on massive stars: asteroseismology, interferometry, and spectropolarimetry
Proceedings IAU Symposium No. 307, 2014
G. Meynet, C. Georgy, J. H. Groh & Ph. Stee, eds.
© International Astronomical Union 2015
doi:10.1017/S1743921314006346

Luminous Infrared Sources in the Local Group: Identifying the Missing Links in Massive Star Evolution

N. Britavskiy[1], A. Z. Bonanos[1] and A. Mehner[2]

[1]IAASARS, National Observatory of Athens, Greece
email: britvavskiy@gmail.com

[2]ESO, Santiago, Chile

Abstract. We present the first systematic survey of dusty massive stars (RSGs, LBVs, sgB[e]) in nearby galaxies, with the goal of understanding their importance in massive star evolution. Using the fact that these stars are bright in mid-infrared colors due to dust, we provide a technique for selecting and identifying dusty evolved stars based on the results of Bonanos *et al.* (2009, 2010), Britavskiy *et al.* (2014), and archival *Spitzer*/IRAC photometry. We present the results of our spectroscopic follow-up of luminous infrared sources in the Local Group dwarf irregular galaxies: Pegasus, Phoenix, Sextans A and WLM. The survey aims to complete the census of dusty massive stars in the Local Group.

Keywords. stars: emission-line, supergiants, galaxies: individual (Pegasus, Phoenix, Sextans A, WLM): local group

1. Introduction

The role of episodic mass loss in massive star evolution is one of the most important open questions of current stellar evolution theory (Smith 2014). Episodic mass loss produces dust and therefore causes evolved massive stars to be very luminous in the mid-infrared and dim at optical wavelengths. We aim to increase the number of investigated luminous mid-IR sources to shed light on the late stages of these objects. To achieve this we employed mid-IR selection criteria to identify dusty evolved massive stars in 4 nearby galaxies.

We selected 4 star-forming dwarf irregular galaxies (dIrr) with existing *Spitzer* photometry, namely Pegasus, Phoenix, Sextans A, and WLM. This work is based on the ESO/VLT proposal 090.D-009 where we observed 99 dusty massive star candidates with the FORS2 MXU spectrograph. The candidate selection was based on mid-IR colors, using 3.6 μm and 4.5 μm photometry from archival *Spitzer* Space Telescope images of nearby galaxies and $J-$band photometry from 2MASS. The targets selection technique was based on the results of Bonanos *et al.* (2009, 2010), which illustrated that each type of dusty massive stars occupies its own region on the color magnitude diagram (CMD). The most luminous targets in mid-IR colors ($M[3.6] < -9$ mag) were observed in 4 dIrr galaxies with exposure time 600 s, a spectral resolution of R $\approx$ 440 and S/N $\approx$ 30.

2. Preliminary results

The spectral type classification revealed 8 new red supergiants (RSGs), 3 new luminous blue variable (LBV)/B[e] supergiant (sgB[e]) candidates, and independently re-identified 3 RSGs in Sextans A (Britavskiy *et al.* 2014) and 4 RSGs in WLM (Levesque & Massey

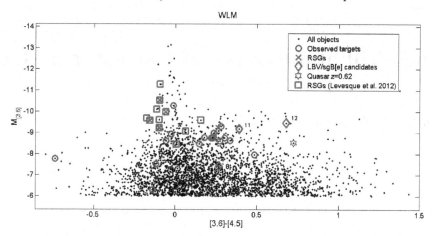

Figure 1. [3.6]-[4.5] color magnitude diagram for one of the observed dwarf irregular galaxies WLM. The positions of our investigated targets are labeled using different symbols. The numbers indicate the emission line objects.

2012). The positions of observed targets on the infrared CMD for WLM are presented in Figure 1.

The most numerous massive dusty objects that we have identified are the RSGs. We have found: 2 new RSGs in Pegasus; 1 in Phoenix; 7 in Sextans A, 3 of them have been independently previously discovered by Britavskiy *et al.* (2014). Also, we found 5 new RSGs in WLM, 4 of them have been identified by Levesque & Massey (2012). The newly discovered RSGs belong to the late spectral types (typically K-M). The positions on the color magnitude diagram of newly discovered and known RSGs for WLM are presented in Figure 1. The region of RSGs is well visible on this diagram, with $[3.6] - [4.5] < 0$. For RSGs for which we have 2 independent observations we will perform spectral type comparison for investigation of the spectral type variability as described in Levesque (2010) and Levesque & Massey (2012). Furthermore, the sample of 4 dIrr galaxies provide to us a good opportunity to investigate dependencies between average spectral types of RSGs vs. metallicities of galaxies. Among the reduced spectra we have identified three emission line spectra. The spectra have prominent hydrogen and iron emission lines that give a hint about the spectral type of these objects, namely B type. Preliminary analysis of the spectra and available photometry suggest the following spectral classification for the objects, which are numbered according to their aperture number on the slit: WLM 8 – an LBV candidate with an HII region, WLM 11 – an LBV candidate, WLM 12 – a sgB[e] candidate. Note, there are no He emission lines, that are typical for LBVs. Massey *et al.* (2007) previously reported that WLM 8 and WLM 12 are hydrogen emission objects.

References

Bonanos, A. Z., Lennon, D. J., Köhlinger, F., *et al.* 2010, *AJ*, 140, 416
Bonanos, A. Z., Massa, D. L., Sewilo, M., *et al.* 2009, *AJ*, 138, 1003
Britavskiy, N. E., Bonanos, A. Z., Mehner, A., *et al.* 2014, *A&A*, 562, A75
Levesque, E. M. 2010, Vol. 425 of *Astronomical Society of the Pacific Conference Series*, p. 103
Levesque, E. M. & Massey, P. 2012, *AJ*, 144, 2
Massey, P., Olsen, K. A. G., Hodge, P. W., *et al.* 2007, *AJ*, 133, 2393
Smith, N. 2014, *ARAA* in press

New windows on massive stars: asteroseismology, interferometry, and spectropolarimetry
Proceedings IAU Symposium No. 307, 2014
G. Meynet, C. Georgy, J. H. Groh & Ph. Stee, eds.

© International Astronomical Union 2015
doi:10.1017/S1743921314006358

Chemical abundances of fast-rotating OB stars

Constantin Cazorla, Thierry Morel, Yaël Nazé and Gregor Rauw

Institut d'astrophysique, géophysique et océanographie, University of Liège, Belgium
email: cazorla@astro.ulg.ac.be

Abstract. Fast rotation in massive stars is predicted to induce mixing in their interior, but a population of fast-rotating stars with normal nitrogen abundances at their surface has recently been revealed (Hunter *et al.* 2009; Brott *et al.* 2011, but see Maeder *et al.* 2014). However, as the binary fraction of these stars is unknown, no definitive statements about the ability of single-star evolutionary models including rotation to reproduce these observations can be made. Our work combines for the first time a detailed surface abundance analysis with a radial-velocity monitoring for a sample of bright, fast-rotating Galactic OB stars to put strong constraints on stellar evolutionary and interior models.

Keywords. stars: abundances, stars: fundamental parameters, stars: rotation.

1. Introduction

By determining the abundances of the key elements expected to be affected by mixing (i.e., He, C, N, O) for a large sample of fast-rotating ($v \sin i > 200 \, \mathrm{km \, s^{-1}}$), bright O 8-B 0 dwarfs in our Galaxy, our project aims at addressing the efficiency of rotational mixing in these objects. Several facilities are used: mainly el TIGRE (HEROS), complemented with archival data from the 1.93 m telescope at the Observatoire de Haute-Provence (SOPHIE, ELODIE), various ESO telescopes (FEROS, UVES), NOT, and AAT. In addition, XMM-*Newton* data will also be used to validate the results obtained in the optical and to give access to elements such as Ne, Si, Mg, and Fe that are not easily measured in the optical.

2. Parameters and CNO abundance determination

Prior to any determination of the atmospheric parameters, radial velocities and projected rotational velocities, $v \sin i$, are estimated. The effective temperature T_{eff}, surface gravity $\log g$, and helium abundance by number $y = N(\mathrm{He})/[N(\mathrm{H}) + N(\mathrm{He})]$, are derived by finding the best match between a set of observed H and He line profiles, and a grid of rotationally-broadened, synthetic profiles. These have been computed using the non-LTE line-formation code DETAIL/SURFACE and Kurucz models. A microturbulence of $10 \, \mathrm{km \, s^{-1}}$ was adopted. An iterative scheme is used: the effective temperature is taken as the value providing the best fit to the He I and He II lines with the same weight given to these two ions, the surface gravity is determined by fitting the wings of the Balmer lines, and the helium abundance is determined by fitting the He I features.

After determining the effective temperature and the surface gravity, CNO abundances are estimated by fitting synthetic profiles to three spectral domains in which the contribution of other elements can be neglected (see Rauw *et al.* 2012).

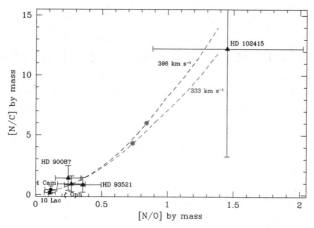

Figure 1. Dependence between the [N/C] and [N/O] abundance ratios (by mass). Except for 10 Lac, solid black symbols denote fast-rotating stars. Solid triangles and squares: archival and el TIGRE data, respectively. Dashed lines show the predictions of Geneva models (Georgy *et al.* 2013) for $15\,M_\odot$, $Z = 0.014$, and two initial rotational velocities. Solid circles indicate the beginning of the red supergiant phase.

3. Results

A slowly-rotating star, 10 Lac (O9 V), was analysed to validate the procedure used to derive the atmospheric parameters and abundances. This star has its parameters and abundances derived from standard, curve-of-growth techniques (see Rauw *et al.* 2012, for another validation test involving two other stars).

As seen in Fig. 1, the N enrichment and C depletion detected in some stars are consistent with the appearance of CNO-cycled material at their surface. The photosphere of HD 102415 appears to be nitrogen overabundant at a level expected for a red supergiant. A single star analysis is thus inappropriate to describe this main-sequence star, raising the possibility of a mass transfer in a binary (see, e.g., Ritchie *et al.* 2012).

4. Future work

New el TIGRE data are being acquired (up to 34 time-resolved spectra per star) in order to enlarge the sample of studied stars. Results will be compared to the predictions of models to investigate the relevance of the conclusions presented by Hunter *et al.* (2009). We will also account for the non-spherical shape due to fast rotation and the resulting gravity darkening. Moreover, a radial-velocity monitoring will be performed.

Acknowledgement

This research was funded through the ARC grant for Concerted Research Actions, financed by the Federation Wallonia-Brussels.

References

Brott, I., Evans, C. J., Hunter, I., *et al.* 2011, *A&A* 530, A116
Georgy, C., Ekström, S., Granada, A., *et al.* 2013, *A&A* 553, A24
Hunter, I., Brott, I., Langer, N., *et al.* 2009, *A&A* 496, 841
Maeder, A., Przybilla, N., Nieva, M.-F., *et al.* 2014, *A&A* 565, A39
Rauw, G., Morel, T., & Palate, M. 2012, *A&A* 546, A77
Ritchie, B. W., Stroud, V. E., Evans, C. J., *et al.* 2012, *A&A* 537, A29

New windows on massive stars: asteroseismology, interferometry, and spectropolarimetry
Proceedings IAU Symposium No. 307, 2014
G. Meynet, C. Georgy, J. H. Groh & Ph. Stee, eds.

© International Astronomical Union 2015
doi:10.1017/S174392131400636X

Massive star archeology in globular clusters

W. Chantereau[1], C. Charbonnel[1,2] and G. Meynet[1]

[1]Departement of Astronomy, University of Geneva, 1290 Versoix, Switzerland
email: william.chantereau@unige.ch

[2]IRAP, CNRS UMR 5277, Université de Toulouse, 31400 Toulouse, France

Abstract. Globular clusters are among the oldest structures in the Universe and they host today low-mass stars and no gas. However, there has been a time when they formed as gaseous objects hosting a large number of short-lived, massive stars. Many details on this early epoch have been depicted recently through unprecedented dissection of low-mass globular cluster stars via spectroscopy and photometry. In particular, multiple populations have been identified, which bear the nucleosynthetic fingerprints of the massive hot stars disappeared a long time ago. Here we discuss how massive star archeology can be done through the lense of these multiple populations.

Keywords. globular clusters: general, stars: evolution, stars: low-mass

1. Introduction and the FRMS scenario

Observations of Globular Clusters (GCs) show strong star-to-star variations of light elements (C, N, O, Na, Mg, Al) while iron and heavier elements stay constant (except in the most massive cases like ωCen, M22, NGC 1851, NGC 2419 and M54). They also reveal multiple structures (main sequences (MS), sub- (SGB), red-(RGB) giant branches) or even extended horizontal branches (HB) in the GCs colour-magnitude diagrams (CMD). This suggests that these stars clusters are composed of at least two populations of stars: the first one has a chemical composition similar to that of halo stars, whereas the second one is characterized by a very peculiar composition. According to the fast-rotating massive stars (FRMS) scenario (Decressin *et al.* 2007; Prantzos & Charbonnel 2006; Krause *et al.* 2013), fast rotating massive stars ($25 - 120 \, M_\odot$) have polluted the intracluster medium with their H-burning products ejecta from which a second generation of stars was born, bearing the fingerprint of the massive stars. In particular, their He content is expected to cover a large range between 0.248 (initial) and 0.8 (for the most O-depleted and Na-enriched ones).

2. The second generation

We compute a grid of standard stellar models (i.e. with neither rotation nor atomic diffusion) for the second generation low-mass stars at low metallicity. The assumed initial chemical composition reflects directly the ejecta pollution from the FRMS (i.e. a higher initial helium content but also the anticorrelation CN, ONa, and MgAl), diluted at various degrees with the intracluster gas. This peculiar initial composition influences greatly the stellar evolution, making these stars hotter and brighter for each evolution phase. Therefore these stars evolve faster; e.g. the age at the turnoff for a $0.8 \, M_\odot$ with an initial He content $Y = 0.248$ is $\sim$13 Gyr when it is only $\sim$161 Myr for a $0.8 \, M_\odot$ with $Y = 0.8$. Additionally the coupled effect of the initial mass and initial helium brings a wide variety of behaviors usually not expected for this range of mass, e.g. the presence

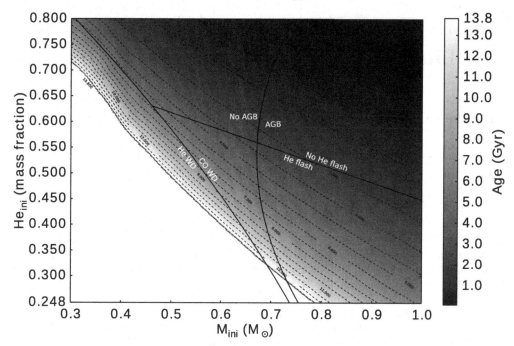

Figure 1. Lifetime and fate of stars versus initial mass and helium content at the metallicity of NGC 6752. The colour code represents the age of the star at the turnoff in Gyr, the blank part represents a time > 13.8 Gyr.

of helium white dwarfs at a lower time than the Hubble one, and exacerbated late hot flashers/AGB-manqué behaviors (see Fig. 1).

3. Direct application

These second generation stars greatly impact the morphology of the current GCs because of their very peculiar composition but also because they outnumber the first generation of stars in GCs (Prantzos & Charbonnel 2006; Carretta *et al.* 2010). For instance Campbell *et al.* (2013) observed that all the Na-rich stars (associated to the second generation, $\sim 70\%$) fail to reach the AGB phase in NGC 6752. A straightforward explanation has been presented by Charbonnel *et al.* (2013) who showed that all the Na-rich stars ([Na/Fe]> 0.4 dex) within the framework of the FRMS scenario do not ascend the AGB, this is the so-called AGB-manqué phenomenon.

References

Campbell, S. W., D'Orazi, V., Yong, D., *et al.* 2013, *Nature* 498, 198
Carretta, E., Bragaglia, A., Gratton, R. G., *et al.* 2010, *A&A* 516, A55
Charbonnel, C., Chantereau, W., Decressin, T., Meynet, G., & Schaerer, D. 2013, *A&A* 557, L17
Decressin, T., Meynet, G., Charbonnel, C., Prantzos, N., & Ekström, S. 2007, *A&A* 464, 1029
Krause, M., Charbonnel, C., Decressin, T., Meynet, G., & Prantzos, N. 2013, *A&A* 552, A121
Prantzos, N. & Charbonnel, C. 2006, *A&A* 458, 135

New windows on massive stars: asteroseismology, interferometry, and spectropolarimetry
Proceedings IAU Symposium No. 307, 2014
G. Meynet, C. Georgy, J. H. Groh & Ph. Stee, eds.

© International Astronomical Union 2015
doi:10.1017/S1743921314006371

Linking 1D Stellar Evolution to 3D Hydrodynamic Simulations

A. Cristini[1], R. Hirschi[1,2], C. Georgy[1], C. Meakin[3], D. Arnett[3] and M. Viallet[4]

[1] Astrophysics group, Keele University, Lennard-Jones Building, Keele, ST5 5BG, UK
email: a.j.cristini@keele.ac.uk

[2] Kavli IPMU (WPI), The University of Tokyo, Kashiwa, Chiba 277-8583, Japan

[3] Department of Astronomy, University of Arizona, Tucson, AZ 85721, USA

[4] Max-Planck-Institut für Astrophysik, Garching, D-85741, Germany

Abstract. In this contribution we present initial results of a study on convective boundary mixing (CBM) in massive stellar models using the GENEVA stellar evolution code (Eggenberger *et al.* 2008). Before undertaking costly 3D hydrodynamic simulations, it is important to study the general properties of convective boundaries, such as the: composition jump; pressure gradient; and 'stiffness'. Models for a $15 M_\odot$ star were computed. We found that for convective shells above the core, the lower (in radius or mass) boundaries are 'stiffer' according to the bulk Richardson number than the relative upper (Schwarzschild) boundaries. Thus, we expect reduced CBM at the lower boundaries in comparison to the upper. This has implications on flame front propagation and the onset of novae.

Keywords. convection, hydrodynamics, stellar dynamics, turbulence, stars: evolution, stars: interiors

---

One of the key properties of a boundary is its 'stiffness'. The 'stiffness' of a convective boundary can be quantified using the bulk Richardson number, Ri_B, this is the ratio of the potential energy for restoration of the boundary to the kinetic energy of turbulent eddies. It is given by

$$\mathrm{Ri}_B = \frac{\Delta B \; L}{v^2/2}$$

$$\text{where} \quad \Delta B(r) = \int_{r_0}^{r} N^2(r')dr'$$

$$\text{and} \quad N^2 = \frac{g\delta}{H_P}\left(\nabla_{\mathrm{ad}} - \nabla + \frac{\phi}{\delta}\nabla_\mu\right)$$

where r is the radius, r_0 the radial coordinate of the convective boundary and L is the length scale that characterises turbulence and is taken to be the pressure scale height at the boundary. The final expression is the Brunt-Väisälä frequency (e.g. see Kippenhahn *et al.* 2013), within the brackets is the familiar Ledoux criterion for convective stability. A 'stiff' boundary will suppress convective boundary mixing (CBM), whereas in the opposite case a 'soft' boundary will be more susceptible to CBM. Typical values of bulk Richardson numbers for 'stiff' and 'soft' boundaries are 10,000 and 10, respectively.

From Fig. 1 (bottom left panel) it can be seen that the bulk Richardson number is larger for the lower convective boundary, implying a 'stiffer' boundary and suppressed convective boundary mixing (CBM). The reason for this is despite the length scale for the lower boundary been slightly smaller, the peak N value is larger for the lower boundary.

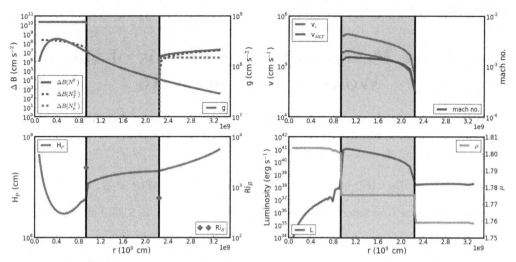

Figure 1. Structure properties of the second C shell burning region as a function of radius (31% of the shell lifetime).
Top left – Buoyancy jump (magenta) and its components, thermal (blue dashed) and compositional (red dashed) and gravitational acceleration (green).
Top right – Convective (red), mixing length theory (green) velocities and Mach number (blue).
Bottom left – Pressure scale height (red) and bulk Richardson number (green diamond).
Bottom right – Luminosity (magenta) and mean molecular weight (cyan). Vertical black lines represent radial positions of convective boundaries and grey areas represent convective regions.

The results presented here are in agreement with 3D simulations e.g. the oxygen burning shell of a 23 $M_\odot$ model by Meakin & Arnett (2007). Suppressed CBM at lower convective boundaries has implications for other areas of astrophysics e.g. Denissenkov *et al.* (2013a) show that the onset of novae is affected by CBM, Denissenkov *et al.* (2013b) and Jones *et al.* (2013) also show that flame front propagation in S-AGB stars is affected by CBM.

Following a preliminary characterisation of the convective boundaries multi-D hydrodynamic simulations of convective nuclear burning shells will commence, using the code PROMPI, a parallelised version of Prometheus. Simulations are planned for the carbon and silicon burning shells of massive stars. These results will be important for the community, in understanding the advanced phases of stellar evolution, and also to produce more accurate 1D pre-supernova progenitor models.

Acknowledgements

The authors acknowledge support from EU-FP7-ERC-2012-St Grant 306901. R.H. acknowledges support from the World Premier International Research Center Initiative (WPI Initiative), MEXT, Japan.

References

Denissenkov, P. A., Herwig, F., Bildsten, L., & Paxton, B. 2013a, *ApJ* 762, 8
Denissenkov, P. A., Herwig, F., Truran, J. W., & Paxton, B. 2013b, *ApJ* 772, 37
Eggenberger, P., Meynet, G., Maeder, A., *et al.* 2008, *Ap&SS* 316, 43
Jones, S., Hirschi, R., Nomoto, K., *et al.* 2013, *ApJ* 772, 150
Kippenhahn, R., Weigert, A., & Weiss, A. 2013, *Stellar Structure and Evolution*
Meakin, C. A. & Arnett, D. 2007, *ApJ* 667, 448

New windows on massive stars: asteroseismology, interferometry, and spectropolarimetry
Proceedings IAU Symposium No. 307, 2014
G. Meynet, C. Georgy, J. H. Groh & Ph. Stee, eds.

© International Astronomical Union 2015
doi:10.1017/S1743921314006383

First Results of the Analysis of the Wolf-Rayet Star WR6

Alex C. Gormaz-Matamala[1] Anthony Hervé[2], André-Nicolas Chené[3], Michel Curé[4] and Ronald Mennickent[1]

[1] Departamento de Astronomía, Universidad de Concepción, Casilla 160-C, Concepción, Chile
email: agormaz@astro-udec.cl

[2] CNRS & Université Montpellier II, Place Eugene Bataillon, F-34095 Montpellier, France
[3] Gemini Observatory, Hawaii, USA.
[4] Instituto de Física y Astronomía, Universidad de Valparaíso. Casilla 5030, Valparaíso, Chile

Abstract. We present the first results of our analysis of the famous variable star, WR6 (HD50896). Using IUE ultraviolet data and an ESPaDOnS spectropolarimetric survey of this star, we plan to determine possible variation of the stellar and wind parameters during the different phases using the radiative transfer code CMFGEN. After the detection of parameter's modifications as a function of the phase, we will analyse deeper the origin of these variability (for example, CIRs?). In the present poster we show the first results of our analysis of the variability and the first step of the stellar parameter determination of the average spectrum of this star.

Keywords. Atmospheres, Mass Loss, Wolf-Rayet.

1. Observational Data and Computational Tools

For optical data, we use ESPaDOnS spectropolarimeter spectra provided by de la Chevrotière *et al.* 2013, whereas for UV data we use IUE spectra from MAST database. For the models, we use the code CMFGEN (Hillier *et al.* 2001).

2. First Results

2.1. *Variability*

Using the formalism of the Temporal Variation Spectrum (TVS: Fullerton *et al.* 1996; Chené 2007; St-Louis *et al.* 2009), together with defining one significant variability level $\sigma_0^2 \chi_N^2 - 1(99\%)$ from the standardised dispersion of our data, we can confirm WR6 line profile variability, as the TVS of the four studied lines are way above this threshold.

2.2. *Stellar Wind Parameter Determinations*

The parameters found and the current best model fit are shown in Table 1 and Figure 1. The terminal velocity was calculated using the CIV 1548 P-Cygni profile: our value coincides with that found by Hamann *et al.* (2006). For temperature, we used the HeII 5411/HeI 5875 ratio (Smith *et al.* 1996).

2.3. *Further Work*

We must check the effects of different β values yet. Also, is necessary to continue testing different abundances (specially carbon and nitrogen) for getting a best fit in those lines.

Table 1. Stellar wind parameters and abundances of our current best fit model (Figure 1).

Parameter	Value	Parameter	Value
$T_{\rm eff}$ [kK]	60	[He/He$_\odot$]	0.525
$\dot{M}[M_\odot\,{\rm yr}^{-1}]$	1.9×10^{-5}	[C/C$_\odot$]	-0.860
v_∞ [km s^{-1}]	1700	[N/N$_\odot$]	1.437
[H/H$_\odot$]	-1.523	[O/O$_\odot$]	0.068

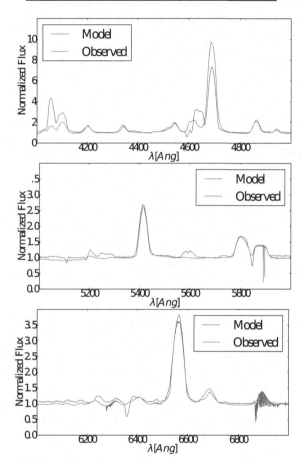

Figure 1. Model compared with the global averaged spectrum.

References

Chené, A.-N. 2007, *Ph.D. thesis*, Université de Montreal (Canada)

de la Chevrotière *et al.*, A. 2013, *ApJ* 764, p.171

Fullerton *et al.*, A. W. 1996, *ApJS* 103, p.475

Hamann *et al.*, W. R. 2006, *A&A* 457, 1015

Hillier *et al.*, J. 2001, *Spectroscopic Challenges of Photoionized Plasmas, ASP Conference Series.* 247, p.343

Smith *et al.*, L. F. 1996, *MNRAS* 281, 163

St-Louis *et al.*, N. 2009, *ApJ* 698, 1951

New windows on massive stars: asteroseismology, interferometry, and spectropolarimetry
Proceedings IAU Symposium No. 307, 2014
G. Meynet, C. Georgy, J. H. Groh & Ph. Stee, eds.

© International Astronomical Union 2015
doi:10.1017/S1743921314006395

Evolution of the rotational properties and nitrogen surface abundances of B-Type stellar populations

A. Granada[1], G. Meynet[1], S. Ekström[1], C. Georgy[2] and L. Haemmerlé[1]

[1] Geneva Observatory, University of Geneva, Maillettes 51, 1290 Sauverny, Switzerland
email: anahi.granada@unige.ch

[2] Astrophysics group, EPSAM, Keele University, Lennard-Jones Labs, Keele, ST5 5BG, UK

Abstract. Stellar evolution models predict that rotation induces the mixing of chemical species, with the subsequent surface abundance anomalies relative to single non-rotating models, even during the main sequence (MS) evolution. The lack of measurable nitrogen surface enrichment in MS rotating stars, such as Be stars, has been interpreted as being in conflict with evolutionary models (e.g. Lennon *et al.* 2005; Hunter *et al.* 2008). In order to have an insight on the kind of ambient we do or we do not expect to find enriched rotating stars, we use our new population synthesis code, to produce synthetic intermediate-mass stellar populations fully accounting for stellar rotation effects, and study their evolution in time.

Keywords. stars: evolution, stars: rotation, stars: abundances.

1. Introduction

We use SYCLIST (SYnthetic CLusters Isochrones & Stellar Tracks, Georgy *et al.* 2014) to study how the fraction of single rotating and nitrogen-enriched B-Type stars evolve, using as inputs the rotating stellar models from Georgy *et al.* (2013) and Granada & Haemmerlé (submitted). We define as B-Type stars those objects with $T_{\rm eff}$ between 10000 and 30000 K. Early B-type have $T_{\rm eff} > 19000$ K. We consider a star as nitrogen-enriched if $(N/H)/(N/H)_{\rm ini} > 2$. Apart from coeval populations, we also computed the fraction of rotating and enriched stars for continuous star formation.

2. Results

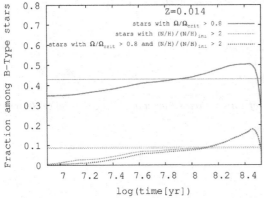

In Fig. 1, the thick lines represent the evolution of the fraction of B-type stars at Z=0.014 with $\Omega/\Omega_{\rm crit} > 0.8$ (solid red), of objects with $(N/H)/(N/H)_{\rm ini} > 2$ (dashed green), and with $\Omega/\Omega_{\rm crit} > 0.8$ and $(N/H)/(N/H)_{\rm ini} > 2$ (dotted blue). The thin lines indicate the same values, for continuous star formation. The values obtained for an age of 25Myr are listed in column 6 of Table 1.

Figure 1. Evolution of different types of B star.

Table 1. Fraction of B and of early-B stars, having properties described in column 1.

Fraction of	Z=0.002		Z=0.006		Z=0.014	
	All B [%]	Early B [%]	All B [%]	Early B [%]	All B [%]	Early B [%]
Population of an age of 25 Myr.						
(1) $\frac{(N/H)}{(N/H)_{ini}}>2$	10.0	47.7	6.6	39.9	4.3	28.5
(2) $\Omega/\Omega_{crit}>0.8$	38.3	26.5	38.0	28.7	38.0	18.4
(1) & (2)	5.1	24.7	3.6	21.4	2.8	15.6
Continuous Star Formation						
(1) $\frac{(N/H)}{(N/H)_{ini}}>2$	42.0	42.7	19.0	25.3	9.0	13.0
(2) $\Omega/\Omega_{crit}>0.8$	49.6	29.5	44.6	20.8	43.3	14.1
(1) & (2)	32.6	22.6	16.5	12.3	8.5	6.0

Fig. 2 shows the Colour Magnitude Diagram (CMD) of the MS of a synthetic cluster of an age of 25Myr with 100 intermediate mass stars, produced by SYCLIST. Black circles indicate late B stars, red squares indicate early-type B stars. Solid symbols indicate rapidly rotating stars and the blue romboid those objects with a large nitrogen enrichment. As expected for this age, around 2% of the B stars are rapidly rotating and nitrogen enriched. Because we can account for inclination effects to build our synthetic clusters, in the example given, the brightest blue square has V sini = 117 km/s and the other V sini = 400 km/s. In terms of the surface velocity, only one of this objects would seem to be rapidly rotating and enriched.

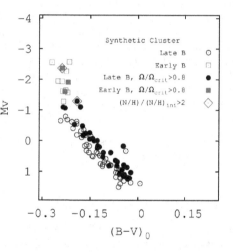

Figure 2. Synthetic CMD, 25 Myr old.

3. Conclusions

• We expect a very low fraction of enriched rotating B-type stars in coeval young populations.

• These fractions are larger among early B-type stars and at smaller Z.

• When continuous star formation is considered, the fractions of rapidly rotating and enriched stars among B-type stars can differ significantly from coeval populations. At Z = 0.014, the fraction is larger among all B stars, but is reduced among early-B stars.

• All these results are dependent on the initial velocity distribution, on the definition of rapidly-rotating and of what we call an enriched stars. We are currently exploring all these dependences.

References

Georgy, C., Ekström, S., Granada, A., *et al.* 2013, *A&A* 553, A24
Georgy, C., Granada, A., Ekström, S., *et al.* 2014, *A&A* 566, A21
Hunter, I., Brott, I., Lennon, D. J., *et al.* 2008, *ApJ (Letters)* 676, L29
Lennon, D. J., Lee, J.-K., Dufton, P. L., & Ryans, R. S. I. 2005, *A&A* 438, 265

New windows on massive stars: asteroseismology, interferometry, and spectropolarimetry
Proceedings IAU Symposium No. 307, 2014
G. Meynet, C. Georgy, J. H. Groh & Ph. Stee, eds.

© International Astronomical Union 2015
doi:10.1017/S1743921314006401

Delta-slow solution to explain B supergiant stars' winds

M. Haucke[1], I. Araya[2], C. Arcos[2],M. Curé[2], L. Cidale[1], S. Kanaan[2], R. Venero[1] and M. Kraus[3]

[1]Facultad de Ciencias Astronómicas y Geofísicas, UNLP, Argentina.
email: mhaucke@fcaglp.unlp.edu.ar

[2]Instituto de Física y Astronomía, UV, Chile.

[3]Astronomický ústav, Akademie věd České Republiky, Ondřejov, Czech Republic.

Abstract. A new radiation-driven wind solution called δ-slow was found by Curé *et al.* (2011) and it predicts a mass-loss rate and terminal velocity slower than the fast solution (m-CAK, Pauldrach *et al.* 1986). In this work, we present our first synthetic spectra based on the δ-slow solution for the wind of B supergiant (BSG) stars. We use the output of our hydrodynamical code HYDWIND as input in the radiative transport code FASTWIND (Puls *et al.* 2005). In order to obtain stellar and wind parameters, we try to reproduce the observed Hα, Hβ, Hγ, Hδ, HeI 4471, HeI 6678 and HeII 4686 lines. The synthetic profiles obtained with the new hydrodynamical solutions are in good agreement with the observations and could give us clues about the parameters involved in the radiation force.

Keywords. stars: mass loss, stars: winds, outflows

1. Introduction

For O supergiant stars (OSGs) and BSGs, the classical theory of radiation-driven winds predicts that their winds' velocity fields can be fitted by a β-law, with a typical value of $\beta = 0.8 - 1$. This describes very well the OSGs winds, but for BSGs, observations show values of $\beta > 1$, indicating a slower outflowing regime at the base of the wind. To explain this behaviour we present here a preliminary analysis, via a line fitting procedure, of the wind of 3 BSGs (HD 52 382, HD 86 440 and HD 91 619) using this new δ-slow solution. To fit the H and He line profiles we used the FASTWIND code, but the input file of the models were done with the code HYDWIND, developed by Curé (2004), which predicts slower and denser winds than the m-CAK theory.

2. Results

The parameters obtained from the line fitting for the observed stars are tabulated in Table 1, and the plots are shown in Figure 1.

• The δ-slow solution fits very well the observations, however for HD 91 619 we need values of $T_{\rm eff}$ and $\log g$ higher than the expected for its spectral type (B7).

• For HD 86 440 and HD 91 619 we obtain a good fit with a δ-slow solution using values of $\delta > 0.4$.

3. Discussion and conclusions

• The δ-slow solution seems to describe the wind structure of the BSGs with $\beta > 1$ at the base of the wind, then it behaves like a $\beta = 0.7$ for an intermediate region

Table 1. Stellar, line force and wind parameters

STAR HD	T_{eff} K	$\log g$ dex	v_{rot} km s^{-1}	$R_\star$ $R_\odot$	α	δ	k	$\dot{M}$ $10^{-6}\,M_\odot\,\text{yr}^{-1}$	v_∞ km s^{-1}	v_{micro} km s^{-1}
52 382	19 000	2.5	100	52	0.52	0.39	0.21	0.47	290	5
86 440	15 000	2.55	60	62	0.52	0.43	0.32	0.15	376	10
91 619	19 000	2.3	40	71	0.52	0.44	0.142	0.5	222	25

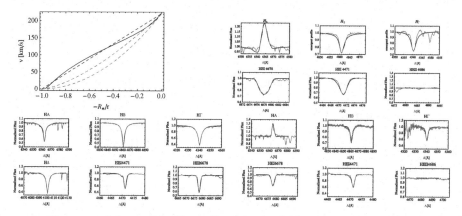

Figure 1. The upper left graphic shows the velocity field of HD 91 619. Black solid line is obtained with δ-slow solution. Black, blue and red dashed lines correspond to β-law with $\beta = 1$, 1.5 and 2, respectively. The other panels are the fittings to H and He lines for HD 52 382 (upper right), HD 86 440 (bottom left) and HD 91 619 (bottom right).

$(1.1 < r/R_\star \lesssim 3.0)$, but finally, in the outer part of the wind, it approaches again to $\beta > 1$ (see Figure 1). The δ-slow solution reproduces well the observed line profiles.

• HD 52 382 is a variable star with no cyclical behaviour (see Morel *et al.* 2004) and this could be the cause of the differences in the stellar and wind parameters found in the literature, for example, $v_\infty = 900\,\text{km s}^{-1}$ (Howarth *et al.* 1997) and $\dot{M} = 0.33 \times 10^{-6}\,M_\odot\,\text{yr}^{-1}$ (Morel *et al.* 2004).

• For HD 86 440 our results of the T_{eff} and $\log g$ are in agreement with the same values obtained by Fraser *et al.* (2010).

• For HD 91 619 we have the same "difficulties" as Markova *et al.* (2008): if we use lower T_{eff} and lower $\log g$ the absorption component of the synthetic profiles is very intense and does not fit the observed one.

• To reproduce some observations we may need δ values > 0.4. These values are greater than $1/3$ which is the value expected for a pure H medium (Puls *et al.* 2000).

References

Curé, M. 2004, *ApJ* 614, 929

Curé, M., Cidale, L., & Granada, A. 2011, *ApJ* 737, 18

Fraser, M., Dufton, P. L., Hunter, I., & Ryans, R. S. I. 2010, *MNRAS* 404, 1306

Howarth, I. D., Siebert, K. W., Hussain, G. A. J., & Prinja, R. K. 1997, *MNRAS* 284, 265

Markova, N., Prinja, R. K., Markov, H., *et al.* 2008, *A&A* 487, 211

Morel, T., Marchenko, S. V., Pati, A. K., *et al.* 2004, *MNRAS* 351, 552

Pauldrach, A., Puls, J., & Kudritzki, R. P. 1986, *A&A* 164, 86

Puls, J., Springmann, U., & Lennon, M. 2000, *A&AS* 141, 23

Puls, J., Urbaneja, M. A., Venero, R., *et al.* 2005, *A&A* 435, 669

New windows on massive stars: asteroseismology, interferometry, and
spectropolarimetry
Proceedings IAU Symposium No. 307, 2014
G. Meynet, C. Georgy, J. H. Groh & Ph. Stee, eds.

© International Astronomical Union 2015
doi:10.1017/S1743921314006413

Massive OB stars at varying Z

A. Herrero[1,2], **M. Garcia**[3], **S. Simón-Díaz**[1,2], **I. Camacho**[1,2],
C. Sabín-Sanjulián[4] and **N. Castro**[5]

[1]Instituto de Astrofísica de Canarias, C/Vía Láctea s/n, E-38200 La Laguna, Spain
email: ahd@iac.es

[2]Departamento de Astrofísica, Universidad de La Laguna, La Laguna, Spain

[3]Centro de Astrobiología, Madrid, CSIC, Spain

[4]Departamento de Física, Universidad de La Serena, Chile

[5]Argelander Institut für Astronomie, Auf dem Hügel 71, 53121 Bonn, Germany

Abstract. Massive stars play a key role in environments with very different metallicities. To interpret the role of massive stars in these systems we have to know their properties at different metallicities. The Local Group offers an excellent laboratory to this aim.

Keywords. stars: fundamental parameters, supergiants, winds; galaxies: Local Group

1. Introduction

Massive stars play a key role in environments with very different metallicities (Z): (a) in the re-ionization of the early Universe (Robertson *et al.* 2010); (b) in the formation of violent phenomena (SNe, GRBs, BH, NS, interacting massive binaries; Langer 2012); (c) in the determination of Starburst properties and Star Formation Rates through UV fluxes or H_α emission (Kennicutt & Evans 2012); or (d) in the chemical evolution of galaxies through stellar winds, supernova explosions and binary interaction. To interpret the role of massive stars in these systems and our observations of starbursts and high-redshift galaxies we have to know their properties at different metallicities. The Local Group offers an excellent laboratory to this aim because we can observe and analyze massive stars individually in varying metallicity conditions. We present here a short summary of our work in some Local Group galaxies, breaking the metallicity frontier of the SMC.

2. OB stars in galaxies of varying Z

2.1. *Sextans A ($Z = 0.07\ Z_\odot$)*

Candidate OB stars in Sextans A were selected following the criteria in Garcia & Herrero (2013) for IC1613. Long-slit spectra were subsequently obtained with OSIRIS@GTC at R = 1000 spectral power in the wavelength interval between 4000 and 5500 Å. Five O stars and eight B stars were observed and classified, and stellar parameters have been obtained for the O stars using the iacob-gbat tool (Simón-Díaz *et al.* 2011). These constitute the largest atlas of individual OB stars at such low metallicities. Final analyses will be presented in Camacho *et al.* (2014, in prep).

2.2. *IC 1613 ($Z = 0.13\ Z_\odot$)*

IC 1613 candidates were selected following the criteria in Garcia & Herrero (2013). Long-slit spectra of IC 1613 OB stars were obtained with OSIRIS@GTC at R = 1000 in the wavelength interval between 4000 and 5500 Å and multi-object spectroscopy was carried out with VIMOS@VLT at R = 1800 covering the interval between 4000 and

6800 Å (Herrero *et al.* 2012). 8 O-stars have been observed with GTC and their parameters were determined using the iacob-gbat tool (Simón-Díaz *et al.* 2011), allowing us to establish the T_{eff} scale of O stars at the up to date lowest metallicity (Garcia & Herrero 2013). The resulting T_{eff} scale is slightly hotter ($\approx$1000 K) than the SMC one.

2.3. LMC: 30 Dor (Z = 0.50 $Z_\odot$)

30 Doradus stars were observed in multi-object, multi-epoch spectroscopy mode with FLAMES@VLT, R = 7000–15000 and λ = [4000, 6800] Å(VFTS project, see Evans *et al.* 2011). 46 OV and 38 Ovz stars were analyzed with the iacob-gbat tool (Simón-Díaz *et al.* 2011) and the results published in Sabín-Sanjulián *et al.* (2014). We clarify the nature of OVz stars (O stars having HeII 4686 stronger than both HeII 4542 and HeI 4471), that were suspected to represent the link between early phases of star formation and the normal, slightly evolved, O dwarfs. Sabín-Sanjulián *et al.* (2014) show that the OVz phenomenon is a natural consequence of the combination of stellar parameters and that OVz stars may lie away from the Zero Age Main Sequence (ZAMS), although it is easier to find them close to it. Effective temperature and wind strength are the primary stellar parameters controlling the OVz phenomenon, while rotational velocity and surface gravity play a secondary role (Sabín-Sanjulián *et al.* 2014).

2.4. Milky Way (Z = 1.0 $Z_\odot$)

We have obtained multi-epoch spectra of 500 Galactic O4-B9 stars with FIES@NOT and HERMES@MERCATOR, with R $\leqslant$ 25000 and λ between 3800 and 9000 Å. Line-broadening parameters were obtained with the iacob-broad tool (Simón-Díaz & Herrero 2014), stellar parameters with the iacob-gbat (Simón-Díaz *et al.* 2011) tool for OB and with the Castro *et al.* (2012) tool for mid and late B. New results for the distributions of vsini and extra broadening effects in OB stars have been presented by Simón-Díaz & Herrero (2014). Among other results, these authors emphasize that a larger number of slowly rotating early B supergiants would be expected as a consequence of angular momentum loss during the evolution of mid O type stars.

3. A comparison of T_{eff} scales

We found no differences between the T_{eff} scales we determined for MW and 30 Dor O dwarfs, in spite of their metallicity difference. Adding data from the literature does not significantly modify this conclusion. The effect is not seen when comparing O stars in IC 1613, Sextans A, the SMC and the MW. The most plausible explanation for this behaviour is that we are facing small numbers statistics (Simón-Díaz *et al.*, in prep.).

References

Castro, N., Urbaneja, M. A., Herrero, A., *et al.* 2012, *A&A* 542, A79
Evans, C. J., Taylor, W. D., Hénault-Brunet, V., *et al.* 2011, *A&A* 530, A108
Garcia, M. & Herrero, A. 2013, *A&A* 551, A74
Herrero, A., Garcia, M., Puls, J., *et al.* 2012, *A&A* 543, A85
Kennicutt, R. C. & Evans, N. J. 2012, *ARA&A* 50, 531
Langer, N. 2012, *ARA&A* 50, 107
Robertson, B. E., Ellis, R. S., Dunlop, J. S., McLure, R. J., & Stark, D. P. 2010, *Nature* 468, 49
Sabín-Sanjulián, C., Simón-Díaz, S., Herrero, A., *et al.* 2014, *A&A* 564, A39
Simón-Díaz, S., Castro, N., Herrero, A., *et al.* 2011, *Journal of Physics Conference Series* 328(1), 012021
Simón-Díaz, S. & Herrero, A. 2014, *A&A* 562, A135

New windows on massive stars: asteroseismology, interferometry, and spectropolarimetry
Proceedings IAU Symposium No. 307, 2014
G. Meynet, C. Georgy, J. H. Groh & Ph. Stee, eds.

© International Astronomical Union 2015
doi:10.1017/S1743921314006425

Massive stars: flare activity due to infalls of comet-like bodies

Subhon Ibadov[1,2] and Firuz S. Ibodov[1]

[1] Lomonosov Moscow State University, SAI, Moscow, Russia
email: ibadovsu@yandex.ru

[2] Institute of Astrophysics, TAS, Dushanbe, Tajikistan

Abstract. Passages of comet-like bodies through the atmosphere/chromosphere of massive stars at velocities more than 600 km/s will be accompanied, due to aerodynamic effects as crushing and flattening, by impulse generation of hot plasma within a relatively very thin layer near the stellar surface/photosphere as well as "blast" shock wave, i.e., impact-generated photospheric stellar/solar flares. Observational manifestations of such high-temperature phenomena will be eruption of the explosive layer's hot plasma, on materials of the star and "exploding" comet nuclei, into the circumstellar environment and variable anomalies in chemical abundances of metal atoms/ions like Fe, Si etc. Interferometric and spectroscopic observations/monitoring of young massive stars with dense protoplanetary discs are of interest for massive stars physics/evolution, including identification of mechanisms for massive stars variability.

Keywords. massive stars, impact-generated stellar flares, comet-like bodies

1. Introduction

High-resolution spectroscopic observations of young stars with dense protoplanetary discs like β Pictoris as well as coronagraphic observations by solar missions indicate the presence of fluxes of comet-like evaporating bodies (FEBs), falling onto Sun/stars (see, e.g. Lagrange *et al.* 1987; Beust *et al.* 1996, http://sungrazer.nrl.navy.mil/).

Disintegration of nuclei of sungrazing comets considered in a vacuum leads to a decrease in their radii to less than 10-20 m prior to impact (Weissman 1983).

Meanwhile, comet-like bodies fall onto massive stars with high orbital velocities $V_* \geqslant V_\odot$ ($V_\odot = 617$ km/s is the parabolic velocity for the Sun). This indicates the possibility of high energy processes in comets (cf. Ibadov 1990, 1996; Ibadov *et al.* 2009; Ibodov *et al.* 2010; Ibodov & Ibadov 2011, 2014). Impulse release of a large energy, up to 10^{32} erg, is possible due to the magnetic reconnection in the active Sun (cf. Somov 1992; Walder *et al.* 2012). We consider physical processes accompanying the passage of star-impacting comet nuclei through the stellar atmosphere, i.e., processes that are capable to lead to massive stars flare activity.

2. Massive stars comet-induced flares

To find, analytically, the law for the rate of the loss of kinetic energy of aerodynamically fully crushed and transversally expanding comet-like body along the height of the star atmosphere, h, we modify the basic equations of the physical theory of meteors (aerodynamic deceleration/evaporation) using the parameter $r = r(h) = \{\exp[(h_* - h)/H]\} - 1$. We use also the comet orbital velocity near the star surface, $V_* = V_\odot (M_* R_\odot / M_\odot R_*)^{1/2}$. Here M_*, $M_\odot$ are the masses of the star and the Sun, R_*, $R_\odot$ are their radii; h_* is the onset height for aerodynamic crushing of comet nucleus, H is the atmosphere height scale.

The energy released around the height of $h = h_m$ (the height of the maximum rate of loss of kinetic energy of the crushed comet nucleus), manifesting as "explosion" in the atmosphere around $r = r_m = r_e$, and the initial temperature of plasma produced within the decelerating/exploding layer with the width, $\Delta h_d = \Delta h_e \approx 0.7H$, are

$$E_e = \frac{\pi \rho_n R_n^3 V_*^2}{3e}, \quad T_0 = \frac{A m_p V_*^2}{12 e k (1 + z + 2x_1/3)}, \tag{2.1}$$

where ρ_n, R_n are the nucleus density and initial radius, A is the mean atomic number, m_p is the proton mass, k is the Boltzmann constant, z is the mean charge multiplicity of plasma ions, x_1 is the mean relative ionization potential; the theory was tested using for the 2013 Chelyabinsk superbolide explosion (cf. Ibadov *et al.* 2009; Ibadov 2012; Ibodov & Ibadov 2011; Grigoryan *et al.* 2013).

Using (2.1) with $\rho_n = 0.5\,\text{g/cm}^3, R_n = 10^5\,\text{cm}, V_* = V_0, A = 20, z = 5, x_1 = 3$ we have $\Delta h_e = 140\,\text{km}, E_e = 7 \cdot 10^{29}\,\text{erg}, T_0 = 7 \cdot 10^6\,\text{K}$. Hence, "superflares" may be due to impacts with comets like comet Hale-Bopp 1995 OI (cf. Ibadov *et al.* 1999; Grigoryan *et al.* 2000; Ibadov *et al.* 2009; Eichler & Mordecai 2012; Ibodov & Ibadov 2014).

3. Conclusions

Impacts of sufficiently large, 100-meter or more, comet-like bodies with stars like the Sun and/or more massive ones will be accompanied by impulse production of a high-temperature plasma, strong "blast" shock wave, shock wave induced ejection of ionized photospheric and cometary matter.

Anomalously intense variable emissions of metal atoms and ions produced during the nucleus "explosion" near the stellar/solar surface may serve as possible indicators of comet impact-generated photospheric stellar/solar flares.

Acknowledgements

The authors are grateful to Prof. G. Meynet for invitation, Dr. G.M. Rudnitskij for useful discussions and V. Vybornov for technical help. The hospitality of the IAUS 307 LOC and MSU is sincerely acknowledged.

References

Beust, H., Lagrange, A.-M., Plazy, F., & Mouillet, D. 1996, *A&A* 310, 181
Eichler, D. & Mordecai, D. 2012, *ApJ (Letters)* 761, L27
Grigoryan, S., Ibadov, S., & Ibodov, F. 2000, *Doklady Physics* 45(9), 463
Grigoryan, S. S., Ibodov, F. S., & Ibadov, S. I. 2013, *Solar System Research* 47, 268
Ibadov, S. 1990, *Icarus* 86, 283
Ibadov, S. 1996, *Physical processes in comets and related objects*
Ibadov, S. 2012, in *IAU Symposium*, Vol. 283 of *IAU Symposium*, pp 392–393
Ibadov, S., Ibodov, F. S., & Grigorian, S. S. 1999, *Romanian Astronomical Journal* 9, 125
Ibadov, S., Ibodov, F. S., & Grigorian, S. S. 2009, in N. Gopalswamy & D. F. Webb (eds.), *IAU Symposium*, Vol. 257 of *IAU Symposium*, pp 341–343
Ibodov, F. S., Grigorian, S. S., & Ibadov, S. 2010, in J. A. Fernandez, D. Lazzaro, D. Prialnik, & R. Schulz (eds.), *IAU Symposium*, Vol. 263 of *IAU Symposium*, pp 269–271
Ibodov, F. S. & Ibadov, S. 2011, in A. Bonanno, E. de Gouveia Dal Pino, & A. G. Kosovichev (eds.), *IAU Symposium*, Vol. 274 of *IAU Symposium*, pp 92–94

110 S. Ibadov & F. S. Ibodov

Ibodov, F. S. & Ibadov, S. 2014, in B. Schmieder, J.-M. Malherbe, & S. T. Wu (eds.), *IAU Symposium*, Vol. 300 of *IAU Symposium*, pp 509–511
Lagrange, A. M., Ferlet, R., & Vidal-Madjar, A. 1987, *A&A* 173, 289
Somov, B. V. (ed.) 1992, *Physical processes in solar flares.*, Vol. 172 of *Astrophysics and Space Science Library*
Walder, R., Folini, D., & Meynet, G. 2012, *Space Sci. Revs* 166, 145
Weissman, P. R. 1983, *Icarus* 55, 448

Melina Bersten

Götz Gräfener

New windows on massive stars: asteroseismology, interferometry, and spectropolarimetry
Proceedings IAU Symposium No. 307, 2014
G. Meynet, C. Georgy, J. H. Groh & Ph. Stee, eds.

© International Astronomical Union 2015
doi:10.1017/S1743921314006437

Study of environment and photosphere of 51 Oph

N. Jamialahmadi, Ph. Berio, B. Lopez, A. Meilland and Ph. Stee

Laboratoire D.-L. Lagrange, UMR 7293 UNS-CNRS-OCA, Boulevard de l'Observatoire,
CS 34229, 06304 NICE Cedex 4, France
email: jami@oca.eu

Abstract. The main objective of this work is to improve our understanding of young fast-rotating stars evolving from the Herbig Ae/Be class to the Vega-like one. We observed with the VEGA instrument on CHARA one object so-called 51 Oph that is probably in such an evolutionary phase, allowing us to measure a mean stellar radius for the first time for this star and to show that the Hα emission was produced in a Keplerian rotating disc. However, additional observations are needed to improve our (u,v) plan coverage in order to measure the flattening of this close-to-critically rotating star and to probe the inner region of its circumstellar gaseous disc. These studies will help to disentangle the gas and dust emission around this late young star and will finally improve our understanding of the planet formation conditions in the inner regions of protoplanetary discs.

Keywords. stars: atmospheres, stars: fundamental parameters, stars: emission-line, stars: Be, stars: kinematics

1. Introduction

51 Ophiuchi (51 Oph) is a rapidly rotating B9.5 Ve star located at 131 pc (Perryman *et al.* 1997) with V magnitude of 4.83 (Mendigutía *et al.* 2012) and age of 0.3×10^6 years (van den Ancker *et al.* 1998) with mass of $\sim 4\,M_\odot$. It appears to be a peculiar source in an unusual transitional state. However, many uncertainties remain on 51 Oph classification in the HR diagram. The spatial structure of the 51 Oph dust disk also remains puzzling. So, this star would be an interesting object for us to figure out because, the gas has not dissipated and may still allow the on going formation and growth of gaseous giant planets.

2. Interferometric observations

51 Oph was observed at medium spectral resolution (R = 5000) with the VEGA instrument (Mourard *et al.* 2009), on CHARA array at Mount Wilson Observatory (California, USA, ten Brummelaar *et al.* 2005). We observed this star in both the continuum and chromospheric spectral line. We selected Hα Balmer line (656.2 nm). We select the calibrators using the SearchCal tool developed at JMMC (Bonneau *et al.* 2006), providing an estimate of the limb-darkened (LD) angular diameter ($\theta_{\rm LD}$).

3. Fundamental parameter and gaseous disk

3.1. *Angular diameter as a fundamental parameter*

The squared visibilities in the continuum around Hα are fitted with a model of UD disk (see Fig.1). We estimated the angular diameter of 0.39 ± 0.001 mas for the first time for this star.

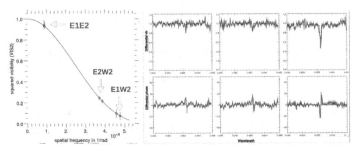

Figure 1. Left: The squared visibility to estimate the angular diameter. The top two points correspond to baseline E1E2, the middle points E2W2 and the down points E1W2. Right: The best fit model for differential visibilities and phase. The red line is model and black line is observational data for Hα line. Clear jumps in the differential phase and differential visibility at the center of the Hα 656.5 nm line are revealed for this star observed at medium spectral resolution.

3.2. *Gaseous disk*

Information related to the position of the photocenter of the gaseous disk can be deduced by measuring the phase of the differential visibility in the narrow spectral bands. This observable has been intensively used to study the kinematic of circumstellar environments. According to Fig.1 (Right), using a keplerian disk model, we succeed to fit our VEGA data. With this fitting, we derived some information for the gaseous disk such as: 1) The gaseous disk is keplerian according to Thi *et al.* (2005) 2) The inclination of the disk is 88° according to Tatulli *et al.* (2008) 3) FWHM in the continuum is 3 stellar diameter and in the line is 10 stellar diameter.

4. Conclusion

By analyzing interferometric measurements derived from CHARA/VEGA data, we have determined the physical extents of the gaseous disk of 51 Oph star. For the first time, we measured the position of the area in the disk where Hα line formed. In addition to this study, we have derived the angular diameter of this star for the first time. More observations with a more complete (u, v) coverage are needed to see whether we can confirm flattening for this fast-rotating star.

References

Bonneau, D., Clausse, J.-M., Delfosse, X., *et al.* 2006, *A&A* 456, 789
Mendigutía, I., Mora, A., Montesinos, B., *et al.* 2012, *A&A* 543, A59
Mourard, D., Clausse, J. M., Marcotto, A., *et al.* 2009, *A&A* 508, 1073
Perryman, M. A. C., Lindegren, L., Kovalevsky, J., *et al.* 1997, *A&A* 323, L49
Tatulli, E., Malbet, F., Ménard, F., *et al.* 2008, *A&A* 489, 1151
ten Brummelaar, T. A., McAlister, H. A., Ridgway, S. T., *et al.* 2005, *ApJ* 628, 453
Thi, W.-F., van Dalen, B., Bik, A., & Waters, L. B. F. M. 2005, *A&A* 430, L61
van den Ancker, M. E., de Winter, D., & Tjin A Djie, H. R. E. 1998, *A&A* 330, 145

New windows on massive stars: asteroseismology, interferometry, and
spectropolarimetry
Proceedings IAU Symposium No. 307, 2014
G. Meynet, C. Georgy, J. H. Groh & Ph. Stee, eds.

© International Astronomical Union 2015
doi:10.1017/S1743921314006449

Line profile variability in spectra of hot massive stars

Alexander Kholtygin, Natallia Sudnik and Viacheslav Dushin

St.Petersburg State University
email: afkholtygin@gmail.com

Abstract. We report the results of our study of the fast line profile variability (LPV) (hours – few days) in the spectra of bright OB and WR stars. All spectra were obtained with 6-m and 1-m telescope of Russian Special Astrophysical Observatory (SAO) and 1.8-m telescope of Bohyunsan Optical Astronomy Observatory, Korea (BOAO). We detected both the stochastic LPV, connected with the formation of small-scale structures in the stellar wind and the regular LPV induced by the large-scale structures in the wind.

Keywords. stars: early-type, line: profiles

1. Observations

Spectral and spectropolarimetric observations of 11 bright OB and 2 WR stars were made in a framework of the program of searching for regular and stochastic LPV in the spectra of OB stars (Kholtygin *et al.* 2003). The observations were obtained in 1997–2011 at SAO and at BOAO. The observations at SAO were made by using 6-m telescope with the Lynx spectrograph (spectral resolution $R = 60000$) and CCD 512×512 pixels, with the NES spectrograph ($R = 60000$) and CCD 1024×1024 and with the MSS spectrograph ($R = 15000$) and CCD $2k \times 2k$, while 1-m telescope observations were made with the CEGS spectrograph ($R = 45000$) and CCD 1242×1152. The reduction of SAO spectra was made with the MIDAS package. The observations at BOAO were performed by using the 1.8-m telescope equipped with the BOES spectrograph ($R = 45000$) and large CCD (2048×4096 pixels). The preliminary reduction of the CCD frames was done with IRAF.

2. Regular LPV

To detect LPV the smTVS analysis (Kholtygin *et al.* 2006) was used. This method allowed us to detect weak variability of the lines of ions Si III, C III, O III, N II, Mg II, S IV in the spectra of the program stars that cannot be detected using TVS as it is shown in the Fig. 1 (left panel). We applied smTVS for left and right polarized light separately. We found that pattern of LPV can differ for left (I_L) and right (I_R) polarized components of the lines in the spectra of all program stars (Fig. 1, right panel). For unblended lines with residual intensities less than 0.9, Fourier analysis was made. The obtained periods of regular LPV appeared in the range from hours to days (see, for example Dushin *et al.* 2013).

3. Stochastic LPV

The stochastic LPV is related with small-scale structures in the stellar wind. The expanding winds of WR stars are believed to be strongly clumped. We illustrate in the Fig. 2 (top panel) the clump contribution into the dynamical spectra of He II $\lambda 5411$ Å

114 A. Kholtygin, N. Sudnik & V. Dushin

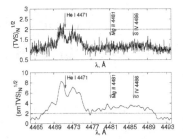 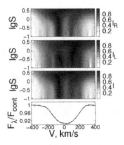

Figure 1. *Left panel*: TVS (top) and smTVS (bottom) normalized by unity of the line
He I $\lambda 4471$ Å in the spectra of λ Cep. Filter width S is 0.2 Å. The horizontal line corresponds to
the significance level 0.001. *Right panel*: Density plot diagram of smTVS for the I_R, I_L and I
components of line profile and the mean profile for the line He I $\lambda 4471$ Å in the spectra of the
star λ Cep (from top to bottom). Darker areas correspond to higher smTVS amplitude.

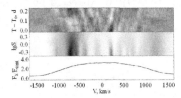 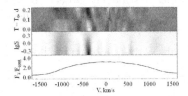

Figure 2. *Left panel*: Dynamical specta of LPV for the line He II $\lambda 5411$ Å in the spectra of
WR 136 taken at July 25/26 (top). Grey scale dynamical plot of the smTVS spectrum as a
function of the filter width S (middle). Nightly mean He II $\lambda 5411$ Å line profile (bottom). *Right
panel*: The same as in the left panel but for the night of July 26/27 (Kholtygin *et al.* 2011).

line. As we can see in the Fig. 2 (middle panel) the regions of the largest amplitude of
the smTVS correspond to the moving bumps in the dynamical spectra of LPV. It means
that smTVS spectra can be used for detecting stochastic line profile variations of small
amplitudes. There is no difference between the pattern of LPV for left and right polarized
components of stellar radiation for 2 studied WR stars. It means that the magnetic field
in the region of the He II line formation is small.

We also found the evidences of the stochastic LPV in the line profiles in the spectra of
O stars using wavelet analysis with MHAT-wavelet as the mother wavelet (e.g. Kholtygin
et al. 2006).

Acknowledgements

AF and NS acknowledge Saint-Petersburg State University (SPbGU) for the research
grant 6.50.1555.2013. AK and VD thank SPbGU for the research project 6.38.18.2013.
NS thank SPbGU for support by the grant 6.41.721.2014.

References

Dushin, V. V., Kholtygin, A. F., Chuntonov, G. A., & Kudryavtsev, D. O. 2013, *Astrophysical
Bulletin* 68, 184

Kholtygin, A. F., Burlakova, T. E., Fabrika, S. N., Valyavin, G. G., & Yushkin, M. V. 2006,
Astronomy Reports 50, 887

Kholtygin, A. F., Fabrika, S. N., Rusomarov, N., et al. 2011, *Astronomische Nachrichten* 332,
1008

Kholtygin, A. F., Monin, D. N., Surkov, A. E., & Fabrika, S. N. 2003, *Astronomy Letters* 29,
175

New windows on massive stars: asteroseismology, interferometry, and spectropolarimetry
Proceedings IAU Symposium No. 307, 2014
G. Meynet, C. Georgy, J. H. Groh & Ph. Stee, eds.

© International Astronomical Union 2015
doi:10.1017/S1743921314006450

Discrete absorption components in the massive LBV Binary MWC 314

A. Lobel[1], C. Martayan[2], M. Corcoran[3], J. H. Groh[4] and Y. Frémat[1]

[1]Royal Observatory of Belgium, Brussels
email: Alex.Lobel@oma.be, alobel@sdf.org

[2]European Southern Observatory, Chile

[3]Goddard Space Flight Center, Greenbelt MD, USA

[4]Geneva Observatory, Versoix, Switzerland

Abstract. We investigate the physical properties of large-scale wind structures around massive hot stars with radiatively-driven winds. We observe Discrete Absorption Components (DACs) in optical He I P Cygni lines of the LBV binary MWC 314 ($P_{\rm orb} = 60.8$ d). The DACs are observed during orbital phases when the primary is in front of the secondary star. They appear at wind velocities between -100 km s^{-1} and -600 km s^{-1} in the P Cyg profiles of HeI $\lambda5875$, $\lambda6678$, and $\lambda4471$, signaling high-temperature expanding wind regions of enhanced density and variable outflow velocity. The DACs can result from wave propagation linked to the orbital motion near the low-velocity wind base. The He I lines indicate DAC formation close to the primary's surface in high-temperature wind regions in front of its orbit, or in dynamical wind regions confined between the binary stars. We observed the DACs with Mercator-HERMES on 5 Sep 2009, 5 May 2012, and 6 May 2014 when the primary is in front of the secondary star. XMM-Newton observations of 6 May 2014 significantly detected MWC 314 in X-rays at an average rate of ~0.015 cts s^{-1}.

Keywords. stars: winds, outflows, emission-line, Be; binaries: eclipsing; X-rays

1. Introduction

Current understanding of the role of the Luminous Blue Variable (LBV) stage in the evolution of the most massive stars is very limited. Important questions are how the eruptions of several $M_\odot$ are triggered, and what role binarity plays in the properties of the LBV stage and in shaping the nebulae observed around LBVs. For example, is most of the mass lost by an LBV star due to a steady radiatively-driven stellar wind, or is the H-rich envelope removed by punctuated eruption-driven mass loss? There are two confirmed LBV binaries: η Car in the Galaxy, and HD 5980 in the SMC. The extremely luminous star MWC 314 has recently been recognized ($P_{\rm orb} = 60.8$ d and $e = 0.23$) as an eclipsing massive binary system (Lobel *et al.* 2013). Indications that MWC 314 is a (dormant) LBV are strong. The star is very luminous ($\log L/L_\odot = 5.9$; see also Miroshnichenko *et al.* 1998) with an optical spectrum and SED nearly identical to the canonical LBV P Cygni. MWC 314 is surrounded by a bipolar Hα nebula that may be the result of an eruption of MWC 314 more than 100,000 years ago (Marston & McCollum 2008). MWC 314 is a single-lined, eclipsing binary system with masses of $40 + 26\,M_\odot$, and radii of $87 + 20\,R_\odot$. The largest diameter of the binary system is ~1.1 AU.

2. Radial velocity curve and DACs in He I P Cyg lines

We measure the radial velocity curve in Fig. 1 from selected absorption lines of the primary star. Two spectra marked with A and C are observed in the quadrature phase

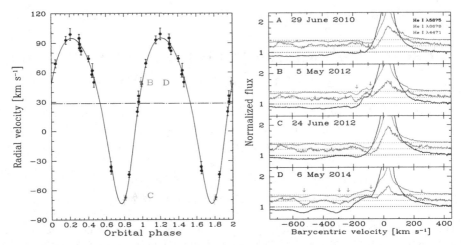

Figure 1. Left-hand panel: orbital radial velocity curve of MWC 314. Two orbital phases B and D when the primary is in front of the secondary ($\phi \sim 0$), and two quadrature phases A and C ($\phi \sim 0.8$) are marked. **Right-hand panels**: the profiles of three He I lines are shown for orbital phases marked A, B, C, and D. DACs are observed in the violet wings of the lines (*marked with arrows*) during the orbital phases B and D. XMM-Newton observed X-rays on 6 May 2014 (D).

close to minimum radial velocity of the primary ($0.75 < \phi < 0.85$). Two spectra observed when the primary is in front of the secondary are marked with B and D ($0.95 < \phi < 1.05$). Three He I lines reveal P Cyg profiles with wind expansion velocities to ~ 1200 km s^{-1}. On 5 May 2012 (B) we observe two discrete absorption components in the extended violet wings of the three lines. The DACs are observed at wind velocities of ~ 100 km s^{-1} and ~ 180 km s^{-1} in the line profiles (*marked with arrows*). The DACs are not clearly observed in the quadrature phases of 29 June 2010 (A) and 24 June 2012 (C). We observe four DACs on 6 May 2014 (D) at wind velocities between ~ 100 km s^{-1} and ~ 600 km s^{-1}.

3. Conclusions

Our high-resolution spectroscopic monitoring program of MWC 314 with Mercator-HERMES during the last 5 years reveals DACs in the orbital phases when the primary is in front of the secondary star on 5 Sep 2009, 5 May 2012, and 6 May 2014. The high-excitation temperatures of the He I lines signal expanding wind regions of enhanced density and variable outflow velocity. The DACs can form close to the primary's surface (of B0-type) in high-temperature and density-enhanced wind regions in front of its orbit. The recurrence of the DACs in orbital phases when the primary is in front of the secondary can result from wave propagation which is physically linked to the orbital motion near the (low-velocity < 150 km s^{-1}) wind base. Alternatively, the DACs can form in dynamical (high-temperature) wind regions confined between the binary stars at the shock interface of a colliding wind region. We expect orbital X-ray variability in MWC 314 (similar to the LBVs η Car and HD 5980) during planned XMM-Newton observations in the quadrature phase of Oct. 2014.

References

Lobel, A., Groh, J. H., Martayan, C., *et al.* 2013, *A&A* 559, A16

Marston, A. P. & McCollum, B. 2008, *A&A* 477, 193

Miroshnichenko, A. S., Frémat, Y., Houziaux, L., *et al.* 1998, *A&AS* 131, 469

New windows on massive stars: asteroseismology, interferometry, and
spectropolarimetry
Proceedings IAU Symposium No. 307, 2014
G. Meynet, C. Georgy, J. H. Groh & Ph. Stee, eds.

© International Astronomical Union 2015
doi:10.1017/S1743921314006462

The mass discrepancy problem in O stars of solar metallicity. Does it still exist?

N. Markova[1] and J. Puls[2]

[1] IANAO, Sofia, Bulgaria
email: nmarkova@astro.bas.bg

[2] Universitäts-Sternwarte, München, Germany

Abstract. Using own and literature data for a large sample of O stars in the Milky Way, we investigate the correspondence between their spectroscopic and evolutionary masses, and try to put constraints on various parameters that might influence the estimates of these two quantities.

Keywords. stars: early type, stars: evolution, stars: fundamental parameters

1. Introduction

In its classical form, the so-called *mass discrepancy* refers to the systematic overestimate of evolutionary masses, M_{evol}^t, compared to spectroscopically derived masses, M_{spec} (e.g., Herrero *et al.* 1992). While continuous improvements in model atmospheres and model evolutionary calculations have reduced the size of the discrepancy (e.g., Repolust *et al.* 2004), however without eliminating it completely (Mokiem *et al.* 2007; Hohle *et al.* 2010; Massey *et al.* 2012), there are also studies (e.g., Weidner & Vink 2010) which argue that, at least for O stars in the Milky Way, the mass discrepancy problem has been solved.

2. Stellar sample and methodology

Our sample consists of 51 Galactic dwarfs, giants and supergiants, with spectral types ranging from O 3 to O 9.7. Forty one of these are cluster/association members; the rest are field stars. For 31 of the sample stars, we used own determinations of stellar parameters, obtained by means of the latest version of the FASTWIND code (Markova *et al.*, in preparation); for the remaining 20, similar data have been derived by Bouret *et al.* (2012) and Martins *et al.* (2012a,b), employing the CMFGEN code instead.

For all sample stars, M_{spec} were calculated from the effective gravities corrected for centrifugal acceleration, whilst M_{evol}^t were determined by interpolation between available tracks along isochrones, as calculated by Ekström *et al.* (2012) and Brott *et al.* (2011). To put constraints on biases originating from uncertain distances and reddening, in parallel to the classical $\log L/L_\odot - \log T_{\mathrm{eff}}$ diagram we also consider a (modified) spectroscopic HRD (sHRD) that is independent of 'observed' stellar radii (for more information, see Markova *et al.* 2014 and Langer & Kudritzki 2014).

3. Results

Our analysis indicates that

i) for objects with $M_{\mathrm{evol}}^{\mathrm{init}} > 35\,M_\odot$, M_{evol}^t are either systematically lower (Ekström models) or roughly consistent (Brott models) with M_{spec}. As $\dot{M}$ scales with $\log L/L_\odot$

(e.g., Vink *et al.* 2000; see also Puls *et al.*, this volume), and as – soon after the ZAMS – the Ekström models with rotation and $M_{\text{evol}}^{\text{init}} \geqslant 40\,M_{\odot}$ become more luminous than the Brott models of the same $M_{\text{evol}}^{\text{init}}$ and T_{eff}, we suggest that the *negative* mass discrepancy established for the Ekström tracks is most likely related to (unrealistically?) high mass-loss rates implemented in these models. (Warning! The good agreement between M_{spec} and M_{evol}^{t} read off the Brott tracks does not necessarily mean that the corresponding mass-loss rates are of the right order of magnitude, see next item).

ii) for objects with $M_{\text{evol}}^{\text{init}} < 35\,M_{\odot}$, M_{evol}^{t} tend to be larger than M_{spec}. As massive hot stars can develop subsurface convection zones (Cantiello *et al.* 2009), and as they can be also subject to various instabilities, we are tempted to speculate that the neglect of turbulent pressure in FASTWIND and CMFGEN atmospheric models might explain the lower M_{spec} compared to M_{evol}^{t}†. Indeed, one might argue that if our explanation was correct a similar discrepancy should be present (but is not observed) for the more massive stars as well. However, such caveat might be easily solved if also the Brott models over-estimate the mass-loss rates, as already suggested by Markova *et al.* (2014), and as also implied from up-to-date comparisons of theoretical and observed $\dot{M}$ (e.g., Najarro *et al.* 2011; Cohen *et al.* 2014)

iii) while for most sample stars the correspondence between M_{spec} and M_{evol}^{t} does not significantly depend on the origin of the latter (HRD or sHRD), there are a number of outliers which, for the case of Brott tracks, demonstrate $M_{\text{evol}}^{t}(\text{sHRD}) > M_{\text{evol}}^{t}(\text{HRD})$, by a factor of 1.5 to 1.8. While specific reasons, such as, e.g., close binary evolution or homogeneous evolution caused by rapid rotation, can in principle explain discrepant masses read off the HRD and sHRD (Langer & Kudritzki 2014), it is presently unclear why this discrepancy does not appear in the Ekström tracks.

iv) the established mass discrepancy does not seem to be significantly biased by uncertain stellar radii; the presence of surface magnetic fields, or systematically underestimated $\log g$-values derived by means of the FASTWIND code (for more information, see Massey *et al.* 2013).

References

Bouret, J.-C., Hillier, D. J., Lanz, T., & Fullerton, A. W. 2012, *A&A* 544, A67
Brott, I., de Mink, S. E., Cantiello, M., *et al.* 2011, *A&A* 530, A115
Cantiello, M., Langer, N., Brott, I., *et al.* 2009, *A&A* 499, 279
Cohen, D. H., Wollman, E. E., Leutenegger, M. A., *et al.* 2014, *MNRAS* 439, 908
Ekström, S., Georgy, C., Eggenberger, P., *et al.* 2012, *A&A* 537, A146
Herrero, A., Kudritzki, R. P., Vilchez, J. M., *et al.* 1992, *A&A* 261, 209
Hohle, M. M., Neuhäuser, R., & Schutz, B. F. 2010, *Astronomische Nachrichten* 331, 349
Langer, N. & Kudritzki, R. P. 2014, *A&A* 564, A52
Markova, N., Puls, J., Simón-Díaz, S., *et al.* 2014, *A&A* 562, A37
Martins, F., Escolano, C., Wade, G. A., *et al.* 2012a, *A&A* 538, A29
Martins, F., Mahy, L., Hillier, D. J., & Rauw, G. 2012b, *A&A* 538, A39
Massey, P., Morrell, N. I., Neugent, K. F., *et al.* 2012, *ApJ* 748, 96
Massey, P., Neugent, K. F., Hillier, D. J., & Puls, J. 2013, *ApJ* 768, 6
Mokiem, M. R., de Koter, A., Evans, C. J., *et al.* 2007, *A&A* 465, 1003
Najarro, F., Hanson, M. M., & Puls, J. 2011, *A&A* 535, A32
Repolust, T., Puls, J., & Herrero, A. 2004, *A&A* 415, 349
Vink, J. S., de Koter, A., & Lamers, H. J. G. L. M. 2000, *A&A* 362, 295
Weidner, C. & Vink, J. S. 2010, *A&A* 524, A98

† By including such a turbulent pressure, one would obtain a spectroscopic $\log g$ that is larger by 0.2 dex, for typical parameters and a turbulent speed of 15 km/s.

New windows on massive stars: asteroseismology, interferometry, and
spectropolarimetry
Proceedings IAU Symposium No. 307, 2014
G. Meynet, C. Georgy, J. H. Groh & Ph. Stee, eds.

© International Astronomical Union 2015
doi:10.1017/S1743921314006474

Investigation of the brightest stars in the Cyg OB2 association

Olga Maryeva[1] and Sergey Parfenov[2]

[1] Special Astrophysical Observatory of the Russian Academy of Sciences
email: olga.maryeva@gmail.com

[2] Ural Federal University

Abstract. We present the results of investigation of most luminous stars belonging to the Cyg OB2 association using quantitative analysis of high-resolution spectra. Physical parameters derived using the CMFGEN and TLUSTY atmospheric codes allow us to estimate the mass and age of these stars.

Keywords. stars: early-type stars: atmospheres stars: winds, outflows stars: evolution

1. Introduction

Cygnus OB2 (Cyg OB2) stellar association, discovered by Munch and Morgan in 1953, is the current leader in number of massive stars among Galactic OB-associations. The interest of researchers in individual stars and the association as a whole is not fading. Numerous articles have investigated the stellar population of Cyg OB2, its interstellar extinction, and star-formation history.

2. Modeling results

E. L. Chentsov and V. G. Klochkova have performed studies of objects belonging to Cyg OB2 for several years using the Russian 6 m telescope of the Special Astrophysical Observatory (SAO RAS). Chentsov *et al.* (2013) published deep spectral description of high resolution spectra of 13 stars in Cyg OB2 and spectral variability of the star #12. We decided to continue the studies and calculate models of these stars.

We used the TLUSTY code by Lanz & Hubeny (2003) to model the atmospheres of the dwarfs (Cyg OB2 #6, #16, and #21) and CMFGEN (Hillier & Miller 1998) for calculation of the atmospheres of supergiants. The results of modeling of the dwarfs are listed in Table 1 and displayed in Figure 1.

As of now, eighteen Galactic O-stars are classified as Ofc (Walborn *et al.* 2010). Four of them are in Cyg OB2. We modeled the spectra of #9 (O 4.5 Ifc) and #11 (O 5.5 Ifc). Table 2 lists the model parameters while the surface abundances of H, He, and CNO are given in Table 3.

Table 1. The resulted parameters and their standard deviations obtained for dwarf stars considered in this study.

	Spec. class	$T_{\rm eff}$ kK	$\log g$ cm s^{-2}	He/H	$V \sin i$ km s^{-1}	$V_{\rm micr}$ km s^{-1}	$M_\star$ $M_\odot$	$\log(L_\star/L_\odot)$	Age Myr
6	O8 V	32.8	3.72	0.11	210	12	24 ± 5	5.1 ± 0.3	5.6 ± 1.3
16	O7.5 V	33.3	3.78	0.16	200	12	24 ± 5	5.1 ± 0.3	5.2 ± 1.4
21	B0 V	31.2	3.97	0.12	25	9	18 ± 2	4.7 ± 0.2	5.1 ± 1.6

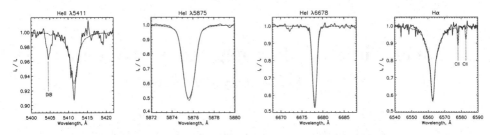

Figure 1. Comparison of the observed spectrum of Cyg OB2 #21 with the best-fit models.

Table 2. Atmospheric parameters of Cyg OB2 #9 and Cyg OB2 #11.

	$T_\star$ kK	$R_\star$ $R_\odot$	$T_{\rm eff}$ kK	$R_{2/3}$ $R_\odot$	$L_\star$ $10^5\,L_\odot$	$\dot{M}_{\rm uncl}$ $10^{-6}\,M_\odot\,{\rm yr}^{-1}$	$\dot{M}_{\rm cl}$ $10^{-6}\,M_\odot\,{\rm yr}^{-1}$	f_∞	V_∞ km s^{-1}	β
#9	37	20	36	21	6.5	3.9	1.1	0.08	1500	1.3
#11	35	22	34.4	22.7	6.5	6	1.7	0.08	2200	1.3

Table 3. The abundances of chemical elements. Solar values are from Asplund *et al.* (2009).

Element	Cyg OB2 #9	Cyg OB2 #11	Sun
H	12	12	12
He	10.7 ± 0.15	10.85 ± 0.15	10.93 ± 0.01
C	8.08 ± 0.05	8.5 ± 0.09	8.39 ± 0.05
N	8.37 ± 0.03	8.28 ± 0.03	7.78 ± 0.06
O	8.18 ± 0.07	8.17 ± 0.07	8.66 ± 0.05

We found He/H $\approx$ 0.1 in the atmospheres of the Ofc stars. Within the errors, it means that their He abundance is equal to initial He abundance of the environment and that He/H did not change during the lifetimes of Cyg OB2 #9 and #11. In Cyg OB2 #11, the N abundance is lower than in other "normal" O stars, while the C abundance is solar. In Cyg OB2 #9 the fraction of N is higher than in #11 and the fraction of C is lower. Cyg OB2 #9 is closer to "normal" O stars than #11. Probably, there is no mixing in atmospheres of Ofc stars that transports the products of CNO cycle from the core towards the stellar surface.

The results discussed here have been published in: Maryeva *et al.* (2014), Maryeva *et al.* (2013) and Maryeva & Zhuchkov (2012).

Olga Maryeva thanks the International Astronomical Union and Dynasty Foundation for grants. The study was supported by the Russian Foundation for Basic Research (projects no. 14-02-31247,14-02-00291).

References

Asplund, M., Grevesse, N., Sauval, A. J., & Scott, P. 2009, *ARA&A* 47, 481
Chentsov, E. L., Klochkova, V. G., Panchuk, V. E., Yushkin, M. V., & Nasonov, D. S. 2013, *Astronomy Reports* 57, 527
Hillier, D. J. & Miller, D. L. 1998, *ApJ* 496, 407
Lanz, T. & Hubeny, I. 2003, *ApJS* 146, 417
Maryeva, O., Zhuchkov, R., & Malogolovets, E. 2014, *Proc. ASA* 31, 20
Maryeva, O. V., Klochkova, V. G., & Chentsov, E. L. 2013, *Astrophysical Bulletin* 68, 87
Maryeva, O. V. & Zhuchkov, R. Y. 2012, *Astrophysics* 55, 371
Walborn, N. R., Sota, A., Maíz Apellániz, J., *et al.* 2010, *ApJ (Letters)* 711, L143

New windows on massive stars: asteroseismology, interferometry, and
spectropolarimetry
Proceedings IAU Symposium No. 307, 2014
G. Meynet, C. Georgy, J. H. Groh & Ph. Stee, eds.

© International Astronomical Union 2015
doi:10.1017/S1743921314006486

OHANA: Eta Carinae's Variability in the Near-IR

Andrea Mehner, Willem-Jan de Wit, Thomas Rivinius and the Paranal VLTI group

European Southern Observatory, Alonso de Cordova 3107, Vitacura, Chile
email: amehner@eso.org

Abstract. Near-IR photometry of η Car since 1972 revealed a long-term trend towards hotter temperatures and a cycloidal behavior of its near-IR colors around periastron passages. Both effects are likely triggered by the companion. We used VLTI AMBER observations from 2004–2014 to investigate η Car's variabilities in the near-IR.

Keywords. stars: atmospheres, (stars:) circumstellar matter, stars: emission-line, Be, stars: individual (eta Carinae), stars: mass loss, stars: winds, outflows

1. Introduction

η Car is one of the key objects for studying the evolution of the most massive stars. Two centuries ago it experienced an eruption in which the star expelled more than $10 \, M_\odot$, creating its bipolar nebula. The recent UV, optical, and X-ray light curves and spectra suggest a physical change in η Car's wind over the last decade (Mehner *et al.* 2010, 2012). The near-IR photometry shows an accelerated brightening since 1998; the star became bluer, matching an apparent increase in blackbody temperature of an optically thick emitting plasma component from about $3\,500$ K to $6\,000$ K over the last 20 years (Mehner *et al.* 2014). This long-term color evolution occurs discontinuously at each periastron passage. In addition, a cycloidal behavior is observed in the colors from -40 to 100 days around periastron (Fig. 1). We investigated the spectral information from 2004–2014 VLTI AMBER data for changes in the near-IR emission lines to complement the recent results from η Car's near-IR photometry.

2. Eta Carinae's Near-IR Flux Spectrum

From 2004–2014 spectro-interferometry of η Car was carried out with VLTI AMBER, delivering both flux and visibility spectra simultaneously. η Car's optically thick wind with an diameter on the order of 4.3 mas (≈ 9 AU) in the K-band was resolved to be asymmetric (Weigelt *et al.* 2007). In early 2014, η Car was observed with AMBER and several baselines as part of the OHANA project, a public Paranal/VLTI Observatory survey at High ANgular resolution of Active OB stars. The survey combines high spatial resolution (4–45 mas) with high spectral resolution (R$\approx$12 000) and covers the He I 2.059 and Brγ lines. Observations over several months with extensive (u,v)-plane coverage allow to investigate temporal and spatial variations. Temporal variations in η Car occur on a timescale of 2–3 weeks and strong small-scale structure can be observed at even the shortest baselines.

η Car has two strong emission lines in the K band: He I 2.059 and Brγ. These contribute $\sim$3.5% to the K band flux at apastron. In the H band, the spectrum is dominated by the hydrogen Bracket series, which contribute $\sim$8% of the total flux. During periastron

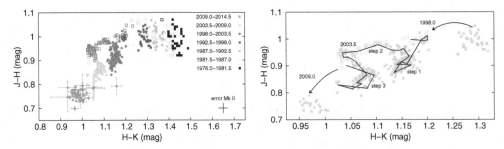

Figure 1. LEFT: Near-IR color-color diagram from 1976–2013. Open squares show data points close to periastron (−40 days to +100 days). RIGHT: Cycloidal pattern for three orbital periods. Filled (open) squares are observations outside (during) periastron passages. The figures are adapted from Mehner *et al.* (2014). See also Whitelock *et al.* (2004) and references therein for details on the η Car near-IR monitoring program at SAAO.

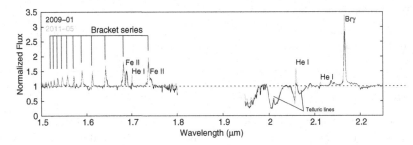

Figure 2. Medium-resolution H and K band VLTI AMBER spectra of η Car during the 2009 periastron and the 2011 apastron when the influence of the companion star is minimal. The telluric bands are not corrected for.

passages the hydrogen lines weaken and only contribute ∼2% of the K band brightness and ∼4% of the H band brightness. These varying hydrogen line contributions are consistent with the H–K ≈ 0.1 mag deviation in the color-color diagram observed close to periastron passages (Fig. 1, step 2). The J band is dominated by Paβ, which contributes ∼10% of the total flux (Smith & Davidson 2001), which is consistent with J–H≈constant during periastron passages.

3. Conclusions

We found no long-term evolution in the near-IR hydrogen and helium lines with respect to the continuum. The cycloidal behavior of the near-IR colors during periastron passages is accounted for by variations in the hydrogen emission line strengths with respect to the continuum. The long-term trend toward a hotter near-IR color is due to a change in the continuum and is triggered at periastron passages.

References

Mehner, A., Davidson, K., Humphreys, R. M., *et al.* 2012, *ApJ* 751, 73

Mehner, A., Davidson, K., Humphreys, R. M., *et al.* 2010, *ApJ (Letters)* 717, L22

Mehner, A., Ishibashi, K., Whitelock, P., *et al.* 2014, *A&A* 564, A14

Smith, N. & Davidson, K. 2001, *ApJ (Letters)* 551, L101

Weigelt, G., Kraus, S., Driebe, T., *et al.* 2007, *A&A* 464, 87

Whitelock, P. A., Feast, M. W., Marang, F., & Breedt, E. 2004, *MNRAS* 352, 447

New windows on massive stars: asteroseismology, interferometry, and spectropolarimetry
Proceedings IAU Symposium No. 307, 2014
G. Meynet, C. Georgy, J. H. Groh & Ph. Stee, eds.

© International Astronomical Union 2015
doi:10.1017/S1743921314006498

Markov Chain Monte-Carlo Models of Starburst Clusters

Jorge Melnick

European Southern Observatory
email: jmelnick@eso.org

Abstract. There are a number of stochastic effects that must be considered when comparing models to observations of starburst clusters: the IMF is never fully populated; the stars can never be strictly coeval; stars rotate and their photometric properties depend on orientation; a significant fraction of massive stars are in interacting binaries; and the extinction varies from star to star. The probability distributions of each of these effects are not *a priori* known, but must be extracted from the observations. Markov Chain Monte-Carlo methods appear to provide the best statistical approach. Here I present an example of stochastic age effects upon the upper mass limit of the IMF of the Arches cluster as derived from near-IR photometry.

Keywords. Methods: statistics; Stars: evolution, atmospheres, rotation.

1. Introduction

The IMF is considered to be fundamental to understand massive star formation (although we have argued elsewhere that the IMF is immanent in the fractal structure of the ISM; eg. Melnick 2009). Starburst clusters are the best places where the IMF can be measured: they contain large numbers of massive stars - important to constrain stochastic effects, and their youth allows to control evolutionary effects. Yet, however young, the stellar populations cannot be strictly coeval, and, I will argue in this paper, even small age differences among the most massive stars have significant effects on the IMF.

A particularly interesting issue that has emerged recently is whether there is an upper limit to how massive stars can be. Crowther *et al.* (2010) found that the most massive stars in 30 Doradus and NGC3603 may have initial masses well above the hitherto canonical mass limit of 120 $M_\odot$. In our photometric study of the Arches cluster (Espinoza *et al.* 2009), we had difficulties matching the most luminous stars to the most massive Geneva models, which left us with the distinct impression that some of these stars could be substantially more massive than 120 $M_\odot$. So I endeavoured to reanalyse our Arches photometry using the most recent Geneva tracks, in particular those including rotation.

Raphael Hirschi kindly provided me with his tracks for masses up to 500 $M_\odot$ (used in Crowther *et al.* 2010), but, although the potential impact of extended atmospheres on the photometric properties of massive stars is well known, there are as yet no comprehensive libraries of such models to be used for population studies. So to compute synthetic colours I used (Castelli & Kurucz 2003) atmospheres. Here I present very preliminary results of comparing this new grid of synthetic photometry with the observations of the Arches cluster.

2. The Arches Starburst cluster

The left panel of Figure 1 shows the ZAMS for the Geneva tracks without rotation and the evolutionary track of a 120 $M_\odot$ star, compared to the best photometric data for

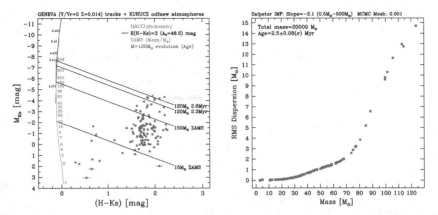

Figure 1. Left. NACO adaptive-optics photometric observations of Arches compared to evolutionary tracks. The ages in the 120 $M_\odot$ track are labelled. The reddening vectors for stars of different masses and ages are shown as solid lines. The dashed line shows the reddening vector for a 1.9 Myr old star of 200 $M_\odot$. **Right.** Rms dispersion versus mass for 10^5 Monte-Carlo draws from a Gaussian age distribution of mean age 2.5 Myr and dispersion 0.08 Myr.

the stars in the Arches cluster from Espinoza *et al.* (2009). This figure illustrates how the inferred initial masses of the most luminous stars depend critically on age: at 2.6 Myr the masses would not be significantly larger than 120 $M_\odot$ whereas at 2.3 Myr, the most luminous stars would be more massive than 200 $M_\odot$.

By definition, *starburst* clusters have ages comparable to the main-sequence life times of their most massive stars: all the massive stars form in a short burst. Barring an act of god, however, this burst cannot be infinitely short, and specifically, cannot be shorter than the free-fall time t_{FF} of parent cloud. For a Virialized cluster t_{FF} is given by $t_{FF} = 1/\sqrt{G\rho} \simeq 0.0243 \times M_{\rm cl}/\sigma_{\rm cl}^3$ Myr where $\sigma_{\rm cl} = 0.0927 \times \sqrt{M_{\rm cl}/R_{\rm cl}}$ km s^{-1}.

2.1. *Results and Future Prospects*

There is some debate about the age of the Arches cluster, but since I am using the same photometric data, here I will use the values of Espinoza *et al.* (2009): age = 2.5 Myr; mass $M_{\rm cl} = 2 \times 10^4\ M_\odot$; radius $R_{\rm cl} = 0.5$ pc, from which $t_{FF} \sim 0.08$ Myr. The right-hand panel of Figure 1 shows Monte-Carlo simulations for a Gaussian distribution of stellar ages with a mean of 2.5 Myr and a dispersion of 0.08 Myr. The average upper mass limit of the IMF is $\sim 125\ M_\odot$ with rms dispersion of 15 $M_\odot$. There is no strong evidence in the photometry, therefore, that the Arches cluster may host stars more massive than 140 $M_\odot$.

The legacy ground-based and HST data sets of 30Dor that are becoming available will allow to use MCMC methods to derive the parent distributions of all the stochastic parameters that define the stellar populations of this iconic starburst cluster: age; extinction; rotation; and binarity. This contribution marks the beginning of such program.

References

Castelli, F. & Kurucz, R. L. 2003, in N. Piskunov, W. W. Weiss, & D. F. Gray (eds.), *Modelling of Stellar Atmospheres*, Vol. 210 of *IAU Symposium*, p. 20P

Crowther, P. A., Schnurr, O., Hirschi, R., *et al.* 2010, *MNRAS* 408, 731

Espinoza, P., Selman, F. J., & Melnick, J. 2009, *A&A* 501, 563

Melnick, J. 2009, in *Revista Mexicana de Astronomia y Astrofisica Conference Series*, Vol. 37 of *Revista Mexicana de Astronomia y Astrofisica Conference Series*, pp 21–31

New windows on massive stars: asteroseismology, interferometry, and
spectropolarimetry
Proceedings IAU Symposium No. 307, 2014
G. Meynet, C. Georgy, J. H. Groh & Ph. Stee, eds.

© International Astronomical Union 2015
doi:10.1017/S1743921314006504

A spectroscopic and photometric study of the interacting binary and double period variable HD 170582

R. E. Mennickent[1], G. Djurašević[2], M. Cabezas[1], A. Cséki[2],
J. Rosales[1], E. Niemczura[3], I. Araya[4] and M. Curé[4]

[1]Departamento de Astronomía, Universidad de Concepción, Chile,
email: rmennick@astroudec.cl

[2]Astronomical Observatory, Belgrade, Serbia, [3]Astronomical Institute, Wrocław University,
Poland, [4]Instituto de Física y Astronomía, Facultad de Ciencias, U. de Valparaíso, Chile

Abstract. We present a spectroscopic and photometric study of the interacting binary and
Double Period Variable HD 170582 based on the analysis of the ASAS V-band light curve and
our high-resolution spectra mostly obtained with CHIRON spectrograph at the 1.5m CTIO
telescope.

Keywords. binaries: eclipsing, binaries: close, stars: mass-loss, stars: fundamental parameters.

1. On the poorly known binary HD 170582

HD 170582 (ASAS ID 183048-1447.5, $\alpha_{2000} = 18 : 30 : 47.5$, $\delta_{2000} = -14 : 47 : 27.8$,
$V = 9.66$ mag, $B - V = 0.41$ mag, spectral type A 9V), is a poorly studied binary star
catalogued semi-detached eclipsing binary with orbital period 16.8599 days in the ASAS
catalogue (Pojmanski 1997). It is located in the region of the cool molecular cloud L379
and is characterized by a long photometric cycle of 536 days. It is the longest-period
member of the Galactic Double Period Variables (DPVs, Fig. 1). DPVs are intermediate
mass interacting binaries showing a long photometric cycle lasting about 33 times the
orbital period, which has been interpreted as cyclic episodes of mass loss (Mennickent
et al. 2003, 2008, 2012a,b; Poleski et al. 2010).

2. Light curve model and spectroscopic analysis

Based on the study of the ASAS V-band light curve we determine an improved orbital
period of 16.87177 ± 0.02084 days and a long period of 587 days. We disentangled the
light curve into an orbital part, determining ephemerides and revealing orbital ellipsoidal
variability with unequal maxima, and a long cycle, showing quasi-sinusoidal changes with
V-band amplitude 0.1 mag. From the analysis of 136 CHIRON/CTIO high-resolution optical spectra, the model of the V-band ASAS light curve and the fit of the spectral energy
distribution, we determine the physical parameters for the stars and the circumprimary
disc, the distance to the system and general system dimensions, reddening and metallicity. For the light curve model we use the code described by Djurašević (1992) solving the
inverse problem for the Roche model with an accretion disk around the more-massive
(hotter) gainer. The best model for the donor star is characterized by $T_2 = 8000 \pm 125$ K,
$\log g_2 = 1.5 \pm 0.25$ and projected rotational velocity $v_{2r} \sin i = 44 \pm 4$ km s^{-1}. Assuming synchronous rotation for the donor we obtain a mass ratio of $q = 0.21$. The disc
contributes about 35% to the system luminosity at the V-band. Two extended regions

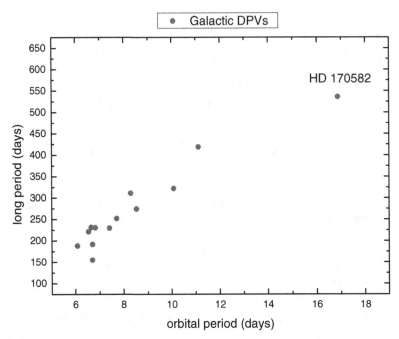

Figure 1. Galactic Double Period Variables according to Mennickent *et al.* (2012a).

located at opposite sides of the disc rim, and hotter than the disc by 67% and 46%, fit the light curve asymmetries. These structures can be attributed to shocks produced by disc gas dynamics and gas stream interaction. The system is seen under inclination 67 degree and it is found at a distance of 238 pc. We discuss the double line nature of He I 5875; two absorption components move in anti-phase during the orbital cycle. A possible origin for one of these component is a place near the stream/disc interaction region. We find that HD 170582 is one of the systems showing a discrepancy between the color excess obtained from diffuse interstellar bands and that obtained from the analysis of the spectral energy distribution.This might be attributed to the influence of circumstellar matter. This study of HD 170582 will help to understand the class of interacting binaries Double Period Variables. A full study has been submitted for publication.

References

Djurašević, G. 1992, *Ap&SS* 197, 17
Mennickent, R. E., Djurašević, G., Kołaczkowski, Z., & Michalska, G. 2012a, *MNRAS* 421, 862
Mennickent, R. E., Kołaczkowski, Z., Djurašević, G., *et al.* 2012b, *MNRAS* 427, 607
Mennickent, R. E., Kołaczkowski, Z., Michalska, G., *et al.* 2008, *MNRAS* 389, 1605
Mennickent, R. E., Pietrzyński, G., Diaz, M., & Gieren, W. 2003, *A&A* 399, L47
Pojmanski, G. 1997, *AcA* 47, 467
Poleski, R., Soszyñski, I., Udalski, A., *et al.* 2010, *AcA* 60, 179

New windows on massive stars: asteroseismology, interferometry, and spectropolarimetry
Proceedings IAU Symposium No. 307, 2014
G. Meynet, C. Georgy, J. H. Groh & Ph. Stee, eds.

© International Astronomical Union 2015
doi:10.1017/S1743921314006516

The Close Binary Frequency of Wolf-Rayet Stars as a Function of Metallicity in M31 and M33

Kathryn F. Neugent and Philip Massey

Lowell Observatory
emails: KNeugent@lowell.edu, massey@lowell.edu

Abstract. Here we investigate whether the inability of the Geneva evolutionary models to predict a large enough WC/WN ratio at high metallicities (while succeeding at lower metallicities) is due to their single star nature. We hypothesize that Roche-lobe overflow in close binary systems may produce a greater number of WC stars at higher metallicities. But, this would suggest that the frequency of close massive binaries is metallicity dependent. We now present our results based on observations of ~100 Wolf-Rayet binaries in the varying metallicity environments of M31 and M33.

Keywords. stars: Wolf-Rayet, binaries: close, stars: evolution

---

Massive star evolutionary models generally predict the correct relative number of WC-type and WN-type Wolf-Rayet (WR) stars at low metallicities, but underestimate the ratio at higher (solar and above) metallicities (Meynet & Maeder 2005; Neugent *et al.* 2012). One possible explanation for this failure is perhaps single-star models are not sufficient and Roche-lobe overflow in close binaries is necessary to produce the "extra" WC stars at higher metallicities. However, this would require the frequency of close massive binaries to be metallicity dependent. Here we test this hypothesis by searching for close WR binaries in the high metallicity environments of M31 and the center of M33 as well as in the lower metallicity environments of the middle and outer regions of M33.

To identify the relative frequency of WR binaries in M31 and M33, we observed 250 candidates 4-6 times across a period of three months. We then looked for stars with statistically significant radial velocity variations by comparing the internal errors (I) with the external scatter (E). Stars with large external radial velocity scatter relative to a small internal error are likely to be close binaries. To account for underestimated internal errors, we adopted an E/I value of > 2 as the dividing line between close binaries and non-binaries (Abt & Levy 1976; Abt 1987). An example of a star with a relatively high E/I value of 4.4 is shown below. Note that the emission comes from the WR (WN5) while the absorption comes from an OB companion (B0).

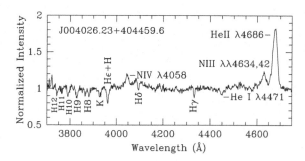

The table below shows the frequency of binaries detected among the M31 and M33 WRs, where we've broken M33 up into three separate regions based on its varying metallicity (see Neugent & Massey 2011). The percentages of close binaries agree across all regions to within a few percent, with the exception of the inner portion of M33. In this region we find that the fraction of close binaries is lower than expected, so this is in the opposite sense of what we would expect if binarity was responsible for the overabundance of WCs. Additionally, we find that the data are consistent with the close binary frequency of WC stars being essentially identical in all four regions.

Region	$\log\frac{O}{H}+12$	Total # Stars	% Binary $\frac{E}{I}>2$	% Binary WNs	WCs
M31 (all)	8.9	106	44 ± 8	57 ± 12	27 ± 9
M33 ($\rho<0.25$)	8.7	44	23 ± 8	20 ± 10	26 ± 13
M33 ($0.25\leqslant\rho<0.50$)	8.4	46	46 ± 12	47 ± 14	40 ± 24
M33 ($\rho\geqslant0.5$)	8.3	54	44 ± 11	43 ± 12	50 ± 27

While studying the relationship between binary frequency and metallicity, we also investigated the nature of our 102 newly identified close WR binaries. With a maximum of 9 observations per star, determining a period was nearly impossible, as shown below where we have phased the radial velocity measurements of J013402.93+305126 by a strongly indicated period of 2.648 days. Clearly, many more observations are needed. Luckily, we have time scheduled on Lowell Observatory's Discovery Channel Telescope this fall to photometrically monitor the most promising binaries to determine periods. Then, we hope to calculate orbits and, finally, masses.

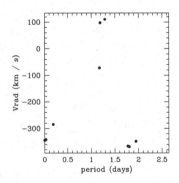

Out of the 250 WRs in M31 and M33, we found 102 stars with $E/I>2$. These close binaries are found throughout M31 and M33, with a binary frequency independent of location, except in the center of M33 where fewer WN binaries were found. Thus, we conclude that a larger binary frequency at high metallicities is not responsible for the discrepancy with the models. Instead, the models seem to either underestimate the life spans of WCs and/or overestimate the life spans for WNs at high metallicities.

Acknowledgements. This work was supported by the NSF under AST-1008020.

References

Abt, H. A. 1987, *ApJ* 317, 353
Abt, H. A. & Levy, S. G. 1976, *ApJS* 30, 273
Meynet, G. & Maeder, A. 2005, *A&A* 429, 581
Neugent, K. F. & Massey, P. 2011, *ApJ* 733, 123
Neugent, K. F., Massey, P., & Georgy, C. 2012, *ApJ* 759, 11

New windows on massive stars: asteroseismology, interferometry, and
spectropolarimetry
Proceedings IAU Symposium No. 307, 2014
G. Meynet, C. Georgy, J. H. Groh & Ph. Stee, eds.

© International Astronomical Union 2015
doi:10.1017/S1743921314006528

Fundamental parameters of B type stars

María-Fernanda Nieva

Institute for Astro- and Particle Physics, Univ. of Innsbruck, Technikerstr. 25/8, 6020
Innsbruck, Austria
Email: Maria-Fernanda.Nieva@uibk.ac.at

Abstract. Fundamental parameters of 26 well-studied sharp-lined single early B-type stars in
OB associations and in the field within a distance of $\leqslant 400$ pc from the Sun are compared to
high-precision data from detached eclipsing binaries (DEBs). Fundamental parameters are derived from accurate and precise atmospheric parameters determined earlier by us from non-LTE
analyses of high-quality spectra, utilising the new Geneva stellar evolution models in the mass-
range ~ 6 to $18\,M_\odot$ at metallicity $Z = 0.014$. Evolutionary masses, radii and luminosities are
determined to better than typically 5%, 10%, and 20% uncertainty, respectively, facilitating the
mass-radius and mass-luminosity relationships to be recovered for single core hydrogen-burning
objects with a similar precision as derived from DEBs. Good agreement between evolutionary
and spectroscopic masses is found. Absolute visual and bolometric magnitudes are derived to
typically $\sim$0.15-0.20 mag uncertainty. Metallicities are constrained to better than 15-20% uncertainty and tight constraints on evolutionary ages of the stars are provided. The spectroscopic
distances and ages of individual sample stars agree with independently derived values for the
host OB associations. The accuracy and precision achieved in the determination of fundamental
stellar parameters from the quantitative spectroscopy of single early B-type stars comes close
(within a factor 2-4) to data derived from DEBs.

Keywords. stars: early-type, stars: evolution, stars: fundamental parameters

1. Introduction

Comprehensive tests of stellar evolution models require as accurate characterisation of
all stellar properties as possible. Besides atmospheric parameters like effective temperature $T_{\rm eff}$ and surface gravity $\log g$ and elemental abundances, a knowledge of fundamental
stellar parameters mass M, radius R and luminosity L, and of age τ is necessary. Primary source of such data are double-lined detached eclipsing binaries (DEBs), which
allow a direct determination of accurate masses and radii at a precision of 1–2%. As
only the $T_{\rm eff}$-ratio of the two stars in a DEB is tightly constrained from light-curve
and radial-velocity-curve analysis, but not $T_{\rm eff}$ of the components, stellar luminosities
remain slightly less constrained. Ages need to be derived by comparison with theoretical
isochrones, with the condition that both components have to be coeval. The most accurate and precise fundamental parameters for massive stars available at present can be
found in the compilation by Torres *et al.* (2010) of early B-type stars.

However, the number of DEBs is limited, and it would be highly valuable if data of
high quality could be obtained for single stars. The best candidate stars are located in
clusters or associations, where distances (and therefore luminosities) and ages can be
constrained from photometry and main-sequence fitting.

We have previously improved the modelling and analysis of the atmospheres of early
B-stars by introducing non-LTE line-formation calculations based on a new generation
of sophisticated model atoms (Nieva & Przybilla 2006, 2007, 2008; Przybilla *et al.* 2008).
Analyses of two samples of stars based on these models (Nieva & Simón-Díaz 2011;

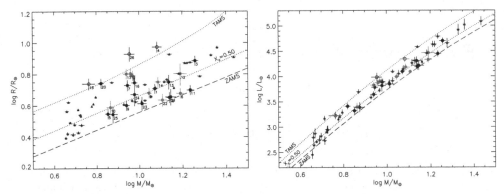

Figure 1. Mass-radius (left panel) and mass-luminosity (right panel) relationships for the sample stars. Black/red dots denote CN-unmixed/mixed chemical composition and open thick circles objects near/beyond core-H exhaustion. Wide circles surrounding the dots mark magnetic stars. Error bars denote 1σ-uncertainties. Data from DEBs (Torres *et al.* 2010) are shown for comparison (small triangles). Abscissa values are evolutionary masses. The ZAMS, 50% core-H depletion and the TAMS are indicated by the dashed/thick/thin-dotted lines, as predicted by the stellar evolution models of Ekström *et al.* (2012).

Nieva & Przybilla 2012), provide a highly accurate and precise atmospheric parameters and chemical abundances. Here, we discuss in particular the fundamental parameters M, R, L of the stars (Nieva & Przybilla 2014). As these are (effectively) single stars, stellar evolution models need to be employed in their derivation. We focus on the grids by Ekström *et al.* (2012), computed for the same metallicity as found for the early B-stars in the solar neighbourhood (Nieva & Przybilla 2012), $Z = 0.014$.

2. Mass-radius and mass-luminosity relationships

Our sample stars consist of 26 apparently slowly-rotating single early B-type stars in OB associations and in the field within a distance of $\leqslant 400\,\mathrm{pc}$ from the Sun. We compare their mass-radius and mass-luminosity relationships with those deduced from DEBs in Fig. 1. We want to draw the attention to the extremely small error bars for these particular DEB data – both components of a DEB have masses and radii determined to $\pm 3\%$, or better. These comprise only data of highest accuracy and precision discussed within the much broader literature on DEBs. Our sample stars fit well into the trends, with error bars coming close to that of the DEB components, typically within a factor 2–4. The precision in luminosity even reaches similar values as that obtained for DEBs. This opens up the possibility to improve on the number statistics and in particular to trace the regions close to the ZAMS in the $M - R$ and $M - L$ relations for the more massive objects, which so far is not covered by DEBs.

References

Ekström, S., Georgy, C., Eggenberger, P., *et al.* 2012, *A&A* 537, A146
Nieva, M. F. & Przybilla, N. 2006, *ApJ (Letters)* 639, L39
Nieva, M. F. & Przybilla, N. 2007, *A&A* 467, 295
Nieva, M. F. & Przybilla, N. 2008, *A&A* 481, 199
Nieva, M.-F. & Przybilla, N. 2012, *A&A* 539, A143
Nieva, M.-F. & Przybilla, N. 2014, *A&A* 566, A7
Nieva, M.-F. & Simón-Díaz, S. 2011, *A&A* 532, A2
Przybilla, N., Nieva, M.-F., & Butler, K. 2008, *ApJ (Letters)* 688, L103
Torres, G., Andersen, J., & Giménez, A. 2010, *A&A Rev.* 18, 67

New windows on massive stars: asteroseismology, interferometry, and spectropolarimetry
Proceedings IAU Symposium No. 307, 2014
G. Meynet, C. Georgy, J. H. Groh & Ph. Stee, eds.

© International Astronomical Union 2015
doi:10.1017/S174392131400653X

A Search for Hot Subdwarf Companions to Rapidly-Rotating Early B Stars

Geraldine J. Peters[1], Douglas R. Gies[2], Luqian Wang[2] and Erika D. Grundstrom[3]

[1]Space Sciences Center & Dept. of Physics & Astronomy, University of Southern California, 835 W. 37th St., Los Angeles, CA 90089-1341, USA
email: gjpeters@mucen.usc.edu

[2]CHARA & Dept. of Physics & Astronomy, Georgia State University, Atlanta, GA 30302-5060, USA

[3]Dept. of Physics & Astronomy, Vanderbilt University, Nashville, TN 37206, USA

Abstract. We continue to search for O-type subdwarf companions in binary systems containing Be primaries. We were not able to confirm an sdO object in π Aqr and HR 2142, even though optical and UV observations suggest their presence. Some possible reasons are enumerated.

Keywords. (stars:) binaries: close, stars: emission-line, Be, (stars:) subdwarfs, (stars:) circumstellar matter, stars: individual (π Aqr, HR 2142)

1. Motivation and method

The rapid rotation seen in some early B stars appears to be the result of angular momentum gain during a prior episode of binary mass transfer where the mass donor is now a sdO object or neutron star. *But just how common is this scenario?* An sdO companion has now been confirmed in three well-known Be stars: ϕ Per (Gies *et al.* 1998), FY CMa (Peters *et al.* 2008), and 59 Cyg (Peters *et al.* 2013). We continue to search for sdO companions by performing a similar analysis of *IUE* HIRES and optical Hα spectra of other bright binary Be stars for which there is evidence that the secondary might be an O subdwarf. Our most recent efforts focus on π Aqr and HR 2142. The method is fully described in Peters *et al.* (2008). Our first task is to find a good set of orbital elements. We look for the signature of the suspected sdO object by cross-correlating the *IUE* spectra with either a template generated from the NLTE model atmospheres of Lanz & Hubeny (2003) or a standard sdO object. We then employ a Doppler tomography algorithm (Bagnuolo *et al.* 1994) to reconstruct the spectrum of the secondary.

2. π Aquarii

π Aqr (B 1 III-IVe, $P = 84.1$ d) is a well known bright Be star that temporarily lost its circumstellar (CS) disk in the mid-1990s. We employed Hα spectra from the Coudé Feed Telescope at the Kitt Peak National Observatory(KPNO), the Be Star Spectra Database (http://basebe.obspm.fr/basebe/), and 22 images from *IUE* to determine good orbital parameters for the primary. Zharikov *et al.* (2013) found that excess Hα emission is located on the side of the disk facing the secondary. Since this is seen in other sdO objects, we searched for a hot companion in the FUV using cross-correlation templates for BD+28 4211 ($T = 82$ kK), BD+75 325 ($T = 53$ kK), and HD 49798 ($T = 48$ kK), but did not find a signature of the secondary.

3. HR 2142

HR 2142 (B 1.5 IV-Vnne, $P = 80.86$ d) has been known as a binary since the early 1970s (Peters 1972, 1983) but the spectrum of the secondary has never been detected. The system displays a two-component shell phase centered on the inferior conjunction of the secondary. The shell lines are red-shifted prior to conjunction, but violet-shifted afterwards. The behavior of the CS material has suggested that HR 2142 is an Algol-type interacting binary undergoing mass transfer (Peters & Gies 2002, Peters 2001, 1983). Several researchers have searched for the secondary in the IR without success, which has led to the suspicion that it is an O subdwarf. We determined a new radial velocity curve for the primary in HR 2142 based upon the combination of 50 measurements from Peters (1983), 87 measurements from the *IUE* cross-correlation functions, 129 measurements of the Hα wings from the KPNO spectra, and BeSS spectra. With a good set of orbital elements for the primary in hand, we searched for an sdO object using our cross-correlation technique, but could not confirm the presence of an O-type subdwarf.

4. Why are there no detections?

Using the same method and software that we employed to detect the subdwarfs in φ Per, FY CMa, and 59 Cyg, we did not find an sdO secondary in the π Aqr and HR 2142 systems, even though the behavior of the CS structures in them suggests that they are similar binaries. Possibly the subdwarfs are cooler or smaller and contribute less than 4% of the light in the FUV, which is about the limit of our ability to detect an sdO object. The lines of the secondary may be broad, or the object immersed in a dense wind. The shell lines in HR 2142 could be the result of a density enhancement due to a shock interface where the wind from the subdwarf collides with the primary's massive CS disk. Another possibility is that the secondary creates a one-armed spiral wake in the disk similar to that predicted for forming planets in a protoplanetary disk (Bate *et al.* 2003, Ogilvie & Lubow 2002). This scenario will be discussed in a forthcoming paper.

The authors appreciate support from NASA grant NNX10AD60G (GJP), NSF grant AST-1009080 (DRG), GSU institutional funding (DRG), and the USC Women in Science and Engineering (WiSE) program (GJP).

References

Bagnuolo, Jr., W. G., Gies, D. R., Hahula, M. E., Wiemker, R., & Wiggs, M. S. 1994, *ApJ* 423, 446

Bate, M. R., Lubow, S. H., Ogilvie, G. I., & Miller, K. A. 2003, *MNRAS* 341, 213

Gies, D. R., Bagnuolo, Jr., W. G., Ferrara, E. C., *et al.* 1998, *ApJ* 493, 440

Lanz, T. & Hubeny, I. 2003, *ApJS* 146, 417

Ogilvie, G. I. & Lubow, S. H. 2002, *MNRAS* 330, 950

Peters, G. J. 1972, *PASP* 84, 334

Peters, G. J. 1983, *PASP* 95, 311

Peters, G. J. 2001, *Publications of the Astronomical Institute of the Czechoslovak Academy of Sciences* 89, 30

Peters, G. J. & Gies, D. R. 2002, in C. A. Tout & W. van Hamme (eds.), *Exotic Stars as Challenges to Evolution*, Vol. 279 of *Astronomical Society of the Pacific Conference Series*, p. 149

Peters, G. J., Gies, D. R., Grundstrom, E. D., & McSwain, M. V. 2008, *ApJ* 686, 1280

Peters, G. J., Pewett, T. D., Gies, D. R., Touhami, Y. N., & Grundstrom, E. D. 2013, *ApJ* 765, 2

Zharikov, S. V., Miroshnichenko, A. S., Pollmann, E., *et al.* 2013, *A&A* 560, A30

New windows on massive stars: asteroseismology, interferometry, and spectropolarimetry
Proceedings IAU Symposium No. 307, 2014
G. Meynet, C. Georgy, J. H. Groh & Ph. Stee, eds.

© International Astronomical Union 2015
doi:10.1017/S1743921314006541

An empirical pipeline for determining the viscosity parameter for Be star disks

Leandro R. Rímulo[1], Alex C. Carciofi[1], Thomas Rivinius[2] and Xavier Haubois[3]

[1]University of São Paulo, Dept. of Astronomy, Brazil
email: lrrimulo@usp.br

[2]ESO, Chile

[3]LESIA, Observatoire de Paris, France

Abstract. Be star phenomenology is strongly associated with their viscous circumstellar disks. Recently, models became available for the temporal evolution of these disks when subject to variable mass ejection rates. In this contribution we will discuss how these dynamical disk models, modeled with the radiative transfer code HDUST, can be used for constraining fundamental disk parameters, such as the α viscosity parameter, and we will report on an ongoing effort to model light curves of a large number of stars.

Keywords. hydrodynamics, radiative transfer, techniques: photometric, stars: mass loss

1. Introduction

Be stars are non supergiant rapid rotating B stars (3 to 15 $M_\odot$), whose spectrum has, or had at some time, Balmer lines in emission, attributed to a geometrically thin circumstellar gaseous decretion disk. Roughly 10% of the B stars in the Galaxy are Be stars, and this percentage grows for the Magellanic Clouds, where metallicity is lower (Martayan *et al.* 2011). The circumstellar disk is created by mass ejected from the rotating stellar surface to the disk, which then diffuses outwards by turbulent viscosity, as observational and theoretical studies have shown. Hence their name "viscous decretion disks" (VDD) (see review by Rivinius *et al.* 2013).

The VDD scenario in the steady state limit has been successfully tested for several Be stars, using the radiative transfer code HDUST (e.g., Carciofi & Bjorkman 2006; Carciofi *et al.* 2007, 2009, 2012). However, there are still several open theoretical questions for the VDD scenario, one of the main ones being the origin and magnitude of the turbulent viscosity. *Measuring* directly (or as directly as possible) the value of Shakura-Sunyaev's α viscosity parameter is of great current interest.

Variability is one of the main observational features of Be stars. They are attributed to a variable mass injection rate from the star to the disk. Using the 1-D time-dependent hydrodynamic code SINGLEBE (Okazaki 2007) associated with the radiative transfer code HDUST (Carciofi & Bjorkman 2006, 2008), it was possible to determine, for the first time, $\alpha = 1.0 \pm 0.2$ for a dissipation phase of the disk of the Be star 28 CMa (Carciofi *et al.* 2012). Currently, using the cited codes, we are developing a simulation based chain of procedures (a pipeline) aimed to estimate the α parameter, for the first time, for hundreds of light curves available from photometric surveys (e.g., OGLE, MACHO, EROS).

2. Modeling dynamical viscous decretion disks

From the α-disk formalism (Shakura & Sunyaev 1973), the equation for the evolution of the distribution of mass in the disk as function of radial distance R and time t is

$$\frac{\partial \Sigma}{\partial t} = \frac{1}{\tau_{\text{vis}}} \left\{ \frac{2}{r} \frac{\partial}{\partial r} \left[r^{\frac{1}{2}} \frac{\partial}{\partial r} \left(r^2 \Sigma \right) \right] + \left(\frac{r_{\text{in}}^2}{r_{\text{in}}^{\frac{1}{2}} - 1} \frac{\delta(r - r_{\text{in}})}{r_{\text{in}}} \right) \Sigma_{\text{in}}(t) \right\} \tag{2.1}$$

where $r = R/R_\star$, $r_{\text{in}} = R_{\text{in}}/R_\star$ (R_s being the radius of mass injection to the disk).

The constant τ_{vis} is given by $\frac{1}{\tau_{\text{vis}}} = \alpha \frac{kT}{\mu m_H (GMR_\star)^{\frac{1}{2}}}$. Therefore, the greater the viscosity parameter α, the faster is the evolution of the disk surface density. The variable $\Sigma_{\text{in}}(t)$ parameter is a more appropriate form of stating the mass injection rate $-\dot{M}_{\text{inj}}(t)$. Both are related by: $2\pi R_\star \Sigma_{\text{in}}(t) \left(\frac{R_\star}{\tau_{\text{vis}}} \right) = -\dot{M}_{\text{inj}}(t) \left(\frac{r_{\text{in}}^{\frac{1}{2}} - 1}{r_{\text{in}}^2} \right)$.

2.1. *The pipeline*

We found that the light curves associated with a disk dissipation following a disk construction that lasted a time Δt (disk build-up time) assume the following empirical law

$$m_\lambda(t) = m_{(t=\infty),\lambda} - (\Delta m_\lambda) e^{-\left(\xi_\lambda \frac{t}{\tau_{\text{vis}}} \right)} \tag{2.2}$$

valid in general, for non-edge-on cases: $i < 60°$ (edge-on cases being observationally identifiable by their behavior in color), where λ denotes a given photometric band.

The limiting magnitude, $m_{(t=\infty),\lambda}$, depends only on the stellar parameters. In order to calibrate empirically ξ_λ and Δm_λ we set up a very large model grid consisting of several spectral types for the central star (from B0 to B9), Σ_{in} (roughly from 0.1 to 4 g cm^2), Δt (from 0.1 to 5 years) and i (from 0° to 65°).

We summarize our pipeline for finding α from light curves as follows:

(*a*) Find a "bump like" light curve of a Be star with known distance and interstellar absorption law, with a few years of coverage, and two color photometry;

(*b*) Exclude the curve if it corresponds to a near edge-on case;

(*c*) Find a good approximation to the stellar parameters $R_\star$, T_{eff} and M by measuring the limiting magnitude $m_{(t=\infty),\lambda}$ in two filters;

(*d*) Measure the build-up interval Δt and the excess Δm_λ directly from the light curve. Then, with our tabulated values, find the asymptotic surface density Σ_{in}. And with it and our tabulated values again, find the ξ_λ parameter;

(*e*) Fit our derived law given by Eq.(2.2) on the dissipation part of the curve and obtain the coefficient of the exponential: $(\xi_\lambda/\tau_{\text{vis}})_{\text{fitted}}$. Then, obtain, the α viscosity parameter: $\alpha = (\xi_\lambda/\tau_{\text{vis}})_{\text{fitted}} / (\xi_\lambda/\alpha\tau_{\text{vis}})_{\text{pipeline}}$.

References

Carciofi, A. C. & Bjorkman, J. E. 2006, *ApJ* 639, 1081
Carciofi, A. C. & Bjorkman, J. E. 2008, *ApJ* 684, 1374
Carciofi, A. C., Bjorkman, J. E., Otero, S. A., *et al.* 2012, *ApJ (Letters)* 744, id. L15
Carciofi, A. C., Magalhães, A. M., Leister, N. V., Bjorkman, J. E., & Levenhagen, R. S. 2007, *ApJ (Letters)* 671, L49
Carciofi, A. C., Okazaki, A. T., Le Bouquin, J.-B., *et al.* 2009, *A&A* 504, 915
Martayan, C., Rivinius, T., Baade, D., Hubert, A.-M., & Zorec, J. 2011, *IAU Circ.* 272, 242
Okazaki, A. T. 2007, *ASP-CS* 361, 230
Rivinius, T., Carciofi, A. C., & Martayan, C. 2013, *A&A Rev.* 21, id.69
Shakura, N. I. & Sunyaev, R. A. 1973, *A&A* 24, 337

New windows on massive stars: asteroseismology, interferometry, and
spectropolarimetry
Proceedings IAU Symposium No. 307, 2014
G. Meynet, C. Georgy, J. H. Groh & Ph. Stee, eds.

© International Astronomical Union 2015
doi:10.1017/S1743921314006553

Westerlund 1 is a Galactic Treasure Chest: The Wolf-Rayet Stars

C. K. Rosslowe and P. A. Crowther

Department of Physics and Astronomy, University of Sheffield
email: chris.rosslowe@shef.ac.uk

Abstract. The Westerlund 1 Galactic cluster hosts an eclectic mix of coeval massive stars. At a modest distance of $4 - 5$ kpc, it offers a unique opportunity to study the resolved stellar content of a young (~ 5 Myr) high mass ($5 \cdot 10^4\ M_\odot$) star cluster. With the aim of testing single-star evolutionary predictions, and revealing any signatures of binary evolution, we discuss on-going analyses of NTT/SOFI near-IR spectroscopy of Wolf-Rayet stars in Westerlund 1. We find that late WN stars are H-poor compared to their counterparts in the Milky Way field, and nearly all are less luminous than predicted by single-star Geneva isochrones at the age of Westerlund 1.

Keywords. stars: early-type, mass loss, Wolf-Rayet - infrared: stars - galaxies: star clusters

1. Introduction

Westerlund 1 (Wd1) is amongst the most massive young clusters in the Galaxy. We witness this coeval collection of stars at an interesting epoch, as Clark *et al.* (2005) have identified a plethora of post-main sequence massive stars, indicating an age of 5±1 Myr and a main sequence turn-off at approximately $30 - 35\ M_\odot$ (O 7V).

A high fraction of dust producing WC stars and coincidental hard X-ray sources amongst the observed Wolf-Rayet (WR) stars suggests a binary fraction approaching unity (Crowther *et al.* 2006). Indeed, Schneider *et al.* (2014) predict that after only a few Myr, the majority of a cluster's most luminous stars are the products of binary interaction. Here we report on preliminary tailored spectral analyses of 15 WR stars in Wd1 from Crowther *et al.* (2006), and discuss how derived parameters compare to single-star and binary evolutionary models.

2. Data & Analysis Method

We obtained NTT/SOFI spectra of 23 WR stars in Wd1 using IJ and HK grisms ($R \sim 1000$), identified by differential narrow-band imaging (Crowther *et al.* 2006).

We have carried out spectral modelling of 15 of these WR stars (neglecting very late WN9-10 or dusty WC) – 2 of which are shown in Fig 1 – using the CMFGEN model atmosphere code (Hillier & Miller 1998) to derive effective temperatures, luminosities, abundances, mass-loss rates and wind velocities.

We constrain luminosities and extinction simultaneously, by requiring model spectral energy distributions to match a combination multi-band photometry and flux calibrated spectra, as shown in Fig. 2.

3. Results

Most WR stars are less luminous than single-star Geneva isochrones covering the expected age of Westerlund 1; older ages are precluded by the presence of high-mass main sequence stars. Late WN stars are generally H-poor compared to their field Milky

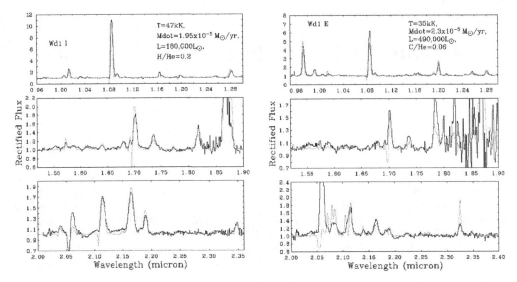

Figure 1. *Left.* NTT/SOFI spectrum (solid) and corresponding CMFGEN model (dotted) for Wd1-I (WN8). *Right* As left for Wd1-E (WC9). CMFGEN parameters are displayed in the upper panels.

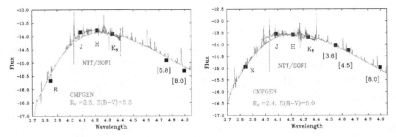

Figure 2. *Left* SED of Wd1-I, with flux calibrated NTT/SOFI spectra and multiband photometry (R:VLT/FORS2, JHK:SOFI, [5.8]-[8.0]:GLIMPSE). *Right* As left for Wd1-E. Model spectra are reddened using a Howarth (1983) extinctin law.

way counterparts. Low luminosity and H deficiency amongst the WN stars is consistent with outcomes of binary evolution (e.g. Eldridge *et al.* 2008). Preliminary mass-loss rates of the H-free early WN stars are systematically lower than those adopted by stellar models for their luminosities. Overall, The WR stars in Westerlund 1 analysed to date display properties that are inconsistent with the current generation of single-star models (Ekström *et al.* 2012). Analysis of the dust producing WC stars is forthcoming.

References

Clark, J. S., Negueruela, I., Crowther, P. A., & Goodwin, S. P. 2005, *A&A* 434, 949
Crowther, P. A., Hadfield, L. J., Clark, J. S., Negueruela, I., & Vacca, W. D. 2006, *MNRAS* 372, 1407
Ekström, S., Georgy, C., Eggenberger, P., *et al.* 2012, *A&A* 537, A146
Eldridge, J. J., Izzard, R. G., & Tout, C. A. 2008, *MNRAS* 384, 1109
Hillier, D. J. & Miller, D. L. 1998, *ApJ* 496, 407
Howarth, I. D. 1983, *MNRAS* 203, 301
Schneider, F. R. N., Izzard, R. G., de Mink, S. E., *et al.* 2014, *ApJ* 780, 117

New windows on massive stars: asteroseismology, interferometry, and spectropolarimetry
Proceedings IAU Symposium No. 307, 2014
G. Meynet, C. Georgy, J. H. Groh & Ph. Stee, eds.
© International Astronomical Union 2015
doi:10.1017/S1743921314006565

Herschel/PACS: Constraining clumping in the intermediate wind region of OB stars

M. M. Rubio-Díez[1], F. Najarro[1], J. O. Sundqvist[2], A. Traficante[3], J. Puls[2], L. Calzoletti[4,5], A. Herrero[6,7], D. Figer[8] and J. Martin-Pintado[1]

[1] Centro de Astrobiología, CSIC-INTA, Madrid, Spain
email: mmrd@cab.inta-csic.es

[2] Universitätssternwarte München, München, Germany

[3] Jodrell Bank Centre for Astrophysics, School of Physics and Astronomy, University of Manchester, Manchester, UK

[4] ASI Science Data Center, Roma, Italy

[5] INAF, Osservatorio Astronomico di Roma, Rome, Italy

[6] Instituto de Astrofísica de Canarias, Tenerife, Spain

[7] Dpto. de Astrofísica, Universidad de La Laguna, Tenerife, Spain

[8] Rochester Institute of Technology, NY, USA

Abstract. At present, it is well established that previously accepted mass-loss rates ($\dot{M}$) of luminous OB stars may be overestimated when clumping is neglected. Our Herschel/PACS Far-Infrared (Far-IR) observations of a set of OB stars allow us to improve our knowledge of clumping stratification, constraining clumping properties in intermediate wind regions. In this work, better sampled clumping structure estimates are provided for ι Ori, ϵ Ori and ξ Per as well as an initial estimate of the clumping properties of the wind from τ Sco. These observations will allow us to obtain reliable mass-loss rates and improve our understanding of the wind physics.

1. Introduction

It is currently accepted that stellar winds of massive stars are not smooth, but show small scale inhomogeneities. Such "wind clumps" have a crucial effect on the standard mass-loss rate ($\dot{M}$) diagnostics: for a given mass loss, *microclumping* (optically thin clumps) increases emission that depends on ρ^2 (H emission, IR/radio free-free emission) while having no effect on processes that depend on ρ (UV resonance lines absorption). On the other hand, *macroclumping* (optically thick clumps) leads to higher $\dot{M}$ estimates if derived from resonance lines. Not taking into account clumping results in inconsistencies between different mass-loss diagnostics and overestimated $\dot{M}$ (Fullerton *et al.* 2006; Sundqvist *et al.* 2010). Therefore, a reliable knowledge of the clumping quantified by the clumping factor ($f_{\rm cl} = f_v^{-1} \equiv < \rho^2 > / < \rho >^2$, Owocki *et al.* 1988) and its radial stratification (Sundqvist & Owocki 2013) is crucial to constrain the "true" $\dot{M}$ of OB stars . We argue that only a consistent treatment of ALL possible diagnostics, scanning different parts of the winds, and analyzed by means of state of the art model atmospheres, will enable us to constrain clumping through the whole wind and $\dot{M}$ itself. To this end, we have assembled a variety of multi-wavelength data (from optical to radio) of a carefully selected sample of 28 O4-B8 stars, including new Far-IR diagnostics of free-free emission. We have used our photometric Herschel/PACS observations at 70, 100 and 160 μm (programs: OT1/OT2_mrubio), which uniquely constrain the clumping properties of the intermediate wind regions, to derive the clumping properties of the entire outflow.

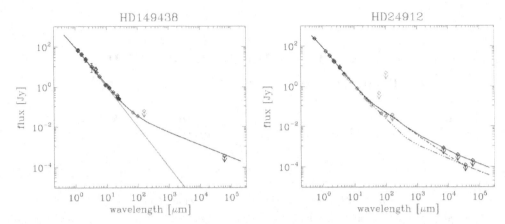

Figure 1. Different computed SED models including microclumping structure of the wind built from IR/radio continuum archival measurements (black and green symbols) and PACS fluxes at 70,100 and 160 μm (red symbols). Arrows denote upper limits. *Left:* τ Sco PACS flux values agree with the new **unclumped model**, $f_{cl}=1$, computed for $\dot{M} = 3.23 \times 10^{-7} M_\odot \, \mathrm{yr}^{-1}$ (black solid line). The blue line represents τ Sco SED for spectral index $\alpha = 2$. *Right:* ξ Per. Clumping models by Puls *et al.* (2006); Solid line: $\dot{M} = 2.3 \times 10^{-6} M_\odot \, \mathrm{yr}^{-1}$ and $f_{clin} = 2.1$, $f_{clmid} = 5$, $f_{clout} = 1$; dotted-dashed line: $\dot{M} = 1.2 \times 10^{-6} M_\odot \, \mathrm{yr}^{-1}$, $f_{clin} = 8$, $f_{clmid} = 20$, $f_{clout} = 1$; dashed-triple dotted line: $\dot{M} = 1.2 \times 10^{-6} M_\odot \, \mathrm{yr}^{-1}$, $f_{clin} = 8$, $f_{clmid} = f_{clout} = 1$. PACS data break the degeneracy between models pointing to lower mass-loss and clumping structure in the intermediate wind region.

2. Remarks

In a first step, we test previous clumping simulations for ι Ori, ϵ Ori, ξ Per (Puls *et al.* 2006 and Najarro *et al.* 2011) against new Herschel/PACS observations at 70, 100 and 160 μm. These results together with the first clumping structure study for τ Sco and previously analyzed objects (Rubio-Díez *et al.* 2013: RD13) make a total subsample of 11 stars (8 O giant/supergiant stars and 3 B supergiant/dwarf stars). When this subsample is considered, we find: **i)** Far-IR observations disentangle the uncertainty in the previous clumping models of 4 O stars of our sample and point to low clumping in the intermediate wind region (e.g.: ξ Per, Fig. 1 *right*; λ^1 Ori and λ Cep, RD13); **ii)** for 2 O stars, our PACS flux measurements confirm the clumping structure derived in previous work (e.g. ζ Puppis, α Cam, see RD13); **iii)** the only PACS detection of ι Ori at 70 μm shows this object as a Rayleigh-Jeans emitter; **iv)** for the two early-B stars analyzed, the measurements at Far-IR wavelengths indicate slight excess emission (τ Sco, Fig. 1 *left*) or a stronger clumped wind than predicted (ϵ Ori); **v)** the uncertainty in the Far-IR flux measurements of CygOB2#7 and #11, due to their complex surroundings (visible in PACS maps), prevents from drawing conclusions about their clumping structure at the intermediate wind region without a parallel spectroscopic analysis. All the sample stars will be subject to a second-stage analysis: a spectroscopic multi-wavelength study with the CMFGEN code (Hillier & Miller 1998). This second phase is mandatory in order to check that the derived clumping properties are consistent with the observed spectral lines, and will provide further insight into the wind physics and mass-loss rates.

References

Fullerton, A. W., Massa, D. L., & Prinja, R. K. 2006, *ApJ* 637, 1025
Hillier, D. J. & Miller, D. L. 1998, *ApJ* 496, 407
Najarro, F., Hanson, M. M., & Puls, J. 2011, *A&A* 535, A32

Owocki, S. P., Castor, J. I., & Rybicki, G. B. 1988, *ApJ* 335, 914

Puls, J., Markova, N., Scuderi, S., *et al.* 2006, *A&A* 454, 625

Rubio-Díez, M. M., Najarro, F., Traficante, A., *et al.* 2013, in *Massive Stars: From alpha to Omega*

Sundqvist, J. O. & Owocki, S. P. 2013, *MNRAS* 428, 1837

Sundqvist, J. O., Puls, J., & Feldmeier, A. 2010, *A&A* 510, A11

William Chantereau

Anahí Granada

New windows on massive stars: asteroseismology, interferometry, and
spectropolarimetry
Proceedings IAU Symposium No. 307, 2014
G. Meynet, C. Georgy, J. H. Groh & Ph. Stee, eds.
© International Astronomical Union 2015
doi:10.1017/S1743921314006577

NGC 3293 revisited by the Gaia-ESO Survey

Thierry Semaan[1], Thierry Morel[1], Eric Gosset[1], Juan Zorec[2], Yves Frémat[3], Ronny Blomme[3] and Alex Lobel[3]

[1] ULg, Institut d'Astrophysique et de Géophysique, Liège, Belgium
email: thierry.semaan@ulg.ac.be

[2] Institut d'Astrophysique de Paris, CNRS (UMR 7095), Paris, France &
Université Pierre et Marie Curie (UMR 7095), Paris, France

[3] Royal Observatory of Belgium, Brussels, Belgium

Abstract. In the framework of the Gaia-ESO survey we have determined the fundamental parameters of a large number of B-type stars in the Galactic, young open cluster NGC 3293. The determination of the stellar parameters is based on medium-resolution spectra obtained with FLAMES/GIRAFFE at ESO-VLT. As a second step, we adopted the accurate parameters to determine the chemical abundances of these hot stars. We present a comparison of our results with those obtained by the 'VLT-FLAMES survey of massive stars' (Evans *et al.* 2005). Our study increases the number of objects analysed and provides an extended view of this cluster.

Keywords. Massive Stars, Fundamental parameters, Chemical abundances

1. Observations, reduction and stars analysed

The observations of NGC 3293 with the multifiber spectrograph FLAMES/GIRAFFE took place from mid-February until beginning of April 2012. During this period, 25 FLAMES/GIRAFFE frames were obtained in the Medusa mode at medium resolution. The GIRAFFE observations were made in 4 grating setups: HR3, HR5A, HR6, and HR14A, which correspond to the wavelength-intervals 4030-4200 Å, 4350-4750 Å and 6300-6700 Å respectively, with a mean resolving power of 20000. The HR3, HR5A, and HR6 domains include several strong helium lines, as well as weak Si_{II}, Si_{III}, C_{II} and O_{II} lines that are very useful for the determination of the fundamental parameters of hot stars. The HR14A domain contains $H\alpha$ and can be used to identify the emission-line stars. The reduction of the GIRAFFE spectra was done by the Cambridge Astronomy Survey Unit (CASU) using a dedicated pipeline. Thanks to the 25 GIRAFFE frames we have more than 500 stars totally covered by the 4 setups. We select only the B stars (no O stars are present in this dataset) thanks to a visual inspection. In a second step we normalised these spectra to the continuum flux level. In our sample we have also detected Be stars, as well as single (SB1) and double-lined binary systems (SB2). The objects firmly identified as Be stars or SB2 systems were excluded from the final sample.

2. Determination of the fundamental parameters

The fundamental parameters ($T_{\rm eff}$, $v \sin i$, $\log g$) and the radial velocities of the B stars are obtained using the GIRFIT code (Frémat *et al.* 2006). This program is based on a least square method and derives the parameters by fitting the observed, normalised spectra with a grid of stellar fluxes computed with the atmospheric code TLUSTY in non-local thermodynamic (NLTE) equilibrium mode. The determination of the fundamental parameters is performed over the whole wavelength domain of 4030-4200 Å and 4350-4750 Å with exceptions for the interstellar bands and the emission components that

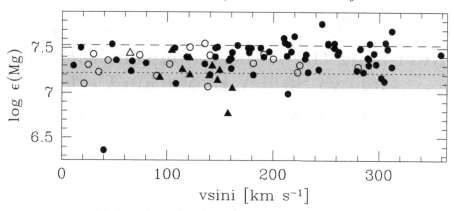

Figure 1. Variation of the Mg abundance, as a function of $v \sin i$. The open symbols represent possible SB2 and all SB1 systems. The horizontal blue strip shows the mean value for this cluster determined by Hunter *et al.* (2009). The horizontal dashed line shows the meteoritic abundance (Asplund *et al.* 2009).

sometimes are filling in the Balmer lines. The microturbulence velocity is fixed in the fitting procedure, but we have computed three different grids for 2, 5 and $10 \, \mathrm{km \, s^{-1}}$. This allows us to determine the best microturbulence velocity among the 3 values for each star.

3. Determination of chemical abundances

After determining T_{eff}, $\log g$, $v \sin i$ and the best microturbulence velocity, we use these parameters to compute the abundances of He, Mg and Si. They are obtained through a best match search between a grid of synthetic spectra and observed line profiles of HeI 4471, HeI 4713, MgII 4481, SiIII 4568, and SiIII 4575 . The grid of synthetic spectra is generated by the NLTE code DETAIL/SURFACE. The variation of the Mg abundance as a function of $v \sin i$ is shown in Fig. 1. Our values are slightly larger than those obtained by Hunter *et al.* (2009) and no dependence on T_{eff}, $\log g$ or $v \sin i$ is detected.

4. Conclusion

This work must be completed with the UVES spectra, which have not yet been analysed. Indeed, this includes some bright stars whose analysis will certainly complete and improve our results. The abundances of some key elements (e.g., CNO) will be determined for the hottest stars of the cluster such that the mixing processes could then be investigated. In the same way, the analysis of Trumpler 14 and NGC 6705 will be completed in the near future.

References

Asplund, M., Grevesse, N., Sauval, A. J., & Scott, P. 2009, *ARA&A* 47, 481
Evans, C. J., Smartt, S. J., Lee, J.-K., *et al.* 2005, *A&A* 437, 467
Frémat, Y., Neiner, C., Hubert, A.-M., *et al.* 2006, *A&A* 451, 1053
Hunter, I., Brott, I., Langer, N., *et al.* 2009, *A&A* 496, 841

New windows on massive stars: asteroseismology, interferometry, and spectropolarimetry
Proceedings IAU Symposium No. 307, 2014
G. Meynet, C. Georgy, J. H. Groh & Ph. Stee, eds.

© International Astronomical Union 2015
doi:10.1017/S1743921314006589

Revisiting the Hunter diagram with the Geneva Stellar Evolution Code

R. Simoniello[1], G. Meynet[2], S. Ekström[2], C. Georgy[3] and A. Granada[2]

[1]Laboratoire AIM, CEA/DSM-CNRS-Université Paris Diderot; CEA, IRFU, SAp, centre de Saclay, F-91191, Gif-sur-Yvette, France
email: rosaria.simoniello@cea.fr

[2]Geneva Observatory, Chemin de Maillettes 51, 1290 Sauverny, Switzerland

[3]Astrophysics group, Lennard-Jones Laboratories, EPSAM, Keele University, Staffordshire ST5 5BG, UK

Abstract. We produced a model grid of rotating main and post-main sequence stars with the Geneva Stellar Evolution Code (GENEC). The initial chemical composition is tailored to compare with observations of early OB type stars in the Large Magellanic Cloud (LMC) and the grid covers stellar masses in the range of $7 \leqslant M/M_\odot \leqslant 15$ and initial velocity between $0\,\mathrm{km\,s^{-1}} \leqslant v\sin(i) \leqslant 300\,\mathrm{km\,s^{-1}}$. The model grid has been used to determine the changes in the surface Nitrogen abundances during the star evolution and the results have been compared with observations.

Keywords. methods: numerical, stars: rotation, stars: abundances

1. Introduction

The rotating evolutionary models for single stars predict that only the fastest rotating stars can be highly enriched in nitrogen (N) while still on the MS. Therefore a particularly strong observational diagnostic for rotational mixing is the N surface abundance of stars undergoing hydrogen fusion through the CNO cycle. A survey of massive stars has been carried out by the Very Large Telescope (VLT) Fiber Large Array Multi-Element Spectrograph (FLAMES) including 750 O and early B type on clusters located in our Galaxy, the Large (LMC) and Small Magellanic Clouds (SMC). A detailed analysis of the relationship between N abundances and the projected rotational velocity ($v\sin i$) has been carried out for early B type stars. The distribution of stars is called the Hunter diagram and it became one of the key diagrams to check the physics of rotational mixing (Hunter *et al.* 2008). It turned out that the current rotating models failed to explain the existence of highly N enriched slow rotators. This lead to intense debates on the unknown causes of the discrepancies and on the identification of other physical mechanisms or ingredients still missing in the description of the theoretical models. Within this context we decided to investigate the efficiency of the rotational mixing on Nitrogen surface abundances in an attempt to quantify the discrepancy with observations.

2. Data analysis

2.1. The model grids

We used **GEN**eva stellar **E**volution **C**ode (GENEC), a one dimensional hydrodynamic code (Eggenberger *et al.* 2008), for the computation of new grids of stellar rotating models for masses between $5 \leqslant M/M_\odot \leqslant 15$, initial velocity between $0 \leqslant v\sin(i) \leqslant 300\,\mathrm{km\,s^{-1}}$

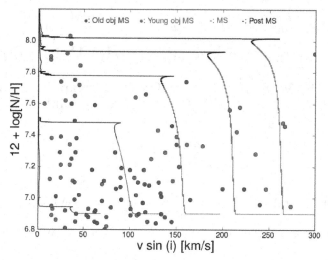

Figure 1. Nitrogen abundance against rotational velocity at LMC metallicity for rotating models at $12\,M_\odot$ (continuous lines) and from observations of the VLT (bullet).

and with initial chemical composition $Z = 0.0047$ matching that observed for the LMC, from which the richest massive star data are obtained.

2.2. *Results*

Figure 1 shows the N surface abundances as function of the projected rotational velocity. The blue and red dots represent the observed sample of early OB type stars in the N11 and NGC2004 LMC cloud downloaded from http://vizier.u-strasbg.fr/viz-bin/VizieR-4 (Hunter *et al.* 2008). They have been split in two groups: young (surface gravity $\log g \geqslant 3.7$) and old ($\log g \leqslant 3.2$) core hydrogen burning objects. The red and black continuous lines represent the amount of nitrogen surface abundances during the evolution for two different evolutionary phase: main and post main sequence stars. The nitrogen abundance is for a $12\,M_\odot$ and for different initial velocity $0 \leqslant vsin(i) \leqslant 300\,\mathrm{km\,s^{-1}}$. As we can see the slowly rotating ($\leqslant 50\,\mathrm{km\,s^{-1}}$) and highly nitrogen enriched ($\geqslant 7.2$) young objects fall off the corresponding young branch of the evolutionary track, while young objects and old but slowly rotating still on the MS can be reproduced by the current rotating models.

3. Conclusion

This model grid constitutes the basis to producing a stellar population synthesis with the SYCLIST code (Georgy *et al.* 2014). This will allow us to get a statistically significant sample covering a specific range of masses and velocities and its comparison with the observed Hunter diagram will tell us the measure of the departure of the theoretical models from the observations.

References

Eggenberger, P., Meynet, G., Maeder, A., *et al.* 2008, *APSS* 316, 43
Georgy, C., Granada, A., Ekström, S., *et al.* 2014, *A&A* 566, A21
Hunter, I., Brott, I., Lennon, D. J., *et al.* 2008, *ApJL* 676, L29

New windows on massive stars: asteroseismology, interferometry, and spectropolarimetry
Proceedings IAU Symposium No. 307, 2014
G. Meynet, C. Georgy, J. H. Groh & Ph. Stee, eds.

© International Astronomical Union 2015
doi:10.1017/S1743921314006590

The properties of single WO stars

F. Tramper[1], S. M. Straal[1], G. Gräfener[2], L. Kaper[1], A. de Koter[1,3], N. Langer[4], H. Sana[5] and J. S. Vink[2]

[1] Anton Pannekoek Institute for Astronomy, University of Amsterdam, PO Box 94249, 1090 GE Amsterdam, The Netherlands
email: F.Tramper@uva.nl

[2] Armagh Observatory, College Hill, BT61 9DG Armagh, Northern Ireland, UK

[3] Instituut voor Sterrenkunde, KU Leuven, Celestijnenlaan 200D, 3001 Leuven, Belgium

[4] Argelander Institut für Astronomie, University of Bonn, Auf dem Hügel 71, D-53121 Bonn, Germany

[5] ESA/Space Telescope Science Institute, 3700 San Martin Drive, Baltimore, MD 21218, USA

Abstract. The enigmatic oxygen sequence Wolf-Rayet (WO) stars represent a very late stage in massive star evolution, although their exact nature is still under debate. The spectra of most of the WO stars have never been analysed through detailed modelling with a non-local thermodynamic equilibrium expanding atmosphere code. Here we present preliminary results of the first homogeneous analysis of the (apparently) single WOs.

Keywords. stars: Wolf-Rayet, stars: fundamental parameters, stars: evolution, stars: individual (WR102, WR142, WR93b, BAT99-123, LH41-1042, DR1), stars: abundances

1. Introduction

The nature of the very rare WO stars is still under debate. Although it is clear that they are highly evolved massive stars, their evolutionary connection to the more common carbon sequence Wolf-Rayet (WC) stars remains unclear. To unravel their nature, we are undertaking a project targeting all the known WO stars. Here we present preliminary results of the first homogeneous analysis of the single WOs.

2. Observational sample and spectral classification

We have obtained the near-ultraviolet to near-infrared spectra of all known WO stars using the X-Shooter spectrograph on ESO's Very Large Telescope (except for the recently discovered WO star in the LMC; Massey *et al.* 2014). The spectra cover a wavelength range from 3 000 to 25 000 Å at a resolving power of $R \sim 8000$. The sample consists of three stars in the Milky Way (MW), two in the Large Magellanic Cloud (LMC), and one in IC 1613. The two WO stars in a binary system have also been observed, but are not analysed in this work. Figure 1 shows the X-Shooter spectra of all WO stars.

The spectral type of the single stars is determined using the classification criteria from Crowther *et al.* (1998). These are based on the equivalent width ratios of Ovi 3811-34 Å / Ov 5590 Å and Ovi 3811-24 Å/ Civ 5801-12 Å, as well as the full width at half maximum of Civ 5801-12 Å. The derived spectral types are given in Table 1.

3. Modelling and conclusions

We model the X-Shooter spectra using CMFGEN (Hillier & Miller 1998) following the method described in Tramper *et al.* (2013). While the observed strong Ovi 3811-34 Å

Table 1. Spectral types and preliminary parameters of the single WO stars.

ID	ID *(figure)*	SpT	log L ($L_\odot$)	T_* (kK)	X_C	X_O
WR102	WO-MW-1	WO2	5.45	210	0.62	0.25
WR142	WO-MW-2	WO2	5.63	200	0.54	0.21
WR93b	WO-MW-3	WO3	5.30	160	0.53	0.18
BAT99-123	WO-LMC-1	WO3	5.26	170	0.55	0.15
LH41-1042	WO-LMC-2	WO4	5.26	150	0.60	0.18
DR1	WO-IC1613-1	WO3	5.68	150	0.46	0.10

Notes: Results for DR1 from Tramper *et al.* (2013). Values for all other stars are preliminary and subject to change.

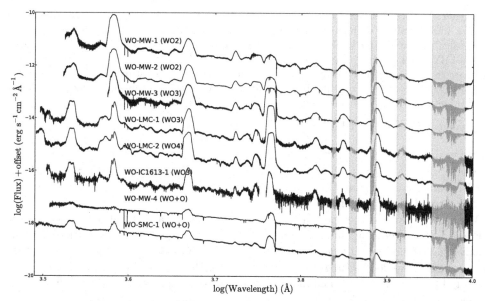

Figure 1. Dereddened, flux-calibrated spectra of the WO stars in the 3 000-10 000 Å wavelength range. The shaded areas indicate regions with telluric features. Labels are defined in Table 1.

emission is not (fully) reproduced, the rest of the spectrum is very well represented by our models. Preliminary parameters derived from the modelling are given in Table 1. Note that the models are currently still being refined, and these values are subject to small changes.

The surface abundances of the WO stars indicate that they have at least burned two-third of the helium in their core, and are expected to explode as type Ic supernovae within 10^5 years. Most evolved is WR102, which may have already exhausted the helium in its core and will likely end its life in less than 10^4 years. Compared to WC stars, the WO stars are hotter and show higher surface abundances of carbon and oxygen.

References

Crowther, P. A., De Marco, O., & Barlow, M. J. 1998, *MNRAS* 296, 367
Hillier, D. J. & Miller, D. L. 1998, *ApJ* 496, 407
Massey, P., Neugent, K. F., Morrell, N., & Hillier, D. J. 2014, *ApJ* 788, 83
Tramper, F., Gräfener, G., Hartoog, O. E., *et al.* 2013, *A&A* 559, A72

New windows on massive stars: asteroseismology, interferometry, and spectropolarimetry
Proceedings IAU Symposium No. 307, 2014
G. Meynet, C. Georgy, J. H. Groh & Ph. Stee, eds.

© International Astronomical Union 2015
doi:10.1017/S1743921314006607

Spectral analysis of LBV stars in M31: AF And and Var 15

A. F. Valeev[1,2], O. Sholukhova[1] and S. Fabrika[1,2]

[1] Special Astrophysical Observatory, Russia
email: azamat@sao.ru

[2] Kazan Federal University, Kremlevskaya 18, 420008 Kazan, Russia

Abstract. We study spectra of two bona fide LBV stars in M31: AF And and Var 15. The spectra were obtained with the 6-m telescope (Russia) from 2005 to 2012. The model spectra were calculated with the CMFGEN code. We have not found strong changes in the spectra of the LBV stars in that time interval, however a certain variability has been detected. We estimate the star and wind parameters, such as luminosity, temperature, raduis, mass loss rate, escape velocity, hydrogen content, and reddening. We study the stars on the Hertzsprung-Russell diagram and find their initial masses using evolutionary tracks by Meynet *et al.* (1994).

Keywords. stars: individual (AF And, Var 15)

The observations were obtained with the 6-m BTA telescope with SCORPIO and MPFS (Integral Field Unit, IFU) spectrographs (Afanasiev & Moiseev 2005). A summary of all observations is presented in Table 1.

We have performed the spectral analysis of two LBV stars in the Andromeda galaxy. AF And has shown no significant changes in its spectrum in 2005–2012. In the M31 Var15 star we have found small changes in the HeII λ4686 indicating a temperature difference between 2005 and 2012. Using the P Cyg profiles in HeI lines of Var15, we have estimated the wind terminal velocity. The model spectra were calculated with the CMFGEN code (Hillier & Miller 1998). In Fig. 2 we present the comparison of the model and observed spectra.

We estimate the star and wind parameters, such as luminosity, temperature, raduis, mass loss rate, escape velocity, hydrogen content, and reddening (Table 2).

The research is partly supported by RFBR grants N 13-02-00885, "Leading Scientific Schools of Russia" 2043.2014.2, grant of the President of RF MK-6686.2013.2, -1699.2014.2. S.F. acknowledges support of the Russian Government Program of Competitive Growth of Kazan Federal University.

Table 1. Summary of observations.

Object	Date	Exposure (sec)	Spectral range (Å)	Spectral resolution (Å)
AF And	2005.10.10	4500 (IFU)	4100–6900	7
	2012.10.14	2 × 900	3850–7200	11.3
	2012.10.15	2 × 900	3850–7200	5.4
Var 15	2005.01.15	3600 (IFU)	4100–6900	7
	2012.10.14	2 × 900	3850–7200	11.8
	2012.10.20	2 × 900	4100–5880	2.3

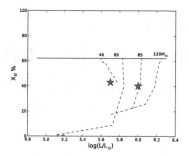

Figure 1. Mass estimates of AF And and Var15 (blue and red asterisks, respectively). The evolution tracks by Meynet *et al.* (1994) for M31 metallicity ($Z = 0.04$) are shown by dashed lines.

Table 2. Stellar parameters.

	AF And	Var 15
T_* [kK]	22.7±0.5	23.1±0.5
v_∞ [km s^{-1}]	$\lesssim$300	300±50
X_H	0.4±0.1	0.43±0.1
$\log(L/L_\odot)$	6.0±0.30	5.7±0.30
$\log(\dot{M}/\sqrt{f})$ [$M_\odot$ yr^{-1}]	-5.0±0.25	-4.7±0.25

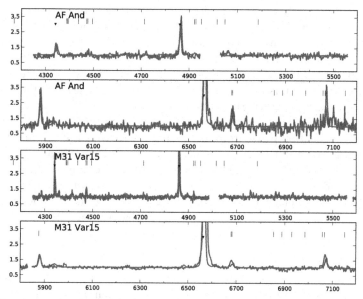

Figure 2. The observed spectra (blue) of AF And and Var15 are presented in comparison with CMFGEN models (red). The red part of Var15 spectrum was obtained with lower resolution. The H lines are marked with triangles and He I lines with vertical bars.

References

Afanasiev, V. L. & Moiseev, A. V. 2005, *Astronomy Letters* 31, 194

Hillier, D. J. & Miller, D. L. 1998, *ApJ* 496, 407

Meynet, G., Maeder, A., Schaller, G., Schaerer, D., & Charbonnel, C. 1994, *A&AS* 103, 97

New windows on massive stars: asteroseismology, interferometry, and spectropolarimetry
Proceedings IAU Symposium No. 307, 2014
G. Meynet, C. Georgy, J. H. Groh & Ph. Stee, eds.

© International Astronomical Union 2015
doi:10.1017/S1743921314006619

Variable C – "a typical" LBV in M 33?

Kerstin Weis[1], Roberta M. Humphreys[2], Birgitta Burggraf[1] and Dominik J. Bomans[1]

[1] Astronomical Institute of the Ruhr-University Bochum Bochum, Germany
email: kweis@astro.rub.de

[2] Minnesota Institute for Astrophysics, University of Minnesota, Minneapolis, USA

Abstract. One of the original Hubble-Sandage variables, Variable C in M33 is thought to be a very typical Luminous Blue Variable (LBV). An observational signature of LBVs is a variable brightness which is coupled to a change in spectral type. We compiled a 110 year long light curve of Var C and a set of spectra covering several decades. Analyzing both data sets, various astonishing changes of Var C, some very recent, emerged. Is Var C a typical or an atypical LBV?

Keywords. stars: variables: other, evolution, mass loss

1. Luminous Blue Variables

LBVs are characterized being luminous evolved massive stars with photometric and spectral variabilities on various timescales and magnitudes. Intrinsic to LBVs is the S Dor variability/cycle in which the spectral type changes from O-B to A-F and back within a few years. V magnitude and colors change, the star being faint in the hot and bright in the cool phase. More violent are the giant eruptions in which the brightness rises spontaneously by several magnitudes and larger amounts of mass are ejected. Giant eruptions have even been mistaken for supernovae (e.g. SN1954J). For further details see Humphreys & Davidson (1994). High mass loss by winds and/or giant eruption forms small ($< 4\,\mathrm{pc}$) circumstellar LBV nebulae, a large fraction ($> 50\%$) being bipolar (Weis 2011). LBVs is a transition phase from main sequence to Wolf-Rayet state. Observations and theoretical stellar evolution models with rotation lowered the mass limit for LBVs to $M_{\mathrm{ini}} \gtrsim 22\,M_\odot$ (Meynet & Maeder 2005). Therefore LBVs are characterized by variability, high mass loss and possibly giant eruptions. However at least temporarily LBVs can appear simply as 'well behaved' normal supergiants! Note that **no** unique classification scheme exists to pinpoint a LBV with just a single observation!

2. Var C in M 33

Compiling several years of photometric and spectroscopic observations with data from the literature, archives and our own observations a lightcurve covering more than 110 years was generated. For several epochs spectral and photometric observations were synchronized. The lightcurve (Fig. 1) reveals variations on various time scales together with a secular brightening. As typical for LBVs the variability appears more irregular, still two prominent maxima (1946 & 1986) are present. Checking for periodicities, Fourier transformation analyses were done, yielding a long term(semi-)periodicity of 40.7 years. The last major maximum was 1986, the next is expected for 2027. A detailed analysis of the lightcurve and spectra of Var C was submitted to A&A by Burggraf *et al.*

What's it's doing now? The 2010 spectrum was dominated by absorption lines (plus some hydrogen lines with P Cygni profiles), and compatible with a B1 to B2 supergiant.

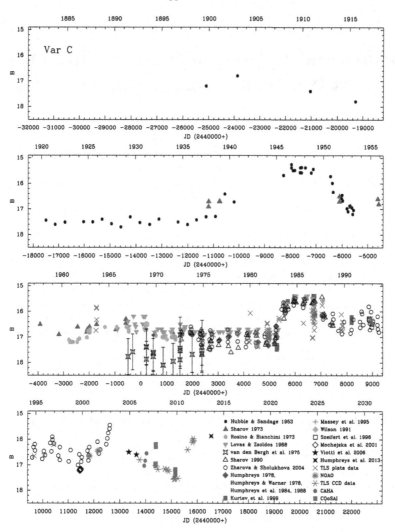

Figure 1. Var C lightcurve in B, starting in 1899

After that Var C started to brighten and a spectrum taken during maximum light in 2013 showed it changed into a late A-type supergiant (confirmed in LBT observations January 2014). It's going through a S Dor cycle! In the last hot phase the stellar wind velocities of Var C were ∼ 30% lower as expected. This manifests either a criteria for the LBV class or that Var C has shed already a large amount of mass. Fur further details on this topic, the data sets and analyses see (Humphreys *et al.* 2013, 2014a,b) and references therein.

References

Humphreys, R. M. & Davidson, K. 1994, *PASP* 106, 1025
Humphreys, R. M., Davidson, K., Gordon, M. S., *et al.* 2014a, *ApJ (Letters)* 782, L21
Humphreys, R. M., Davidson, K., Grammer, S., *et al.* 2013, *ApJ* 773, 46
Humphreys, R. M., Weis, K., Davidson, K., Bomans, D. J., & Burggraf, B. 2014b, *ApJ* 790, 48
Meynet, G. & Maeder, A. 2005, *A&A* 429, 581
Weis, K. 2011, in C. Neiner, G. Wade, G. Meynet, & G. Peters (eds.), *IAU Symposium*, Vol. 272 of *IAU Symposium*, pp. 372–377

New windows on massive stars: asteroseismology, interferometry, and spectropolarimetry
Proceedings IAU Symposium No. 307, 2014
G. Meynet, C. Georgy, J. H. Groh & Ph. Stee, eds.

© International Astronomical Union 2015
doi:10.1017/S1743921314006620

Variational approach for rotating-stellar evolution in Lagrange scheme

Nobutoshi Yasutake[1] and Shoichi Yamada[2]

[1]Physics Department, Chiba Institute of Technology,
Shibazono 2-1-1, Narashino, Chiba, 275-0023, Japan
email: nobutoshi.yasutake@p.chibakoudai.jp

[2]Advanced Research Institute for Science and Engineering, Waseda University,
Okubo 3-4-1, Shinjuku, Tokyo, 169-8555, Japan

Abstract. We have developed an entirely new formulation to obtain self-gravitating, axisymmetric configurations in permanent rotation. It is based on the Lagrangian variational principle and, as a consequence, will allow us to apply it to stellar evolution calculations rather easily. We adopt a Monte Carlo technique, which is analogous to those employed in other fields, e.g. nuclear physics, in minimizing the energy functional. We also present the analogies between the study on rotating stellar configurations and the one on deformed nuclei. Possible applications are not limited to main sequence stars but will be extended to e.g. compact stars, proto-stars and planets. We believe that our formulation will be a major break-through then.

Keywords. stars: rotation— stars: evolution — stars: protostars

1. Introduction-Edges and nodes in Lagrange variational principle

Studies on evolutions of spherically symmetric stars have been advanced exponentially over the years by many researchers (Maeder & Meynet 2000; Woosley *et al.* 2002) after the innovation of Henyey method (Henyey *et al.* 1964). This method is only applicable to spherical (non rotating) topics, and most of the stellar evolution calculations with this method are based on the Lagrange scheme, which is very easy to be applied to stellar evolution calculations. While, in the last ten years, rotation has become an essential part of stellar models comparing with observational results such as Be-stars, there is no appropriate Lagrange scheme for fully multi-dimensional calculations unfortunately.

In this study, we introduce a novel formulation in Lagrange method made of nodes and edges, in which the Lagrange variables, such as mass, angular momentum, entropy, and chemical fractions, are conserved on each node. Our scheme is also based on the variational principle, hence we can get optimal stellar structures finding the minimal energy (N. Yasutake, T. Noda, H. Sotani, T. Maruyama, T. Tatsumi 2013). Although detailed validations will be presented in the sequel, we show an example using our technique.

2. Deformed-star/pasta (deformed nuclear structures) duality

We adopt a Monte Carlo technique in variational principle, which is analogous to those employed in other fields, e.g. nuclear physics, in minimizing the energy functional, which is evaluated on a triangulated mesh. We can, actually, find out many analogies between deformed-stars with rotation and deformed nuclear structures; e.g. while deformed nuclear structures, known as *"pasta structures"*, exist under the balance between Coulomb

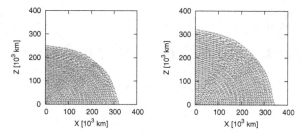

Figure 1. stellar structures in equilibrium shown by edges and nodes. Left panel is the barotropic case, and right one the baroclinic case, where entropy is added to the pole area.

interaction and surface tension (N. Yasutake, T. Noda, H. Sotani, T. Maruyama, T. Tatsumi 2013), the deformed rotating-stars do between gravitation and rotation.

3. Applicational result and discussion

In Fig. 1, we show an examples of rotating structures for stellar evolutions. We assume that the stars are composed of hydrogen alone and employ an ideal gas EOS. Left panel shows the barotropic case, where the stellar mass is $0.6M_\odot$, and the entropy is set as $15.0\,k_B$ on each node. Here k_B is the Boltzmann's constant. We adopt a spherically symmetric configuration as an initial-guess structure, and add to it the angular momentum, which satisfies the following law on the angular velocity: $\Omega = \Omega_0 X_0^2/(X^2 + X_0^2)$. Here X_0 and Ω_0 are model parameters and the former is set to the radius of the spherical star whereas the latter is $\Omega_0 = 10^{-6}\pi \cdot 5000\,\mathrm{rad\,s^{-1}}$, which is roughly 5000 times the solar angular velocity. Note that the resultant configuration of the angular velocity is given when the initial structure changes to the optimal structure as shown in the left panel.

Right panel shows the baroclinic case, where the entropy has been artificially changed from the barotropic case according to the relation: $K_i = K_0(1 + \theta_i/\pi)$ where K_0 is the constant of the ideal-gas EOS $P = K_0\rho^{5/3}$, and θ_i is the latitude of node i.

Note that we can follow up the trajectories of nodes completely; we can see from where to where each node moves in the change of entropy distribution. Although I adopt a simple entropy change here, we can apply realistic nuclear-reactions instead of it. Hence, we can easily apply our technique to stellar evolution calculations. This is one of the strong points on our method.

Another point is that our method is useful to check the instability of stellar structures since our method is based on the variational principle, which is useful for evaluations of mixing regions. We will show it in the nearly future.

We have just assumed a main sequence star here, but our method is applicable to e.g. compact stars, proto-stars and planets. We believe that our formulation will be a major break-through then. We have already checked that our result is consistent with the one by Hachisu method (Hachisu 1986).

References

Hachisu, I. 1986, *ApJS* 61, 479

Henyey, L. G., Forbes, J. E., & Gould, N. L. 1964, *ApJ* 139, 306

Maeder, A. & Meynet, G. 2000, *ARA&A* 38, 143

N. Yasutake, T. Noda, H. Sotani, T. Maruyama, & T. Tatsumi 2013, *Recent Advances in Quarks Research*, Chapt. Thermodynamical description of hadron-quark phase transition and its implications on compact-star phenomena, Nova Publishers

Woosley, S. E., Heger, A., & Weaver, T. A. 2002, *Reviews of Modern Physics* 74, 1015

New windows on massive stars: asteroseismology, interferometry, and spectropolarimetry
Proceedings IAU Symposium No. 307, 2014
G. Meynet, C. Georgy, J. H. Groh & Ph. Stee, eds.

© International Astronomical Union 2015
doi:10.1017/S1743921314006632

Wolf-Rayet stars from Very Massive Stars

Norhasliza Yusof

Department of Physics, University of Malaya, 50603 Kuala Lumpur, Malaysia
email: `norhaslizay@um.edu.my`

Abstract. Many studies focused on very massive stars (VMS) within the framework of Pop. III stars, because this is where they were thought to be abundant. In this work, we focus on the evolution of VMS in the local universe following the discovery of VMS in the R136 cluster in the Large Magellanic Cloud (LMC). We computed grids of VMS evolutionary tracks in the range $120 - 500\,M_\odot$ with solar, LMC and Small Magellanic Cloud metallicities. All models end their lives as Wolf-Rayet (WR) stars of the WC (or WO) type. We discuss the evolution and fate of VMS around solar metallicity with particular focus on the WR phase. For example, we show that a distinctive feature that may be used to disentangle Wolf-Rayet stars originating from VMS from those originating from lower initial masses is the enhanced abundances of Ne and Mg at the surface of WC stars.

Keywords. stars: evolution, rotation, Wolf-Rayet

1. Introduction

In this work we discuss very massive stars (VMS) in the local Universe. Although there is high interest within Pop. III framework (Heger & Woosley 2002; Yoon *et al.* 2012), evidence from observations of main sequence stars (Crowther *et al.* 2010) and superluminous supernovae (Gal-Yam *et al.* 2009) suggest that VMS exist in the local Universe. Our main interest is the evolution of VMS, in particular during their Wolf-Rayet (WR) phases before the end of their lives. Here we discuss the WR stars originating from VMS at solar metallicity ($Z = 0.014$). As we show below, Wolf-Rayet stars from VMS have distinctive features that can be used to distinguish them from those evolving from less massive stars (less than $100\,M_\odot$).

2. Wolf Rayet stars from Very Massive Stars

We have computed the evolution of very massive stars from $120\,M_\odot$ to $500\,M_\odot$ for both rotating and non-rotating models (Yusof *et al.* 2013). The models extend at least until the end of He-burning and, in most cases, until the start of O-burning. All the computed models evolves into WR stars, which is assumed here to happen when the surface mass fraction of hydrogen is less than 0.3 and when the effective temperature is greater then 10^4 K. In this present work, VMS go through the following phases; O–late WN– early WN –WC/WO. This is similar to the M_{OWR} evolution scenario reported in Meynet & Maeder (2005).

In Fig. 1 we present the evolution of the surface abundances of 60 and $150\,M_\odot$ models as a function of total mass. From the figure, one can observed that mass loss effects and internal mixing change the models surface composition. Qualitatively, there are very small differences between the 60 and $150\,M_\odot$ models. In the $150\,M_\odot$ model, the star has larger cores, and the transition to the various WR stages occurs at higher total masses compared to the $60\,M_\odot$ model. This confirms the general idea that a more massive WR star originates from a more massive O-type star. Figure 1 also shows that all surface

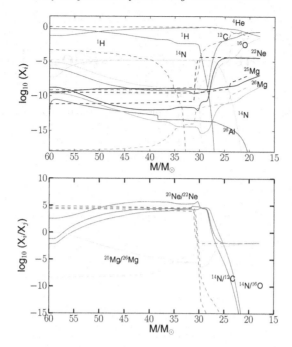

Figure 1. *Top:* Surface abundances of 150 $M_\odot$ (dashed) and 60 $M_\odot$ models at solar metallicity (dotted) as a function of the total mass. *Bottom:* Idem, but for surface abundance ratios.

abundances and abundance ratios are very similar for a given WR phase. It is therefore not possible to distinguish a WR originating from a VMS based on its surface chemical composition.

Although it is difficult to distinguish WRs originating from massive stars to those from VMS, we can examine the structure at the final timestep of these models. From our analysis, we find that VMS produce higher ^{20}Ne abundance. This is due to higher central temperatures during the core He-burning phase, when the reaction ^{16}O$(\alpha, \gamma)^{20}$Ne is more active and, thus, more ^{20}Ne is produced. Producing large amounts of ^{20}Ne implies also enhanced ^{24}Mg. We obtain layers rich in ^{20}Ne and ^{24}Mg in the 500 $M_\odot$ model. This implies that strong overabundances of these two isotopes at the surface of WC stars can be taken as a signature for an initially very massive stars as the progenitor of that WC star.

Acknowledgements

N. Yusof acknowledge support from the Fundamental Research Grant Scheme (FRGS/FP003-2013A) under Ministry of Education of Malaysia and IAU for travel grant.

References

Crowther, P. A., Schnurr, O., Hirschi, R., *et al.* 2010, *MNRAS* 408, 731
Gal-Yam, A., Mazzali, P., & Ofek, E. O. 2009, *Nature* 462, 624
Heger, A. & Woosley, S. E. 2002, *ApJ* 567, 532
Meynet, G. & Maeder, A. 2005, *A&A* 429, 581
Yoon, S.-C., Dierks, A., & Langer, N. 2012, *A&A* 542, A113
Yusof, N., Hirschi, R., Meynet, G., *et al.* 2013, *MNRAS* 433, 1114

New windows on massive stars: asteroseismology, interferometry, and spectropolarimetry
Proceedings IAU Symposium No. 307, 2014
G. Meynet, C. Georgy, J. H. Groh & Ph. Stee, eds.

© International Astronomical Union 2015
doi:10.1017/S1743921314006644

Massive Star Asteroseismology in Action

Conny Aerts[1,2]

[1]Institute of Astronomy, KU Leuven, Celestijnenlaan 200 D, 3001 Leuven, Belgium
email: `Conny.Aerts@ster.kuleuven.be`

[2]Department of Astrophysics/IMAPP, Radboud University Nijmegen, 6500 GL Nijmegen, the Netherlands

Abstract. After highlighting the principle and power of asteroseismology for stellar physics, we briefly emphasize some recent progress in this research for various types of stars. We give an overview of high-precision high duty-cycle space photometry of OB-type stars. Further, we update the overview of seismic estimates of stellar parameters of OB dwarfs, with specific emphasis on convective core overshoot. We discuss connections between pulsational, rotational, and magnetic variability of massive stars and end with future prospects for asteroseismology of evolved OB stars.

Keywords. asteroseismology, stars: oscillations (including pulsations), stars: interiors, stars: evolution, stars: rotation, methods: data analysis, methods: statistical, magnetic fields, waves, turbulence

1. Progress in Asteroseismology

Asteroseismology is a fairly new and powerful way of studying stellar physics, including important excursions to exoplanetary science and the study of galactic structure and evolution. Even in the context of stellar interiors alone, there are many aspects to this research field (e.g., Aerts *et al.* 2010a, for an extensive monograph). A single snapshot picture capturing the basic idea of seismic modelling is provided in Fig. 1, where X, Z, M, τ are the initial hydrogen fraction and metallicity, the mass, and the age, respectively. In this sketch, we assumed the model computations to be based on the simplistic (but practical!) one-dimensional time-independent mixing-length theory of convection described by the two free parameters $\alpha_{\rm MLT}$ and $\alpha_{\rm ov}$, which are the mixing length and convective overshoot parameters, both expressed in units of the local pressure scale height. For core-hydrogen burning massive stars, the choice of $\alpha_{\rm MLT}$ is of limited importance as long as it is within reasonable limits (e.g., between 1.5 and 2.0), while the poorly known amount of convective overshooting expressed by $\alpha_{\rm ov}$ plays a pivotal role in their structure, in all of the evolutionary phases. This $\alpha_{\rm ov}$ is thus a key parameter to estimate and asteroseismology is a relatively new way to do so, as we will discuss further on.

With the high-precision space photometry from the CoRoT and *Kepler* missions at hand for thousands of FGK-type stars, priority was given to the determination of basic seismic observables and stellar parameters of such stars. The most popular observables are the so-called large and small frequency separations, which are the inverse of twice the sound travel time between the centre and the surface of the star and a measure for the sound-speed gradient in the stellar interior, respectively. The frequency of maximum power is also an important basic observable. Stochastically-excited modes are easy to identify as they form specific ridges in so-called échelle diagrams, in which the detected frequencies are reduced modulo the large frequency separation. Moreover, the seismic diagnostics are easily interpretable in terms of the fundamental parameters of the stars by means of scaling relations (e.g., Kallinger *et al.* 2010; Huber *et al.* 2011; Hekker *et al.*

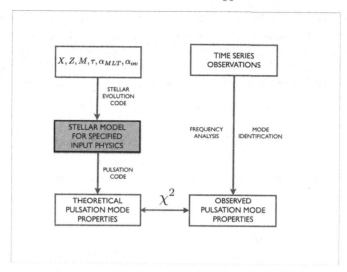

Figure 1. The principle of asteroseismic modelling in a snapshot, where the dependency on the chosen input physics of the equilibrium models is highlighted in shaded grey. Figure courtesy of Dr. Katrijn Clémer.

2011). The scheme represented in Fig. 1 hence delivers high-precision stellar parameters that are typically an order of magnitude more precise than those derived from classical ground-based snapshot spectroscopy or multicolour photometry (e.g., Chaplin *et al.* 2014), even taking into account uncertainties in the input physics. The seismic parameters derived from scaling relations are particularly welcomed for host stars of exoplanets (e.g., Gilliland *et al.* 2011; Huber *et al.* 2013; Chaplin *et al.* 2013; Van Eylen *et al.* 2014; Lebreton & Goupil 2014) and for galactic clusters and population studies (e.g., Corsaro *et al.* 2012; Miglio *et al.* 2013; Stello *et al.* 2013; Casagrande *et al.* 2014).

The discovery of dipole mixed modes in low-mass evolved stars (Beck *et al.* 2011) allowed to go beyond the use of simple scaling relations and led to the derivation of their nuclear burning phase and hence evolutionary stage (Bedding *et al.* 2011). Such information is not accessible from classical data because core-helium burning and hydrogen-shell burning red giants have the same surface properties. Moreover, after two years of monitoring with the *Kepler* satellite, the detected rotational splitting of dipole mixed modes led to the derivation of their interior rotational properties, with core rotation typically only a factor 5 to 20 faster than envelope rotation (Beck *et al.* 2012; Mosser *et al.* 2012; Deheuvels *et al.* 2014), pointing to major shortcomings in our understanding of the angular momentum distribution inside stars (Eggenberger, these proceedings).

Many remarkable seismic studies of other types of stars are not mentioned here, due to space constraints and for that reason, we also limit the rest of the paper to massive stars. Their variability is far more diverse than that of low-mass stars due to phenomena like fast rotation, mass loss, close binarity, magnetism, etc. Hence, the case of massive OB star asteroseismology is challenging. A specific challenge is connected with the nature of the oscillations, i.e., the majority of the detected oscillation modes in massive stars are self-driven by a heat mechanism, which operates along with a yet unknown mode selection mechanism (e.g., Aerts *et al.* 2010a). This creates sparse low-order mode frequency spectra roughly in the range of 30 to 200 μHz and/or dense high-order gravity-mode spectra at low frequencies, typically below 20 μHz. Moreover, the rotation of most massive stars is such that their rotationally split pulsation mode multiplets get merged in frequency spectra, preventing pattern recognition as for high-order pressure modes in low-mass stars from échelle diagrams. In addition to these inherent difficulties based on the physics of the pulsations of such stars, OB-type targets have been avoided as much as possible in the fields of the CoRoT and *Kepler* exoplanet programmes, to maximise the efficiency of the search for transiting planets. Despite this limitation, progress in the asteroseismology of OB stars the past decade is large and is our point of focus in the remainder of this paper.

2. Asteroseismic Data of OB Stars in the Space Age

OB stars have been part of the space photometry revolution since a decade now, with observations from the MOST, CoRoT, and *Kepler* satellites. We review the observational studies based on space photometry and refer to the mentioned papers for the exquisite light curves and the accompanying Fourier spectra of the oscillation frequencies.

2.1. *Core-Hydrogen Burning Massive Stars*

MOST was a pioneering mission for massive stars. Although it could only monitor stars for typically up to 6 weeks, resulting in limited frequency precision, and with relatively low brightness precision (typically 0.1 mmag), it allowed to detect at least twice as many modes than known from the ground for selected β Cep stars (Aerts *et al.* 2006b; Handler *et al.* 2009), to detect pulsations in Be stars (Saio *et al.* 2007), and to discover new Slowly Pulsating B stars (Aerts *et al.* 2006a; Cameron *et al.* 2008; Gruber *et al.* 2012).

The asteroseismology programme of the CoRoT mission consisted of long runs of five months (frequency precision $\sim 0.1\,\mu$Hz) as well as short runs of a few weeks, dedicated to bright stars (visual magnitude between 6 and 9) resulting in amplitude precision of roughly 10 μmag. This programme included detailed studies of carefully selected B stars and showed that the majority of them is pulsating in numerous modes, some of which with clear period spacings (Degroote *et al.* 2010a; Pápics *et al.* 2012) while others have less structured dense frequency spectra (Degroote *et al.* 2009b; Thoul *et al.* 2013). Some of the B stars turned out to be rotational variables without pulsations (Degroote *et al.* 2011; Pápics *et al.* 2011). One short run of CoRoT was dedicated to the monitoring of O stars during three weeks. These data showed diversity in the variability patterns with one β Cep-type pulsator (Briquet *et al.* 2011), one primary of a long-period binary with stochastic modes (Degroote *et al.* 2010b), three stars with power excess at low frequency of unknown origin (Blomme *et al.* 2011) and one complex O-type binary, also known as Plaskett's star (Mahy *et al.* 2011). In addition, the CoRoT asteroseismology programme contained a number of Be stars (Neiner *et al.* 2009; Gutiérrez-Soto *et al.* 2009; Diago *et al.* 2009; Huat *et al.* 2009; Desmet *et al.* 2010; Emilio *et al.* 2010; Neiner *et al.* 2012a). These Be stars also revealed quite diverse variability patterns, including an outburst measured in real time for the B0.5IVe star HD 49330 due to the beating of non-radial pulsation modes, stochastically excited gravito-inertial modes, rotational modulation, and accretion phenomena in a close binary.

Numerous B dwarfs were also found in the CoRoT exoplanet programme (Degroote *et al.* 2009a), after variability classification by means of multivariate Gaussian mixtures (Debosscher *et al.* 2009; Sarro *et al.* 2009) supported by ground-based follow-up spectroscopy (Sarro *et al.* 2013). Similar studies have been done for the *Kepler* field (Debosscher *et al.* 2011), along with manual searches for B pulsators (Balona *et al.* 2011), resulting in the detection of variability with similar precision in amplitude than with CoRoT but with a frequency resolution $\sim 0.01\,\mu$Hz following the four years of monitoring. It is thanks to this excellent frequency precision that the detection of rotational splitting in slow rotators became a reality (Pápics *et al.* 2014), with major progress and future potential in the seismic modelling of massive stars, as outlined below for two recently discovered pulsators.

Despite these observational studies for hundreds of OB dwarfs, only 16 of them could be modelled according to the scheme in Fig. 1 was achieved, *the* major obstacle being the lack of identification of the pulsation modes. However, several promising new cases based on four years of *Kepler* data are currently emerging. As a side remark, the detection of the numerous pulsation modes at low frequency in so many OB dwarfs, which was not possible from ground-based data, prompted the need for additional excitation mechanisms because the classical heat mechanism is not able to explain all the detected oscillations. One way to solve this is to increase the opacity, either globally in the star or in the excitation layer (Salmon *et al.* 2012; Walczak *et al.* 2013). A promising new excitation mechanism concerns stochastic gravity modes triggered by core convection (Belkacem *et al.* 2010; Samadi *et al.* 2010; Saio 2011; Shiode *et al.* 2013), although it remains unclear what their amplitudes are at the stellar surface.

2.2. *Evolved Massive Stars*

The case of asteroseismology of evolved OB stars is far more challenging, but accordingly more interesting, with large future potential. The challenge is not only due to the longer oscillation periods (which can reach up to several months) but also because the oscillations are influenced by various poorly understood physical processes which take place in the outer atmosphere and wind of such objects, affecting the boundary conditions that are of importance for the theoretical predictions needed to perform the scheme in Fig. 1.

Seismic data of evolved OB stars are scarce, even in the space age of asteroseismology. Also on the front of supergiant photometric data from space, MOST delivered the first interesting case with a 37 d light curve of the B2Ib/II star HD 163899. This revealed the simultaneous excitation of pressure and gravity modes with frequencies below 30 μHz and amplitudes of a few mmag (Saio *et al.* 2006). A MOST campaign of 28 d dedicated to the B8Ia supergiant Rigel was too short to uncover all the gravity modes found from six years of spectroscopic monitoring with frequencies as low as 0.2 μHz and up to 12 μHz (Moravveji *et al.* 2012). Of a completely different nature is the 137 d CoRoT light curve of the B6I star HD 50064, which, combined with spectroscopy, led to the conclusion that periodic mass-loss episodes due to a variable large-amplitude mode seem to occur (Aerts *et al.* 2010b). Finally, a 26 d CoRoT light curve of the B8Ib star HD 46769 led to the detection of low-amplitude rotational modulation with a frequency of 2.4 μHz and an amplitude of some 100 μmag rather than pulsations (Aerts *et al.* 2013).

The only evolved massive star observed by the *Kepler* mission is the bright eclipsing binary V380 Cyg, which required a customized mask (Tkachenko *et al.* 2012). Its *Kepler* light curve, along with extensive time-resolved spectroscopy led to the detection of rotational modulation along with low-amplitude stochastic variability at low frequency in the primary (Tkachenko *et al.* 2014). This variability is of similar nature than the patterns detected in the three hottest CoRoT O dwarfs (Blomme *et al.* 2011) and is compatible with theoretical predictions for stochastic gravity waves excited by the convective core.

The detailed theoretical interpretation of the periodic oscillations in supergiants remains to be done. While Godart *et al.* (2009) showed that mass loss and large overshooting reduce the extent of an otherwise occurring intermediate convective zone, preventing the excitation of gravity modes due to a heat mechanism acting in the metal opacity bump, Daszyńska-Daszkiewicz *et al.* (2013) do find such modes irrespective of an intermediate convection zone or not. Theoretical work on the improvement of stellar structure models of evolved massive stars will benefit from the seismic and spectroscopic data, following the scheme in Fig. 1.

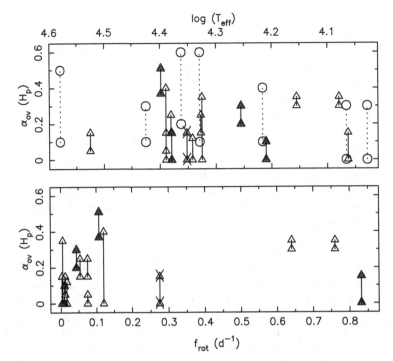

Figure 2. Core overshoot parameters from asteroseismology of OB stars as a function of the rotational frequency (lower panel) and effective temperature (upper panel), where single stars are indicated by open triangles and spectroscopic binaries by filled triangles; the magnetic pulsator is indicated with an additional cross. The estimates for the primaries of eclipsing binaries (Claret 2007) are indicated in the upper panel by circles connected with dotted lines.

3. Seismic Modelling of OB Dwarfs

Unlike solar-like stars and red giants, where ensemble asteroseismology can be achieved from scaling relations, the seismic modelling of massive stars requires a star-by-star treatment of the scheme in Fig. 1 and is thus immensely work-intensive. Aerts (2013) provided a compilation of the seismically derived stellar parameters according to Fig. 1, including the core overshoot value assuming the Schwarzschild criterion of convection and a fully mixed overshoot region, for 8 single and 3 binary non-emission OB dwarfs, among which one magnetic pulsator. The majority of these studies was based on multisite campaigns. In addition, seismic modelling was achieved for the late Be stars HD 181231 and HD 175869 (Neiner *et al.* 2012b). New studies since then were done for the β Cep stars γ Peg (Walczak *et al.* 2013) and σ Sco (Tkachenko *et al.* 2014), as well as for the B8.3V star KIC 10526294 (Pápics *et al.* 2014). The latter star is a young slowly rotating ($P_{rot} = 190$ d) B star with 19 rotationally split and quasi-equally spaced dipole gravity modes detected in the 4-year *Kepler* light curve. Frequency inversion of its triplets leads to counter-rotation in its envelope (Triana *et al.*, submitted). This star, along with the A-type pulsator KIC 11145123 whose envelope rotates slightly faster than its core, as revealed by rotationally split triplets and quintuplets (Kurtz *et al.* 2014), will undoubtedly give rise to future improvement of the input physics for stars of intermediate mass, because the current models cannot explain the observed pulsational properties.

A major finding from asteroseismology and of relevance for stellar evolution of massive stars is the need for extra mixing, enlarging the core masses of OB dwarfs, already from their early life. This was also hinted at long ago from modelling of the turn-off point of clusters, but this method has limited predictive power due to observational biases. Isochrone fitting of eclipsing binaries is a more powerful method to pinpoint α_{ov} (Claret 2007; Torres *et al.* 2014) independently from asteroseismology. In Fig. 2 we provide a compilation of the α_{ov} values derived from asteroseismology for 16 OB dwarfs and from isochrone fitting for 7 unevolved eclipsing binaries. Only 9 of the 23 stars are compatible with the absence of core overshoot. Other than that, no obvious relation emerges. The same conclusion is found when comparing α_{ov} with the mass.

4. Spin-offs of Massive Star Asteroseismology

4.1. *Macroturbulence*

The wings of the profiles of metal lines of a large fraction of OB stars cannot be explained by 1-dimensional atmosphere models, unless a macroturbulent velocity field is added (Símon-Díaz, these proceedings). Independently of whether one uses an isotropic or a radial-tangential Gaussian (Gray 2005) to describe this macroturbulence, supersonic speeds are often needed to bring the data in agreement with the model predictions, particularly for O stars and supergiants.

Aerts *et al.* (2009) suggested that the collective pulsational velocity broadening due to gravity modes could be a viable physical explanation for macroturbulence in massive stars. However, while B dwarfs are known to undergo pulsations with accompanying periodic line-profile variations (e.g., De Cat & Aerts 2002; Aerts & De Cat 2003), pulsations with the required frequencies and velocity amplitudes have yet to be firmly established for large samples of O stars and B supergiants. Indeed, the samples of OB stars observed by Simón-Díaz & Herrero (2014) and Markova *et al.* (2014) reveal a diversity of time-variability in terms of line broadening, with the dwarfs having clear line-profile variations but the evolved stars not necessarily so while their line wings are strongly broadened. It therefore seems that convective velocities in the outer envelope are a more plausible explanation than pulsations for evolved stars (Símon-Díaz, these proceedings).

4.2. *Multivariate Statistical Analysis of Oscillations, Rotation, and Magnetic Field*

In the spirit of this symposium of considering an integrated approach, Aerts *et al.* (2014) made a multivariate study by combining spectropolarimetric, asteroseismic and spectroscopic observables derived from high-precision data for 64 galactic dwarfs. This 10-dimensional data set is complete in $\log T_{\rm eff}$, $\log g$, and $v \sin i$ and is more than 94% complete in the rotational frequency $f_{\rm rot}$ and the magnetic field strength. An estimate of the nitrogen abundance is available for 59% of the targets while 32% of the stars have gravity-mode oscillation frequencies and 32% have pressure-mode oscillation frequencies. The stars without oscillations all have tight upper limits on the amplitudes.

One of the major aims of this study was to search for correlations between the nitrogen abundance and the other nine observables, from multivariate data analysis. This was achieved by applying a statistical technique called *multiple imputation* for the missing values, followed by linear regression from both backward and forward selection.

No significant correlation was found between the rotational observables $v \sin i$ or $f_{\rm rot}$ and the nitrogen abundance. The latter did correlate with $\log T_{\rm eff}$ and with the dominant acoustic oscillation frequency of the stars. Also a correlation between the dominant gravity-mode frequency and the magnetic field strength was found, but none of these two observables beared any relation with the nitrogen abundance.

While the sample may be prone to biases in the spectroscopic observables (cf. Przybilla, these proceedings), the study showed that rotation cannot be the only physical process to cause mixing in stellar interiors of galactic dwarfs and that magnetic fields are not a good alternative explanation. The role of other phenomena causing chemical mixing, such as oscillation modes or internal gravity waves (Mathis, these proceedings) could be a viable alternative or additional explanation.

5. Conclusions and Future Prospects

The past five years, photometric asteroseismic data improved in precision from mmag to μmag and in duty cycle from tens of percents to more than 90%. Moreover, data became available for thousands of stars in almost all evolutionary stages, instead of being limited to a handful of solar-like stars, young B stars, subdwarfs and white dwarfs. As a consequence, asteroseismology opened a new window for stellar physics in such a way that stellar structure and evolution studies are currently observationally driven.

The discovery of gravity-mode oscillations in thousands of CoRoT and *Kepler* targets allows to probe the physics in and near stellar cores in both red giants (Eggenberger, these proceedings) and in massive stars. A major finding is that the internal stellar rotation, and along with it the angular momentum distribution inside stars, is not at all what current theories predict it should be. While the deviation between theory and observations in terms of the interior rotation profile in the A-type star KIC 11145123 and in the B-type star KIC 10526294 remains modest and can qualitatively be explained by the act of internal gravity waves (Mathis, these proceedings), the discrepancy for red giants is two orders of magnitude and requires at least one hitherto omitted strong coupling mechanism between the stellar core and the envelope (see also Maeder, these proceedings).

While asteroseismology of massive stars has been in action since more than a decade now, after the first seismic probing of the core overshooting (Aerts *et al.* 2003), the number of stars with detailed tuning of the internal physics is far too low and, moreover, restricted to dwarfs. This is first of all due to the limited number of such stars in the fields of CoRoT and *Kepler* (some tens compared to many thousands of giants), but also has to do with the lack of mode identification for the majority of OB-type stars observed in white-light space photometry. The problem of mode identification can only be overcome from extensive data sets of multicolour photometry or high-resolution high-precision spectroscopy of bright stars (Aerts *et al.* 2010b, Chapter 6), from the detection of period spacings of high-order gravity modes, (e.g., Degroote *et al.* 2010b; Pápics *et al.* 2012), from the occurrence of rotationally split multiplets (e.g., Aerts *et al.* 2003; Pamyatnykh *et al.* 2004; Briquet *et al.* 2007), or, ideally, a combination of the latter two cases (Pápics *et al.* 2014). The detection of rotationally split multiplets of high-order gravity modes was only achieved for two dwarfs with a convective core so far, and only after analysing four years of *Kepler* data (Kurtz *et al.* 2014; Pápics *et al.* 2014). These two stars will certainly be modelled in more detail in the coming months, with attention to improvement of the input physics of models, because the current theory cannot explain the details of the observed seismic behaviour (Moravveji, these proceedings).

A new opportunity in the coming months and years, are several hundreds of massive OB stars to be observed by the *Kepler* 2-wheel mission, a.k.a. K2 (Howell *et al.* 2014). It remains to be seen how many of those OB-type targets turn out to be pulsators with rotational splitting, given the rather limited time base of three months of continuous monitoring, but even a low percentage of the hundreds of submitted OB stars will open new avenues for massive star seismology including numerous evolved OB stars.

Another near-future avenue for improved seismic modelling is the addition of an accurate distance, and by implication a model-independent luminosity and radius estimate, from the ESA cornerstone mission Gaia (launched 19 December 2013 and currently in its commissioning phase). A direct accurate angular diameter measurement for bright stars from interferometry (van Belle, these proceedings), combined with an accurate parallax determination, would have similar capacity. The combination of asteroseismic data and a model-independent radius estimate implies a serious reduction in the uncertainties of the physical quantities of the stars from seismic inference, as outlined in detail by Cunha *et al.* (2007) and Huber *et al.* (2012) for stars with stochastically-excited oscillations. Given the current lack of accurate distances for massive stars, and of good calibrators for a high-precision interferometric radius determination of pulsating OB stars, progress in this area is to be expected in the next years.

Finally, the recently approved ESA M3 mission PLATO (Rauer *et al.* 2014), to be launched in 2024, will measure numerous OB-type stars in its very wide Field-of-View ($2250\,\mathrm{deg}^2$). This will include high-cadence (2.5 s) two-colour measurements of stars with visual brightness between 4 and 8 to be obtained by its two bright-star telescopes with a 95% duty cycle for a minimum of 2 years and, possibly, fainter OB stars in white light with a cadence of 32 s observable with all of the 32 normal telescopes, during the step-and-stare phase of the mission.

References

Aerts, C. 2013, in *EAS Publications Series*, Vol. 64, pp 323–330

Aerts, C., Christensen-Dalsgaard, J., & Kurtz, D. W. 2010a, *Asteroseismology, Astronomy and Astrophsyics Library, Springer Berlin Heidelberg*

Aerts, C. & De Cat, P. 2003, *Space Sci. Revs* 105, 453

Aerts, C., De Cat, P., Kuschnig, R., *et al.* 2006a, *ApJ (Letters)* 642, L165

Aerts, C., Lefever, K., Baglin, A., *et al.* 2010b, *A&A* 513, L11

Aerts, C., Marchenko, S. V., Matthews, J. M., *et al.* 2006b, *ApJ* 642, 470

Aerts, C., Molenberghs, G., Kenward, M. G., & Neiner, C. 2014, *ApJ* 781, 88

Aerts, C., Puls, J., Godart, M., & Dupret, M.-A. 2009, *A&A* 508, 409

Aerts, C., Simón-Díaz, S., Catala, C., *et al.* 2013, *A&A* 557, A114

Aerts, C., Thoul, A., Daszyńska, J., *et al.* 2003, *Science* 300, 1926

Balona, L. A., Pigulski, A., Cat, P. D., *et al.* 2011, *MNRAS* 413, 2403

Beck, P. G., Bedding, T. R., Mosser, B., *et al.* 2011, *Science* 332, 205

Beck, P. G., Montalban, J., Kallinger, T., *et al.* 2012, *Nature* 481, 55

Bedding, T. R., Mosser, B., Huber, D., *et al.* 2011, *Nature* 471, 608

Belkacem, K., Dupret, M. A., & Noels, A. 2010, *A&A* 510, A6

Blomme, R., Mahy, L., Catala, C., *et al.* 2011, *A&A* 533, A4

Briquet, M., Aerts, C., Baglin, A., *et al.* 2011, *A&A* 527, A112

Briquet, M., Morel, T., Thoul, A., *et al.* 2007, *MNRAS* 381, 1482

Cameron, C., Saio, H., Kuschnig, R., *et al.* 2008, *ApJ* 685, 489

Casagrande, L., Silva Aguirre, V., Stello, D., *et al.* 2014, *ApJ* 787, 110

Chaplin, W. J., Basu, S., Huber, D., *et al.* 2014, *ApJS* 210, 1

Chaplin, W. J., Sanchis-Ojeda, R., Campante, T. L., *et al.* 2013, *ApJ* 766, 101

Claret, A. 2007, *A&A* 475, 1019

Corsaro, E., Stello, D., Huber, D., *et al.* 2012, *ApJ* 757, 190

Cunha, M. S., Aerts, C., Christensen-Dalsgaard, J., *et al.* 2007, *A&A Rev.* 14, 217

Daszyńska-Daszkiewicz, J., Ostrowski, J., & Pamyatnykh, A. A. 2013, *MNRAS* 432, 3153

De Cat, P. & Aerts, C. 2002, *A&A* 393, 965

Debosscher, J., Blomme, J., Aerts, C., & De Ridder, J. 2011, *A&A* 529, A89

Debosscher, J., Sarro, L. M., López, M., *et al.* 2009, *A&A* 506, 519

Degroote, P., Acke, B., Samadi, R., *et al.* 2011, *A&A* 536, A82

Degroote, P., Aerts, C., Baglin, A., *et al.* 2010a, *Nature* 464, 259

Degroote, P., Aerts, C., Ollivier, M., *et al.* 2009a, *A&A* 506, 471

Degroote, P., Briquet, M., Auvergne, M., *et al.* 2010b, *A&A* 519, A38

Degroote, P., Briquet, M., Catala, C., *et al.* 2009b, *A&A* 506, 111

Deheuvels, S., Doğan, G., Goupil, M. J., *et al.* 2014, *A&A* 564, A27

Desmet, M., Frémat, Y., Baudin, F., *et al.* 2010, *MNRAS* 401, 418

Diago, P. D., Gutiérrez-Soto, J., Auvergne, M., *et al.* 2009, *A&A* 506, 125

Emilio, M., Andrade, L., Janot-Pacheco, E., *et al.* 2010, *A&A* 522, A43

Gilliland, R. L., McCullough, P. R., Nelan, E. P., *et al.* 2011, *ApJ* 726, 2

Godart, M., Noels, A., Dupret, M.-A., & Lebreton, Y. 2009, *MNRAS* 396, 1833

Gray, D. F. 2005, *The Observation and Analysis of Stellar Photospheres, 3rd Edition, Cambridge University Press*

Gruber, D., Saio, H., Kuschnig, R., *et al.* 2012, *MNRAS* 420, 291

Gutiérrez-Soto, J., Floquet, M., Samadi, R., *et al.* 2009, *A&A* 506, 133

Handler, G., Matthews, J. M., Eaton, J. A., *et al.* 2009, *ApJ (Letters)* 698, L56

Hekker, S., Elsworth, Y., De Ridder, J., *et al.* 2011, *A&A* 525, A131

Howell, S. B., Sobeck, C., Haas, M., *et al.* 2014, *PASP* 126, 398

Huat, A.-L., Hubert, A.-M., Baudin, F., *et al.* 2009, *A&A* 506, 95

Huber, D., Bedding, T. R., Stello, D., *et al.* 2011, *ApJ* 743, 143

Huber, D., Chaplin, W. J., Christensen-Dalsgaard, J., *et al.* 2013, *ApJ* 767, 127

Huber, D., Ireland, M. J., Bedding, T. R., *et al.* 2012, *ApJ* 760, 32

Kallinger, T., Mosser, B., Hekker, S., *et al.* 2010, *A&A* 522, A1

Kurtz, D. W., Saio, H., Takata, M., *et al.* 2014, *MNRAS*, in press (arXiv1405.0155)

Lebreton, Y. & Goupil, M.-J. 2014, *A&A* , in press (arXiv1406.0652)

Mahy, L., Gosset, E., Baudin, F., *et al.* 2011, *A&A* 525, A101

Markova, N., Puls, J., Simón-Díaz, S., *et al.* 2014, *A&A* 562, A37

Miglio, A., Chiappini, C., Morel, T., *et al.* 2013, *MNRAS* 429, 423

Moravveji, E., Guinan, E. F., Shultz, M., Williamson, M. H., & Moya, A. 2012, *ApJ* 747, 108

Mosser, B., Goupil, M. J., Belkacem, K., *et al.* 2012, *Astronomy & Astrophysics* 548, A10

Neiner, C., Floquet, M., Samadi, R., *et al.* 2012a, *A&A* 546, A47

Neiner, C., Gutiérrez-Soto, J., Baudin, F., *et al.* 2009, *A&A* 506, 143

Neiner, C., Mathis, S., Saio, H., *et al.* 2012b, *A&A* 539, A90

Pamyatnykh, A. A., Handler, G., & Dziembowski, W. A. 2004, *MNRAS* 350, 1022

Pápics, P. I., Briquet, M., Auvergne, M., *et al.* 2011, *A&A* 528, A123

Pápics, P. I., Briquet, M., Baglin, A., *et al.* 2012, *A&A* 542, A55

Pápics, P. I., Moravveji, E., Aerts, C., *et al.* 2014, *A&A* , in press (arXiv1407.2986)

Rauer, H., Catala, C., Aerts, C., *et al.* 2014, *Experimental Astronomy* , in press (arXiv1310.0696)

Saio, H. 2011, *MNRAS* 412, 1814

Saio, H., Cameron, C., Kuschnig, R., *et al.* 2007, *ApJ* 654, 544

Saio, H., Kuschnig, R., Gautschy, A., *et al.* 2006, *ApJ* 650, 1111

Salmon, S., Montalbán, J., Morel, T., *et al.* 2012, *MNRAS* 422, 3460

Samadi, R., Belkacem, K., Goupil, M. J., *et al.* 2010, *Ap&SS* 328, 253

Sarro, L. M., Debosscher, J., Aerts, C., & López, M. 2009, *A&A* 506, 535

Sarro, L. M., Debosscher, J., Neiner, C., *et al.* 2013, *A&A* 550, A120

Shiode, J. H., Quataert, E., Cantiello, M., & Bildsten, L. 2013, *MNRAS* 430, 1736

Simón-Díaz, S. & Herrero, A. 2014, *A&A* 562, A135

Stello, D., Huber, D., Bedding, T. R., *et al.* 2013, *ApJ (Letters)* 765, L41

Thoul, A., Degroote, P., Catala, C., *et al.* 2013, *A&A* 551, A12

Tkachenko, A., Aerts, C., Pavlovski, K., *et al.* 2014, *MNRAS* 442, 616

Tkachenko, A., Aerts, C., Pavlovski, K., *et al.* 2012, *MNRAS* 424, L21

Torres, G., Vaz, L. P. R., Sandberg Lacy, C. H., & Claret, A. 2014, *AJ* 147, 36

Van Eylen, V., Lund, M. N., Silva Aguirre, V., *et al.* 2014, *ApJ* 782, 14

Walczak, P., Daszyńska-Daszkiewicz, J., Pamyatnykh, A. A., & Zdravkov, T. 2013, *MNRAS* 432, 822

Discussion

CHIEFFI: Can you provide the mass of the convective core of the stars when you do seismic modelling?

AERTS: Sure, since we so far considered the Schwarzschild criterion and added a fully mixed overshoot region, we could have taken the full core mass as a quantity, rather than the overshoot parameter. We will take it up when we redo a homogeneous remodelling of all 16 stars mentioned in Sect. 3, because these results have been assembled over the past decade and not all the studies contain this information. Also, the modelling has been done with improved input physics over the years.

MEYNET: In massive stars, the convective core decreases in mass as a function of time, so to provide a constraint to the stellar models, you need to determine the size of the core, as well as the age and initial mass of the star.

AERTS: Correct. That is precisely what we do in seismic modelling (cf. Aerts 2013): we start from a huge grid of stellar models computed from the Hayashi track up the H-shell burning, with the following parameters: M, initial X and Z, age, and core overshoot parameter. The grid is so dense that the pulsation frequencies from one model to the next on each evolutionary track, differs less than the precision of the detected oscillation frequencies. In this huge grid, we select the models that best fit the seismic data according to Fig. 1. In that way, we get high-precision estimates of the four parameters, for the adopted input physics. Point taken that the theoreticians want us to provide not only the mass and the age, but also the size of the core and we shall document it in future papers (cf. point by Alessandro Chieffi).

SAIO: Is there any correlation between the core overshooting parameter and the rotational frequency of the star?

AERTS: Cf. Fig. 2 in the paper: not really. The diversity of this parameter for massive stars is much larger than anticipated before, which points out that the extra mixing necessary to explain the seismic data is probably caused by various mixing phenomena whose efficiency differs from star to star. This is in line with the few results available from isochrone fitting of massive eclipsing binaries, where one relies on the input physics in the same way as for seismic modelling. These have now been added in the plot, for comparison with the seismic estimates.

NOELS: When using an inversion procedure for the SPB KIC 10526294, you already need to have a good estimate of the density distribution. With a counter-rotating envelope, the density is likely different so it seems to me that an iteration process should be made in order to achieve the "true" rotation profile.

AERTS: Completely agreed that our result cannot be but a 1st step and we now need to plug in the profile to see what kind of effect it would have on the stellar structure, in particular on the mixing and on the angular momentum distribution it will imply. For the moment, we did not yet do such an iteration. Where we are today, is that we found stellar models that explain the 19 equally spaced zonal modes satisfactorily, all the best models have kernels with a strongly differing probing power in the stellar envelope, and the inversion gives consistent results independently of which of the better models we take to compute the kernels. Moreover, we took a broad range in the smoothing parameter that occurs in the inversion procedure to make sure the result is independent of the choice. In addition, we did forward modelling based on the computed $\Omega(r)$ to check if the observed splittings, including their asymmetries, are reconstructed and that is indeed the case. We will move on to the next steps in this subject, keeping in mind that *Kepler* also delivered other types of main-sequence pulsators which do not lead to a rotation profile as expected from theory. As far as I am aware, only internal gravity waves can explain the observations of the newly discovered SPB KIC 10526294.

DE KOTER: The Hunter diagram shows that there are groups of stars that we do not understand at all, e.g., slowly rotating N-enriched objects. I am surprised that your statistical study shows that there is a correlation between N and T_{eff}. Your stars cover the entire main sequence and therefore age enters because age and N are coupled through rotational mixing. Stars in the Hunter diagram tend to be near-TAMS stars.

AERTS: Our sample indeed covers the entire main sequence (cf. plot in our paper). The sample selection was purely based on the availability of high-precision observables and our explicit aim was *not* to rely on any theoretical predictions. Note that Rivero González *et al.* (2012, A&A, 537, A79) came to the same conclusion from a completely independent study of LMC O stars, for which they found that mixing occurs already early on in the core-hydrogen burning phase for a considerable fraction of O stars and hence cannot be due to rotational mixing alone.

Conny Aerts

New windows on massive stars: asteroseismology, interferometry, and
spectropolarimetry
Proceedings IAU Symposium No. 307, 2014
G. Meynet, C. Georgy, J. H. Groh & Ph. Stee, eds.

© International Astronomical Union 2015
doi:10.1017/S1743921314006656

Asteroseismology of red giants to constrain angular momentum transport

P. Eggenberger

Geneva Observatory, University of Geneva, Maillettes 51, 1290, Sauverny, Switzerland
email: patrick.eggenberger@unige.ch

Abstract. Asteroseismic data obtained by the *Kepler* spacecraft have led to the recent detection
and characterization of rotational frequency splittings of mixed modes in red-giant stars. This has
opened the way to the determination of the core rotation rates for these stars, which is of prime
importance to progress in our understanding of internal angular momentum transport. In this
contribution, we discuss which constraints can be brought by these asteroseismic measurements
on the modelling of angular momentum transport in stellar radiative zones.

Keywords. stars: evolution, stars: interiors, stars: oscillations, stars: rotation

1. Introduction

The inclusion of rotational effects basically changes all outputs of stellar models (e.g.
Maeder 2009). A coherent physical description of rotational effects has to take into account simultaneously the transport of chemical elements and angular momentum. The
efficiency of angular momentum transport is then found to have an important impact on
the properties of rotating models. To progress in our understanding of internal angular
momentum transport, one needs direct observational constraints on the internal rotation
of stars, which are now available thanks to seismic data.

Helioseismic measurements indicate a nearly flat rotation profile in the radiative zone of
the Sun (Brown *et al.* 1989; Elsworth *et al.* 1995; Kosovichev *et al.* 1997; Couvidat *et al.*
2003; García *et al.* 2007), while solar models including only shellular rotation predict
a rapidly rotating core (Pinsonneault *et al.* 1989; Chaboyer *et al.* 1995; Eggenberger
et al. 2005; Turck-Chièze *et al.* 2010). This is a first indication that an undetermined
mechanism for the angular momentum transport is at work in the solar interior.

In the continuity of helioseismic measurements, solar-like oscillations have been observed for a large number of stars. In particular, rotational frequency splittings have
been recently detected and characterized for red giants (Beck *et al.* 2012; Deheuvels
et al. 2012; Mosser *et al.* 2012; Deheuvels *et al.* 2014). In this contribution, we briefly discuss the implications of these new observational constraints on the modelling of angular
momentum transport in stellar radiative zones.

2. Mixed oscillation modes in the red giant KIC 8366239

The first detection and characterization of rotational frequency splittings of mixed
modes has been obtained for the red giant KIC 8366239 (Beck *et al.* 2012). These measurements show that oscillation modes at the center of the dipole forests have lower
values of rotational splitting than modes in the wings of the dipole forests. This directly
indicates that differential rotation is present in the interior of this red giant, with a core
rotating more rapidly than the surface (Beck *et al.* 2012). At first sight, this seems to
be in good agreement with stellar models, which predict differential rotation during the

red-giant phase because of the rapid contraction of the central layers after the main sequence (e.g Palacios *et al.* 2006; Eggenberger *et al.* 2010). However, by computing models including rotation for this target and comparing the theoretical values for the rotational splittings with the observed ones, we conclude that rotating models predict too steep internal rotation profiles compared to asteroseismic measurements (Eggenberger *et al.* 2012; Marques *et al.* 2013). This clearly indicates that an additional mechanism for the internal angular momentum transport is at work during the post-main sequence phase.

In order to study how this unknown transport mechanism can be constrained by asteroseismic data, Eggenberger *et al.* (2012) introduced an additional viscosity ν_{add} corresponding to this process in the equation describing the internal transport of angular momentum. By computing models of KIC 8366239 for different values of ν_{add} and of the initial velocity on the zero-age main sequence (ZAMS), one finds that the mean value for the efficiency of this unknown transport mechanism is strongly constrained by asteroseismic measurements to the value of $\nu_{\mathrm{add}} = 3 \cdot 10^4 \, \mathrm{cm^2 \, s^{-1}}$ for this $1.5 \, M_\odot$ red giant (Eggenberger *et al.* 2012).

3. KIC 7341231: mixed modes in a low-mass red giant

In the preceding section, we have seen that an additional mechanism for the internal transport of angular momentum is needed during the post-main sequence evolution of a $1.5 \, M_\odot$. This star exhibits a convective core during the main sequence and one can then wonder whether such a mechanism is also needed for a low-mass star with a radiative core during the main sequence. Rotational splittings have been obtained for KIC 7341231, a red giant with a mass of about $0.84 \, M_\odot$ (Deheuvels *et al.* 2012). For this star with a radiative core during the main sequence, one also finds that an efficient additional mechanism for the internal transport of angular momentum is needed in order to correctly reproduce the asteroseismic data (Ceillier *et al.* 2012, 2013).

As for the more massive red giant KIC 8366239, the efficiency of this unknown transport process can be determined for KIC 7341231 by computing models including an additional viscosity ν_{add}. In this case, we find a lower value of about $3 \cdot 10^3 \, \mathrm{cm^2 \, s^{-1}}$ for the additional viscosity. The model of KIC 7341231 has a convective enveloppe during the main sequence and the initial velocity on the ZAMS is thus found to depend on the adopted efficiency for the associated surface magnetic braking during the main sequence. Interestingly, the value of the additional viscosity ν_{add} can be precisely determined independently from the uncertainties on the adopted magnetic braking during the main sequence. The differences in ν_{add} between KIC 8366239 and KIC 7341231 suggests that the efficiency of the undetermined additional process for the transport of angular momentum decreases when the mass of the star decreases.

4. Rotational splittings for a sample of six subgiants and young red giants

Asteroseismic measurements have been recently obtained for a sample of six *Kepler* young red giants (Deheuvels *et al.* 2014). Estimates of the core and surface rotation rates have then been deduced for these stars. These results suggest that the degree of differential rotation increases when evolution proceeds, with an increase of the core rotation rate and a decrease of the surface rotation rate when gravity decreases (Deheuvels *et al.* 2014).

Interestingly, rotational splittings suggest that for two of these targets a discontinuous internal rotation profile is favored compared to a smooth rotation profile. The determination of the location of such a discontinuity suggests that it corresponds to the location of the hydrogen-burning shell. These results seem to indicate that the efficiency of the additional mechanism of internal angular momentum transport should strongly decrease in a region of large chemical gradients. Such a sharp discontinuity in the rotation profile seems to be difficult to reproduce with fossil magnetic fields and could favor diffusive processes with a reduced efficiency related to the presence of steep chemical gradients.

By comparing the deduced core and surface rotation rates with rotating models including an additional viscosity associated to the unkown transport process, one can precisely determined the efficiency of this process for the six targets. This also suggests an increase of ν_{add} with the mass in good agreement with the preceding result obtained by comparing KIC 8366239 and KIC 7341231. Interestingly, the simultaneous determination of the core and surface rotation rates for these stars can also constrain the efficiency of the magnetic braking associated to the presence of a convective envelope during the main-sequence phase. One also finds that the change of the efficiency of the transport of angular momentum in the radiative zone has a negligible impact on the surface rotation rates during the red-giant phase (see dashed lines in Fig. 1). This is obvious in view of the difference in the moment of inertia between the core and the enveloppe of a red giant. It is however interesting to recall this point in the context of red giants exhibiting a rapid surface rotation velocity, which origin is more related to binary interaction or planet engulfment than to an efficient transport of angular momentum in the radiative interior.

5. Core rotation rates for a large sample of red giants

Mosser *et al.* (2012) have determined mean core rotation rates from rotational splittings for a large sample of red giants observed with the *Kepler* spacecraft. Interestingly, this study suggests that core rotation rates slightly decrease during the evolution on the red-giant branch.

As shown in Fig. 1, a model including only shellular rotation (dotted line) predicts a very rapidly rotating core during the red giant phase. Moreover, it predicts an increase of the core rotation rate when the star evolves on the red-giant branch, which is also in contradiction with asteroseismic data. When an additional constant viscosity of $3 \cdot 10^4 \, \mathrm{cm}^2 \, \mathrm{s}^{-1}$ is included (continuous line in Fig. 1), the model correctly reproduces the asteroseismic data available for the red giant KIC 8366239 and also predicts a slight increase in the core rotation rates of red giants with a radius lower than about $5 \, R_\odot$. This is in good agreement with core rotation rates deduced from asteroseismic data. However, for higher values of the radius, this model predicts an increase of the core rotation rates when evolution proceeds. This discrepancy with the trend deduced from asteroseismic measurements indicates that the efficiency of the unknown transport mechanism of angular momentum must increase when the star evolves along the red-giant branch.

6. Comparison with mechanisms already included in stellar evolution codes

In the preceding sections, we have seen that according to asteroseismic data an unknown mechanism is needed in addition to meridional circulation and the shear instability for the transport of angular momentum in the radiative zone of red giants. Asteroseismic data also bring valuable constraints on the efficiency of such a process for red giants of

168   P. Eggenberger

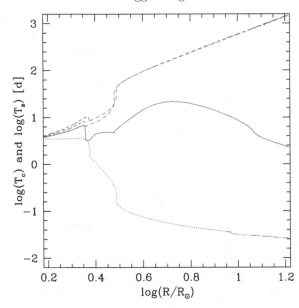

Figure 1. Core (continuous and dotted lines) and surface (dashed lines) rotation periods for two rotating models of $1.5\,M_\odot$ with an initial velocity of $20\,\mathrm{km\,s^{-1}}$ on the ZAMS. The dotted line (blue) indicates a model computed with shellular rotation only. The continuous line (red) corresponds to the same model except for the inclusion of an additional viscosity of $3 \times 10^4\,\mathrm{cm^2\,s^{-1}}$.

different masses and at different evolutionary stages. One can then wonder whether these seismic constraints can be correctly reproduced by mechanisms already included in some evolution codes such as magnetic fields and internal gravity waves.

Internal gravity waves are found to be able to efficiently transport angular momentum in the radiative zone of solar-type stars (see e.g. Zahn *et al.* 1997; Talon & Charbonnel 2003; Charbonnel & Talon 2005; Mathis *et al.* 2013). It would then be interesting to see whether rotating models of red giants including the transport of angular momentum by internal gravity waves are able to correclty reproduce the asteroseismic constraints now available for the internal rotation of red giants.

Concerning magnetic fields, Cantiello *et al.* (2014) recently compared rotating models of red giants including the Tayler-Spruit dynamo (Spruit 1999, 2002) to asteroseismic measurements. They find that models including the transport of angular momentum by the Tayler-Spruit dynamo are in better agreement with asteroseismic determinations of core rotation rates of red giants than models including only rotational effects. However, models including the Tayler-Spruit dynamo still predict core rotation rates which are at least ten times higher than the observed values.

7. Summary

Asteroseismic constraints on the internal rotation of red giants are now available to help us progress in the modelling of the different transport processes at work in stellar radiative zones. The goal is now to use these new valuable constraints to obtain a coherent global physical description of transport processes in radiatives zones that should be valid for low and massive stars, from the pre-main sequence to advanced stages of evolution.

Preliminary results of comparison of these asteroseismic measurements with stellar models clearly show that an undetermined mechanism for the internal transport of angular momentum is needed during the post-main sequence evolution. This is the case for low-mass red giants with a radiative core during the main sequence as well as for more massive stars with a convective core during the main sequence. Thanks to seismic data, the efficiency of this unknown transport process can be precisely determined. Asteroseismic measurements suggest an increase of the efficiency of this process when the mass increases as well as when the star evolves on the red-giant branch.

Concerning the physical nature of this unknown process, it will be interesting to compare the rotation profiles of red-giant models including angular momentum transport by internal gravity waves to the asteroseismic constraints now available. Different clues may suggest that transport of angular momentum by magnetic instabilities could be a promising candidate. While a first study of the impact of the Tayler-Spruit dynamo indicates that the efficiency of this process is not sufficient to account for the low core rotation rates deduced from seismic data (Cantiello *et al.* 2014), it will be particularly interesting to consider the effects of the azimuthal magnetorotational instability in this context (e.g. Rüdiger *et al.* 2014, and see the contribution by M. Gellert in this volume).

References

Beck, P. G., Montalban, J., Kallinger, T., *et al.* 2012, *Nature* 481, 55

Brown, T. M., Christensen-Dalsgaard, J., Dziembowski, W. A., *et al.* 1989, *ApJ* 343, 526

Cantiello, M., Mankovich, C., Bildsten, L., Christensen-Dalsgaard, J., & Paxton, B. 2014, *ApJ* 788, 93

Ceillier, T., Eggenberger, P., García, R. A., & Mathis, S. 2012, *Astronomische Nachrichten* 333, 971

Ceillier, T., Eggenberger, P., García, R. A., & Mathis, S. 2013, *A&A* 555, A54

Chaboyer, B., Demarque, P., & Pinsonneault, M. H. 1995, *ApJ* 441, 865

Charbonnel, C. & Talon, S. 2005, *Science* 309, 2189

Couvidat, S., García, R. A., Turck-Chièze, S., *et al.* 2003, *ApJ (Letters)* 597, L77

Deheuvels, S., Doğan, G., Goupil, M. J., *et al.* 2014, *A&A* 564, A27

Deheuvels, S., García, R. A., Chaplin, W. J., *et al.* 2012, *ApJ* 756, 19

Eggenberger, P., Maeder, A., & Meynet, G. 2005, *A&A* 440, L9

Eggenberger, P., Miglio, A., Montalban, J., *et al.* 2010, *A&A* 509, A72

Eggenberger, P., Montalbán, J., & Miglio, A. 2012, *A&A* 544, L4

Elsworth, Y., Howe, R., Isaak, G. R., *et al.* 1995, *Nature* 376, 669

García, R. A., Turck-Chièze, S., Jiménez-Reyes, S. J., *et al.* 2007, *Science* 316, 1591

Kosovichev, A. G., Schou, J., Scherrer, P. H., *et al.* 1997, *Sol. Phys.* 170, 43

Maeder, A. 2009, *Physics, Formation and Evolution of Rotating Stars*, Springer Berlin Heidelberg

Marques, J. P., Goupil, M. J., Lebreton, Y., *et al.* 2013, *A&A* 549, A74

Mathis, S., Decressin, T., Eggenberger, P., & Charbonnel, C. 2013, *A&A* 558, A11

Mosser, B., Goupil, M. J., Belkacem, K., *et al.* 2012, *A&A* 548, A10

Palacios, A., Charbonnel, C., Talon, S., & Siess, L. 2006, *A&A* 453, 261

Pinsonneault, M. H., Kawaler, S. D., Sofia, S., & Demarque, P. 1989, *ApJ* 338, 424

Rüdiger, G., Gellert, M., Schultz, M., Hollerbach, R., & Stefani, F. 2014, *MNRAS* 438, 271

Spruit, H. C. 1999, *A&A* 349, 189

Spruit, H. C. 2002, *A&A* 381, 923

Talon, S. & Charbonnel, C. 2003, *A&A* 405, 1025

Turck-Chièze, S., Palacios, A., Marques, J. P., & Nghiem, P. A. P. 2010, *ApJ* 715, 1539

Zahn, J.-P., Talon, S., & Matias, J. 1997, *A&A* 322, 320

Discussion

MORAVVEJI: Can we take the lessons learned about viscosity on RGB phase to other stellar evolution phases?

EGGENBERGER: Yes, the idea is to use all available observational constraints to progress in the modelling of transport processes in stellar radiative zones. In this context, the asteroseismic data obtained for red giants are particularly valuable. A coherent physical description of these processes should then be valid for low-mass stars as well as for massive stars and from the pre-main sequence to advanced stages of evolution.

HIRSCHI: Can you confirm that a process transporting angular momentum all the way to the centre of the star is needed to reproduce observations?

EGGENBERGER: Yes, core rotation rates of red giants deduced from asteroseismic measurements clearly show that an efficient process for the transport of angular momentum is needed in addition to meridional circulation and the shear instability to extract angular momentum from the central layers.

Patrick Eggenberger

Michalis Kourniotis

New windows on massive stars: asteroseismology, interferometry, and
spectropolarimetry
Proceedings IAU Symposium No. 307, 2014
G. Meynet, C. Georgy, J. H. Groh & Ph. Stee, eds.

© International Astronomical Union 2015
doi:10.1017/S1743921314006668

Photometric Variability of OB-type stars as a New Window on Massive Stars

M. Kourniotis[1,2], A. Z. Bonanos[1], I. Soszyński[3], R. Poleski[3,4], G. Krikelis[2] and the OGLE team

[1]IAASARS, National Observatory of Athens, GR-15236 Penteli, Greece
email: mkourniotis@astro.noa.gr, bonanos@astro.noa.gr

[2]Section of Astrophysics, Astronomy and Mechanics, Faculty of Physics, University of Athens, Panepistimiopolis, GR15784 Zografos, Athens, Greece

[3]Warsaw University Observatory, Al. Ujazdowskie 4, 00-478 Warszawa, Poland

[4]Department of Astronomy, Ohio State University, 140 W. 18th Ave., Columbus, OH 43210, USA

Abstract. We present the first systematic study of 4646 spectroscopically confirmed early-type massive stars in the Small Magellanic Cloud (SMC), using variability as a tool to confine the physics of OB-type massive stars. We report the discovery of ~ 100 massive eclipsing systems which are useful for the accurate determination of the fundamental parameters of massive stars and we evaluate the frequency of multiplicity. In addition, we explore the occurrence of the Oe/Be phenomenon and provide a large number of candidate non-radial pulsators, which can be further studied via asteroseismology. The results of this work (Kourniotis *et al.* 2014) will contribute to a better understanding of the role of metallicity in triggering processes associated to matter ejections and/or disk formation, which in turn affect mass loss and stellar rotation.

Keywords. stars: emission-line, Be, (stars:) binaries: eclipsing, galaxies: individual (SMC), stars: variables: other

1. Introduction

Photometric variability has been acclaimed to be a unique tool to constrain the physics of hot massive stars. Extended photometric surveys have already yielded large catalogs of eclipsing binaries (EBs) and pulsating variables in the Galaxy and nearby galaxies. Eruptive events due to strong photometric processes have been monitored, allowing a deep insight on the nature of massive variables such as Be stars. The occurrence and intensity of the Be phenomenon has been already reported to be higher in low metallicity environments owing to the lower mass loss rates that result in lower angular momentum losses (Vink *et al.* 2001). Consequently, massive stars reach a high Ω/Ω_c ratio, triggering the formation of an equatorial disk, the modification of which results in photometric and spectroscopic variability (Rivinius *et al.* 2013).

The low metallicity of the SMC provides an ideal laboratory in the context of evaluating the frequency of Be stars among environments of different metallicity. Motivated by the availability of a large photometric database with long-term lightcurves provided by the wide-field monitoring of the OGLE-III survey, as well as of an extended catalog of 5324 massive stars with known spectral types in the SMC (Bonanos *et al.* 2010), we aim to exploit the time domain in order to identify photometrically candidate Be stars, report newly discovered eclipsing binaries and hot, candidate pulsators and correlate variability with the spectral type of massive stars in the SMC.

2. Input Catalogues and Analysis

The input catalog of 5324 massive stars in the SMC provided by Bonanos *et al.* (2010), consists of 12 Wolf Rayet, 3983 OB-type stars and 1329 stars of later spectral types, with a typical accuracy to one luminosity class and one spectral sub-type. The OGLE-III observations were conducted between 2001 and 2009 with the 1.3-m Warsaw telescope at Las Campanas Observatory in Chile that covered 14 square degrees in the direction of the SMC (Udalski *et al.* 2008). High-quality light curves were provided in the Cousins $I-$band and Johnson V photometric band and matched with the stars of the input catalogue. Very bright and thus saturated stars as well as stars that were located out of the OGLE-III field were excluded, resulting in a final list of 4646 stars with $I-$band magnitudes between $12.6 - 19.4$ mag.

As a first step, the light curves were studied with respect to their mean magnitude and standard deviation. We used the σ_I vs. I plane and imposed an empirical curve based on the OGLE accuracy, following the method proposed by Graczyk & Eyer (2010) to distinguish between real variables and non-variables (Fig. 1). A second, shifted curve was added to the plane to further confine stars with high-amplitude fluctuations mainly attributed to the Be mechanism. We then flagged as high-amplitude variables those stars located above our suggested curve, low-amplitude those found between the two curves and "constant" stars those that lie below the primary, empirical curve. Eclipsing binaries were identified following the method introduced by Graczyk & Eyer (2010) based on the skewness and the kurtosis of the light curves. We focused on the region of the skewness-kurtosis plane where we expect EBs to lie and searched the light curves for periodic signals using the Analysis of Variance (AoV) algorithm (Schwarzenberg-Czerny 1989). We then folded the light curves to the period that corresponds to the frequency with the highest peak in the power spectrum, yielding ~ 200 EBs and a large number of sinusoidally varying periodic stars representative of pulsators and rotational variables.

3. Results

Following the previously described periodic search, we identified 205 eclipsing binaries, 101 of which are newly reported for their eclipsing nature. We assign the spectral type to 59 known EBs and derive periods, which in many cases are more precise than those reported in the literature owing to their long-term monitoring. Among the identified EBs, we further report rare cases of transient EBs, double-periodic variables considered to be systems surrounded by a circumbinary disk and binaries with evident indication of apsidal motion.

A total of 249 periodic stars present sinusoidal behaviour with equal minima, corresponding to pulsators and rotating variables. We classified these stars as "short" and "long period" variables, when displaying periods shorter and longer than 3 days, respectively. We then flagged variables as "periodic with extra variability" when their periodic signal varies in amplitude or baseline by more than 0.05 mag (see Fig. 8, Kourniotis *et al.* 2014). The remaining periodic variables consist of 126 classical Cepheids and 50 stars flagged as "rotating variables" being sinusoidally periodic variables with unequal minima which implies a strong indication of tidal distortion effects for an ellipsoidal star or starspot modulation.

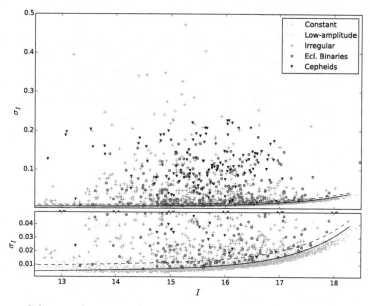

Figure 1. σ_I vs. I diagram for our studied stars. The solid curve is proposed to differentiate real from spurious variables (Graczyk & Eyer 2010). Low-amplitude variables are located between the proposed and our suggested (dashed) curve. A zoomed region is presented in the lower panel.

Our high-amplitude, stochastic variables flagged as "irregular", on their majority display a combination of short or long-term outbursts and/or trends. We distinguished 4 modes of variability similar to the modes presented in Keller *et al.* (2002) for classifying the blue variables in the LMC, that consist of high-amplitude, long-term outbursts ("bumper" events), short-term and low-amplitude, sudden outbursts ("flicker" events), monotonic trends across the 8-year time domain ("monotonic") and decline of magnitude that lasts up to 1000 days ("fading" events) (see Fig. 9, Kourniotis *et al.* 2014). Of our 443 irregular variables, $\sim$76% are newly reported as variables in the present work.

4. Discussion

We proceed to study the variability of 4646 stars as defined in the previous section, with respect to the spectral type. Fig. 2 presents 4 pie charts showing the distribution of variability for O, early B, late B and AFG-type stars. The most prominent feature is the decline of variability –particularly of "irregular" type– beyond the B2 spectral type. Short-term periodic stars are concentrated in early-type spectral bins whereas eclipsing binaries are mostly frequent in O-type stars, in agreement with the high binarity frequency reported among stars of this type (Sana *et al.* 2012).

The majority of the irregular variables exhibit reddening, as shown in Fig. 3, which presents a color-magnitude diagram (CMD) for our studied stars with respect to the type of variability, whereas low-amplitude variability in O-type stars is mainly restricted to the blue regime possibly owing to wind variability. We propose irregular and low-amplitude variability for B-type stars to be a photometric criterion for defining a candidate Be star. Applying this criterion to our stars, the fraction of candidate early Be stars is found to be $30 \pm 1\%$, consistent with the respective fractions derived in the literature using different methods.

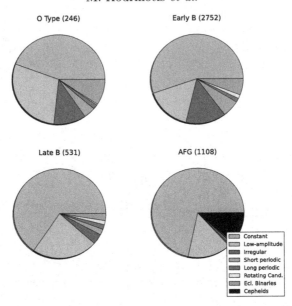

Figure 2. Pie charts displaying the distribution of variability for our studied O, early B, late B and AFG-type stars.

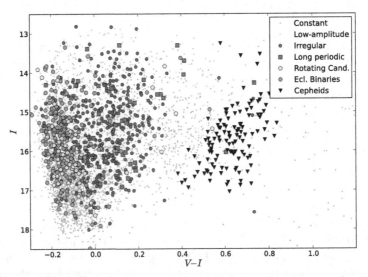

Figure 3. Color-magnitude diagram of our studied stars for the type of variability.

Of our spectroscopically confirmed Be stars, $78 \pm 4\%$ exhibit prominent variability on the 8-year time domain of OGLE-III i.e. irregular, low-amplitude or short-term periodic variability. We caution that since the fraction of short-term periodic variables is incomplete and that low-amplitude variables may have a smaller chance of being detected with Hα emission during spectrscopy than irregular variables, the derived fraction of photometrically variable Be stars corresponds to a lower limit.

Early-type short-term periodic variables of mainly low-amplitude, occupy distinctive regions in the CMD where the known Slowly Pulsating B stars (SPBs) and non-radially pulsating Be variables of the SMC are located (see Fig. 17, Kourniotis *et al.* 2014). They thus provide ground for further examination of the properties of the metal opacity bump in low metallicity environments (Salmon *et al.* 2012). Having examined 4 pulsating Be stars with extra variability as defined in Sec. 3, we find that they are rotating close to their critical speed, likely enhancing matter ejections that cause the observed additional variability.

References

Bonanos, A. Z., Lennon, D. J., Köhlinger, F., *et al.* 2010, *AJ* 140, 416
Graczyk, D. & Eyer, L. 2010, *AcA* 60, 109
Keller, S. C., Bessell, M. S., Cook, K. H., Geha, M., & Syphers, D. 2002, *AJ* 124, 2039
Kourniotis, M., Bonanos, A. Z., Soszyński, I., *et al.* 2014, *A&A* 562, A125
Rivinius, T., Carciofi, A. C., & Martayan, C. 2013, *A&A Rev.* 21, 69
Salmon, S., Montalbán, J., Morel, T., *et al.* 2012, *MNRAS* 422, 3460
Sana, H., de Mink, S. E., de Koter, A., *et al.* 2012, *Science* 337, 444
Schwarzenberg-Czerny, A. 1989, *MNRAS* 241, 153
Udalski, A., Soszyński, I., Szymański, M. K., *et al.* 2008, *AcA* 58, 329
Vink, J. S., de Koter, A., & Lamers, H. J. G. L. M. 2001, *A&A* 369, 574

Discussion

FOSSATI: The fraction of rotating variables among O-type stars appears to be quite small. Could that be due to selection effects?

KOURNIOTIS: We denoted as candidate rotating variables those sinusoidally periodic stars with unequal minima as this implies indication of starspot modulation. The fraction of candidate rotating variables is incomplete, since "constant" stars were not searched for periodic signals and hence, we discard from our study rotating variables with very low amplitude. Rotating variables shall be also found among our short/long periodic variables. Therefore, the particular fraction shall not be considered as intrinsic but rather, as a lower limit.

New windows on massive stars: asteroseismology, interferometry, and spectropolarimetry
Proceedings IAU Symposium No. 307, 2014
G. Meynet, C. Georgy, J. H. Groh & Ph. Stee, eds.

Behaviour of Pulsations in Hydrodynamic Models of Massive Stars

C. C. Lovekin[1] and J. A. Guzik[2]

[1]Department of Physics, Mount Allison University, Sackville, NB, Canada
email: `clovekin@mta.ca`

[2]XTD-NTA, Los Alamos National Laboratory, Los Alamos, NM 87545

Abstract. We have calculated the pulsations of massive stars using a nonlinear hydrodynamic code including time-dependent convection. The basic structure models are based on a standard grid published by Meynet *et al.* (1994). Using the basic structure, we calculated envelope models, which include the outer few percent of the star. These models go down to depths of at least 2 million K. These models, which range from 40 to 85 solar masses, show a range of pulsation behaviours. We find models with very long period pulsations (> 100 d), resulting in high amplitude changes in the surface properties. We also find a few models that show outburst-like behaviour. The details of this behaviour are discussed, including calculations of the resulting wind mass-loss rates.

Keywords. stars: oscillations, stars: variables: other

1. Introduction

S Doradus variables, or Luminous Blue Variables (LBVs), show photometric variability on several different timescales. The shortest, the "microvariability" has timescales of weeks to months. For example, van Genderen *et al.* (1998) show the photometric variability of R85, which is a B5Iae α Cyg variable and quite possibly also an LBV. The light curve varies by about 0.2 to 0.3 magnitudes on timescales of about 100 days. Their best fit solution found two periods, one at 390 d and one at 83.5 d (van Genderen *et al.* 1998).

These stars periodically undergo larger outbursts, during which the visual magnitude of the stars increases while the bolometric magnitude stays constant. This process recurs on either short (< 10 years) or long (> 20 years) timescales. The mechanism that produces these events is unclear, although it is thought to be related to their proximity to the Humphreys-Davidson (HD) limit (Humphreys & Davidson 1994). As these hot massive stars evolve towards the red side of the HR diagram, they encounter the HD limit and become unstable, then undergo a period of mass loss before settling back on the stable (hot) side of the HD limit. This process repeats for a period of a few times 10^4 years.

LBVs can also undergo giant eruptions as observed in η Carinae. These events can eject large quantities of mass (10 $M_\odot$ or more during the Great Eruption of η Car), and produce an increase in both bolometric and visual luminosity (Smith *et al.* 2003).

In this work, we investigate the possibility that the interaction between radial pulsations and time-dependent convection can drive the long-period variability observed in these stars. A few of our models undergo outburst-like events that may be related to the outbursts seen in S Doradus variables. We also present preliminary estimates of the mass-loss rates in these stars.

2. Models

Our models are based on the stellar evolution model grid calculated by Meynet *et al.* (1994). These models include enhanced mass-loss rates on the post-main sequence, but do not include rotation. In this work, we focus on the 60 and 85 $M_\odot$ models at metallicities of $Z = 0.004$, 0.008, 0.02, and 0.04. We selected models along each track ranging from approximately the middle of the main sequence through the start of core helium burning and used the current mass, effective temperature, luminosity, and composition at the surface and in the core to calculate an envelope model. Our envelope models contain 60 zones, including the outer few percent of the mass, and go down to a depth of $2 \cdot 10^6$ K, which ensures that we have completely captured the damping and driving regions. For more details on the model calculations, refer to Lovekin & Guzik (2014a,b).

We use the DYNSTAR hydrodynamic code, which includes time-dependent convection (TDC, Ostlie 1990). This is a more realistic approximation than the standard mixing length model when the timescales of the pulsations are similar to the convective timescale, as is the case here. Because the convective motions increase gradually after a region becomes convectively unstable, energy can be trapped in the lower layers, causing the radiative luminosity in these regions to surpass the Eddington limit. In models with convective regions, we expect to see longer periods in the hydrodynamic model, as well as other behaviour not predicted by the linear non-adiabatic pulsation calculations. Our model of TDC is described more fully in Lovekin & Guzik (2014a), and builds on previous work (see Guzik & Lovekin 2012).

We calculate the pulsation periods using a linear non-adiabatic pulsation code and DYNSTAR. The pulsation periods are extracted from the nonlinear results as described in Lovekin & Guzik (2014a). A comparison of the resulting periods shows that the nonlinear periods can be longer than the linear periods by factors of 50 or more, while few models have nonlinear periods that are longer by a factor of 1000 or more. Models with convection zones show larger period enhancements than models without convection, suggesting that the interaction between pulsation and TDC produces longer periods than predicted by the linear model alone.

3. Long Period Pulsations

Plotting in the Hertzsprung-Russell (HR) diagram the location of models with periods > 50 days shows an instability strip that is consistent with the location of the HD limit in the HR diagram. The models in this region are typically near the end of their main-sequence lifetime and are mainly located to the red of the HD limit. The resulting variability is typically multiperiodic, with a dominant period on the order of 100 days. One of the longest lightcurves calculated is over 15 000 days, although simulation times of a few hundred days, capturing 2–3 pulsation cycles are more typical, as shown in Fig. 1. The peak-to-peak amplitude of this pulsation is on the order of 30 km s^{-1}. Given the long pulsation period, the expansion and contraction produces huge changes in effective temperature ($\sim 9\,000$ K) and radius ($\sim 600\,R_\odot$). The resulting magnitude changes, shown in the bottom half of Fig. 1, are typically on the order of 0.2 magnitudes, consistent with those seen in R85 (van Genderen *et al.* 1998).

4. Outbursts

A few of our models show outburst-like events during which the surface expands rapidly in the first few days of the simulation. We divide these events into two classes, based on the peak expansion speed of the surface. Minor outbursts have peak expansion velocities of $20 - 30$ km s^{-1}, while the major outbursts have peaks between 50 and 80 km s^{-1}. The radial velocity curve of a typical example of each class is shown in Fig. 2. Even in the major outbursts, the expansion velocities are not high enough to eject mass, but the resulting variation in surface parameters will likely drive increased mass loss. Unfortunately, our hydrodynamic calculations do not allow us to remove mass from the star during the calculation, so we cannot reliably follow the behaviour of the star after the outburst.

In our models, outbursts occur preferentially in models with lower metallicity. The model showing the major outburst illustrated in Fig. 2 has $Z = 0.004$, and the other outbursting models occur all either at $Z = 0.004$ or $Z = 0.008$. This is a consequence of the high mass-loss rates assumed by the Meynet et al. (1994) models, as higher metallicity models do not reach the region of the HR diagram where outbursts are common.

4.1. *Minor Outbursts*

Minor outbursts have peak expansion velocities of $20 - 30$ km s^{-1}, which is comparable to the long-period pulsation amplitudes. Minor outbursts typically occur after approximately 10 days of simulation time. As the envelope expands and contracts during pulsation, the interior becomes convectively unstable. Because of the TDC, the convective velocities, and hence the convective luminosities increase gradually, with the increase in luminosity lagging behind the increase in velocity by about 0.5–1.0 day. This pattern is first seen in the deepest zone we followed (about 1/2 way into the envelope), and then gradually propagates out to the surface. When the peak in the convective luminosity reaches the surface, this drives a sudden jump in the expansion rate of the surface. As a result, the effective temperature and radius increase by 7 000 K and 200 $R_\odot$ respectively. During this time the bolometric magnitude remains approximately constant.

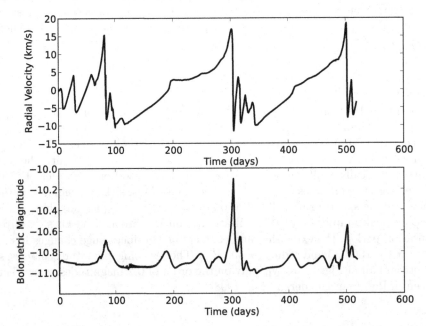

Figure 1. *Top:* The radial velocity variation for a typical model pulsating with a long period. This model is an 85 $M_\odot$ model with metallicity $Z = 0.004$ near the end of the main sequence. *Bottom:* The bolometric magnitude variation. Variations are typically about 0.2 magnitudes, as seen in R85 and other S Dor variables.

4.2. *Major Outbursts*

Major outbursts occur earlier in the simulation than the minor outbursts, typically around 5 days. In these models, the envelope is initially convective, but then the convection turns off as the star pulsates. When the convection turns off, the radiative luminosity spikes, jumping up to 20–25 times the Eddington luminosity. This sharp increase in the radiative luminosity drives the expansion of the star, reaching peak expansion velocities of up to 80 km s^{-1}. This pattern results in major changes in the temperature and radius. During the spike in luminosity, the temperature jumps by more than 10 000 K, then drops by nearly 20 000 K as the star expands. The spike in luminosity corresponds to a brightening of factor of 3 in bolometric magnitude.

5. Mass-Loss Rates

We have used the surface properties as a function of time to calculate mass-loss rates using the Vink *et al.* (2001) mass-loss prescription. These results are highly uncertain, as the Vink rates are based on calculations of line-driven winds in main sequence stars, while our models include post-main sequence stars and the winds are likely to be continuum-driven. Given the rapid variation in the surface parameters of our models however, it is not clear that any simple prescription is valid, so we have chosen the Vink rates for ease of use. It is our expectation that while the absolute values of the resulting mass-loss rates are inaccurate, the relative changes in rate are reasonable.

For the long period model shown in Fig. 1, the mass-loss rate varies between 10^{-8} and $10^{-12}\ M_\odot$/yr, a variation of 4 orders of magnitude. The mass-loss rates calculated for this model are considerably lower than expected for an 85 $M_\odot$ star, which is an indication

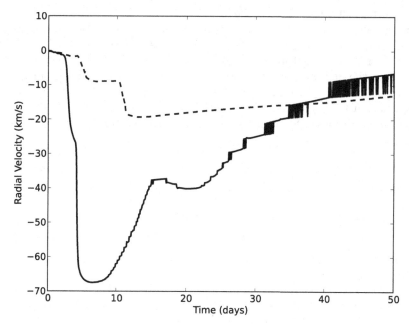

Figure 2. The radial velocity curves for a major (solid) and minor (dashed) outburst. Minor outbursts occur later in the simulation than major outbursts, and have significantly lower peak expansion velocities. The calculation after $\sim$ 15–20 days is uncertain, as our simulations do not remove mass from the envelope. Although the expansion velocities are significantly lower than the escape speed for these stars, the outbursts are expected to drive an increase in mass loss (see Section 5).

that the mass-loss rates we use here are not a good choice. This particular model has $(\log L/L_\odot, \log T_{\mathrm{eff}}/\mathrm{K}) = (6.151, 4.498)$, which places it just past the end of the main sequence, where the Vink mass-loss rates are not expected to be accurate.

For the major outburst shown in Fig. 2, we find that the mass-loss rate increases dramatically during the sharp spike in radiative luminosity. The mass-loss rate increases by nearly 5 orders of magnitude, from $\sim 10^{-8}\ M_\odot/\mathrm{yr}$ to $\sim 10^{-3}\ M_\odot/\mathrm{yr}$ before dropping as the star expands. The peak mass-loss rate in this model is in very good agreement with the observed mass-loss rates during S Doradus type outbursts. However, this model is slightly cooler than the long-period model discussed above, with $(\log L/L_\odot, \log T_{\mathrm{eff}}) = (6.195, 4.168)$, putting this star even further past the main sequence. As a result, the Vink mass-loss rates are not expected to be appropriate for this star, and we should only consider the changes in mass-loss rate to be reasonable.

6. Conclusions

We have performed hydrodynamic calculations of massive pulsating stars including the effects of time-dependent convection (TDC). We have found that TDC interacts with the pulsation to significantly alter the pulsation characteristics, producing periods that are tens or even thousands of times longer than predicted by linear non-adiabatic calculations. Even at relatively low amplitudes $(10 - 15\ \mathrm{km\,s}^{-1})$ this variability can act over periods of hundreds of days to produce very large changes in the radius, luminosity, and effective temperature of the star. The resulting magnitude changes are typically 0.2–3.0 magnitudes, similar to those observed in hot massive stars such as R85. Simple mass loss calculations have shown that the mass-loss rate in these stars could vary by as much as 4 orders of magnitude over the pulsation cycle.

A small subset of our models show outburst-like behaviour, classified based on the peak expansion velocity as either major $(50 - 80\ \mathrm{km\,s}^{-1})$ or minor $(20 - 30\ \mathrm{km\,s}^{-1})$. Even the highest peak speed in the major outbursts is well below the escape speed for these stars, but the mechanism that produces the outburst also produces a large increase in luminosity and temperature. As a result, the star brightens by 3 magnitudes for about a day before the star begins to expand. The peak mass-loss rate at this point is comparable to the mass-loss rates in S Dor variables, although the duration in our models is shorter.

References

Guzik, J. A. & Lovekin, C. C. 2012, *The Astronomical Review* 7(3), 13
Humphreys, R. M. & Davidson, K. 1994, *PASP* 106, 1025
Lovekin, C. & Guzik, J. 2014a, *MNRAS, submitted*
Lovekin, C. & Guzik, J. 2014b, *MNRAS, in press*
Meynet, G., Maeder, A., Schaller, G., Schaerer, D., & Charbonnel, C. 1994, *A&AS* 103, 97
Ostlie, D. A. 1990, in J. R. Buchler (ed.), *Numerical Modelling of Nonlinear Stellar Pulsations Problems and Prospects*, p. 89
Smith, N., Gehrz, R. D., Hinz, P. M., *et al.* 2003, *AJ* 125, 1458
van Genderen, A. M., Sterken, C., & de Groot, M. 1998, *A&A* 337, 393
Vink, J. S., de Koter, A., & Lamers, H. J. G. L. M. 2001, *A&A* 369, 574

Discussion

LOBEL: The microvariability of LBVs observed in V is typically a few tenths of a magnitude with quasi-periods from days to months. Do the hydro-models show that they can be attributed to pulsations or time-dependent convection, or both?

LOVEKIN: The pulsations seen in our models are produced by an interaction between pulsations and time-dependent convection. Without the convection, the periods would be much shorter. The periods we find are typically larger than expected for the LBV microvariability, but that's somewhat artificial. For this work, we used an arbitrary cut-off of 20 days and only considered models with larger periods. It would be interesting to go back and look at the shorter period models in more detail.

DE KOTER: When the LBVs inflate on a years-decades timescale they all end up with $T_{\rm eff} \sim 8000\,$K, regardless their luminosity. This implies that the radius inflation factor is mass (or temperature) dependent. Do your predictions also show such a mass (or $T_{\rm eff}$) dependent inflation?

LOVEKIN: This is something we haven't looked at. This could be very interesting and I'll definitely a look into this.

Catherine Lovekin

New windows on massive stars: asteroseismology, interferometry, and spectropolarimetry
Proceedings IAU Symposium No. 307, 2014
G. Meynet, C. Georgy, J. H. Groh & Ph. Stee, eds.

© International Astronomical Union 2015
doi:10.1017/S1743921314006681

Asteroseismic Diagnostics for Semi-Convection in B Stars in the Era of K2

Ehsan Moravveji†

Instituut voor Sterrenkunde, KU Leuven, Celestijnenlaan 200D, B-3001 Leuven, Belgium
email: Ehsan.Moravveji@ster.kuleuven.be

Abstract. Semi-convection is a slow mixing process in chemically-inhomogeneous radiative interiors of stars. In massive OB stars, it is important during the main sequence. However, the efficiency of this mixing mechanism is not properly gauged yet. Here, we argue that asteroseismology of β Cep pulsators is capable of distinguishing between models of varying semi-convection efficiencies. We address this in the light of upcoming high-precision space photometry to be obtained with the *Kepler* two-wheel mission for massive stars along the ecliptic.

Keywords. asteroseismology, stars: oscillations (including pulsations), stars: interiors, stars: evolution, stars: rotation, variables: others

1. Introduction

Non-radial pulsation is a common phenomenon among B dwarfs. Early-type B stars - widely known as β Cep stars - are pulsationally unstable against low-order, low-degree radial and non-radial pressure (p-) and gravity (g-) modes. Their mass ranges from ~ 8 to $20\,M_\odot$ (see Aerts *et al.* 2010, for details). Contrary to their fully mixed convective cores, the mixing of species in their radiative interior occurs on a long - yet unconstrained - time scale. There are several mixing mechanisms that operate (simultaneously) in radiative zones, among which rotational mixing and semi-convection. In this paper, we limit ourselves to slowly-rotating B stars.

The pulsation frequencies of stars are highly sensitive to their internal structure, and can be used as a proxy to test different input physics. Miglio *et al.* (2008) already showed the effect of extra mixing induced by, e.g., rotation, atomic diffusion, and convective overshooting on the period spacing of g-modes in heat-driven pulsators; their conclusions can be extended to include semi-convection as an extra mixing mechanism.

In the near future, the *Kepler* two-wheel mission, (hereafter K2, Howell *et al.* 2014) will provide high-precision space photometry of a handful of late-O and early B-type pulsators in the ecliptic plane. We emphasise that K2 will conduct pioneering observations, since such space photometry is scarce for massive stars, particularly for objects more massive than $8\,M_\odot$.

In this paper, we put forward asteroseismic diagnostics to probe semi-convective mixing in massive main-sequence stars. Following on Miglio *et al.* (2008), we address the possibility of constraining the efficiency of semi-convection in massive stars in light of the upcoming high-precision data to be assembled by the K2 mission and already present in the CoRoT archive.

† Postdoctoral Fellow of the Belgian Science Policy Office (BELSPO), Belgium

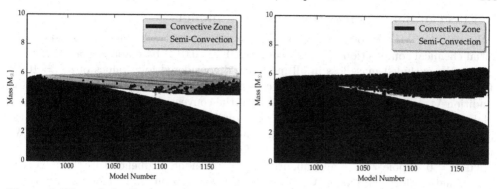

Figure 1. Kippenhahn diagrams showing the evolution of 15 $M_\odot$ stellar models with $\alpha_{sc} = 10^{-1}$ (left) and $\alpha_{sc} = 1$ (right). The convective zones are shown in blue, and the semi-convective zones are shown in purple.

2. Semi-Convective Mixing

Semi-convection is a slow mixing process believed to operate in the chemically inhomogeneous parts of radiative zones, where the g-modes are oscillatory (Schwarzschild & Härm 1958; Kato 1966; Langer *et al.* 1983; Noels *et al.* 2010). In other words, semi-convection acts in those layers of the star where the radiative temperature gradient ∇_{rad} takes values in between the adiabatic ∇_{ad} and the Ledoux ∇_L gradients, i.e., $\nabla_{ad} < \nabla_{rad} < \nabla_L$. Here, $\nabla_L = \nabla_{ad} + \varphi/\delta\, \nabla_\mu$, with $\varphi = (\partial \ln \rho/\partial \ln T)_{P,\mu}$, $\delta = (\partial \ln \rho/\partial \ln \mu)_{P,T}$ and $\nabla_\mu = d \ln \mu/d \ln P$. Semi-convection occurs due to the stabilizing effect of the composition gradient ∇_μ against the onset of convection in the chemically inhomogeneous layers on top of the receding convective core. This mixing process is believed to be present in stars with $M \gtrsim 15\,M_\odot$ (Langer *et al.* 1985; Langer 1991). The reason is the increasing effect of the radiation pressure with mass, and the local increase of ∇_{rad} with respect to ∇_{ad} outside the convective core. Therefore, semi-convection is not expected in intermediate to late B-type stars. For this reason, β Cep stars are optimal candidates to investigate if such a mixing can leave observable footprints. The semi-convective mixing is typically described in a diffusion approximation (Langer *et al.* 1983). See Section 6.2 in Maeder (2009) for an overview of different mixing schemes.

The stellar evolution code MESA (Paxton *et al.* 2011, 2013) follows the prescription by Langer *et al.* (1985), where the semi-convective diffusion coefficient D_{sc} is defined according to Kato (1966) and Langer *et al.* (1983):

$$D_{sc} = \alpha_{sc} \frac{\kappa_r}{6 c_p \rho} \frac{\nabla - \nabla_{ad}}{\nabla_L - \nabla}, \tag{2.1}$$

where $\kappa_r = 4acT^3/3\kappa\rho$ is the radiative conductivity, c_p the specific heat at constant pressure, and ρ denotes the density. The efficiency of semi-convective mixing is controlled by the free parameter α_{sc}, which determines the length scale and time scale of the vibrational mixing associated to semi-convective zones. The parameter α_{sc} is not calibrated from observations, Langer *et al.* (e.g., 1985) favoured $\alpha_{sc} = 10^{-1}$, while later on Langer (1991) preferred $0.01 \leqslant \alpha_{sc} \leqslant 0.04$. Note that the value of α_{sc} depends on the choice of the opacity tables and on the numerical scheme to compute D_{sc} according to Eq. (2.1). We aim at constraining α_{sc} using K2 data.

3. Effect of Semi-Convection on β Cep models

We used the MESA code to calculate four evolutionary tracks for a $15\,M_\odot$ star with initial chemical composition $(X, Y, Z) = (0.710, 0.276, 0.014)$ based on Nieva & Przybilla (2012). We excluded mass loss in all models and considered four values of $\alpha_{sc} = 10^{-6}$, 10^{-4}, 10^{-2}, and 1. We used the Ledoux criterion of convection, and made sure that the condition $\nabla_{rad} = \nabla_{ad} = \nabla_L$ was satisfied from the convective side of the core boundary (Gabriel *et al.* 2014); see also Noels (these proceedings). On each evolutionary track, we stored an equilibrium model for a central hydrogen abundance of $X_c = 0.10$. We subsequently used the GYRE pulsation code (Townsend & Teitler 2013) to calculate radial $\ell = 0$, dipole $\ell = 1$ and quadrupole $\ell = 2$ mode frequencies in the adiabatic approximation, for each model. We restricted the comparison of the mode behaviour to low-order modes, i.e., $-5 \leqslant n_{pg} \leqslant +3$, where $n_{pg} = n_p - n_g$.

Figure 1 shows Kippenhahn diagrams for models with $\alpha_{sc} = 10^{-2}$ (left) and $\alpha_{sc} = 1$ (right). For relatively limited semi-convective mixing (i.e., $\alpha_{sc} = 10^{-2}$), an extended semi-convective zone (grey points) develops and continues at roughly the same mass coordinate. On the other hand, for efficient semi-convective mixing (i.e. $\alpha_{sc} = 1$) the former zones are identified as convective and an extended intermediate convective zone (hereafter ICZ) develops. In our models, the ICZ encapsulates $\sim 1\,M_\odot$. The formation and presence of an ICZ largely impacts the later evolution of the star. From an asteroseismic point of view, models harbouring an ICZ have smaller radiative zones (measured from the surface), hence a smaller cavity for g-mode propagation. As a result, it is expected that the presence of an ICZ, which in turn results from different semi-convective efficiencies, affects the adiabatic frequencies of low-order low-degree p- and g-modes.

A word of caution about matching detected frequencies of unidentified modes from a grid of asteroseismic models is worthwhile to be made here. Figure 2a shows that the frequency of the radial fundamental mode $\ell = 0, n_{pg} = 1$ is quite close to that of an $\ell = 2, n_{pg} = -2$ non-radial g-mode. Without robust mode identification — which by itself is intricate — the interpretation of detected pulsation frequencies can be quite misleading. Therefore, the photometric light curves to be assembled by K2 must be complemented with ground-based multi-colour photometry and/or high-resolution spectroscopy to identify at least one of the detected modes. Non-adiabatic computations might help to identify modes, although there are still severe disagreements between the observed modes in OB stars and those predicted to be excited in the sense that we detect many more modes than foreseen by theory. It is therefore safer not to rely on excitation computations when identifying detected frequencies.

Figure 2a shows the frequencies of $\ell = 0, 1$ and 2 p- and g-modes for different values of α_{sc}. For better visibility, few modes with identical ℓ and n_{pg} are connected by lines. Clearly, the change in α_{sc} shifts most of the frequencies to slightly lower/higher values. The frequency difference of the modes with fixed ℓ and n_{pg} increases with decreasing α_{sc}. An important question is whether we are able to capture such subtle differences observationally from the K2 space photometry.

To answer that, we take the model with $\alpha_{sc} = 1$ as the *reference* model, hence its frequencies are $f_i^{(ref)}$. We compare the relative frequency change $\delta f_i^{(th)}$ of each theoretical frequency f_i with respect to the reference frequency as

$$\delta f_i^{(th)} = \frac{|f_i^{(ref)} - f_i|}{f_i^{(ref)}}. \tag{3.1}$$

Here, $f_i^{(ref)}$ and f_i are both calculated using GYRE. From an observational point of view, the frequency precision Δf depends on the total observation time base ΔT. A

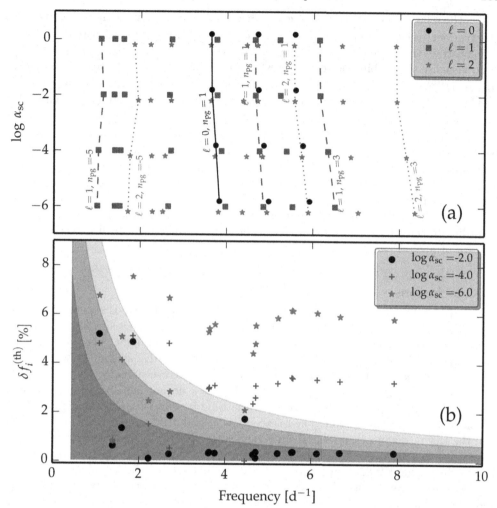

Figure 2. Adiabatic low radial order $-5 \leqslant n_{\rm pg} \leqslant +3$ frequencies of $15\,M_\odot$ models with $Z = Z_\odot$ and $X_c = 0.10$. Both panels share the same frequency range on the abscissa. (a) The ordinate is the logarithm of $\alpha_{\rm sc}$ (Eq. 2.1). Circles are radial $\ell = 0$ modes, squares are non-radial dipole $\ell = 1$ p- and g-modes, and stars are non-radial quadrupole $\ell = 2$ p- and g-modes. (b) Comparison of the relative theoretical frequency change $\delta f^{\rm (th)}$ with the relative observed frequency precision $\delta f^{\rm (obs)}$ as a function of frequency; see Eqs. (3.1) and (3.2). Dark grey, grey and light grey show 1σ, 2σ and 3σ precision levels, respectively. Circles, plus marks, and stars correspond to models with $\alpha_{\rm sc} = 10^{-2}$, 10^{-4}, and 10^{-6}, respectively. The reference frequencies are taken from the $\alpha_{\rm sc} = 1$ model. Consult the color version of this figure in electronic format.

conservative estimate by Loumos & Deeming (1978) is $\Delta f \approx 2.5/\Delta T$. For K2, the planned ΔT is approximately 75 days. We also define the relative observed frequency precision $\delta f_i^{\rm (obs)}$

$$\delta f_i^{\rm (obs)} = \frac{\Delta f}{f_i^{\rm (ref)}}, \tag{3.2}$$

where $\delta f_i^{\rm (obs)}$ stands for a 1σ uncertainty level, and measures the observational frequency precision required to capture the effect of a specific feature — in our case the presence/absence of the ICZ. If by varying a stellar structure free parameter — here

$\alpha_{\rm sc}$ — the relative theoretical frequency change is significantly larger than the estimated observational relative frequency precision, i.e. $\delta f_i^{\rm (th)} \gtrsim 3\,\delta f_i^{\rm (obs)}$, then asteroseismology can constrain the value of that parameter. Note the arbitrary choice of 3σ here.

To visualise the probing power of the K2 data, Figure 2b shows the relative frequency change $\delta f_i^{\rm (th)}$ with respect to the reference model ($\alpha_{\rm sc} = 1$) for models with $\log \alpha_{\rm sc} = -2$ (circles), -4 (plus marks) and -6 (stars), respectively. In the background of the same plot, we show $\delta f_i^{\rm (obs)}$ (dark grey), $2 \times \delta f_i^{\rm (obs)}$ (grey) and $3 \times \delta f_i^{\rm (obs)}$ (light grey), respectively. The distribution of the circles is roughly inside the 1σ zone, which makes the distinction between models with $\alpha_{\rm sc} = 1$ and $\alpha_{\rm sc} = 10^{-2}$ quite challenging. However, plus marks that compare $\log \alpha_{\rm sc} = 10^{-4}$ models with those of the reference model, or stars that compare $\log \alpha_{\rm sc} = 10^{-6}$ models with the reference model are more than 3σ away from $\delta f^{\rm (obs)}$. For the latter, K2 holds the potential to constrain the efficiency of semi-convection as an extra mixing mechanism, provided that we can find good seismic models of the stars according to the scheme outlined in Aerts (these proceedings) to which we then add the concept of semi-convective mixing.

4. Conclusions

Unfortunately, the *Kepler* mission observed no O-type and early-B stars so far. Thus, we are yet unable to estimate the richness of the frequency spectrum of the K2 light curves for massive dwarfs. The public release of the K2 field 0 light curves is scheduled for September 2014; since there are several OB pulsators on K2 silicon, we will soon be able to gauge the quality of K2 data for massive star asteroseismology.

In our comparisons, we assumed that the K2 data do not suffer from instrumental effects, hence that $\delta f^{\rm (obs)}$ depends only on the time base of the K2 campaigns. This is of course an idealised situation. The telescope jitter and drift can easily increase $\delta f^{\rm (obs)}$. Yet, the seismic diagnostic potential stays valid since the $\delta f^{\rm (th)}$ can exceed the 3σ level for several modes, and hence the effect of semi-convective mixing can hopefully be detected and studied using K2 light curves.

Based on Figure 2, there is no real preference between modes of different order $n_{\rm pg}$ and degree ℓ in providing asteroseismic diagnostics for semi-convection; in other words, all low-order p- and g-modes possess the same potential to provide a constraint on $\alpha_{\rm sc}$. The semi-convection free parameter was varied in a broad range, from 10^{-6} to 1, in our exercise. It is of course easier to discriminate between models with large differences in $\alpha_{\rm sc}$. According to Figure 2b, the distinction between models with $\alpha_{\rm sc} = 10^{-6}$ and 1 is more within reach than distinguishing between models with $\alpha_{\rm sc} = 10^{-2}$ and 1.

The main message of our work is to emphasis the importance of semi-convection along with the shrinking convective cores in massive OB stars, when the observed frequencies of β Cep stars are compared with theoretical frequencies. Consequently, semi-convection — as one of the extra mixing mechanisms in inhomogeneous layers of the stellar radiative interior — could be employed as an extra dimension when modelling stars based on grid calculations coupled to asteroseismic forward modelling (Briquet *et al.* 2007).

References

Aerts, C., Christensen-Dalsgaard, J., & Kurtz, D. W. 2010, *Asteroseismology, Astronomy and Astrophsyics Library, Springer Berlin Heidelberg*
Briquet, M., Morel, T., Thoul, A., *et al.* 2007, *MNRAS* 381, 1482
Gabriel, M., Noels, A., Montalban, J., & Miglio, A. 2014, *ArXiv e-prints*
Howell, S. B., Sobeck, C., Haas, M., *et al.* 2014, *PASP* 126, 398

Kato, S. 1966, *PASJ* 18, 374

Langer, N. 1991, *A&A* 252, 669

Langer, N., El Eid, M. F., & Fricke, K. J. 1985, *A&A* 145, 179

Langer, N., Fricke, K. J., & Sugimoto, D. 1983, *A&A* 126, 207

Loumos, G. L. & Deeming, T. J. 1978, *Ap&SS* 56, 285

Maeder, A. 2009, *Physics, Formation and Evolution of Rotating Stars*

Miglio, A., Montalbán, J., Noels, A., & Eggenberger, P. 2008, *MNRAS* 386, 1487

Mowlavi, N. & Forestini, M. 1994, *A&A* 282, 843

Nieva, M.-F. & Przybilla, N. 2012, *A&A* 539, A143

Noels, A., Montalban, J., Miglio, A., Godart, M., & Ventura, P. 2010, *Ap&SS* 328, 227

Pápics, P. I., Moravveji, E., Aerts, C., *et al.* 2014, *A&A* , in press (arXiv1407.2986)

Paxton, B., Bildsten, L., Dotter, A., *et al.* 2011, *ApJS* 192, 3

Paxton, B., Cantiello, M., Arras, P., *et al.* 2013, *ApJS* 208, 4

Schwarzschild, M. & Härm, R. 1958, *ApJ* 128, 348

Townsend, R. H. D. & Teitler, S. A. 2013, *MNRAS* 435, 3406

Discussion

GEORGY: First a comment: in Georgy *et al.* (2014), we showed that using the Ledoux criterion improves the fit of the observations of supergiants. However, we did not claim that this is the unique solution.

GEORGY: It seems that your best fit points systematically towards the highest Z you tried for the SPB KIC 10526294. Did you try still higher Z?

MORAVVEJI: As shown in Pápics *et al.* (2014), there exist strong correlations between the different grid parameters — like mass, Z, X_c, and overshooting — so that you always find smaller χ^2 by increasing Z. But we have to stop somewhere, and we have difficulty fine-tuning Z without constraints from spectroscopy.

HIRSCHI: What is the significance of small χ^2 differences when the χ^2 values are so high?

MORAVVEJI: This means that a limited range of parameters can almost equivalently reproduce the observed frequencies. However, in general when performing seismic modelling of a massive star, the χ^2_{red} of the best models from a grid including semi-convection and no overshooting is to be compared with that of the best model(s) from a grid taking only overshooting and no semi-convection. This is our aim for the future K2 data.

NOELS: In a $3\,M_\odot$ star computed with the Schwarzschild criterion, there is no semi-convective zone surrounding the convective core. On the other hand, the use of the Ledoux criterion must give exactly the same extent of convective cores, whatever the mass of the star. In particular, no semi-convective zone is present in models of $3\,M_\odot$ computed with the Ledoux criterion. I suspect a misplaced convective boundary, which leads you to interpret the layers outside that boundary as semi-convective layers instead of convective ones. Have you checked that the radiative temperature gradient is exactly equal to the adiabatic one at the boundary? This should necessarily be the case, with a radiative temperature gradient lower than the adiabatic one outside the convective core.

MORAVVEJI: We are aware that in the current version of MESA, $\nabla_{\mathrm{rad}} > \nabla_{\mathrm{ad}}$ from the convective side of the boundary when using Ledoux criterion. We have fixed this short-coming, and then we observe almost identical behaviour of the core boundary whether Schwarzschild or Ledoux criterion is used. Yet, we need to improve some minor aspects of our fix, and then will publish it for use by the community.

New windows on massive stars: asteroseismology, interferometry, and spectropolarimetry
Proceedings IAU Symposium No. 307, 2014
G. Meynet, C. Georgy, J. H. Groh & Ph. Stee, eds.

© International Astronomical Union 2015
doi:10.1017/S1743921314006693

Are the stars of a new class of variability detected in NGC 3766 fast rotating SPB stars?

S. J. A. J. Salmon[1], J. Montalbán[1], D. R. Reese[3], M.-A. Dupret[1] and P. Eggenberger[2]

[1] Département d'Astrophysique, Géophysique et Océanographie, Université de Liège, Allée du 6 Août 17, 4000 Liège, Belgium
email: salmon@astro.ulg.ac.be

[2] Observatoire de Genève, Université de Genève, Chemin des Maillettes 51, 1290 Sauverny, Switzerland

[3] School of Physics and Astronomy, University of Birmingham, Edgbaston, Birmingham, B15 2TT, UK

Abstract. A recent photometric survey in the NGC 3766 cluster led to the detection of stars presenting an unexpected variability. They lie in a region of the Hertzsprung-Russell (HR) diagram where no pulsation are theoretically expected, in between the δ Scuti and slowly pulsating B (SPB) star instability domains. Their variability periods, between $\sim$0.1–0.7 d, are outside the expected domains of these well-known pulsators. The NCG 3766 cluster is known to host fast rotating stars. Rotation can significantly affect the pulsation properties of stars and alter their apparent luminosity through gravity darkening. Therefore we inspect if the new variable stars could correspond to fast rotating SPB stars. We carry out instability and visibility analysis of SPB pulsation modes within the frame of the traditional approximation. The effects of gravity darkening on typical SPB models are next studied. We find that at the red border of the SPB instability strip, prograde sectoral (PS) modes are preferentially excited, with periods shifted in the 0.2–0.5 d range due to the Coriolis effect. These modes are best seen when the star is seen equator-on. For such inclinations, low-mass SPB models can appear fainter due to gravity darkening and as if they were located between the δ Scuti and SPB instability strips.

Keywords. stars: rotation, stars: variables: other, stars: early-type.

1. Introduction

Intermediate-mass stars can exhibit various types of pulsation during the main sequence; mid- to late B-type stars present high-order gravity (g) modes with periods $\gtrsim$ 1 d (SPB pulsators), whereas late A- and early F-type stars show low-order pressure (p) and g modes with periods between 0.3 h and 6 hr (δ Scuti stars). These pulsations are driven by the κ mechanism due to the iron-group and HeII opacity bumps, respectively. In early A-type stars, none of these two opacity bumps fit the conditions to efficiently activate the κ mechanism (e.g. Pamyatnykh 1999), so that they are not expected to pulsate.

However, Mowlavi *et al.* (2013, hereafter Mo13) recently detected a significant number (36) of unknown variable stars in NGC 3766. These stars span over 3 magnitudes and are all located in the HR diagram between those identified as SPB or δ Scuti candidates. The amplitudes of these new variable stars are typically lower by a factor 2 to 3. Moreover, their periods are between $\sim$0.1 d and 0.7 d, distinguishing them from SPB or δ Scuti modes. The existence of variable stars with such properties, often referred as Maia stars, is a recurrent debate (e.g. De Cat *et al.* 2007). Yet, low-amplitude variable stars observed in the field of the CoRoT mission (Degroote *et al.* 2009) and in NGC 884 (Saesen *et al.* 2010) appeared as possible new such candidates, with properties similar to those found in NGC 3766 by Mo13. Quite recently, Moździerski *et al.* (2014) reported the detection of a very similar population of variable stars in the NGC 457 cluster, rising once again the question about the possible origin of this variability.

Mo13 already suggested the role of rotation, since B stars of the NGC 3766 cluster are known to be fast rotators with typical rotational velocities half or more their break-up velocities (McSwain *et al.* 2008). Using the traditional approximation of rotation (TAR), we study whether rotation can both shift periods to an unusual range and reproduce the properties of the variability amplitudes. Furthermore, the centrifugal distortion induced by rotation can also affect luminosity and effective temperature of stars (von Zeipel 1924). Hence the observed flux will depend on the stellar inclination (e.g. Maeder & Peytremann 1972). We analyse for different rotational velocities and inclinations whether gravity darkening can displace SPBs towards fainter and cooler regions in the HR diagram.

The paper is structured as follows; in the second and third sections, we present respectively the stability analysis and visibility computations of two SPB models. In Sec. 4, we show how the visual properties of these models are affected by gravity darkening. The paper ends with our concluding remarks.

2. Instability domain in the traditional approximation framework

The TAR is based on the assumption that the rotational frequency, Ω, is moderate in comparison to the critical rotation rate of the star. Here, we assume a solid-body rotation and adopt $\Omega_{\mathrm{crit}} = (GM/R_e^3)^{1/2}$ as the critical rotation rate, where G is the gravitational constant, M the mass of the star, and R_e the radius at the equator.

Townsend (2005a) showed that periods of classic SPB g modes can be significantly shifted with rotation whilst the limit of their instability strip is barely shifted towards cooler temperatures. In addition, Savonije (2005) and Townsend (2005b) studied modes that have no counterpart in the case with no rotation. These modes are known as Yanai and Rossby waves and present a hybrid character, since their restoring force is a mix between the Coriolis force and buoyancy. The Yanai modes can present periods well below 1 d and could explain low period modes in late B-type stars as it was advanced by Savonije (2005). However, these modes might be difficult to observe as we will show in Sec. 3.

The limit of validity for the TAR is not clearly determined. Ballot *et al.* (2012) showed in a comparison with non-spherical models that the TAR underestimates the periods of modes up to 4% for $\eta \simeq 5$, where $\eta = 2\Omega/\omega_{\mathrm{co}}$ is the spin parameter and ω_{co} the angular pulsation frequency. Since we look at effects that could shift periods from ~ 1 to ~ 0.3 d, that is by 70%, the TAR computations should nevertheless remain of sufficient precision.

With help of the CLES code (Scuflaire *et al.* 2008) and adopting the solar chemical mixture, we compute two SPB models of 2.9 and 3.2 $M_\odot$, representative of the red border of the SPB instability strip. We perform a stability analysis with the MAD code including the TAR (Bouabid *et al.* 2013) for the models of about 20 Myr, in agreement with the age estimated for NGC 3766 (Aidelman *et al.* 2012). Periods in the inertial frame (P_{in}) of the modes found to be excited are given in Table 1, where modes with azimuthal order $m < 0$ are prograde.

Among the classic SPB g modes, i.e. those presenting a counterpart in the case with no rotation, the prograde sectoral modes (PS; $m = -\ell$), are preferentially excited at the red border of the instability strip (see details in Townsend 2005a). Moreover, their P_{in}

Table 1. Inertial periods (in days) of modes found to be excited in the 2.9 and 3.2 $M_\odot$ aged of $\sim$ 20 Myr, and for various rotation frequencies. Figures between brackets give the number of excited modes.

Mass ($M_\odot$)	$\Omega/\Omega_{\rm crit}$	$\ell = 1$ $m = -1$	$\ell = 2$ $m = -2$	$\ell = 1$ $m = 0$	Yanai $m = 1$	Yanai $m = 2$
2.9	0.20	0.43 - 0.53 (7)	–	–	5.09 - 8.63 (6)	0.63 - 0.64 (8)
	0.40	0.29 - 0.33 (7)	–	–	–	0.298 - 0.304 (5)
	0.60	0.22 - 0.24 (7)	–	–	–	–
3.2	0.20	0.43 - 0.58 (10)	0.23 - 0.28 (8)	0.57 - 0.84 (10)	4.05 - 9.57 (11)	0.64 - 0.66 (10)
	0.40	0.30 - 0.35 (9)	0.15 - 0.18 (9)	0.46 - 0.58 (7)	1.16 - 1.84 (9)	0.30 - 0.32 (11)
	0.60	0.22 - 0.25 (9)	0.12 - 0.13 (9)	0.40 -0.45 (4)	0.64 - 0.81 (7)	0.19 - 0.20 (9)

are shifted downwards 0.23 - 0.58 d and 0.12 - 0.25 d for 0.20 and 0.60 $\Omega_{\rm crit}$, respectively. Axisymmetric modes ($m = 0$) are only excited in the 3.2 $M_\odot$ case with periods around 0.5 d but tend to stabilise as Ω increases. The other classic g modes are not presented in Table 1 because they are stable in almost every cases, excepting a low number of them in the 3.2 $M_\odot$ model rotating at 0.20 $\Omega_{\rm crit}$ (see more details in Salmon, S. J. A. J. *et al.* 2014).

Yanai modes with $m = 1$ present periods in the usual SPB domain, though they decrease below 1 d as rotation increases (0.60 $\Omega_{\rm crit}$). Clearly, $m = 2$ Yanai modes present periods quite smaller than 1 d, as in the case of PS g modes.

3. Visibilities of the unstable modes

Computing the amplitudes of non-radial pulsations is a difficult non-linear problem, which cannot be carried out presently. We instead compute mode visibilities that we normalise by the highest visibility reached by one of the modes. Initially suggested by Savonije (2013), this approach makes it possible for a relative comparison between the modes.

We determine the visibilities (in the CoRoT visible passband) following Townsend

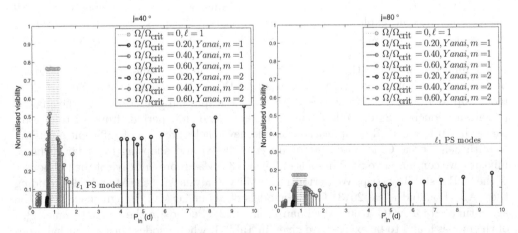

Figure 1. Normalised visibilities (see text) as a function of the inertial periods of all the Yanai modes found to be unstable in the 3.2 $M_\odot$ model aged $\sim$ 20 Myr. Left (resp. right) panels correspond to an observer viewing angle of 40° (resp. 80°). The horizontal grey lines represent the normalised visibilities of the PS ℓ_1 modes, which present the same value at 0.20, 0.40 and 0.60 $\Omega_{\rm crit}$.

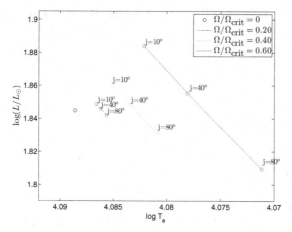

Figure 2. Luminosity and effective temperature of the $2.9\,M_\odot$ model aged of ~ 20 Myr in the cases of no rotation and rotation at 0.20, 0.40 and 0.60 $\Omega_{\rm crit}$, for inclinations j from $10°$ to $80°$.

(2003) and adopting a normalisation as presented above (see details in Salmon, S. J. A. J. *et al.* 2014). The visibilities of classic SPB g modes found to be excited in Table 1 were presented in Salmon, S. J. A. J. *et al.* (2014). We found that the PS $\ell = 1$ (ℓ_1 hereafter) modes are the most visible when the star is seen towards the equator, with visibilities typically $\sim$30-40 % that of SPB modes in a case with no rotation. Therefore, the ℓ_1 PS modes reproduce qualitatively the observed ratio between the amplitudes of the new variables and SPB stars of Mo13.

The visibilities of the Yanai $m = 1$ and $m = 2$ modes are shown in Fig. 1, where the horizontal lines depict those of the PS ℓ_1 modes. As they present no equatorial confinement, Yanai modes are the most visible when the star is seen towards mid-latitudes ($j = 40°$), $m = 1$ modes reaching up to 50 % of the visibility of SPB modes with no rotation. However, their visibilities drop to ~ 20 % and ~ 10 % when the star is seen close to the pole ($j = 10°$, not presented) and close to the equator ($j = 80°$), respectively. In this latter case, their visibilities are ~ 3 times smaller than those of PS ℓ_1 modes. The visibility of $m = 2$ Yanai modes never exceeds 10 %. As a consequence, Yanai modes with periods < 0.65 d should be difficult to detect in comparison with the PS modes. Hence, these latter appear as good candidates to explain both the periods and amplitudes of the new variable stars.

4. Gravity darkening due to rotation

We now determine whether the effect of rotation on the visual properties of SPB stars can make them appear outside their instability strip in the HR diagram. We assume the star is in solid-body rotation and adopt the Roche model to describe the centrifugal distortion. We then follow Georgy *et al.* (2014) to compute the effects of limb- and gravity darkening on the apparent effective temperature (T_e) and luminosity (L), depending on the inclination and rotation rate.

Figure 2 illustrates these effects for a Geneva model (Eggenberger *et al.* 2008) of $2.9\,M_\odot$ with the same properties as the analog CLES model considered in the previous sections. The larger the rotation rate is, the cooler the stars appears. Meanwhile, the more the star is seen close to the equator ($j = 80°$), the more it appears cooler and fainter. It results that SPB stars at the red border of the instability strip can be displaced outside the strip and appears as if they were between the SPB and δ Scuti instability domains (see Salmon, S. J. A. J. *et al.* 2014). In particular, the PS modes are the most visible when the star is highly inclined, i.e. when the star appears shifted at most towards low T_e and L.

5. Conclusion

The presence of fast rotators in NGC 3766 has led us to include rotation in our study, resulting in a scenario able to account for the variable stars detected in the cluster. Using the TAR, we have shown that PS modes of fast rotating SPB stars are good candidates to explain this new kind of variability. At the red border of the classic SPB instability strip, the periods of excited modes are shifted from 0.7–1.2 d (no rotation) to 0.15–0.6 d or 0.12–0.45 d depending on the rotation rate. In particular, the PS modes are there preferentially excited. They behave as Kelvin waves, which are confined to the equator, making them visible only in highly inclined stars (low latitudes facing the observer).

Moreover, this combination of a fast rotation rate and high inclination coincide with large shifts towards cooler T_e and fainter L due to gravity darkening. It leads to a displacement of these pulsators outside the classic SPB instability strip. For such inclinations, the visibilities of ℓ_1 PS modes are approximately 2 to 3 times lower than in a non-rotating star. These values correspond qualitatively to the ratio between the variability amplitudes of the SPB and new variable stars observed by Mo13. The Yanai modes with the lowest periods ($\lesssim 0.65$ d, $m = 2$) present low visibilities, clearly smaller than those of PS modes. However, some of the new variable stars close to the SPBs detected by Mo13 present modes of both SPB- and new variability type; they might correspond to stars less inclined and presenting $m = 1$ Yanai or axisymmetric classic g modes.

This scenario might also explain the similar new variable stars detected by CoRoT (Degroote *et al.* 2009), and in the NGC 884 and NGC 457 clusters (Saesen *et al.* 2010; Moździerski *et al.* 2014). In the future, spectroscopic campaigns devoted to these stars could in particular confirm the role of rotation.

References

Aidelman, Y., Cidale, L. S., Zorec, J., & Arias, M. L. 2012, *A&A* 544, A64
Ballot, J., Lignières, F., Prat, V., Reese, D. R., & Rieutord, M. 2012, in H. Shibahashi, M. Takata, & A. E. Lynas-Gray (eds.), *Progress in Solar/Stellar Physics with Helio- and Asteroseismology*, Vol. 462 of *ASP Conf. Ser.*, p. 389
Bouabid, M.-P., Dupret, M.-A., Salmon, S., *et al.* 2013, *MNRAS* 429, 2500
De Cat, P., Briquet, M., Aerts, C., *et al.* 2007, *A&A* 463, 243
Degroote, P., Aerts, C., Ollivier, M., *et al.* 2009, *A&A* 506, 471
Eggenberger, P., Meynet, G., Maeder, A., *et al.* 2008, *Ap&SS* 316, 43
Georgy, C., Granada, A., Ekström, S., *et al.* 2014, *A&A* 566, A21
Maeder, A. & Peytremann, E. 1972, *A&A* 21, 279
McSwain, M. V., Huang, W., Gies, D. R., Grundstrom, E. D., & Townsend, R. H. D. 2008, *ApJ* 672, 590
Mowlavi, N., Barblan, F., Saesen, S., & Eyer, L. 2013, *A&A* 554, A108
Moździerski, D., Pigulski, A., Kopacki, G., Kołaczkowski, Z., & Stęślicki, M. 2014, *AcA* 64, 89
Pamyatnykh, A. A. 1999, *AcA* 49, 119
Saesen, S., Carrier, F., Pigulski, A., *et al.* 2010, *A&A* 515, A16
Salmon, S. J. A. J., Montalbn, J., Reese, D. R., Dupret, M.-A., & Eggenberger, P. 2014, *A&A* 569, A18
Savonije, G. J. 2005, *A&A* 443, 557
Savonije, G. J. 2013, *A&A* 559, A25
Scuflaire, R., Théado, S., Montalbán, J., *et al.* 2008, *Ap&SS* 316, 83

Townsend, R. H. D. 2003, *MNRAS* 343, 125
Townsend, R. H. D. 2005a, *MNRAS* 360, 465
Townsend, R. H. D. 2005b, *MNRAS* 364, 573
von Zeipel, H. 1924, *MNRAS* 84, 665

Discussion

MEYNET: In case faster rotating stars are more frequent at low metallicity, can we expect more stars showing this type of variability?

SALMON: I expect, for a lower metallicity, a lower number of prograde sectoral modes found to be excited, since their driving mechanism is dependent on the metal content of the star. Similarly, excited axisymmetric modes should only appear to be excited in slightly more massive SPB stars than at larger metallicity.

ANDERSON: Since you argue that the new variables in NGC 3766 are seen equator-on, have you deduced whether this is supported by random inclinations in that sample of stars?

SALMON: We plan in the future to carry out kinds of Monte-Carlo simulations of stellar populations in which the rotation rates and inclinations would vary randomly. We would then check whether the fraction of stars in these simulations that fits the conditions to present prograde sectoral modes agrees with the observations. Yet, the fact that some new variable stars close to the SPBs detected in NGC 3766 present modes with periods both close to 1 day and 0.3 day could correspond to less-slanted stars, showing axisymmetric or Yanai modes and being hence less shifted due to the gravity darkening.

SAIO: If the traditional approximation is not used, some modes are damped due to mode interactions.

SALMON: I agree with that comment. Lee (2008) computed pulsations in a 2D model, taking into account centrifugal distortion and showed that the driving of retrograde modes was less efficient, in part due to coupling between modes. On the contrary, the driving of prograde modes seemed to be amplified.

Sébastien Salmon

New windows on massive stars: asteroseismology, interferometry, and spectropolarimetry
Proceedings IAU Symposium No. 307, 2014
G. Meynet, C. Georgy, J. H. Groh & Ph. Stee, eds.
© International Astronomical Union 2015
doi:10.1017/S174392131400670X

Asteroseismology of OB stars with hundreds of single snapshot spectra
(and a few time-series of selected targets)

S. Simón-Díaz[1,2]

[1]Instituto de Astrofísica de Canarias, 38200 La Laguna, Tenerife, Spain
email: ssimon@iac.es

[2]Departamento de Astrofísica, Universidad de La Laguna, 38205 La Laguna, Tenerife, Spain

Abstract. Imagine we could do asteroseismology of large samples of OB-type stars by using just one spectrum per target. That would be great! But this is probably a crazy and stupid idea. Or maybe not. Maybe we have the possibility to open a new window to investigate stellar oscillations in massive stars that has been in front of us for many years, but has not attracted very much our attention: the characterization and understanding of the so-called macroturbulent broadening in OB-type stars.

Keywords. stars: early-type, stars: oscillations (including pulsations), line: profiles, techniques: spectroscopic

1. Introduction

1.1. *From macroturbulent broadening to pulsations (a bit of context)*

First references to macroturbulent broadening in O-type stars and early-B supergiants can be found in Struve (1952), Slettebak (1956), Conti & Ebbets (1977), Penny (1996), and Howarth *et al.* (1997). Based on the lack of this type of stars with sharp absorption lines, these authors proposed that rotation was not the only broadening mechanism shaping their line-profiles. This hypothesis was definitely confirmed with the advent of high-quality spectroscopic observations and its analysis by means of adequate techniques (e.g., Ryans *et al.* 2002; Simón-Díaz & Herrero 2007, 2014).

Modern measurements of the non-rotational broadening component soon allowed to discard that it was produced by any type of large scale turbulent motion (Simón-Díaz *et al.* 2010). In this context, Aerts *et al.* (2009) revived an alternative scenario to explain its physical origin: the *pulsational hypothesis*.

1.2. *The pulsational view of macroturbulent broadening (and vice versa)*

The basic idea proposed by Aerts *et al.* (also previously indicated by Lucy 1976, and Howarth 2004) is that the collective pulsational velocity broadening due to gravity modes could be a viable physical explanation for macroturbulence in massive stars. The presence of an important (variable) pulsational broadening component is firmly established in B dwarfs and giants located in the β Cep and SPB instability domains (e.g., Aerts *et al.* 2014, and references therein); however, the macroturbulent-pulsational broadening connection in O-type stars and B supergiants (B Sgs), located in a region of the HR diagram which to-date is (by far) less explored and understood from an asteroseismic point of view, required further (observational) confirmation.

As discussed by Simón-Díaz *et al.* (2012), this hypothesis might also open the possibility to use macroturbulent broadening as a spectroscopic indicator of the occurrence of a certain type of stellar oscillations in massive stars. Obviously, this alternative strategy will never be able to compete with a detailed asteroseismic study; however, if at some point the macroturbulent – pulsational broadening connection is confirmed, this spectroscopic feature could become a cheap, single-snapshot way to detect and investigate pulsations in massive stars from a complementary perspective. *Reward may be juicy, worth a shot!*

1.3. *The macro-pulsa project*

Motivated by the characterization of the macroturbulent broadening in the whole O and B-type star domain, and the investigation of its postulated pulsational origin, we have compiled during the last 6 years a large high-resolution spectroscopic dataset comprising more than 3500 spectra of about 500 Galactic O4-B9 stars (including all luminosity classes). In this talk I highlight the most important results obtained to-date in this context from the exploitation of this unique spectroscopic dataset.

2. Line-broadening in OB stars: a single-snapshot overview

2.1. *Observations, methods, and some results*

The spectroscopic observations are drawn from the *IACOB spectroscopic database of Northern Galactic OB star*. Last described in Simón-Díaz & Herrero (2014, SDH14), the database now also include new† HERMES spectra of bright B2–B9 stars. We have discarded from the initial sample all stars detected as SBx (x $\geqslant$ 2) and applied the methodology described in SDH14 to disentangle the rotational ($v \sin i$) and macroturbulent (v_{mac}, using a radial-tangential prescription) broadening components from the O III λ 5592, Si III λ 4552, Mg II λ 4481 or C II λ 4267 line-profiles. A first estimation of the effective temperature (T_{eff}) and gravity ($\log g$) in the O- and B-star samples was obtained by means of updated versions of the grid-based automatized tools described in Simón-Díaz *et al.* (2011) and Castro *et al.* (2012), respectively.

Fig. 1 summarizes the results of the line-broadening analysis of our final sample in the $v \sin i - v_{\mathrm{mac}}$ diagram. Stars with $v \sin i \geqslant 180 \, \mathrm{km \, s^{-1}}$ are excluded from this discussion (see SDH14). We differentiate six regions in the diagram depending of the relative contribution of the rotational and macroturbulent broadenings to the line-profile. The most important result are (1) the strong correlation found in stars with $v_{\mathrm{mac}} \geqslant v \sin i$, and (2) that stars with similar ($v \sin i, v_{\mathrm{mac}}$) combinations present almost identical profiles independently of their spectral type and luminosity class.

Stars with and without a clear macroturbulent broadening component (the former grouped in the shadowed gray region in Fig. 1) are located separately in the $\log T_{\mathrm{eff}} - \log g$ diagram in Fig. 2. The two diagrams are complemented with state-of-the-art evolutionary tracks and computations of high-order g-mode instability domains. Interestingly, there is a clear separation between the regions in the HR diagram where the stars with/without an important contribution of the macroturbulent broadening are located. However, there does not seem to be any clear correlation between these regions and the predicted instability domains for high order g-modes. Especially challenging for the pulsational hypothesis (in terms of g-modes) are the O-stars and the late-B Sgs.

---

† The *IACOB database* initially comprised FIES@NOT spectra of bright O- and early B-type stars. Given the similar capabilities and performance of HERMES@MERCATOR, we have decided to join together the IACOB spectra with those gathered as part of the IACOB-sweG (P.I. Negueruela) and macro-pulsa projects (P.I. Simón-Díaz), mainly using HERMES.

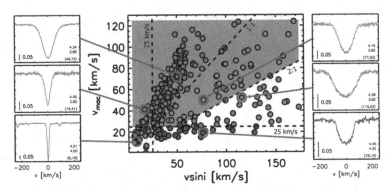

Figure 1. $v \sin i - v_{\mathrm{mac}}$ diagram including 100 O- and 200 B-type stars. The diagram is divided into six regions depending of the relative contribution of the rotational and macroturbulent broadenings to the global shape of the line-profiles (see also notes in SDH14). The three regions with a clear contribution of the macroturbulent broadening are shadowed in gray. Panels to the left and right show one illustrative profile per zone. Numbers quoted in each panel are the (normalized) flux scale, $\log T_{\mathrm{eff}}$, $\log g$, and the measured $v \sin i$ and v_{mac} (both in $\mathrm{km\,s}^{-1}$).

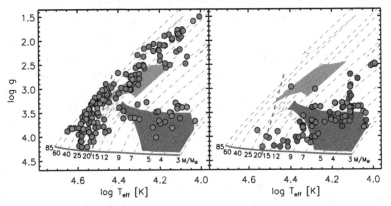

Figure 2. $\log T_{\mathrm{eff}} - \log g$ diagrams with the analyzed stars, separated by cases with (right) and without (left) a clear macroturbulent broadening component. The red tilted line separates O- and B-stars. In the background: evolutionary tracks from Ekström *et al.* (2012) and high-order g-mode instability strips from Miglio *et al.* (2007, OPAL, GN93, MS, $M \leqslant 18\,M_\odot$; dark gray) and (Godart 2011, OPAL, GN93, Vink mass-loss rates, MS and post-MS, $M \geqslant 10\,M_\odot$; light gray).

3. Macroturbulent broadening and line-profile variability

3.1. *Observations, methods, and some results*

During the last 6 years we have been compiling spectroscopic time-series (with both FIES and HERMES) of a selected subsample of 10 O stars and 11 B Sgs. All targets (except three: one O-dwarf, one B-dwarf, and one late-B Sg) were selected as having a dominant macroturbulent broadening component. For a few of them we already have more than 150 spectra, for the rest we count on a few dozens of spectra. The main bulk of the observations are characterized by a cadence of 4-8 spectra per night separated by 0.5-1.5 h. The typical exposure time is between a few seconds and 15 min. The length of each separated run is typically a few nights (being 10 nights the longest one). We are investigating line-profile variability (LPV) using the moment method (Aerts *et al.* 1992), and line-broadening variability using the IACOB-BROAD tool (SDH14).

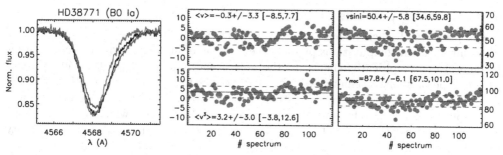

Figure 3. Illustrative example of the type of line-profile variability found in the sample of O stars and B supergiants for which we have obtained spectroscopic time-series. [Right panels] Variability in the first and third moments of the line-profile (center), and the two parameters defining the line-broadening (right). Quoted numbers indicate the mean value, standard deviation, minimum and maximum values, respectively. All quantities in km s^{-1} except for $\langle v^3 \rangle$ (in 10^4 km^3 s^{-3}). [Left panel] Three characteristic profiles having $\langle v^3 \rangle = 0$ (black) and maximum negative/positive skewness (red and blue, respectively).

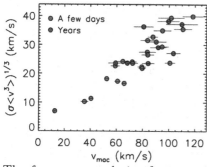

Figure 4. Observational correlation found between the size of the macroturbulent broadening and the amplitude of variation of the skewness of the line-profile. Horizontal lines indicate the range of variability of v_{mac}. Close-by red and blue symbols correspond to the same stars but quantities computed covering different temporal scales (a few days and years, respectively).

The frequency analysis of our spectroscopic time-series (prior to any mode identification) is a complicated task due to the non-optimal observational cadence and the complexity of the line-profile variability detected in this type of stars (C. Aerts, these proceedings). However, we can already highlight some first observational results which must be taken into account by any attempt to explain the physical origin of the macroturbulent broadening:

• Line-profiles of all the studied stars show LPVs (with similar characteristics as those illustrated in Fig. 3). Despite the variability, the global characteristic *V-shape* of the profiles remains *roughly* constant. Indeed, a similar shape can be found in very short ($\sim$ a few seconds) and much longer (e.g. co-adding all the spectra obtained during 6 years) exposures.

• The typical peak-to-peak amplitude of variability of the first moment ($\langle v \rangle$, centroid) is $\sim$ a few km s^{-1}. We also find small variability ($\leqslant 10\%$) in the $v \sin i$ and v_{mac} measurements provided by IACOB-BROAD.

• We find strong correlations between v_{mac} and the amplitude of variability of the third moment ($\langle v^3 \rangle$, skewness); this was shown for the first time in Simón-Díaz *et al.* (2010) and is now confirmed with a larger sample and a longer time-span (see Fig. 4).

• Although the precise identification of the frequencies of line-profile variability is difficult, we find some systematic hits of all stars showing multi-periodic variability in the high-order g-mode frequency domain ($\sim$ a few hours to several days). However, the possibility of non-strictly-periodic variability cannot be discarded.

4. Concluding remarks

The presence of macroturbulent broadening in the whole O-type star domain, where instability domain computations predict no excitation of high-order gravity mode, seems to be a strong empirical challenge to the pulsational hypothesis (at least in reference to this type of stellar oscillations). The same occurs for the late-B Sgs. However, the detected line-profile variability in all studied stars with macroturbulent broadening, and the empirical correlations found between v_{mac} and $\sigma(\langle v^3 \rangle)$ points towards to existence of a connection between the physical origin of this spectroscopic feature and stellar variability phenomena occurring in the photosphere of stars with masses above 15 $M_\odot$.

The understanding of the macroturbulent broadening is still an open issue, but the new observational material and the strategy proposed here will certainly help us to find the solution of a long-standing question in the field of massive stars. Even if the pulsational hypothesis may be rejected at some point, any other proposed scenario must fulfill the empirical (single snapshot and time-dependent) constraints described in Sects. 2 and 3. In this context, it is important to have also present other related studies proving complementary empirical constraints (e.g., Sundqvist et al. 2013; Markova et al. 2014), or alternative scenarios to the origin of the macroturbulent broadening (e.g., Cantiello et al. 2009), and maintain the initiated synergies between this research line and Asteroseismology. A new window on massive stars is now fully open.

Acknowledgements

Although they are not included in the co-author list of this proceeding, I would like to thank many colleagues for interesting discussions during the development of this project. In particular to A. Herrero, C. Aerts, P. Degroote, J. Puls, M. Godart, N. Markova, and E. Moravveji. Special thanks to N. Castro for providing me with the stellar parameters of the B star sample, and to a long list of observing colleagues who are making the compilation of the observational data needed for this project more manageable: I. Negueruela, R. Dorda, I. Camacho, K. Rübke, P. de Cat, S. Triana, C. González, A. González, E. Niemczura, D. Drobek. Last, but not least, the NOT and MERCATOR staff for their high competence and always useful assistance. This work has been funded by the Spanish Ministry of Economy and Competitiveness under the grants AYA2010-21697-C05-04, and Severo Ochoa SEV-2011-0187, and by the Canary Islands Government under grant PID2010119.

References

Aerts, C., de Pauw, M., & Waelkens, C. 1992, A&A 266, 294
Aerts, C., Puls, J., Godart, M., & Dupret, M.-A. 2009, A&A 508, 409
Aerts, C., Simon-Diaz, S., Groot, P. J., & Degroote, P. 2014, ArXiv e-prints
Cantiello, M., Langer, N., Brott, I., et al. 2009, A&A 499, 279
Castro, N., Urbaneja, M. A., Herrero, A., et al. 2012, A&A 542, A79
Conti, P. S. & Ebbets, D. 1977, ApJ 213, 438
Ekström, S., Georgy, C., Eggenberger, P., et al. 2012, A&A 537, A146
Godart, M. 2011, Ph.D. thesis, University of Liege, Belgium
Howarth, I. D. 2004, in A. Maeder & P. Eenens (eds.), Stellar Rotation, Vol. 215 of IAU Symposium, p. 33
Howarth, I. D., Siebert, K. W., Hussain, G. A. J., & Prinja, R. K. 1997, MNRAS 284, 265
Lucy, L. B. 1976, ApJ 206, 499
Markova, N., Puls, J., Simón-Díaz, S., et al. 2014, A&A 562, A37
Miglio, A., Montalbán, J., & Dupret, M.-A. 2007, Communications in Asteroseismology 151, 48
Penny, L. R. 1996, ApJ 463, 737
Ryans, R. S. I., Dufton, P. L., Rolleston, W. R. J., et al. 2002, MNRAS 336, 577

Simón-Díaz, S., Castro, N., Herrero, A., *et al.* 2012, in L. Drissen, C. Rubert, N. St-Louis, & A. F. J. Moffat (eds.), *Proceedings of a Scientific Meeting in Honor of Anthony F. J. Moffat*, Vol. 465 of *Astronomical Society of the Pacific Conference Series*, p. 19

Simón-Díaz, S., Castro, N., Herrero, A., *et al.* 2011, *Journal of Physics Conference Series* 328(1), 012021

Simón-Díaz, S. & Herrero, A. 2007, *A&A* 468, 1063

Simón-Díaz, S. & Herrero, A. 2014, *A&A* 562, A135

Simón-Díaz, S., Herrero, A., Uytterhoeven, K., *et al.* 2010, *ApJ (Letters)* 720, L174

Slettebak, A. 1956, *ApJ* 124, 173

Struve, O. 1952, *PASP* 64, 117

Sundqvist, J. O., Petit, V., Owocki, S. P., *et al.* 2013, *MNRAS* 433, 2497

Discussion

HENRICHS: Why did this very nice study only use one spectral line for O stars? The reason I ask is that we found that nearly all H and He lines in our O-star sample show significant variations beyond $v \sin i$, a signature of an additional mechanism that exclude pulsational origin.

SIMÓN-DÍAZ: Thanks for your comment, Huib. For the moment we are concentrating in metal lines, where macroturbulent broadening is more easily disentangled from other broadening agents and for which we expect to be less affected by other mechanisms leading to line-profile variability, such as, for example, wind variability. But at some point, we plan to extend our study to more lines. We will certainly account for the results of your sample at this point.

ANDERSON: Have you considered the effect of changing radial and tangential contributions on the macroturbulence values you derive? If these are dependent, they may help to explain some of the red noise you observe for the pulsating stars.

SIMÓN-DÍAZ: This is something we have considered to explore at some point; however, we are aware this is a difficult task since the line-broadening characterization then becomes multi-parametric and highly degenerated. But we keep it in mind, thanks.

LANDSTREET: The "radial-tangential" line component used to separate macroturbulent motions from rotational broadening and convective motions (microturbulence) is *not* a model of any actual hydrodynamic motions or of pulsation. It is simply a convenient mathematical form used to model the very triangular line profiles found in many stars, which cannot be fit with any combination of Gaussian (microscopic) and rotational profiles. This "macroturbulence" line profile component may be well-correlated with the presence of pulsations at the stellar surface, but it is *not* a model of these motions, and so tuning parameters of this line component (e.g. ratio of "radial" to "tangential" components) does not provide useful insight. Radial-tangential macroturbulence is most useful to identify stellar atmospheres in which certain types of motions occur, which can be studied in more detail.

SIMÓN-DÍAZ: Thanks, John. I completely agree.

*New windows on massive stars: asteroseismology, interferometry, and
spectropolarimetry*
Proceedings IAU Symposium No. 307, 2014
G. Meynet, C. Georgy, J. H. Groh & Ph. Stee, eds.
© International Astronomical Union 2015
doi:10.1017/S1743921314006711

Probing high-mass stellar evolutionary models with binary stars

A. Tkachenko†

Instituut voor Sterrenkunde, KU Leuven, Celestijnenlaan 200D, B-3001 Leuven, Belgium
email: Andrew.Tkachenko@ster.kuleuven.be

Abstract. Mass discrepancy is one of the problems that is pending a solution in (massive) binary star research field. The problem is often solved by introducing an additional near core mixing into evolutionary models, which brings theoretical masses of individual stellar components into an agreement with the dynamical ones. In the present study, we perform a detailed analysis of two massive binary systems, V380 Cyg and σ Sco, to provide an independent, asteroseismic measurement of the overshoot parameter, and to test state-of-the-art stellar evolution models.

Keywords. asteroseismology, star: oscillations (including pulsations), line: profiles, methods: data analysis, techniques: photometric, techniques: spectroscopic, (stars:) binaries: spectroscopic, stars: fundamental parameters, stars: individual (V380 Cyg, σ Sco)

1. Introduction

One of the major problems that is currently pending a solution in binary star research field is the so-called mass discrepancy problem. It stands for the difference between the component masses inferred from binary dynamics (hereafter, dynamical masses) and those obtained from spectral characteristics of stars and evolutionary models (hereafter, theoretical masses). The mass discrepancy problem observed in massive O- and B-type stars has been known for more than 20 years already and has been discussed in detail by Herrero *et al.* (1992). Hilditch (2004) pointed out that the discrepancy does not disappear when the effects of rotation are included into the models.

A remarkable mass discrepancy has been reported by Guinan *et al.* (2000) for the primary components of the V380 Cyg system. The authors showed that large amount of core overshoot ($\alpha_{ov} = 0.6$ pressure scale height) can account for the difference between dynamical and theoretical mass of the primary component. This large amount of overshoot contradicts the typical value of $\alpha_{ov} <= 0.2\,H_P$ observed for single stars of similar mass (see e.g. Aerts 2013, Aerts *et al.* 2003, 2011; Briquet *et al.* 2011). Moreover, the second largest value after V380 Cyg of $\alpha_{ov} \sim 0.45$ has also been measured in a binary system, for the $8\,M_\odot$ primary component of the θ Ophiuchi system (Briquet *et al.* 2007). Recently, Garcia *et al.* (2014) found that convective overshoot $\alpha_{ov} > 0.35$ is required to reproduce observed absolute dimensions of both components of the V578 Mon system. Thus, there seems to be a tendency of measuring larger core overshoot in binary stars than in single objects, within the same stellar mass range. The reason is that this parameter often effectively accounts for the above mentioned mass discrepancy, but we need to investigate the feasibility of using core overshoot alone to explain such a complex problem as discrepancy between dynamical and theoretical masses in binary stars.

In this paper, we present a study of two massive binary star systems, V380 Cyg and σ Sco, and aim at measuring core overshoot parameter for pulsating components using asteroseismic methods.

† Postdoctoral Fellow of the Fund for Scientific Research (FWO), Flanders, Belgium

Table 1. Key orbital, physical, and atmospheric parameters of the V380 Cyg and σ Sco systems, as derived from our photometric and/or spectroscopic data.

Parameter		V380 Cyg		σ Sco	
		Primary	Secondary	Primary	Secondary
Period,	(day)	12.425719		33.016±0.012	
Periastron passage time,	(HJD)	24 54 602.888±0.007		24 34 886.11±0.04	
Periastron long.,	(degree)	138.4±0.4		288.1±0.8	
eccentricity,		0.2261±0.0004		0.383±0.008	
RV semi-amplitude,	(km s^{-1})	93.54±0.07	152.71±0.22	30.14±0.35	47.01±0.98
Mass,	($M_\odot$)	11.43±0.19	7.00±0.14	14.7±4.5	9.5±2.9
Radius,	($R_\odot$)	15.71±0.13	3.819±0.048	9.2±1.9	4.2±1.0
$T_{\rm eff}$,	(K)	21 700±300	22 700±1 200	25 200±1 500	25 000±2 400
$\log g$,	(dex)	3.104±0.006	4.120±0.011	3.68±0.15	4.16±0.15
v sin i,	(km s^{-1})	98±2	38±2	31.5±4.5	43.0±4.5

2. V380 Cyg

V380 Cyg is a bright ($V = 5.68$) double-lined spectroscopic binary (SB2, Hill & Batten 1984) consisting of two B-type stars residing in an eccentric 12.4 day orbit. The primary component is an evolved star, whereas the secondary just started its life on the main-sequence. Pavlovski *et al.* (2009) revisited the *U, B, V* light curves obtained by Guinan *et al.* (2000) and collected about 150 high-resolution échelle spectra using several telescopes. The authors presented a revised orbital solution, and used spectral disentangling technique (Simon & Sturm 1994) formulated in Fourier space (Hadrava 1995), as implemented in the FDBINARY code (Ilijić *et al.* 2004), to determine disentangled spectra of both binary components. Similar to the results of Guinan *et al.* (2000), a remarkable mass discrepancy was found for the primary component of V380 Cyg. Moreover, Pavlovski *et al.* (2009) came to the same conclusion as Hilditch (2004) did, namely that the effects of rotation included into evolutionary models could not fully account for the observed discrepancy.

The discovery of seismic signal in the primary component of V380 Cyg (Tkachenko *et al.* 2012) opened up an opportunity of obtaining an independent measurement of the overshoot parameter for this binary system. We base our analysis on about 560 days long time-series of high-precision *Kepler* photometry, and about 400 high-resolution, high signal-to-noise ratio (S/N) spectra obtained with HERMES spectrograph (Raskin *et al.* 2011) attached to 1.2-meter Mercator telescope (La Palma, Canary Islands).

The effects of binarity in the *Kepler* light curve were modelled using the JKTWD wrapper (Southworth *et al.* 2011) of the 2004 version of the Wilson-Devinney code (Wilson & Devinney 1971; Wilson 1979). The time-series of spectra were analysed with the FDBINARY code to determine spectroscopic orbital solution and disentangled spectra of both stellar components. We found that light curve yields poor constraints on the shape of the orbit, because of strong correlation between eccentricity e and longitude of periastron ω. Since the quantities $e \cos\omega$ and $e \sin\omega$ are respectively well constrained from photometry and spectroscopy, we constrained the orbital shape by iterating between the two analyses: the light curve was used to determine best fit ω for a given e, and the spectral disentangling to find the best e for a given ω. Analysis of the disentangled spectra delivered accurate atmospheric parameters and individual abundances for both binary components. Table 1 lists some key orbital, physical, and atmospheric properties of this binary system; for more details, reader is referred to Tkachenko *et al.* (2014a).

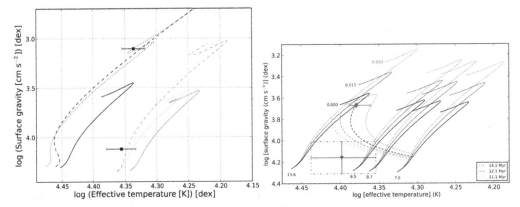

Figure 1. Left: Location of the primary and secondary components of V380 Cyg in the $T_{\mathrm{eff}} - \log g$ diagram. Solid, long- and short-dashed tracks correspond to models 1, 2 and 3 in Table 2, respectively. T_{eff} and $\log g$ values are those from Table 1. **Right:** Position of the primary (circle) and the secondary (triangle) of the σ Sco system in the $T_{\mathrm{eff}} - \log g$ diagram, along with the MESA evolutionary tracks. The initial masses as well as the overshoot parameter f_{ov} are indicated in the plot. The atmospheric parameters T_{eff} and $\log g$ are those from Table 3. The isochrones corresponding to the age of the system of 12.1 Myr and its error bars deduced from seismology of the primary are indicated by the dashed lines.

Table 2. Evolutionary model parameters for both components of the V380 Cyg system. α_{ov} and v stand for the overshoot parameter and initial rotation rate, respectively.

Parameter		Primary			Secondary		
		Model 1	Model 2	Model 3	Model 1	Model 2	Model 3
M,	$(M_\odot)$	11.43	12.00	12.90	7.00	7.42	7.42
Z,	(dex)	0.014	0.012	0.012	0.014	0.012	0.012
α_{ov},	(H_P)	0.2	0.6	0.3	0.2	0.6	0.2
v,	$(\mathrm{km\,s^{-1}})$	0	241	243	0	0	0
Age,	(Myr)	–	21.5	18	–	21.5	18

Photometric residuals obtained after the subtraction of our best fit model were subjected for frequency analysis. The majority of the frequencies are variable both in appearance and amplitude, in agreement with the conclusions made by Tkachenko et al. (2012) about stochastic nature of the signal. The variability consistent with the expected rotation period of the primary component has been detected on top of the binarity and stochastic oscillation signals in the star. We speculate that this signal comes from rotational modulation of spot-like chemical abundance or temperature structures on the surface of the primary component. To verify this hypothesis, spectral line profiles of the primary has been examined for spot signatures, after subtracting the contribution of the companion star from the composite spectra of V380 Cyg. We found a remarkable variability in all observed silicon lines of more massive binary component, which could not be detected in spectral lines of other chemical elements and was found to be consistent with the rotational period of the star. INVERS8 (Piskunov & Rise 1993) code was used to perform Doppler Imaging analysis based on several prominent lines of doubly ionized silicon found in the spectrum of the primary component. The obtained results suggest the presence of two high-contrast stellar surface abundance spots which are located either at the same latitude or longitude.

Finally, we compare our revised fundamental stellar parameters of both components of the V380 Cyg system with the state-of-the-art evolutionary models computed with

Table 3. Fundamental parameters of both components of the σ Sco, after seismic modelling of the primary. Parameters determined spectroscopically are highlighted in boldface.

Parameter		Primary	Secondary
Mass,	$(M_\odot)$	$13.5^{+0.5}_{-1.4}$	$8.7^{+0.6}_{-1.2}$
Radius,	$(R_\odot)$	$8.95^{+0.43}_{-0.66}$	$3.90^{+0.58}_{-0.36}$
Luminosity (log (L)),	$(L_\odot)$	$4.38^{+0.07}_{-0.15}$	$3.73^{+0.13}_{-0.15}$
Age of the system,	(Myr)		12.1
Overshoot (f_{ov}),	(H_P)	$0.000^{+0.015}$	–
T_{eff},	(K)	$23\,945^{+500}_{-990}$	**$25\,000^{+2\,400}_{-2\,400}$**
log g,	(dex)	$3.67^{+0.01}_{-0.03}$	**$4.16^{+0.15}_{-0.15}$**

the MESA code (Paxton *et al.* 2011, 2013). Figure 1 (left) shows the location of both components of V380 Cyg in the T_{eff} − log g diagram along with the evolutionary tracks. The two models that fit the positions of both stars in the diagram, taking into account the error bars, are illustrated by long- and short-dashed lines (models 2 and 3 in Table 2, respectively). The dynamical mass models for both binary components are shown by solid lines (model 1 in Table 2). The MESA model predictions clearly point to mass discrepancy for the primary component, in agreement with the findings by Guinan *et al.* (2000) and Pavlovski *et al.* (2009). **We conclude that present-day single-star evolutionary models are inadequate for this particular binary system, and lack a serious amount of near-core mixing.**

3. σ Sco

Sigma Scorpii is a double-lined spectroscopic binary in a quadruple system. Two components are early B-type stars, residing in an eccentric, 33 day period orbit. Though the star was a subject of numerous photometric and spectroscopic studies in the middle of 20th century, its double-lined nature was discovered by Mathias *et al.* (1991). So far, the studies by Mathias *et al.* (1991), Pigulski (1992), and North *et al.* (2007) have been the most extensive ones focusing on orbital and physical properties of the system.

Besides the σ Sco system is a spectroscopic binary, its evolved primary component is known to be unstable to β Cep-type stellar pulsations. Moreover, according to Kubiak (1980), the amplitude of the dominant, radial pulsation mode of the primary is comparable to the orbital semi-amplitude K_1 of this star. This fact was not taken into account in either of the previous studies focusing on orbital solution, but is taken into consideration in our study.

Our analysis is based on some 1000 high-resolution spectra collected with the CORALIE spectrograph attached to the 1.2-meter Euler Swiss Telescope (La Silla, Chile). Orbital parameters of the system were initially derived based on an iterative approach, and further on refined using the method of spectral disentangling. The spectral disentangling was applied to a couple of dozen carefully selected spectra and corresponding to a zero pulsation phase (unperturbed profile), since the method assumes no variability intrinsic to stellar components forming a binary system. For more details on the procedure, reader is referred to Tkachenko *et al.* (2014b). The disentangled spectra were used to determine accurate atmospheric parameters and chemical composition of both stars. The final set of orbital parameters as well as the spectroscopically derived values of T_{eff} and log g are given in Table 1. The masses and radii listed in this table were determined from our orbital parameters and interferometric value of the orbital inclination angle reported by North *et al.* (2007).

We further made use of the fact that the primary component of σ Sco is a radial mode pulsator, and performed asteroseismic analysis of this star. Evolutionary models were computed with the MESA code, while p- and g-mode eigenfrequencies for mode degrees $l = 0$ to 3 have been calculated in the adiabatic approximation with the GYRE stellar oscillation code (Townsend & Teitler 2013). The addition of the seismic constraints to the modelling implied a drastic reduction in the uncertainties of the fundamental parameters, and provided an age estimate (see Table 3). Figure 1 (right) shows the position of both components of the σ Sco system in the $T_{\rm eff} - \log g$ diagram, along with the evolutionary tracks. The error bars are those obtained from the evolutionary models and 3σ spectroscopic uncertainties for the primary and secondary, respectively. **Though we make an a priori assumption in our seismic modelling that the models are appropriate for the primary component, similar to the case of V380 Cyg, we still find mass discrepancy for the main-sequence secondary component of the σ Sco system.**

References

Aerts, C. 2013, *EAS Publications Series* 64, 323
Aerts, C., Thoul, A., Daszyńska, J., *et al.* 2003, *Science* 300, 1926
Aerts, C., Briquet, M., Degroote, P., *et al.* 2011, *A&A* 534, A98
Briquet, M., Morel, T., Thoul, A., *et al.* 2007, *MNRAS* 381, 1482
Briquet, M., Aerts, C., Baglin, A., *et al.* 2011, *A&A* 527, A112
Garcia, E. V., Stassun, K. G., Pavlovski, K., *et al.* 2014, *AJ* 148, 39
Guinan, E. F., Ribas, I., Fitzpatrick, E. L., *et al.* 2000, *ApJ* 544, 409
Hadrava, P. 1995, *A&AS* 114, 393
Herrero, A., Kudritzki, R. P., Vilchez, J. M., *et al.* 1992, *A&A* 261, 209
Hilditch, R. W. 2004, *ASP Conference Series* 318, 198
Hill, G. & Batten, A. H. 2004, *A&A* 141, 39
Ilijić, S., Hensberge, H., Pavlovski, K., & Freyhammer, L. M. 2004, *ASP Conference Series* 318, 111
Kubiak, M. 1980, *AcA* 30, 41
Mathias, P., Gillet, D., & Crowe, R. 1991, *A&A* 252, 245
North, J. R., Davis, J., Tuthill, P. G., *et al.* 2007, *MNRAS* 380, 1276
Pavlovski, K., Tamajo, E., Koubský, P., *et al.* 2009, *MNRAS* 400, 791
Paxton, B., Bildsten, L., Dotter, A., *et al.* 2011, *ApJS* 192, 3
Paxton, B., Cantiello, M., Arras, P., *et al.* 2013, *ApJS* 208, 4
Pigulski, A. 1992, *A&A* 261, 203
Piskunov, N. E. & Rice, J. B. 1993, *PASP* 105, 1415
Raskin, G., van Winckel, H., Hensberge, H., *et al.* 2011, *A&A* 526, A69
Simon, K. P. & Sturm, E. 2009, *A&A* 281, 286
Southworth, J., Zima, W., Aerts, C., *et al.* 2011, *MNRAS* 414, 2413
Tkachenko, A., Aerts, C., Pavlovski, K., *et al.* 2012, *MNRAS* 424, L21
Tkachenko, A., Degroote, P., Aerts, C., *et al.* 2014a, *MNRAS* 438, 3093
Tkachenko, A., Aerts, C., Pavlovski, K., *et al.* 2014b, *MNRAS* 442, 616
Townsend, R. H. D. & Teitler, S. A. 2013, *MNRAS* 435, 3406
Wilson, R. E. 1979, *ApJ* 234, 1054
Wilson, R. E. & Devinney, E. J. 1971, *ApJ* 166, 605

Discussion

MAEDER: The overshooting parameter is just a fitting parameter on the core mass. The fact you get higher overshoot parameters for some binaries could be due to other effects, for example shears created by tidal interactions in binaries.

TKACHENKO: Agreed. This is exactly why we do this kind of research. We want to answer the question: is it true that the overshoot parameter is different for components in binary systems from single stars of similar mass? If so, why?

Andrew Tkachenko

Sergio Simón-Díaz

Ehsan Moravveji

New windows on massive stars: asteroseismology, interferometry, and
spectropolarimetry
Proceedings IAU Symposium No. 307, 2014
G. Meynet, C. Georgy, J. H. Groh & Ph. Stee, eds.

© International Astronomical Union 2015
doi:10.1017/S1743921314006723

Rotation and the Cepheid Mass Discrepancy

Richard I. Anderson[1], Sylvia Ekström[1], Cyril Georgy[2], Georges Meynet[1], Nami Mowlavi[1] and Laurent Eyer[1]

[1]Observatoire de Genève, Université de Genève, 51 Ch. des Maillettes, 1290 Sauverny,
Switzerland
email: `richard.anderson@unige.ch`

[2]Astrophysics group, Lennard-Jones Laboratories, EPSAM, Keele University, Staffordshire,
ST5 5BG, UK

Abstract. We recently showed that rotation significantly affects most observable Cepheid quantities, and that rotation, in combination with the evolutionary status of the star, can resolve the long-standing Cepheid mass discrepancy problem. We therefore provide a brief overview of our results regarding the problem of Cepheid masses. We also briefly mention the impact of rotation on the Cepheid period-luminosity(-color) relation, which is crucial for determining extragalactic distances, and thus for calibrating the Hubble constant.

Keywords. stars: evolution, stars: rotation, supergiants, Cepheids, distance scale

1. Introduction

Classical Cepheids are evolved intermediate-mass stars observed during brief phases of stellar evolution that render them highly precise standard candles. They are furthermore excellent laboratories of stellar structure and evolution thanks to their variability and location in the Hertzsprung-Russell diagram. Despite the adjectives *classical* and *standard*, Cepheids are all but sufficiently well understood. A key symptom of this is the 45-year-old Cepheid mass discrepancy problem (Christy 1968; Stobie 1969a,b,c) that has been estimated until recently to be in the range of $10 - 20\%$ (Bono *et al.* 2006) and has motivated much research into convective core overshooting (e.g. Prada Moroni *et al.* 2012) and enhanced mass-loss (Neilson & Lester 2008).

2. Rotation to the Rescue

While convective core overshooting is successful in increasing core size and thereby increasing luminosity at fixed mass, it cannot fully explain the mass discrepancy, since high values ($\geqslant 20\%$ of pressure scale height) of convective core overshooting also suppress the appearance of blue loops at the low-mass end, cf. Anderson *et al.* (2014, Fig. 1). This is a problem, since the majority of Cepheids are understood to reside on blue loops and have short periods, i.e., they originate from relatively low ($\sim 5\,M_\odot$) mass B-stars.

We recently presented the first detailed investigation of the effect of rotation on populations of classical Cepheids (Anderson *et al.* 2014) based on the latest Geneva stellar evolution models (Ekström *et al.* 2012; Georgy *et al.* 2013) that incorporate a homogeneous treatment of rotation over a large range of masses. We found that rotation, together with evolutionary status (i.e., identification of the instability strip (IS) crossing) can resolve the mass discrepancy, and mass-luminosity relations (MLRs) of models for typical initial rotation rates agree better with observed Cepheid masses than models without rotation, see Fig. 1. Furthermore, rotation does not suppress the appearance of

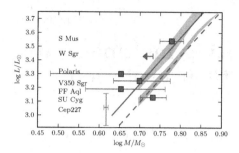
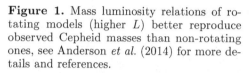

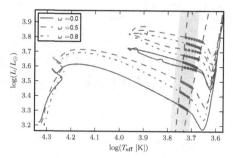

Figure 1. Mass luminosity relations of rotating models (higher L) better reproduce observed Cepheid masses than non-rotating ones, see Anderson *et al.* (2014) for more details and references.

Figure 2. Evolutionary tracks for a $7\,M_\odot$ model with Solar metallicity and different initial rotation rates quantified by $\omega = \Omega/\Omega_{\rm crit}$. The effect of rotation during the main sequence carries through to the later evolutionary stages.

blue loops (cf. Fig. 2) and is thus in better agreement with observations than models invoking high overshooting values.

3. Implications

An important consequence of rotation is that no unique MLR applies to all stars. The farther a star evolves along the main sequence, the larger this difference tends to become. The difference in main sequence turn off luminosity between models of different rotation rates carries over into the more advanced evolutionary stages. For Cepheids, luminosity also tends to increase between the 2nd and 3rd IS crossings, adding further complexity. To estimate a Cepheid's mass given the luminosity, its evolutionary status must therefore be taken into account. Measured rates of period change provide empirical measurements of the IS crossings, and are furthermore sensitive to initial rotation.

Finally, we point out that rotation can lead to intrinsic scatter in the period-luminosity relation (PLR) and the period-luminosity-color-relation (PLCR). The PLCR follows from inserting an MLR into the pulsation equation ($P \propto 1/\sqrt{\bar\rho}$, Ritter 1879). As there is no unique MLR (cf. above), there cannot be a unique PLCR. This finding has potentially important implications for the accuracy of Cepheid distances and thus for the distance scale. Further investigation in this direction is in progress.

References

Anderson, R. I., Ekström, S., Georgy, C., *et al.* 2014, *A&A* 564, A100
Bono, G., Caputo, F., & Castellani, V. 2006, *MemSAIt* 77, 207
Christy, R. F. 1968, *QJRAS* 9, 13
Ekström, S., Georgy, C., Eggenberger, P., *et al.* 2012, *A&A* 537, A146
Georgy, C., Ekström, S., Granada, A., *et al.* 2013, *A&A* 553, A24
Neilson, H. R. & Lester, J. B. 2008, *ApJ* 684, 569
Prada Moroni, P. G., Gennaro, M., Bono, G., *et al.* 2012, *ApJ* 749, 108
Ritter, A. 1879, *Wiedemanns Annalen* VIII, 173
Stobie, R. S. 1969a, *MNRAS* 144, 461
Stobie, R. S. 1969b, *MNRAS* 144, 485
Stobie, R. S. 1969c, *MNRAS* 144, 511

New windows on massive stars: asteroseismology, interferometry, and
spectropolarimetry
Proceedings IAU Symposium No. 307, 2014
G. Meynet, C. Georgy, J. H. Groh & Ph. Stee, eds.

© International Astronomical Union 2015
doi:10.1017/S1743921314006735

Tidal interactions in rotating multiple stars and their impact on their evolution

P. Auclair-Desrotour[1], S. Mathis[2] and C. Le Poncin-Lafitte[3]

[1]IMCCE, Observatoire de Paris, UMR 8028 du CNRS, UPMC,
77 Av. Denfert-Rochereau, 75014 Paris, France

[2]Laboratoire AIM Paris-Saclay, CEA/DSM - CNRS - Université Paris Diderot, IRFU/SAp
Centre de Saclay, F-91191 Gif-sur-Yvette Cedex, France

[3]SYRTE, Observatoire de Paris, UMR 8630 du CNRS, UPMC,
77 Av. Denfert-Rochereau, 75014 Paris, France

email: `pierre.auclair-desrotour@obspm.fr`, `stephane.mathis@cea.fr`,
`christophe.leponcin@obspm.fr`

Abstract. Tidal dissipation in stars is one of the key physical mechanisms that drive the evolution of binary and multiple stars. As in the Earth oceans, it corresponds to the resonant excitation of their eigenmodes of oscillation and their damping. Therefore, it strongly depends on the internal structure, rotation, and dissipative mechanisms in each component. In this work, we present a local analytical modeling of tidal gravito-inertial waves excited in stellar convective and radiative regions respectively. This model allows us to understand in details the properties of the resonant tidal dissipation as a function of the excitation frequencies, the rotation, the stratification, and the viscous and thermal properties of the studied fluid regions. Then, the frequencies, height, width at half-height, and number of resonances as well as the non-resonant equilibrium tide are derived analytically in asymptotic regimes that are relevant in stellar interiors. Finally, we demonstrate how viscous dissipation of tidal waves leads to a strongly erratic orbital evolution in the case of a coplanar binary system. We characterize such a non-regular dynamics as a function of the height and width of resonances, which have been previously characterized thanks to our local fluid model.

Keywords. hydrodynamics, waves, turbulence, (stars:) binaries (including multiple): close, (stars:) planetary systems, stars: rotation, stars: oscillations (including pulsations), stars: interiors, stars: evolution

---

Many stars of our galaxy are components of close binary or multiple stellar systems. In such multiple stars, tides play a key role to modify the rotational evolution of the components and of their orbit (de Mink *et al.* 2013, Langer in this volume). The same applies to the case of star-planet systems (Albrecht *et al.* 2012). In this framework, Zahn (1977) and Ogilvie & Lin (2007) demonstrated that the dissipation of the kinetic energy of the low-frequency eigenmodes of oscillation of the stars excited by tides (i.e. tidal inertial waves driven by the Coriolis acceleration in convection zones and gravito-inertial waves driven by the buoyancy restoring force and the Coriolis acceleration in stably stratified radiation zones) drives tidal migration, the circularization of the orbits, and the synchronization and alignment of the spins (Barker & Ogilvie 2009).

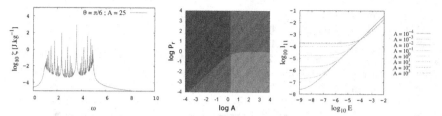

Figure 1. *Left:* Example of the frequency-spectra of the viscous dissipation of tidal gravito-inertial waves. *Middle:* the four asymptotic regimes for tidal waves in stars. *Right:* Example of scaling law for the width at mid-height of the resonance $\{m, n\} \equiv \{1, 1\}$ as a function of E and A.

1. Tidal waves in a box: a reduced model to understand tidal dissipation in stars

In this study, we consider a two-body system. The central body A, which has a mass M_A and a radius R_A, rotates with an angular velocity Ω. The punctual-mass tidal perturber B (M_B) has an orbit relative to A with a semi-major axis a and a mean motion $\tilde{n}$. We compute the viscous dissipation of tidal gravito-inertial waves, which are driven by the buoyancy and the Coriolis acceleration, using a local Cartesian model that generalizes the one of Ogilvie & Lin (2004). We thus consider a box of characteristic length L inclined with a co-latitude θ relatively to the rotation axis. The fluid has a density ρ, a kinematic viscosity ν, a thermal diffusivity κ, and a Brunt-Väisälä frequency N (with $N \approx 0$ in convective regions). The control parameters of the system are: i) the Froude number $A = [N/(2\Omega)]^2$ that compares the buoyancy force and the Coriolis acceleration, ii) the Ekman number $E = (2\pi^2\nu)/(\Omega L^2)$ that compares the viscous diffusion time to $t_\Omega = (2\Omega)^{-1}$, and iii) the thermal number $K = (2\pi^2\kappa)/(\Omega L^2)$ that compare the thermal diffusion time to t_Ω. The viscous dissipation (Fig. 1, left) strongly depends on the tidal frequency $\sigma = 2(\tilde{n} - \Omega)$ and of the fluid parameters (Ω, A, E, K). This complex behavior must be taken into account when studying the dynamical evolution of the system (Mathis & Le Poncin-Lafitte 2009). Four asymptotic regimes (see Fig. 1, middle) are identified. In convection zones ($A \leqslant 0$), we identify tidal inertial waves dissipated by viscous (in blue; $P_r = \nu/\kappa$ is the Prandt number) or thermal (in purple) diffusions. Respectively, in radiation zones ($A > 0$), we identify tidal gravito-inertial waves dissipated by thermal (in orange) or viscous (in red) diffusions. In these regimes, we have derived analytic scaling laws for the eigenfrequencies (ω_{mn}; m and n correspond to the horizontal and vertical wave numbers) of resonances, their width at half-height (l_{mn}) (Fig. 1, right), their height (h_{mn}), their number ($N_{\rm kc}$), and the amplitude of the non-resonant background ($H_{\rm bg}$) that corresponds to the so-called equilibrium tide (e.g. Remus *et al.* 2012). Finally, we demonstrated that in the case of such resonant tidal dissipation spectra, the orbital evolution becomes erratic (Witte & Savonije 1999; Auclair-Desrotour *et al.* 2014). This behavior is observed through jumps in semi-major axis that scale as $\Delta a/a \propto lH^{1/4}$, where the width at mid-height l and height H of resonances are characterized thanks to obtained scaling laws.

References

Albrecht, S., Winn, J. N., Johnson, J. A., *et al.* 2012, *ApJ* 757, 18

Auclair-Desrotour, P., Le Poncin-Lafitte, C., & Mathis, S. 2014, *A&A* 561, L7

Barker, A. J. & Ogilvie, G. I. 2009, *MNRAS* 395, 2268

de Mink, S. E., Langer, N., Izzard, R. G., Sana, H., & de Koter, A. 2013, *ApJ* 764, 166

Mathis, S. & Le Poncin-Lafitte, C. 2009, *A&A* 497, 889
Ogilvie, G. I. & Lin, D. N. C. 2004, *ApJ* 610, 477
Ogilvie, G. I. & Lin, D. N. C. 2007, *ApJ* 661, 1180
Remus, F., Mathis, S., & Zahn, J.-P. 2012, *A&A* 544, A132
Witte, M. G. & Savonije, G. J. 1999, *A&A* 350, 129
Zahn, J.-P. 1977, *A&A* 57, 383

From left to right Raphael Hirschi, Patrick Eggenberger and Andrea Miglio

From left to right Matteo Cantiello, Artemio Herrero and Francisco Najarro

New windows on massive stars: asteroseismology, interferometry, and spectropolarimetry
Proceedings IAU Symposium No. 307, 2014
G. Meynet, C. Georgy, J. H. Groh & Ph. Stee, eds.

© International Astronomical Union 2015
doi:10.1017/S1743921314006747

Constraints on stellar evolution from white dwarf asteroseismology

Agnès Bischoff-Kim

Penn State Worthington Scranton
email: axk55@psu.edu

Abstract. High mass and low mass stars follow a similar evolution until the inert core phase that follows the end of the core helium burning stage. In particular, one common phase of stellar evolution is the alpha capture reaction that turns carbon into oxygen in the core. We can obtain constraints on this reaction rate by studying the remnants of low mass stars, as this is the ultimate reaction that occurs in their core. We also present results that allow us to test the time dependent calculations of diffusion in dense interiors.

Keywords. dense matter, diffusion, nuclear reactions, stars: AGB and post-AGB, stars: interiors, stars: oscillations, white dwarfs

1. Introduction

A small subset of white dwarfs pulsate in non-radial, g-mode oscillations. These stars have been used to study a number of fundamental physical processes (e.g Bischoff-Kim *et al.* 2008; Bischoff-Kim 2008; Córsico *et al.* 2013), including the emission of weakly interacting particles. Pulsating white dwarfs are useful because we have a good set of observed frequencies for a dozen of them and they are relatively simple to model (neglible nuclear burning, no extended atmospheres, non-contracting, well determined equations of states...). With white dwarf asteroseismology, we can test stellar evolution calculations. We focus here on two processes: i) the $^{12}C(\alpha, \gamma)^{16}O$ nuclear reaction rate (section 2) and ii) the diffusion of elements over time in stellar interiors (section 3).

In white dwarf asteroseismology, we parameterize the core chemical profiles and allow them to take the shape required to have the models' frequencies match the observed frequencies. This is in contrast with stellar evolution based asteroseismology, where the chemical profiles are calculated based on nuclear reaction rates, diffusion, convection and other other relevant physical processes. While perhaps less physical, such modelling ties internal structure more directly to the observed frequencies. Results of white dwarf asteroseismology can then be confronted with what stellar evolution calculations yield. For more details on the models and methods used in this study, we refer the reader to Bischoff-Kim & Østensen (2011).

2. The $^{12}C(\alpha, \gamma)^{16}O$ rate

Two white dwarfs for which we have pulsation spectrums of high enough quality for detailed asteroseismic studies are GD 358 and CBS 114, both helium atmosphere white dwarfs. Metcalfe (2003) determined what the rates of the $^{12}C(\alpha, \gamma)^{16}O$ nuclear reaction needed to be in order to produce the carbon and oxygen abundance profiles infered from the asteroseismic study of GD 358 and CBS 114. For both stars, the rates found were consistent with the NACRE nuclear reaction rates (Angulo *et al.* 1999).

211

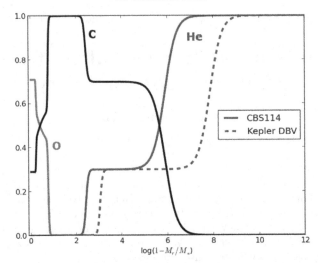

Figure 1. Composition profiles for helium atmosphere pulsating white dwarfs. Each profile is the result of asteroseismic fits. The basic shape of the profiles was borrowed from results of time-dependent diffusion calculations (Dehner & Kawaler 1995) and parameterized. The carbon and oxygen profiles are for CBS 114. For clarity, the carbon and oxygen profiles for the Kepler DBV are not shown.

3. Diffusion of helium

Another example of the use of white dwarf asteroseismology to test stellar evolution calculations is the diffusion of helium. For helium atmosphere white dwarfs, helium is the lightest element present. Time dependent calculations predict that over time, helium floats up to the surface (e.g. Dehner & Kawaler 1995). The asteroseismic study of white dwarfs in different stages in their evolution qualitatively confirm this. Asteroseismic studies of CBS 114 (Metcalfe *et al.* 2005) and KIC 8626021, a helium atmosphere white dwarf (DBV) found in the original *Kepler* field, show that the helium layer in CBS 114 is thicker than the helium layer in the Kepler DBV. The Kepler DBV is hotter than CBS 114, meaning it is younger. This result is shown in Fig. 1.

4. Summary and Future Work

White dwarfs hold fossilized in their core, the results of nuclear synthesis on the AGB. Some white dwarfs pulsate and we can probe their interior structure , allowing us to test these nuclear reaction rates. We also observe a trend in the thicknesses of their helium envelopes, consistent with the outward diffusion of helium as white dwarfs cool. This trend needs to be quantitatively checked against time dependent diffusion calculations.

References

Angulo, C., Arnould, M., Rayet, M., *et al.* 1999, *Nuclear Physics A* 656, 3

Bischoff-Kim, A. 2008, *Communications in Asteroseismology* 154, 16

Bischoff-Kim, A., Montgomery, M. H., & Winget, D. E. 2008, *ApJ* 675, 1512

Bischoff-Kim, A. & Østensen, R. H. 2011, *ApJl* 742, L16

Córsico, A. H., Althaus, L. G., García-Berro, E., & Romero, A. D. 2013, *Journal of Cosmology and Astroparticle Physics* 6, 32

Dehner, B. T. & Kawaler, S. D. 1995, *ApJl* 445, L141

Metcalfe, T. S. 2003, *ApJl* 587, L43

Metcalfe, T. S., Nather, R. E., Watson, T. K., *et al.* 2005, *A& A* 435, 649

New windows on massive stars: asteroseismology, interferometry, and spectropolarimetry
Proceedings IAU Symposium No. 307, 2014
G. Meynet, C. Georgy, J. H. Groh & Ph. Stee, eds.

© International Astronomical Union 2015
doi:10.1017/S1743921314006759

Radiative Levitation in Massive Stars: A self-consistent approach

Durand D'souza and Achim Weiss

Max Planck Institute for Astrophysics, Garching, Germany
email: durand@mpa-garching.mpg.de

Abstract. In B stars, the transportation of metals such as iron due to radiative levitation is thought to trigger sub-surface convection, which may lead to significant stellar pulsations through the iron-induced kappa mechanism (Pamyatnykh *et al.* 2004). The main goal of this work is to model the evolution of these stars by treating radiative levitation in an accurate and efficient manner in order to enable rapid asteroseismic analysis using the Opacity Project data and codes (Seaton 2005) with the GARching STellar Evolution Code (Weiss & Schlattl 2008).

Keywords. stars: abundances, stars: evolution, stars: interiors, stars: oscillations (including pulsations), diffusion, atomic processes

Asteroseismic diagnostics of massive stars requires the consideration of the various pulsation modes in detail. In massive O stars, the opacity peak in iron at $\log(T) \approx 5.3$ and the large luminosities involved naturally give rise to a sub-surface convection zone in the radiative envelope of the star. These sub-surface convective zones have also been linked to non-radial pulsations, such as in β Cephei and slowly pulsating B stars (Cantiello *et al.* 2009). In less massive B stars with similar metallicity, the reduced luminosity means that a larger opacity is necessary to trigger convection. This increase is hypothesised to be due to radiative levitation of elements, especially iron.

For consistency, we utilise the Opacity Project (OP) data and codes (Seaton 2005) within the GARching STellar Evolution Code (Weiss & Schlattl 2008) to calculate the Rosseland mean opacity at each point in the star on-the-fly. Different methods of calculating the Rosseland mean opacity produce somewhat different results (Fig. 1). As the Opacity Project does not consider the effects of high temperatures and molecular opacities at low temperatures, we need to interpolate with the OPAL tables within the relevant temperature domain. Composition changes due to diffusion and radiative acceleration, amongst other processes, are treated by solving the Burgers equations exactly (Thoul *et al.* 1994). The radiative acceleration profile for a $5\,M_\odot$ star may be seen in Fig. 2. We intend to apply the Single-Valued Parameter (SVP) approximation developed by LeBlanc & Alecian (2004), which has been shown to produce similar radiative acceleration profiles to exact methods at a substantially reduced computational cost.

References

Cantiello, M., Langer, N., Brott, I., *et al.* 2009, *A&A* 499, 279
LeBlanc, F. & Alecian, G. 2004, *MNRAS* 352, 1329
Pamyatnykh, A. A., Handler, G., & Dziembowski, W. A. 2004, *MNRAS* 350, 1022
Seaton, M. J. 2005, *MNRAS* 362, L1
Thoul, A. A., Bahcall, J. N., & Loeb, A. 1994, *ApJ* 421, 828
Weiss, A. & Schlattl, H. 2008, *Ap&SS* 316, 99

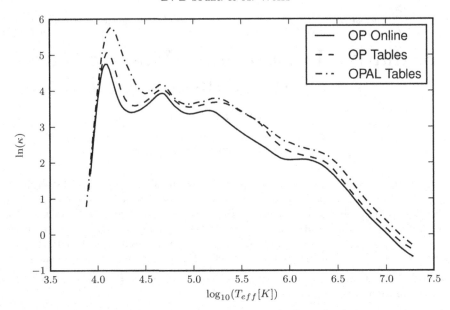

Figure 1. Differences between methods used to calculate the Rosseland mean opacity can lead to different results. In this plot, Rosseland mean opacities from on-the-fly calculations (OP Online) are compared with those interpolated from precomputed tables (OP Tables and OPAL Tables) for a $1.7\,M_\odot$ star. While the Online version is self-consistent, it does not include modifications for high temperatures or for low temperatures, where molecular opacities play a big role.

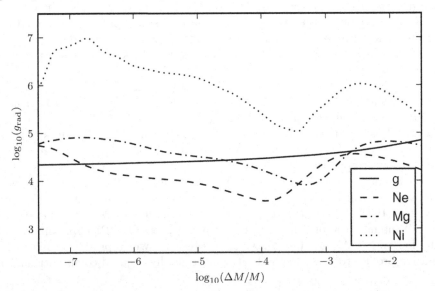

Figure 2. A plot showing the difference in radiative acceleration behaviour between various elements as a function of position in the star. The solid line represents gravitational acceleration. Neon has a radiative acceleration profile that is generally weaker than gravitational acceleration so it will gravitationally settle while Nickel will radiatively levitate. In general, higher mass stars have higher radiative accelerations due to the larger contribution from the radiation field.

New windows on massive stars: asteroseismology, interferometry, and spectropolarimetry
Proceedings IAU Symposium No. 307, 2014
G. Meynet, C. Georgy, J. H. Groh & Ph. Stee, eds.

© International Astronomical Union 2015
doi:10.1017/S1743921314006760

Leaky-wave-induced disks around Be stars: a pulsational analysis on their formation

Melanie Godart[1], Hiromoto Shibahashi[1] and Marc-Antoine Dupret[2]

[1]Dept. of Astronomy, The University of Tokyo, Japan
email: melanie.godart@gmail.com

[2]Dept. of Astrophysics, Geophysics and Oceanography, University of Liège, Belgium

Abstract. Be stars are B-type stars near the main sequence which undergo episodic mass loss events detected by emission lines, whose line shape and intensity vary with a timescale of the order of decades. Spectroscopic observations show a large rotation velocity such that one of the prevailing scenarios for the formation of the equatorial disk consists in an increasing equatorial rotation velocity to the break-up limit where gravity is challenged by the centrifugal force. We investigate here a new scenario recently suggested by Ishimatsu & Shibahashi (2013), in which the transport of angular momentum through the photosphere would be achieved by leaky waves, keeping the rotation velocity still below the break-up limit.

Keywords. stars: emission-line, Be, oscillations, rotation

1. Leaky-wave-induced disks – scenario

The emission lines in the spectrum of Be stars are now widely accepted to be due to episodic mass-loss from the equatorial region of the star, occurring quasi-periodically and forming a cool circumstellar disk (e.g. Porter & Rivinius 2003). One of the prevailing scenarios for the formation of the equatorial disk consists in an increasing equatorial rotation velocity to the break-up limit where gravity is challenged by the centrifugal force. Whether Be stars rotate at the critical break-up limit or not (e.g. Ekström *et al.* 2012; Granada *et al.* 2012) remains unclear although their high rotation places Be stars in the center of discussions about angular momentum (AM) transport. Non-radial oscillations have then been investigated as another mechanism to transport the AM and increase the rotation up to the break-up velocity (e.g. Neiner *et al.* 2013). Recently, Ishimatsu & Shibahashi (2013); Shibahashi (2014) suggested a new scenario in which the transport of AM through the photosphere would be achieved by leaky waves, keeping the rotation velocity still below the break-up limit. In this scenario, g-modes (excited by the κ-mechanism operating in the iron opacity bump) redistribute the AM but the wave leakage through the photosphere prevents a large deposition of AM at the stellar surface keeping a lower surface velocity. The waves are not fully reflected at the stellar surface and loose therefore their standing wave characrisctic. A reflection occurs when the frequency of the mode is larger than a critical frequency. This latter one increases with the rotation velocity and the reflective boundary may no longer occur for a high rotation rate, allowing the wave leakage and the transport of AM to the circumstellar envelope to form a disk. The wave loses then its energy and the oscillations are damped. Episodic mass loss would occur in a timescale of the order of the mode lifetime.

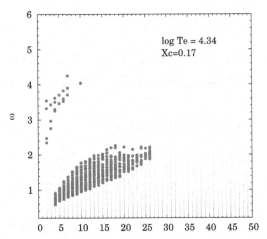

Figure 1. Dimensionless angular frequency range for excited modes computed for $l = 1$ to 50 in a $10\,M_\odot$ model with $X_c = 0.17$. Red dots stand for the stable modes while solid red circles stand for the unstable modes.

2. Stellar and pulsational models

Stellar evolution models were computed with CLES (Scuflaire *et al.* 2008). We adopted the AGS05 metal mixture (Asplund *et al.* 2005), the corresponding OP opacity tables (Badnell *et al.* 2005), $X = 0.070$ and $Z = 0.014$ and a convective core overshooting of 0.1. Non-adiabatic computations have been performed with MAD (Dupret 2002), in which the traditional approximation for rotation has been implemented (Bouabid *et al.* 2013) to perform modes for $1 \leqslant l \leqslant 4$. The efficiency of wave leakage (given by the integral over the surface of the wave kinetic energy) should increase with the rotation as the critical frequency increases. We derive an increase by 75% of the wave leakage between $\Omega/\Omega_c = 0.37$ and $\Omega/\Omega_c = 0.55$ showing, indeed, an increasing efficiency of the leaking effect with rotation. These results are however preliminary and further investigation need to be carry on. The lack of observational evidence of non-radial oscillations in all Be stars is not invalidating the scenario. Indeed, high degree l oscillation modes are expected to be excited in the Be star mass range. Figure 1 shows the range of dimensionless angular frequencies for g-modes ($l = 1$ to 50) in a $10\,M_\odot$ model (without rotation) near the end of the main sequence. $l = 4$ to 26 modes are found unstable. The lifetimes of such modes are in good agreement with the quasi-periodic Be phenomenom (9.45 year for a mode of $l = 15$, $\omega = 1.25$, $P = 12$h).

Acknowledgements

This research has been funded by the Japanese Society for Promotion of Science (JSPS).

References

Asplund, M., Grevesse, N., & Sauval, A. J. 2005, in T. G. Barnes, III & F. N. Bash (eds.), *Cosmic Abundances as Records of Stellar Evolution and Nucleosynthesis*, Vol. 336 of *Astronomical Society of the Pacific Conference Series*, p. 25

Badnell, N. R., Bautista, M. A., Butler, K., *et al.* 2005, *MNRAS* 360, 458

Bouabid, M.-P., Dupret, M.-A., Salmon, S., *et al.* 2013, *MNRAS* 429, 2500

Dupret, M.-A. 2002, *Non-radial non-adiabatic oscillations of near main sequence variable stars*, Vol. 71

Ekström, S., Georgy, C., Granada, A., Wyttenbach, A., & Meynet, G. 2012, in R. Capuzzo-Dolcetta, M. Limongi, & A. Tornambè (eds.), *Advances in Computational Astrophysics: Methods, Tools, and Outcome*, Vol. 453 of *Astronomical Society of the Pacific Conference Series*, p. 353

Granada, A., Ekström, S., Georgy, C., *et al.* 2012, in A. C. Carciofi & T. Rivinius (eds.), *Circumstellar Dynamics at High Resolution*, Vol. 464 of *Astronomical Society of the Pacific Conference Series*, p. 117

Ishimatsu, H. & Shibahashi, H. 2013, in H. Shibahashi & A. E. Lynas-Gray (eds.), *Astronomical Society of the Pacific Conference Series*, Vol. 479 of *Astronomical Society of the Pacific Conference Series*, p. 325

Neiner, C., Mathis, S., Saio, H., & Lee, U. 2013, in H. Shibahashi & A. E. Lynas-Gray (eds.), *Astronomical Society of the Pacific Conference Series*, Vol. 479 of *Astronomical Society of the Pacific Conference Series*, p. 319

Porter, J. M. & Rivinius, T. 2003, *PASP* 115, 1153

Scuflaire, R., Théado, S., Montalbán, J., *et al.* 2008, *Ap&SS* 316, 83

Shibahashi, H. 2014, in J. A. Guzik, W. J. Chaplin, G. Handler, & A. Pigulski (eds.), *IAU Symposium*, Vol. 301 of *IAU Symposium*, pp 173–176

Mélanie Godart

New windows on massive stars: asteroseismology, interferometry, and spectropolarimetry
Proceedings IAU Symposium No. 307, 2014
G. Meynet, C. Georgy, J. H. Groh & Ph. Stee, eds.

© International Astronomical Union 2015
doi:10.1017/S1743921314006772

Time Resolved Photometric and Spectroscopic Analysis of Chemically Peculiar Stars

Santosh Joshi[1], Gireesh C. Joshi[2], Y. C. Joshi[1] and Rahul Aggrawal[1]

[1] Aryabhatta Research Institute of Observational Sciences (ARIES), Nainital -263129, India
email: `santosh@aries.res.in`

[2] Center of Advanced Study, Department of Physics, D. S. B. Campus, Kumaun University, Nainital-263002, India

Abstract. Here we present the report on the "Nainital–Cape survey" research project aiming to search for and study the pulsational variability of main-sequence chemically peculiar (CP) stars. For this study, the time-series photometric observations of the sample stars were carried out at the 1.04 m ARIES telescope (India), while the high-resolution spectroscopic and spectropolarimetric observations were carried out at the the 6.0 m Russian telescope. Under this project, we have recently found clear evidence of photometric variability in the Am star HD 73045, which is likely to be pulsating in nature with a period of about 36 min, hence adding a new member to the family of the δ Scuti pulsating variables that have peculiar abundances.

Keywords. stars: chemically peculiar, variables : δ Scuti, techniques : photometric, spectroscopic, polarimetric

1. Introduction

Chemically peculiar (CP) stars show strong and/or weak absorption lines of certain elements in their optical spectrum in comparison to normal stars of similar spectral type. The two classes of CP stars known as metallic-line (Am) and A-peculiar(Ap) stars are important tools for asteroseismological studies because some of them exhibit multi-periodic pulsational variability. Am stars are non-magnetic and generally found in a binary system. Many of them are now known to exhibit low-amplitude pulsational variability with a period range similar to those of δ-Scuti variables. Conversely, Ap stars are magnetic with field strengths of the order of kG. One of the sub-group of this class is known as rapidly oscillating Ap (roAp) stars which pulsate with periods between 5 and 23 min. The pulsations of Am and Ap stars are important astrophysical tools to study the complex relationship between stellar pulsations and magnetic fields in the presence of atmospheric abundance anomalies. To study these phenomena, we have initiated a survey project at ARIES (India) to search for the pulsational variability in Ap and Am stars.

2. Sample selection, observations and data reduction

The criteria of sample selection was to select those A- and F-type stars having Strömgren indices similar to those of known roAp stars (Joshi *et al.* 2009). The photometric observations were carried out at the 1.04 m telescope of ARIES in the high-speed photometric mode with a continuous 10 s integrations through a Johnson B filter and 30″ diaphragm aperture. The data reduction process comprises of visual inspection of the light curve to identify and remove bad data points, correction for coincident counting losses, subtraction of the interpolated sky background, and correction for the mean

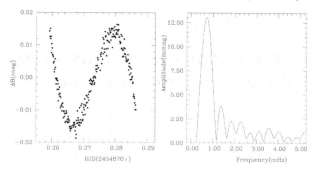

Figure 1. The light curve (left panel) and corresponding amplitude spectrum (right panel) of the Am star HD 73045.

atmospheric extinction. The resulting time-series were then analyzed using Discrete Fourier Transform (Joshi *et al.* 2006).

3. Results

The "Nainita-Cape" survey is running at ARIES since the last fifteen years, during which we found many interesting results. Firstly, we discovered a new roAp star (HD 12098) at ARIES. This is a magnetic star pulsating with a period of 7.6 min (Girish *et al.* 2001). Secondly, δ Scuti–type pulsations were discovered in 6 CP stars, namely HD 13038, HD 13079 (Martinez *et al.* 2001), HD 98851, HD 102480 (Joshi *et al.* 2003), HD 113878, and HD 118660 (Joshi *et al.* 2006). Among these pulsating variables, HD 98851 and HD 102480 are classified as unusual pulsators due to the presence of alternating high and low maxima. We have also performed time-resolved photometric, high-resolution spectroscopic and spectro-polarimetric analysis of the Ap star HD 13038 and the Am star HD 207561 (Joshi *et al.* 2010, 2012). Apart from these discoveries, we have recently detected the pulsational variability of HD 73045 with a period of ~ 36.23 min. This star belongs to the Praesepe open star cluster (Fossati *et al.* 2008). It has been classified as an Am star, in which Ca and Sc are underabundant and the Fe-peak elements are overabundant (Fossati *et al.* 2007). Here we present the preliminary light curve of HD 73045 from ARIES, Nainital obtained on HJD 2451943 (Fig. 1, left panel), while the amplitude spectrum of the time-series is shown on the right panel. A detail asteroseismic study of this star will be presented elsewhere. The prospect of our survey looks bright in the light of the upcoming observational facilities at Devasthal (ARIES), Nainital.

Acknowledgements

SJ acknowledge the International Astronomical Union for providing the travel grant to attend the international symposium IAU 307. Part of this work was done under the Indo-Russian RFBR project INT/RFBR/P-118 supported by DST, Govt. of India and Russian Academy of Science, Russia.

References

Fossati, L., Bagnulo, S., Landstreet, J., *et al.* 2008, *A&A* 483, 891
Fossati, L., Bagnulo, S., Monier, R., *et al.* 2007, *A&A* 476, 911
Girish, V., Seetha, S., Martinez, P., *et al.* 2001, *A&A* 380, 142
Joshi, S., Girish, V., Sagar, R., *et al.* 2003, *MNRAS* 344, 431
Joshi, S., Mary, D. L., Chakradhari, N. K., Tiwari, S. K., & Billaud, C. 2009, *A&A* 507, 1763
Joshi, S., Mary, D. L., Martinez, P., *et al.* 2006, *A&A* 455, 303
Joshi, S., Ryabchikova, T., Kochukhov, O., *et al.* 2010, *MNRAS* 401, 1299
Joshi, S., Semenko, E., Martinez, P., *et al.* 2012, *MNRAS* 424, 2002
Martinez, P., Kurtz, D. W., Ashoka, B. N., *et al.* 2001, *A&A* 371, 1048

New windows on massive stars: asteroseismology, interferometry, and
spectropolarimetry
Proceedings IAU Symposium No. 307, 2014
G. Meynet, C. Georgy, J. H. Groh & Ph. Stee, eds.

© International Astronomical Union 2015
doi:10.1017/S1743921314006784

Stochastic excitation of gravity waves in rapidly rotating massive stars

S. Mathis[1,2] and C. Neiner[2]

[1]Laboratoire AIM Paris-Saclay, CEA/DSM - CNRS - Université Paris Diderot, IRFU/SAp
Centre de Saclay, F-91191 Gif-sur-Yvette Cedex, France
email: stephane.mathis@cea.fr

[2]LESIA, Observatoire de Paris, CNRS UMR 8109, UPMC, Univ. Paris-Diderot,
5 place Jules Janssen, 92195 Meudon, France

Abstract. Stochastic gravity waves have been recently detected and characterised in stars thanks to space asteroseismology and they may play an important role in the evolution of stellar angular momentum. In this context, the observational study of the CoRoT hot Be star HD 51452 suggests a potentially strong impact of rotation on stochastic excitation of gravito-inertial waves in rapidly rotating stars. In this work, we present our results on the action of the Coriolis acceleration on stochastic wave excitation by turbulent convection. We study the change of efficiency of this mechanism as a function of the waves' Rossby number and we demonstrate that the excitation presents two different regimes for super-inertial and sub-inertial frequencies. Consequences for rapidly rotating early-type stars and the transport of angular momentum in their interiors are discussed.

Keywords. hydrodynamics, turbulence, waves, stars: oscillations, stars: rotation, stars: interiors

1. Stochastic gravity waves in massive stars

Thanks to asteroseismology using high-resolution photometry in space, our knowledge of stellar structure, rotation, and oscillations is currently undergoing a revolution (e.g. Aerts *et al.* 2010, Aerts, this volume). Until now, our understanding of the excitation of gravity modes in massive stars was mainly based on the κ-mechanism paradigm. However, these stars have a convective core on the main-sequence as well as a possible sub-surface convective zone that also contribute to waves excitation (e.g. Browning *et al.* 2004; Cantiello *et al.* 2009; Samadi *et al.* 2010). The stochastically excited waves can then reach the surface of the star where they become visible (Shiode *et al.* 2013) as in the case of the CoRoT hot Be star HD 51452 observed by Neiner *et al.* (2012). Many massive stars are rapidly rotating, as is HD 51452. Therefore, it becomes important to study the impact of (rapid) rotation on the stochastic excitation of gravity modes, which become gravito-inertial waves (hereafter GIWs) because of the action of the Coriolis acceleration, to provide a correct interpretation of asteroseismic data for such stars. Indeed, both the observation of HD 51452 and recent numerical simulations (Rogers *et al.* 2013) showed a trend in increase of the efficiency of the stochastic excitation mechanism with rotation that must be understood.

2. The impact of rotation on stochastic excitation

In Mathis *et al.* (2014), we have formally demonstrated that rotation, through the Coriolis acceleration, modifies the stochastic excitation of gravity waves and GIWs, the control

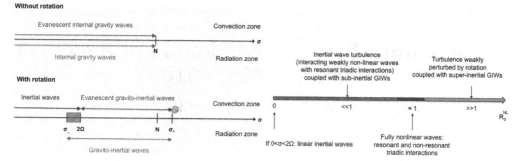

Figure 1. *Left:* low-frequency spectrum of waves in a non-rotating ($\Omega = 0$, top) and in a rotating (bottom) star (N is the usual Brunt-Väisälä frequency). The red box corresponds to sub-inertial GIWs where the cavities of gravity and inertial waves are coupled. *Right:* wave-turbulence couplings as a function of the non-linear Rossby number $\left(R_{\mathrm{o}}^{\mathrm{NL}}\right)$. Resonant excitation is obtained for $R_{\mathrm{o}}^{\mathrm{NL}} \approx R_{\mathrm{o}}$. (Taken from Mathis *et al.* (2014), courtesy *Astronomy & Astrophysics*).

parameters being the wave's Rossby number $R_{\mathrm{o}} = \sigma/2\Omega$ (σ and Ω are the wave frequency and the rotation rate respectively) and the non-linear Rossby number $R_{\mathrm{o}}^{\mathrm{NL}} = t_{\Omega}/t_{\mathrm{c}}$ of convective turbulent flows (t_{c} is the convective turn-over time and $t_{\Omega} = (2\Omega)^{-1}$). On one hand, in the super-inertial regime ($\sigma > 2\Omega$, *i.e.* $R_{\mathrm{o}} > 1$), the evanescent behaviour of GIWs in convective regions becomes increasingly weaker as the rotation speed grows until $R_{\mathrm{o}} = 1$. Simultaneously, the turbulent energy cascade towards small scales is slowed down. The coupling between super-inertial GIWs and given turbulent convective flows is then amplified. On the other hand, in the sub-inertial regime ($\sigma < 2\Omega$, *i.e.* $R_{\mathrm{o}} < 1$), GIWs become propagative inertial waves in convection zones. In the case of rapid rotation, turbulent flows, which become strongly anisotropic, result from their non-linear interactions. Sub-inertial GIWs that correspond to propagating inertial waves in convection zones are then intrinsically and strongly coupled to rapidly rotating convective turbulence. These two different regimes are summarised in Fig. 1.

These results are of great interest for asteroseismic studies since GIW amplitudes are thus expected to be larger in rapidly rotating stars, a conclusion that is supported by recent observations and numerical simulations. For example, until recently, stochastic gravity waves were thought to be of too low amplitude to be detected even with space missions such as CoRoT (Samadi *et al.* 2010). The discovery of stochastically excited GIWs in the rapid rotator HD 51452 proved that such waves can be detected. The interpretation of observed pulsational frequencies in rapidly rotating massive stars should take this into account for example in Be and Bn stars. Finally, the transport of angular momentum by stochastic GIWs may have important consequences in active massive stars such as Be stars, in which it may be at the origin of observed outbursts (Lee *et al.* 2014).

References

Aerts, C., Christensen-Dalsgaard, J., & Kurtz, D. W. 2010, *Asteroseismology*, Springer
Browning, M. K., Brun, A. S., & Toomre, J. 2004, *ApJ* 601, 512
Cantiello, M., Langer, N., Brott, I., *et al.* 2009, *A&A* 499, 279
Lee, U., Neiner, C., & Mathis, S. 2014, *ArXiv e-prints*
Mathis, S., Neiner, C., & Tran Minh, N. 2014, *A&A* 565, A47
Neiner, C., Floquet, M., Samadi, R., *et al.* 2012, *A&A* 546, A47
Rogers, T. M., Lin, D. N. C., McElwaine, J. N., & Lau, H. H. B. 2013, *ApJ* 772, 21
Samadi, R., Belkacem, K., Goupil, M. J., *et al.* 2010, *Ap&SS* 328, 253
Shiode, J. H., Quataert, E., Cantiello, M., & Bildsten, L. 2013, *MNRAS* 430, 1736

New windows on massive stars: asteroseismology, interferometry, and spectropolarimetry
Proceedings IAU Symposium No. 307, 2014
G. Meynet, C. Georgy, J. H. Groh & Ph. Stee, eds.

© International Astronomical Union 2015
doi:10.1017/S1743921314006796

An attempt of seismic modelling of β Cephei stars in NGC 6910

D. Moździerski, Z. Kołaczkowski and E. Zahajkiewicz

Asstronomical Institute, University of Wrocław, Kopernika 11, 51-622 Wrocław
email: mozdzierski@astro.uni.wroc.pl

Abstract. We present preliminary results of seismic modeling of β Cephei-type stars in NGC 6910 based on simultaneous photometric and spectroscopic observations carried out in 2013 in Białków (photometry) and Apache Point (spectroscopy) observatories.

Keywords. stars: oscillations, open clusters and associations: NGC 6910, stars: early-type, stars: fundamental parameters

1. Introduction

The recent seismic modeling of bright β Cephei stars (see, e.g. Aerts *et al.* 2003, Pamyatnykh *et al.* 2004, Dupret *et al.* 2004, Daszyńska-Daszkiewicz & Walczak 2010) has brought some constraints on the convective overshooting of the core and an indication that massive star cores rotate faster than their envelopes. The perspectives for doing asteroseismology of these stars are therefore promising.

A new way of using asteroseismology is to simultaneously model many stars of the same pulsation type, which is called ensemble asteroseismology. This type of asteroseismology can be done e.g. for members of an open cluster. A prerequisite of a successful asteroseismology study is mode identification and determination of some global stellar parameters, which can be done much easier for stars in open clusters than for field objects. This is because cluster membership makes that some stellar parameters (distance, reddening, age, chemical composition) can be safely assumed to be the same for all stars, while other parameters (e.g. masses and radii) are strictly related. One of the best candidates for ensemble asteroseismology is the young open cluster NGC 6910 in which Kołaczkowski *et al.* (2004) discovered four β Cephei variables. A new campaign focused on this cluster allowed to detect at least eight β Cephei-type members (Pigulski 2008). The frequency spectra of β Cephei stars in this cluster, arranged according to the decreasing brightness (i.e. mass), show a very interesting progress of frequencies of excited modes. This is exactly what one could expect for p modes in stars located at the same isochrone in a cluster. This is a strong argument for trying ensemble asteroseismology of this cluster.

2. Observations and Results

New observations of NGC 6910 were made in 2013. The photometric observations were obtained in Białków Observatory (Poland) during 21 nights. These observations were carried out with a 60-cm reflecting telescope and the attached CCD camera covering $13' \times 12'$ field of view. About 4000 CCD frames were acquired using the B, V, and I_C filters of the Johnson-Kron-Cousins photometric system. The spectroscopic observations were carried out at the Apache Point Observatory (APO) ARC 3.5-m telescope and the ARC Echelle Spectrograph (ARCES) during five nights. In total, we have taken 36 spectra of NGC 6910-18 and single-epoch spectra of two other β Cep-type stars: NGC 6910-14

Table 1. Atmospheric parameters of analysed β Cep stars in NGC 6910.

Star	$\log(T_{\mathrm{eff}}/\mathrm{K})$ [1]	$\log(T_{\mathrm{eff}}/\mathrm{K})$ [2]	$\log L/L_\odot$	$v \sin i$ [km s^{-1}]
NGC 6910-14	4.447	4.443	4.182	125
NGC 6910-16	4.447	—	—	149
NGC 6910-18	4.398	4.400	4.025	94

Notes:
[1] Based on our spectroscopy.
[2] Based on Strömgren photometry.

and NGC 6910-16. The spectra have a resolving power of 31500 and cover a range between 3200 and 10000 Å. Photometric observations were calibrated in a standard way. For each frame, we calculated aperture and profile magnitudes of the stars using the DAOPHOT II package (Stetson 1987), and then derived differential magnitudes. Spectroscopic observations were reduced with standard IRAF routines.

We determined T_{eff} and $v \sin i$ of the observed stars (NGC 6910-14, -16, -18) using our spectra and the BSTAR2006 grid of non-LTE model atmospheres of Lanz & Hubeny (2007) and the ROTIN3 program. We used also the UVBYBETA code and literature u, v, b, y, and β magnitudes to obtain $\log L/L_\odot$ and $\log T_{\mathrm{eff}}$ of NGC 6910-14 and NGC 6910-18. The results are shown in Table 1. In the case of NGC 6910-18, we found that the amplitudes of the two dominating modes with frequencies $f_1 = 6.1549 \, \mathrm{d}^{-1}$ and $f_2 = 6.3890 \, \mathrm{d}^{-1}$ remained almost unchanged in comparison to 2005–2007 observations. We performed mode identification for this star with the methods developed by Daszyńska-Daszkiewicz *et al.* (2005) for five stellar models using B, V, and I_C time-series photometry. The evolutionary tracks were computed with the Warsaw-New Jersey evolutionary code adopting the OP opacities, the solar mixture, rotational velocity $V_{\mathrm{rot}} = 100 \, \mathrm{km \, s}^{-1}$, hydrogen abundance $X = 0.7$, metallicity parameter $Z = 0.015$ and no overshooting from the convective core. We identified f_1 as an $l = 3$ mode whereas f_2 can be identified as $l = 0$, 1 or 2.

This work was supported by the NCN grant No. 2012/05/N/ST9/03898 and has received funding from the EC Seventh Framework Programme (FP7/2007-2013) under grant agreement no. 269194. The work is based on observations obtained with the Apache Point Observatory 3.5-meter telescope, which is owned and operated by the Astrophysical Research Consortium. Some calculations have been carried out in Wrocław Centre for Networking and Supercomputing (http://www.wcss.wroc.pl), grant No. 219. We thank Dr. J. Jackiewicz for making available telescope time on the APO ARC 3.5-m telescope and for his support during observations. We thank A. Pigulski, P. Bruś, G. Kopacki and P. Śródka for making some observations of NGC 6910 in Białków.

References

Aerts, C., Thoul, A., Daszyńska, J., *et al.* 2003, *Science* 300, 1926
Daszyńska-Daszkiewicz, J., Dziembowski, W. A., & Pamyatnykh, A. A. 2005, *A&A* 441, 641
Daszyńska-Daszkiewicz, J. & Walczak, P. 2010, *MNRAS* 403, 496
Dupret, M.-A., Thoul, A., Scuflaire, R., *et al.* 2004, *A&A* 415, 251
Kołaczkowski, Z., Pigulski, A., Kopacki, G., & Michalska, G. 2004, *AcA* 54, 33
Lanz, T. & Hubeny, I. 2007, *ApJS* 169, 83
Pamyatnykh, A. A., Handler, G., & Dziembowski, W. A. 2004, *MNRAS* 350, 1022
Pigulski, A. 2008, *Journal of Physics Conference Series* 118(1), 012011
Stetson, P. B. 1987, *PASP* 99, 191

New windows on massive stars: asteroseismology, interferometry, and
spectropolarimetry
Proceedings IAU Symposium No. 307, 2014
G. Meynet, C. Georgy, J. H. Groh & Ph. Stee, eds.

© International Astronomical Union 2015
doi:10.1017/S1743921314006802

Pulsation Period Change & Classical Cepheids: Probing the Details of Stellar Evolution

Hilding R. Neilson[1], Alexandra C. Bisol[2], Ed Guinan[2] and Scott Engle[2]

[1] East Tennessee State University
email: neilsonh@etsu.edu

[2] Villanova University

Abstract. Measurements of secular period change probe real-time stellar evolution of classical Cepheids making these measurements powerful constraints for stellar evolution models, especially when coupled with interferometric measurements. In this work, we present stellar evolution models and measured rates of period change for two Galactic Cepheids: Polaris and l Carinae, both important Cepheids for anchoring the Cepheid Leavitt law (period-luminosity relation). The combination of previously-measured parallaxes, interferometric angular diameters and rates of period change allows for predictions of Cepheid mass loss and stellar mass. Using the stellar evolution models, We find that l Car has a mass of about 9 $M_\odot$ consistent with stellar pulsation models, but is not undergoing enhanced stellar mass loss. Conversely, the rate of period change for Polaris requires including enhanced mass-loss rates. We discuss what these different results imply for Cepheid evolution and the mass-loss mechanism on the Cepheid instability strip.

Keywords. stars: individual (l Carinae, Polaris), stars: mass loss, Cepheids

1. Introduction

Classical Cepheids are powerful probes of stellar structure thanks to their pulsation periods. Measurements of the change of pulsation period directly constrain stellar evolution models (Eddington 1919). This has consequences for understanding the transition from hot, blue main sequence stars to the end points of asymptotic giant branch evolution and supernovae progenitors. But, measuring the rate of period change requires decades of time-domain observations, spanning about one century. For l Carinae, $P = 35.5$ days, and Polaris, $P = 4.97$ days, we measure $\dot{P} = 4.46 \pm 1.46$ and 23.7 ± 6.5 s yr^{-1}, respectively.

2. Rates of Period Change

Period change is measured from time-series observations of the Cepheid light curve and computing an O-C diagram that plots the period measured at some time minus a reference period. A parabolic structure indicates that the period is changing and that change is constant (Percy 2007). However, determining the evolutionary state of Cepheid from period change is not obvious as there are three crossings of the instability strip. Two crossings occur when the star is evolving redwards and period change is positive. The second crossing from the red giant stars to hotter effective temperatures corresponds to negative rates of period change. Hence, we require another constraint, one provided by interferometry. For instance, assuming parallaxes from *Hipparcos* and HST (van Leeuwen *et al.* 2007; Benedict *et al.* 2007), the radius of Polaris and l Car are 43.5 ± 0.8 and 159.9 ± 16.6 $R_\odot$, respectively (Kervella *et al.* 2006; Mérand *et al.* 2006). This permits

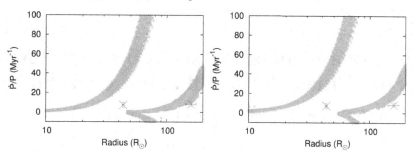

Figure 1. Rates of period change relative to pulsation period from stellar evolution models of Cepheids assuming 10^{-9} (left) and 10^{-6} $M_\odot$ yr^{-1} (right) during the Cepheid phase of evolution. The red points represent measured rates for Polaris and l Car. There are three sequences of period change, large positive correspond to the first crossing of the instability strip while the smaller positive sequence is the third crossing. Negative rates represent the second crossing.

the comparison of period changes and radii for these stars to constrain evolution and fundamental properties. We plot synthetic rates of period change in Fig. 1 based on stellar evolution models computed using the Bonn code (Neilson *et al.* 2011, 2012a,b) for two cases: the first assuming a Cepheid mass-loss rate of 10^{-9} and 10^{-6} $M_\odot$ yr^{-1}.

3. Outlook

Based on the models, Polaris is poorly represented. We suggest that Polaris is undergoing enhanced mass loss at the rate of order 10^{-7} - 10^{-6} $M_\odot$ yr^{-1}, but stellar evolution models do not evolve across the instability strip at those rates making it challenging to reproduce the observed properties (Neilson 2014). The period change for l Car is consistent only with stellar evolution models with smaller mass-loss rates $< 10^{-7}$ $M_\odot$ yr^{-1}.

Based on that comparison, we determine fundamental parameters such as mass, $M = 8.6 \pm 0.5$ $M_\odot$, effective temperature $T_{\rm eff} = 4960 \pm 280$ K and luminosity $\log L/L_\odot = 4.04 \pm 0.09$. These results provide valuable insights into the evolution of stars at the mass threshold between core-collapse supernovae and white dwarf formation. Interestingly, the findings hint at a possibility that, counterintuitively, massive Cepheids have weaker winds than lower-mass Cepheids. This conjecture is based on the combination of time-domain astronomy and optical interferometric observations constraining stellar evolution models.

Acknowledgements

We acknowledge funding from NSF grants AST-0807664 and AST-0507542 and NASA grants GO-12302 and GO-13019.

References

Benedict, G. F., McArthur, B. E., Feast, M. W., *et al.* 2007, *AJ* 133, 1810
Eddington, A. S. 1919, *The Observatory* 42, 338
Kervella, P., Mérand, A., Perrin, G., & Coudé du Foresto, V. 2006, *A&A* 448, 623
Mérand, A., Kervella, P., Coudé du Foresto, V., *et al.* 2006, *A&A* 453, 155
Neilson, H. R. 2014, *A&A* 563, A48
Neilson, H. R., Cantiello, M., & Langer, N. 2011, *A&A* 529, L9
Neilson, H. R., Engle, S. G., Guinan, E., *et al.* 2012a, *ApJ (Letters)* 745, L32
Neilson, H. R., Langer, N., Engle, S. G., Guinan, E., & Izzard, R. 2012b, *ApJ (Letters)* 760, L18
Percy, J. R. 2007, *Understanding Variable Stars*, Cambridge University Press
van Leeuwen, F., Feast, M. W., Whitelock, P. A., & Laney, C. D. 2007, *MNRAS* 379, 723

New windows on massive stars: asteroseismology, interferometry, and
spectropolarimetry
Proceedings IAU Symposium No. 307, 2014
G. Meynet, C. Georgy, J. H. Groh & Ph. Stee, eds.

© International Astronomical Union 2015
doi:10.1017/S1743921314006814

Pulsations of massive stars beyond TAMS: effects of mass loss, diffusion, overshooting

Jakub Ostrowski and Jadwiga Daszyńska-Daszkiewicz

Instytut Astronomiczny, Uniwersytet Wrocławski, ul. Kopernika 11, 51-622 Wrocław, Poland
email: ostrowski@astro.uni.wroc.pl

Abstract. We present an influence of mass loss, element diffusion and convective overshooting on instability areas of SPBsg stars with masses of $13 - 18\,M_\odot$. Evolutionary phases before and after helium core ignition are considered. We discuss how these effects affect the structure of the blue loops and hence a coverage of instability areas.

Keywords. stars: early-type, stars: supergiants, stars: oscillations, stars: evolution

1. Introduction

The discovery of g-mode pulsations in B-type post main sequence star HD 163899 (B2 Ib/II; Klare & Neckel 1977; Schmidt & Carruthers 1996) has led to a new class of pulsating variables named Slowly Pulsating B-type supergiants (SPBsg; Saio *et al.* 2006). Despite efforts of a few different groups (Saio *et al.* 2006; Godart *et al.* 2009; Daszyńska-Daszkiewicz *et al.* 2013) these objects are still very poorly known. There is no agreement even on the evolutionary status of these stars and the formation and properties of the blue loops are one of the biggest uncertainties related to the precise determination of it. Here, we investigate the influence of element diffusion, mass loss and overshooting on emerging of the blue loops and the coverage of instability areas.

2. Theoretical models

The models with masses $M = 13 - 18\,M_\odot$ have been calculated with MESA evolution code (Paxton *et al.* 2011, 2013) and non-adiabatic pulsation code of Dziembowski (1977). We use OPAL opacity tables (Iglesias & Rogers 1996) computed for the AGSS09 mixture (Asplund *et al.* 2009). Ledoux criterion for convective instability and exponential model of overshooting (Herwig 2000) are used. Pulsation calculations cover spherical harmonic degrees $\ell = 0, 1, 2$. All effects of rotation are neglected. We use the mass-loss models by Vink *et al.* (2001) for $\log T_{\rm eff} > 4.0$ and de Jager *et al.* (1988) for $\log T_{\rm eff} < 4.0$. If element diffusion is applied, Burglers equations are solved with the method and diffusion coefficients of Thoul *et al.* (1994).

3. Instability areas

We calculated instability areas with various parameters. Inward overshooting from the outer convective zone on the RGB phase is included in all presented models because it appeared to be indispensable for the emergence of the blue loops in the studied range of stellar masses. In the left panel of Fig. 1 we depict reference models (OPAL tables, $Z = 0.015$) without mass loss and element diffusion. The central panel shows models with element diffusion and models with mass loss are depicted in the right panel. The instability areas on the blue loops have similar size for reference models and models

226

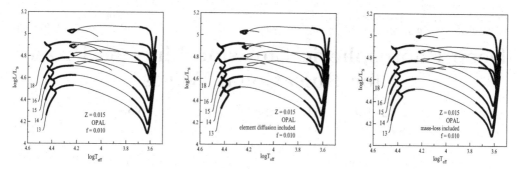

Figure 1. The HR diagrams with the evolutionary tracks for masses $M = 13-18\,M_\odot$ calculated until the end of the core helium burning with OPAL tables and metallicity $Z = 0.015$. Instability areas are shown with thick lines. In the left panel reference models without additional effects are shown, the central panel shows models with element diffusion and the right panel – models with mass loss included.

with diffusion. The presence of mass loss leads to shrinking of the blue loop and hence to decreasing the instability area. The instability areas during hydrogen-shell burning phase are very similar for all three cases.

4. Conclusions

Overshooting, element diffusion and mass loss affect pulsations mainly through their influence on the blue loops. Their influence on the instability areas during hydrogen-shell burning phase is negligible. The most important effect is the inward overshooting from outer convective zone on the RGB phase because it is essential for the development of the blue loops in the studied models. The presence of mass loss leads to smaller or entirely vanishes the instability areas on the blue-loops whereas the effects of element diffusion on the blue loops and hence the shape of the instability strip are very small.

Acknowledgements

The work was financially supported by the Polish NCN grants 2013/09/N/ST9/00611, 2011/01/M/ST9/05914 and 2011/01/B/ST9/05448.

References

Asplund, M., Grevesse, N., Sauval, A. J., & Scott, P. 2009, *ARA&A* 47, 481
Daszyńska-Daszkiewicz, J., Ostrowski, J., & Pamyatnykh, A. A. 2013, *MNRAS* 432, 3153
de Jager, C., Nieuwenhuijzen, H., & van der Hucht, K. A. 1988, *A&AS* 72, 259
Dziembowski, W. 1977, *AcA* 27, 95
Godart, M., Noels, A., Dupret, M.-A., & Lebreton, Y. 2009, *MNRAS* 396, 1833
Herwig, F. 2000, *A&A* 360, 952
Iglesias, C. A. & Rogers, F. J. 1996, *ApJ* 464, 943
Klare, G. & Neckel, T. 1977, *A&AS* 27, 215
Paxton, B., Bildsten, L., Dotter, A., *et al.* 2011, *ApJS* 192, 3
Paxton, B., Cantiello, M., Arras, P., *et al.* 2013, *ApJS* 208, 4
Saio, H., Kuschnig, R., Gautschy, A., *et al.* 2006, *ApJ* 650, 1111
Schmidt, E. G. & Carruthers, G. R. 1996, *ApJS* 104, 101
Thoul, A. A., Bahcall, J. N., & Loeb, A. 1994, *ApJ* 421, 828
Vink, J. S., de Koter, A., & Lamers, H. J. G. L. M. 2001, *A&A* 369, 574

New windows on massive stars: asteroseismology, interferometry, and spectropolarimetry
Proceedings IAU Symposium No. 307, 2014
G. Meynet, C. Georgy, J. H. Groh & Ph. Stee, eds.

© International Astronomical Union 2015
doi:10.1017/S1743921314006826

Deep Photospheric Emission Lines as Probes for Pulsational Waves

Th. Rivinius[1], M. Shultz[1,2,3] and G. A. Wade[3,2]

[1]ESO - European Organisation for Astronomical Research in the Southern Hemisphere, Chile
email: triviniu@eso.org

[2]Dept. of Physics, Engineering Physics and Astronomy, Queen's University, Canada

[3]Dept. of Physics, Royal Military College of Canada, Canada

Abstract. Weak line emission originating in the photosphere is well known from O stars and widely used for luminosity classification. The physical origin of the line emission are NLTE effects, most often optical pumping by far-UV lines. Analogous lines in B stars of lower luminosity are identified in radially pulsating β Cephei stars. Their diagnostic value is shown for radially pulsating stars, as these lines probe a much larger range of the photosphere than absorption lines, and can be traced to regions where the pulsation amplitude is much lower than seen in the absorption lines.

Keywords. stars: oscillations, stars: atmospheres, stars: individual (ξ^1 CMa, BW Vul)

1. Introduction

Weak line emission originating in the photosphere is well known from luminous O stars. The physical origin of the line emission is NLTE effects, most often optical pumping by far-UV lines. In early B hypergiants FeIII emission lines were identified to be pumped by HeI lines (Wolf & Stahl 1985). Rivinius *et al.* (1997) found these lines to have the least negative radial velocity and the least degree of variability. This means they are seated deeply in the photosphere. Analogous lines can be found in early B-type β Cephei stars of lower luminosity. Their diagnostic value is shown for radially pulsating stars, as these lines probe a much larger range of the photosphere than absorption lines.

2. Observations

Observations of ξ^1 CMa (116 proprietary intensity spectra) and BW Vul (434 archival intensity spectra) were carried out with ESPaDOnS/CFHT on Mauna Kea ($R = 68\,000$, 370 to 1000 nm, $S/N > 100$ for all observations). Data were usually obtained by continuously observing the targets over several hours.

For ξ^1 CMa the formation of the weak emission lines in the photosphere is corroborated by their magnetic signature, which is identical in shape, but inverted in sign w.r.t. the absorption lines. This is the expected behaviour for photospherically formed emission. (See Shultz *et al.* (this volume) for a more in-depth discussion of the magnetic properties of ξ^1 CMa).

For most lines, the pumping or fluorscence mechanisms were not yet identified. However, for several of the emission lines that could be identified, the upper levels coincide with the upper levels of strong potassium resonance lines below about 100 nm. The lines in which emission is seen were tentatively identified as

- **SiII:** $\lambda\lambda$ 6347, 6371.
- **SiIII:** $\lambda\lambda$ 6462 (or CII), 7463, 7467, 8103.

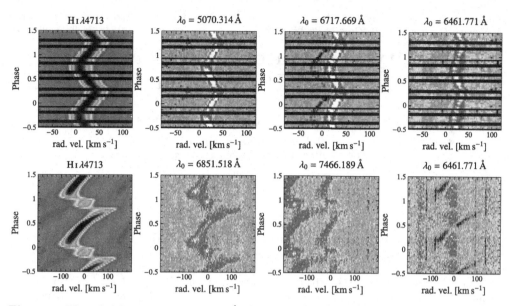

Figure 1. Phased β Cephei-variability in ξ^1 CMa (top) and BW Vul (bottom) in selected lines. Absorption lines, like the rightmost, all have similar amplitudes. The lower amplitude in some emission lines (three leftmost panels in each row) is evident.

- FeIII: λ 5879.
- CII: $\lambda\lambda$ 6151, 6462 (or SiIII)
- **Unidentified:** $\lambda\lambda$ 5056, 5070, 6587, 6718, 6777, 6804, 6806, 6852, 6928, 7113, 7116, 7388, 7552, 7512, 7852, 8236, 8287, 8293, 8630, 9826, 9855.

3. Discussion

Fig. 1 shows that the pulsation amplitude in the emission lines is either similar or smaller than in the absorption lines. It is known for radial pulsators that there is an amplitude gradient in the photosphere. For high amplitude pulsators, the amplitude increases with height (see Fig. 2 of Fokin *et al.* 2004, for models).

However, the typical change of amplitude is rather small over the range of the photospheric absorption lines. Fig. 6 of Nardetto *et al.* (2013) indicates changes of up to 10% between weak and strong absorption lines. In the case of ξ^1 CMa, the emission lines with the lowest peak-to-peak amplitudes have an amplitude of only 50% of the amplitude of the absorption lines, while for BW Vul the amplitude seems even to approach zero in the most extreme case (see lower row of Fig. 1, HeI λ4713 vs. $\lambda_0 = 6461.771$ Å).

References

Fokin, A., Mathias, P., Chapellier, E., Gillet, D., & Nardetto, N. 2004, *A&A* 426, 687
Nardetto, N., Mathias, P., Fokin, A., *et al.* 2013, *A&A* 553, A112
Rivinius, T., Stahl, O., Wolf, B., *et al.* 1997, *A&A* 318, 819
Wolf, B. & Stahl, O. 1985, *A&A* 148, 412

New windows on massive stars: asteroseismology, interferometry, and
spectropolarimetry
Proceedings IAU Symposium No. 307, 2014
G. Meynet, C. Georgy, J. H. Groh & Ph. Stee, eds.

© International Astronomical Union 2015
doi:10.1017/S1743921314006838

Stability boundaries for massive stars in the sHR diagram

Hideyuki Saio[1], Cyril Georgy[2] and Georges Meynet[3]

[1] Tohoku University, Japan
email: `saio@astr.tohoku.ac.jp`

[2] Keele University, UK

[3] Geneva observatory, Switzerland

Abstract. Stability boundaries of radial pulsations in massive stars are compared with positions of variable and non-variable blue-supergiants in the spectroscopic HR (sHR) diagram (Langer & Kudritzki 2014), whose vertical axis is $4 \log T_{\rm eff} - \log g \, (= \log L/M)$. Observational data indicate that variables tend to have higher L/M than non-variables in agreement with the theoretical prediction. However, many variable blue-supergiants are found to have values of L/M below the theoretical stability boundary; i.e., surface gravities seem to be too high by around 0.2-0.3 dex.

Keywords. stars: evolution, stars: oscillations, stars: mass loss, Hertzsprung-Russell diagram

Recently, a spectroscopic HR (sHR) diagram was introduced by Langer & Kudritzki (2014). Its horizontal axis is $\log T_{\rm eff}$, common to the ordinary HR diagram, while the vertical axis is $4 \log T_{\rm eff} - \log g$ (in solar units). One of the differences from the ordinary HRD is that no distance information is needed for plotting stars in the sHRD, although accurate estimates of the surface gravity $\log g$ are needed. Theoretically, the vertical axis is equal to $\log(L/M)$ (in solar units), which is affected by mass loss in the post-main-sequence evolution more sensitively than $\log L$ in the ordinary HRD. Because of these properties, sHRD is useful, in addition for the properties discussed by Langer & Kudritzki (2014), for comparing pulsational stability boundaries with observed positions of blue supergiants (BSG).

A massive star becomes a BSG (BSG1) after the main-sequence evolution, and evolves to the red supergiant (RSG) stage. After losing significant mass in the RSG stage, the star becomes a BSG again (BSG2) (Ekström *et al.* 2012). Saio *et al.* (2013) found that radial pulsations in the BSG region are excited by the strange mode instability only in BSG2 but not in BSG1 (Fig. 1). Since the luminosity is not very different between BSG1 and BSG2 for a given mass, the distributions of BSG1s and BSG2s (and hence the distribution of variable and non-variable BSGs) are similar and mixed in the HR diagram. In the sHRD diagram, however, the evolutionary track is affected more strongly by mass loss, and BSG2 is located significantly higher than BSG1, so that the pulsational instability region, which is governed mainly by L/M, is separated roughly irrespective to the mass (right panel of Fig. 1).

Fig. 2 compares BSGs in NGC 300 (left panel) and BSGs in the Milky Way (right panel) with theoretical stability boundaries in the sHRD. Generally, variable BSGs are located above the non-variable BSGs (with some exceptions), which is consistent with the theoretical prediction. However, the observational non-variable/variable boundary (for MW) seems lower than the theoretical stability boundary of pulsation; i.e., observational values of $\log g$ seem somewhat too high. The reason for the discrepancy is not clear at the moment.

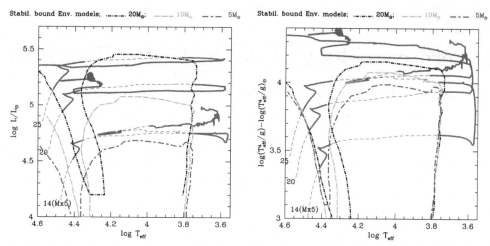

Figure 1. Stability boundaries of radial pulsations for envelope models of $20\,M_\odot$, $10\,M_\odot$, and $5\,M_\odot$ are shown in the ordinary HR diagram (left panel) and sHR diagram (right panel). Parts of thick solid lines in evolutionary tracks indicate where at least one radial modes are excited.

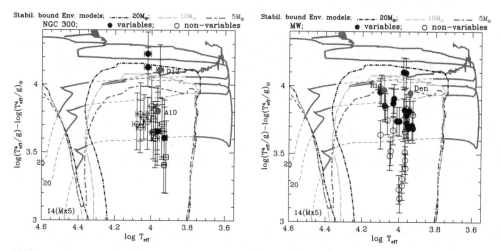

Figure 2. BSGs in NGC 300 (left panel; Bresolin *et al.* 2004; Kudritzki *et al.* 2008) and BSGs in the Milky Way (right panel; Firnstein & Przybilla 2012) are plotted in the sHR diagram. Filled (open) circles indicate variable (non-variable) BSGs. Stability boundaries shown in Fig. 1 (right panel) are also shown for comparison. The stars D12 and A10 in NGC 300 (left panel) show regular light curves which look consistent with radial pulsations.In the right panel 'Rig' and 'Den' stand for Rigel and Deneb, respectively.

References

Bresolin, F., Pietrzyński, G., Gieren, W., *et al.* 2004, *ApJ* 600, 182
Ekström, S., Georgy, C., Eggenberger, P., *et al.* 2012, *A&A* 537, A146
Firnstein, M. & Przybilla, N. 2012, *A&A* 543, A80
Kudritzki, R.-P., Urbaneja, M. A., Bresolin, F., *et al.* 2008, *ApJ* 681, 269
Langer, N. & Kudritzki, R. P. 2014, *A&A* 564, A52
Saio, H., Georgy, C., & Meynet, G. 2013, *MNRAS* 433, 1246

New windows on massive stars: asteroseismology, interferometry, and spectropolarimetry
Proceedings IAU Symposium No. 307, 2014
G. Meynet, C. Georgy, J. H. Groh & Ph. Stee, eds.
© International Astronomical Union 2015
doi:10.1017/S174392131400684X

Asteroseismology of the SPB star HD 21071

Wojciech Szewczuk and Jadwiga Daszyńska-Daszkiewicz

Instytut Astronomiczny, Uniwersytet Wrocławski, Wrocław, Poland
email: szewczuk@astro.uni.wroc.pl

Abstract. We perform mode identification for the frequency peaks detected in the light variation of HD 21071 including the effects of rotation via the traditional approximation. We find the angular numbers (ℓ, m) for all observed frequencies and limit the range of the rotational velocity. In the next step, we make an attempt towards seismic modelling of the star in order to constrain its global parameters.

Keywords. stars: oscillations, early-type, rotation

1. Introduction

HD 21071 (V576 Per) is a slowly pulsating B-type star with the brightness of $V = 6.08$ mag, the spectral type of B5IV or B7V (Reed 2005), rotating at a speed of at least 50 km/s (Abt *et al.* 2002). The effective temperature $T_{\rm eff} = 4.164(7)$ and metallicity $Z = 0.0082(16)$ was derived from the IUE ultraviolet spectra by Niemczura (2003). The luminosity of $\log L/L_\odot = 2.444(76)$, needed to put the star on the H-R diagram, was determined in this paper. Using the Geneva time-series photometry, De Cat *et al.* (2007) found four pulsational frequencies: $\nu_1 = 1.18843$, $\nu_2 = 1.14934$, $\nu_3 = 1.41968$ and $\nu_4 = 0.95706$ c/d. The dominant one ν_1 was also found in the spectroscopic data (De Cat 2002).

2. Mode identification and frequency fitting

In order to make mode identification we apply the method proposed by Daszyńska-Daszkiewicz *et al.* (2008), which takes into account the Coriolis force in the framework of the traditional approximation. The results are presented in Tab. 1. We considered the models from the centre and four edges of the error box. For ν_4 in the model #4 we were unable to find mode identification in the common range of $V_{\rm rot}$ with other frequencies. They are indicated by a question mark. The similar situation take place for the scenario $\nu_4(\ell = 1, m = 0)$ in the model #5. Depending on the model parameters, the rotation velocity of HD21071 ranges from about 150 to 250 km s^{-1}.

3. Frequency fitting

Fitting the two (ν_1, ν_2) and three (ν_1, ν_2, ν_3) frequencies was performed for models with $\log L/L_\odot$ and $\log T_{\rm eff}$ from the 3σ error box. We considered the rotational velocities, $V_{\rm rot}$, in the range from 150 to 255 km s^{-1}. In Fig. 1 the results are shown for $V_{\rm rot} = 160$ km s^{-1}. There are six combinations of the radial orders that fit two frequencies and only one (g_{13}, g_{14}, g_9) that fits three frequencies for the depicted value of $V_{\rm rot}$.

Table 1. Mode identification for models from the centre and edges of the error box.

ν [c/d]	(ℓ, m) [$V_{\mathrm{rot}}^{\min}$ km s^{-1} − $V_{\mathrm{rot}}^{\max}$ km s^{-1}]				
	#1	#2	#3	#4	#5
1.18843	(1,0) [211-212]	(1,0) [148-170]	(1,0) [174-200]	(1,0) [249-255]	(1,0) [229-245] (1,1) [107-110]
1.14934	(1,0) [211-212]	(1,0) [148-170]	(1,0) [174-200]	(1,0) [249-255]	(1,0) [229-245] (1,1) [107-110]
1.41968	(1,0) [211-212]	(1,0) [148-170]	(1,0) [174-200]	(1,0) [249-255]	(1,0) [229-245] (1,1) [107-110] (2,0) [108-110]
0.95706	(2,-1) [211-212]	(1,-1) [148-170] (2,-1) [148-170] (2,-2) [148-170]	(1,-1) [174-200] (2,-1) [174-200] (2,-2) [174-200]	(1,0) [50-180]? (2,-1) [50-189]? (2,-2) [180-189]?	(1,0) [50-215]? (1,0) [107-110] (2,-1) [107-110]

Notes: **#1:** $\log T_{\mathrm{eff}} = 4.164$, $\log L/L_\odot = 2.444$; **#2:** $\log T_{\mathrm{eff}} = 4.171$, $\log L/L_\odot = 2.368$; **#3:** $\log T_{\mathrm{eff}} = 4.157$, $\log L/L_\odot = 2.368$; **#4:** $\log T_{\mathrm{eff}} = 4.171$, $\log L/L_\odot = 2.520$; **#5:** $\log T_{\mathrm{eff}} = 4.157$, $\log L/L_\odot = 2.520$

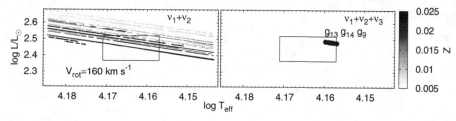

Figure 1. The seismic models of HD 21071 on the H-R diagram for $V_{\mathrm{rot}} = 160\,\mathrm{km\,s^{-1}}$. Only models with unstable modes are shown. Grayscale codes metallicity Z. The left panel: the seismic models fitting ν_1 and ν_2. The right panel: the seismic models fitting ν_1, ν_2 and ν_3.

4. Conclusions

Mode identification in the framework of the traditional approximation is slightly depended on the model. Despite of this fact, we succeeded in determining the angular numbers (ℓ, m) of the frequencies observed in HD 21071. The simultaneously determined value of the rotational velocity is more sensitive to the adopted stellar parameters and we could constrain its value to the range of about $(150 - 250)\,\mathrm{km\,s^{-1}}$.

In the next step, we try to find rotating models which reproduce the well identified frequencies. There are models which fit two and three observed frequencies in the determined range of $V_{\mathrm{rot}} \in (150 - 250)\,\mathrm{km\,s^{-1}}$. When only unstable modes are considered, the seismic models fitting ν_1 and ν_2 become more luminous and their effective temperatures increase with the rotation rate. Above $V_{\mathrm{rot}} = 180$ km/s there are no models with unstable modes fitting the three frequencies.

Acknowledgements

WS was financially supported by the Polish NCN grant DEC-2012/05/N/ST9/03905 and JDD by the Polish NCN grants 2011/01/M/ST9/05914, 2011/01/B/ST9/05448.

References

Abt, H. A., Levato, H., & Grosso, M. 2002, *ApJ* 573, 359

Daszyńska-Daszkiewicz, J., Dziembowski, W. A., & Pamyatnykh, A. A. 2008, *Journal of Physics Conference Series* 118(1), 012024

De Cat, P. 2002, in *IAU Colloq. 185: Radial and Nonradial Pulsationsn as Probes of Stellar Physics*, Vol. 259 of *Astronomical Society of the Pacific Conference Series*, p. 196

De Cat, P., Briquet, M., Aerts, C., *et al.* 2007, *A&A* 463, 243

Niemczura, E. 2003, *A&A* 404, 689

Reed, B. C. 2005, *AJ* 130, 1652

Participants during the CERN visit

New windows on massive stars: asteroseismology, interferometry, and
spectropolarimetry
Proceedings IAU Symposium No. 307, 2014
G. Meynet, C. Georgy, J. H. Groh & Ph. Stee, eds.
© International Astronomical Union 2015
doi:10.1017/S1743921314006851

Spectral Effects of Pulsations in Blue Supergiants

S. Tomić, M. Kraus and M. E. Oksala

Astronomický Ústav AVČR, Fričova 298, 25165 Ondřejov, Czech Republic
email: sanja@sunstel.asu.cas.cz

Abstract. We have been spectroscopically monitoring a number of blue supergiants, focusing
on several strategic photospheric and wind lines. Our aim is to detect line profile variability, and
to determine its origin. Here, we present preliminary results for ρ Leo and ϵ Ori. We conduct an
asteroseismic analysis of HeI $\lambda6678$. We find in each star multiple periods raging from hours to
several days. In addition, we observe strong, night to night variability in Hα.

Keywords. line: profiles, stars: oscillations, supergiants, stars: early-type

1. Introduction

Blue supergiants (BSGs) are luminous, evolved massive stars. Their photospheric lines
are broadened by both stellar rotation and so-called macroturbulence. The nature of this
often large macroturbulence (Simón-Díaz & Herrero 2014) is not clear yet, but has been
suggested to be connected to stellar pulsations (Aerts *et al.* 2009). So far, only a few
BSGs have been reported to pulsate (Lefever *et al.* 2007; Saio *et al.* 2006; Kraus *et al.*
2012). However, BSGs are well known to display strong variability of their wind lines
such as Hα, which might indicate a connection between pulsations and variable wind
conditions.

We initiated an observing campaign to monitor spectroscopically a sample of northern
Galactic BSGs. Our goal is to identify pulsation periods and their possible link to variable
mass-loss. For this study we focus on two objects, ρ Leo and ϵ Ori. Both are known to
show variability of their optical line profiles.

2. Observations, data analysis and preliminary results

With the Perek 2m telescope at the Ondřejov Observatory, we obtained 394 spectra of
ϵ Ori and 254 spectra of ρ Leo with SNR > 300 in the period 2014 February 6 to March
14. Spectral time series were obtained for both stars during the final seven nights of the
observing run. The observations include Hα and HeI $\lambda6678$.

Clearly, both lines display strong profile variability. Hα changes from night to night in
both shape and strength (Tomić *et al.*, in preparation). We focus on the HeI $\lambda6678$ line
and compute the first three moments of its profile. The first and third moment vary in
phase (Fig. 1) indicating pulsations. We apply the Lomb-Scargle method for the period
analysis. For both stars, the identified frequencies, their amplitudes, and their phases are
listed in Table 1. For ρ Leo, we identified 9 periods ranging from $\sim$5 hours to $\sim$6 days.
For ϵ Ori, we find 13 periods ranging from $\sim$2 hours to $\sim$3 days. We prewhitened the
data with each identified period and made a final fitting using a least-square method.

Acknowledgements

This work was supported by GAČR (14-21373S), MŠMT (LG14013) and RVO:67985815.

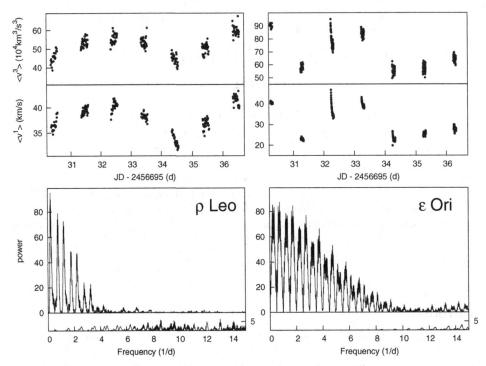

Figure 1. Top: Variability in the first and third moment during seven consecutive nights. Bottom: Lomb-Scargle periodograms of the first moments and their residuals.

Table 1. Lists of identified frequencies

ρ Leo			ε Ori		
Frequency 1/d	amplitude km/s	phase	Frequency 1/d	amplitude km/s	phase
0.79	6.43	4.26	3.75	-2.89	3.45
1.25	2.45	4.77	2.81	-8.09	5.36
0.16	5.98	1.48	1.35	-9.13	7.08
2.22	-3.02	-5.06	4.70	1.97	-0.53
2.78	1.47	-1.02	1.64	22.97	2.75
0.37	0.91	0.80	0.44	4.77	9.12
0.89	5.77	5.27	0.62	-30.28	1.67
1.87	2.79	1.83	2.20	11.35	11.35
4.73	0.44	4.73	4.19	6.99	-4.41
			3.09	6.10	3.25
			5.77	2.97	1.15
			5.13	1.57	6.05
			10.99	-1.07	3.47

References

Aerts, C., Puls, J., Godart, M., & Dupret, M.-A. 2009, *A&A* 508, 409

Kraus, M., Tomić, S., Oksala, M. E., & Smole, M. 2012, *A&A* 542, L32

Lefever, K., Puls, J., & Aerts, C. 2007, *A&A* 463, 1093

Saio, H., Kuschnig, R., Gautschy, A., *et al.* 2006, *ApJ* 650, 1111

Simón-Díaz, S. & Herrero, A. 2014, *A&A* 562, A135

New windows on massive stars: asteroseismology, interferometry, and
spectropolarimetry
Proceedings IAU Symposium No. 307, 2014
G. Meynet, C. Georgy, J. H. Groh & Ph. Stee, eds.
© International Astronomical Union 2015
doi:10.1017/S1743921314006863

Is λ Cep a pulsating star?

J.M. Uuh-Sonda[1], P. Eenens[1] and G. Rauw[2]

[1]Departamento de Astronomía, Universidad de Guanajuato,
Apdo. Postal 144, 36000, Guanajuato, Guanajuato, Mexico
email: juuh@astro.ugto.mx, eenens@ugto.mx

[2]Institut d'Astrophysique et de Géophysique, Université de Liège,
Allée du 6 Août, 400 Liège, Belgium
email: rauw@astro.ulg.ac.be

Abstract. It has been proposed that the variability seen in absorption lines of the O6Ief star λ Cep is periodical and due to non-radial pulsations (NRP). We have obtained new spectra during six campaigns lasting between five and nine nights. In some datasets we find recurrent spectral variations which move redward in the absorption line profile, consistent with perturbations on the stellar surface of a rotating star. However the periods found are not stable between datasets, at odds with the NRP hypothesis. Moreover, even when no redward trend is found in a full dataset of an observing campaign, it can be present in a subset, suggesting that the phenomenon is short-lived, of the order of a few days, and possibly linked to transient magnetic loops.

Keywords. stars: early-type, stars: oscillations, stars: spots, stars: rotation, stars: individual (λ Cep), stars: variables: general

1. Introduction

In O-type and O-supergiant stars, line profile variability (LPV) is a widespread phenomenon (Fullerton *et al.* 1996). However, the origin of such variability is still uncertain and debatable. It is usually attributed to pulsations, rotational modulations, magnetic fields, structures in the stellar wind, or a mix of these phenomena. The presence of non-radial pulsations (NRP) has been reported mainly in late-O stars (e.g. HD 93521, Rauw *et al.* 2008), while in the case of early-O stars the determination of genuine NRP has been more complicated, since their spectra are dominated by the so-called "red-noise" (e.g. HD 46223, HD 46150 and HD 46966, Blomme *et al.* 2011).

In 1999, de Jong *et al.*reported the presence of low-order NRP with $P_1 = 12.3$ hrs ($\ell = 3$) and $P_2 = 6.6$ hrs ($\ell = 5$) in the O6 Ief star, λ Cep (HD 210839). This study was conducted by analyzing the HeI $\lambda4713$ line observed for five nights in a multi-site campaign. Since other Oef stars have displayed a strong epoch-dependence in their variability (e.g. Rauw *et al.* 2003), the results of de Jong *et al.* (1999) needed confirmation.

2. Observations, analysis and results of our data

We have collected 495 spectra during six observing campaigns, four in the Observatoire de Haute Provence (OHP, France) and two in the National Astronomical Observatory of San Pedro Martir (SPM, Mexico). To analyze our datasets, we make use of the tools used by Rauw *et al.* (2008). The results of the variability and Fourier analysis, for our HeI $\lambda4471$ and HeII $\lambda4542$ absorption lines, are described in Uuh-Sonda *et al.* (2014).

3. NRP? Long-term instability of variations

NRP are expected to produce a pattern of alternating absorption excesses and deficits that cross the line profile from blue to red as the star rotates (Telting & Schrijvers

1997). This results in a progressive variation of the phase of the modulation across the line profile. In our data, only a subset of the frequencies identified yields a monotonic progression of the phase across those parts of the line profile where the amplitude of the variation is large. However, such frequencies behave differently for different epochs or lines, i.e. there is no long-term stability in the frequencies found. This suggests that we might not be observing the same mode at different epochs. Also we could not find the frequencies reported by de Jong *et al.* (1999). It appears thus that there is no evidence for persistent NRP in λ Cep (Uuh-Sonda *et al.* 2014).

4. Co-rotating magnetic bright spots: short-term variability

Among the possible causes of LPV, the idea of "co-rotating magnetic bright spots" at the stellar surface has been gaining strength. Such magnetic spots would be created by magnetic fields generated in a sub-surface convective layer of massive stars (Cantiello *et al.* 2009), and their lifetime would be only of a few hours (Cantiello & Braithwaite 2011). It has also been proposed that the co-rotating magnetic bright spots are at the base of the so-called "stellar prominences", which cause cyclical variation in the wind of massive stars (e.g. for λ Cep, Henrichs & Sudnik 2013). As these magnetic bright spots are rotating across the line of sight, they would mimic the line profile variations usually attributed to NRP, but on short scales. Recently, Ramiaramanantsoa *et al.* (2014) reported the first convincing case of co-rotating bright spots on an O-star, ξ Per (O7.5III(n)((f))).

Our analysis reveals that, in some of our campaigns, there is a persistent variability moving redward in the absorption line profile, consistent with perturbations on the stellar surface of a rotating star. However, in other campaigns such variability is less noticeable. Even during the same campaign the variability can behave differently from day to day. This suggests that there is a genuine variability but it manifests itself only on short timescales, i.e. the phenomenon that causes such variability is transient.

If the variability observed in λ Cep is produced by short-lived bright spots, some patterns could be detected in shorter datasets better than in longer ones. In order to test this hypothesis, we have analyzed sub-sets of our campaigns. This is indeed the case for example for our OHP-June-2010 HeII λ4542 line. A frequency of 0.87 d^{-1} is found in the analysis of both the complete seven-day dataset and of a subset of only the first six nights. However only the latter shows the related redward behavior across the line profile. This suggests a transient phenomenon that faded away during the seventh night, just as bright spots would.

We are inclined to the idea that this transitional phenomenon is actually the presence of numerous magnetic bright spots rotating with the stellar surface (Uuh-Sonda *et al.* 2014). Indeed, this idea has also been suggested by Ramiaramanantsoa *et al.* (2014). However, this hypothesis still needs to be confirmed.

References

Blomme, R., Mahy, L., Catala, C., *et al.* 2011, *A&A* 533, A4
Cantiello, M. & Braithwaite, J. 2011, *A&A* 534, A140
Cantiello, M., Langer, N., Brott, I., *et al.* 2009, *A&A* 499, 279
de Jong, J. A., Henrichs, H. F., Schrijvers, C., *et al.* 1999, *A&A* 345, 172
Fullerton, A. W., Gies, D. R., & Bolton, C. T. 1996, *ApJS* 103, 475
Henrichs, H. F. & Sudnik, N. P. 2013, *ArXiv:1310.5264*
Ramiaramanantsoa, T., Moffat, A. F. J., Chené, A.-N., *et al.* 2014, *MNRAS* 441, 910
Rauw, G., De Becker, M., van Winckel, H., *et al.* 2008, *A&A* 487, 659
Rauw, G., De Becker, M., & Vreux, J.-M. 2003, *A&A* 399, 287
Telting, J. H. & Schrijvers, C. 1997, *A&A* 317, 723
Uuh-Sonda, J. M., Rauw, G., Eenens, P., *et al.* 2014, *Rev. Mexicana AyA* 50, 67

New windows on massive stars: asteroseismology, interferometry, and spectropolarimetry
Proceedings IAU Symposium No. 307, 2014
G. Meynet, C. Georgy, J. H. Groh & Ph. Stee, eds.

© International Astronomical Union 2015
doi:10.1017/S1743921314006875

Seismic analysis of the massive β Cephei star 15 Canis Majoris

Przemysław Walczak and Gerald Handler

Nicolaus Copernicus Astronomical Center, Polish Academy of Sciences
email: pwalczak@camk.edu.pl

Abstract. 15 Canis Majoris is a quite massive ($M \sim 14\,M_\odot$) main sequence pulsator of the β Cephei type. Recent photometric (Handler 2014) and spectroscopic (Saesen & Briquet, priv. comm.) observations confirm four pulsational frequencies and indicate possible additional modes. We calculated models fitting two frequencies identified as radial and dipole modes. Our analysis indicates rather effective overshooting from the convective core as well as a strong dependence of the minimal required overshooting parameter ($\alpha_{\mathrm{ov,min}}$) on the metallicity, Z ($\alpha_{\mathrm{ov,min}} \sim -2.5Z$).

When incorporating the non-adiabatic f-parameter (Daszyńska-Daszkiewicz & Walczak 2009), defined as the ratio of the bolometric flux changes to the radial displacement, significant differences between the opacity tables were obtained. The comparison of the models derived with different codes is also interesting. We used two evolutionary codes: Warsaw-New Jersey (Pamyatnykh *et al.* 1998) and MESA (Paxton *et al.* 2011) and some systematic differences were found.

Keywords. stars: individual 15CMa, stars: variables: β Cephei, stars: interiors

1. Introduction

The comparison between the observational and theoretical amplitude ratios of the light changes in the Strömgren uvy filters gave us identifications of the mode degree, ℓ, for four pulsational modes of 15 CMa. We derived $\ell = 1$ for $\nu_1 = 5.418522(3)$ c/d, $\ell = 0$ for $\nu_2 = 5.183250(6)$ c/d, $\ell = 1$ or 3 for $\nu_3 = 5.308302(8)$ c/d and $\ell = 1$ for $\nu_4 = 5.52139(2)$ c/d.

The comparison between the empirical and theoretical values of the f-parameter for the radial mode ν_2 indicates that this mode is most probably the fundamental one. The method of determination of the empirical values of the f-parameter can be found in Daszyńska-Daszkiewicz *et al.* (2003, 2005).

2. Asteroseismic models

Seismic model fitting of two well-identified frequencies, ν_2 (the $\ell = 0$ mode) and ν_4 (the $\ell = 1$ mode), is presented in Fig. 1. Since we do not know the azimuthal order of the dipole modes, we assumed $m = 0$ for ν_4. For this reason, our analysis should be treated with caution. It is interesting, however, that when we assumed ν_1 being the centroid dipole mode ($m = 0$) we could not find any seismic model fitting both ν_1 and ν_2. It may indicate that our assumption is correct, since ν_1 and ν_4 seem to belong to the same rotationally split triplet.

Models shown in Fig. 1 were calculated with the Warsaw-New Jersey evolutionary code (WNJ). In the left panel we used the OP opacity tables (Seaton 2005) and in the right panel the OPAL data (Iglesias & Rogers 1996). Modes ν_2 and ν_4 are in general unstable in almost all presented models. Only the OPAL models with small Z and high α_{ov} are stable (upper left part of the right panel).

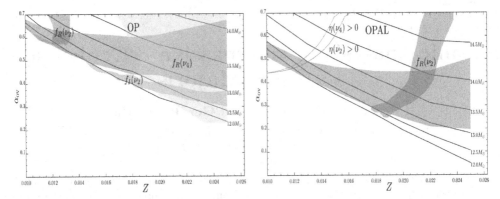

Figure 1. Seismic models that fit two frequencies: ν_2 (the $\ell = 0$ mode) and ν_4 (the $\ell = 1$ mode). On both panels we overplot lines of constant mass (black solid). Grey areas indicate models lying inside the observational error box of the effective temperature ($T_{\rm eff}$) and surface gravity ($\log g$) of 15 CMa, and their error estimates were taken from Shobbrook *et al.* (2006). Regions labelled with $f_R(\nu)$ and $f_I(\nu)$ mark models that fit the real and imaginary parts of the empirical values of the f-parameter, respectively.

Unfortunately, we were unable to find seismic models fitting both the real and imaginary parts of the empirical f-parameter, neither for ν_2 nor ν_4. This may indicate problems with the opacity coefficient. For B-type pulsators, the f seismic tool probes in particular stellar metallicity and opacities. Although in case of 15 CMa there is no clear preference towards any opacity table, we can notice that the OPAL models can not fit the imaginary part of the f-parameter, neither for ν_2 nor ν_4. There is also a lack of models fitting the real part of f for ν_4 while f_R for ν_2 indicates much higher metallicity.

In general, the compatibility between WNJ and MESA models is rather good (MESA models are not shown), although the f-parameter of WNJ models indicates slightly larger metallicity than in case of MESA models. MESA models have also higher instability parameter, η.

Acknowledgments. We gratefully thank Sophie Saesen and Maryline Briquet for providing the spectroscopic data and Anne Thoul for helpful discussions. Calculations have been carried out using resources provided by Wroclaw Centre for Networking and Supercomputing (http://wcss.pl), grant No. 265. This study is supported by the Polish National Science Centre grants No DEC-2013/08/S/ST9/00583 and 2011/01/B/ST9/05448.

References

Daszyńska-Daszkiewicz, J., Dziembowski, W., & Pamyatnykh, A. 2003, *A&A* 407, 999
Daszyńska-Daszkiewicz, J., Dziembowski, W., & Pamyatnykh, A. 2005, *A&A* 441, 641
Daszyńska-Daszkiewicz, J. & Walczak, P. 2009, *MNRAS* 398, 1961
Handler, G. 2014, IAU Symposium 301, p. 417
Iglesias, C. & Rogers, F. 1996, *ApJ* 464, 943
Pamyatnykh, A., Dziembowski, W., Handler, G., & Pikall, H. 1998, *A&A* 333, 141
Paxton, B., Bildsten, L., Dotter, A., *et al.* 2011, *ApJS* 192, 3
Seaton, M. 2005, *MNRAS* p. 1
Shobbrook, R., Handler, G., Lorenz, D., & Mogorosi, D. 2006, *MNRAS* 369, 171

New windows on massive stars: asteroseismology, interferometry, and
spectropolarimetry
Proceedings IAU Symposium No. 307, 2014
G. Meynet, C. Georgy, J. H. Groh & Ph. Stee, eds.

© International Astronomical Union 2015
doi:10.1017/S1743921314006887

An interferometric journey
around massive stars

Anthony Meilland and Philippe Stee

Laboratoire Lagrange UMR7293 OCA - CNRS - UNS - Bd de l'Observatoire, Nice, France
email: ame@oca.eu

Abstract. Since the construction in the late nineties of modern facilities such as the VLTI
or CHARA, interferometry became a key technique to probe massive stars and their often-
complex circumstellar environments. Over the last decade, the development of a new generation
of beam combiners for these facilities enabled major breakthroughs in the understanding of
the formation and evolution of massive stars. In this short review, we will present few of these
advances concerning young stellar objects, binarity, mass loss, and stellar surfaces.

Keywords. techniques: interferometric, stars: binaries (including multiple): close, stars: circumstellar matter, stars: early-type, stars: emission-line, Be, stars: mass loss, stars: imaging

1. Introduction

Interferometry is the only observational technique allowing to spatially resolved structures smaller than the milliarcsecond, and that is not going to change soon. Actually,
even the upcoming ELTs would have only one tenth of the current spatial resolution of
CHARA (and one quarter of the VLTI).

Nevertheless, even if interferometry is the technique providing the highest spatial resolution, one interferometric observation only gives a very sparse knowledge of an object
through the measurement of one or a few spatial frequencies of its Fourier transform.
Thus, the more complex an object is, the more measurements are needed and, in that
sense, interferometry is indeed a very time-consuming technique.

Another major issue with interferometric observations is the random wavefront perturbation stemming from the atmosphere. Such time-dependent perturbation makes it
impossible to estimate the absolute interferometric phase. Unfortunately, this phase contains most of the information on a object as shown in Fig. 1 where either the phase
or the modulus of Fourier transform is replaced by a random distribution. Despite this
impossibility to directly measure the phase, partial information can be retrieve using
a linear combination of the measured phases on three telescopes, called closure phase,
which cancels the atmospheric contributions.

The recent increase in number of telescopes of the interferometric arrays and the
improve capability of the new instruments that can simultaneously combine the light
from more telescopes, with a current record of six for the MIRC instrument installed on
CHARA, partly solved these issues. It allows to simultaneously measure more visibilities (up to 15) and closure phases (up to 20). With such an improvement compared to
previous facilities or instruments working with two or three telescopes, better and faster
sampling of the object Fourier transform can be obtained, and finally, image reconstruction techniques can be used efficiently.

In the following we will present a short review of some of the last-decade major results
obtained using optical interferometry on the formation (Sect. 2), the binarity (Sect. 3),
the circumstellar environment (Sect. 4), and the surface (Sect. 5) of massive stars.

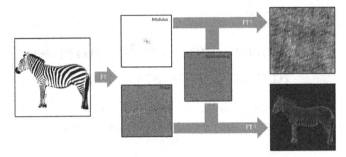

Figure 1. Example illustrating that most of the information on an object geometry is contained in the phase part of its Fourier transform and not in its modulus.

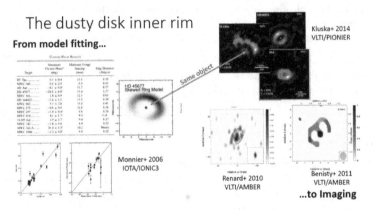

Figure 2. The inner rim of dusty disk around Herbig Ae/Be stars seen by optical interferometry. From Monnier *et al.* (2006), Renard *et al.* (2010), Benisty *et al.* (2011), and Kluska *et al.* (2014).

2. The formation of massive stars

The gain in sensibility of modern optical interferometers made it possible to observe the circumstellar environment of young stellar objects (YSO). Many studies were conducted in the near-infrared to constrain the dust sublimation radius and in the mid-infrared to probe colder material further from the central star. From PTI, IOTA, and Keck aperture masking measurements, Monnier & Millan-Gabet (2002) presented the first statistical study of the near-IR extension of YSO. This work was then completed by Keck-I measurements (Monnier *et al.* 2005) and presented in a more complete review in Dullemond & Monnier (2010). The relation they derived was compatible with simple dusty disk models with an inner rim at an expected dust sublimation temperature.

However, to discriminate between different disk models, a more detailed view of the dusty disk inner rim was needed. For instance, the measure of inner-rim skewness, that can only be done using closure phases, can help to put constraints on the inner-rim height and the disk opacity. First measurements of this skewness was done by Monnier *et al.* (2006) using IOTA. More recent facilities such as VLTI/AMBER and VLTI/PIONIER were used to obtain images of the inner-rim (Renard *et al.* 2010, Benisty *et al.* 2011, and Kluska *et al.* 2014). These results are summarized in Fig. 2.

Similar studies were conducted in the mid-infrared, mainly with the VLTI/MIDI (Leinert *et al.* 2004, Preibisch *et al.* 2006 ,di Folco *et al.* 2009), to probe the larger scale environment and constrain the disk density distribution, flaring, temperature law, and chemistry thanks to MIDI high-enough spectral resolution, that allows to resolve the large PAH and silicate spectral features.

On the other hand, a higher spectral resolution was needed to resolve the hydrogen emission lines and study the geometry and kinematics of the circumstellar gas. Such studies were made possible with the VLTI/AMBER whose resolution is R = 1 500 in MR mode and R = 12 000 in HR mode. The first study was done by Malbet *et al.* (2007) on MWC 297. They showed that, for this object, the Brγ emission was coming from a region larger than the dusty-disk inner rim and that it was expanding, ruling out the possibility of an emission coming from an inner gaseous disk. This object was observed again by Weigelt *et al.* (2011) with a higher spectral resolution and they managed to fully model their observations with a disk-wind model. For an other object, HD 104237, Tatulli *et al.* (2007) found that the extension of the gaseous emission was of the order of the dusty disk inner rim.

The first small survey on the gaseous emission of Herbig stars was published in Kraus *et al.* (2008). They have shown that the gaseous emission can stem from various regions in the circumstellar environment. Of the five stars in their survey, two had emission originating from compact regions compatible with the inner disk. For the three others, the emission was at least as extended as the disk inner rim, favoring the hypotheses of a disk-wind or a very extended stellar wind.

Finally, the VEGA instrument, working in the visible and installed in 2008 on the CHARA array, made it possible to study the gas in the Hα emission line. Two objects were studied : AB Aur by Rousselet-Perraut *et al.* (2010) and HD 200775 by Benisty *et al.* (2013). They both show an extended gaseous emission compatible with a wind model, but highly flattened in the case of AB Aur.

All these previously-cited studies concern stars up to about 10 $M_{\odot}$. The formation processes of the more massive stars are still high debated. Actually, two main scenarios are competing to explain their formation. One is a scaled-up version of Herbig stars scenario, an accretion disk phase followed by a blow-up of the remaining material by the stellar wind. The second one called "competitive accretion", involves a high stellar density (cluster cores), which attracts and forces accretion onto the most massive stars. In this scenario all massive stars should belong to multiple systems. The first very massive young stars observed with the VLTI/AMBER (Kraus *et al.* 2010) showed no clue of binarity and a disk-like structure compatible with the first scenario (see Fig. 3). However, such result need to be confirmed on a larger sample of massive young stellar object, and the binary fraction of massive stars should also be measured.

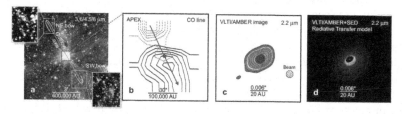

Figure 3. Zoom in on the massive star IRAS 13481-6124 (Kraus *et al.* 2010). From left to right: Spitzer and APEX images showing a bipolar outflow, VLTI/AMBER image and corresponding best-fit radiative model showing a disk like structure perpendicular to the outflow.

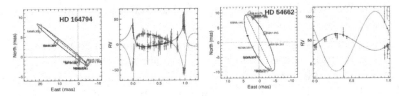

Figure 4. Examples of two massive stars 3D-orbit constrained using interferometric and radial velocity measurements. From yet unpublished VLTI/PIONIER data.

3. The multiplicity of massive stars

Optical interferometry is also a suitable technique for studying stellar multiplicity, especially when it comes to systems with separation of the order of the milliarcsecond. For instance, in an attempt to determine the binary ratio of massive stars, a joint program using the VLTI/PIONIER for small separation (1 to 45 mas) and the VLT/NACO for larger separation (30 mas to 8") was initiated. 173 southern O stars were observed during this Southern Massive stArs at High Angular Resolution (SMASH+) program. The results of this survey, published in Le Bouquin (2014), showed a very high binary fraction of the sample, 0.86, with an average number of companions of 1.83 ± 0.32. They conclude that almost all massive stars are in multiple systems.

Optical interferometry can also be used to determine the visual orbit of binary systems (i.e. projected onto the sky-plane). Combined with spectroscopic measurements of radial velocity, the full three-dimensional orbit of the system can be constrained. This allows to derive the total mass of the system and of the masses of its components with a precision down to a few percents. Many works has been done on massive binary systems, for instance: Sana *et al.* (2013) on the triple O system HD 150136 using the VLTI/PIONIER, Kraus *et al.* (2009) on the binary YSO θ^1 Ori C with the VLTI/AMBER, or Monnier *et al.* (2011) of the Wolf-Rayet binary WR 104 using CHARA/CLASSIC and IOTA/IONIC3. Figure 4 presents the case of two massive stars recently observed with the VLTI/PIONIER.

During their evolution, stars in binary systems can interact with each other through their circumstellar or circumbinary environment. Such interactions, that can be once again well constrained using optical interferometry, are mainly due to gravitational effects or particle collisions (mainly in stellar winds).

The Be-binary system β Lyr is a good example of what can be done on interacting systems using interferometry . This semidetached binary is composed of two B-type stars, a primary supergiant filling its Roche lobe, and a secondary surrounded by an accretion disk fed by the primary material leaking in through the system's first Lagrangian point. Near-infrared reconstructed-images during a complete orbital period were obtained by Zhao *et al.* (2008) using the CHARA/MIRC combiner (see Fig. 5). On these images two components are seen, one unresolved, the primary, and one elongated, the secondary surrounded by the accretion disk. To go further in this study, and disentangle the stellar and circumstellar emission, the system was also observed in $H\alpha$ with the CHARA/VEGA instrument (Bonneau *et al.* 2011). The authors found that the emission was coming from a structure larger than the binary separation, most probably a circumbinary disk with possible contributions from polar jets.

Another interacting binary system discovered and studied using interferometry is the yellow hypergiant HR 5171A. Chesneau *et al.* (2014) showed that this star, one of the largest known star, i.e. $R \sim 1300 R_\odot$, is in contact with a previously unknown companion, and that the system is probably in a common envelope phase.

Due to a lack of sensibility, only a few studies of wind collisions were conducted using interferometry. The main results were obtained with the Keck aperture masking

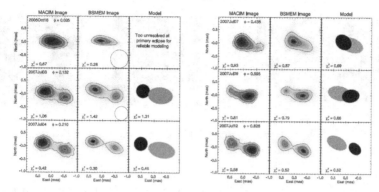

Figure 5. The Be-binary system β Lyr as seen by CHARA/MIRC during one full orbital period. From Zhao *et al.* (2008).

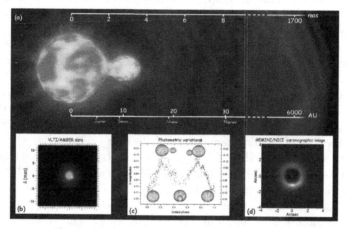

Figure 6. VLTI/AMBER observations and long term photometry revealed the binary nature and common envelope phase of the yellow hypergiant HR 5171A. From Chesneau *et al.* (2014).

experiment (Tuthill *et al.* 1999, Harries *et al.* 2004 and Monnier *et al.* 2007a) on imaging of dust around Wolf-Rayet stars. They found out that the dust was located in a pinwheel nebulae, i.e. a spiral-like pattern tracing the orbital motion of the wind-collision layer. The first Wolf-Rayet stars observed with long-baseline optical interferometry were γ^2 Vel (Millour *et al.* 2007) and WR 118 (Millour *et al.* 2009b). In both papers, the authors manage to resolve the system, and in that second one, they found closure phase signal compatible with a pinwheel model.

4. Mass loss and circumstellar environments

As a natural tracer of mass loss, the study of massive stars circumstellar environment is crucial to improve their evolution models. To characterize not only the total mass loss but its possible anisotropy, one has to put some constraints on the geometry and kinematics of the close surrounding of these massive stars. Interferometry and in particular spectro-interferometry are indeed powerful techniques to probe these environments whose extensions are usually of the order of few millarscseconds in the visible and near-IR and tens of milliarseconds in the mid-IR.

Chesneau *et al.* (2010) used the CHARA/VEGA instrument to study the wind of two blue supergiants: Deneb and Rigel. He managed to model these high-spectral resolution

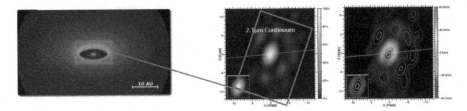

Figure 7. The A[e] supergiant seen by the VLTI instruments MIDI (left) and AMBER, in the 2.1 μm continuum (center) and the Brγ line (right). From Meilland *et al.* (2010) and Millour *et al.* (2011).

(R=30000) observations centered on Hα using the CMFGEN radiative transfer code. The authors also found a strong asymmetry in Deneb wind.

Using VLTI/AMBER and VLTI/VINCI measurements Groh *et al.* (2010) probed the heart of η Car circumstellar environment. They studied possible effects from fast-rotation of the primary and wind collision to explain the break of symmetry of the mass loss and the formation of the Homonculus nebula around these very massive objects. The interferometric study of η Car is still going on and new results are presented further in this book.

Actually, the mass-loss break of symmetry, its origin and its consequences are major issues in the understanding of stellar evolution. Binarity and rotation are the most common physical phenomena that can cause departure from a spherically symmetric mass loss, affecting the loss of angular momentum of the star itself.

Among the objects concerned by this break of symmetry are the stars showing the B[e] phenomenon, i.e. presence of forbidden lines produced in a diluted and highly illuminated medium and of strong IR-excess due to the presence of dust produced in a dense and cool medium. Both rotation through the bi-stable model and binarity were proposed to explain the formation of an equatorial structure dense enough to allow dust formation in their circumstellar environment.

These B[e] stars were perfect targets for the VLTI instruments MIDI and AMBER and the brightest of them were studied in the last decade : CPD -57°2874 (Domiciano de Souza *et al.* 2007), Hen 3-1191 (Lachaume *et al.* 2007), HD 87643 (Millour *et al.* 2009a), HD 62623 (Meilland *et al.* 2010, Millour *et al.* 2011), HD 50138 (Borges Fernandes *et al.* 2011), MWC 300 (Wang *et al.* 2012), HD 327083 (Wheelwright *et al.* 2012), V921 Sco (Kreplin *et al.* 2012), and HD 85567 (Wheelwright *et al.* 2013). Most of these study modeled either with simple geometric model or radiative transfer code such as MC3D, conclude that the dust was located in a circumstellar disk and that the gaseous emission was coming from structures smaller than the dusty-disk inner rim.

In Fig. 7 we present the results obtained on the most extensively studied object with both MIDI and AMBER : the A[e] supergiant HD 62623. For this object, the authors found that both the gas and dust formed a stratified disk in Keplerian rotation, and that the stellar velocity was too small for the rotation to explain the break of symmetry of the environment but too high for a supergiant. They concluded that this star has probably gone through a spin-up phase due to interaction with a spectroscopically detected companion. However, the detailed mechanisms for the disk formation remain unknown.

Classical Be stars were already favored targets for interferometers in the nineties. However, these studies were quite limited by the data lack of accuracy and the small number of telescopes of the interferometric arrays. In the last decade, many objects were observed with modern facilities, and data were analyzed using either simple geometric or kinematic models (Tycner *et al.* 2004, Tycner *et al.* 2006, Meilland *et al.* 2008, Delaa *et al.* 2011,

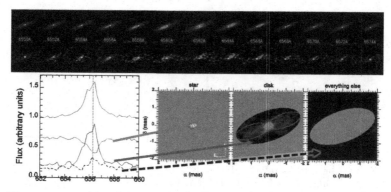

Figure 8. Narrow-spectral band images in the Hα line (top) and extracted spectrum for the star and disk (bottom) for the Be-shell star φ Per. From CHARA/VEGA measurement to be published in Mourard *et al.* (2014).

Stee *et al.* 2012, Kraus *et al.* 2012) or radiative transfer codes such as SIMECA (Meilland *et al.* 2007a, Meilland *et al.* 2007b), BEDSIK (Tycner *et al.* 2011), or HDUST (Štefl *et al.* 2009, Carciofi *et al.* 2009). In all cases the authors found that the circumstellar emission stems from a thin equatorial disk. The more detailed studies, using spectral resolution allowed to probe the kinematics and show that disks were dominated by rotation, probably close to the Keplerian one. The disk density law was also constrained, showing that the observations were compatible with the viscous excretion disk model.

Two Be binary stars were extensively studied to understand the effect of binarity on the Be phenomenon : δ Sco (Tycner *et al.* 2011, Meilland *et al.* 2011, Meilland *et al.* 2013) and Achernar (Kervella & Domiciano de Souza 2006, Kanaan *et al.* 2008, Kervella *et al.* 2009). Finally, some statistical studies of Be stars environments were conducted to constrain the general properties of their disks (Gies *et al.* 2007, Meilland *et al.* 2009, Touhami *et al.* 2013) and also to derive a mean rotational rate of Be stars showing that they were very close to their critical rotation (Meilland *et al.* 2012).

Fig. 8 shows an example of the possibility of Be stars observations with modern spectro-interferometric instruments. In this forthcoming paper, Mourard *et al.* (2014) manage to reconstruct narrow-spectral band images in the Hα line probing the disk geometry and kinematics of this Be-shell star. As shown in the figure, the technique is similar to integral-field spectroscopy but with a forty-times higher spatial resolution.

5. From diameters measurements to stellar imaging

Stellar radius is one of the main input parameters of stellar interior models. It is often used to derive the star mass, age and other physical parameters. With no direct angular measurements, the radius is generally inferred from photometry but with some important biases and poor accuracy. Modern interferometers allow to obtain accurate angular diameters, and using Hipparcos parallaxes, to derive more accurate linear radii.

Accurate angular diameters are also needed to calibrate the surface-brightness relation for some distance estimators such Cepheids and eclipsing binaries. Up to recently, such relation was poorly constrained for early type stars. Recently, significant improvements (see Fig. 9) were obtained using the two CHARA visible beam-combiners, PAVO (Maestro *et al.* 2013) and VEGA (Challouf *et al.* 2014).

Actually, the first star measured with interferometry almost a hundred years ago, the red supergiant Betelgeuse, is still being observed with modern facilities. Haubois *et al.* (2009) obtained the first reconstructed image of its stellar surface. He managed

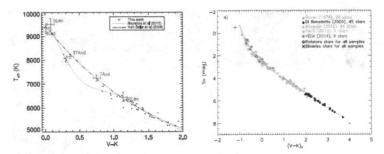

Figure 9. Surface brightness relation for early type stars determined with CHARA/PAVO (Maestro *et al.* 2013, left) and CHARA/VEGA (Challouf *et al.* 2014,right).

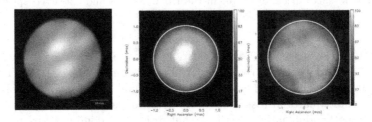

Figure 10. Reconstructed images of red supergiants: Betelgeuse (left), RS Per (center) and T Per (right) . From Haubois *et al.* (2009) and Baron *et al.* (2014).

to detect giant "spots" on the surface that are compatible with convection cells (see Fig. 10). Two other red supergiants were imaged by (Baron *et al.* 2014) in the K-band continuum with CHARA/MIRC whereas the yellow supergiant Canopus was resolved by the VLTI/AMBER (Domiciano de Souza *et al.* 2012).

On the other hand, using the VLTI/AMBER spectro-interferometric capability, Ohnaka *et al.* (2009) resolved the 2.3 μm CO bands of Betelgeuse. These observations allowed to study the geometry and kinematics, showing evidence of asymmetry and convection in a 1.3 R_* extended atmosphere. New observations published in Ohnaka *et al.* (2011) showed significant changes in the kinematics and geometry. Similar results were found for an other red supergiant, i.e. Antares (Ohnaka *et al.* 2013).

Stellar surfaces of five intermediate-mass fast-rotators were imaged in the K band using the CHARA/MIRC instrument (Monnier *et al.* 2007b, Zhao *et al.* 2009, Che *et al.* 2011). The reconstructed images, presented in Fig. 11, allowed to constrain the photosphere flattening and gravity darkening induced by the Von Zeipel effect. Actually, the authors found that their observations are better fitted by a β exponent of 0.22, i.e. slightly different from the 0.25 value derived in Von Zeipel work for radiative stars. Finally, using the ESTER model for interior of fast rotating stars, Espinosa Lara & Rieutord (2011) found values compatible to the one measured by MIRC.

However, Delaa *et al.* (2013) showed that the measured value of the β parameter can be biased in case of a latitudinal differential rotation. Such degeneracy on the gravity darkening measurement could be removed using spectro-interferometry in photospheric lines as show in Fig. 12.

Finally, the differential phase measurements can also be used to measure the stellar diameters of unresolved fast-rotators as shown by Domiciano de Souza *et al.* (2012) using VLTI/AMBER high-resolution measurements of the close-to-critical rotator Achernar. Using the same technique Hadjara *et al.* (2014) even managed to derive the radius for slower rotators.

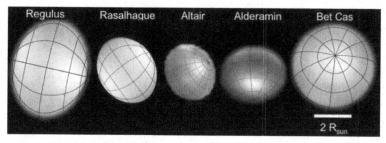

Figure 11. Fast rotators as seen by CHARA/MIRC. From Monnier *et al.* (2007b), Zhao *et al.* (2009), and Che *et al.* (2011).

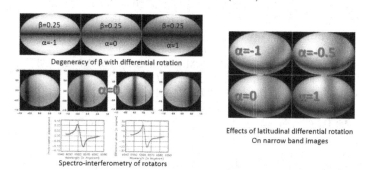

Figure 12. Stellar rotation probed by spectro-interferometry. Top left: Intensity map in the continuum showing the degeneracy of the β parameters for different values of the latitudinal-differential rotation parameter α. Bottom left: Narrow band images at different wavelengths through a photopheric line for $\alpha = 0$ and corresponding differential phase signal. Narrow band images at one wavelength for different values of the α parameter. From Delaa *et al.* (2013).

References

Baron, F., Monnier, J. D., Kiss, L. L., *et al.* 2014, *ApJ* 785, 46

Benisty, M., Perraut, K., Mourard, D., *et al.* 2013, *A&A* 555, A113

Benisty, M., Renard, S., Natta, A., *et al.* 2011, *A&A* 531, A84

Bonneau, D., Chesneau, O., Mourard, D., *et al.* 2011, *A&A* 532, A148

Borges Fernandes, M., Meilland, A., Bendjoya, P., *et al.* 2011, *A&A* 528, A20

Carciofi, A. C., Okazaki, A. T., Le Bouquin, J.-B., *et al.* 2009, *A&A* 504, 915

Challouf, M., Nardetto, N., Mourard, D., *et al.* 2014, *Submitted to A&A*

Che, X., Monnier, J. D., Zhao, M., *et al.* 2011, *ApJ* 732, 68

Chesneau, O., Dessart, L., Mourard, D., *et al.* 2010, *A&A* 521, A5

Chesneau, O., Meilland, A., Chapellier, E., *et al.* 2014, *A&A* 563, A71

Delaa, O., Stee, P., Meilland, A., *et al.* 2011, *A&A* 529, A87

Delaa, O., Zorec, J., Domiciano de Souza, A., *et al.* 2013, *A&A* 555, A100

di Folco, E., Dutrey, A., Chesneau, O., *et al.* 2009, *A&A* 500, 1065

Domiciano de Souza, A., Driebe, T., Chesneau, O., *et al.* 2007, *A&A* 464, 81

Domiciano de Souza, A., Hadjara, M., Vakili, F., *et al.* 2012, *A&A* 545, A130

Dullemond, C. P. & Monnier, J. D. 2010, *ARA&A* 48, 205

Espinosa Lara, F. & Rieutord, M. 2011, *A&A* 533, A43

Gies, D. R., Bagnuolo, Jr., W. G., Baines, E. K., *et al.* 2007, *ApJ* 654, 527

Groh, J. H., Madura, T. I., Owocki, S. P., Hillier, D. J., & Weigelt, G. 2010, *ApJ (Letters)* 716, L223

Hadjara, M., Domiciano de Souza, A., Vakili, F., *et al.* 2014, *Submitted to A&A*

Harries, T. J., Monnier, J. D., Symington, N. H., & Kurosawa, R. 2004, *MNRAS* 350, 565

Haubois, X., Perrin, G., Lacour, S., *et al.* 2009, *A&A* 508, 923

Kanaan, S., Meilland, A., Stee, P., *et al.* 2008, *A&A* 486, 785

Kervella, P. & Domiciano de Souza, A. 2006, A&A 453, 1059

Kervella, P., Domiciano de Souza, A., Kanaan, S., et al. 2009, A&A 493, L53

Kluska, J., Malbet, F., Berger, J.-P., et al. 2014, in M. Booth, B. C. Matthews, & J. R. Graham (eds.), IAU Symposium, Vol. 299 of IAU Symposium, pp 117–118

Kraus, S., Hofmann, K.-H., Benisty, M., et al. 2008, A&A 489, 1157

Kraus, S., Hofmann, K.-H., Menten, K. M., et al. 2010, Nature 466, 339

Kraus, S., Monnier, J. D., Che, X., et al. 2012, ApJ 744, 19

Kraus, S., Weigelt, G., Balega, Y. Y., et al. 2009, A&A 497, 195

Kreplin, A., Kraus, S., Hofmann, K.-H., et al. 2012, A&A 537, A103

Lachaume, R., Preibisch, T., Driebe, T., & Weigelt, G. 2007, A&A 469, 587

Le Bouquin, J.-B. 2014, Submitted to A&A

Leinert, C., van Boekel, R., Waters, L. B. F. M., et al. 2004, A&A 423, 537

Maestro, V., Che, X., Huber, D., et al. 2013, MNRAS 434, 1321

Malbet, F., Benisty, M., de Wit, W.-J., et al. 2007, A&A 464, 43

Meilland, A., Delaa, O., Stee, P., et al. 2011, A&A 532, A80

Meilland, A., Kanaan, S., Borges Fernandes, M., et al. 2010, A&A 512, A73

Meilland, A., Millour, F., Kanaan, S., et al. 2012, A&A 538, A110

Meilland, A., Millour, F., Stee, P., et al. 2007a, A&A 464, 73

Meilland, A., Millour, F., Stee, P., et al. 2008, A&A 488, L67

Meilland, A., Stee, P., Chesneau, O., & Jones, C. 2009, A&A 505, 687

Meilland, A., Stee, P., Spang, A., et al. 2013, A&A 550, L5

Meilland, A., Stee, P., Vannier, M., et al. 2007b, A&A 464, 59

Millour, F., Chesneau, O., Borges Fernandes, M., et al. 2009a, A&A 507, 317

Millour, F., Driebe, T., Chesneau, O., et al. 2009b, A&A 506, L49

Millour, F., Meilland, A., Chesneau, O., et al. 2011, A&A 526, A107

Millour, F., Petrov, R. G., Chesneau, O., et al. 2007, A&A 464, 107

Monnier, J., Millan-Gabet, R., & Billmeier, R. 2005, ApJ 624, 832

Monnier, J. D., Berger, J.-P., Millan-Gabet, R., et al. 2006, ApJ 647, 444

Monnier, J. D. & Millan-Gabet, R. 2002, ApJ 579, 694

Monnier, J. D., Tuthill, P. G., Danchi, W. C., Murphy, N., & Harries, T. J. 2007a, ApJ 655, 1033

Monnier, J. D., Zhao, M., Pedretti, E., et al. 2011, ApJ (Letters) 742, L1

Monnier, J. D., Zhao, M., Pedretti, E., et al. 2007b, Science 317, 342

Mourard, D., Monnier, J., Che, X., Meilland, A., & Millour, F. 2014, Submitted to A&A

Ohnaka, K., Hofmann, K.-H., Benisty, M., et al. 2009, A&A 503, 183

Ohnaka, K., Hofmann, K.-H., Schertl, D., et al. 2013, A&A 555, A24

Ohnaka, K., Weigelt, G., Millour, F., et al. 2011, A&A 529, A163

Preibisch, T., Kraus, S., Driebe, T., van Boekel, R., & Weigelt, G. 2006, A&A 458, 235

Renard, S., Malbet, F., Benisty, M., Thiébaut, E., & Berger, J.-P. 2010, A&A 519, A26

Rousselet-Perraut, K., Benisty, M., Mourard, D., et al. 2010, A&A 516, L1

Sana, H., Le Bouquin, J.-B., Mahy, L., et al. 2013, A&A 553, A131

Stee, P., Delaa, O., Monnier, J. D., et al. 2012, A&A 545, A59

Tatulli, E., Isella, A., Natta, A., et al. 2007, A&A 464, 55

Touhami, Y., Gies, D. R., Schaefer, G. H., et al. 2013, ApJ 768, 128

Tuthill, P. G., Monnier, J. D., & Danchi, W. C. 1999, Nature 398, 487

Tycner, C., Ames, A., Zavala, R. T., et al. 2011, ApJ (Letters) 729, L5

Tycner, C., Gilbreath, G. C., Zavala, R. T., et al. 2006, AJ 131, 2710

Tycner, C., Hajian, A. R., Armstrong, J. T., et al. 2004, AJ 127, 1194

Štefl, S., Rivinius, T., Carciofi, A. C., et al. 2009, A&A 504, 929

Wang, Y., Weigelt, G., Kreplin, A., et al. 2012, A&A 545, L10

Weigelt, G., Grinin, V. P., Groh, J. H., et al. 2011, A&A 527, A103

Wheelwright, H. E., de Wit, W. J., Oudmaijer, R. D., & Vink, J. S. 2012, A&A 538, A6

Wheelwright, H. E., Weigelt, G., Caratti o Garatti, A., & Garcia Lopez, R. 2013, A&A 558, A116

Zhao, M., Gies, D., Monnier, J. D., *et al.* 2008, *ApJ (Letters)* 684, L95
Zhao, M., Monnier, J. D., Pedretti, E., *et al.* 2009, *ApJ* 701, 209

Anthony Meilland

New windows on massive stars: asteroseismology, interferometry, and
spectropolarimetry
Proceedings IAU Symposium No. 307, 2014
G. Meynet, C. Georgy, J. H. Groh & Ph. Stee, eds.

© International Astronomical Union 2015
doi:10.1017/S1743921314006899

Basics of Optical Interferometry:
A Gentle Introduction

Gerard T. van Belle

Lowell Observatory
email: gerard@lowell.edu

Abstract. The basic concepts of long-baseline optical interferometery are presented herein.

Keywords. stars: fundamental parameters, techniques: interferometers, techniques: high angular resolution, etc.

1. Introduction: Interferometry is Inevitable

Stars are, from an angular size standpoint, small. This rather qualitative, relative statement can actually be easily quantified in a back-of-the-envelope fashion. Using our sun – an object of 30 minutes of arc in angular diameter† – as a prototype, we can rapidly derive the size regime in which we must work to be directly examining the sizes, shapes, and ultimately surface morphologies of stars. Since the sun delivers a (mildly staggering) apparent magnitude of $V \approx -26$, with the next nearest star (in terms of brightness) at about $V \approx 0$, we see that from $V_\odot - V_{\text{star}} = -2.5 \log(I_\odot / I_{\text{star}})$ that the difference in intensity is roughly a factor of 10 billion. Under this back-of-the-envelope approach, with all stars having identical surface brightnesses, intensity scales simply with disk area, and we estimate angular diameter θ:

$$\frac{I_\odot}{I_{\text{star}}} = \frac{A_\odot}{A_{\text{star}}} = \left(\frac{\theta_\odot}{\theta_{\text{star}}}\right)^2 \tag{1.1}$$

From our value of 30' for the sun, we arrive at a size of 12 milliarcseconds (mas) for the nearest, brightest stars; this number diminishes rapidly into the sub-mas regime for sample sizes greater than one or two dozen.‡

Given that conventional ground-based telescopes are limited by the atmosphere to resolutions of roughly 0.25-0.50 arcseconds (under ideal observing conditions that occur less than 10% of the time), already this regime is out of reach. Adaptive optics ameliorates the atmospheric resolution limit somewhat, but aperture size still imposes its limits - eg. the Keck 10-m telescopes can reach a limiting resolution of $1.22\lambda/D \approx 30$ mas (assuming AO-enabled J-band observations). Again, an angular resolution that is insufficient for the task of stellar surface imaging.

This problem becomes even more daunting when considering things we may wish to accomplish in the not-so-distant future. The notion of surface imaging of extrasolar planets is one that provokes the 'giggle factor' response in many, and yet the ravenous hunger of the field for increased knowledge of these objects will ultimately lead to considering solutions for this challenge. A 10,000km object at a distance of 10 parsecs is roughly 10

---

† About the size of your thumbnail at arm's length.

‡ As an aside, much of the concepts in this article are presented in far greater detail – and far more precisely – in the summer school writeups such as that of Lawson (2000).

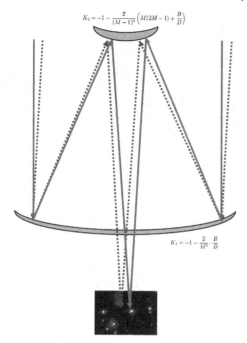

$$K_2 = -1 - \frac{2}{(M-1)^3}\left(M(2M-1) + \frac{B}{D}\right)$$

$$K_1 = -1 - \frac{2}{M^3} \cdot \frac{B}{D}$$

Figure 1. Path lengths through a 'conventional' single aperture telescope. All paths for the on-axis star (solid lines) from the star, through the atmosphere, and through the optical system, to the detector, are equal to optical tolerances ($\sigma_d < 100$nm); the same is true of the off-axis star (dotted line). Although only the edge light rays from both stars are show, this applies to the entire ensemble of rays across the entrance aperture. Meeting this pathlength equality condition dictates the specific shapes of the primary and secondary mirrors, relative to the sky and relative to each other.

microarcseconds (μas) in size – a regime where clearly the highest angular resolution capabilities will be needed. The need for angular resolution at this scale is intimidating but any future astrophysics roadmap ultimately arrives there: *interferometry is inevitable*.

2. Interferometry is Already Here

Although you, gentle reader, are now faced with the prospect of interferometry being inevitable, do not despair! For one main unrealized lesson of many is that **all telescopes are interferometers** – the technique of interferometry is already here throughout astronomy. Illustrated in Fig. 1 is this little-acknowledged fact, which can be best thought of in this context as a question of path length: all paths through the instrument, from stars in the field of view to detector, must be equal to optical tolerances. If this condition is not met, the telescope will not be diffraction limited.

The best example of failing to meet this condition is the original Hubble Space Telescope: the shape of the primary, rather than meeting its prescription to its specified tolerance of roughly $\lambda/100$, actually came in at $10 \times \lambda$. However, since the error in the primary was a smooth deviation, corrective optics could be introduced into its optical path to return the overall system to a condition where the pathlength equality condition was satisfied†.

A second instructive example here are the largest telescopes currently available to astronomy – those based upon giant segmented mirrors, such as the existing Keck-1 and Keck-2 telescopes, or the planned E-ELT, TMT, or GMT telescopes. These facilities have primary mirrors (and in the case of the next generation, secondary mirrors as well) that are built up of many individual elements. As a whole, the general shape of the resulting mirrors still satisfy the pathlength equality condition, albeit with slight gaps between the individual segments. If one were to increase the size of the gaps between the segments, while maintaining the pathlength equality condition, a useful diffraction-limited signal

† You can think of it as an adaptive optics system with an actuator response rate of 10 nHz.

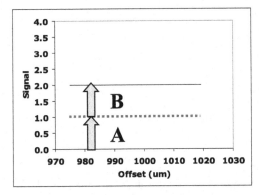 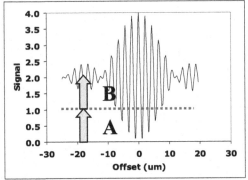

Figure 2. Photometric signal as described in §3 for a two telescope interferometer observing a point source, combined without matching pathlengths (left), and as the zero-path condition is met (right). The peaked appearance of the right-hand signal leads to the moniker 'fringes'.

would still fall upon the detector. In this fashion one could leverage the benefits of such a telescope being an interference device – an interferometer – while increasing its resolving power.

In the end, a long-baseline optical interferometer (LBOI) is simply such a device: the space between its mirrors segments is large – very large – but it is still meeting this important pathlength equality condition. The additional wrinkle introduced with LBOI is that the segments have been replaced by telescopes themselves; the practical impact of this wrinkle is that the on-sky field-of-view of a LBOI is much smaller than that for a conventional telescope.

3. What Does a Simple Interferometer 'See'?

For the simplest case of a two-telescope interferometer, one may think of the light collected from the two telescopes – let's call them 'A' and 'B' – passing through the optical system, meeting up with each other at a 50% transmission / 50% reflection beam splitter, and one of those two combined beams then being concentrated upon a single-element photodetector†.

If the light from 'A' and 'B' arrives at the photodetector without the paths through each of the interferometer's arm, through the atmosphere, and back to the star, being exactly equal‡, all that the detector will register is the amount of light from telescope 'A' in combination with 'B' (Fig. 2, left). However, if care is taken to match the pathlengths from the two telescopes, a much different signal is seen – the signal will appear to fluctuate above and below the A+B level (Fig. 2, right). This is the phenomenon of constructive and destructive interference happening as a zero-path differential is achieved.¶

If our maximum constructive (V_+) and destructive (V_-) signals are measured, we can characterize the amplitude of these 'fringes' simply as:

$$V = \frac{V_+ - V_-}{V_+ + V_-} \tag{3.1}$$

† This is in fact the *exact* architecture for some of the early facilities and/or 'first light' instrumentation of later facilities: IRMA (Dyck *et al.* 1993), IOTA (Dyck *et al.* 1995), CHARA (ten Brummelaar *et al.* 2005).

‡ Well, 'exactly' here means to within a fraction of operational wavelength.

¶ A frequently asked question: "Is this 'creating' signal on the detector?" No. Bear in mind that the 50/50 beam splitter has *two* outputs; the other of the two combined beams shows an equal but opposite signal from Figure 2, right – conservation of energy applies even here.

which is often referred to as the **visibility**. For a perfect optical system (including an atmosphere that does not degrade the signal) observing an unresolved star, $V = 1$.

This is all good and well, but where things get interesting is with a resolved star (Figure 3). A resolved star can be thought of as sending light through one's optical system from each of two halves. Each side of the star individually is unresolved, but the angular separation upon the star corresponds to a slightly different zero-path condition through the optical system. Since the interference patterns of each half is seen simultaneously, they smear each other out, reducing the contrast or viability of the star's signal . The reason why this is interesting is that there is a direct relationship between the amount of reduction in visibility and the angular size of a star:

$$V = \frac{2J_1(x)}{x}, \text{ where } x = \frac{\theta_{UD}\pi B}{\lambda} \qquad (3.2)$$

with J_1 is the first-order Bessel function, θ_{UD} is the angular diameter of an equivalent uniformly illuminated disk†, B is the *projected* baseline between the two telescopes, π is simply the familiar numerical constant, and λ is the operational wavelength.

As such, in the most straightforward case‡ a measurement of V can be used to directly establish the angular diameter of a star. In practice, rigorously measuring V is complicated by an uncooperative atmosphere, necessitating short integration times (typically shorter than an atmospheric coherence time, ∼a few milliseconds), and non-perfect optical systems, necessitating unresolved calibrator-science target-calibrator observational interleaves to explicitly measure the true point-response of the instrument.¶

The underlying physical description of what is being observed is provided by the **van Cittert-Zernike theorem** (van Cittert 1934; Zernike 1938); for a LBOI, the effective implication of this theorem is that the visibility measurements we make in this context correspond to the *real* components of the Fourier transform of the image of the object being observed.

4. Fundamental Stellar Parameters from Interferometry

Angular sizes of stars are, by themselves, not terribly useful. It is in conjunction with *ancillary* data products where such data reveal their true strength. The most immediate fundamental stellar parameter – and the necessary ancillary data product – is fairly obvious, namely **linear size**. If we know the distance to a star, the linear size is realized immediately from $R = \pi\theta$, where R is the linear radius‖, and π is the parallax. In many cases, the real challenge is determination of π – not always readily accomplished, even in this age of *Hipparcos* and *Gaia*.

The second immediate fundamental parameter, **effective temperature**, comes from the *definition* of luminosity, L:

$$L = 4\pi\sigma R^2 T_{eff}^2 \rightarrow T_{eff} \propto \left(\frac{F_{bol}}{\theta^2}\right)^{1/4} \qquad (4.1)$$

when we divide by distance on both sides: L becomes the bolometric flux, F_{bol}, and R

---

† A poor but ultimately serviceable zeroth-order approximation of the intensity distribution across the face of a star.

‡ The star isn't *over*-resolved, is a reasonably uniform disk, isn't non-spherical, etc. etc.

¶ One interesting aspect of measuring V, typically at the 1-5% level, to arrive at θ in Equation 3.2: the resolution limit is in this specific regime is not 1.22 λ/D but more like $0.25\lambda/D$. Hence the spatial resolution of LBOI is typically better than people naïvely expect.

‖ Historically, angular sizes have always been in terms of 'angular diameter', whereas linear sizes have been in terms of 'linear radius'. Another one of the many idiosyncrasies of astronomy.

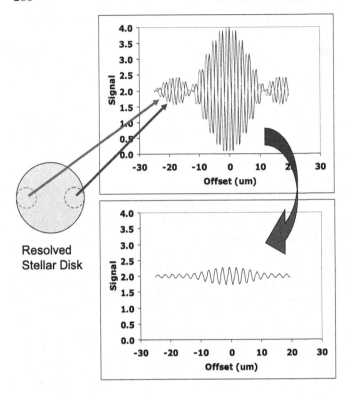

Figure 3. An optical interferometer has sufficient resolving power that, for each of the two halves of a star that is sufficiently large, there is a slight path length difference. Each half creates 'perfect' fringes, but the two halves combine incoherently, effectively smearing the resulting interference signal and reducing the amplitude of the fringes.

becomes θ. Here we see one of the major strengths of this approach: determination of effective temperature – an essentially *macroscopic* quantity – is being accomplished by macroscopic means, in contrast to 'microscopic' means such as spectroscopy. As with linear size, the real challenge in many cases is not the determination of θ but the determination of $F_{\rm BOL}$. Direct observation of the entire flux budget arriving here at Earth for a star is complicated by unobservable windows, and estimates of losses due to interstellar reddening are typically somewhat model dependent.

5. What Does a More Complicated Interferometer 'See'?

Modern facilities such as VLTI, CHARA and NPOI are building upon these fundamentals by incorporating more than 2 telescopes feeding light to the back end instrumentation – for the former, up to 4 beams can be combined simultaneously, while for the latter two facilities, up to 6 beams can be combined at once. Each pair of telescopes can in theory produce a single visibility measurement of the object being observed, with rapid gains as the number of telescopes N_T is increased – the number of visibility points N_V goes as

$$N_V = \frac{N_T(N_T - 1)}{2} \qquad (5.1)$$

However, where a 3-plus element interferometer gains is in its ability to observe the **closure phase** of an object. As illustrated in Figure 4, consider the case where 3 telescope are each pointed at the same object. The true path length back to the star for a given telescope, seen in Figure 4a can be represented by an absolute phase value ϕ. Returning to the formality of the van Cittert-Zernike theorem, the phase corresponds to the *imaginary* part of a component in the Fourier transform of the image of the object being observed.

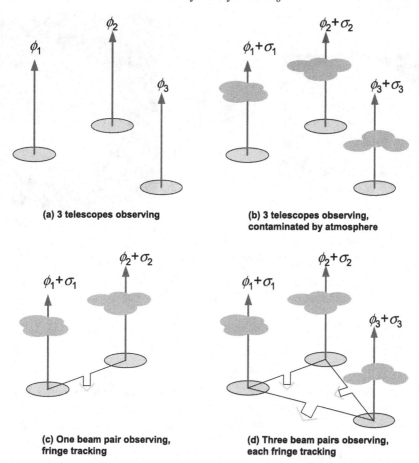

(a) 3 telescopes observing

(b) 3 telescopes observing, contaminated by atmosphere

(c) One beam pair observing, fringe tracking

(d) Three beam pairs observing, each fringe tracking

Figure 4. An illustration of observing closure phase, as discussed in §5.

Unfortunately, the atmosphere induces for each telescope an unknown about of 'piston' error (Figure 4b), which is a time-variable amount of pathlength error, typically many times greater than the operational optical wavelength†. For even single-baseline observing with a pair of telescopes, this necessitates some form of fringe tracking, such that the position of a delay line joining the two apertures can account for the atmospheric piston error $\sigma_1 + \sigma_2$ (Figure 4c). Unfortunately, in the single-baseline case, positive acquisition and tracking of interference fringes between the two telescopes does not provide any information on the phases - the single observable, that of the necessary delay line position (Φ_{12}) to obtain fringes - is insufficient to solve for the two unknowns σ_1 and σ_2 and arrive at some knowledge of the object's phases $\{\phi_1, \phi_2\}$.

However, in the case of 3 telescopes, the landscape alters in a subtle but significant way. In this situation, the 3 unknowns $\{\sigma_1, \sigma_2, \sigma_3\}$ are matched by 3 observables $\{\Phi_{12}, \Phi_{23}, \Phi_{31}\}$. Mathematically, we can see this by summing around the closure phase triangle:

$$\Phi_{12} = (\phi_1 + \sigma_1) - (\phi_2 + \sigma_2) \qquad (5.2)$$

$$\Phi_{23} = (\phi_2 + \sigma_2) - (\phi_3 + \sigma_3) \qquad (5.3)$$

† For example, for visible light or near-infrared observing, piston error is on the order of $\sim$1-10μm, on $\sim$1-10ms timescales.

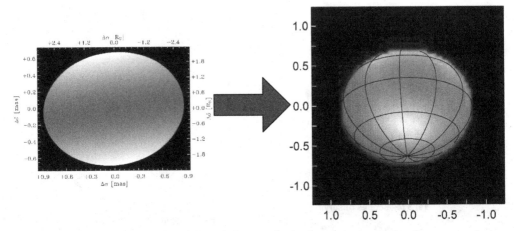

Figure 5. An illustration of the utility of closure phase: the appearance of α Cep as seen in van Belle *et al.* (2006) (left; only visibility data was available) and Zhao *et al.* (2009) (right; visibility and closure phase data). The overall scale of the object and its oblateness is well captured in both – a characteristic of visibility data – but the asymmetric surface brightness characteristic of the non-edge-on von Zeipel effect is missed in the earlier study.

$$\Phi_{31} = (\phi_3 + \sigma_3) - (\phi_1 + \sigma_1) \tag{5.4}$$

which leads to

$$\Phi_{123} = \Phi_{12} + \Phi_{23} + \Phi_{31} = \phi_1 + \phi_2 + \phi_3 \tag{5.5}$$

Although the closure phase Φ_{123} is not an individual absolute phase measurement (such as ϕ_1), we see in Equation 5.5 that it is related to the combined true phase information of the observed object and is free of atmospheric noise. In the case of a single triple of telescopes, 1/3 of the 'true' image information is thus recovered. As with visibility points, the number of closure phases rapidly increases with number of telescopes:

$$N_{CP} = \frac{(N_T - 2)}{N_T} \tag{5.6}$$

In the case of 4 telescopes, 50% of the object phase information is recovered; for 6, that number is 66%.

In general terms, the closure phase information is related to the imaginary part of the Fourier transform of the image upon the sky; the visibility is the real part. A more practical short-hand way of thinking of closure phase is that it is related to the degree of asymmetry of the object's intensity distribution upon the sky. Without closure phase, significant morphology features can be difficult to recover from interferometric data (Figure 5).

6. Summary

The new 4- and 6-way closure phase instruments that are available or soon to be deployed are enabling a second revolution in optical interferometry's impact upon astronomy through high resolution (Monnier 2007). This includes MIRC on CHARA†; MATISSE, GRAVITY and PIONIER on VLTI; and Classic and VISION on NPOI.

Overall, the technique of optical interferometry is one that has achieved a significant degree of maturity since the simple beginnings of its modern era, roughly 25 years ago.

† VEGA and PAVO are also notable new CHARA instruments but are only 3T devices.

The modern instruments and the supporting infrastructure of data analysis tools are significantly extending its community accessibility and scientific reach. Although the technique is viewed by some as still somewhat esoteric, grappling with the basic concepts of fringe **visibility** (essentially, the *size* of a interference signal) and fringe **phase** (essentially, the *position* of a interference signal) help de-mystify the underlying concepts.

References

Boyajian, T. S., van Belle, G., & von Braun, K. 2014, *AJ* 147, 47

Dyck, H. M., Benson, J. A., Carleton, N. P., *et al.* 1995, *AJ* 109, 378

Dyck, H. M., Benson, J. A., & Ridgway, S. T. 1993, *PASP* 105, 610

Lawson, P. R. (ed.) 2000, *Principles of Long Baseline Stellar Interferometry*

Monnier, J. D. 2007, *New Astron. Revs* 51, 604

ten Brummelaar, T. A., McAlister, H. A., Ridgway, S. T., *et al.* 2005, *ApJ* 628, 453

van Belle, G. T. 1999, *PASP* 111, 1515

van Belle, G. T., Ciardi, D. R., ten Brummelaar, T., *et al.* 2006, *ApJ* 637, 494

van Belle, G. T., Thompson, R. R., & Creech-Eakman, M. J. 2002, *AJ* 124, 1706

van Belle, G. T. & van Belle, G. 2005, *PASP* 117, 1263

van Cittert, P. H. 1934, *Physica* 1, 201

Whitelock, P. A., van Leeuwen, F., & Feast, M. W. 1997, in R. M. Bonnet, E. Høg, P. L. Bernacca, L. Emiliani, A. Blaauw, C. Turon, J. Kovalevsky, L. Lindegren, H. Hassan, M. Bouffard, B. Strim, D. Heger, M. A. C. Perryman, & L. Woltjer (eds.), *Hipparcos - Venice '97*, Vol. 402 of *ESA Special Publication*, pp 213–218

Zernike, F. 1938, *Physica* 5, 785

Zhao, M., Monnier, J. D., Pedretti, E., *et al.* 2009, *ApJ* 701, 209

Discussion

AERTS: We made several efforts to get interferometric radii of B-type pulsators (β Crucis, σ Scorpii, β Centauri) but in each case we faced the problem of systematic uncertainties due to lack of good enough calibrators resulting in highly inaccurate estimates (incompatible with other data from spectroscopy and asteroseismology). How can this be solved or improved? Can you do, e.g., 12 Lac, HD 180642 and try to get the radius of these bright B-type pulsators as test case?

VAN BELLE: Using the crude $\{V, K\}$ angular size predictor in van Belle (1999) (updated nicely in Boyajian *et al.* 2014), we can examine the rough angular sizes we should expect for these stars, which reveals some of the answer:

Star	Spectral Type	V (mag)	K (mag)	theta (mas)
β Cru	B1 IV	1.25	1.99	1.05
σ Sco	B1 III	2.89	2.40	1.15
β Cen	B1 IV	0.60	1.28	1.5
12 Lac	B1 III	5.23	5.62	0.22
HD 180642	B1 II	8.29	7.79	0.10

The key here is that these stars are *small* - angular sizes measures of less than 1.5mas are very hard to do and require facilities with the highest amounts of angular resolution; since these are all southern hemisphere objects I presume that the attempts have been made with VLTI, which would find these angular scales challenging. For the first three, a facility such as CHARA or NPOI should be able to resolve the northern hemisphere counterparts of these objects (and a corresponding set of sufficiently small calibration objects, as described in a suitable reference such as van Belle & van Belle 2005). NPOI has greater spatial resolution by virtue of the fact it operates at $\sim 3\times$ shorter wavelengths on similar baselines to VLTI; CHARA has $\sim 3\times$ longer baselines (and is starting to operate at $\sim 3\times$ shorter wavelengths for additional spatial resolution).

The latter two stars are particularly small, but more significantly, they are very faint relative to the capabilities of modern facilities which is an additional challenge.

MASON: Regarding the relative magnitude limit of Gaia and interferometers – will the fundamental limit on diameters (eg. nearby stars) be set by Hipparcos parallax errors?

VAN BELLE: It's an interesting question where the 'fundamental limit' may be connected not so much to specific missions, but to basic astrophysics or particular objects. Cool evolved stars tend to have a certain intrinsic luminosity (eg. very roughly, $10^4 - 10^5 L_\odot$) which means that at a given distance, they'll have a given angular size which 'competes' with its parallax. What this translates to in slightly more quantitative terms of a specific example is that, for Mira variable stars, the angular size for a given star is typically $3\times$ greater than the parallax, so to make a parallax measurement, one is trying to measure substellar diameter photocenter shifts. Coupling this fact to the expectation that these objects will have surface morphologies and rotational periods that both change on month to year time scales – which matches the cadence of any parallax program – and you can see that there are challenges in obtaining a parallax for these objects that are above and beyond objects with smaller angular sizes (this is discussed in more detail in §3.4 of van Belle *et al.* 2002). This is a mission independent phenomenon. A sufficiently large ensemble of data can lead to meaningful results (Whitelock *et al.* 1997) but this does mean that, for individual stars, distances remain an unsolved problem.

For the hot stars of interest for this symposium, though, the relationship between their angular sizes and parallaxes mean that sub-photocenter measures are not being attempted, and this specific problem is not a concern.

On the point of Gaia: Gaia is going to be not simply an amazing mission but a truly revolutionary one; however, the specific performance in the regime of its bright limit remains to be seen. It is possible that its distances for (the relatively bright) stars of interest to interferometry will not improve substantially upon the Hipparcos results.

Gerard van Belle

New windows on massive stars: asteroseismology, interferometry, and spectropolarimetry
Proceedings IAU Symposium No. 307, 2014
G. Meynet, C. Georgy, J. H. Groh & Ph. Stee, eds.
© International Astronomical Union 2015
doi:10.1017/S1743921314006905

The photosphere and circumstellar environment of the Be star Achernar

Daniel M. Faes[1,2], Armando Domiciano de Souza[2], Alex C. Carciofi[1] and Philippe Bendjoya[2]

[1]Instituto de Astronomia, Geofísica e Ciências Atmosféricas, Universidade de São Paulo, Rua do Matão 1226, Cidade Universitária, 05508-900, São Paulo, SP, Brazil
email: moser@usp.br

[2]Lab. J.-L. Lagrange, UMR 7293 - Observatoire de la Côte d'Azur (OCA), Univ. de Nice-Sophia Antipolis (UNS), CNRS, Valrose, 06108 Nice, France

Abstract. Achernar is a key target to investigate high stellar rotation and the Be phenonemon. It is also the hottest star for which detailed photospheric information is available. Here we report our results to determine the photospheric parameters of Achernar and evaluate how the emission of a Viscous Decretion Disk (VDD) around it would be observable. The analysis is based on interferometric data (PIONIER and AMBER at ESO-VLTI), complemented by spectroscopy and polarimetry for the circumstellar emission. For the first time fundamental parameters of a Be photosphere were determined. The presence of a residual disk at the quiescent phase and some characteristics of the new formed disk (2013 activity) are also discussed. This is rare opportunity to precisely determine the stellar brightness distribution and evaluate the evolution of a just formed Be disk.

Keywords. stars: individual (Achernar), stars: fundamental parameters, techniques: interferometric, circumstellar matter, stars: emission-line, Be

1. Introduction

Be stars are known to be rapid rotators and this property is believed to be fundamentally linked to the existence of the disk. Other mechanisms, such as magnetic activity, non-radial pulsations and stochastic processes at the photosphere are also necessary agents to explain the Be phenomenon. The investigation of the photosphere of Be stars, and of their circumstellar (CS) disks, has entered a new phase during last years, both observationally and theoretically. See Rivinius *et al.* (2013b) for a review.

Optical and infrared interferometry is a technique capable of bringing qualitatively new information by resolving the stars and their disks at the milliarcsecond (mas) level. Due to the relatively strict magnitude limits of current interferometers, bright nearby Be stars are among the most popular and most frequently observed targets.

2. An interferometric review of Achernar

Achernar (α Eridani, HD10144) is the closest and brightest Be star in the sky. It is also the hottest and more massive star from which we have detailed photospheric information (van Belle 2012). Moreover, among them, it is the one of the highest rotation rate (Domiciano de Souza *et al.* 2014).

The successive generations of beam combiners of ESO-VLTI (Very Large Telescope Interferometer) have being used to study Achernar. This section summarize these early interferometric studies. All parameters derived from interferometry strongly depend on modeling (i.e., accurate brightness distribution), which is wavelength dependent. Nevertheless, fundamental physical properties from both photosphere and CS environment could be constrained based on the high-angular resolution information available.

2.1. *First oblateness determination with VINCI*

Domiciano de Souza *et al.* (2003) determined for the first time the oblateness of Achernar (and of a Be star) with K-band visibility data from VINCI. It was also the first estimation of the on-sky orientation of the star. The diameter ratio between the equator and the pole $2a/2b = 1.56$ was found adjusting to each projected position angle (PA) of the observations the size of an uniform disc.

Although undoubtedly indicating a high rotation rate for the star, that was not necessarily implying a rotation above the Roche limit: the radii ratio depends on the underlying stellar model (the fit of multiple uniform disks is just a first approximation of the stellar brightness distribution). Knowing that the star had a little activity at the time of the observations (e.g. Vinicius *et al.* 2006), the authors have also explicitly ignored the contribution of the CS component in the data since it was a much weaker emission than the stellar one ($< 5\%$).

2.2. *On a polar emission in the interferometric signal*

Kervella & Domiciano de Souza (2006) analyzed Achernar's VINCI data from both K and H bands. They have done a single fit on them adopting a uniform ellipse brightness distribution for the star. The diameter ratio between the equator and the pole was then $2a/2b = 1.41$.

In this scenario, the addition of a elongated CS envelope with Gaussian brightness distribution superimposed to the stellar model significantly improved the quality of the fitting. This CS component had $\approx 4.5\%$ of the total flux and removed the trend that appeared in the residuals of the uniform ellipse in PAs around the polar direction. This study was made using only the VINCI data as constraint, and did not relate the CS emission with other observables.

2.3. *The first multi-technique analyzes*

Carciofi *et al.* (2008) and Kanaan *et al.* (2008) have independently done a multi-technique fitting of Achernar at the time of VINCI observations. They have used SED and Hα line profiles as constraints, and linked them with polarimetric estimates. Although taking into account the brightness distribution by the Von Zeipel effect and the Roche geometry due to the high rotation of the star, they found distinct results: while Carciofi *et al.* argue that the presence of a residual CS disk in Achernar is sufficient to explain the observed quantities with a near-critical rotating star, Kanaan *et al.* conclude that at the time of the VINCI observations Achernar had either a small or no CS disk, but it did have a polar, stellar wind.

2.4. *Proof of concept: what information is contained on diff. phases*

Domiciano de Souza *et al.* (2012a) have done the first consistent determination of multiple photospheric parameters of Achernar (radius, rotational velocity, inclination angle and on-sky orientation). This work was based on AMBER high resolution spectro-interferometry along the H Brackett γ line. The photospheric parameters were found using the CHARRON code (Domiciano de Souza *et al.* 2002, 2012b) fitting the differential phases. The derived equatorial angular diameter was compatible with previous values from visibilities, and the diameter ratio was below the Roche limit ($2a/2b = 1.45$). This

Table 1. Physical parameters of Achernar derived from the fit of the RVZ model (CHARRON code) to VLTI/PIONIER H band data using the MCMC method (emcee code).

Model free parameters fitted	Values and uncertainties
Equatorial radius ($R_\odot$)	9.16 (+0.23; -0.23)
Equatorial rotation velocity ($km\,s^{-1}$)	298.8 (+6.9; -5.5)
Rotation-axis inclination angle ($^\circ$)	60.6 (+7.1; -3.9)
Gravity-darkening coefficient (β)	0.166 (+0.012; -0.010)
Position angle of the visible pole ($^\circ$)	216.9 (+0.4; -0.4)

time, no positive detection of circumstellar activity was found at the time of observations (2009-2010), from both polarimetry and spectroscopy.

This study shows that differential phases are capable of providing useful information, but multiple parameter fitting using only a differential quantity is tricky (see Domiciano de Souza *et al.* 2014, for a discussion).

3. The photospheric characterization with PIONIER

A quantitative jump was given to the characterization of Achernar by Domiciano de Souza *et al.* (2014). Their analysis include observations from PIONIER 4 beam combiner, obtaining high precision visibility measurements with closure phase information. Its photosphere was studied in detail since no important circumstellar activity was detected from 2009 to 2011. The employed data include photometry, spectroscopy, polarimetry and interferometry (squared visibilities, closure and differential phases) all considered in a single analysis.

The adopted model for the stellar photosphere is the Roche model (rigid rotation and mass concentrated in the stellar center) with a generalized form of the von Zeipel's gravity darkening (RVZ model). The model-fitting was performed using the emcee code (Foreman-Mackey *et al.* 2013), an implementation of a Markov Chain Monte Carlo (MCMC) method. A summary of the Achenar's parameters is in Table 1.

The stellar diameter ratio $2a/2b$ is about 1.352, which corresponds to a not so high rotation rate. Still, it is the highest rotation rate observed in a angularly resolved star so far. The determined β parameter supports the model of Espinosa Lara & Rieutord (2011), which is a promising explanation to gravity darkening variation in rotating stars.

Independently of the parametric fitting, interferometric reconstruction image was performed using the MIRA software (Thiébaut 2008). The difference between the brightness distribution of the two methods is such that it excludes any circumstellar emission above 1.5% from the total flux in the considered epoch of the observations.

4. The recent disk activity

In the beginning of 2013 Achernar started a new Be active phase after about 7 years of quiescence, in which was mostly a normal B star. Our campaign aims to characterize the new formed disk using VLTI spectro-interferometry and its radiative transfer modeling using the HDUST code (Carciofi & Bjorkman 2006, 2008). Specifically, we aim to characterize for the first time the mass loss rate $\dot{M}$ and the viscous diffusion coefficient α to the newly formed Be disk based on the VDD model (c.f. Rivinius *et al.* 2013b, and references therein). According to this model, the CS disk surface density $\Sigma(r)$ can be a function of time according to the Eq. 4.1.

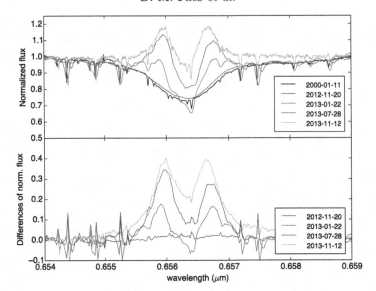

Figure 1. Hα line profile of Achernar.

$$\frac{\partial \Sigma}{\partial t} = \frac{1}{\tau_{\mathrm{vis}}} \left(\frac{2}{r}\frac{\partial}{\partial r} \left[r^{1/2}\frac{\partial}{\partial r}(r^2\Sigma) \right] + \Sigma_{\mathrm{in}}(t)\frac{r_{\mathrm{in}}^2}{r_{\mathrm{in}}^{1/2}-1}\frac{1}{r}\delta(r-r_{\mathrm{in}}) \right) \qquad (4.1)$$

The new activity has been report by Napoleão & Marcon (priv. comm.) looking at the Hα line profile. The Hα evolution in 2013 can be seen in Fig. 1. As an example of phenomenon associated with the stellar activity is the Achernar's spin-up discovered by Rivinius *et al.* (2013a). So, this is a rare opportunity to evaluate the process of the evolution of the just formed Be disk high-angular resolution measurements: its importance is to unveil the spin unknown mass-loss mechanism(s) in Be stars and the process governing the circumstellar material in the close vicinity of the stellar photosphere in such a nearby star.

References

Carciofi, A. C. & Bjorkman, J. E. 2006, *ApJ* 639, 1081

Carciofi, A. C. & Bjorkman, J. E. 2008, *ApJ* 684, 1374

Carciofi, A. C., Domiciano de Souza, A., Magalhães, A. M., Bjorkman, J. E., & Vakili, F. 2008, *ApJ (Letters)* 676, L41

Carciofi, A. C., Magalhães, A. M., Leister, N. V., Bjorkman, J. E., & Levenhagen, R. S. 2007, *ApJ (Letters)* 671, L49

Domiciano de Souza, A., Hadjara, M., Vakili, F., *et al.* 2012a, *A&A* 545, A130

Domiciano de Souza, A., Kervella, P., Faes, D. M., *et al.* 2014, *A&A* accepted

Domiciano de Souza, A., Kervella, P., Jankov, S., *et al.* 2003, *A&A* 407, L47

Domiciano de Souza, A., Vakili, F., Jankov, S., Janot-Pacheco, E., & Abe, L. 2002, *A&A* 393, 345

Domiciano de Souza, A., Zorec, J., & Vakili, F. 2012b, in S. Boissier, P. de Laverny, N. Nardetto, R. Samadi, D. Valls-Gabaud, & H. Wozniak (eds.), *SF2A-2012: Proceedings of the Annual meeting of the French Society of Astronomy and Astrophysics*, pp 321–324

Espinosa Lara, F. & Rieutord, M. 2011, *A&A* 533, A43

Foreman-Mackey, D., Hogg, D. W., Lang, D., & Goodman, J. 2013, *PASP* 125, 306

Kanaan, S., Meilland, A., Stee, P., *et al.* 2008, *A&A* 486, 785

Kervella, P. & Domiciano de Souza, A. 2006, *A&A* 453, 1059

Rivinius, T., Baade, D., Townsend, R. H. D., Carciofi, A. C., & Štefl, S. 2013a, *A&A* 559, L4

Rivinius, T., Carciofi, A. C., & Martayan, C. 2013b, *A&A Rev.* 21, 69

Thiébaut, E. 2008, *MIRA: an effective imaging algorithm for optical interferometry*

van Belle, G. T. 2012, *A&A Rev.* 20, 51

Vinicius, M. M. F., Zorec, J., Leister, N. V., & Levenhagen, R. S. 2006, *A&A* 446, 643

Discussion

RIVINIUS: Can you comment on the presence (or absence) of a polar wind above Achernar?

FAES: We have no reason to believe this feature was present in Achernar at the time of our observations. As we have shown in Domiciano de Souza *et al.* (2014) about the non-presence of a residual disk in the data, the hydrogen emission of such wind would be detected in linear polarization and spectroscopic line profiles before having an inter-ferometric significance on IR.

STEE: There was a clear detection of a polar wind by Kervella & Domiciano (2006) confirmed by Kanaan (2008) of about 5% in the K band. Do you have any evidence of such wind in your data? If you add a 5% extended contribution along the polar axis, is it still compatible with your data?

FAES: As said, there is no evidence of such feature. We have even tried to minimize the photospheric parameters of Achernar in addition with another Gaussian emission, having it free position, size and intensity. An additional component was unable to improve data fitting, comprising the characteristics of such polar axis emission. Furthermore, Achernar's reconstructed image differs from the data in a maximum of 1.5% in flux, and this difference is not extensive in the polar direction.

PETERS: Have you attempted to incorporate the observed FUV photometric variability into your models? On the 1980-90's a 30% flux variability with a period of ~ 0.7 day was observed in the FUV by instrumentation on Galileo [satellite]. A lesser variability at the same period was seen in IUE flux data. It seems to me that α Eri had a hot spot that rotated into/out of our line-of-sight.

FAES: This is something that we have not done yet that is to compare its historical measured variability and our modeling. For example, Carciofi *et al.* (2007) detected linear polarization variations on timescales as short as 1 hr and as long as several weeks. Their modeling strongly suggested that short-term variations can originate from discrete mass ejection events occurring at the photosphere-disk interface. To analyze the evolution of Achernar in light of the new photospheric parameters is the next step of our research. Thank you to remind us about theses UV records.

DOMICIANO DE SOUZA: UV variations of 0.7 day period should be compared to the new photospheric model parameters. There is no indication of polar wind in the PIONIER data, but there is a residual flux of 1%. The VINCI data show a decrease of visibility at low spatial frequency. This is an observational result. Other interpretations different from the polar wind can be tested in combination with the new data.

PULS: Just one comment regarding the "polar wind", that has been mentioned now a couple of times. Let me remind you that we have already discussed this in Paris (IAUS "Active OB-Stars"), and that a typical wind from a B-type would have a very low mass-loss rate ($\sim 10^{-8} M_\odot \, \mathrm{yr}^{-1}$). The corresponding IR-emission would be very low, with $\tau \approx 1$ very close to the photosphere. So, what has been observed (whatever it is) is completely inconsistent with wind theory, whether prolate or not.

FAES: I completely agree. As a suggestion, we could refer to this feature as *interferometric polar signal* instead of "polar wind" since there is no reason to relate this signal to a wind from the theoretical point of view.

Daniel Faes

André Maeder and Corinne Charbonnel

New windows on massive stars: asteroseismology, interferometry, and
spectropolarimetry
Proceedings IAU Symposium No. 307, 2014
G. Meynet, C. Georgy, J. H. Groh & Ph. Stee, eds.

© International Astronomical Union 2015
doi:10.1017/S1743921314006917

Zooming into Eta Carinae with interferometry

Jose H. Groh

Geneva Observatory, Geneva University, Chemin des Maillettes 51, CH-1290 Sauverny,
Switzerland
email: jose.groh@unige.ch

Abstract. Shaped by strong mass loss, rapid rotation, and/or the presence of a close companion,
the circumstellar environment around the most massive stars is complex and anything but
spherical. Here we provide a brief overview of the high spatial resolution observations of Eta
Carinae performed with the Very Large Telescope Interferometer (VLTI). Special emphasis is
given to discuss VLTI/AMBER and VLTI/VINCI observations, which directly resolve spatial
scales comparable to those where mass loss originates. Studying scales as small as a few milliarcseconds allows us to investigate kinematical effects of rotation and binarity in more detail
than ever before.

Keywords. stars: atmospheres stars: individual (Eta Carinae) stars: mass-loss stars: rotation
stars: variables: general supergiants

Recent advances in long-baseline optical interferometry allow one to directly resolve astrophysical objects at the milli-arcsecond (mas) spatial scale. For Galactic massive stars,
this approaches scales comparable to those where mass loss originates, stellar surfaces
and winds become distorted by rotation, and the presence of close companions can be
directly diagnosed. Mass-loss rates generally increase as evolution proceeds, causing the
outer hydrostatic layers of massive stars to eventually become hidden by an optically-
thick stellar wind. Thus, as massive stars evolve, their appearance is increasingly affected
by mass loss – a phenomenon that can be directly probed with optical interferometry. In
addition, the interplay between mass loss and rotation will also shape the circumstellar
environment around massive stars in a complex manner (Owocki et al. 1996). Likewise,
if a massive companion star is present, strong disturbances in the stellar wind structure
may occur, in particular if the wind of the companion has enough momentum to cause
the presence of a wind-wind collision zone between the two stars (Stevens et al. 1992).

Because of its large brightness and unique nature among massive stars, Eta Carinae
has been a natural target for optical interferometers. Most, if not all groups working
on Eta Car, support a massive binary scenario with an eccentric orbit, as suggested by
Damineli et al. (1997). The orbital period is 2022.7±1.3 d (Damineli et al. 2008b), and the
eccentricity $e \geqslant 0.9$ (Corcoran 2005). The orbital orientation has been controversial (see
discussion in Madura et al. 2012), but the first determination in 3-D space suggests an
orbital inclination angle $i \approx 130°$ to $145°$, longitude of periastron $\omega \approx 240°$ to $285°$, and
an orbital axis projected on the sky at a position angle $\mathrm{PA}_z \approx 302°$ to $327°$ from North
to East (Madura et al. 2012). The combined mass of the system is larger than 110 solar
masses (Hillier et al. 2001), and only loose constraints on the individual masses exist.
The combined luminosity is $\simeq 5 \cdot 10^6 \ L_\odot$ (Davidson & Humphreys 1997), and thought
to be accounted for mainly by the primary star (hereafter η_A), since it dominates the
ultraviolet, optical, and near-infrared spectrum (Hillier et al. 2001, 2006; Groh et al. 2012,
hereafter G12). The most recent spectroscopic analysis suggests that η_A has an effective

temperature of $\sim 9,400$ K, mass-loss rate of $8.5 \cdot 10^{-4} M_\odot \mathrm{yr}^{-1}$, and wind terminal velocity of ~ 420 km s^{-1} (G12; see also Hillier *et al.* 2001, 2006). Concerning the companion star (hereafter η_B), only indirect constraints exist on its temperature ($T_{\mathrm{eff}} \simeq 36,000 - 41,000$ K, Mehner *et al.* 2010; see also Verner *et al.* 2005; Teodoro *et al.* 2008) and luminosity ($10^5 L_\odot \leqslant L_\star \leqslant 10^6 L_\odot$; Mehner *et al.* 2010). These come from ionization studies of the ejecta surrounding Eta Car, as η_B has never been observed directly. X-ray studies have constrained the wind properties of η_B ($v_\infty \sim 3000$ km s^{-1} and $\dot{M} \sim 1.4 \cdot 10^{-5}$ $M_\odot \mathrm{yr}^{-1}$; e. g., Parkin *et al.* 2011).

1. A close view of Eta Carinae with the VLTI

Here we review Eta Car VLTI observations obtained in the K-band with the beam-combiner instruments VINCI and AMBER. VINCI (now decommissioned) operated as a broadband instrument and combined light from 2 telescopes delivering visibilities. These are the amplitude of the Fourier transform of an object's brightness distribution on the sky, and visibilities are directly related to *sizes*. In contrast to VINCI, AMBER combines light from up to 3 telescopes and has 3 spectral resolution modes: low- (LR, $R \sim 30$), medium- (MR, $R \sim 1500$), and high-resolution (HR, $R \sim 12000$). In addition to visibilities, AMBER delivers differential phases, which are related to photocenter shifts, and closure phases, which measures how point-symmetric the brightness distribution is.

Eta Carinae was observed with VINCI from 2001 November until 2004 January, using mainly the siderostats. These dataset have been discussed in van Boekel *et al.* (2003) and Kervella (2007). AMBER observations with the UTs in MR and HR, obtained between 2004–2005, were reported by Weigelt *et al.* (2007).

VINCI provided our first milli-arcsecond view of Eta Car (van Boekel *et al.* 2003) and two main breakthroughs: the K-band (pseudo) photosphere was spatially resolved and the brightness distribution was non-spherical in 2003. van Boekel *et al.* (2003) reported visibility measurements of Eta Car for a range of baseline position angles (P.A.) on the sky. These data were analyzed by these authors assuming that the brightness distribution of Eta Car can be represented by a two-dimensional Gaussian. Under this assumption, the VINCI dataset can be fitted by an ellipsoid with a ratio of major to minor axis of 1.25 ± 0.05, with the major axis aligned at P.A.= $134 \pm 7°$ East of North. The inferred Gaussian full-width at half maximum (FWHM) diameters are between $\sim 6 - 8.5$ mas depending on P.A. This elongation of the K-band emission has been explained as being due to the presence of a dense polar wind generated by the rapid rotation of η_A (van Boekel *et al.* 2003). However, any influence from η_B was not considered at this point, which is worth mentioning given that the 2003 VINCI data were obtained at orbital phase $\phi = 10.93$, i.e. relatively close to periastron passage of η_B. Here we assume the ephemeris from Damineli *et al.* (2008a), the orbital cycle labeling from Groh & Damineli (2004), and that periastron occurs at $\phi = 0$.

AMBER revolutionized again our view of Eta Car, as it became possible to resolve the object within spectral lines and, thus, obtain kinematical information. Weigelt *et al.* (2007) were able to resolve Eta Car in multiple spectral channels within Bracket γ, Heɪ 2.058 μm, and K-band continuum, finding 50% encircled-energy diameters of 9.6, 6.5, and 4.2 mas, respectively. The different values of the K-band continuum diameters found with VINCI and AMBER are consistent when accounting for the fact that a Gaussian approximation is not adequate for the brightness distribution of Eta Car (Weigelt *et al.* 2007; Kervella 2007). The AMBER dataset also supports a similar elongation of the K-band emission along P.A.= $120 \pm 15°$. The Hillier *et al.* (2001) model reproduces the K-band continuum visibilities, indicating that the dense stellar wind of η_A is indeed being resolved with AMBER. At least during 2004–2005, it does not seem that a significant amount of hot dust is present within the field-of-view of the UTs (~ 70 mas). The extension of the Br γ emitting region is also roughly reproduced by the spherical Hillier *et al.* (2001) model.

The AMBER data shows significant signals in the differential and closure phases within the Br γ and HeI 2.058 μm lines. These quantities reveal a complex pattern of photocenter shifts and image asymmetries as a function of velocity within the line. The Br γ behavior has been interpreted in the context of a dense polar wind with the rotation axis tilted at $\sim 41°$ from the observer. A geometrical model has been developed by Weigelt *et al.* (2007) to explain the relative photocenter shifts seen in the blue and red wings of Br γ.

AMBER also shed light on the binarity of Eta Car, although light from η_B was not detected. A lower limit for the K-band brightness ratio of η_A to η_B of ~ 50 has been determined (Weigelt *et al.* 2007), which is consistent with the brightness ratio of 200 expected for an O star companion with $T_{eff} \simeq 34,000$ K and $L_\star = 10^6 L_\odot$ (Hillier *et al.* 2006). Given the findings of Mehner *et al.* (2010) that $L_\star = 10^6 L_\odot$ is likely an upper limit, and that η_B could have $L_\star$ as low as $10^5 L_\odot$, it may well be that one needs to be sensitive to brightness ratios of 500 or higher to directly detect η_B.

2. An elongated K-band photosphere: rotation or binary induced?

Given the ubiquitous evidence for binarity in Eta Car, one may wonder how much the presence of η_B would affect the interferometric observables described above, and the conclusion that a rapidly rotating η_A causes the deformation in the K-band photosphere. With the goal of constraining the rotational velocity (v_{rot}) of η_A and probing the influence of η_B on the wind of η_A, Groh *et al.* (2010) analyzed the aforementioned VINCI and AMBER interferometric measurements of Eta Car with 2D radiative transfer models based on the Busche & Hillier (2005) code.

Ignoring the presence of η_B, rotation and the spatial orientation of the rotation axis of η_A can be investigated based on the effects of rotation on the wind density structure, which determines the geometry of the K-band emitting region. Prolate wind models with a ratio of v_{rot} to the critical velocity for break-up (v_{crit}) of $W = 0.77 - 0.92$ (assuming the prescription from Owocki *et al.* 1996), inclination angle of the rotation axis of $i = 60° - 90°$, and position angle on the sky of P.A. $= 108° - 142°$ reproduce

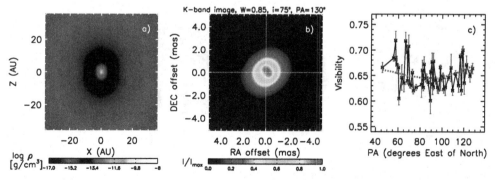

Figure 1. *(a):* Density structure of the latitude-dependent wind models of η_A in the x-z plane (i. e., equator-on). *(b):* K-band image projected on the sky. *(c):* VINCI visibilities for the 24 m baseline as a function of baseline P.A. (connected black asterisks) compared to the respective model prediction (red dotted line). This model assumes a prolate wind with $W = 0.85$, $i = 75°$, P.A. $= 130°$. Adapted from Groh *et al.* (2010).

simultaneously K-band continuum visibilities from VLTI/VINCI and closure phase measurements with VLTI/AMBER (see Fig. 1). Interestingly, oblate models with $W = 0.73 - 0.90$ and $i = 80° - 90°$ produce similar fits to the interferometric data, but require P.A. $= 210° - 230°$. Therefore, both prolate and oblate models suggest that the rotation axis of the primary star is not aligned with the Homunculus polar axis. While a prolate wind is thought to arise in a gravity-darkened, fast-rotating star with a radiative envelope (Owocki et al. 1996, 1998), an oblate wind can be produced by a fast-rotating star when gravity-darkening is not important (Bjorkman & Cassinelli 1993; Owocki et al. 1994, 1996, 1998).

η_A has been routinely referred to as the prototype of a massive star with a fast, dense polar wind created by rapid stellar rotation, even though the system is believed to contain a massive companion, η_B. What happens when the influence of η_B is taken into account? Three-dimensional hydrodynamical simulations have shown that the geometry of the wind of η_A is severely affected by the wind of η_B, which creates a low-density WWC cavity in the wind of η_A and a thin, dense wind-wind interacting region between the two winds (Pittard & Corcoran 2002; Okazaki et al. 2008; Parkin et al. 2009, 2011; Madura et al. 2012). Groh et al. (2010) presented an extension of the 2D radiative transfer code of Busche & Hillier (2005) to handle massive binary systems (see also Groh et al. 2012 for further details on the implementation). Interestingly, when the aforementioned density effects arising due to the presence of η_B are taken into account, an otherwise spherical primary wind carved by the companion can explain the observed VINCI and AMBER data (Groh et al. 2010). Figure 2a presents the assumed density structure of η_A's carved wind, with a distance of the apex of the WWC to η_A of $d_{\mathrm{apex}} = 10$ AU, half-opening angle of the cavity of $\alpha = 54°$, and width of the shocked walls of $\delta\alpha = 3°$. The best-fit orbital orientation is sketched in Fig. 2b, with $i = 41°$, $\omega = 270°$, and P.A. $= 40°$. Note that a similar good fit (Fig. 2d) would be also obtained with the orbital inclination suggested by Madura et al. (2012) ($i = 138°$).

Therefore, assuming the standard orbital and wind parameters of Eta Car, even if η_A has a spherical wind, its inner density structure can be sufficiently disturbed by η_B, mimicking the effects of a prolate/oblate latitude-dependent wind in the available interferometric observables in the K-band continuum. Therefore, fast rotation may not be the only explanation for the interferometric observations (Groh et al. 2010).

We have presented a brief overview of the high spatial resolution investigations of the Eta Car massive binary system reported in the literature. We hope the reader is

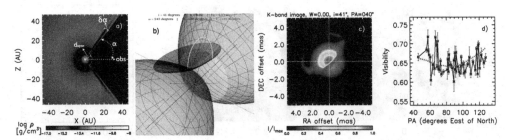

Figure 2. *Panels a,c,d:* Similar to Figure 1, but for the η_A model with a spherical wind including a cavity and compressed walls created by the wind of η_B. The model is appropriate for the VINCI observations ($\phi = 10.93$). See text for model parameters. *Panel b:* Sketch of the binary orbit (yellow) and the shock cone orientation at $\phi = 10.93$ relative to the Homunculus (not to scale), assuming a counterclockwise motion of η_B on the sky. The orbital plane is assumed to be in the skirt plane with the orbital axis (blue) aligned to the Homunculus axis of symmetry. North is up and East is to the left. From Groh et al. (2010).

convinced that zooming into Eta Car significantly increases the amount of information available and the constraints that can be put on hydrodynamic and radiative transfer models. The joint effort of many observers and theoreticians have shown that the binary interaction in Eta Car is complex, and no firm constraints on the individual masses exist yet. We encourage efforts into this direction, since this may present a rare opportunity to weight one of the most massive stars in the Galaxy.

Acknowledgements

I would like to thank my collaborators in studies of high spatial resolution observations of Eta Car: T. Madura, T. R. Gull, S. P. Owocki, D. J. Hillier, G. Weigelt, O. Absil, J.-P. Berger, H. Sana, J.-B. Le Bouquin, and M. de Becker.

References

Bjorkman, J. E. & Cassinelli, J. P. 1993, *ApJ* 409, 429

Busche, J. R. & Hillier, D. J. 2005, *AJ* 129, 454

Corcoran, M. F. 2005, *AJ* 129, 2018

Damineli, A., Conti, P. S., & Lopes, D. F. 1997, *New. Astron.* 2, 107

Damineli, A., Hillier, D. J., Corcoran, M. F., *et al.* 2008a, *MNRAS* 386, 2330

Damineli, A., Hillier, D. J., Corcoran, M. F., *et al.* 2008b, *MNRAS* 384, 1649

Davidson, K. & Humphreys, R. M. 1997, *ARA&A* 35, 1

Groh, J. H. & Damineli, A. 2004, *Information Bulletin on Variable Stars* 5492, 1

Groh, J. H., Hillier, D. J., Madura, T. I., & Weigelt, G. 2012, *MNRAS* 423, 1623

Groh, J. H., Madura, T. I., Owocki, S. P., Hillier, D. J., & Weigelt, G. 2010, *ApJ (Letters)* 716, L223

Hillier, D. J., Davidson, K., Ishibashi, K., & Gull, T. 2001, *ApJ* 553, 837

Hillier, D. J., Gull, T., Nielsen, K., *et al.* 2006, *ApJ* 642, 1098

Kervella, P. 2007, *A&A* 464, 1045

Madura, T. I., Gull, T. R., Owocki, S. P., *et al.* 2012, *MNRAS* 420, 2064

Mehner, A., Davidson, K., Ferland, G. J., & Humphreys, R. M. 2010, *ApJ* 710, 729

Okazaki, A. T., Owocki, S. P., Russell, C. M. P., & Corcoran, M. F. 2008, *MNRAS* 388, L39

Owocki, S. P., Cranmer, S. R., & Blondin, J. M. 1994, *ApJ* 424, 887

Owocki, S. P., Cranmer, S. R., & Gayley, K. G. 1996, *ApJ (Letters)* 472, L115

Owocki, S. P., Cranmer, S. R., & Gayley, K. G. 1998, *Ap&SS* 260, 149

Parkin, E. R., Pittard, J. M., Corcoran, M. F., & Hamaguchi, K. 2011, *ApJ* 726, 105

Parkin, E. R., Pittard, J. M., Corcoran, M. F., Hamaguchi, K., & Stevens, I. R. 2009, *MNRAS* 394, 1758

Pittard, J. M. & Corcoran, M. F. 2002, *A&A* 383, 636

Stevens, I. R., Blondin, J. M., & Pollock, A. M. T. 1992, *ApJ* 386, 265

Teodoro, M., Damineli, A., Sharp, R. G., Groh, J. H., & Barbosa, C. L. 2008, *MNRAS* 387, 564

van Boekel, R., Kervella, P., Schöller, M., *et al.* 2003, *A&A* 410, L37

Verner, E., Bruhweiler, F., & Gull, T. 2005, *ApJ* 624, 973

Weigelt, G., Kraus, S., Driebe, T., *et al.* 2007, *A&A* 464, 87

Discussion

PULS: Two comments. 1. When transforming the wind shape to the rotational speed, one has to be careful, because a direct relation is only provided if the ionization structure does not change from pole to equator. 2. For such high mass-loss rates, also optically-thick clumping effects (porosity and vorosity) might need to be considered.

GROH: I totally agree with your remarks. We used the Owocki *et al.* parametrization that is indeed more appropriate for O stars. However, regardless of the relationship used, the strong deformation of the K-band photosphere points out to a relatively high ratio of rotational over critical speed.

DE KOTER: You find that the axis of symmetry of the Homunculus is offset from the rotational axis of the star (though they are aligned as found by van Boekel *et al.* 2003). Do you have ideas as to what may have caused this?

GROH: Great question. Van Boekel *et al.* assumed that the rotational axis and the Homunculus symmetry axes were aligned, since they could only measure the azimuthal angle (i.e. the position angle on the sky), but not the inclination angle. Our results suggest that the inclination angle is offset by at least 20 degrees. One possibility is that the angular momentum loss was not axisymmetric during the formation of the Homunculus, which would have caused a recoil of the rotational axis of the star. Another possibility is that the secondary star ejected the Homunculus, so the Homunculus symmetry axis would be related to the rotational axis of the secondary star, and not that of the primary which is what we measure.

WEIS: I have a comment to the previous question by A. de Koter. You also have to create an equatorial plane/disk. My question: you showed a spheroidal structure in the K-band from Eta Car that represents the shape and reveals the rotation and inclination of the rotational axis. It is however known that LBVs in their cool phases (which can be multiple) do inflate their radius and form pseudo-photospheres, mimicking a cool star. Do the K-band observation probe this pseudo-photosphere or can we look into the deeper layers?

GROH: Unfortunately we cannot look deeper that the surface where the optical depth is roughly 2/3, so the K-band observations probe the pseudo-photosphere at this wavelength. This is a physical limit, as we can only observe the surface where the photons are thermalized and finally decouple from the matter.

José Groh

New windows on massive stars: asteroseismology, interferometry, and spectropolarimetry
Proceedings IAU Symposium No. 307, 2014
G. Meynet, C. Georgy, J. H. Groh & Ph. Stee, eds.

© International Astronomical Union 2015
doi:10.1017/S1743921314006929

Evidences for a large hot spot on the disk of Betelgeuse (α Ori)

M. Montargès[1], P. Kervella[1], G. Perrin[1], A. Chiavassa[2] and J. B. Le Bouquin[3]

[1]LESIA, Observatoire de Paris, CNRS, UPMC, Université Paris-Diderot, 5 place Jules Janssen, 92195 Meudon, France
email: miguel.montarges@obspm.fr

[2]Laboratoire Lagrange, UMR 7293, Université de Nice Sophia-Antipolis, CNRS, Observatoire de la Côte d'Azur, BP. 4229, 06304 Nice Cedex 4, France

[3]UJF-Grenoble 1 /CNRS-INSU, Institut de Planétologie et dAstrophysique de Grenoble (IPAG) UMR 5274, Grenoble, France

Abstract. Massive evolved stars contribute to the chemical enrichment of the Galaxy. When they die as supernova but also through their mass loss during the several thousands of years of their red supergiant (RSG) phase. Unfortunately the mass loss mechanism remains poorly understood. Detailed study of the CSE and photosphere of nearby RSGs is required to constrain this scenario.

Betelgeuse is the closest RSG (197 pc) and therefore has a large apparent diameter (~ 42 mas) which makes it a very interesting target. For several years, our team has lead a multi-wavelength and multi-scale observing program to characterize its mass loss. We will review here our recent results in near-infrared interferometry.

Keywords. stars: individual: Betelgeuse, supergiants, stars: late-type, techniques: interferometric, infrared: stars

1. Introduction

Massive stars contribute to the chemical evolution during the later stages of their existence through their spectacular supernova and in the quieter RSG phase. Material is moving away from the RSG thus it is cooling and forming molecules and dust that can reach the interstellar medium (ISM). Yet, several processes remain poorly understood. In particular we still do not know how the mass loss of RSG is triggered as these stars are not experiencing flares or pulsations like AGB stars. Josselin & Plez (2007) suggested that huge convection cells could locally lower the effective gravity and levitate material in combination with radiative pressure on molecular lines. With the detection of a magnetic field around α Ori (Aurière *et al.* 2010), the hypothesis of the triggering of the mass loss by the dissipation of Alfvén waves (Airapetian *et al.* 2010) remains also an explanation.

As the closest RSG, Betelgeuse is the ideal target for observers to constrain the mass loss models. Hot spots were observed on this star by Haubois *et al.* (2009) with a reconstructed image in the H band from IOTA interferometric data and interpreted as a convective pattern by Chiavassa *et al.* (2010). More recently Ohnaka *et al.* (2009, 2011) observed upward and downward motions of CO in the MOLsphere with K band interferometry at high spectral resolution.

Table 1. Best fitted value of the VLTI/AMBER data for the uniform disk (UD) and limb-darkened disk (LDD)

Model	θ (mas)	α	χ^2
UD	40.9 ± 0.52	-	110
LDD	41.8 ± 0.57	0.10 ± 0.02	55

2. VLTI/AMBER observations

AMBER (Petrov *et al.* 2007, Astronomical Multi-BEam combineR,) is installed at the ESO Very Large Telescope Interferometer (VLTI). It is combining the light from three telescopes in the J, H or K band with spectral dispersion. Therefore it produces three observables : the visibilities, the differential and the closure phases. For a more detailed presentation of the interferometric techniques, see Van Belle's contribution to this conference.

We observed Betelgeuse with VLTI/AMBER with the Auxiliary Telescopes (ATs) of 1.8 m of diameter on January 1st, 2nd and 3rd and February 17th 2011. We used the medium spectral resolution of the instrument (R = 1500) in the H band and in the K band (centered at 2.3 μm). The data reduction and the analysis of the CO and H_2O absorption domain ($\lambda \in [2.245; 2.45\ \mu m]$) are presented in detail in Montargès *et al.* (2013). The data were reduced using `amdlib` 3.0.3, the AMBER reduction package†. The stars HR 1543, HR 2275, HR 2469, HR 2508 and HR 3950 were observed as interferometric calibrators.

Here we will focus on the K band continuum data ($\lambda \in [2.1; 2.245\ \mu m]$). The data were fitted with two analytical models. First we used a uniform disk (UD) for which $I = I_0$ on the stellar disk and a power law limb-darkened disk (Hestroffer 1997) where $I = I_0 \mu^\alpha$ with $\mu = \cos\theta$, the cosinus of the angle between the line of sight and the center of the disk. The results of these fits are presented in Table 1. The visibilities together with the best fitted analytical model are plotted on Fig. 1.

If these analytical model reproduce well the first and second lobe of the visibility function (the χ^2 falls to 5.27 and 4.28 for the UD and the LDD respectively when considering only those spatial frequencies while the best fitted parameters stay in their error bars), from Fig. 1 it appears that these simple models cannot reproduce the high spatial frequency signal which reveals the presence of smaller structures on the star photosphere.

To analyse the higher order lobes (spatial frequencies > 60 arcsec^{-1}) of the visiblity function and the closure phases, we used radiative hydrodynamics (RHD) simulations from the CO^{5}BOLD code (Freytag *et al.* 2012). The non-gray model st35gm03n13 was picked up (Chiavassa *et al.* 2011). The resolution of the grid was 235^3 with a step of 8.6 $R_\odot$. The parameters of the model are compared to those of Betelgeuse in Table. 2.

From an initial homogeneous state the simulation is evolving. When the convection stationary regime is in place, several temporal snapshots are saved, each one becoming a realization of the convective pattern of the star. Intensity maps are produced with the 3D pure-LTE radiative transfer code OPTIM3D (Chiavassa *et al.* 2009) in each spectral channel of AMBER between 2.1 and 2.245 μm. As we do not know the orientation of the simulation relatively to the real star on the sky, we rotate each snapshot with a step of 10°. Thus we have a simulation grid of snapshots and rotation angle. The interferometric observables are then extracted from these maps (Chiavassa *et al.* 2010).

† Available at `http://www.jmmc.fr/amberdrs`

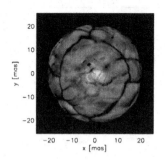
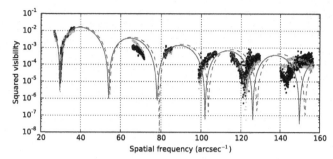

Figure 1. Best fitted snapshot of the RHD simulation. *Left:* Intensity map of the best fitted snapshot. *Right:* The black dots represent the squared visibilities of the VLTI/AMBER observations. The grey dots correspond to the best fitted snapshot. For comparison, the best fitted UD model is represented by the dashed line and the best fitted LDD model by the continuous line.

Table 2. Comparison between the physical caracteristics of Betelgeuse and of the RHD simulation.

Parameter	Betelgeuse	Model
M ($M_\odot$)	11.6 (a)	12.0
L ($L_\odot$)	6.31×10^4 (b)	8.94×10^4
R ($R_\odot$)	645 (b)	846
T_{eff}	3640 (c)	3430
$\log(g)$	-0.300 (d)	-0.354

Notes:
(a): Neilson *et al.* (2011), (b): Perrin *et al.* (2004) , (c) Levesque *et al.* (2005), (d) Harper *et al.* (2008).

We fitted the observed visibilities with the ones we derived from the RHD simulation. The best fitted snapshot and rotation angle give $\chi^2 = 7.47$ (See Fig. 1). Therefore the RHD simulation reproduces our VLTI/AMBER continuum observations better than the analytical UD and LDD model. However our closure phases are poorly reproduced (the χ^2 goes up to 17 if we include the is observable). This means that the convective pattern we observe on our RHD simulation does not correspond exactly to the features on Betelgeuse's photosphere but that they match a convective signature. In other words, the structures observed with VLTI/AMBER in the K band on the photosphere of the star are clearly caused by convective cells but because of the uncertainties on their number, shape and position, we are unable to reproduce exactly their distribution within the actual observations and the use of this simulation. Only such statistical approach can validate the measured signal.

From the intensity map of the best fitted snapshot (Fig. 1), we notice that we do not observe giant convection cells or hot spots as observed by Haubois *et al.* (2009). The detailed study has been submitted to *A&A* (Montargès *et al.* 2014).

3. VLTI/PIONIER monitoring of the photosphere

We continued to observe Betelgeuse at the VLTI the following years. We gathered data with the new VLTI/PIONIER instrument on January 31th 2012, February 09th 2013 and January 11th 2014. PIONIER (Le Bouquin *et al.* 2011, Precision Integrated-Optics Near-infrared Imaging ExpeRiment,) is a 4-telescopes instrument located at the Paranal observatory. It gives access to six visibilities and 3 independent closure phases

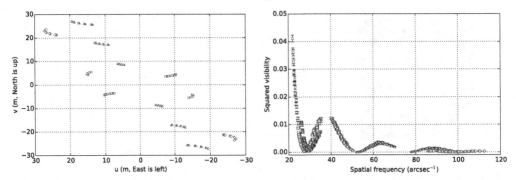

Figure 2. VLTI/PIONIER data of Betelgeuse of February 09th 2013 (see the online version of this proceeding for a color version of this figure). *Left:* (u, v) plane of the observations with a symbol-coded PA. *Right:* Visibilities with matching symbols.

in one observation. It is a low spectral resolution instrument. We used it in the H band. Betelgeuse was observed with the four ATs in their compact array configuration. The data were reduced and calibrated using the `pndrs` package provided by the PIONIER team.

In 2012, Sirius was used as interferometric calibrator but is located 27° away from Betelgeuse. As the interferometric transfer function of PIONIER is mainly sensitive to the distance between the science target and its calibrators and to avoid systematics we change our calibrators in 2013 and 2014 and used 56 Ori (6° from Betelgeuse), b Ori (7°) and LTT 11688 (5°). Betelgeuse is too bright at infrared wavelengths, diaphragms were used in 2012 to lower the incoming flux. In 2013 we used a neutral density filter instead and in 2014 we lowered the exposure time by acquiring the data on three pixels of the H band (usually seven pixels are read)

For the three epochs the first zero of the visbility function occurs at different spatial frequencies along two directions of the (u, v) plane (see Fig. 2 for an example on the 2013 epoch). This could indicate that the star has not the same size depending on the azimuthal direction probed. For the 2013 epoch, we can fit both direction on which we probe the first lobe by the power law limb-darkened disk described in Sect. 2. Counting 0° in the North direction and 90° in the East one, we have $\theta_{\mathrm{LDD}} = 48.14 \pm 0.19$ mas for $\chi^2 = 46.89$ in the 32° azimuthal direction (North-East) and $\theta_{\mathrm{LDD}} = 44.54 \pm 0.32$ mas for $\chi^2 = 114.62$ at 113° (South-East). Considering that Betelgeuse is 197 pc away (Harper *et al.* 2008) this would cause a difference of stellar radius of $\sim 85\,R_\odot$ or 7% $R_\star$ if one consider the mean stellar radius. With the long rotational period of the star (Uitenbroek *et al.* 1998, 17 yr,) the elliptical shape of the star is most likely unreal, moreover, it fails to reproduce the higher order lobes and gives a significant greater mean angular diameter than the 2011 observations (see Sect. 2).

Therefore, we adopted another model to fit these data : we used a power law limb-darkened disk as described earlier on which we added a gaussian hot spot characterized by its center ($x_{\mathrm{gauss}}^{\mathrm{center}}$, $y_{\mathrm{gauss}}^{\mathrm{center}}$) its full width at half maximum (FWHM) and its flux weight relatively to the stellar disk flux. The results of these fits are presented on Table 3 and Fig. 3. This model reproduces the double first zero of the visibility function along the two direction of the (u, v) plane and allows conserving a single angular diameter for Betelgeuse as expected (it also ensures continuity with the 2011 value).

Table 3. Derived parameters from the fit of the limb-darkened disk and gaussian hot spot model to the VLTI/PIONIER data.

Epoch	2012	2013	2014
$\theta_{\rm LDD}$ (mas)	42.75 ± 0.05	43.47 ± 0.05	45.65 ± 0.08
$\alpha_{\rm LDD}$	0.13 ± 0.01	0.17 ± 0.01	0.37 ± 0.02
Weight$_{\rm LDD}$	0.88 ± 0.29	0.82 ± 0.24	0.91 ± 0.42
$x_{\rm gauss}^{\rm center}$ (mas)	19.0 ± 0.2	15.16 ± 0.23	18.71 ± 0.53
$y_{\rm gauss}^{\rm center}$ (mas)	-7.23 ± 0.23	-2.79 ± 0.21	-20.37 ± 0.41
FWHM$_{\rm gauss}$ (mas)	20.92 ± 0.51	24.73 ± 0.31	21.88 ± 0.87
Weight$_{\rm gauss}$	0.12 ± 0.04	0.18 ± 0.05	0.09 ± 0.04
χ^2	61.13	86.48	97.22

Notes:
The $x_{\rm gauss}^{\rm center}$ and $y_{\rm gauss}^{\rm center}$ are offsets from the star center. North and East are counted as positive offsets.

Figure 3. Intensity map of the best fitted model of the VLTI/PIONIER data by a limb-darkened disk and a gaussian spot. *Left:* January 2012. *Center:* February 2013. *Right:* January 2014.

4. Conclusion

VLTI near infrared data enabled us to monitor the photosphere of Betelgeuse, the closest RSG : we detected the signature of the convection with VLTI/AMBER in 2011 and monitor a hot spot (certainly a giant convection cell) with VLTI/PIONIER in 2012, 2013 and 2014. The VLTI/PIONIER model fitting will shortly be complemented by a projected spot on a sphere model (Baron *et al.* 2014) a RHD simulation analysis. These observations allow following the morphology and dynamics of the photosphere. Thanks to imaging at larger scale at different wavelengths (UV, radio), we will have a consistent view of Betelgeuse's CSE and a complete simultaneous snapshot of its mass loss which will strongly constrain the models.

Acknowledgements

This work is based on AMBER and PIONIER observations made with ESO Telescopes at the Paranal Observatory under programmes ID 086.D-0351, 286.D-5036(A), 288.D-5035(A), 090.D-0548(A), 092.D-0366(A) and 092.D-0366(B). This research received the support of PHASE, the high angular resolution partnership between ONERA, Observatoire de Paris, CNRS and University Denis Diderot Paris 7. We acknowledge financial support from the Programme National de Physique Stellaire (PNPS) of CNRS/INSU, France.

References

Airapetian, V., Carpenter, K. G., & Ofman, L. 2010, *ApJ* 723, 1210

Aurière, M., Donati, J.-F., Konstantinova-Antova, R., *et al.* 2010, *A&A* 516, L2

Baron, F., Monnier, J. D., Kiss, L. L., *et al.* 2014, *ApJ* 785, 46

Chiavassa, A., Freytag, B., Masseron, T., & Plez, B. 2011, *A&A* 535, A22

Chiavassa, A., Haubois, X., Young, J. S., *et al.* 2010, *A&A* 515, A12

Chiavassa, A., Plez, B., Josselin, E., & Freytag, B. 2009, *A&A* 506, 1351

Decin, L., Cox, N. L. J., Royer, P., *et al.* 2012, *A&A* 548, A113

Freytag, B., Steffen, M., Ludwig, H.-G., *et al.* 2012, *Journal of Computational Physics* 231, 919

Harper, G. M., Brown, A., & Guinan, E. F. 2008, *AJ* 135, 1430

Haubois, X., Perrin, G., Lacour, S., *et al.* 2009, *A&A* 508, 923

Hestroffer, D. 1997, *A&A* 327, 199

Josselin, E. & Plez, B. 2007, *A&A* 469, 671

Le Bouquin, J.-B., Berger, J.-P., Lazareff, B., *et al.* 2011, *A&A* 535, A67

Levesque, E. M., Massey, P., Olsen, K. A. G., *et al.* 2005, *ApJ* 628, 973

Montargès, M., Kervella, P., Perrin, G., & Ohnaka, K. 2013, in P. Kervella, T. Le Bertre, & G. Perrin (eds.), *EAS Publications Series*, Vol. 60 of *EAS Publications Series*, pp 167–172

Montargès, M., Kervella, P., Perrin, G., *et al.* 2014, *A&A*, Submitted

Neilson, H. R., Lester, J. B., & Haubois, X. 2011, in S. Qain, K. Leung, L. Zhu, & S. Kwok (eds.), *Astronomical Society of the Pacific Conference Series*, Vol. 451, p. 117

Ohnaka, K., Hofmann, K.-H., Benisty, M., *et al.* 2009, *A&A* 503, 183

Ohnaka, K., Weigelt, G., Millour, F., *et al.* 2011, *A&A* 529, A163

Perrin, G., Ridgway, S. T., Coudé du Foresto, V., *et al.* 2004, *A&A* 418, 675

Petrov, R. G., Malbet, F., Weigelt, G., *et al.* 2007, *A&A* 464, 1

Uitenbroek, H., Dupree, A. K., & Gilliland, R. L. 1998, *AJ* 116, 2501

Discussion

ARNETT: How many zones does the 3D simulation have? Numerical resolution makes a difference and even heroic calculations can be unresolved.

MONTARGÈS: The resolution of the simulation is 235^3. Indeed, as with every simulation the resolution constrains a lot the result. Increasing the numerical resolution brings more details in the structure but the overall convection is still there (ie., the large granules). We can try to increase to very high number the resolution but, for sure, we are still far from fully resolving all the convective structures from small to large cells.

DOMICIANO DE SOUZA: Maybe to guide you in the model comparisons for Betelgeuse you could use analytical models with spots to have an idea of how the star looks like.

MONTARGÈS: That's what we are trying to do : constraining the biggest spot with improved model fitting (spot on the limb of a sphere instead of a limb-darkened disk + gaussian spot model) and then, use this best fitted model to constrain the "first order" of the RHD simulation which will provide in addition the smaller structures. This is not an easy task though. It works in principle if the model spots have the right shape and if the number of spots is correct. The difficulty is that the high frequency data cannot be fully filtered as closure phases are a mix of different ranges of frequency. Rejecting all closure phases with a frequency larger than a threshold leaves with fewer low frequencies than what has been measured by the interferometer. Hence the difficulty sometimes to compare the low frequency fit with the image reconstructed using all the set of data.

LOBEL: Could you give us an estimate of the formation height of the H-band continuum in alpha Ori's extended atmosphere? Do you expect to find more surface spots when looking in the V-band, or at the formation level of photospheric absorption lines? Note that the UV sport observed in 1998 was imaged in the "chromosphere" which extends over 100 times the photospheric radius. Would you think that the spots you see in the H-band are physically linked to the features observed in the UV, for example due to plumes that protrude the chromosphere?

MONTARGÈS: We usually consider that the H band continuum is forming at $\tau = 1$, meaning at the angular diameter measured for the star. Concerning the convection observed in the V band, the strong atomic and molecular electronic absorption should increase the opacity thus stimulating the convection. But spots are also created by temperature inhomogeneities and we should keep in mind that M stars have a lower flux in the visible. The counter part of the convection in the chromosphere is one of the motivation of our executed HST/STIS imaging program but to establish a strong link between the different scales, one would need to perform a long survey at different scales and wavelengths. Note that Chiavassa *et al.* (2011) suggested a strong contribution of the convection at shorter wavelengths.

MEYNET: Betelgeuse is a runaway star. I do not know whether the rapid movement of the star may be a cause for the differences in radius you obtain from different directions ? Could for instance the movement change the wind near the star and make some anisotropies in some wavelengths?

MONTARGÈS: The bow shock observed by Decin *et al.* (2012) is several arcmin away from the star. At the local scale ($\sim$ tens of mas for the photosphere) such motion has poor or even no effect. The dynamics is dominated by the mass loss and the gravity of the star.

STEE: Since you have hydro-simulations at small spatial resolution, have you tried to convolve your high-res images with the "beam" of the interferometer to see if you can reproduce the observations in the first visibility lobe?

MONTARGÈS: Indeed the simulations do not reproduce such a big hot spot on the RSG photosphere but the convolution should not solve the hot spot issue as the first lobe is well resolved by the interferometer (the VLTI compact configuration alone allows to explore the visibility function up to the fifth lobe)

Miguel Montargès

New windows on massive stars: asteroseismology, interferometry, and spectropolarimetry
Proceedings IAU Symposium No. 307, 2014
G. Meynet, C. Georgy, J. H. Groh & Ph. Stee, eds.

© International Astronomical Union 2015
doi:10.1017/S1743921314006930

On the atmospheric structure and fundamental parameters of red supergiants

M. Wittkowski[1], B. Arroyo-Torres[2], J. M. Marcaide[2,3], F. J. Abellan[2], A. Chiavassa[4], B. Freytag[5], M. Scholz[6], P. R. Wood[7] and P. H. Hauschildt[8]

[1] ESO Garching, Germany, email: mwittkow@eso.org

[2] University of València, Spain

[3] DIPC, Donostia-San Sebastian, Spain

[4] University of Nice Sophia-Antipolis, France

[5] Uppsala University, Sweden

[6] University of Heidelberg, Germany and University of Sydney, Australia

[7] Australian National University, Canberra, Australia

[8] Hamburg Observatory, Germany

Abstract. We present near-infrared spectro-interferometric studies of red supergiant (RSG) stars using the VLTI/AMBER instrument, which are compared to previously obtained similar observations of AGB stars. Our observations indicate spatially extended atmospheric molecular layers of water vapor and CO, similar as previously observed for Mira stars. Data of VY CMa indicate that the molecular layers are asymmetric, possibly clumpy. Thanks to the spectro-interferometric capabilities of the VLTI/AMBER instrument, we can isolate continuum band-passes, estimate fundamental parameters of our sources, locate them in the HR diagram, and compare their positions to recent evolutionary tracks. For the example of VY CMa, this puts it close to evolutionary tracks of initial mass $25 - 32\,M_\odot$. Comparisons of our data to hydrostatic model atmospheres, 3d simulations of convection, and 1d dynamic model atmospheres based on self-excited pulsation models indicate that none of these models can presently explain the observed atmospheric extensions for RSGs. The mechanism that levitates the atmospheres of red supergiant is thus a currently unsolved problem.

Keywords. convection, techniques: interferometric, stars: atmospheres, stars: evolution, stars: fundamental parameters, stars: individual (VY CMa, KW Sgr, UY Sct, AH Sco, V602 Car, HD 95687), stars: mass loss, supergiants

1. Introduction

Asymptotic giant branch (AGB) and red supergiant (RSG) stars are located in the Hertzsprung-Russell-Diagram (HRD) at cool effective temperatures between about 2500 K and 4500 K and cover a large range of luminosities depending on their initial mass. Levesque *et al.* (2005) showed that observed HRD positions of RSGs are not as cool as previously thought and that they are consistent with the red edge of stellar evolutionary tracks.

Due to the low temperatures of AGB and RSG stars, molecules and dust can form, and are subsequently expelled into the insterstellar medium via a stellar wind with widely overlapping mass-loss rates of $(4 \cdot 10^{-8} - 8 \cdot 10^{-5})\,M_\odot/\text{yr}$ (AGB stars) and $(2 \cdot 10^{-7} - 3 \cdot 10^{-4})\,M_\odot/\text{yr}$ (RSGs) (De Beck *et al.* 2010). The mechanism for this mass-loss process has been established for carbon-rich AGB stars (e.g.; Wachter *et al.* 2002; Mattsson *et al.*

2010), but are less well understood for oxygen-rich AGB stars (Woitke 2006; Höfner 2008; Bladh *et al.* 2013) and even less for RSGs (Josselin & Plez 2007). Both AGB stars and RSGs are affected by pulsation and convection. Variable RSGs have pulsation amplitudes of about 3 times less than AGB stars (Wood *et al.* 1983).

Interferometry is well suited to provide direct measurements of angular diameters of AGB and RSG stars, to directly estimate their effective temperatures based on measured angular diameters and measured bolometric fluxes, to constrain the stratification of their atmospheres and of molecular layers, and to probe surface inhomogeneities. Most powerful is the technique of spectro-interferometry, which allows us to separate continuum and molecular bands. For instance, Perrin *et al.* (2004, 2005) used narrow-band interferometry at the IOTA interferometer to confirm the "molecular layer scenario" for AGB and RSG stars, respectively.

2. AGB stars

Spectro-interferometry using the VLTI/AMBER instrument has proven to be a very powerful tool to study the atmospheres and close molecular layers of AGB stars, starting with the observations of the Mira S Ori by Wittkowski *et al.* (2008). The spectro-interferometric capabilities of AMBER allow us to observe the source at near-continuum and molecular bands simultaneously, resulting in a diagnostic curve of visibility versus wavelength. This curve shows a "bumpy" shape, which is a signature of the presence of molecular layers lying above the atmosphere: At near-continuum wavelengths, the molecular opacity is low, the target appears relatively small, and the observed visibilities are relatively high. At other wavelengths, the molecular opacity –in the near-IR range most importantly of water vapor and CO– is larger, the target appears larger, and the visibilities are smaller. For comparison, cool giants on the first giant branch that do not exhibit extensions of molecular layers, have a smooth curve of visibility versus wavelength.

Follow-up observations of a number of Miras using the medium spectral resolution mode of AMBER by Wittkowski *et al.* (2011) as well as studies by Woodruff *et al.* (2009) and Hillen *et al.* (2012) showed that observed visibilities are well consistent with predictions by the latest 1d dynamic model atmosphere series based on self-excited pulsation models of oxygen-rich Mira (**CODEX** models, Ireland *et al.* 2008, 2011). Closure phase data show deviations from point symmetry in molecular bands, possibly caused by clumpy extended molecular layers, but which affect the visibility moduli by less than a few percent. Best-fit parameters based on the **CODEX** models are consistent with independent estimates. In the dynamic model atmospheres, shock fronts reach the atmospheric layers, which leads to a geometric extension of the atmosphere of a few stellar radii.

3. The red supergiant VY CMa

The parameters of the red supergiant VY CMa have been controversially discussed during the last decade with effective temperatures between 2700 K and 3650 K, radii between $600 R_\odot$ and $3000 R_\odot$, and initial masses between $12 M_\odot$ and $40 M_\odot$. A previous interferometric angular diameter was obtained with a broad filter in the K-band (Monnier *et al.* 2004), which may be contaminated by molecular and dusty circumstellar emission.

Wittkowski *et al.* (2012) observed VY CMa with the VLTI/AMBER instrument using the low and medium spectral resolution modes. The visibility data at low spectral resolution indicate again a "bumpy" curve, resembling those of Miras as discussed above in Sect. 2, which is indicative of atmospheric layers of CO and H_2O. The closure phases, which are indicative of deviations from point symmetry, show small deviations from symmetry near the near-continuum bandpass at $2.25\,\mu$m and strong asymmetries in the CO and H_2O bands. Medium resolution data show in addition the presence of strong CO bandheads in the flux and visibility spectra. A comparison with hydrostatic PHOENIX model atmospheres shows that molecular features of water vapor and CO are consistent with the PHOENIX models in the integrated *flux* spectra, but these models do not reproduce the strong observed features in the *visibility* spectra. This means that, though molecular opacities are included in these model atmospheres, they are clearly too compact compared to our observations.

However, although molecular layers are observed at certain bandpasses, we could also separate two near-continuum bandpasses around $1.7\,\mu$m (in the H-band) and $2.25\,\mu$m (in the K-band) that are little contaminated by molecular emission. The visibility curves versus spatial frequency reach up to the first minimum of the visibility function and beyond, and are consistent with a UD or a PHOENIX atmosphere model. They give consistently a Rosseland angular diameter of 11.3 ± 0.3 mas. Together with the recently improved distance estimate to VY CMa and the well probed bolometric flux, this interferometric measurement corresponds to a Rosseland radius of $1420 \pm 120\,R_\odot$, an effective temperature of 3490 ± 90 K and a luminosity of $(2.7 \pm 0.4) \cdot 10^5\,L_\odot$.

Compared to the ranges discussed above, this places VY CMa near the larger effective temperature of 3650 K by Massey *et al.* (2006), but with a radius twice as large as that of Massey *et al.* and at the same time less than half as large as the largest radii discussed so far. The luminosity estimate is improved by the improved distance. With these values, VY CMa is confirmed to be located close to the red limit of evolutionary tracks of initial masses $25 \pm 10\,M_\odot$.

4. More red supergiants

Arroyo-Torres *et al.* (2013) and Arroyo-Torres *et al.* (in prep.) used the VLTI/AMBER instrument in a similar way to observe a larger sample of red supergiants covering cool spectral types between M3 and M4–5, including AH Sco (M4–5), UY Sct (M4), KW Sgr(M4), V602 Car (M3), and HD 95687 (M3). These sources show CO features in their flux spectra that are consistent with synthetic spectra based on PHOENIX models. However, as in the case of VY CMa, all of these sources also show much stronger drops of the visibility in the CO bandheads than predicted by the PHOENIX models. This confirms for a larger sample of RSGs that, though the opacities of CO are included in the PHOENIX models, the CO layers are observed to be much more geometrically extended than predicted. Some of these sources also show indications of geometrically extended layers of water vapor. Uniform disk diameters in the water band are increased by up to 25% and in the CO band-heads by up to 50%.

As a simple characterization of the observed extensions of the CO layers, Arroyo-Torres *et al.* (in prep.) computed the ratio of the observed visibilities in the near-continuum band just before the first CO bandhead (average between $2.27\,\mu$m and $2.28\,\mu$m) and in the first (2-0) CO line at $2.29\,\mu$m. Figure 1 shows the obtained values for red supergiants compared to those of Miras (Wittkowski *et al.* 2011) and red giants (Arroyo-Torres *et al.* 2014). For the red giants, the ratio is close to unity and consistent with the PHOENIX model atmospheres. The RSGs and Miras show similar ratios that lie significantly above unity. This illustrates that these sources show similarly large extensions of the CO layer, which are not predicted by the hydrostatic PHOENIX models. Furthermore, Fig. 1 suggests a correlation between the visibility ratio of our RSGs and the luminosity, contrary to the Miras. This may suggest that, unlike for Miras, RSGs develop large atmospheric

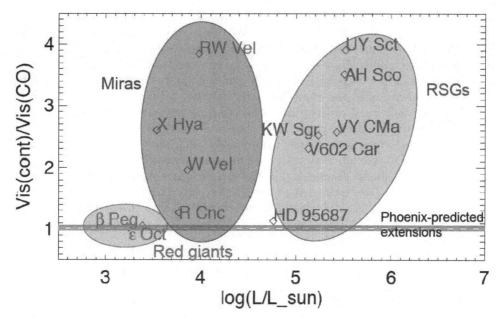

Figure 1. Atmospheric extensions in the CO 2-0 line at 2.29 μm, measures as the ratio of the visibilities in the nearby continuum and in the line, versus luminosity. Shown are red giants, Miras, and red supergiants of our sample.

extensions only for high luminosities of above about $10^5 L_\odot$ and that a radiatively driven process might (partly) explain their atmospheric extensions.

5. Comparisons to 1d pulsation and 3d convection models

In the case of Miras, `CODEX` dynamic model atmospheres based on self-excited pulsation models have been successful to describe interferometric observations, in particular their extended atmospheric molecular layers (cf. Sect. 2). As the observed visibility spectra and atmospheric extensions of our sample of RSGs are similar to those of Miras (Sect. 4), one may speculate whether similar models could also explain the atmospheric extensions of RSGs despite the different fundamental parameters. Doubts have been expressed because of the lower variability of RSGs compared to Miras (Josselin & Plez 2007), but a detailed study of pulsation models for RSGs has not yet been available. Indeed, variability amplitudes of variable RSGs are typically lower by a factor of 2-3 compared to Miras (Wood *et al.* 1983). Arroyo-Torres *et al.* (in prep) calculated a pulsation model based on stellar parameters that are typical for an RSG star and in particular similar to those of V602 Car of our sample. The amplitude of the photospheric radius variation was about 10% with radial velocities of up to about 5 km/s, which reproduces the amplitude of the visual lightcurve of V602 Car and typical observed long-term velocities. Whilst shock fronts enter the stellar atmosphere in a typical CODEX model of a Mira variable at or below optical depth 1, leading to a geometric extension of the stellar atmosphere of the order a few Rosseland radii, it turned out that no shock fronts reach at any phase the atmospheric layers in case of the RSG model. As a result, the model produces a compact atmosphere with extensions similar to those of the hydrostatic `PHOENIX` models.

Photospheric convection has been discussed as a possible explanation of high velocities on short time scales that were observed in atmospheres of RSGs (Gray 2000), and possibly as a mechanism to levitate the atmospheres to radii where dust can form. Arroyo-Torres (in prep) thus compared the visibility data of RSGs to 3d radiative hydro-dynamical (RHD) simulations of stellar convection. We used two simulations (st35gm03n07, st35gm03n13) from Chiavassa *et al.* (2011) and references therein. These models take into account the Doppler shifts occurring due to convective motions. Using the method explained in detail in Chiavassa *et al.* (2009) we computed azimuthally averaged intensity profiles and averaged the monochromatic intensity profiles to match the spectral channels of the individual observations. The resulting intensity profiles show that the intensity in the CO bandhead is lower by a factor of about 2 compared to the intensity in the continuum, which is consistent with observed flux spectra. The detailed surface structure in the CO line intensity map appears less corrugated and the details such as intergranular lanes almost disappear. The CO line surface is slightly more extended, but only by few percent (7%, estimated at the 0% intensity radius). The model-predicted visibility curves of the 3d RHD simulation are very similar to those of the hydrostatic PHOENIX models at the AMBER resolution. This means that also these can not explain the large observed atmospheric extensions of RSG stars.

In summary, comparisons of our RSG data to hydrostatic model atmospheres, 1d dynamic model atmospheres based on self-excited pulsation models, and 3d simulations of convection can all not explain the observed large atmospheric extensions of red supergiants. The mechanism that levitates the atmospheres of red supergiant is thus a currently unsolved problem. The observed correlation of atmospheric extension with luminosity may hint toward a radiatively driven levitation process as suggested by Josselin & Plez (2007).

References

Arroyo-Torres, B., Marti-Vidal, I., Marcaide, J. M., *et al.* 2014, *A&A* 566, A88
Arroyo-Torres, B., Wittkowski, M., Marcaide, J. M., & Hauschildt, P. H. 2013, *A&A* 554, A76
Bladh, S., Höfner, S., Nowotny, W., Aringer, B., & Eriksson, K. 2013, *A&A* 553, A20
Chiavassa, A., Freytag, B., Masseron, T., & Plez, B. 2011, *A&A* 535, A22
Chiavassa, A., Plez, B., Josselin, E., & Freytag, B. 2009, *A&A* 506, 1351
De Beck, E., Decin, L., de Koter, A., *et al.* 2010, *A&A* 523, A18
Gray, D. F. 2000, *ApJ* 532, 487
Hillen, M., Verhoelst, T., Degroote, P., Acke, B., & van Winckel, H. 2012, *A&A* 538, L6
Höfner, S. 2008, *A&A* 491, L1
Ireland, M. J., Scholz, M., & Wood, P. R. 2008, *MNRAS* 391, 1994
Ireland, M. J., Scholz, M., & Wood, P. R. 2011, *MNRAS* 418, 114
Josselin, E. & Plez, B. 2007, *A&A* 469, 671
Levesque, E. M., Massey, P., Olsen, K. A. G., *et al.* 2005, *ApJ* 628, 973
Massey, P., Levesque, E. M., & Plez, B. 2006, *ApJ* 646, 1203
Mattsson, L., Wahlin, R., & Höfner, S. 2010, *A&A* 509, A14
Monnier, J. D., Millan-Gabet, R., Tuthill, P. G., *et al.* 2004, *ApJ* 605, 436
Perrin, G., Ridgway, S. T., Mennesson, B., *et al.* 2004, *A&A* 426, 279
Perrin, G., Ridgway, S. T., Verhoelst, T., *et al.* 2005, *A&A* 436, 317
Wachter, A., Schröder, K.-P., Winters, J. M., Arndt, T. U., & Sedlmayr, E. 2002, *A&A* 384, 452
Wittkowski, M., Boboltz, D. A., Driebe, T., *et al.* 2008, *A&A* 479, L21
Wittkowski, M., Boboltz, D. A., Ireland, M., *et al.* 2011, *A&A* 532, L7
Wittkowski, M., Hauschildt, P. H., Arroyo-Torres, B., & Marcaide, J. M. 2012, *A&A* 540, L12
Woitke, P. 2006, *A&A* 460, L9
Wood, P. R., Bessell, M. S., & Fox, M. W. 1983, *ApJ* 272, 99
Woodruff, H. C., Ireland, M. J., Tuthill, P. G., *et al.* 2009, *ApJ* 691, 1328

Discussion

ARNETT: Are the pulsation simulations 1d and the radiative-hydrodynamics simulations 3d? Are the 3d simulations sufficiently resolved?

WITTKOWSKI: The pulsation model that I showed is 1d. The 3d RHD simulations in principle include pulsations if these are excited. They are small for RSG stars and larger for AGB stars. It would be interesting to inject the velocities from the 1d pulsation model into the 3d RHD model as an initial condition. The resolution of the 3d models is still limited. We clearly should try to go for a better resolution of the RSG models. However, we presently do not expect the atmospheric velocity fields (due to convection and pulsation alone) to grow enough to remotely reach the amplitude necessary to give a molsphere extension of a few stellar radii as observed.

PULS: You argued that the extension might be related to a wind. Have you checked whether with typical mass-loss rates the wind can reach a typical optical depth of unity around the sonic point?

WITTKOWSKI: We have not yet checked this, but it would be interesting to calculate.

MEYNET: Just a comment. Actually, the position in the HRD of red giants and supergiants depends, among other things, on the value of the mixing length. For a given physics, changing ℓ/H_P will move to redder (lower value) or hotter values (higher values) allowing to fit observed positions. So a good fit can actually be obtained by a given choice of this free parameter.

WITTKOWSKI: Thank you for the comment. The Lagarde *et al.* grid fits our positions of red giants in the HRD better than the Ekström *et al.* grid. There might be more differences between these grids than the treatment of thermohaline mixing.

Markus Wittkowski

New windows on massive stars: asteroseismology, interferometry, and spectropolarimetry
Proceedings IAU Symposium No. 307, 2014
G. Meynet, C. Georgy, J. H. Groh & Ph. Stee, eds.

© International Astronomical Union 2015
doi:10.1017/S1743921314006942

Amplitude Modulation of Cepheid Radial Velocity Curves as a Systematic Source of Uncertainty for Baade-Wesselink Distances

Richard I. Anderson

Observatoire de Genève, Université de Genève, 51 Ch. des Maillettes, 1290 Sauverny, Switzerland
email: `richard.anderson@unige.ch`

Abstract. I report on the recent discovery of modulation in the radial velocity curves in four classical Cepheids. This discovery may enable significant improvements in the accuracy of Baade-Wesselink distances by revealing a not previously considered systematic source of uncertainty.

Keywords. techniques: radial velocities, stars: individual ℓ Carinae, stars: oscillations, supergiants, Cepheids, distance scale

1. Introduction

Cepheids are excellent standard candles thanks to a relation between the logarithm of the period and their intrinsic brightness (Leavitt & Pickering 1912). This period-luminosity relation (PLR) finds application on Galactic and extragalactic scales and is crucial to precisely measure the expansion rate of the universe (Riess *et al.* 2011).

Baade-Wesselink (BW) type methods determine geometric distances by comparing angular and linear radius variations due to pulsation. They have been applied both to Cepheids in the Galaxy and the Magellanic Clouds (see Gieren *et al.* 2013, and references therein). Hence, different metallicities are probed using a homogeneous method, which is of great value regarding a possible metallicity dependence of the PLR. However, the accuracy of BW-type distances has suffered from certain difficulties usually ascribed to the projection (p-)factor (e.g. Nardetto *et al.* 2009), which is used to translate measured radial velocity into pulsation velocity.

2. Radial Velocity Amplitude Modulations

Using high-precision radial velocities (RVs) obtained with the *Coralie* spectrograph mounted to the Swiss Euler telescope at La Silla, Anderson (2014) was recently able to show that Cepheid radial velocities are subject to modulation. This was demonstrated for two short-period, as well as two long-period Cepheids. Figure 1 exemplifies this phenomenon for the long-period ℓ Carinae.

3. A Systematic Uncertainty for BW distances

BW-type methods determine distance from the ratio of linear (ΔR) and angular ($\Delta\theta$) radius variations due to pulsation. ΔR is computed as the integral of the pulsational velocity curve and therefore depends on the epoch of observation, if modulation is present. In consequence, combining measurements of $\Delta\theta$ from one epoch with ΔR from another epoch can lead to systematic errors in the derived distance, reaching up to $5-15\%$

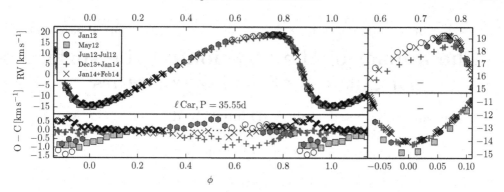

Figure 1. The modulated radial velocity curve of ℓ Carinae. Figure from Anderson (2014).

(Anderson 2014). It is thus crucial to measure angular and linear radius variations contemporaneously (equivalent pulsation cycles). While the impact of modulation on the distance estimate may average out over very long timescales (if the mean RV curve is stable), it appears likely that modulation explains a good portion of what has thus far limited the accuracy of BW distances.

4. The way forward

There are several possible explanations for the origin of RV modulations. From the fact that modulations in short-period Cepheids appear steady over longer timescales (years), while long-period Cepheids exhibit seemingly stochastic (cycle-to-cycle) modulation, it appears likely that different effects are responsible for the observed phenomenon. A longer observational baseline is needed to investigate possible periodicity and the frequency of the phenomenon.

Observed period jitter in the *Kepler* Cepheid V1154 Cygni (Derekas *et al.* 2012) and *flickering* observed in two Cepheids with *MOST* (Evans *et al.* 2014) suggests that RV modulations have a photometric counterpart. However, no Cepheid has as of yet been found to exhibit both RV modulation and photometric flickering. Only the long-period Cepheid RS Puppis exhibits both strong stochastic period variations and RV modulation. Both phenomena are obvious in the unprecedented *Coralie* RV data set.

In conclusion, significant improvements in BW distance accuracy may be achieved by determining linear and angular radius variations contemporaneously, if RV curve modulations directly correlate with modulations in angular diameter. Dense contemporaneous time-series of high-quality spectroscopic, photometric, and interferometric measurements are required to investigate this exciting possibility.

References

Anderson, R. I. 2014, *A&A* 566, L10

Derekas, A., Szabó, G. M., Berdnikov, L., *et al.* 2012, *MNRAS* 425, 1312

Evans, N. R., Szabó, R., Szabados, L., *et al.* 2014, in J. A. Guzik, W. J. Chaplin, G. Handler, & A. Pigulski (eds.), *IAU Symposium*, Vol. 301 of *IAU Symposium*, pp 55–58

Gieren, W., Storm, J., Nardetto, N., *et al.* 2013, in R. de Grijs (ed.), *IAU Symposium*, Vol. 289 of *IAU Symposium*, pp 138–144

Leavitt, H. S. & Pickering, E. C. 1912, *Harvard College Observatory Circular* 173, 1

Nardetto, N., Gieren, W., Kervella, P., *et al.* 2009, *A&A* 502, 951

Riess, A. G., Macri, L., Casertano, S., *et al.* 2011, *ApJ* 730, 119

New windows on massive stars: asteroseismology, interferometry, and spectropolarimetry
Proceedings IAU Symposium No. 307, 2014
G. Meynet, C. Georgy, J. H. Groh & Ph. Stee, eds.

© International Astronomical Union 2015
doi:10.1017/S1743921314006954

The impact of the rotation on the surface brightness of early-type stars

M. Challouf[1,2], N. Nardetto[1], A. Domiciano de Souza[1], D. Mourard[1], H. Aroui[2], P. Stee[1] and A. Meilland[1]

[1]Laboratoire Lagrange, UMR7293, UNS/CNRS/OCA, 06300 Nice, France
email: mounir.challouf@oca.eu

[2]Lab. Dyn. Moléculaire et Matériaux Photoniques, UR11ES03, UT/ESSTT, Tunis, Tunisie

Abstract. The surface brightness colors (SBC) relation is a very important tool to derive the distance of extragalactic eclipsing binaries. However, for early-type stars, this SBC relation is critically affected by the stellar environment (wind, circumstellar disk, etc...) and/or by the fast rotation. We calculated 6 models based on the code of Domiciano de Souza *et al.* (2012) considering different inclinations and rotational velocities. Using these results, we quantify for the first time the impact of the rotation on the SBC relation.

Keywords. stars: early-type, stars: rotation, techniques: interferometric, methods: numerical

1. Introduction

Detached eclipsing double-lined spectroscopic binaries offer a unique opportunity to measure directly, and very accurately the distance (Graczyk *et al.* 2011; Wyrzykowski *et al.* 2003; Macri *et al.* 2001) to nearby galaxies.

The early-type stars, very bright, are currently easily detected in the LMC (Pietrzyński *et al.* 2009; Pawlak *et al.* 2013) but there is no accurate SBCR available for these stars. Using the unique capabilities of the Visible spEctroGraph and polArimeter (VEGA) (Mourard *et al.* 2009) in this field of research, Challouf *et al.*(2014) have improved the surface brightness (with a σ=0.17 mag) by determining the angular diameter of a dozen of OBA stars with a precision of 1.5%.

An empirical SBCR is already well established for late-type stars (Di Benedetto 2005) coming from accurate determinations of stellar angular diameters by interferometry. But for early-type stars the interferometric study is a bit difficult because the majority of these stars have a stellar environment (wind, circumstellar disk, etc...) or they are fast rotators.

2. Impact of the rotation

For improving the surface brightness colors relationship for this type of star, we use numerical tool CHARRON (Code for High Angular Resolution of Rotating Objects in Nature) (Domiciano de Souza *et al.* 2012, 2002) code to model the rapid rotation and to better estimate the interferometric parameters. The stellar photospheric shape is given by the commonly adopted Roche approximation (rigid rotation and mass concentrated in the stellar center), which is well adapted for non-degenerate, fast rotating stars.

The effective temperature $T_{\rm eff}$ at the surface for fast rotators is non uniform (dependend on the co-latitude θ) due to the decreasing effective gravity $g_{\rm eff}$ (gravitation plus centrifugal acceleration) from the poles to the equator (gravity darkening effect). We model the gravity darkening as a generalized form of the von Zeipel law (von Zeipel

1924): $T_{\text{eff}}(\theta) = Kg_{\text{eff}}^{\beta}(\theta)$ where β is the gravity darkening coefficient, and K is the proportionality constant between T_{eff} and g_{eff}, which depends on the stellar physical parameters.

Once $T_{\text{eff}}(\theta)$ and $g_{\text{eff}}(\theta)$ are defined, we use the spectral synthesis code SYNSPEC (Hubeny & Lanz 2011) and the ATLAS9 stellar atmosphere models (Kurucz 1979) to compute specific intensity maps of the star. The surface brightness relations can be directly obtained from these intensity maps and from the input parameters of the model, in particular equatorial radius and distance.

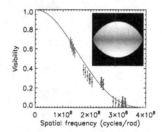

model	M1	M2	M3	M4	M5	M6
$V-K$	−0.7	−0.5	−0.3	0	0.3	0.5
T_{eff} [K]	23701	18418	13660	9794	8353	7822
$\log g$ [dex]	4	3.5	4	4	4	4
$M[M_{\odot}]$	13.61	5.59	4.11	2.52	2.17	1.99
$R[R_{\odot}]$	10	5.01	2.88	2.39	2.23	1.99

Table 1. Parameters of our sample based on Worthey & Lee (2011)

Figure 1. Modeled intensity distributions and the fourier transform of the intensity map at 720 nm for 95% of V_c and $i = 90°$

We used the intensity maps constructed by the CHARRON code (Fig. 1) describing the distribution of surface brightness of the stars from a number of parameters that depends on the type of objects studied (see Tab. 1). In a second step, we calculate the complex visibility from the Fourier transform (FT thereafter) of these maps in accordance with the theorem of Zernike-Van Cittert. We obtained thus the complex visibility. The module of the complex visibility provides the evolution of fringe contrast as a function of spatial frequency. In the case of a star deformed by the rotation , the diameter measurement must take into account the orientation of the interferometer in order to be interpreted correctly. In this case, the overall shape of the flattened star is determined by performing visibility measurements for different orientations of the interferometer (coverage of plane (u,v)) in order to probe different parts of the FT module. The visibility curves correspond to the component of the FT module along baselines that are then adjusted by a model of uniform disk (Fig. 1) using the similar configuration of the W1W2E2 and S2W2 VEGA/CHARA telescopes for 5 inclinations (0°, 25°, 50°, 75°, and 90°) and 5 rotational velocity (0.0Vc, 0.25Vc, 0.50Vc, 0.75Vc, and 0.95Vc). With these diameters we can calculate the surface brightness for each star and by comparing the theoretical results and the simulation we estimate that the error caused by rotation is 0.12 mag and the error due to the inclination is of 0.10 mag.

Acknowledgments: The research leading to these results has received funding from the European Community's Seventh Framework Programme under Grant Agreement 312430.

References

Di Benedetto, G. P. 2005, *MNRAS* 357, 174

Domiciano de Souza, A., Hadjara, M., Vakili, F., *et al.* 2012, *A&A* 545, A130

Domiciano de Souza, A., Vakili, F., & Jankov, S. e. a. 2002, *A&A* 393, 345

Graczyk, D., Soszyński, I., Poleski, R., *et al.* 2011, *AcA* 61, 103

Hubeny, I. & Lanz, T. 2011, *Synspec*, Astrophysics Source Code Library

Kurucz, R. L. 1979, *ApJS* 40, 1

Macri, L. M., Stanek, K. Z., Sasselov, D. D., & et, a. 2001, *AJ* 121, 870

290 M. Challouf *et al.*

Mourard, D., Clausse, J. M., Marcotto, A., *et al.* 2009, *A&A* 508, 1073
Pawlak, M., Graczyk, D., Soszyński, I., *et al.* 2013, *AcA* 63, 323
Pietrzyński, G., Thompson, I. B., Graczyk, D., *et al.* 2009, *ApJ* 697, 862
von Zeipel, H. 1924, *MNRAS* 84, 665
Worthey, G. & Lee, H.-c. 2011, *ApJS* 193, 1
Wyrzykowski, L., Udalski, A., Kubiak, M., *et al.* 2003, *AcA* 53, 1

Thomas Rivinius

Katie Gordon

New windows on massive stars: asteroseismology, interferometry, and spectropolarimetry
Proceedings IAU Symposium No. 307, 2014
G. Meynet, C. Georgy, J. H. Groh & Ph. Stee, eds.

© International Astronomical Union 2015
doi:10.1017/S1743921314006966

The circumstellar environment of the B[e] star GG Car: an interferometric modeling

A. Domiciano de Souza[1] †, M. Borges Fernandes[2], A. C. Carciofi[3] and O. Chesneau[1]

[1] Laboratoire Lagrange, UMR 7293, OCA, UNS, CNRS, CS 34229, 06304 Nice, France
email: Armando.Domiciano@oca.eu

[2] Observatório Nacional/MCTI, R. Gal. José Cristino 77, 20921-400 São Cristovão, RJ, Brazil

[3] IAG, USP, R. do Matão 1226, Cidade Universitária, São Paulo, SP - 05508-900, Brazil

Abstract. The research of stars with the B[e] phenomenon is still in its infancy, with several unanswered questions. Physically realistic models that treat the formation and evolution of their complex circumstellar environments are rare. The code HDUST (developed by A. C. Carciofi and J. Bjorkman) is one of the few existing codes that provides a self-consistent treatment of the radiative transfer in a gaseous and dusty circumstellar environment seen around B[e] supergiant stars. In this work we used the HDUST code to study the circumstellar medium of the binary system GG Car, where the primary component is probably an evolved B[e] supergiant. This system also presents a disk (probably circumbinary), which is responsible for the molecular and dusty signatures seen in GG Car spectra. We obtained VLTI/MIDI data on GG Car at eight baselines, which allowed to spatially resolve the gaseous and dusty circumstellar environment. From the interferometric visibilities and SED modeling with HDUST, we confirm the presence of a compact ring, where the hot dust lies. We also show that large grains can reproduce the lack of structure in the SED and visibilities across the silicate band. We conclude the dust condensation site is much closer to the star than previously thought. This result provides stringent constraints on future theories of grain formation and growth around hot stars.

Keywords. stars: emission-line, Be, stars: individual (GG Car), supergiants, techniques: high angular resolution, techniques: interferometric, spectroscopic, photometric

---

1. Introduction

The B[e] phenomenon, observed in different classes of stars, is characterized in particular by strong emission lines (permitted and forbidden) and strong IR excess due to hot circumstellar dust (e.g. Lamers *et al.* 1998). Among the stars presenting the B[e] phenomenon, the supergiants (sgB[e] stars) are a challenge to models of stellar evolution (they are not predicted by them) and of dust formation around hot, massive stars. Fast-rotation of the central star and binarity seem to be intimately related to the sgB[e].

In the last decade, optical/IR long-baseline interferometry (OLBI) became an important tool to deeply and directly study the circumstellar environment (CSE) of the brightest B[e] stars. We used the ESO-VLTI/MIDI beam combiner to study the CSE of binary system GG Car, where the primary component is probably a sgB[e] (Marchiano *et al.* 2012; Kraus *et al.* 2013). However, its distance is very uncertain, ranging from 1 to 5 kpc (Borges Fernandes, private comm.; Marchiano *et al.* 2012). Recent works (based on spectropolarimetry, CO emission lines) suggest the presence of a rotating disk-like structure, probably a narrow circumbinary ring (Pereyra *et al.* 2009; Kraus *et al.* 2013).

† Based on observations performed at the European Southern Observatory, Chile under ESO Programs 074.D-0101 and 076.D-0663.

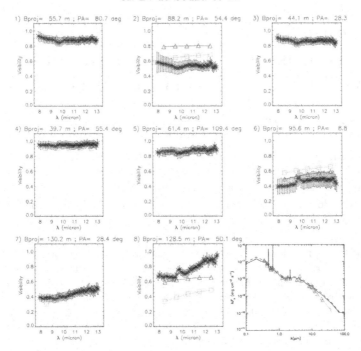

Figure 1. Observed MIDI visibilities and SED (several literature data) measured on GG Car. The model was calculated with the HDUST code (see text for further details).

2. Results

GG Car was observed between Dec. 2004 and Dec. 2005 with MIDI, the mid-IR, two-telescope interferometric beam-combiner of ESO-VLTI. As a preliminary study to interpret the observations we used the grid of HDUST sgB[e] models from Domiciano de Souza & Carciofi (2012). The central star is modeled as a B type star of $T_{\rm eff} \sim 20\,000$ K, $R \sim 10\,R_\odot$, and $L \sim 10^4\,L_\odot$. Figure 1 compares the MIDI visibilites and SED with our HDUST model. Our preliminary models can mostly describe the VLTI/MIDI data and SED. However, they are not a perfect representation of the circumstellar environment of GG Car; for example some line profiles (not shown here) are not well reproduced.

As an important result our analysis suggests that, despite of the uncertainty in the distance, the hot-dust site is ring shaped and somewhat compact, with the dust being formed at distances of $\lesssim 100\,R$ (much closer to the star than previously thought). We also find that large grains (sizes from 1 to 50 μm) are required if amorphous silicate are used (as in our models) to reproduce the lack of mid-IR structure in the SED and visibilities.

References

Domiciano de Souza, A. & Carciofi, A. C. 2012, in T. Carciofi, A. C. & Rivinius (ed.), *Astronomical Society of the Pacific Conference Series*, Vol. 464 of *Astronomical Society of the Pacific Conference Series*, p. 149

Kraus, M., Oksala, M. E., Nickeler, D. H., *et al.* 2013, *A&A* 549, A28

Lamers, H. J. G. L. M., Zickgraf, F.-J., de Winter, D., Houziaux, L., & Zorec, J. 1998, *A&A* 340, 117

Marchiano, P., Brandi, E., Muratore, M. F., *et al.* 2012, *A&A* 540, A91

Pereyra, A., de Araújo, F. X., Magalhães, A. M., Borges Fernandes, M., & Domiciano de Souza, A. 2009, *A&A* 508, 1337

New windows on massive stars: asteroseismology, interferometry, and spectropolarimetry
Proceedings IAU Symposium No. 307, 2014
G. Meynet, C. Georgy, J. H. Groh & Ph. Stee, eds.

© International Astronomical Union 2015
doi:10.1017/S1743921314006978

Angular Diameters of O- and B-type Stars

Kathryn Gordon,[1] Douglas Gies[1] and Gail Schaefer[2]

[1]Department of Physics and Astronomy, Georgia State University, Atlanta, GA 30302, USA
email: kgordon@chara.gsu.edu
[2]The CHARA Array of Georgia State University, Mt. Wilson, CA 91023, USA

Abstract. We are observing a sample of 10 O-type stars and 60 B-type stars to determine angular diameters using the Center for High Angular Resolution Astronomy (CHARA) Array, the foremost optical long baseline interferometer in the world. Our goal is to establish accurate stellar parameters to test modern theories of stellar evolution that include rotation. We will combine our stellar angular diameter measurements with flux and line measurements from spectroscopy, projected rotational velocities, and distances to determine radius, effective temperature, luminosity, equatorial rotational velocity, and evolutionary mass. Knowing these properties will allow us to place the stars in a Hertzsprung-Russell diagram and obtain estimates for the age and evolutionary state.

Keywords. stars: fundamental parameters (classification, colors, luminosities, masses, radii, temperatures, etc.), techniques: interferometric

1. The CHARA Array

The CHARA Array is a state of the art optical/infrared interferometer operated by Georgia State University at the Mt. Wilson Observatory in California (ten Brummelaar *et al.* 2005). The Array consists of six 1-m diameter telescopes arranged in a Y configuration, and it provides the highest resolution of any telescope at visible and near-infrared wavelengths. It is capable of resolving details as small as 200 micro-arcseconds. The light from the telescopes is sent through light pipes across the mountain to be combined in the beam combining lab. Moving carts with mirrors in the lab allow the path length difference of the beams to be adjusted so that the interference fringes may be observed. There are six beam combiners in use with the Array allowing observations from the *R*-band with PAVO to the *H*- and *K*-bands with CLASSIC and CLIMB.

2. Fringe Visibility Data

The PAVO beam combiner (Ireland *et al.* 2008) operates in the *R*-band and splits light up into several spectral channels with tiny prisms. Figure 1 shows example PAVO data of the B9 V star HD 11502 from three brackets of observations. The angular sizes and physical radii (from Hipparcos distances) are listed in the figure. A bracket consists of observations of a calibrator star, followed by the target, followed by another calibrator star. The ideal calibrator will be as small as possible so as to be unresolved, but still bright enough to be close in magnitude to the target. The data are fit to obtain an angular size measurement. The CLIMB beam combiner (Sturmann *et al.*

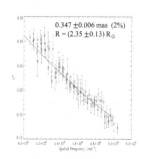

Figure 1. Fringe squared visibility data from PAVO of HD 11502.

2010) operates in the H- and K-bands and can use up to three telescopes at once. CLIMB has a resolution limit, operating on longest possible baselines, of about 0.5 mas in the H-band, while PAVO has a resolution limit of about 0.2 mas in the R-band.

3. A New Companion for λ Ori

The O star λ Ori currently has four known companions with the closest being about 4 arcsec away. Our visibility observations of λ Ori show an oscillation in visibility with spatial frequency that is generally associated with the presence of a close binary companion. Preliminary fits of the visibility data indicate an angular separation of 1 to 3 mas which would put it at 0.3 to 1.0 AU with a period of 10 to 54 days. These values are calculated using a distance of 324 pc from Hipparcos and a typical mass of 23 $M_\odot$ for an O 8 III star and 20 $M_\odot$ for the companion (based on a flux ratio of $f_2/f_1 = 0.8$). Because we only have a single baseline of coverage for each night of observation, more modeling and observations with multiple baselines are necessary to further constrain the separation and orbit.

4. Results and Future Work

We currently have data on 25 stars from several observing runs at the CHARA Array, starting in 2012. More runs are planned through 2015. Priority is given to O stars and those B stars that have measured parallaxes in the Hipparcos catalog with errors less than 10% and/or that are members of clusters. We have closure phase data from CLIMB which will allow us to look for any light asymmetry caused by rotation or nearby companions. We also have good (u, v) coverage from multiple baselines with CLIMB and PAVO that will allow us to detect an elliptical stellar shape caused by rapid rotation.

We obtain $T_{\rm eff}$ by combining the angular diameter measurement with the integrated flux from the relationship $F_{\rm bol} = (1/4)\theta^2 \sigma T_{\rm eff}^4$. The physical radii of our stars are found by combining the angular diameter with a known distance from parallax or cluster membership. Once $T_{\rm eff}$ and the radius are known we can use a star's position in the H-R diagram to obtain mass and age estimates from theoretical evolutionary tracks. Our measurements of stellar properties, such as radius, temperature, mass, and age will provide fundamental reference data for nearby massive stars and will be of key importance for testing new stellar evolutionary models (Brott *et al.* 2011; Ekström *et al.* 2012).

References

Brott, I., de Mink, S. E., Cantiello, M., *et al.* 2011, *A&A* 530, A115
Ekström, S., Georgy, C., Eggenberger, P., *et al.* 2012, *A&A* 537, A146
Ireland, M. J., Mérand, A., ten Brummelaar, T. A., *et al.* 2008, in *Society of Photo-Optical Instrumentation Engineers (SPIE) Conference Series*, Vol. 7013 of *Society of Photo-Optical Instrumentation Engineers (SPIE) Conference Series*
Sturmann, J., ten Brummelaar, T., Sturmann, L., & McAlister, H. A. 2010, in *Society of Photo-Optical Instrumentation Engineers (SPIE) Conference Series*, Vol. 7734 of *Society of Photo-Optical Instrumentation Engineers (SPIE) Conference Series*
ten Brummelaar, T. A., McAlister, H. A., Ridgway, S. T., *et al.* 2005, *ApJ* 628, 453

*New windows on massive stars: asteroseismology, interferometry, and
spectropolarimetry*
Proceedings IAU Symposium No. 307, 2014
G. Meynet, C. Georgy, J. H. Groh & Ph. Stee, eds.

© International Astronomical Union 2015
doi:10.1017/S174392131400698X

Binarity of the LBV HR Car

Th. Rivinius[1], H. M. J. Boffin[1], W. J. de Wit[1], A. Mehner[1], Ch. Martayan[1], S. Guieu[2] and J.-B. Le Bouquin[2]

[1]ESO - European Organisation for Astronomical Research in the Southern Hemisphere, Chile
email: triviniu@eso.org

[2]Univ. Grenoble Alpes, CNRS, IPAG, France

Abstract. VLTI/AMBER and VLTI/PIONIER observations of the LBV HR Car show an interferometric signature that could not possibly be explained by an extended wind, more or less
symmetrically distributed around a single object. Instead, observations both in the Brγ line
and the *H*-band continuum are best explained by two point sources (or alternatively one point
source and one slightly extended source) at about 2 mas separation and a contrast ratio of about
1:5. These observations establish that HR Car is a binary, but further interpretation will only be
possible with future observations to constrain the orbit. Under the assumption that the current
separation is close to the maximum one, the orbital period can be estimated to be of the order
of 5 years, similar as in the η Car system. This would make HR Car the second such LBV binary.

Keywords. stars: individual (HR Car)

1. Introduction

Luminous Blue Variables (LBVs) are a brief phase in the evolution of massive stars,
but a very important one. The giant eruption remains enigmatic, but the discovery of
the flagship LBV η Car to be a five-year highly eccentric binary put focus on possible
binarity induced mechanism for the giant outbursts, and prompted binarity searches
among LBVs.

So far, however, while several wide LBV binaries were identified, LBV systems similar
to η Car (relatively close and eccentric) have not been found, with the possible exception
of the LBV candidate MWC 314 (Lobel *et al.* 2013; see as well the preliminary summary
by Martayan *et al.* 2012). This is rather surprising as it is thought that given their very
high multiplicity rate, more than 70% of all massive stars will exchange mass with a
companion (Sana *et al.* 2012).

2. Observations

2.1. *AMBER/VLTI OHANA Data*

The LBV HR Car was observed as part of the OHANA sample (spectrally resolved 3-
beam interferometry of Brγ with AMBER at the Very Large Telescope Interferometer,
VLTI; see Rivinius *et al.*, this volume). The OHANA data obtained for HR Car showed
a clear and temporally stable (over the months of observation) phase signature across
the blue part of the emission line, but little to no visibility signature. This marks a
photocentre displacement of the emission line with respect to the continuum. Such a
displacement is hard to explain with the more-or-less symmetric, but variable wind of a
single supergiant star. Ad-hoc explanations are:

- **A binary**, and the emission is associated with the secondary, not the primary;
- **A binary**, where part of the emission is formed at the location of the secondary,
part at the location of the primary, and part in a wind collision zone;

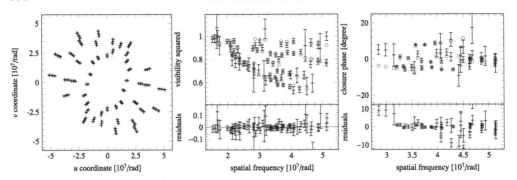

Figure 1. PIONIER observations of HR Car, showing *uv*-coverage, visibilities, and closure phases. Crosses mark the observed data, open circles the model, residuals are shown in the lower respective panels (computed with LITpro, see Tallon-Bosc *et al.* 2008).

- **A single star** with a dense nebular structure nearby that forms the hydrogen emission (possibly ejected in a previous eruption).

2.2. *PIONIER Data*

The available OHANA data would not allow further distinction of these hypotheses, so HR Car was observed with PIONIER. PIONIER is a 4-beam interferometric instrument working in the *H*-band continuum (Le Bouquin *et al.* 2011), and thus is not as sensitive to nebular contributions as OHANA observations.

Analysis of the obtained PIONIER data firmly establishes the presence of two different sources with a contrast ratio of about 1:5 and a separation of about 2 mas (Rivinius *et al.*, in prep.). Neither source is extended in itself, or only marginally so, i.e., the data strongly support the binary hypothesis (see Fig. 1).

Judging by the strength of the emission lines in various hydrogen lines and the appearance of the visual spectrum in general, a binary with a wind collision zone seems to be the most attractive. While this interpretation is work in progress, the pure proof of binarity was already delivered by the PIONIER observations.

2.3. *Conclusions*

At the estimated distance of HR Car ($\sim$5000 pc), observations indicate that the two components have a projected separation of $\sim$ 10 AU. Assuming a total mass of the system of $\sim$ 40 $M_\odot$, this separation would correspond roughly to an orbital period of about 5 years. A final value will depend on the eccentricity and inclination of the system. However, unless HR Car is a wide system seen at a very unfavourable (and unlikely) projection, it would be the second known η Car-like LBV binary. The proposed scenario of a wind-wind effect being responsible for the AMBER signature is similar to the model for the LBV candidate binary MWC 314 by Lobel *et al.* (2013).

References

Le Bouquin, J.-B., Berger, J.-P., Lazareff, B., *et al.* 2011, *A&A* 535, A67
Lobel, A., Groh, J. H., Martayan, C., *et al.* 2013, *A&A* 559, A16
Martayan, C., Lobel, A., Baade, D., *et al.* 2012, in A. C. Carciofi & T. Rivinius (eds.), *Circumstellar Dynamics at High Resolution*, Vol. 464 of *Astronomical Society of the Pacific Conference Series*, p. 293
Sana, H., de Mink, S. E., de Koter, A., *et al.* 2012, *Science* 337, 444
Tallon-Bosc, I., Tallon, M., Thiébaut, E., *et al.* 2008, Vol. 7013 of *Society of Photo-Optical Instrumentation Engineers (SPIE) Conference Series*, article ID 70131J

New windows on massive stars: asteroseismology, interferometry, and spectropolarimetry
Proceedings IAU Symposium No. 307, 2014
G. Meynet, C. Georgy, J. H. Groh & Ph. Stee, eds.

© International Astronomical Union 2015
doi:10.1017/S1743921314006991

AMBER/VLTI Snapshot Survey on Circumstellar Environments

Th. Rivinius[1], W.J. de Wit[1], Z. Demers[2,3], A. Quirrenbach[3] and the VLTI Science Operations Team

[1] ESO - European Organisation for Astronomical Research in the Southern Hemisphere, Chile
email: triviniu@eso.org

[2] Department of Physics and Astronomy, The University of Western Ontario, London, Canada

[3] Landessternwarte Königstuhl, D-69117, Heidelberg, Germany

Abstract. OHANA is an interferometric snapshot survey of the gaseous circumstellar environments of hot stars, carried out by the VLTI group at the Paranal observatory. It aims to characterize the mass-loss dynamics (winds/disks) at unexplored spatial scales for many stars. The survey employs the unique combination of AMBER's high spectral resolution with the unmatched spatial resolution provided by the VLTI. Because of the spatially unresolved central OBA-type star, with roughly neutral colour terms, their gaseous environments are among the easiest objects to be observed with AMBER, yet the extent and kinematics of the line emission regions are of high astrophysical interest.

Keywords. stars: winds, outflows, circumstellar matter

1. Introduction

The **O**bservatory survey at **H**igh **AN**gular resolution of **A**ctive OB stars (OHANA) combines high spectral with high spatial resolution across the Brγ and He$\iota\lambda$2.056 lines to characterize the dynamics of winds and disks. It is carried out by the VLTI group at the Paranal observatory with the three-beam combining instrument AMBER (Petrov *et al.* 2007). The survey was designed to make use of the observing time not requested by other programs, usually due to bad weather or unsuitable local sidereal time slots.

2. Observations and Data Reduction

The survey targets consist of twelve bright Be stars, thirteen O and B type supergiants, and one interacting binary (see Table 1). Almost 300 observations were obtained. By design, namely targeting quantities relative to the adjacent continuum, no calibrators were observed. However, in some nights calibrators, taken for technical purpose or other programs using the same setup, are available. These have been added to the database.

Basic data reduction was performed with amdlib, v3.0.6 (Tatulli *et al.* 2007; Chelli *et al.* 2009), and then processed further with idl. In particular:

• The pixel shifts between the spectral channels were a matter of concern, and seem not to be entirely stable. Whether this is a real effect or a consequence of noise affecting the determination of the shift is under investigation.

• Since the program aims for relative quantities, which do not suffer from degradation of absolute visibility, 100% of the frames were selected for display in this work. However, the final reduction includes several lower selection ratios as well.

• In the case of continuous observations of more than 30 minutes, (u, v) points were merged into 30 minute bins.

Table 1. Observed targets, spectral types, and data obtained. For each spectral line, the number of observations on the small, intermediate, and large telescope configurations (s–i–l) are given.

Target	Sp. type	Brγ s–i–l	HeIλ2.056 s–i–l	Target	Sp. type	Brγ s–i–l	HeIλ2.056 s–i–l
Be Stars				**OBA Supergiants**			
μ Cen	B2 Vnpe	2–5–3	0–1–0	η Car	LBV	22–16–5	3–1–0
χ Oph	B2 Vne	0–0–1	0–0–0	HR Car	LBV	4–4–1	2–0–0
ζ Tau	B2 IVe-sh	2–1–0	1–0–0	ζ Pup	O4 If	4–6–1	1–2–2
δ Cen	B2 IVne	3–5–2	1–1–1	ι Ori	O9 III	1–2–0	1–0–0
ϵ Cap	B3 Ve-sh	1–5–4	0–2–0	ζ Ori	O9.7 Iab	2–1–2	1–0–1
β^1 Mon A	B3 Ve	6–8–0	2–1–0	ϵ Ori	B0 Iab	1–1–0	0–0–0
β^1 Mon B	B3 ne	2–1–0	0–0–0	κ Ori	B0 Iab	3–2–0	2–0–0
β^1 Mon C	B3 e	2–1–0	0–0–0	ζ^1 Sco	B0.5 Ia+	0–1–3	0–0–1
P Car	B4 Vne	6–5–2	2–3–1	γ Ara	B1 Ib	0–1–0	0–1–0
β Psc	B6 Ve	1–4–4	0–2–0	HR 6142	B1 Ia	0–0–1	0–0–1
η Tau	B7 IIIe	0–0–0	1–0–0	ϵ CMa	B2 Iab	4–3–1	1–1–1
Electra	B8 IIIe	0–0–0	1–0–0	HD 53 138	B3 Ia	11–16–3	2–5–1
Interacting Binary				V 533 Car	A6 Iae	3–5–2	1–1–1
SS Lep	A1 V + M6 II	7–1–2	1–1–2				

- Intensity spectra were extracted, and the absolute wavelength scale corrected using telluric lines. The flux continuum was normalized to unity.
- Visibilities in the continuum were normalized to unity, phases in the continuum to zero. If calibrators were taken, these were used to check for and eventually remove instrumental ripples.
- RMS in the continuum was measured for each quantity to estimate data quality.

The raw data have become public immediately, and the results of the final reduction of the Brγ observations will be made public as soon as they are complete. The reduction of the HeIλ2.056 observations is pending.

3. Data Description and First Impressions

Due to the snapshot/backup/filler nature of the program, the data quality is inhomogeneous. Typical values for a good data set are an uncertainty of the visibility (normalized to unity) of about ±0.05, and of the phase ±2°, at a SNR of the combined spectrum of above 100. Selected data sets of the target stars are shown in Figs. 1 and 2. For each of the four targets, four baselines are shown, taken from two observations. The uppermost panels for each target show the flux spectra, then subpanels a-d show visibility and phase (upper and lower resp. profiles), while the centered panel show the (u, v) plane covered by the four baselines shown.

3.1. Be Stars

Visual inspection of the Be star observations shows them to be compatible with the canonical picture, namely a circumstellar decretion disk. The targets span all inclinations (equatorial to pole-on) and spectral subtypes. For some of the brighter stars the disk is already well resolved in the intermediate configuration (typical baseline lengths 30–70m), and overly resolved in the large configuration (typical baseline lengths 80–130m) Data for β^1 Mon and μ Cen are shown in Fig. 1. μ Cen shows a broad shallow ramp-type wing in the line, which is reflected in the phase. This may be the signature of freshly ejected material closer to the star than the bulk of the disk.

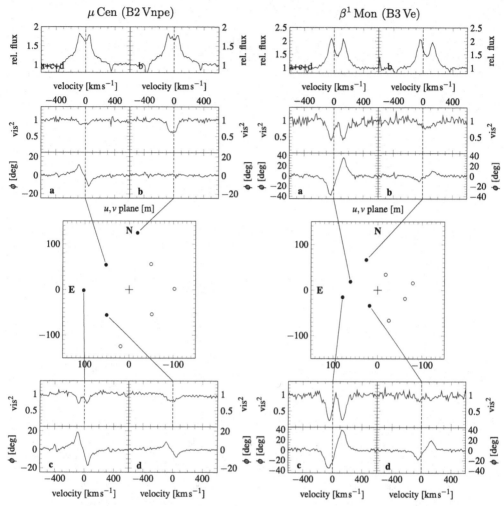

Figure 1. Example of OHANA data for the Be stars μ Cen and β^1 Mon.

3.2. *OB Supergiants*

The OBA supergiants can be divided into three distinct sets, namely LBVs and the remaining ones along the wind bi-stability at about $T_{\text{eff}} = 22\,000$ K. For the LBVs η Car and HR Car, dedicated contributions to this volume are presented by Mehner *et al.* and Rivinius *et al.* The O- and early B-supergiants do not show any obvious signature in visibility or phase, meaning their winds are too small to detect. In turn, the later B- and A-type supergiants do show such signatures in visibility, but again very little in phase. This is obvious for the hypergiant ζ^1 Sco, indicating an extended, but largely symmetric wind (Fig. 2, left). For HD 53 138 this is less obvious in Fig. 2, right, but comparison with calibrator data shows the visibility signature to be real, not instrumental. For HD 53 138 spectral emission variability goes together with a changing visibility and phase signature, which may indicate a variable, asymmetric wind. Further analysis of the data at hand may provide constraints on the size and clumping of these winds, while further observations may be able to trace variability, in particular for the slower winds with flow times of up to several weeks.

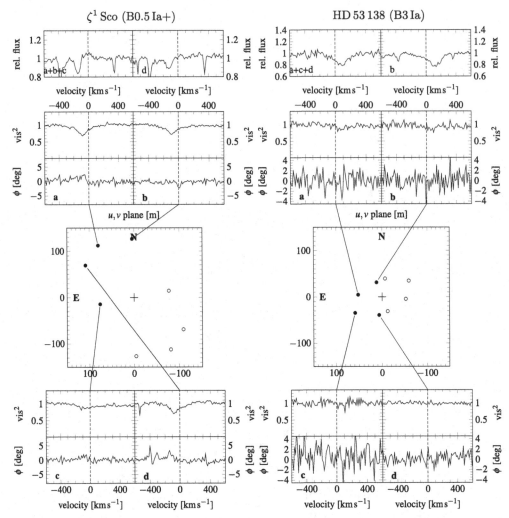

Figure 2. Example of OHANA data for the B-type supergiants ζ^1 Sco and HD 53 138.

3.3. *Interacting Binaries*

The only observed interacting binary was SS Lep, with barely any interferometric signature. Small wiggles seen in the visibility of some longer baselines need to be verified.

4. Conclusions

The OHANA survey provided interferometric data of the circumstellar environments of Be stars and OBA supergiants. The raw data is publicly available, the reduced data will become so as soon as the final reduction has passed quality control tests. The reduced data will be made available from `http://activebstars.iag.usp.br/index.php/34-ohana`.

References

Chelli, A., Utrera, O. H., & Duvert, G. 2009, *A&A* 502, 705
Petrov, R. G., Malbet, F., Weigelt, G., *et al.* 2007, *A&A* 464, 1
Tatulli, E., Millour, F., Chelli, A., *et al.* 2007, *A&A* 464, 29

New windows on massive stars: asteroseismology, interferometry, and spectropolarimetry
Proceedings IAU Symposium No. 307, 2014
G. Meynet, C. Georgy, J. H. Groh & Ph. Stee, eds.

© International Astronomical Union 2015
doi:10.1017/S1743921314007005

Recent highlights of spectropolarimetry applied to the magnetometry of massive stars

J. H. Grunhut

European Southern Observatory, Karl-Schwarzschild-Str. 2, D-85748 Garching, Germany
email: jgrunhut@eso.org

Abstract. Spectropolarimetry is a powerful tool used to probe fundamental properties of stars that cannot typically be measured in any other way. A new generation of high-resolution spectropolarimeters (ESPaDOnS at the Canada-France-Hawaii telescope, Narval at the Télescope Bernad Lyot, and HARPSpol at the 3.6-m ESO telescope) and dedicated observing campaigns (such as the Magnetism in Massive Stars (MiMeS) project) have led to significant improvements in both our observational and theoretical understanding of the underlying physics governing massive stars. In this article I review recent advances in the field of stellar magnetism of massive stars acquired using spectropolarimetry.

Keywords. polarization, surveys, stars: magnetic fields, stars: pre–main-sequence, stars: early-type, stars: emission-line, Be, stars: late-type, stars: Wolf-Rayet

---

1. Introduction

As discussed by Landstreet (these proceedings), the most direct method to investigate stellar magnetism is through spectropolarimetry. In particular, obtaining circularly polarized Stokes V spectra, which measures the difference between left- and right-hand circular polarization, presents the easiest way to measure the exceptionally small wavelength splitting caused by Zeeman splitting, and, is in practice, only limited by the achievable signal-to-noise ratio (SNR) of an observation. Ever since the first magnetic detection in the chemically peculiar Ap star 78 Vir (Babcock 1947), this has effectively been the primary method used to detect and diagnose magnetic fields.

The study of magnetism in massive stars ($M \gtrsim 8\,M_\odot$) is still a relatively new field. While the first magnetic detection in the archetype magnetic B2V star σ Ori E goes all the way back to the pivotal work of Landstreet & Borra (1978), only a handful of new magnetic detections have been found in the years to follow (e.g. Borra & Landstreet 1979; Landstreet 1982; Thompson & Landstreet 1985; Bohlender *et al.* 1987). In fact, most of our knowledge regarding magnetism in higher-mass stars ($M \gtrsim 1.5\,M_\odot$) comes from studies of the chemically peculiar A- and B-type stars. This largely reflects the fact that, unlike intermediate-mass magnetic stars that show strong and distinctive photospheric chemical peculiarities compared to non-magnetic A- and B-type stars (and therefore can be used as a proxy for magnetism), the strong, radiatively-driven outflows of more massive stars generally inhibit these chemical peculiarities, making them difficult to identify. Furthermore, the relatively weak fields of these stars coupled with relatively few spectral lines from which to directly diagnose the presence of magnetism in the optical spectra of massive stars meant that these fields remained undetected by previous generations of instrumentation.

Over the last 15 years a new generation of instruments [ESPaDOnS at the Canada-France-Hawaii telescope (CFHT), Narval at the Télescope Bernad Lyot (TBL), and

HARPSpol at the ESO's 3.6-m telescope] and new techniques, such as Least-Squares Deconvolution (LSD; Donati *et al.* 1997), have been at the centre of studies that have significantly contributed to our understanding of magnetism in massive stars. Their large wavelength coverage (ideal for multi-line techniques) in addition to their high efficiency and their mounting on medium-sized telescopes (which allow us to reach high SNR in a reasonable amount of time for these stars), have provided us with the necessary tools to really study the magnetic properties of massive stars.

2. Magnetism in main sequence stars

It is well-known that magnetic fields are detected in a small population of main sequence (MS) massive stars. The majority of the magnetically detected massive stars are high-mass extensions of the chemically-peculiar intermediate-mass stars that show He peculiarities (e.g. Bohlender *et al.* 1987). It was only about 15 years ago that the first confirmed detections of magnetic fields in stars more massive than $\sim$10 $M_\odot$ (earlier than B2) were made, e.g. the classical B1III pulsator β Cep (Henrichs *et al.* 2000; Donati *et al.* 2001) and the young massive O7V star θ^1 Ori C (Donati *et al.* 2002). In fact, to give the reader a sense of how young the field really is, prior to 2009 only 3 magnetic O-type stars were even known (θ^1 Ori C; the Of?p star HD 191612 (Donati *et al.* 2006); the evolved O9.7Ib binary star ζ Ori A (Bouret *et al.* 2008)), and 9 of the 13 of the most massive stars (earlier than B2) were discovered between 2005 and 2009.

The majority of what we knew about magnetism in higher-mass stars is derived from observations of intermediate-mass stars ($1.5 < M < 8\,M_\odot$). Compared to the magnetic fields found in low-mass stars ($M \lesssim 1.5\,M_\odot$), the incidence and characteristics of magnetic fields found in these stars are significantly different. While magnetism in low-mass stars is essentially ubiquitous, magnetic fields are only detected in 5-10% of the population of intermediate-mass stars (e.g. Bagnulo *et al.* 2006). Observations suggest that the majority of intermediate-mass stars present globally-ordered, topologically simple magnetic fields, often characterized by a dipole or a low-order multipole (e.g. Aurière *et al.* 2007) that are effectively *frozen* into the star - the fields appear to be stable over timescales of at least decades and any observed variability is a consequence of rotational modulation (e.g. Silvester *et al.* 2012).

While strong, detectable surface magnetic fields in massive stars appear to be rather rare, they nevertheless have an important impact. Magnetic fields can confine and modulate the highly ionized radiatively driven winds of massive stars (e.g. ud-Doula & Owocki 2002), which is likely responsible for the production of periodic high-energy X-ray emission (e.g. Gagne *et al.* 1997; Babel & Montmerle 1997) and other line-profile-variability that is commonly observed in hot OB stars (e.g. Grunhut *et al.* 2012a,b). Furthermore, the interaction of the magnetic field with the stellar wind may enhance the shedding of angular momentum, and therefore have a significant impact on the rotation rates of OB stars (e.g. Weber & Davis 1967; ud-Doula *et al.* 2008). These effects have far-reaching consequences on the evolution of massive stars, which has been shown to be strongly influenced by stellar rotation (e.g. Meynet *et al.* 2011). Therefore, it is necessary that the magnetic field characteristics be well understood for massive stars. To date, the largest and most productive investigation of the magnetic properties of massive stars has been the Magnetism in Massive Stars (MiMeS) Project.

The MiMeS Project was formed by an international team of recognized researchers with the goal to better understand the complex and puzzling field of magnetism of massive stars. In 2008, the MiMeS Project was awarded "Large Program" (LP) status (640 hours) by Canada and France (PI G. Wade) with ESPaDOnS from late 2008 through

2012. This was shortly followed by additional LPs with Narval (a total of ~500 hours, PI C. Neiner) and with HARPSpol (a total of ~300 hours, PI E. Alecian). In addition to these LPs, the MiMeS Project was and still is supported by numerous PI programs from such observatories as the Anglo-Australia Telescope, Chandra, Dominion Astrophysical Observatory, HST, MOST, SMARTS, and the VLT.

One aspect of the MiMeS Project includes a "Survey Component" (SC), the main goal of which was to discover newly magnetic stars and to provide critical missing information about the incidence and statistical properties of large-scale magnetic fields in a statistically-significant sample of massive stars. Over 4800 high-precision (median signal-to-noise ratio (S/N) of ~800 per pixel), high-resolution ($R \sim 68000$), broad-band (370-1050 nm) Stokes V spectra were analyzed within the context of the MiMeS Project, corresponding to approximately 550 OB stars with spectral types between B9 and O5. In order to increase the SNR and enhance our sensitivity to weak magnetic Zeeman signatures, the LSD technique, as implemented by Kochukhov *et al.* (2010), was used to produce mean profiles from the unpolarized light (Stokes I), circularly polarized light (Stokes V) and also from the diagnostic spectrum (N; which diagnoses the presence of spurious signatures resulting from instrumental or other first-order systematic effects). The detection of any signal in the LSD mean V and N profiles was diagnosed using a standard χ^2 analysis (Donati *et al.* 1997).

Of the ~550 stars observed within the context of the MiMeS Project, approximately 65 show evidence for magnetic fields. Of these detected targets, about 30 were previously established as magnetic stars prior to the survey and are therefore ignored in the following statistics. The bulk incidence (i.e. the total number of previously-unknown magnetic stars in the sample relative to the total sample) is $7 \pm 1\%$. Approximately 420 B-type stars in the sample were previously unknown to host magnetic fields, and of this sample, 32 stars were found to be magnetic. This translates into a magnetic incidence fraction of $7 \pm 1\%$. Of particular interest to this conference are the most massive stars in the sample - the O-type stars. Approximately 95 O-type stars in the survey were previously unknown to host magnetic fields, and 6 were found to be magnetic, for an incidence of $6 \pm 1\%$. The MiMeS Project has carried out that largest and most sensitive investigation of the magnetic properties of O-type stars, and in the end has now tripled the number of previously known magnetic O-type stars.

The detected magnetic stars (e.g. Grunhut *et al.* 2012a,b; Wade *et al.* 2012a) exhibit periodic variability of their Stokes V profiles (with periods in the range of 0.5 d to many years), as a result of the rotational modulation of the projected magnetic field. These detections were systematically found to present organized magnetic fields with important dipole components. The surface polar field strengths range from several hundred G up to 20 kG. In general, the magnetic characteristics of detected B-type stars and O-type stars are very similar. Of particular note, the MiMeS Project is responsible for discovering the most rapidly-rotating, magnetic, early-type stars (HR 7355 (Oksala *et al.* 2010; Rivinius *et al.* 2010, 2013); HR 5907 (Grunhut *et al.* 2012a)) and the most strongly magnetic O-type star (NGC 1624-2 (Wade *et al.* 2012b)) known to date.

Based on the statistics learned from the MiMeS Project, it can be concluded that the incidences and characteristics of large-scale magnetic fields in B- and O-type stars are essentially the same. The physical and statistical properties of their fields are qualitatively identical to those of intermediate-mass stars with spectral types between F0 to A0 on the MS. The MiMeS survey therefore establishes that the basic physical characteristics of magnetism in stars with radiative envelopes remains unchanged across more than 1.5 decades of stellar mass, from mid-F to early O-type stars ($M \sim 1.5 \, M_\odot$ to $\sim 50 \, M_\odot$).

Other than providing critical missing information regarding the statistical properties of magnetic fields in massive stars, another goal of the survey was to investigate the connection between magnetism and observed phenomena. Listed here are some of these results.

The MiMeS Project has observed all of the known Galactic stars of the Of?p class, and detected or confirmed the presence of a magnetic field in each of these stars (see e.g. Wade *et al.* 2012b, and references therein). The Of?p classification pertains to O-type stars that exhibit CIII λ4650 in emission with a strength comparable to the neighbouring NIII lines (Walborn 1972; Walborn *et al.* 2010). It can therefore be concluded that the Of?p stars represent a magnetic class of O-type stars, and the peculiar spectral properties are likely a consequence of the interaction of their winds with the magnetic fields (e.g. Sundqvist *et al.* 2012). This is the first established magnetic class of O-type stars (similar to the chemically-peculiar Ap/Bp stars), and as there are now 5 known extra-Galactic Of?p stars (Walborn *et al.* 2010; Massey *et al.* 2014), this therefore implies the identification of the first magnetic stars outside of the Galaxy.

About 100 pulsating β Cep and slowly-pulsating B-type (SPB) stars were observed as part of the MiMeS SC. It has recently been suggest (Hubrig *et al.* 2006, 2009) that there is a high magnetic incbidence fraction (∼50%) among these stars, which has far-reaching consequences regarding the connection between pulsations and magnetism. However, the MiMeS SC results suggest a significantly different result with a magnetic incidence fraction of $10 \pm 3\%$. This value is fully consistent with the larger sample of O- and B-type stars and suggests that pulsating B-type stars are not more likely to be magnetic than non-pulsating B-type stars. Furthermore, a monitored investigation of a sample of claimed magnetic detections with characterized magnetic models by Hubrig *et al.* (2011) have also been shown to be inconsistent with the MiMeS data (Shultz *et al.* 2012). It can therefore be concluded that there is no connection between magnetism and pulsations in B-type stars.

In a sample of about 90 classical Be stars, MiMeS has failed to unambiguously detect a magnetic field in any of these stars. The sample of classical Be stars were identified according to the classification of Porter & Rivinius (2003) - as early-type emission line stars with Keplerian disks - to distinguish these stars from other emission line stars whose emission is of a different origin (e.g. accretion disks from Herbig stars). Be stars are also known to be rapidly-rotating (e.g. Rivinius *et al.* 2013), yet despite this fact, MiMeS has achieved a magnetic sensitivity similar to that of the larger sample and has successfully detected fields in other rapidly-rotating stars (e.g. Oksala *et al.* 2010; Grunhut *et al.* 2012a) and therefore expected to detect fields in ∼10 stars. The majority of the Be sample was chosen to exclude stars whose optical spectra were dominated by emission lines, but most still showed significant emission. However, the lack of detections is not a result of this emission as magnetic fields have been detected in other emission-line stars (e.g. Herbig Be stars or Of?p stars). Hence it can be concluded that the lack of detected magnetic classical Be stars, indicates that decretion Keplerian Be disks are not of magnetic origin (Neiner *et al.* 2012).

3. Magnetism in pre-main sequence stars

The pre-main sequence (PMS) phase of evolution is defined as the period of stellar formation after the proto-stellar phase, but before the MS phase where a star is undergoing core-hydrogen burning. At the beginning of the PMS phase the star is on the birth-line – the locus of points in the HR diagram where the star is, for the first time, observable at optical wavelengths after shedding the majority of its proto-stellar envelope and is no longer undergoing significant accretion (Stahler 1983). At this point, nuclear-burning has not yet started in the star's core and slow gravitational collapse is the main source of energy. Just before the end of the PMS phase, nuclear reactions begin, contributing more and more energy to the luminosity of the star. As the star evolves closer to the zero age main sequence (ZAMS), gravitational contraction slows, eventually coming to a stop.

As previously discussed, main sequence, magnetic higher-mass stars all show *stable* large-scale fields, at least over the lifetime observations. The leading hypothesis suggests that these magnetic fields are *fossil* fields – the *frozen* remnants of the Galactic field accumulated and possibly enhanced by a dynamo field generated in an earlier phase of evolution (e.g. Cowling 1945; Mestel 2001; Moss 2001). This has been supported by numerical and analytical studies that find that surviving fields can relax into configurations exhibiting long-term stability (Braithwaite 2009; Duez & Mathis 2010) and can approximately explain the field topology and other general characteristics of magnetism in higher-mass stars. However, prior to about 2005, the global magnetic properties of HAeBe stars were unknown, and therefore direct evidence linking the magnetic properties of MS stars with PMS stars was completely lacking.

To address this issues, a high-resolution spectropolarimetric survey of 128 Herbig Ae/Be (HAeBe) stars (Herbig 1960; Vieira *et al.* 2003) was carried out using ESPaDOnS and Narval. A sample of field HAeBe stars were selected from the catalogue of The *et al.* (1994) and Vieira *et al.* (2003), while an additional sample were selected from members of the following young clusters: NGC 2244 (Park & Sung 2002), NGC 2264 (Sung *et al.* 1997), and NGC 6611 (de Winter *et al.* 1997). LSD was used to increase the sensitivity to detect weak Zeeman signatures. This survey obtained magnetic detections in 8 stars (e.g. Wade *et al.* 2005; Alecian *et al.* 2008b) and found a magnetic incidence fraction of ∼6%, similar to the incidence that is found for intermediate-mass MS stars and now MS massive stars. Spectropolarimetric monitoring of many of the detected stars was also carried out in order to characterize the strength and structure of their surface fields. The results show that the magnetic properties of HAeBe star are similar to what is found on the MS - the fields are mainly dipolar, are strong (surface field strengths of 300 G to 4 kG), and stable over many years (e.g. Alecian *et al.* 2008a, 2009). This survey has therefore established a direct link between the magnetic properties of stars found on the PMS and on the MS. It can therefore be concluded that the magnetic properties of A/B stars must have been shaped before the HAeBe phase of stellar evolution (Alecian *et al.* 2013).

4. Magnetism in evolved massive stars

As a massive star completes its MS evolution, its envelope rapidly grows while its surface temperature cools until it eventually forms a cool supergiant. Characterized by a helium-burning core and a deep convective hydrogen-burning envelope, the internal structure of a supergiant is in strong contrast to its MS OB progenitor, which is characterized by a convective hydrogen-burning core and a radiative envelope. The most massive stars (initial mass $> 25\,M_\odot$) go through a phase of extreme mass-loss and eject their envelope in the Wolf-Rayet (WR) phase before ending their lives in a dramatic Type Ib supernova explosion, while the less massive stars (initial mass between about 8 and $25\,M_\odot$) evolve into red supergiants that explode as Type II supernovae. Ultimately, the end product of a massive star is a neutron star or black hole (e.g. Crowther *et al.* 1995; Eldridge 2008).

It is well known that magnetic fields exist in at least some fraction of massive stars and that magnetic flux conservation of the fields found in MS A- and B-type stars is in relative agreement with the fluxes observed in their ultimate descendants: white dwarfs and neutron stars (e.g. Tout *et al.* 2004; Ferrario & Wickramasinghe 2006; Walder *et al.* 2012). It therefore is unsurprising to expect that at least some fraction of stars found in an intermediate phase of evolution should be magnetic.

4.1. *Wolf-Rayet stars*

While it has only recently been established by the MiMeS collaboration, and now confirmed by the B-fields in OB stars (BOB) collaboration (see e.g. Morel *et al.*, these proceedings), that magnetic fields are systematically detected in about 5-10% of stars with radiative envelopes, WR stars, which are the direct descendants of the most massive OB stars, are known to show evidence of corotating interaction regions (CIRs), which some have postulated to be the result of magnetic fields (Harries 2000). Motivated by this and the previously discussed matter, de la Chevrotière *et al.* (2013, 2014) (as part of the MiMeS collaboration) have recently attempted to constrain the magnetic field characteristics of these stars.

However, unlike the previously discussed results for MS and PMS stars, the diagnosis and detection of magnetic fields in WR stars is complicated by the fact that the majority of the visible optical spectral lines are in emission, a consequence of their formation in a strong stellar wind. Not only does this make the application of multi-line techniques extremely difficult to be useful, but this also has the potential to fundamentally change the Stokes V profile shape relative to the signatures that are formed in the stellar photosphere – the typical large-scale field in higher-mass stars is generally well-described by a dipole; however, in the presence of a strong stellar wind, or at large distances from the photosphere (where the wind energy density is greater than the magnetic energy density), the field lines are stretched open and better resemble a split monopole (ud-Doula & Owocki 2002) – and therefore makes characterization of these lines challenging. Gayley & Ignace (2010) have investigated this issue and have developed a model of the expected shape of a Zeeman signature formed in stellar wind. As expected, the shape of the Zeeman signature formed in an outflowing wind with an underlying split monopole configuration is considerably different from that of a Zeeman signature formed in a hydrostatic atmosphere with an underlying mainly-dipole configuration: the latter can usually be well-described by an *s-shaped* profile, while the former reflects a *heartbeat* shape to its profile (see Fig. 3 of Gayley & Ignace 2010 for an example).

Taking all this into consideration, de la Chevrotière *et al.* (2013, 2014) have carried out the most sensitive and detailed search for the presence of large-scale magnetic fields in 12 WR-stars. None of their observations resulted in the unambiguous detection of any Zeeman signature in their data. However, using Bayesian inference methods, they were able to derive upper limits for these non-detections and, in some stars, find marginal evidence favouring a non-zero magnetic model. They find an average upper limit of the allowed field strength of ∼500 G in the wind is consistent with the non-detections in their sample and that 3 of the 12 stars show marginal evidence for magnetic fields with likely field strengths of ∼100-200 G, or not exceeding ∼2 kG at their most upper limits permitted by their Bayesian assessment. If these marginal detections are taken at face value, then this would imply a considerably higher magnetic incidence fraction (∼25%) among WR stars than their progenitors (the OB stars with a ∼7% magnetic incident rate). However, robust conclusions cannot be derived from such small number statistics, as also pointed out by the authors.

5. Magnetism in cool supergiants

Active late-type supergiants are luminous in X-rays and show emission in chromospheric ultraviolet (UV) Si IV lines, with some stars showing evidence of flaring phenomena usually associated with the presence of a corona and magnetic reconnection generated via dynamo fields (e.g. Tarasova 2002; Ayres 2005). Additionally, a class of 'non-coronal' or inactive supergiants – that exhibit weak or no X-ray emission and have UV spectra containing mostly low-temperature or relatively weaker flux chromospheric lines compared to 'normal' supergiants – are known to exist, and can be explained by the presence of magnetic loops submerged within their thick extended chromospheres (Ayres 2005). Furthermore, observations of Betelgeuse have revealed evidence for an extended chromosphere (Carpenter & Robinson 1997) and surface brightness fluctuations that can be described by giant convective cells (e.g. Chiavassa *et al.* 2010, and references therein).

These observational phenomena have hinted that significant magnetic fields should be present in supergiants. The first confirmation of this was the detection of Zeeman signatures found in the mean LSD profiles obtained with high-resolution spectropolarimeters corresponding to a magnetic field in Betelgeuse (Aurière *et al.* 2010) and the systematic detection of magnetic fields in a larger population of supergiants (Grunhut *et al.* 2010). The original sample presented by Grunhut *et al.* (2010) has more than doubled in size and now contains about 70 stars. New statistics derived from this sample indicate that approximately $\gtrsim$30-40% of supergiant stars are found to have detected Zeeman signatures in their mean LSD profiles (an example is provided in Fig. 1). An HR-diagram showing the position of these detected stars is presented in Fig. 1. The detected fields have globally weak surface-averaged magnetic fields, as inferred from their weak longitudinal magnetic field measurements on the order of $\sim$ G (compared to the 100-1000s of G in their OB progenitors). The characteristics of these profiles suggest that many of these stars host topologically complex fields that significantly depart from the large-scale dipoles typically found in their OB progenitors. If we just include stars that are cooler than the granulation boundary that suggests the onset of a significant convective envelope (Gray 1989) (as shown in Fig. 1), then the incidence rate is somewhat higher at $\sim$50-60%. For stars hotter than the granulation boundary, the incidence rate drops significantly to $\sim$10%. Based on these results we can make some basic conclusions: the significantly higher incidence fraction of magnetic fields among cool supergiants suggests these stars cannot be the descendants of the magnetic OB stars. The higher incidence fraction of fields among stars with deep convective envelopes suggest that these fields are likely drive by dynamo mechanisms (e.g. Parker 1955); however, the extremely slow rotation of the coolest supergiants (such as Betelgeuse) would argue for an *exotic* dynamo model for at least some of these stars, as they are rotating too slow for the solar-like $\alpha - \omega$ dynamo models to be efficient. Numerical models of Dorch (2004) have already shown that magnetic fields can be generated from large convective cells.

6. Conclusions

It is now well-established that large-scale magnetic fields, detected from the induced circularly polarized signatures reflecting Zeeman splitting in the mean line profiles of high-resolution spectra, are detected in at least some fraction of stars at every phase of stellar evolution of massive stars – from the pre-MS all the way to the latest stages of evolution before the star violently explodes. The similar statistics and magnetic properties of massive stars during the pre-MS and MS phases suggests that the physics of magnetism in stellar radiative zones remains unchanged over millions of years of evolution, and

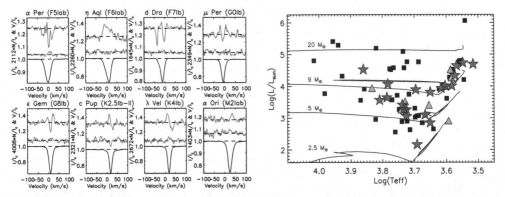

Figure 1. Left: Example mean Stokes V (top), diagnostic null (middle) and Stokes I line profiles obtained from cool supergiants. The profiles in the upper panels are likely a consequence of topologically complex surface magnetic fields in these stars. Right: HR diagram showing the position of all supergiants observed as part of a survey of cool supergiants. Evolutionary tracks (sold lines) are also included. Black squares correspond to stars with non detection of a Zeeman signature, orange triangles reflect stars with a possibly detected Zeeman signature, and red stars represent stars with an unambiguously detected Zeeman signature.

that the detected fields are formed and fossilized in the star prior to the pre-MS phase. With only a small number of WR stars investigated in any detail and no unambiguous detection of a field in these stars, no conclusions can be drawn about their potential relationship to possible magnetic OB star progenitors. Lastly, as the magnetic incidence fraction of magnetism in cool supergiants greatly exceeds similar measurements in MS massive stars, it stands to reason that the majority of supergiants are not the descendants of MS magnetic stars. This suggests that the detected fields are not fossil remnants of an earlier phase of evolution and are therefore likely generated contemporaneously via some dynamo process.

References

Alecian, E., Catala, C., Wade, G. A., *et al.* 2008a, *MNRAS* 385, 391
Alecian, E., Wade, G., Catala, C., *et al.* 2008b, *A&A* 481, L99
Alecian, E., Wade, G. A., Catala, C., *et al.* 2009, *MNRAS* 400, 354
Alecian, E., Wade, G. A., Catala, C., *et al.* 2013, *MNRAS* 429, 1001
Aurière, M., Donati, J.-F., Konstantinova-Antova, R., *et al.* 2010, *A&A* 516, L2
Aurière, M., Wade, G. A., Silvester, J., *et al.* 2007, *A&A* 475, 1053
Ayres, T. R. 2005, *ApJ* 618, 493
Babcock, H. W. 1947, *ApJ* 105, 105
Babel, J. & Montmerle, T. 1997, *A&A* 323, 121
Bagnulo, S., Landstreet, J. D., Mason, E., *et al.* 2006, *A&A* 450, 777
Bohlender, D., Landstreet, J., Brown, D., & Thompson, I. 1987, *ApJ* 323, 325
Borra, E. F. & Landstreet, J. D. 1979, *ApJ* 228, 809
Bouret, J.-C., Donati, J.-F., Martins, F., *et al.* 2008, *MNRAS* 389, 75
Braithwaite, J. 2009, *MNRAS* 397, 763
Carpenter, K. G. & Robinson, R. D. 1997, *ApJ* 479, 970
Chiavassa, A., Haubois, X., Young, J. S., *et al.* 2010, *A&A* 515, A12
Cowling, T. G. 1945, *MNRAS* 105, 166
Crowther, P. A., Hillier, D. J., & Smith, L. J. 1995, *A&A* 293, 172
de la Chevrotière, A., St-Louis, N., & Moffat, A. F. J., the MiMeS Collaboration 2013, *ApJ* 764, 171

de la Chevrotière, A., St-Louis, N., & Moffat, A. F. J., the MiMeS Collaboration 2014, *ApJ* 781, 73

de Winter, D., Koulis, C., The, P. S., *et al.* 1997, *A&AS* 121, 223

Donati, J.-F., Babel, J., Harries, T., *et al.* 2002, *MNRAS* 333, 55

Donati, J.-F., Howarth, I., Bouret, J.-C., *et al.* 2006, *MNRAS* 365, L6

Donati, J.-F., Semel, M., Carter, B., Rees, D., & Collier Cameron, A. 1997, *MNRAS* 291, 658

Donati, J.-F., Wade, G., Babel, J., *et al.* 2001, *MNRAS* 326, 1265

Dorch, S. B. F. 2004, *A&A* 423, 1101

Duez, V. & Mathis, S. 2010, *A&A* 517, A58

Eldridge, J. J. 2008, *Royal Society of London Philosophical Transactions Series A* 366, 4441

Ferrario, L. & Wickramasinghe, D. 2006, *MNRAS* 367, 1323

Gagne, M., Caillault, J.-P., Stauffer, J. R., & Linsky, J. L. 1997, *ApJL* 478, L87

Gayley, K. G. & Ignace, R. 2010, *ApJ* 708, 615

Gray, D. F. 1989, *PASP* 101, 832

Grunhut, J. H., Rivinius, T., Wade, G. A., *et al.* 2012a, *MNRAS* 419, 1610

Grunhut, J. H., Wade, G. A., Hanes, D. A., & Alecian, E. 2010, *MNRAS* 408, 2290

Grunhut, J. H., Wade, G. A., Sundqvist, J. O., *et al.* 2012b, *MNRAS* 426, 2208

Harries, T. J. 2000, *MNRAS* 315, 722

Henrichs, H. F., de Jong, J. A., Donati, D.-F., *et al.* 2000, in Y. V. Glagolevskij & I. I. Romanyuk (eds.), *Magnetic Fields of Chemically Peculiar and Related Stars*, pp 57–60

Herbig, G. H. 1960, *ApJS* 4, 337

Hubrig, S., Briquet, M., De Cat, P., *et al.* 2009, *Astronomische Nachrichten* 330, 317

Hubrig, S., Briquet, M., Schöller, M., *et al.* 2006, *MNRAS* 369, L61

Hubrig, S., Ilyin, I., Schöller, M., *et al.* 2011, *ApJL* 726, L5

Hussain, G. A. J. & Alecian, E. 2014, in *IAU Symposium*, Vol. 302 of *IAU Symposium*, pp 25–37

Kochukhov, O., Makaganiuk, V., & Piskunov, N. 2010, *A&A* 524, A5

Landstreet, J. & Borra, E. 1978, *ApJL* 224, L5

Landstreet, J. D. 1982, *ApJ* 258, 639

Massey, P., Neugent, K. F., Morrell, N., & Hillier, D. J. 2014, *ApJ* 788, 83

Mestel, L. 2001, in G. Mathys, S. Solanki, & D. Wickramasinghe (eds.), *Magnetic Fields Across the Hertzsprung-Russell Diagram*, Vol. 248 of *Astronomical Society of the Pacific Conference Series*, p. 3

Meynet, G., Eggenberger, P., & Maeder, A. 2011, *A&A* 525, L11

Moss, D. 2001, in G. Mathys, S. Solanki, & D. Wickramasinghe (eds.), *Magnetic Fields Across the Hertzsprung-Russell Diagram*, Vol. 248 of *Astronomical Society of the Pacific Conference Series*, p. 305

Neiner, C., Grunhut, J. H., Petit, V., *et al.* 2012, *MNRAS* 426, 2738

Oksala, M., Wade, G., Marcolino, W., *et al.* 2010, *MNRAS* 405, L51

Park, B.-G. & Sung, H. 2002, *AJ* 123, 892

Parker, E. N. 1955, *ApJ* 122, 293

Porter, J. M. & Rivinius, T. 2003, *PASP* 115, 1153

Rivinius, T., Carciofi, A. C., & Martayan, C. 2013, *A&A Rev.* 21, 69

Rivinius, T., Szeifert, T., Barrera, L., *et al.* 2010, *MNRAS* 405, L46

Shultz, M., Wade, G. A., Grunhut, J., *et al.* 2012, *ApJ* 750, 2

Silvester, J., Wade, G. A., Kochukhov, O., *et al.* 2012, *MNRAS* 426, 1003

Stahler, S. W. 1983, *ApJ* 274, 822

Sundqvist, J., ud-Doula, A., Owocki, S., *et al.* 2012, *MNRAS* p. L433

Sung, H., Bessell, M. S., & Lee, S.-W. 1997, *AJ* 114, 2644

Tarasova, T. N. 2002, *Astronomy Reports* 46, 474

The, P. S., de Winter, D., & Perez, M. R. 1994, *A&AS* 104, 315

Thompson, I. B. & Landstreet, J. D. 1985, *ApJL* 289, L9

Tout, C. A., Wickramasinghe, D. T., & Ferrario, L. 2004, *MNRAS* 355, L13

ud-Doula, A. & Owocki, S. 2002, *ApJ* 576, 413

ud-Doula, A., Owocki, S., & Townsend, R. 2008, *MNRAS* 385, 97

310   J. H. Grunhut

Vieira, S. L. A., Corradi, W. J. B., Alencar, S. H. P., *et al.* 2003, *AJ* 126, 2971
Wade, G., Grunhut, J., Gräfener, G., *et al.* 2012a, *MNRAS* 419, 2459
Wade, G. A., Drouin, D., Bagnulo, S., *et al.* 2005, *A&A* 442, L31
Wade, G. A., Maíz Apellániz, J., Martins, F., *et al.* 2012b, *MNRAS* 425, 1278
Walborn, N. R. 1972, *AJ* 77, 312
Walborn, N. R., Sota, A., Maíz Apellániz, J., *et al.* 2010, *ApJL* 711, L143
Walder, R., Folini, D., & Meynet, G. 2012, *Space Science Reviews* 166, 145
Weber, E. & Davis, Jr., L. 1967, *ApJ* 148, 217

Discussion

KHALACK: Could you please comment on the possibility of detecting magnetic fields in the Horizontal-Branch (HB) stars?

GRUNHUT: There are currently no plans to observe HB stars. As the HB stars are typically found (and also identified) in Globular clusters, these stars tend to be too faint to acquire high-resolution spectropolarimetry with the current generation of instruments and the telescopes they are mounted on. Perhaps this is possible with low-resolution instruments that are mounted on larger telescopes (such as FORS on the VLT).

MAEDER: Are you planning to also observe magnetic fields in evolutionary stages earlier that the Herbig Ae/Be phase, such as stars that live in the Hayashi or FU Orionis phase?

GRUNHUT: Evelyne Alecian and Gaitee Hussain are already leading a project that aims to observe the earliest phases of intermediate-mass evolution that is permitted by optical spectropolarimetric studies (Hussain & Alecian 2014). At earlier evolutionary phases the star is still enshrouded in its cocoon and therefore infrared spectropolarimeters are required to observe stars at these phases. Therefore, one of the primary goals for the next generation of spectropolarimeters (e.g. SPIRou at the CFHT, CRIRES+ at the VLT) is to probe the magnetic properties of these young stars in the early stages of their formation.

Jason Grunhut

*New windows on massive stars: asteroseismology, interferometry, and
spectropolarimetry*
Proceedings IAU Symposium No. 307, 2014
G. Meynet, C. Georgy, J. H. Groh & Ph. Stee, eds.

© International Astronomical Union 2015
doi:10.1017/S1743921314007017

Basics of spectropolarimetry

John D. Landstreet

University of Western Ontario, London, Canada and
Armagh Observatory, Armagh, Northern Ireland
email: jlandstr@uwo.ca

Abstract. Many astronomical sources of radiation emit polarised radiation, for example because
of the presence of a disk which produces linear polarisation by scattering some photospheric radiation, or because of the presence of a magnetic field, which leads to circular and sometimes
linear polarisation of spectral line profiles. Measuring the wavelength dependence of the polarisation of radiation from such sources can reveal valuable and interesting constraints on the
nature of the objects observed. This paper summarises the basic ideas of spectropolarimetry
and describes some of the information it can provide.

Keywords. Polarisation, instrumentation: polarimeters, techniques: polarimetric, stars: magnetic fields, stars: circumstellar matter

1. Introduction

Normal stellar spectroscopy starts with measurement of the intensity of a beam of
starlight as a function of wavelength. It is well known that such data provide an astonishing range of valuable information about the light source. However, spectroscopy does
not exhaust the available information in the light beam. The light may be *linearly and/or
circularly polarised*. Measurement of the polarisation state of the light as a function of
wavelength, known as spectropolarimetry, can provide new and valuable constraints on
the geometric structure of the light source (for example, on the shape and state of a
circumstellar disk), or reveal the strength and structure of a magnetic field present in
the star.

This paper is intended to provide a simple introduction to the subject of spectropolarimetry. I will qualitatively describe how polarisation of starlight can arise from the
geometry of the source or a magnetic field. I will then discuss some basic methods of
measuring the wavelength dependence of polarisation, and in particular explain how it
is possible to reliably detect polarisation as low as one part in 10^5. Finally I will discuss
how important it is to be both careful and critical in measuring such tiny effects.

2. Linear polarisation due to scattering

Everyone is familiar with some of the phenomena of linear polarisation. If you look
at sunlight reflected obliquely from a puddle of water through polarising sunglasses, the
intensity of the light reflected by the puddle is diminished relative to the brightness of
the surroundings. If you take off the polarising sunglasses and look through them as you
rotate them around an axis normal to one lens, the brightness of the reflected light varies.
The brightness is near its minimum value with the glasses in their normal orientation,
and maximum when the glasses are at 90° to this orientation.

Light is a transverse wave of electricity and magnetism, and if the direction of the
electric vector is not distributed randomly around the direction of propagation, the light
is said to be *linearly polarised*. The direction of the average electric vector is the plane of

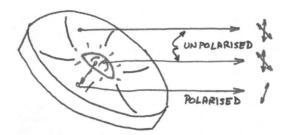

Figure 1. Polarisation of light from a star-disk system. The light emitted by both the stellar photosphere and by the hot circumstellar disk is essentially unpolarised, but the photospheric light that is *scattered* by free electrons in the disk is strongly polarised in the plane perpendicular to the plane containing the incident and scattered light beams.

polarisation, and the difference between the light intensity in that plane and the intensity in the orthogonal plane, divided by the total intensity, is the fractional polarisation.

In the puddle reflection observation described above, the reflection of light polarised parallel to the plane of reflection is less efficient than reflection of light polarised parallel to the surface of the puddle, so the reflected light has net polarisation parallel to the plane of the surface of the puddle.

This experiment illustrates the basic ideas of astronomical polarisation. Natural phenomena can preferentially emit, reflect, or scatter light with a preferred plane of plarisation. The polarisation of the light then carries information about the source or the reflection phenomenon. This information can be extracted by measuring the direction and amplitude of the polarisation. This is done by using a filter that passes one direction of polarisation but not the orthogonal direction (the sunglasses, whose lenses transmit vertically polarised light but not horizontally polarised light), and measuring the *intensity* of the transmitted light in each of the two orthogonal polarisation planes.

A typical astronomical situation leading to linearly polarised light is shown in Fig. 1. A star is surrounded by an accretion disk (a Herbig AeBe star, a mass transfer binary, etc.) or a decretion disk (a classical Be star, etc.) which is heated by the central star. The distant observer sees (1) light coming directly from the stellar photosphere, (2) light emitted thermally by the hot disk material, and (3) photospheric light that has been scattered by the free electrons in the disk. Components (1) and (2) are unpolarised, but component (3) is strongly linearly polarised because the scattering material in the disk is not distributed around the star in a spherically symmetric way.

Detection of linear polarisation from a star showing evidence of circumstellar material immediately allows one to conclude that the distribution of material around the star is not spherically symmetric, but is somehow flattened. This usually means that the matter is in a disk of some kind. Observing the *spectrum* of the linear polarisation of such a system can provide much further information about the relative importance of emission and scattering phenomena. For example, scattered (and thus polarised) photons may have to pass through part of the disk on their way to the observer, and so strong absorption in the disk can reduce the number of scattered photons escaping. Because of this, the wavelength dependence of polarisation can be used to constrain such model parameters as disk inclination to the line of sight, disk density, and disk thermal structure.

Some recent articles discussing the use of spectropolarimetry to understand the circumstellar material around hot stars include Vink *et al.* (2005); Davies *et al.* (2007); Carciofi *et al.* (2007); Halonen & Jones (2013).

There are some basic points to keep in mind about linear spectropolarimetry of stars.

• The observed fractional polarisation p due to scattering is of the order of $p \sim$ (fraction of scattered photons)·(geometric factor). Observation of non-zero linear polarisation, so that (geometric factor) $\neq 0$, reveals immediately that the system departs from spherical symmetry.

• Fractional polarisation values are often small in astrophysical situations. Circumstellar disks frequently lead to linear polarisation of order 1% or less. This implies that linear polarisation spectra usually require *a lot of photons*, of order $10^6 - 10^{10}$ photons per final (possibly binned or smoothed) pixel, in order to have measurement uncertainties that are small compared to the expected polarisation.

• The scattered photons usually require some detailed modelling in order to extract quantitatively useful information about a system.

3. Zeeman splitting and polarisation in spectral lines due to magnetic fields

One of the most widely exploited areas of spectropolarimetry is the use of (mostly circularly) polarised spectra to study magnetic fields in stars. This kind of measurement relies on polarisation produced by the Zeeman effect in spectral lines, which allows detection and (often) modelling of even very weak magnetic fields (down to the level of ~ 1 Gauss $\sim 10^{-4}$ Tesla).

The Zeeman effect arises from the fact that each energy level (state) of an atom has a magnetic moment, of the order of $e\hbar/2mc$, which is aligned with the total angular momentum vector **J**. Thus when an atom is placed in a magnetic field, each single state of unperturbed energy E_{i0} splits into $2J+1$ closely spaced states having energies (in cgs Gaussian units)

$$E_i = E_{i0} + g_i(e\hbar/2mc)Bm_J, \qquad (3.1)$$

where e, m, c and $\hbar$ are the electron charge and mass, the speed of light, and Planck's constant divided by 2π, B is the magnetic field strength, m_J is the magnetic quantum number (the projection of **J** on the field direction), and g_i is the "Landé factor", usually between 0 and 3, that varies from one energy level to another. See Eisberg & Resnick (1985) for more details

As a result, when an atom is placed in a magnetic field, a single transition betwen two energy levels E_i and E_f splits into a number of closely spaced spectral lines produced by all the allowed transitions ($m_{Ji} - m_{Jf} = 0$ or ± 1) between the "magnetic sublevels" of the two states. This is illustrated in Fig. 2 for the $\lambda4574$ Å line of SiIII, whose lower level (with $J = 1$) splits into three levels, while the upper level (with $J = 0$) does not split at all. Transitions with $\Delta m_J = \pm 1$ are called σ components, while those with $\Delta m_J = 0$ are π components.

Eq. (3.1) can easily be used to show that the wavelength shifts of the components produced by Zeeman splitting of a line at λ_0 are given by

$$\Delta\lambda_{ij} = \frac{e\lambda_0^2}{4\pi mc^2}(g_i m_{Ji} - g_j m_{Jj}) = 4.67\,10^{-13}\lambda_0^2 B(g_i m_{Ji} - g_j m_{Jj}), \qquad (3.2)$$

where in the second form of the equation, λ is in Å units and B is in Gauss. In the case of the $\lambda4574$ Å line of SiIII, for which the upper level has $J = 0$ and hence $m_{Jj} = 0$, and for which the lower level has $g_i = 2.0$, the σ components (due to lower levels with $m_{Ji} = \pm 1$) are separated from the π component (with $m_{Ji} = 0$) by $\Delta\lambda = \pm 2.0 \cdot 4.67\,10^{-13} \cdot 4574^2 \cdot B$. Thus measurement of Zeeman splitting can be used to deduce magnetic field strength.

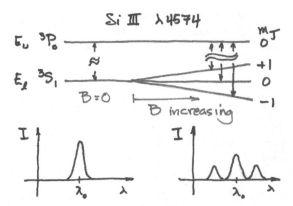

Figure 2. Splitting of the levels involved in the transition producing the 4574 Å line of Si III. The upper part of the figure shows how the energy levels split as the magnetic field B is increased; the two panels below sketch the appearance of this line (in emission) with $B = 0$ (left) and $B \neq 0$ (right).

A further important feature of Zeeman splitting is the *polarisation* properties of the subcomponents of a spectral line formed in a magnetic field. Look again at Fig. 2. If the field splitting the line is parallel to the line of sight (a *longitudinal field*), the spectral line component (or components) arising from transitions in which m_J changes by $+1$ are *circularly polarised* in one sense, the transitions with $\Delta m_J = -1$ are circularly polarised in the opposite sense, and the component with $\Delta m_J = 0$ have zero intensity. In a *transverse field*, the line components formed from transitions with $\Delta m_J = \pm 1$ are *linearly polarised* normal to the field, while the components formed from $\Delta m_J = 0$ transitions are linearly polarised parallel to the field.

These effects of a magnetic field on atomic transitions provide us with tools for measuring the strength of a magnetic field present in a star's atmosphere. This is because the splitting of spectral lines, and the polarisation properties of the various components, alter the absorption lines in the stellar spectrum. If the field is so large that the separation of $\Delta m_J = \pm 1$ components (called σ components) from the central $\Delta m_J = 0$ (π components) is larger than any of the other broadening mechanisms affecting the spectral line, the magnitude of the magnetic field strength $\langle B \rangle$, averaged over the visible hemisphere of the star, can be determined by using Eq 3.1. Such splitting is illustrated for the Si III multiplet (2) triplet in Fig. 3.

The polarisation produced by the Zeeman effect is also illustrated in this figure. The lower spectrum, showing the net circular polarisation spectrum (intensity of right circularly polarised light minus the intensity of left circularly polarised light, divided by total intensity, as a function of wavelength), shows a large spike of circular polarisation in each of the outer (σ) components of the split spectral line, produced by transitions with $\Delta m_J = \pm 1$. Note that the net circular polarisation has opposite sign in the two σ groups. Because circular polarisation in the σ line components is produced only by the longitudinal component of the magnetic field, the intensity of the circular polarisation spikes allows us to estimate the strength of the line-of-sight (or longitudinal) component $\langle B_z \rangle$ of the magnetic field, averaged over the visible hemisphere.

However, in most cases the magnetic field is too weak, and the rotational broadening too strong, to produce clear Zeeman splitting of spectral lines. But even in stars in which the weak field does not lead to obvious splitting or at least deformation of the line intensity, the circular polarisation is often quite readily detected and measured. This is possible because the Zeeman effect effectively leads to an absorption line as seen in

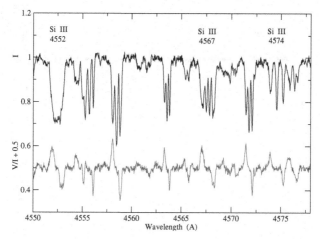

Figure 3. The upper spectrum shows Zeeman splitting in an intensity spectrum of the very strongly magnetic Bp star HD 215441. The lower spectrum shows the net circular polarisation (shifted by +0.5 for clarity). Notice that the two outer Zeeman components of each split line in the upper spectrum are strongly circularly polarised in the lower spectrum, while the central components show almost no circular polarisation.

one circular polarisation having a mean wavelength that is slightly different than the mean wavelength of the same line observed in the opposite sense of circular polarisation. Because we can measure wavelength (radial velocity) differences between spectral lines with an uncertainty that is much smaller than the width of the line, we can still detect the circular polarisation signature of a magnetic field in at much smaller strengths than those needed to produce visible splitting. This effect is illustrated in Fig. 4.

In the case illustrated in Fig. 4, the circular polarisation signature of a field is visible in each spectral line, but it is clear that if the field were ten or twenty times smaller, the signature would be lost in the noise. We can nevertheless often detect such tiny polarisation signatures by noticing the obvious similarity of the polarisation signal in the different lines. This makes it possible to average both intensity and polarisation profiles of many spectral lines to bring up a detectable signal. This process is often called "least squares deconvolution", or LSD.

A more complete discussion of magnetic field meaurement using the Zeeman effect is provided by Landstreet (2009) and by Donati & Landstreet (2009).

4. Measuring the polarisation

4.1. *The Stokes vector*

We now turn to the practicalities of actually measuring polarisation. We have seen that polarisation of light means that the direction of oscillation of the electric field in a beam of light has some preferential orientation and/or some net rotation. We can describe the polarisation state of the light "completely" by reporting the results of four experiments: measurements of (1) the total intensity of the beam; (2) the difference between the intensity of beam as measured through a perfect linear polariser oriented (say) vertically (0°) and one oriented at 90° to the orientation of the first measurement; (3) the difference between the intensity observed when the polariser is oriented at 45° and when it is at 135°; and (4) the difference between the intensity measured through a circular polariser that passes only right circularly polarised light and one that passes only left circularly polarised light. The last three measurements are often normalised to the total beam

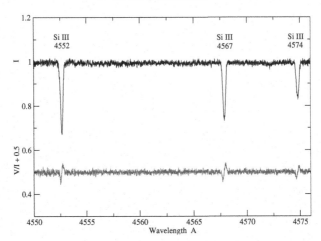

Figure 4. The upper spectrum shows an intensity spectrum of the B2p magnetic star HD 96446, with no obvious Zeeman splitting. The lower spectrum shows the net circular polarisation (shifted by +0.5 for clarity). Although no Zeeman splitting is visible in the intensity spectrum, the circular polarisation signature of a field is obvious in each spectral line.

intensity and expressed as percentages or decimal fractions. The three measurements of polarisation compare "orthogonal" polarisation state to one another.

The results of these four measurements are usually written as a vector $\{I, Q, U, V\}$, known as the Stokes vector.

The quantities in the Stokes vector usually vary with wavelength. The first component, $I(\lambda)$, is of course simply the usual intensity spectrum of a star. The other three components report how the polarisation of that light varies with wavelength. For example, a circular polarisation measurement using ESPaDOnS at the CFHT or FORS at ESO, to measure a stellar magnetic field, will result in two spectra: $I(\lambda)$ (the usual intensity spectrum) and $V(\lambda)/I(\lambda)$ (the normalised circular polarisation spectrum; see Figs 3 and 4).

4.2. *How polarisation is measured*

In practice, polarisation spectra – $Q(\lambda)$, $U(\lambda)$ or $V(\lambda)$ – are measured using normal spectrographs which have been modified by the addition of some specialised optical elements that allow one to separately meaure the intensity of the beam in two orthogonal polarisation states. These two resulting spectra are then summed to form $I(\lambda)$, and the difference provides $Q(\lambda)$, $U(\lambda)$, or $V(\lambda)$. It would require more space than is available in this brief review to describe how these polarising optics work, and how they are actually fabricated, so I will simply describe operationally what they do. Anyone interested in more details should consult the really excellent little book "Polarized Light" by Shurcliff & Ballard, a text from the 1960s US Commission on College Physics series.

A simple system for measuring a linear polarisation spectrum would be to place a polarizing beam splitter known as a Wollaston prism in the collimated beam of a single-order (low-dispersion) spectrograph. A Wollaston prism splits an incident light beam into two orthogonal linearly polarised components, which leave the prism in different directions. If the beam splitter splits the beam perpendicularly to the dispersion direction, the result would be two spectra side by side. One would register the intensity $I_0(\lambda)$ of starlight polarised parallel to an axis at (say) 0° on the sky, and the other would register the intensity $I_{90}(\lambda)$ of starlight in the orthogonal direction. From these two spectra we could obtain the $I(\lambda) = I_0 + I_{90}$ and $Q(\lambda) = I_0 - I_{90}$ spectra.

Notice that we want to measure the intensities I_0 and I_{90} *simultaneously* in order to avoid having to compare spectra taken at different times (for example with a single rotating polariser just behind the entrance slit), with different guiding, seeing and transparency conditions, and perhaps different flexure. Such sequential measurements could easily lead to difference spectra of very low accuracy.

However, the scheme I have just described is not very practical. It requires rotating the whole spectrograph by 45° around the telescope axis in order to measure the Stokes U component! Instead, we make use in practice of the properties of retarding wave plates to simplify the operation, increase the capabilities, and reduce the cost of our polarimeter.

The important feature of retarder wave-plates is that they allow us to convert polarisation in one of the Stokes components (say linear polarisation parallel to the 0° direction) into some other polarisation form that might be easier to measure. They also allow us to exchange polarisation states so that we can have, for example, I_0 in the upper spectrum on the CCD and I_{90} in the lower spectrum, and then exchange the two beams so that I_0 is in the lower CCD spectrum and I_{90} in the upper. If we then compute the polarisation spectrum as the average of meaurements made in these two settings, a number of kinds of systematic errors cancel out.

It is easy to understand how wave plates affect the polarisation state of a light beam. Basically, a wave plate is a device that resolves an electromagnetic wave into two components parallel to two orthogonal axes on the face of the waveplate. As the wave travels through the wave plate, the components of the incoming wave along these two axes travel at different speeds. The phase of one component is thus shifted with respect to the other, and this changes its polarisation state.

An example can make this clear. Suppose a plane-polarised wave with its plane of polarisation at 45° to each of the two wave plate axes enters the wave plate. Where the wave enters, the wave plate resolves the electric vector into components along each of the two principal axes. Since the wave is linearly polarised at 45° to these axes, the two wave components oscillate in phase. If the wave plate retards one component by 1/4-wave relative to the other, when the wave exits the wave plate, the maximum of the electric vector will rotate steadily in a circle from one axis to the other. The outcoming wave will now be completely circularly polarised.

If the retardation is 1/2-wave, the components along the two principal axies will emerge 180° out of phase. The wave will emerge linearly polarised, but aligned at −45° to the two axes instead of +45°. Effectively a 1/2-wave plate reflects the plane of polarisation around one of the two axes, thus rotating its plane of polarisation. This property allows us to easily exchange which of two orthogonal wave components falls onto each of the two spectra on the detector.

Thus a polarimetric analyser that can measure all the polarisation components Q, U and V can be constructed by using rotatable and interchangeable 1/4– and 1/2–wave plates followed by a beam-splitting polariser. This system is usually inserted into the spectrograph beam as far up the beam as possible, often before the light reaches the spectrograph entrance slit.

A much fuller discussion of the theory of polarisation measurement, and of data treatment, is provided by Bagnulo *et al.* (2009).

5. Precision, accuracy, and caution

As discussed above, polarisation is now almost always determined by computing the difference in intensity of two beams whose intensities are proportional to the two orthogonal polarisation states being measured (e.g. right and left circularly polarisation), which are recorded simultaneously, usually on a CCD or similar detector. Then the association of each orthogonal polarisation state with a light beam is reversed with the aid of a rotatable wave plate: if the upper CCD spectrum measured the intensity of right circularly polarised light in the first measurement, in the second it measures the intensity of left circular polarised light. By averaging the two difference spectra (after changing the sign of one), most kinds of systematic error due to different transmissivities of the two beams cancel out. The result is a measurement of one Stokes polarisation component.

If a small slit or aperture is involved (as it usually is in high resolution spectropolarimetry), seeing and guiding variations may still lead to small variations in the differences between the overall continuum level of one beam relative to the other, but the relative wavelength scales of the two beams are very well determined. Thus when the continuum polarisation is forced to zero, the *difference* between line profiles in the two orthogonal polarisation states, particularly any small wavelength shifts of one profile relative to the other, can be determined with much higher precision than the actual wavelength position or shape of either line profile separately. In practice, it is possible to detect and measure polarisations of substantially less than 0.01% if the data have sufficiently high signal-to-noise ratios. (Note that measuring such small polarisations requires very high photon counts per pixel, either in the original spectrum or at least in the LSD averaged spectral line.)

However, although such high precisions can be obtained with current facility spectropolarimeters, as in any precision work it is essential to be cautious about taking measured results at face value. It is extremely important to carry out polarisation observations of suitable standards such as stars showing no polarisation or known polarisation, and to compute various "check sums" which, if all is working correctly, will result in "null polarisation" spectra showing no significant signal. Only by studying the behaviour of the spectropolarimeter carefully and critically can one determine the actual precision that can be achieved.

An example of the problems that can arise from acceptance of computed precisions at face value is shown by data obtained with the FORS spectrograph at the ESO VLT. This instrument can be used as a very efficient low-resolution spectropolarimeter which can reliably detect magnetic fields of a few hundred G in hot stars of $m_V \sim 10$ or even fainter. However, Bagnulo *et al.* (2012, 2013) have shown by careful examination of the entire data set of magnetic measurements obtained with FORS1 that the actual polarimetric precision that can be achieved with this instrument is not quite as high as one would expect from the photon counts. Excess noise arises from several sources, particularly from seeing fluctuations in very short exposures, guiding variations, instrument flexures, and cosmic rays.

The presence of these extra noise sources have led to reports of numerous magnetic field detections at the $3 - 5\sigma$ level in large surveys of Be stars (Hubrig *et al.* 2009b), of β Cep and SPB pulsating B stars (Hubrig *et al.* 2009a) of O stars (Hubrig *et al.* 2008), and of hot subdwarfs (O'Toole *et al.* 2005). Reanalysis of the same data using different reduction algorithms and more conservative signal-to-noise estimates (Bagnulo *et al.* 2012; Landstreet *et al.* 2012), as well as reobservation of a number of the stars with reported fields with the ESPaDOnS spectropolarimeter at CFHT (Shultz *et al.* 2012), have shown that more than 80% of the magnetic field detections reported by these FORS1 surveys are spurious. The revised data show magnetic field detections in a much smaller fraction (or even none) of the observed stars than originally reported. Note, however, that with more conservative treatment and interpretation of the data, the results from these surveys are still extremely useful.

The moral is quite simple: actually achieving extremely high precision requires extreme care and repeated verification.

To obtain the most accurate and precise magnetic field measurements of sharp line stars, there is a big advantage to using a static, stabilised high-dispersion spectrographs such as ESPaDOnS at the CFHT, NARVAL at the TBL, or HARPSpol at ESO La Silla. Such spectropolarimeters take full advantage of the precision that can be obtained from narrow line profiles, and generally lack at least some of the instabilities of Cassegrain instruments. With these instruments it is possible to confidently detect and measure magnetic fields $\langle B_z \rangle$ in early B stars of the order of $50 - 100$ G or even less if enough photons are available.

For stars with broad lines (say $v \sin i \geqslant 200$ km s^{-1}), low-resolution spectropolarimeters, such as ESO's FORS or ISIS at the WHT, are equally powerful choices, and the advantage brought by the large collecting area of a VLT can be very useful. For such instruments, the practical field detection limit is 200–300 G.

With today's facility instruments, it is completely practical for anyone to make spectropolarimetric observations. However, as with any unfamiliar technology, it is essential to learn about the problems and limitations of the techniques before pushing to the very highest precision limits of the instrument.

References

Bagnulo, S., Fossati, L., Kochukhov, O., & Landstreet, J. D. 2013, *A&A* 559, A103

Bagnulo, S., Landolfi, M., Landstreet, J. D., *et al.* 2009, *PASP* 121, 993

Bagnulo, S., Landstreet, J. D., Fossati, L., & Kochukhov, O. 2012, *A&A* 538, A129

Carciofi, A. C., Magalhães, A. M., Leister, N. V., Bjorkman, J. E., & Levenhagen, R. S. 2007, *ApJ (Letters)* 671, L49

Davies, B., Vink, J. S., & Oudmaijer, R. D. 2007, *A&A* 469, 1045

Donati, J.-F. & Landstreet, J. D. 2009, *ARA&A* 47, 333

Eisberg, R. & Resnick, R. 1985, *Quantum Physics of Atoms, Molecules, Solids, Nuclei, and Particles, 2nd Edition*

Halonen, R. J. & Jones, C. E. 2013, *ApJ* 765, 17

Hubrig, S., Briquet, M., De Cat, P., *et al.* 2009a, *Astronomische Nachrichten* 330, 317

Hubrig, S., Schöller, M., Savanov, I., *et al.* 2009b, *Astronomische Nachrichten* 330, 708

Hubrig, S., Schöller, M., Schnerr, R. S., *et al.* 2008, *A&A* 490, 793

Landstreet, J. D. 2009, in C. Neiner & J.-P. Zahn (eds.), *EAS Publications Series*, Vol. 39 of *EAS Publications Series*, pp 1–20

Landstreet, J. D., Bagnulo, S., Fossati, L., Jordan, S., & O'Toole, S. J. 2012, *A&A* 541, A100

O'Toole, S. J., Jordan, S., Friedrich, S., & Heber, U. 2005, *A&A* 437, 227

Shultz, M., Wade, G. A., Grunhut, J., *et al.* 2012, *ApJ* 750, 2

Vink, J. S., Harries, T. J., & Drew, J. E. 2005, *A&A* 430, 213

Discussion

NIEVA: Will the high precision be biased towards sharp-lined stars?

LANDSTREET: Yes, this is a consequence of how the measurements are made. The field of a star with many sharp and deep lines can be measured more precisely than the field of a similar star with few broad and shallow lines, just as abundances or radial velocities can be more precisely measured for the sharp-line star.

AERTS: Could you explain how to unravel line-profile shapes due to pulsations and due to a magnetic field if they act together? *i.e.* what is the effect of pulsation on your magnetic modelling?

LANDSTREET: This is discussed in the talk of Coralie Neiner.

HERRERO: Instruments and telescopes introduce additional polarisation through reflections. Is it better to use standards to calibrate the observations or theoretical knowledge about the instrument+telescope behaviour?

LANDSTREET: One should use all available means to understand instrumental polarisation and phase shifts. However, most polarimetric analysers today are at Cassegrain focus, where the axial symmetry and nearly normal reflections from primary and secondary mirrors introduce almost no instrumental polarisation or phase shifts. After the polarisation analyser optics, the polarisation measurement becomes a comparison of two *intensities*, and any polarising properties of the optical train are unimportant.

PULS: Could you comment on potential depolarisation effects by the ISM (inhomogeneities, B-fields, etc) for the stars we are interested in?

LANDSTREET: This is an interesting possibility, but I do not know of any work that has considered it.

John Landstreet

New windows on massive stars: asteroseismology, interferometry, and spectropolarimetry
Proceedings IAU Symposium No. 307, 2014
G. Meynet, C. Georgy, J. H. Groh & Ph. Stee, eds.

© International Astronomical Union 2015
doi:10.1017/S1743921314007029

Magnetic Field - Stellar Winds Interaction

Asif ud-Doula

Penn State Worthington Scranton, Dunmore, PA 18512, USA
email: asif@psu.edu

Abstract. As per the recent study by the MiMeS collaboration, only about 10% of massive stars possess organized global magnetic fields, typically dipolar in nature. The competition between such magnetic fields and highly non-linear radiative forces that drive the stellar winds leads to a highly complex interaction. Such an interplay can lead to a number of observable phenomena, e.g. X-ray, wind confinement, rapid stellar spindown. However, due to its complexity, such an interaction cannot usually be modeled analytically, instead numerical modeling becomes a necessary tool. In this talk, I will discuss how numerical magnetohydrodynamic (MHD) simulations are employed to understand the nature of such magnetized massive star winds.

Keywords. stars: early-type, stars: magnetic fields, stars: mass loss, stars: rotation, stars: winds, outflows, X-rays: stars

1. Introduction

Massive hot stars are so luminous that their light can impart enough momentum to lift material off the stellar surface leading to a high mass loss. Such a mechanism is well described by the pioneering work of Castor *et al.* (1975, known as CAK). They have shown that scattering of photons off lines of partially ionized metals can lead to an outward force that is stronger than the inward pull of gravity.

Such outflows are expected to be steady, smooth and spherically symmetric. However, there is evidence that massive star winds have extensive structures and variability on a range of spatial and temporal scales. Relatively small-scale, stochastic structure seems most likely a natural result of the strong, intrinsic instability of the line-driving mechanism itself (Owocki 1994; Feldmeier 1995). But larger-scale structure – e.g. as evidence by explicit UV line profile variability in even low signal-to-noise IUE spectra (Kaper *et al.* 1996; Howarth & Smith 1995) – seems instead likely to be the consequence of wind perturbation by processes occurring in the underlying star. For example, the photospheric spectra of many hot stars show evidence of radial and/or non-radial pulsation, and in a few cases there is evidence linking this with observed variability in UV wind lines (Telting *et al.* 1997; Mathias *et al.* 2001).

An alternate scenario is that, in at least some hot stars, surface magnetic fields could perturb, and perhaps even channel, the wind outflow, leading to rotational modulation of wind structure that is diagnosed in UV line profiles, and perhaps even to magnetically confined wind-shocks with velocities sufficient to produce the relatively hard X-ray emission seen in some hot-stars.

However, the interaction of stellar winds with magnetic fields is highly complex, and it cannot be easily modeled analytically. As such, numerical tools must be employed. In what follows, I briefly discuss how numerical computations are used to study magnetized line-driven winds.

2. Magneto-hydrodynamic Equations

Massive star winds are mainly composed of fully ionized hydrogen and helium, and trace amounts of partially ionized metals ($\sim 10^{-4}$). These winds are dense enough that the mean free paths between collisions are relatively small, and the plasma can be treated as a single fluid. The time-dependent MHD (magnetohydrodynamic) equations governing the system include the conservation of mass,

$$\frac{D\rho}{Dt} + \rho \nabla \cdot \mathbf{v} = 0, \tag{2.1}$$

and the equation of motion

$$\rho \frac{D\mathbf{v}}{Dt} = -\nabla p + \frac{1}{4\pi}(\nabla \times \mathbf{B}) \times \mathbf{B} - \frac{GM\hat{\mathbf{r}}}{r^2} + \mathbf{g}_{\text{external}}, \tag{2.2}$$

where ρ, p, and $\mathbf{v}$ are the mass density, gas pressure, and velocity of the fluid flow, and $D/Dt = \partial/\partial t + \mathbf{v} \cdot \nabla$ is the advective time derivative. The gravitational constant G and stellar mass M set the radially directed ($\hat{\mathbf{r}}$) gravitational acceleration. The term $\mathbf{g}_{\text{external}}$ represents the total external force that may include the centrifugal force. In this case it is the line-force and the force due to scattering of the stellar luminosity L by the free electron opacity κ_e. The magnetic field $\mathbf{B}$ is constrained to be divergence free

$$\nabla \cdot \mathbf{B} = 0, \tag{2.3}$$

and, under the assumption of an idealized MHD flow with infinite conductivity, its inductive generation is described by

$$\frac{\partial \mathbf{B}}{\partial t} = \nabla \times (\mathbf{v} \times \mathbf{B}). \tag{2.4}$$

In addition we solve an explicit equation for conservation of energy,

$$\rho \frac{D}{Dt} \left(\frac{e}{\rho} \right) = -p \nabla \cdot v + \Lambda_{\text{cool}}. \tag{2.5}$$

Here the ideal gas law is used to compute the pressure, p, and Λ_{cool} is radiative cooling.

3. The Wind Magnetic Confinement Parameter

In an interplay between magnetic field and stellar wind, the dominance of the field is determined how strong the field is relative to the wind. In order to understand the competition between these two, let us first define a characteristic parameter for the relative effectiveness of the magnetic fields in confining and/or channeling the wind outflow. Specifically, consider the ratio between the energy densities of field vs. flow as discussed by ud-Doula & Owocki (2002),

$$\eta(r, \theta) \equiv \frac{B^2/8\pi}{\rho v^2/2} \approx \frac{B^2 r^2}{\dot{M} v}$$
$$= \left[\frac{B_*^2(\theta) R_*^2}{\dot{M} v_\infty} \right] \left[\frac{(r/R_*)^{-2n}}{1 - R_*/r} \right], \tag{3.1}$$

where the latitudinal variation of the surface field has the dipole form given by $B_*^2(\theta) = B_0^2(\cos^2\theta + \sin^2\theta/4)$. In general, a magnetically channeled outflow will have a complex flow geometry, but for convenience, the second equality in eq. (3.1) simply characterizes the wind strength in terms of a spherically symmetric mass loss rate $\dot{M} = 4\pi r^2 \rho v$. The third equality likewise characterizes the radial variation of outflow velocity in terms of the phenomenological velocity law $v(r) = v_\infty(1 - R_*/r)$, with v_∞ the wind terminal speed; this equation furthermore models the magnetic field strength decline as a power-law in radius, $B(r) = B_*(R_*/r)^{(n+1)}$, where, e.g., for a simple dipole $n = 2$.

With the spatial variations of this energy ratio thus isolated within the right square bracket, we see that the left square bracket represents a dimensionless constant that characterizes the overall relative strength of field vs. wind. Evaluating this in the region of the magnetic equator ($\theta = 90°$), where the tendency toward a radial wind outflow is in most direct competition with the tendency for a horizontal orientation of the field, one can thus define a equatorial 'wind magnetic confinement parameter',

$$\eta_* \equiv \frac{B_*^2(90°)R_*{}^2}{\dot{M}v_\infty} = 0.4\,\frac{B_{100}^2\,R_{12}^2}{\dot{M}_{-6}\,v_8}. \tag{3.2}$$

where $\dot{M}_{-6} \equiv \dot{M}/(10^{-6}\,M_\odot/\text{yr})$, $B_{100} \equiv B_0/(100\ \text{G})$, $R_{12} \equiv R_*/(10^{12}\ \text{cm})$, and $v_8 \equiv v_\infty/(10^8\ \text{cm/s})$. As these stellar and wind parameters are scaled to typical values for an OB supergiant, e.g. ζ Pup, the last equality in eq. (3.2) immediately suggests that for such winds, significant magnetic confinement or channeling should require fields of order ~ 100 G. By contrast, in the case of the sun, the much weaker mass loss ($\dot{M}_\odot \sim 10^{-14}\,M_\odot/\text{yr}$) means that even a much weaker global field ($B_0 \sim 1$ G) is sufficient to yield $\eta_* \simeq 40$, implying a substantial magnetic confinement of the solar coronal expansion. But in Bp stars the magnetic field strength can be of order kG with $\dot{M}_\odot \sim 10^{-10}\,M_\odot/\text{yr}$ leading $\eta_* \leqslant 10^6$. Thus, the confinement in Bp stars is very extreme.

It should be emphasized that $\dot{M}$ used in the above formalism is predicted theoretical value obtained for a spherically symmetric non-magnetic wind.

3.1. *Alfvén Radius*

What determines the extent of the effectiveness of magnetic, is the Alfvén radius, R_A, where flow and Alfvén velocities are equal. This can be derived from eq. 3.1 where the second square bracket factor shows the overall radial variation; n is the power-law exponent for radial decline of the assumed stellar field, e.g. $n = 2$ for a pure dipole, and β is the velocity-law index, with typically $\beta \approx 1$. For a star with a non-zero field, we have $\eta_* > 0$, and so given the vanishing of the flow speed at the atmospheric wind base, this energy ratio always starts as a large number near the stellar surface, $\eta(r \rightarrow R_*) \rightarrow \infty$. But from there outward it declines quite steeply, asymptotically as r^{-4} for a dipole, crossing unity at the Alfvén radius defined implicitly by $\eta(R_A) \equiv 1$.

For a canonical $\beta = 1$ wind velocity law, explicit solution for R_A along the magnetic equator requires finding the appropriate root of

$$\left(\frac{R_A}{R_*}\right)^{2n} - \left(\frac{R_A}{R_*}\right)^{2n-1} = \eta_* , \tag{3.3}$$

which for integer $2n$ is just a simple polynomial, specifically a quadratic, cubic, or quartic for $q = 1$, 1.5, or 2. Even for non-integer values of $2n$, the relevant solutions can be approximated (via numerical fitting) to within a few percent by the simple general expression,

$$\frac{R_A}{R_*} \approx 1 + (\eta_* + 1/4)^{1/(2n)} - (1/4)^{1/(2n)} . \tag{3.4}$$

For weak confinement, $\eta_* \ll 1$, we find $R_A \rightarrow R_*$, while for strong confinement, $\eta_* \gg 1$, we obtain $R_A \rightarrow \eta_*^{1/(2n)} R_*$. In particular, for the standard dipole case with $n = 2$, we expect the strong-confinement scaling $R_A/R_* \approx \eta_*^{1/4}$.

Clearly R_A represents the radius at which the wind speed v exceeds the local Alfvén speed V_A. It also characterizes the maximum radius where the magnetic field still dominates over the wind. For chemically peculiar Ap/Bp stars where stellar fields are of order kG, the $\eta_* \gg 1$, e.g. for σ Ori E it is about 10^7, implying an Alfvén radius $\sim 60R_*$. Thus, in Bp (and Ap) stars wind is trapped to large radii creating extensive magnetospheres.

4. MHD Simulations

The initial magnetohydrodynamic (MHD) simulations by ud-Doula & Owocki (2002) assumed, for simplicity, that radiative heating and cooling would keep the wind outflow nearly isothermal at roughly the stellar effective temperature. The simulations studied the dynamical competition between field and wind by evolving MHD simulations from an initial condition when a dipole magnetic field is suddenly introduced into a previously relaxed, one-dimensional spherically symmetric wind.

Immediately after the introduction of the field, the dynamic interplay between the wind and the field leads to two distinct regions. Along the polar region, the wind streams radially outward freely, stretching the field lines into a radial configuration, as can be inferred from the left panel of illustrative Fig. 1. If the field is strong enough, around the magnetic equator a region of closed magnetic loops is formed wherein the flow from opposite hemispheres collides to make strong shocks, quite similar to what was predicted in the semi-analytic, fixed-field models of Babel & Montmerle (1997a). The shocked material forms dense disk-like structure which is opaque to line-driving. But, its support against gravity by the magnetic tension along the convex field lines is inherently unstable, leading to a complex pattern of fall back along the loop lines down to the star, again as suggested by the left panel of Fig. 1.

Note that even for weak field models with moderately small confinement, $\eta_* \leqslant 1/10$, the field still has a noticeable global influence on the wind, enhancing the density and decreasing the flow speed near the magnetic equator.

5. X-ray Emission from Massive Stars

But to model the actual X-ray emission from shocks that form from the magnetic channeling and confinement, subsequent efforts (see ud-Doula 2003; Gagné et al. 2005) have relaxed this simplification to include a detailed energy equation that follows the radiative cooling of shock-heated material. In particular, building upon the initial suggestion by Babel & Montmerle (1997b) that such *Magnetically Confined Wind Shocks* (MCWS) might explain the relatively hard X-ray spectrum observed by *Rosat* for the O7V star θ^1 Ori C, Gagné et al. (2005) have applied such energy-equation, MHD simulations toward a detailed, dynamical model of the more extensive *Chandra* X-ray observations of this star. Based on the spectropolarimetric measurement (see Donati et al. 2002) of a ca. 1100 G field for θ^1 Ori C, combined with empirical and theoretical estimates of the wind momentum and stellar radius, the simulations assumed a moderately large magnetic confinement parameter, $\eta_* \approx 10$.

Overall, the associated X-ray emission of this MHD model matches quite well the key properties of the *Chandra* observations for θ^1 Ori C (Gagné et al. 2005), including: the relatively hard X-ray spectrum that arises from the high post-shock temperatures $T \sim 20 - 30$ MK; the relative lack of broadening or blue-shift from X-ray lines emitted from the nearly static, shock-heated material; and the X-ray light curve eclipse that stems from the moderate source radius $r \approx 1.5R_*$ for the bulk of the X-ray emission.

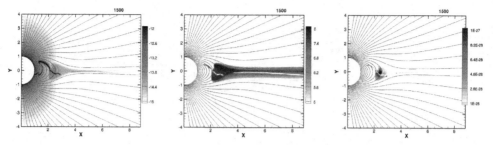

Figure 1. Color plots of log density (left) and log temperature (middle) for arbitrary snapshot of structure in the standard model with $\eta_* = 100$. The right panel plots the proxy X-ray emission XEM_{T_x} (weighted by the radius r), on a *linear* scale for a threshold X-ray temperature $T_x = 1.5\,\mathrm{MK}$.

5.1. *Shock Retreat*

In a further development ud-Doula *et al.* (2014) examine the effects of radiative cooling on X-ray emission from MCWS in massive stars with radiatively driven stellar winds. They show that for stars with high enough mass loss to keep the shocks radiative, the MHD simulations indicate a linear scaling of X-ray luminosity with mass-loss rate; but for lower luminosity stars with weak winds, X-ray emission is reduced and softened by a *shock retreat* resulting from the larger post-shock cooling length, which within the fixed length of a closed magnetic loop forces the shock back to lower pre-shock wind speeds. The basic idea behind such a shock retreat is illustrated in Fig. 2.

Through a parameter study of cooling, they show that overall scalings of time-averaged X-rays in the numerical MHD simulations are well matched by the L_X computed from a semi-analytic "X-ray Analytic Dynamic Magnetosphere" (XADM) model. However, the values of L_X are about a factor 5 lower in the MHD models, mostly likely reflecting an overall inefficiency of X-ray emission from the repeated episodes of dynamical infall.

6. 3D MHD

To get a full picture of the wind-field interaction, one needs ultimately full 3D MHD simulations. Using the specific parameters chosen to represent the prototypical slowly rotating magnetic O star θ^1 Ori C, for which centrifugal and other dynamical effects of rotation are negligible ud-Doula *et al.* (2013) have computed the first fully 3D MHD model of its wind. The computed global structure in latitude and radius resembles that found in previous 2D simulations, with unimpeded outflow along open field lines near the magnetic poles, and a complex equatorial belt of inner wind trapping by closed loops near the stellar surface, giving way to outflow above the Alfvén radius. In contrast to this previous 2D work, the 3D simulation described here now also shows how this complex structure fragments in azimuth, forming distinct clumps of closed loop infall within the Alfvén radius, transitioning in the outer wind to radial spokes of enhanced density with characteristic azimuthal separation of $15° - 20°$. Applying these results in a 3D code for line radiative transfer, they show that emission from the associated 3D 'dynamical magnetosphere' (DM) matches well the observed Hα emission seen from θ^1 Ori C, fitting both its dynamic spectrum over rotational phase and the observed level of cycle-to-cycle stochastic variation. Comparison with previously developed 2D models for the Balmer emission from a dynamical magnetosphere generally confirms that time averaging over 2D snapshots can be a good proxy for the spatial averaging over 3D azimuthal wind

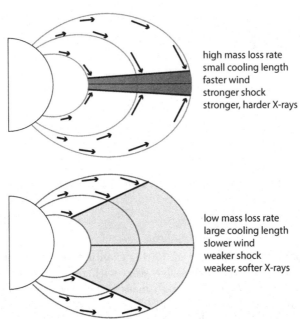

high mass loss rate
small cooling length
faster wind
stronger shock
stronger, harder X-rays

low mass loss rate
large cooling length
slower wind
weaker shock
weaker, softer X-rays

Figure 2. Schematic illustration of the "shock retreat" from inefficient cooling associated with a lower mass loss rate $\dot{M}$, showing a hemispheric, planar slice of a stellar dipole magnetic field. Wind outflow driven from opposite foot-points of closed magnetic loops is channeled into a collision near the loop top, forming magnetically confined wind shocks (MCWS). For the high $\dot{M}$ case in the upper panel, the efficient cooling keeps the shock-heated gas within a narrow cooling layer, allowing the pre-shock wind to accelerate to a high speed and so produce strong shocks with strong, relatively hard X-ray emission. For the low $\dot{M}$ case in the lower panel, the inefficient cooling forces a shock retreat down to lower radii with slower pre-shock wind, leading to weaker shocks with weaker, softer X-ray emission.

structure, as illustrated in Fig. 3. Nevertheless, fully 3D simulations will still be needed to model the emission from magnetospheres with non-dipole field components.

7. Role of Rotation

As noted above, MHD simulations of ud-Doula & Owocki (2002); ud-Doula (2003) indicate that a dipole magnetic field can confine the flow within closed loops that extend out to about the Alfvèn radius, $R_A \approx \eta_*^{1/4} R_*$. In rotating models such closed loops tend also to keep the outflow in rigid-body rotation with the underlying star, and so R_A also roughly represents the radius of maximum rotational spin-up of the wind azimuthal speed.

To further characterize such rotational effects, one can define a *Keplerian corotation radius* R_K at which rigid-body rotation would yield an equatorial centrifugal acceleration that just balances the local gravitational acceleration from the underlying star,

$$\frac{GM}{R_K^2} = \frac{v_\phi^2}{R_K} = \frac{V_{\rm rot}^2 R_K}{R_*^2}, \tag{7.1}$$

where $V_{\rm rot}$ is the stellar surface rotation speed at the equator. This can be solved to yield

$$R_K = W^{-2/3} R_*, \tag{7.2}$$

where $W \equiv V_{\rm rot}/V_{\rm crit}$, with $V_{\rm crit} \equiv \sqrt{GM/R_*}$ the critical rotation speed.

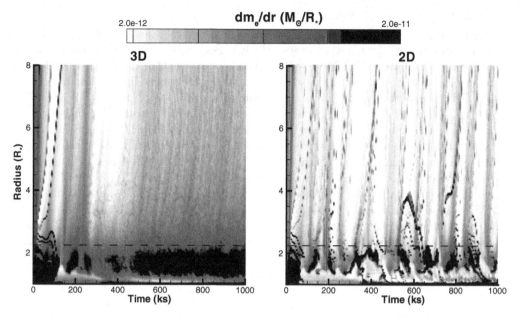

Figure 3. Equatorial mass distribution dm_e/dr (in units $M_\odot/R_*$) for the azimuthally averaged 3D model (left) and for a corresponding 2D model (right), plotted versus radius and time, with the dashed horizontal line showing the Alfvén radius $R_A \approx 2.23R_*$. Following a similar adjustment to the initial condition, the long-term evolution of the 2D models is characterized by a complex pattern of repeated strings of infall, whereas the 3D azimuthally averaged model settles into a relatively smooth asymptotic state characterized by an enhanced mass near and below the Alfvén radius.

The Alfvén radius R_A essentially defines the extent of the magnetosphere. If the star is a slow rotator and its $R_K > R_A$ then all the material within co-rotating region does not have enough centrifugal support to remain in orbit. As such, they rain down back onto the stellar surface leading to dynamical magnetosphere with specific implications about, for instance, Hα emission (for more details see Sundqvist *et al.* 2012).

On the other hand, if $R_A > R_K$ as is the case for fast rotating chemically peculiar magnetic Bp stars, the material trapped within above R_K is forced into co-rotation leading to *rigidly rotating magnetosphere*. The prototypical Bp star σ Ori E is a prime example of such a case. However, relatively low mass loss rate ($\dot{M} \sim 10^{-10} M_\odot$ yr^{-1} coupled with extremely strong magnetic field ($B \sim 10^4$ G), leads to very high magnetic confinement ($\eta_* \sim 10^7$). This in turn implies very high Alfvén speed, and thus very short Courant timestep, needed to preserve numerical stability, rendering direct MHD modelling impractical. As such, other methods must be employed to model such stars. These are discussed in the next two sections.

One other obvious impact of magnetic field in stellar wind is the angular momentum loss. The simulations done in a series of papers (see ud-Doula *et al.* 2006, 2008, 2009) reveal that magnetic field plays a significant role in braking and the timescale for such braking for a star with an aligned-dipole field can characterized by $\tau_{\rm brake} \approx 0.15\tau_{\rm mass}\eta_*^{-1/2}$, where $\tau_{\rm mass}$ is the mass-loss timescale (see ud-Doula *et al.* 2009). Nearly 80% of the angular momentum is carried away by stresses in the magnetic field, confirming that magnetic braking is an efficient process.

8. Rigidly Rotating Magnetosphere

Townsend & Owocki (2005) present a semi-analytic method for modelling the circumstellar environment of early-type stars, which is essentially a generalized extension of Babel and Montmerle model discussed above. They assume that the magnetic field lines remain rigid and corotate with the star without being influenced by the dynamics of the wind. This is essentially equivalent to the assumption of $\eta_* \to \infty$.

This RRM model can be applied to an arbitrary field geometry and tilt angle between the field and rotation axes. There are no dynamic forces involved in this model, instead effective gravitational + centrifugal potential is calculated based on the constrained motion of the plasma. Accumulation surfaces are assumed to lie along the location of minimum effective potential. Typically, for a tilted-dipole field case, such a surface will take the form of a geometrically thin warped disk that lies between magnetic and rotational axes.

When the accumulation process (in most cases stellar wind) is known, one can evaluate the density throughout the circumstellar environment allowing us to calculate observational diagnostics such as emission-line spectra. Indeed, Townsend and Owocki (2005) Townsend & Owocki (2005) show that RRM model can applied to explain Hα emission from the strongly magnetic Bp star σ Ori E remarkably well.

9. Arbitrary Rigid-Field Hydrodynamics

As a further improvement to the RRM model, Townsend *et al.* (2007) introduced a new Rigid-Field Hydrodynamics method to modelling the circumstellar environments of strongly magnetic massive stars as defined above. Just like in the RRM method, the field lines are treated as rigid, but now the flow along the lines is computed self-consistently using hydrodynamical equations including line force due to radiation. In this ansatz, flow along each field line is considered to be an independent 1D flow. They perform a large number of such 1D calculations for differing field lines, then piece them together to build up a time-dependent 3D model of a magnetosphere.

Since the flow along each field line can be solved independently of other field lines, the computational cost of this approach is a fraction of an equivalent full MHD simulation. However, this method suppresses any coupling between the flow on adjacent field lines which in practice may have some important ramifications. But overall RFHD model provides a good match with observations of Bp star magnetospheres.

This method is now superseded by *Arbitrary* Rigid-Field Hydrodynamics (ARFHD) method which now allows any configuration of self-consistent magnetic topology. It also improves some of the numerical algorithms and radiative cooling and includes thermal conduction (see Chris Bard's contribution in this volume).

10. Conclusions

Science is a lifelong journey, but unfortunately for some of us that journey gets interrupted abruptly. As a small tribute to two of our beloved colleagues and friends, Stanislav Štefl and Olivier Chesneau, who recently left us quite abruptly, I too end this report abruptly.

Acknowledgements

This work has been partially supported by NASA through Chandra Award numbers TM4-15001A issued by the Chandra X-ray Observatory Center which is operated by the Smithsonian Astrophysical Observatory for and behalf of NASA under contract

NAS8-03060. This work was also carried out with partial support by NASA ATP Grants NNX11AC40G to the University of Delaware.

References

Babel, J. & Montmerle, T. 1997a, *ApJ (Letters)* 485, 29
Babel, J. & Montmerle, T. 1997b, *A&A* 323, 121
Castor, J. I., Abbott, D. C., & Klein, R. I. 1975, *ApJ* 195, 157
Donati, J.-F., Babel, J., Harries, T. J., *et al.* 2002, *MNRAS* 333, 55
Feldmeier, A. 1995, *A&A* 299, 523
Gagné, M., Oksala, M. E., Cohen, D. H., *et al.* 2005, *ApJ* 628, 986
Howarth, I. D. & Smith, K. C. 1995, *ApJ* 439, 431
Kaper, L., Henrichs, H. F., Nichols, J. S., *et al.* 1996, *A&AS* 116, 257
Mathias, P., Aerts, C., Briquet, M., *et al.* 2001, *A&A* 379, 905
Owocki, S. P. 1994, *Ap&SS* 221, 3
Sundqvist, J. O., ud-Doula, A., Owocki, S. P., *et al.* 2012, *MNRAS* 423, L21
Telting, J. H., Aerts, C., & Mathias, P. 1997, *A&A* 322, 493
Townsend, R. H. D. & Owocki, S. P. 2005, *MNRAS* 357, 251
Townsend, R. H. D., Owocki, S. P., & ud-Doula, A. 2007, *MNRAS*, 382, 139
ud-Doula, A. 2003, *Ph.D. thesis*, University of Delaware
ud-Doula, A., Owocki, S., Townsend, R., Petit, V., & Cohen, D. 2014, *MNRAS* 441, 3600
ud-Doula, A. & Owocki, S. P. 2002, *ApJ* 576, 413
ud-Doula, A., Owocki, S. P., & Townsend, R. H. D. 2008, *MNRAS* 385, 97
ud-Doula, A., Owocki, S. P., & Townsend, R. H. D. 2009, *MNRAS* 392, 1022
ud-Doula, A., Sundqvist, J. O., Owocki, S. P., Petit, V., & Townsend, R. H. D. 2013, *MNRAS* 428, 2723
ud-Doula, A., Townsend, R. H. D., & Owocki, S. P. 2006, *ApJ (Letters)* 640, L191

Discussion

LOBEL: Thank you for the nice presentation. In your simulations with confined magnetic fields, do you find large-scale coherent density or velocity structures that build up and stay stable in the wind beyond the Alfvén radius? I am hinting at CIRs or structures that can cause DACs in spectra of hot massive stars.

UD-DOULA: That is not something entirely obvious. In 2D simulations, by design, structures within Alfvén radius are coherent azimuthally but in 3D simulations, as shown in ud-Doula *et al.* (2013), such structures fragment in azimuth, forming distinct clumps of closed loop infall within the Alfvén radius, transitioning in the outer wind to radial spokes of enhanced density with characteristic azimuthal separation of 15°–20°. Now, rotation will add centrifugal ejections of some of these clumps to the mix. Will they indeed form CIRs? This is something that will need further investigation.

Asif ud-Doula

New windows on massive stars: asteroseismology, interferometry, and spectropolarimetry
Proceedings IAU Symposium No. 307, 2014
G. Meynet, C. Georgy, J. H. Groh & Ph. Stee, eds.
© International Astronomical Union 2015
doi:10.1017/S1743921314007030

The BinaMIcS project: understanding the origin of magnetic fields in massive stars through close binary systems

E. Alecian[1,2], C. Neiner[2], G. A. Wade[3], S. Mathis[4,2], D. Bohlender[5], D. Cébron[6], C. Folsom[7], J. Grunhut[8], J.-B. Le Bouquin[1], V. Petit[9], H. Sana[10], A. Tkachenko[11], A. ud-Doula[12] and the BinaMIcS collaboration

[1] UJF-Grenoble 1/CNRS-INSU, Institut de Planétologie et d'Astrophysique de Grenoble (IPAG) UMR 5274, 38041 Grenoble, France, email: evelyne.alecian@obs.ujf-grenoble.fr

[2] LESIA, Observatoire de Paris, CNRS UMR 8109, UPMC, Université Paris Diderot, France

[3] Dept. of Physics, Royal Military College of Canada, Kingston, ON, Canada K7K 0C6

[4] Laboratoire AIM Paris-Saclay, CEA/DSM – CNRS – Université Paris Diderot, IRFU/SAp Centre de Saclay, 91191 Gif-sur-Yvette Cedex, France

[5] DAO, CNRC, Victoria, BC V9E 2E7, Canada

[6] Université Grenoble Alpes, CNRS, ISTerre, Grenoble, France

[7] LATT – CNRS/Université de Toulouse, 14 Av. E. Belin, Toulouse F-31400, France

[8] ESO, Garching bei München, Germany

[9] Bartol Research Institute, University of Delaware, Newark, DE 19716, USA

[10] ESA/Space Telescope Science Institute, Baltimore, MD 21218, USA

[11] Instituut voor Sterrenkunde, KU Leuven, Celestijnenlaan 200D, B-3001 Leuven, Belgium

[12] Penn State Worthington Scranton, Dunmore, PA 18512, USA

Abstract. It is now well established that a fraction of the massive ($M > 8\,M_\odot$) star population hosts strong, organised magnetic fields, most likely of fossil origin. The details of the generation and evolution of these fields are still poorly understood. The BinaMIcS project takes an important step towards the understanding of the interplay between binarity and magnetism during the stellar formation and evolution, and in particular the genesis of fossil fields, by studying the magnetic properties of close binary systems. The components of such systems are most likely formed together, at the same time and in the same environment, and can therefore help us to disentangle the role of initial conditions on the magnetic properties of the massive stars from other competing effects such as age or rotation. We present here the main scientific objectives of the BinaMIcS project, as well as preliminary results from the first year of observations from the associated ESPaDOnS and Narval spectropolarimetric surveys.

Keywords. stars: binaries: close, stars: magnetic fields

1. Introduction

The Binarity and Magnetic Interactions in various classes of stars (BinaMIcS) project proposes to bring new constraints on the physical processes of magnetic stars and of binary systems, by studying the interplay between magnetism and binarity. Binary stars have many advantages compared to single stars. Thanks to their gravitational interaction, model-independent measurements of the stellar fundamental properties (such as masses or radii) can be made. Binary systems also represent laboratories of novel physical processes through star-star interactions (e.g. tidal deformation and flows, mutual heating, wind-wind collisions, magnetospheric coupling, pulsation excitation). Finally, they bring the possibility to investigate the origin of magnetic fields in stars with similar formation and evolution scenarios, and the same age.

In the last decade, magnetic fields have been intensely studied in stars of various ages and masses. We now have a clear global picture of stellar magnetism: convective dynamos in the upper envelope of cool stars (below ∼6500 K) generate complex and variable magnetic fields, while in the hotter stars with radiative outer envelopes and masses above $1.5\,M_\odot$, fossil fields are observed in only 5-10% of them (see the review of Donati & Landstreet 2009; Wade *et al.* 2013). Details on the origin and evolution of stellar magnetism of binary systems are not yet well understood. The BinaMIcS project† proposes to bring new constraints by addressing how star-star interaction in close binary systems can affect or be affected by magnetic fields. With this aim we have gathered a large international consortium of about 80 expert scientists in stellar magnetism and/or binary/multiple systems, observers, modellers, and theoreticians, and have proposed to address four scientific objectives.

2. Scientific objectives

2.1. *How do tidally-induced internal flows impact fossil or dynamo fields ?*

An initially eccentric binary system with non-synchronised, non-aligned components will tend to an asymptotic state with circular orbits, synchronised components, and aligned spins (e.g. Hut 1980). Such an evolution is possible thanks to a dissipation of the kinetic energy of internal flows into heat. During this process, three types of flows are generated in the stellar interiors: the equilibrium tide, a 3D-large scale flow induced by hydrostatic adjustment of a star in response to the perturbing gravitational field of the companion (e.g. Remus *et al.* 2012); the dynamical tide corresponding to the excitation of eigenmodes of the stellar interior oscillations by the tidal potential (e.g. Ogilvie & Lin 2007); and the elliptical instability, a 3D-turbulent flow that can be excited in a rotating fluid of ellipsoidal shape (e.g. Cébron & Hollerbach 2014). One of the objectives of the BinaMIcS project will be to address how such flows can affect the fossil fields of hot massive stars, or the dynamo fields in cool stars.

2.2. *How do magnetospheric Star-Star interactions modify stellar activity ?*

In hot or cool close systems, with two magnetic components, we expect important magnetic interactions between the components' magnetospheres. Intense magnetic reconnection phenomena may develop during the motion of the secondary (and its magnetosphere) through the magnetosphere of the primary, and vice-versa (e.g. Gregory *et al.* 2014). We therefore aim at understanding the modification of stellar activity by such events.

2.3. *What is the magnetic impact on angular momentum exchanges ?*

In low-mass solar-type stars, stellar winds originate from the magnetic coronae (e.g. Parker 1960). In massive stars, winds are driven by spectral line absorption of the stellar radiation field ("line-driving", Castor *et al.* 1975). In both cases, these winds carry away angular momentum and slow down the rotation of the surface layers of the stars (e.g. ud-Doula *et al.* 2009). This phenomenon is amplified if the winds are magnetised (e.g. Roxburgh 1983). The combination of tidal friction and magnetised winds may have a significant impact on the evolution of binary systems (e.g. Barker & Ogilvie 2009). We will bring observational constraints on the proposed impacts.

† http://lesia.obspm.fr/BinaMIcS/

2.4. *What is the impact of magnetic fields on stellar formation, and vice-versa ?*

In hotter higher-mass stars ($M > 1.5\,M_\odot$), the fossil magnetic fields stored into their upper radiative zones are believed to be the results of fields accumulated or amplified during the star formation (e.g. Borra *et al.* 1982). While this hypothesis explains well most of the properties of the fields of high-mass stars (Moss 2001), one of the biggest challenges is to understand why only a small fraction (5-10%) of high-mass stars host fossil fields (e.g. Wade *et al.* 2013). A natural explanation would be the existence of fundamental differences in the initial conditions (IC) of star forming regions (e.g. local density, local magnetic strength, etc.). An excellent way to test this hypothesis is to study the magnetic fields in close binary systems, which contain two stars formed at the same time and with the same IC. Such a study will help us to disentangle IC effects from other (e.g. evolutionary) effects.

3. Observing strategies and resources

To address all the objectives exposed above, we focus on close binary systems (with orbital periods lower than 20 d), i.e. binary systems in which we expect a significant mutual interaction via tidal or magnetospheric interaction. The two components of such close binary systems are also very likely to have been formed from the same condensed molecular core, and therefore to share a similar history (e.g. Bonnell & Bate 1994), which is essential to study fossil field history.

We aim at acquiring high-resolution spectropolarimetric datasets of close binary systems of a sufficient quality to get sensitive measurements of the magnetic fields at the surface of both components. The origin of the surface magnetic fields in cool and hot stars being different, the observing strategy we have chosen is different. In the case of cool stars, magnetic fields are ubiquitous, and we have selected about 40 close binary systems with cool components in which indirect signs of magnetic activity ensure the detection of the magnetic fields in our data. Each system will be monitored and a magnetic map at the surface of the two components will be obtained.

In the case of hotter higher-mass binary systems, fossil fields are only expected in a small fraction of them. We have therefore built two samples. The first one, the Survey Component (SC), gathers about 220 close ($P_{\rm orb} < 20$ d) binary systems of magnitude V brighter than 8 mag in which we will look for magnetic fields (see the primary spectral types distribution in Fig. 1). We used three different catalogues to built this sample: the 9th catalogue of spectroscopic binary orbits (SB9 Pourbaix *et al.* 2009), the CHARA catalogue (Taylor *et al.* 2003), and a catalogue of O-type spectroscopic binaries (Sana & Evans 2011). Whenever a magnetic field is detected in this sample, it is transferred to the second sample. The second Targeted Component (TC) sample contained 6 systems at the start of the project, and has been build by isolating the previously selected SC systems that were reported to have at least one magnetic component. The purpose of this sample is to perform dense monitoring of each object in order to map the magnetic fields at the surface of the stars. Both samples will allow us to establish the magnetic properties (incidence, nature, geometry, strength) in close binary systems and compare them to single stars, already analysed in previous programs (e.g. MiMeS, Wade *et al.* 2013).

The observational resources are mainly based on two high-resolution spectropolarimetric large programmes (LP) of ∼600 h and ∼150 h respectively allocated on ESPaDOnS (Canada-France-Hawaii Telescope) and Narval (Telescope Bernard Lyot, France). Data acquisition started in early 2013 and will end in late 2016. In order to better constrain the

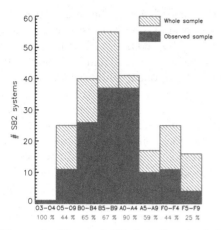

Figure 1. Observed (filled, red) and whole (dashed, blue) BinaMIcS samples as a function of the spectral type of the primary component. The fractions of targets already observed are indicated per spectral type at the bottom.

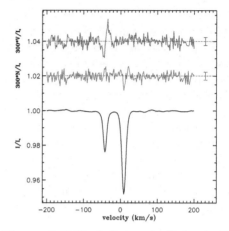

Figure 2. LSD I (bottom), V (top), N (middle) profiles of the F4+F5 binary system HD 160922. The V and N profiles have been shifted in the Y-axis and amplified for display purpose. Note the Zeeman signature in the secondary (left) component only.

orbital parameters (inclination and flux ratio), or the stellar activity of the cool systems of our sample, we have started or proposed additional programs using complementary instrumentation: a PIONIER/VLTI program (PI: J.-B. Le Bouquin) is acquiring interferometric data of the magnetic BinaMIcS targets ; a CHANDRA proposal (PI: C. Argiroffi) has been submitted to analyse the X-ray activity of an eccentric pre-main-sequence system along the orbit, simultaneously with its magnetic activity ; photometric observations of cool systems will be obtained to improve the surface mapping of the components.

4. First look analysis and preliminary results

We will focus here only on the results we obtained so far in high-mass systems.

We have determined the exposure times of each SC target in order to reach a detection limit of 150 G or lower. These exposure times take into account the magnitude and binarity of the system, and the $v \sin i$ and spectral type of the stars. All these quantities can affect the amplitude of the detection of the magnetic field in the polarised spectra, hence the detection limit. The data have been reduced using the fully automatic Libre-Esprit reduction tool available at both telescopes. Libre-Esprit provides us with optimally reduced intensity I and circularly polarised V spectra. A null N spectrum is also computed in a way such that polarisation is cancelled, allowing us to check for spurious polarisation in our data (Donati *et al.* 1997).

In order to increase the signal-to-noise ratio (SNR) of our data we applied the Least-Squares Deconvolution (LSD, Donati *et al.* 1997) technique to all our data. This technique provides us with the mean I, V, and N profiles of the spectral lines with SNRs increased by a factor of 3 to 30 (depending on the spectral type) compared to SNRs in the individual spectral lines. This technique allows us to detect longitudinal magnetic fields as low as a few gauss with a reasonable exposure time.

Of the 220 SC targets, we have observed 151 of them. We have detected only one magnetic field so far, in the secondary component of HD 160922 (Fig. 2, Neiner & Alecian 2013). The spectral types of the primary and secondary components of HD 160922 are respectively F4 V and F5 V. A spectral type of F5 corresponds to the generally accepted limit below which the surface magnetic field is no longer likely to be fossil, and is of dynamo origin – while this limit still needs to be explored in more details. It is therefore not yet clear what the nature of the field detected in HD 160922 is. We will therefore assume that our number of fossil field detections, today, is between 0 and 1. If we include the 6 previously known TC targets with one magnetic component, and take into account that each system contains 2 stars, we now have gathered sensitive magnetic measurements in 314 high-mass stars belonging to close binary systems. 6 or 7 of them host a fossil field, leading to a fossil field incidence lower than 2%. This is 3 to 5 times lower than what is observed in high-mass single stars. We can therefore conclude that fossil magnetic fields seem to be more absent in close binary systems than in single stars. This result is consistent with the work of Carrier *et al.* (2002) based on a much smaller sample, and only on A-type stars. Here our larger sample allows us to extend this result to higher masses, i.e. to all stars with radiative envelopes, and to quantify the differences between close binaries and single stars (Alecian *et al.* in prep.).

5. Perspectives

This very first result, while preliminary, is interesting to analyse in the context of magnetic impact on star formation. Recent theoretical developments have shown that weakly magnetised molecular cores can fragment, while the fragmentation is highly reduced in the case of strongly magnetised cores (e.g. Commerçon *et al.* 2011). In a very simple view, if initial conditions can be retained throughout the stellar formation process until the final stage of a binary star, then we would expect that single stars are more magnetised than binary stars, which is what we observe. However, detailed calculation of magnetised collapse tend to show that initial conditions are lost at the end of the first collapse (Masson *et al.*, priv. comm.). Therefore, the properties of fossil fields in single and close binary systems might not come from the earliest stages of star formation, indicating that we still need to understand the discrepancy between single stars and close binary systems. Progress in our understanding of the magnetic impact in the later stages of star formation (from the second collapse to the Hayashi phase) appear crucial to understand the origin and evolution of fossil fields.

References

Barker, A. J. & Ogilvie, G. I. 2009, *MNRAS* 395, 2268
Bonnell, I. A. & Bate, M. R. 1994, *MNRAS* 271, 999
Borra, E. F., Landstreet, J. D., & Mestel, L. 1982, *ARA&A* 20, 191
Castor, J. I., Abbott, D. C., & Klein, R. I. 1975, *ApJ* 195, 157
Cébron, D. & Hollerbach, R. 2014, *ApJ (Letters)* 789, L25
Commerçon, B., Hennebelle, P., & Henning, T. 2011, *ApJ (Letters)* 742, L9
Donati, J.-F. & Landstreet, J. D. 2009, *ARA&A* 47, 333
Donati, J.-F., Semel, M., Carter, B. D., Rees, D. E., & Collier Cameron, A. 1997, *MNRAS* 291, 658
Gregory, S. G., Holzwarth, V. R., Donati, J.-F., *et al.* 2014, in *European Physical Journal Web of Conferences*, Vol. 64 of *European Physical Journal Web of Conferences*, p. 8009
Hut, P. 1980, *A&A* 92, 167
Moss, D. 2001, in G. Mathys, S. K. Solanki, & D. T. Wickramasinghe (eds.), *Magnetic Fields Across the Hertzsprung-Russell Diagram*, Vol. 248 of *Astronomical Society of the Pacific Conference Series*, pp 305–+
Neiner, C. & Alecian, E. 2013, in *EAS Publications Series*, Vol. 64 of *EAS Publications Series*, pp 75–79
Ogilvie, G. I. & Lin, D. N. C. 2007, *ApJ* 661, 1180

Parker, E. N. 1960, *ApJ* 132, 175

Pourbaix, D., Tokovinin, A. A., Batten, A. H., *et al.* 2009, *VizieR Online Data Catalog* 1, 2020

Remus, F., Mathis, S., & Zahn, J.-P. 2012, *A&A* 544, A132

Roxburgh, I. W. 1983, in J. O. Stenflo (ed.), *Solar and Stellar Magnetic Fields: Origins and Coronal Effects*, Vol. 102 of *IAU Symposium*, pp 449–459

Sana, H. & Evans, C. J. 2011, in C. Neiner, G. Wade, G. Meynet, & G. Peters (eds.), *IAU Symposium*, Vol. 272 of *IAU Symposium*, pp 474–485

Taylor, S. F., McAlister, H. A., & Harvin, J. A. 2003, in *American Astronomical Society Meeting Abstracts*, Vol. 35 of *Bulletin of the American Astronomical Society*, p. 1342

ud-Doula, A., Owocki, S. P., & Townsend, R. H. D. 2009, *MNRAS* 392, 1022

Wade, G. A., Grunhut, J., Petit, V., *et al.* 2013, in *Massive Stars: From alpha to Omega*

Discussion

KHALACK: Can you say which part of your stars in the selected binary systems have slow axial rotation ($v \sin i < 20$ km.s^{-1})?

ALECIAN: Our spectra will be able to provide this information. We have not retrieved it yet.

Evelyne Alecian

New windows on massive stars: asteroseismology, interferometry, and spectropolarimetry
Proceedings IAU Symposium No. 307, 2014
G. Meynet, C. Georgy, J. H. Groh & Ph. Stee, eds.

© International Astronomical Union 2015
doi:10.1017/S1743921314007042

Revealing the Mass Loss Structures of Four Key Massive Binaries Using Optical Spectropolarimetry

Jamie R. Lomax

Homer L. Dodge Department of Physics and Astronomy, University of Oklahoma, Norman, OK, 73019, USA
email: Jamie.R.Lomax@ou.edu

Abstract. The majority of massive stars are members of binary systems. However, in order to understand their evolutionary pathways, mass and angular momentum loss from these systems needs to be well characterized. Self-consistent explanations for their behavior across many wavelength regimes need to be valid in order to illuminate key evolutionary phases. I present the results of linear spectropolarimetric studies of three key binaries (β Lyrae, V356 Sgr, V444 Cyg, and WR 140) which reveal important geometric information about their circumstellar material. β Lyrae exhibits a repeatable discrepancy between secondary eclipse in the total and polarized light curves that indicates an accretion hot spot has formed on the edge of the disk in the system. The existence of this hot spot and its relationship to bipolar outflows within the system is important in the understanding of mass transfer dynamics in Roche-lobe overflow binaries. Preliminary work on V356 Sgr suggests the system maybe surrounded by a common envelope. V444 Cyg shows evidence that its shock creates a cone with a large opening angle of missing material around the WN star. This suggests the effects of radiative inhibition or braking, can be significant contributors to the location and shape of the shock within colliding wind binaries. The intrinsic polarization component of WR 140 is likely due to the formation of dust within the system near periastron passages. Continued work on these and additional objects will provide new and important constraints on the mass loss structures within binary systems.

Keywords. techniques: photometric, (stars:) binaries: eclipsing, stars: individual (beta Lyr, V356 Sgr, V444 Cyg, WR 140)

1. Introduction

Podsiadlowski *et al.* (1992) showed that between 30 and 50% of massive binaries have initial periods that are short enough (under 1500 days) for the individual stars to interact within a system. Additional work by Sana *et al.* (2012) suggests that up to 70% of all massive stars are members of a binary at some point during their lives. Given the importance of mass loss on massive star evolution, it is imperative that we better understand how binary systems exchange and lose material.

Polarization studies of massive binaries can give important, geometric information about the mass loss structures that form from strong stellar winds and during phases of Roche lobe overflow. This is particularly useful for understanding mass loss in systems which remain unresolved. Below I describe the results of polarimetric studies of four massive binary systems: two Roche lobe overflow objects and two colliding-wind binaries.

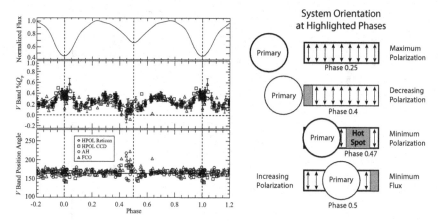

Figure 1. *V* Band Polarization of β Lyrae (Lomax *et al.* 2012). The top left panel is the normalized *V* band light curve from Harmanec *et al.* (1996). The middle and bottom left panels display the intrinsic polarization and position angle versus phase. Dashed vertical lines indicated eclipse locations. The solid curve in the middle panel represents the Fourier fit to the data. The solid horizontal line in the bottom panel indicates the error-weighted mean position angle of the system. Point styles indicate the source of the data. The right panel shows the system orientation including the location of the hot spot at different phases as indicated.

2. Roche-Lobe Overflow

2.1. β Lyr

β Lyrae is a semi-detached, eclipsing binary. The primary, B6-8 II, giant star is losing material via Roche lobe overflow to the secondary B0.5 main sequence star (Hubeny & Plavec 1991). A thick accretion disk has formed around the secondary which almost completely obscures its interior. Bipolar outflows are also known to exist in the system (Harmanec *et al.* 1996; Hoffman *et al.* 1998).

Figure 1 shows the polarization behavior of β Lyrae with phase (Lomax *et al.* 2012). The average position angle of the polarization, excluding the behavior around secondary eclipse, is consistent with the orientation of the disk and system on the sky, indicating that the broad band polarization is caused by scattering in the disk. Three distinct features appear in the $\% Q_p$. The first is the increase in polarization at primary eclipse caused by a decrease in the unpolarized flux due to the disk blocking the star losing material. Second, there are local maxima at quadurature phases due to light scattering off the edge of the disk. The last feature is a secondary eclipse that occurs at phase 0.481 ± 0.001 that is associated with a position angle rotation. This is consistent with a hot spot forming on the edge of the accretion disk due to the impact of the mass stream (Lomax *et al.* 2012) and has been confirmed through modeling of β Lyr's light curve (Mennickent & Djurašević 2013). Size estimates derived from the polarimetric data indicated that the hot spot has a width of approximately $22 - 33\,R_\odot$ across the disk edge (Lomax *et al.* 2012, the disk diameter is $60\,R_\odot$ for comparison).

2.2. V356 Sgr

There are two major differences between the β Lyr system and V356 Sgr that should be kept in mind when comparing the two systems. The first is the disk in β Lyr is eclipsed at phase 0.5 while it is eclipses at phase 0.0 in V356 Sgr. This is because V356 Sgr has an A2 supergiant that is losing mass to its brighter, primary B star (Wilson & Caldwell 1978). Additionally, the accretion disk that has formed around the primary star is optically thin (Wilson & Caldwell 1978), where as β Lyr's is optically thick.

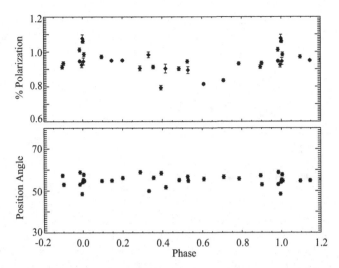

Figure 2. *V* Band Polarization of V356 Sgr (Malatesta *et al.* 2013 AAS proceedings). The top panel displays the total polarization (interstellar polarization has not been removed) versus phase while the bottom panel shows the position angle versus phase.

Figure 2 displays preliminary polarization curves of V356 Sgr (Malatesta *et al.* 2013 AAS proceedings). The relative lack of eclipse effects compared to β Lyr and the lack of polarization features at quadrature phases, which if they did exist would be due to scattering off the disk edge, indicate that the scattering region in V356 Sgr might be circumbinary material, such as a common envelope, or a combination of circumbinary material and the disk within the system.

3. Colliding Wind Binaries

3.1. *V444 Cyg*

V444 Cyg is an eclipsing WR+O system with a short 4.2 day period. Evidence for colliding winds exists in a variety data from *IUE, ASCA, ROSAT, Einstein,* and *XMM-Newton* despite a naive estimate (based on mass-loss rates and terminal velocities) that the wind of the WR star should impact the surface of the O star instead of forming a shock (Moffat *et al.* 1982; Koenigsberger & Auer 1985; Pollock 1987; Shore & Brown 1988; Corcoran *et al.* 1996; Maeda *et al.* 1999; Fauchez *et al.* 2011). Theoretical work has suggested that the situation is more complex; radiative braking and radiative inhibition may play an important roll in the location of the shock (Owocki & Gayley 1995; Stevens & Pollock 1994).

Figure 3 shows the polarization behavior of V444 Cyg's HeII 4686 with phase. The Stokes $\%U_p$ parameter (the p indicates that the standard Stokes parameters have been rotated so that the error-weighted mean $\%Q$ is zero) has a different polarization behavior around secondary eclipse compared to other phases, where as the Stokes $\%Q_p$ parameter appears to scatter about 0% more equally. To show this I have calculated the error weighted mean for each parameter in the 0.3 to 0.75, and the 0.75 to 1.3 phase regions. Figure 3 shows that the error weighted mean $\%U_p$ is more negative around secondary eclipse than at other phases at a 2σ significance. This type of behavior is very similar to the behavior of modeled polarization for a single degenerate Type Ia supernovae where the companion carves a cavity out of the ejecta (Kasen *et al.* 2004). In that scenario viewing angles close to the cavity produce a more negative polarization than viewing

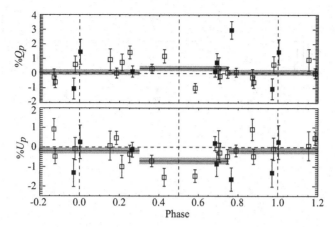

Figure 3. He<small>II</small> 4686 Polarization of V444 Cyg. The top panel displays the Stokes Q parameter and the bottom panel displays the Stokes U parameter. Solid filled squares indicate data taken with HPOL at Ritter Observatory while open symbols are HPOL data from the Pine Bluff Observatory. Horizontal dashed lines indicate zero points for each parameter, while vertical dashed lines make eclipse phases. Grey boxes represent the 1σ uncertainty on the error-weighted mean (solid line inside the box) Stokes parameter for the phase range over which it is plotted.

angles farther away. For V444 Cyg, the cavity is carved out of the WN-star wind by the shock and O-star wind while viewing angles close to the cavity occur around secondary eclipse.

3.2. *WR 140*

WR 140 is the canonical long period, high eccentricity WR+O binary system. Periastron passage sparks the production of dust, which quickly expands, dissipates, and is destroyed (Monnier *et al.* 2002; Taranova & Shenavrin 2011). Previous studies of WR 140 have indicated that the wavelength dependent polarization signatures observed from the system are consistent with dust. Figure 4 shows the polarization light curve of the system. The points show large variability after the 1985 periastron passage that does not exist before the 1993 periastron passage. This is also consistent with dust being the primary scattering region in the system.

4. Future Work

The results of these studies show that polarization is a powerful tool for revealing the mass loss structures around binary systems. Despite β Lyr and V356 Sgr being very similar systems, their polarization light curves are very different. This suggests that a large scale polarimetric survey of Roche lobe overflow systems is needed to better characterize their circumstellar and circumbinary material. The difference in mean polarization of V444 Cyg at secondary eclipse compared to other phases is currently only suggestive at a 2σ detection level. Work is ongoing to obtain more observations to fill in the light curve and better characterize this behavior. Polarimetric monitoring of WR 140 has historically been poor. Large gaps between epochs of data exist at crucial dust production phases. Better observational cadence, particularly through periastron passage, could potentially reveal new information about the formation and destruction of dust in the system.

Acknowledgements

This research includes contributions from the following collaborators: Jennifer L

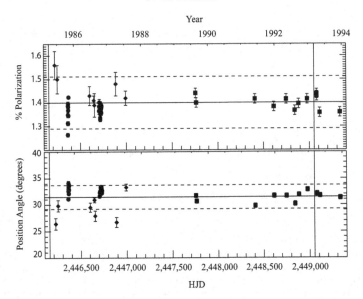

Figure 4. Polarization of WR 140. Diamonds indicate data taken with the Lyot instrument at Pine Bluff Observatory, squares are data from HPOL at the Pine Bluff Observatory, and circles indicate archival data from Robert *et al.* (1989).

Hoffman (The University of Denver), James Davidson (University of Toledo), John Wisniewski (University of Oklahoma), and Michael Malatesta (University of Oklahoma).

References

Corcoran, M. F., Stevens, I. R., Pollock, A. M. T., *et al.* 1996, *ApJ* 464, 434

Fauchez, T., De Becker, M., & Nazé, Y. 2011, *Bulletin de la Societe Royale des Sciences de Liege* 80, 673

Harmanec, P., Morand, F., Bonneau, D., *et al.* 1996, *A&A* 312, 879

Hoffman, J. L., Nordsieck, K. H., & Fox, G. K. 1998, *AJ* 115, 1576

Hubeny, I. & Plavec, M. J. 1991, *AJ* 102, 1156

Kasen, D., Nugent, P., Thomas, R. C., & Wang, L. 2004, *ApJ* 610, 876

Koenigsberger, G. & Auer, L. H. 1985, *ApJ* 297, 255

Lomax, J. R., Hoffman, J. L., Elias, II, N. M., Bastien, F. A., & Holenstein, B. D. 2012, *ApJ* 750, 59

Maeda, Y., Koyama, K., Yokogawa, J., & Skinner, S. 1999, *ApJ* 510, 967

Mennickent, R. E. & Djurašević, G. 2013, *MNRAS* 432, 799

Moffat, A. F. J., Firmani, C., McLean, I. S., & Seggewiss, W. 1982, in C. W. H. De Loore & A. J. Willis (eds.), *Wolf-Rayet Stars: Observations, Physics, Evolution*, Vol. 99 of *IAU Symposium*, pp 577–581

Monnier, J. D., Tuthill, P. G., & Danchi, W. C. 2002, *ApJ (Letters)* 567, L137

Owocki, S. P. & Gayley, K. G. 1995, *ApJ (Letters)* 454, L145

Podsiadlowski, P., Joss, P. C., & Hsu, J. J. L. 1992, *ApJ* 391, 246

Pollock, A. M. T. 1987, *ApJ* 320, 283

Robert, C., Moffat, A. F. J., Bastien, P., & Drissen, L., St.-Louis, N. 1989, *ApJ* 347, 1034

Sana, H., de Mink, S. E., de Koter, A., *et al.* 2012, *Science* 337, 444

Shore, S. N. & Brown, D. N. 1988, *ApJ* 334, 1021

Stevens, I. R. & Pollock, A. M. T. 1994, *MNRAS* 269, 226

Taranova, O. G. & Shenavrin, V. I. 2011, *Astronomy Letters* 37, 30

Wilson, R. E. & Caldwell, C. N. 1978, *ApJ* 221, 917

Discussion

PETERS: Eclipse of V356 Sgr in the FUSE FUV definitely shows the presence of a jet structure that emits strongly in OVI (implies a 300,000 K plasma). Could this structure produce a polarization signature?

LOMAX: Yes. In β Lyr we see a polarization signature from the jets in the UV continuum and the emission lines in the optical. However, we don't see similar signatures in V356 Sgr.

DAVID-URAZ: On the WR 140 polarization curve there doesn't seem to be as much variability after the 1993 periastron; is that just lack of data?

LOMAX: I do think we've missed a lot of variability because of lack of observations, particularly in that periastron passage, but maybe even at other times as well.

DAVID-URAZ: Were there any polarization observations during the 2010 periastron campaign?

LOMAX: No. There is an unpublished observation from 2013, but nothing that occurred during the 2010 periastron passage.

WADE: What are the systematic contributions to scatter in β Lyr or V356 Sgr? In other words, how close to the photon limit are your uncertainties?

LOMAX: Systematic uncertainties for the HPOL data are small and about $\pm 0.02\%$ in the broadbands. A paper currently in prep. (Davidson *et al.*) quotes this in a more complete manner with more exact numbers and a comparison to how our systematic uncertainties change with time. I am uncertain about the other data because it was all archival and I haven't been able to find information for those instruments.

Jamie Lomax

New windows on massive stars: asteroseismology, interferometry, and spectropolarimetry
Proceedings IAU Symposium No. 307, 2014
G. Meynet, C. Georgy, J. H. Groh & Ph. Stee, eds.

© International Astronomical Union 2015
doi:10.1017/S1743921314007054

The B Fields in OB Stars (BOB) Survey

T. Morel,[1] N. Castro,[2] L. Fossati,[2] S. Hubrig,[3] N. Langer,[2]
N. Przybilla,[4] M. Schöller,[5] T. Carroll,[3] I. Ilyin,[3] A. Irrgang,[6]
L. Oskinova,[7] F. R. N. Schneider,[2] S. Simon Díaz,[8,9] M. Briquet,[1]
J. F. González,[10] N. Kharchenko,[11] M.-F. Nieva,[4,6] R.-D. Scholz,[3]
A. de Koter,[12,13] W.-R. Hamann,[7] A. Herrero,[8,9] J. Maíz Apellániz,[14]
H. Sana,[15] R. Arlt,[3] R. Barbá,[16] P. Dufton,[17] A. Kholtygin,[18]
G. Mathys,[19] A. Piskunov,[20] A. Reisenegger,[21] H. Spruit,[22]
and S.-C. Yoon[23]

[1]Institut d'Astrophysique et de Géophysique, Liège, Belgium

[2]Argelander-Institut für Astronomie, Bonn, Germany

[3]Leibniz-Institut für Astrophysik Potsdam (AIP), Potsdam, Germany

[4]Institute for Astro- and Particle Physics, University of Innsbruck, Austria

[5]European Southern Observatory, Garching, Germany

[6]Dr. Remeis Observatory & ECAP, Bamberg, Germany

[7]Institut für Physik und Astronomie der Universität Potsdam, Germany

[8]Instituto de Astrofísica de Canarias, La Laguna, Spain

[9]Universidad de La Laguna, Dpto. de Astrofísica, La Laguna, Spain

[10]Instituto de Ciencias Astronomicas, de la Tierra, y del Espacio, San Juan, Argentina

[11]Main Astronomical Observatory, Kiev, Ukraine

[12]Astronomical Institute Anton Pannekoek, Amsterdam, The Netherlands

[13]Instituut voor Sterrenkunde, Universiteit Leuven, Leuven, Belgium

[14]Instituto de Astrofísica de Andalucía-CSIC, Granada, Spain

[15]ESA/Space Telescope Science Institute, Baltimore, USA

[16]Departamento de Física, La Serena, Chile

[17]Astrophysics Research Centre, Belfast, UK

[18]Chair of Astronomy, Saint-Petersburg State University, Russia

[19]European Southern Observatory, Santiago, Chile

[20]Institute of Astronomy of the Russian Acad. Sci., Moscow, Russia

[21]Pontificia Universidad Católica de Chile, Santiago, Chile

[22]Max-Planck-Institut für Astrophysik, Garching, Germany

[23]Department of Physics and Astronomy, Seoul National University, Seoul, Republic of Korea

Abstract. The B fields in OB stars (BOB) survey is an ESO large programme collecting spectropolarimetric observations for a large number of early-type stars in order to study the occurrence rate, properties, and ultimately the origin of magnetic fields in massive stars. As of July 2014, a total of 98 objects were observed over 20 nights with FORS2 and HARPSpol. Our preliminary results indicate that the fraction of magnetic OB stars with an organised, detectable field is low. This conclusion, now independently reached by two different surveys, has profound implications for any theoretical model attempting to explain the field formation in these objects. We discuss in this contribution some important issues addressed by our observations (e.g., the lower bound of the field strength) and the discovery of some remarkable objects.

Keywords. magnetic fields, stars: early-type, stars: magnetic fields, stars: individual (HD 164492C, CPD –57° 3509, HD 54879, β CMa, ε CMa)

1. Magnetic fields in OB stars

Magnetic fields are one of the key factors affecting the evolution and properties of massive stars. Yet it is only very recently that the number of magnetic OB stars known has reached a level that allows us to evaluate the field incidence, examine the properties of the fields, and critically test the various models that were proposed for their creation. The picture now emerging is that relatively strong fields (above, say, 100-200 G at the surface) are only found in a small fraction of all massive stars and that the field topology is rather simple (dipolar, or, in some rare cases, low-order multipolar). Moreover, the field strength is not directly linked to the stellar parameters. These characteristics are similar to those presented by the chemically-peculiar Ap/Bp stars (Donati & Landstreet 2009). This suggests a similar origin of the field.

Despite the dramatic progress made over the last few years, the answers to some important questions are still eluding us, e.g., the effects of fields on the internal rotational profile and on the transport of the chemical species. Even how the field is created is not completely settled. The magnetic field permeating the interstellar medium (ISM) is amplified during star formation and may naturally relax into a large-scale, mostly poloidal field emerging at the surface (e.g., Braithwaite & Spruit 2004). The similarity between the magnetic properties of OB and Ap/Bp stars suggests that today we observe the remnant of the field frozen-in in the ISM. However, in view of the significant fraction of OB stars that may suffer a merger or a mass-transfer event during their evolution (Sana *et al.* 2012), it cannot be ruled out that fields are created through such processes (e.g., Wickramasinghe *et al.* 2014). Finally, a dynamo operating in subsurface convection layers could produce short-lived, spatially localised magnetic structures (Cantiello & Braithwaite 2011) that are, however, much more challenging to detect.

A better understanding of the origin and effects of magnetic fields requires the identification of additional magnetic objects (e.g., only in about ten O stars has a field been firmly detected). This has motivated us to initiate the B fields in OB stars (BOB) survey.

2. The BOB survey

A total of 35.5 observing nights were allocated during P91-P96 as part of an ESO large programme (191.D-0255; PI: Morel). About 20 nights are dedicated to obtain snapshot observations of a large number of OB stars, while the remaining nights are devoted to confirm the field detection for the candidate magnetic stars and to better characterise the field properties for those that are firmly identified as being magnetic. Two different state-of-the-art instruments with circular polarisation capabilities were used (with low and high spectral resolution, respectively): FORS2 at the VLT for the fainter targets and HARPSpol at the 3.6-m telescope at La Silla for the brighter ones. About two thirds of the nights are scheduled on HARPSpol. As of July 2014, 20 nights of observations were completed. Only one night (with FORS2) was lost because of bad weather.

Previously known magnetic OB stars on average appear to have rotation speeds significantly lower than the rest of the population. We therefore mostly targeted stars with $v \sin i$ below $60 \, \mathrm{km \, s^{-1}}$ to enhance the probability of detecting fields. Contrary to the MiMeS survey (Grunhut *et al.*, these proceedings), we concentrate on normal, main-sequence OB stars and do not consider, e.g., Of?p, Be or Wolf-Rayet stars. The sample is composed in roughly equal parts of O ($\sim$40%) and B ($\sim$60%) stars. The vast majority are late O- and early B-type stars. BOB and MiMeS can be viewed as two complementary surveys in the sense that there are very few targets in common.

One important aspect of our survey is that the data reduction and analysis are carried out completely independently by two groups (one from the Argelander-Institut für Astronomie in Bonn and the other from the Leibniz-Institut für Astrophysik in Potsdam) to ensure that the results are robust. The two groups process both the FORS2 and HARPSpol data separately, and employ different tools and analysis techniques (for details, see Hubrig *et al.* 2014).

3. The occurrence of magnetic fields in massive stars

The results obtained by the only other large-scale survey of this kind up to now (MiMeS; Wade *et al.* 2013) indicate that about 7% of massive stars host a magnetic field detectable with current instrumentation ($\gtrsim 100\,\mathrm{G}$). We have so far observed 98 targets and only unambiguously detected five magnetic stars. For all the stars, the detection is not only confirmed by the two groups (Bonn and Potsdam), but the field measurements also systematically agree within the errors. In addition, the field is detected at a high significance level with both FORS2 and HARPSpol.

Therefore, our results tend to support those independently obtained by MiMeS, and confirm that the incidence rate of strong magnetic fields is low in massive stars and is similar to that inferred for intermediate-mass stars. It should be emphasised, however, that a number of candidate magnetic stars are still being followed up and that the preliminary incidence rate that we obtain (~5%) may eventually be revised upwards.

Regardless of the exact figures, the scarcity of strong fields has far-reaching implications, from the interpretation of the statistical properties of young stellar populations (e.g., impact of magnetic braking on the rotational velocities, X-ray characteristics) to the fate of massive stars as degenerate objects following the supernova explosion (e.g., as magnetars).

4. The first magnetic stars discovered in the course of the survey

4.1. *A magnetic field in a multiple system in the Trifid nebula*

One of the aims of our survey is to uncover magnetic stars with specific and unusual characteristics that would allow us to discriminate between the various channels that could lead to the field formation. An interesting discovery is therefore the detection of a magnetic field in a multiple system in the Trifid Nebula (Hubrig *et al.* 2014), which is a very young and active site of star formation. We first observed the three brightest components in the central part of the nebula (A, C, and D; Kohoutek *et al.* 1999) with FORS2 and clearly detected a circularly polarised signal in component C (HD 164492C).

Further observations on two consecutive nights with HARPSpol confirmed the existence of a field with a longitudinal strength ranging from 400 to 700 G. These high-resolution observations reveal complex and variable line profiles pointing towards a multiple system (made up of at least two early B-type stars). The situation is complicated further by the possible existence of chemical patches on the surface of some components. We will keep monitoring this system to establish whether only one or more components are magnetic. A complete characterisation of this peculiar system may provide valuable information about the interplay between binarity and magnetic fields in massive stars.

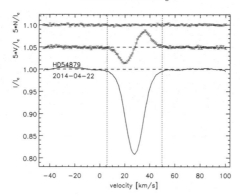

Figure 1. Stokes I (*black*), V (*red*), and diagnostic null N (*blue*) profiles of HD 54879 obtained with HARPSpol through least-squares deconvolution (LSD). From Castro *et al.*, in preparation.

4.2. *A new magnetic, helium-rich star with a tight age constraint*

The variability of the rare magnetic, helium-rich stars (the prototype is σ Ori E) arises from a dipolar field that is tilted with respect to the rotational axis. The competition between radiative levitation and gravitational settling in the presence of a stellar wind leads to photospheric abundance anomalies. Some stars undergo rapid rotational braking (e.g., Mikulášek *et al.* 2008), which provides an unique opportunity to study virtually in real time the poorly-understood effects of angular momentum loss through magnetically channeled, line-driven winds (e.g., ud-Doula *et al.* 2009).

Our observations of the B1 star CPD –57° 3509 in the young ($\sim$ 10 Myr) open cluster NGC 3293 with FORS2 and HARPSpol reveal a strong and rapidly varying field (by up to 900 G for the longitudinal component between two consecutive nights). The field is found to change polarity, which shows that both magnetic hemispheres are visible as the star rotates. The polar field exceeds 3 kG assuming a dipole geometry. A preliminary NLTE spectral analysis indicates that the star is helium rich ($\sim$ 3 times solar) and has evolved throughout about one third of its main-sequence lifetime (Przybilla *et al.*, in prep.). This is one of the most evolved He-rich stars with a tight age constraint, promising to provide crucial information on the evolution of stars with magnetically-confined stellar winds.

4.3. *A non-peculiar magnetic O star*

The few magnetic O stars known are very often peculiar. Their strong magnetic fields are believed to give rise to spectral peculiarities and/or to drive periodic line-profile variations (e.g., the Of?p stars or θ^1 Ori C). In contrast, we have discovered a narrow-lined O 9.7 V star (HD 54879) hosting a strong field (with a dipole strength above 2 kG; Fig.1), yet displaying no evidence in the few optical spectra taken over five years for spectral peculiarity or variability. Only the broad and emission-like Hα profile is variable. This might be related to the presence of a centrifugal magnetosphere (Castro *et al.*, in prep.). Further observations are necessary to confirm the lack of spectral peculiarities and, if so, to understand the distinct behaviour with respect to other strongly magnetised O stars. A parallel investigation of the magnetic variability also needs to be undertaken.

4.4. *Weak fields in OB stars*

One of the most intriguing properties of magnetic stars of intermediate mass is the bimodal nature of the fields that are either strong (above 300 G) or extremely weak ($\lesssim$ 1 G). The lack of objects with ordered fields of intermediate strength appears to reveal a real dichotomy (e.g., Lignières *et al.* 2014). Investigating the origin of this "magnetic

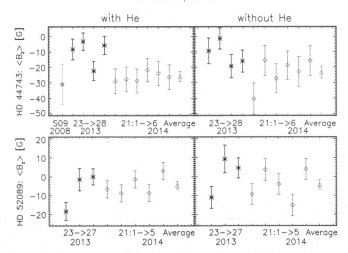

Figure 2. Time series of the longitudinal field measurements of β CMa (top panels) and ε CMa (bottom panels) using a line mask which contains (left panels) or does not contain (right panels) helium lines. Black crosses: observations carried out on four different nights in December 2013. Red rhombs: values obtained from consecutive observations on 21 April 2014 (the blue triangle is the average value obtained on that night). Green circle: ESPaDOnS measurement of β CMa carried out in 2008 by Silvester *et al.* (2009). From Fossati *et al.*, in preparation.

desert" is essential to understand the origin and evolution of fields in stars that cannot support a dynamo acting in a deep, outer convective envelope.

To estimate the lower bound of the field in more massive stars, we have obtained very high-quality observations with HARPSpol of two very bright, early B-type stars (β CMa and ε CMa), and repeatedly detected for both stars a weak Zeeman signature across the line profiles. Interestingly, the field appears to be constant within the errors and relatively weak in both cases. The longitudinal components are at most $\sim$ 30 G in modulus (Fig.2), which translates into a polar strength of $\sim$ 150 G assuming a dipolar geometry. Although all the available measurements of β CMa are consistent with a field of that magnitude, there is some indication in the literature for a stronger field in ε CMa (Hubrig *et al.* 2009; Bagnulo *et al.* 2012). Therefore, the case for a weak field is much stronger in β CMa.

Detecting fields of that magnitude in fainter targets is very challenging. Are the weak fields found in these two bright objects only the tip of the iceberg? Is there a large population of stars with weaker fields or that are even truly non-magnetic (see Neiner *et al.* 2014)? It seems conceivable that weak fields are considerably more widespread than the data currently available would suggest. Fields that are remnants of the star formation are prone to decay on evolutionary time-scales (see Landstreet *et al.* 2008). A different field strength distribution for intermediate- and high-mass stars thus raises the issue of a mass-dependent time-scale for the field decay (Fossati *et al.*, in prep.).

Acknowledgements

TM acknowledges financial support from Belspo for contract PRODEX GAIA-DPAC. LF acknowledges financial support from the Alexander von Humboldt Foundation. MB is F.R.S.-FNRS Postdoctoral Researcher, Belgium.

References

Bagnulo, S., Landstreet, J. D., Fossati, L., & Kochukhov, O. 2012, *A&A* 538, A129
Braithwaite, J. & Spruit, H. C. 2004, *Nature* 431, 819

Cantiello, M. & Braithwaite, J. 2011, *A&A* 534, A140
Donati, J.-F. & Landstreet, J. D. 2009, *ARA&A* 47, 333
Hubrig, S., Briquet, M., De Cat, P., *et al.* 2009, *Astronomische Nachrichten* 330, 317
Hubrig, S., Fossati, L., Carroll, T. A., *et al.* 2014, *A&A* 564, L10
Kohoutek, L., Mayer, P., & Lorenz, R. 1999, *A&AS* 134, 129
Landstreet, J. D., Silaj, J., Andretta, V., *et al.* 2008, *A&A* 481, 465
Lignières, F., Petit, P., Aurière, M., Wade, G. A., & Böhm, T. 2014, *ArXiv e-prints (1402.5362)*
Mikulášek, Z., Krtička, J., Henry, G. W., *et al.* 2008, *A&A* 485, 585
Neiner, C., Monin, D., Leroy, B., Mathis, S., & Bohlender, D. 2014, *A&A* 562, A59
Sana, H., de Mink, S. E., de Koter, A., *et al.* 2012, *Science* 337, 444
Silvester, J., Neiner, C., Henrichs, H. F., *et al.* 2009, *MNRAS* 398, 1505
ud-Doula, A., Owocki, S. P., & Townsend, R. H. D. 2009, *MNRAS* 392, 1022
Wade, G. A., Grunhut, J., Alecian, E., *et al.* 2013, *ArXiv e-prints (1310.3965)*
Wickramasinghe, D. T., Tout, C. A., & Ferrario, L. 2014, *MNRAS* 437, 675

Discussion

LANDSTREET: Possible detection of a σ Ori E analogue is very interesting. Have you looked for indications of a trapped magnetosphere with detailed comparisons of Balmer line profiles as observed with model profiles, to identify weak emission or shell absorption?

MOREL: This is certainly something we plan to do, but we have still not investigated this in detail. Dedicated follow-up observations are needed to fully characterise the behaviour of the Balmer lines. We have recently been granted time on UVES to monitor this object. These spectra can be used for that purpose.

WADE: The detection of V signatures in $\beta + \varepsilon$ CMa rely on magnetic precision of ~ 5 G. For how many other stars is a comparable precision achieved? (Since this will tell us about the frequency of such apparently weak fields.)

MOREL: A comparable precision was only achieved for another star, and in that case a field was not detected. Such high-quality data can only be obtained for very bright stars, and it cannot be ruled out that weak fields also exist in the fainter targets.

AERTS: Did you try to fold your polarimetric data of β CMa on the seismic rotation period to check if they are consistent, as is the case for V2052 Oph and HD 43317?

MOREL: We still have too few measurements to investigate whether the magnetic period is compatible with the rotational one determined by Mazumdar *et al.* (2006, A&A, 459, 589). However, it has to be kept in mind that the weakness of the magnetic signal may hamper a clear detection of the variability.

Thierry Morel

New windows on massive stars: asteroseismology, interferometry, and
spectropolarimetry
Proceedings IAU Symposium No. 307, 2014
G. Meynet, C. Georgy, J. H. Groh & Ph. Stee, eds.
© International Astronomical Union 2015
doi:10.1017/S1743921314007066

Unraveling the variability of σ Ori E

M. E. Oksala[1], O. Kochukhov[2], J. Krtička[3], M. Prvák[3] and Z. Mikulášek[3]

[1] Astronomical Institute, Academy of Sciences of the Czech Republic, Fricova 298, 251 65
Ondřejov, Czech Republic
email: meo@udel.edu

[2] Department of Physics and Astronomy, Uppsala University, Box 516, Uppsala 75120, Sweden

[3] Institute of Theoretical Physics and Astrophysics, Masaryk University, 611 37 Brno, Czech
Republic

Abstract. σ Ori E (HD 37479) is the prototypical helium-strong star shown to harbor a strong
magnetic field, as well as a magnetosphere consisting of two clouds of plasma. The observed optical ($ubvy$) light curve of σ Ori E is dominated by eclipse features due to circumstellar material,
however, there remain additional features unexplained by the Rigidly Rotating Magnetosphere
(RRM) model of Townsend & Owocki (2005). Using the technique of magnetic Doppler imaging
(MDI), spectropolarimetric observations of σ Ori E are used to produce maps of both the magnetic field topology and various elemental abundance distributions. We also present an analysis
utilizing these computed MDI maps in conjunction with non-local thermodynamical equilibrium
TLUSTY models to study the optical brightness variability of this star arising from surface inhomogeneities. It has been suggested that this physical phenomena may be responsible for the
light curve inconsistencies between the model and observations.

Keywords. stars: magnetic fields, stars: rotation, stars: early-type, stars: circumstellar matter,
stars: individual: HD 37479, techniques: spectroscopic, techniques: polarimetric, ultraviolet: stars

1. Introduction

The B2Vp star σ Ori E (HD 37479) has long been known as the prototypical He-strong
magnetic Bp star. Because of its status, it has been one of the most well studied of
such stars, exhibiting numerous types of observed variability all modulated on the stellar
rotation period. The combination of high rotation speed ($\sim 40\%$ of the critical speed)
and a strong, global magnetic field ($B_p \sim 10\,\mathrm{kG}$) makes σ Ori E an excellent laboratory
to study the interaction of magnetism, rotation, and mass-loss; three essential parameters to determine the stellar evolutionary course. With this enigmatic object in mind,
Townsend & Owocki (2005) developed their Rigidly Rotating Magnetosphere (RRM)
model to describe the circumstellar environment of an oblique magnetic rotator in which
rapid rotation keeps magnetically trapped material supported against gravity via centrifugal forces, with material accumulating at the intersection between the rotational and
magnetic equators.

This analytical model, applied to the specific case of σ Ori E by Townsend *et al.* (2005),
produced a prediction of two co-rotating "clouds" of plasma, thought to be responsible
for several observed variations, including H_α emission and what appear to be eclipses
in the optical photometric light curve. Good agreement was found between observations
and model computed data, mainly because of the implementation of an offset dipolar
magnetic configuration to achieve an asymmetry in the size of the two plasma clouds
to better fit the observations. Regardless, the fit to the photometric light curve by this
purely circumstellar model was still unable to reproduce one major feature, a brightening

of the star directly after the second cloud passes across the stellar surface. This discrepancy suggests important physics are missing from the model, and led to a mass effort to obtain new, current data to determine the origin of this feature. New and historical photometry demonstrate that the stellar rotation rate is decreasing due to the influence of the magnetic field on angular momentum of the system (Townsend *et al.* 2010). Photometry with the *MOST* satellite revealed a precise, stable brightness variation, confirming the asymmetry of the eclipses and the presence of additional brightening unfit by the RRM model (Townsend *et al.* 2013). A set of high resolution spectropolarimetric data of σ Ori E was analyzed by Oksala *et al.* (2012) to determine line profile variability and to more precisely calculate the variation of the longitudinal magnetic field. These authors also showed that the de-centered dipole invoked by Townsend *et al.* (2005) is not compatible with the observed circular polarization signatures.

In light of these substantial changes in our view of σ Ori E, the constraints and parameters used in the RRM model must also be revisited. Ultimately our goal is to achieve a model such that we properly understand the separate contributions to the variability from the star and the magnetically confined material. In this work, we present two first steps towards this aim, magnetic Doppler imaging (MDI) and a synthetic light curve analysis.

2. Magnetic Doppler Imaging

We obtained a total of 18 high-resolution (resolving power $R = 65000$) broadband (370–1040 nm) spectra of σ Ori E in November 2007 with the Narval spectropolarimeter on the 2.2m Bernard Lyot telescope (TBL) at the Pic du Midi Observatory in France and in February 2009 with the spectropolarimeter ESPaDOnS on the 3.6-m Canada-France-Hawaii Telescope (CFHT), as part of the Magnetism in Massive Stars (MiMeS) Large Program (Wade *et al.* 2011).

Magnetic Doppler imaging (MDI), previously employed to study Ap/Bp stars, investigates both the surface inhomogeneities of various elements and the magnetic field topology assuming rotational variation of line profiles with time. We applied the MDI code INVERS10, developed by Piskunov & Kochukhov (2002), to our set of spectra to determine, for σ Ori E, the stellar magnetic field and surface abundance features of a number of different elements.

2.1. Magnetic Field

The Stokes I and Stokes V line profiles of He I 5867 Å and 6678 Å were used to determine the best fit magnetic topology for σ Ori E. With the longitudinal magnetic field curve computed by Oksala *et al.* (2012) as an additional constraint in addition to the multipolar regularization, the derived magnetic field configuration suggests a magnetic field with both a dipole with polar strength $B_d = 7.434$ kG and orientation angles $\beta_d = 47.1°$ and $\gamma_d = 97.5°$, and a quadrupole component with strength $B_q = 3.292$ kG. The quadrupole axis is offset from the dipole where the positive and negative poles are clearly not separated by 180°.

2.2. Surface Abundances

With the derived magnetic field topology set as a fixed parameter, INVERS10 was then used to compute the distribution of various chemical elements on the surface of the star. The set of high resolution time-series spectra of σ Ori E contains lines suitable for modeling the abundances of Fe, Si, C, and He. Other chemical elements are present in the spectrum, but their line strength was too weak for accurate determination of any

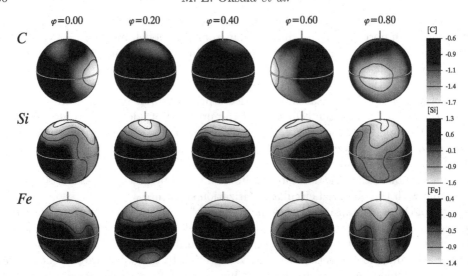

Figure 1. The chemical abundance distribution of σ Ori E derived from Stokes I and V profiles of the C, Fe and Si lines. The star is shown at 5 equidistant rotational phases viewed at the inclination angle, $i = 75°$ and $v\sin i = 140$ km s^{-1}. The scale gives abundance as ϵ_{Elem} corresponding to $\log(\text{N}_{\text{Elem}}/\text{N}_{\text{tot}})$ for metals. The rotation axis is vertical. The contour step size is 0.5 dex.

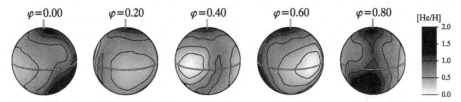

Figure 2. The chemical abundance distribution of σ Ori E derived from the 4713 Å He I line. See Fig.1 for details. The scale gives abundance as $\log(\text{He/H})$.

surface structure. He lines required a slightly more rigorous treatment, and were fit with INVERS13 (for more details see Kochukhov *et al.* 2012, 2013).

The Stokes I spectral lines were fit to good agreement between the observations and the model fits for all of these lines. Figure 1 displays spherical maps of the abundance distribution for Fe, Si, and C, with He displayed in Fig. 2. The maps indicate that the metals all show a similar pattern over the stellar surface. The minimum abundances are found at rotational phase 0.8, in both a spot at the equator and at the top pole. An equatorial spot at phase 0.5 has the maximum value on the surface. The He abundance map indicates a large spot of overabundance at phase 0.8, located in the lower hemisphere. The minimum abundance is located at phase 0.6, with a normal, solar level of He. The spot of enhanced He does not appear to be correlated with the location of the magnetic poles. Note the anti-correlation between the strength of metals and He, previously established by Oksala *et al.* (2012) from equivalent width variations.

3. Synthesis of Strömgren photometric light curves

The output from the MDI abundance analysis produces tables of a grid of values over the stellar surface for various latitudes, which can then be used as input to determine the brightness variation due to the presence of inhomogeneous spots. Krtička *et al.* (2007)

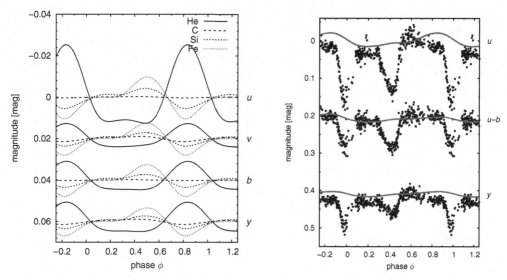

Figure 3. *Left:* Predicted light variations of σ Ori E in the Strömgren photometric system calculated using abundance maps of one element only. The abundance of other elements was fixed. *Right:* The observed Strömgren photometric light curves of σ Ori E (Hesser *et al.* 1977, black filled dots). Over-plotted is the predicted light variations (red, solid lines) computed taking into account helium, silicon, carbon, and iron surface abundance distributions.

have used this technique for the case of the He-strong star HD 37776, and find excellent agreement between synthetic and observed light curves. The output of MDI for σ Ori E was used as input for a similar treatment, in which grids of TLUSTY model atmospheres (Hubeny & Lanz 1995; Lanz & Hubeny 2007) and SYNSPEC synthetic spectra were computed, using a fixed $T_{\rm eff}$ and $\log g$, but with varying abundance values.

The radiative flux in a band c, at a distance D from the star, with radius $R_\star$ is (Mihalas 1978)

$$f_c = \left(\frac{R_\star}{D}\right)^2 \int\limits_{\substack{\text{visible}\\\text{surface}}} I_c(\theta, \Omega)\cos\theta\,\mathrm{d}\Omega. \tag{3.1}$$

The specific band intensity $I_c(\theta, \Omega)$ is obtained by interpolating between intensities $I_c(\theta, \varepsilon_{\rm He}, \varepsilon_{\rm Si}, \varepsilon_{\rm Cr}, \varepsilon_{\rm Fe})$ at each surface point, calculated from the grid of synthetic spectra. The magnitude difference in a given band is defined as

$$\Delta m_c = -2.5 \log\left(\frac{f_c}{f_c^{\rm ref}}\right), \tag{3.2}$$

where f_c is calculated from Eq. 3.1 and $f_c^{\rm ref}$ is the reference flux obtained under the condition that the mean magnitude difference over the rotational period is zero.

To obtain predicted light curves, the magnitude differences were computed for each individual rotational phase. For each element, an individual light curve was computed, with the results plotted in the left side of Fig. 3. Finally, a light curve representative of the photospheric contribution to the brightness variation is produced by combining the effects from all element contributions (right side of Fig. 3). He dominates this final produce due to its extreme abundance compared to the relatively smaller abundance changes in metals. As He has its maximum brightness at phase 0.8, it is clear that the photosphere is not responsible for the excess brightness in the observed photometric light curve seen at phase 0.6.

4. Conclusions

We have used new spectropolarimetric observations to study the elemental surface structure and magnetometry of the prototypical Bp star σ Ori E. The analysis and results presented in Section 3 clearly demonstrate that photospheric inhomogeneities on the stellar surface are not responsible for the light curve feature not fit by the RRM model. This conclusion indicates that there are remaining physics missing, likely in the treatment of the circumstellar model, which at present is quite simple. However, the model for comparison in this case lacks the proper magnetic model determined in Section 2.1. Currently, work is progressing on updating the RRM model to be more consistent with the observed physical properties of the star. This work will allow for a more accurate determination of whether the model needs to include additional physical processes, such as scattering.

References

Hesser, J. E., Ugarte, P. P., & Moreno, H. 1977, *ApJ (Letters)* 216, L31
Hubeny, I. & Lanz, T. 1995, *ApJ* 439, 875
Kochukhov, O., Mantere, M. J., Hackman, T., & Ilyin, I. 2013, *A&A* 550, A84
Kochukhov, O., Wade, G. A., & Shulyak, D. 2012, *MNRAS* 421, 3004
Krtička, J., Mikulášek, Z., Zverko, J., & Žižňovský, J. 2007, *A&A* 470, 1089
Lanz, T. & Hubeny, I. 2007, *ApJS* 169, 83
Mihalas, D. 1978, *Stellar atmospheres, 2nd edition*, W. H. Freeman and Co., San Fransisco
Oksala, M. E., Wade, G. A., Townsend, R. H. D., *et al.* 2012, *MNRAS* 419, 959
Piskunov, N. & Kochukhov, O. 2002, *A&A* 381, 736
Townsend, R. H. D., Oksala, M. E., Cohen, D. H., Owocki, S. P., & ud-Doula, A. 2010, *ApJ (Letters)* 714, L318
Townsend, R. H. D. & Owocki, S. P. 2005, *MNRAS* 357, 251
Townsend, R. H. D., Owocki, S. P., & Groote, D. 2005, *ApJ (Letters)* 630, L81
Townsend, R. H. D., Rivinius, T., Rowe, J. F., *et al.* 2013, *ApJ* 769, 33
Wade, G. A., Alecian, E., Bohlender, D. A., *et al.* 2011, in C. Neiner, G. Wade, G. Meynet, & G. Peters (eds.), *IAU Symposium*, Vol. 272 of *IAU Symposium*, pp 118–123

Discussion

KHALACK: Could you please comment about possible errors in the simulated light curve caused by taking into account just contribution from horizontal stratification only four elements?

OKSALA: We consider all element contributions from lines that we are able to detect a significant spectral feature with variability. As He dominates the light curve structure, any elements with minimal abundances will not contribute any significant amount.

TKACHENKO: When do you the light curve synthesis, do you assume a global model atmosphere, or you have a grid of models with each model in it being specific to a given surface element?

OKSALA: A grid of models are computed for each combination of varied abundance for all four elements. Each "grid" surface element is correlated with the appropriate model based on the output MDI maps.

New windows on massive stars: asteroseismology, interferometry, and spectropolarimetry
Proceedings IAU Symposium No. 307, 2014
G. Meynet, C. Georgy, J. H. Groh & Ph. Stee, eds.

© International Astronomical Union 2015
doi:10.1017/S1743921314007078

Constraining general massive-star physics by exploring the unique properties of magnetic O-stars: Rotation, macroturbulence & sub-surface convection

Jon O. Sundqvist

Universitätssternwarte, Scheinerstr. 1, D-81679 Müenchen; Germany
email: mail@jonsundqvist.com

Abstract. A quite remarkable aspect of non-interacting O-stars with detected surface magnetic fields is that they all are very slow rotators. This paper uses this unique property to first demonstrate that the projected rotational speeds of massive, hot stars, as derived using current standard spectroscopic techniques, can be severely overestimated when significant "macroturbulent" line-broadening is present. This may, for example, have consequences for deriving the statistical distribution of rotation rates in massive-star populations. It is next shown how such macroturbulence (seemingly a universal feature of hot, massive stars) is present in all but one of the magnetic O-stars, namely NGC 1624-2. Assuming then a simple model in which NGC 1624-2's exceptionally strong, large-scale magnetic field suppresses atmospheric motions down to layers where the magnetic and gas pressures are comparable, first empirical constraints on the formation depth of this enigmatic hot-star macroturbulence is derived. The results suggest it originates in the thin sub-surface convection zone of massive stars, consistent with a physical origin due to, e.g., stellar pulsations excited by the convective motions.

Keywords. stars: early-type, stars: rotation, stars: magnetic fields, convection

1. Introduction

Over the past decade, new generations of spectropolarimeters and large survey programs have revealed that roughly ∼10 % of all massive main-sequence stars harbor large-scale, organized surface magnetic field, quite similar to intermediate-mass ApBp stars (see, e.g., Wade *et al.* 2012; Grunhut, this Volume). The fields are strong, typically on the order of kG, and their fundamental origin is basically unknown, although recent observations of Herbig pre-main sequence stars point toward surviving fossils from early phases of stellar formation (Alecian *et al.* 2013). A particularly neat property of these magnetic massive stars is that they are *oblique rotators* (meaning their magnetic and rotation axes are offset), so that their rotation periods can be readily measured from the observed variation of the line-of-sight field (e.g., Borra & Landstreet 1980) or from photometric/spectral variations caused by their circumstellar magnetospheres (e.g., Howarth *et al.* 2007). This paper focuses on (non-interacting) magnetic O-stars, which all have very long measured rotation periods (likely because they have been spun down through magnetic braking by their strong stellar winds, e.g. Petit *et al.* 2013). By means of high-quality spectra collected within the Magnetism in Massive Stars project (MiMeS, Wade *et al.* 2012), I use these unique properties to examine:

- The accuracy of standard methods for inferring rotation rates of massive stars.
- General origin (and magnetic inhibition of) "macroturbulence" in hot stars.

Table 1. Stellar and magnetic parameters for the sample O-stars, including $v \sin i$ as implied from the measured rotation periods and macroturbulent velocities θ (assuming here isotropic macroturbulence, θ_G, see text). Table adapted from Sundqvist *et al.* (2013a).

Star	Spec. type	$T_{\rm eff}$ [kK]	$\log g$ [cgs]	$B_{\rm pole}$ [kG]	$P_{\rm rot}$ [d]	$v \sin i$ [km s^{-1}]	θ_G [km s^{-1}]
NGC 1624-2	O6.5-O8 f?cp	35	4.0	20	158	0	$2.2 \pm {}^{0.9}_{2.2}$
HD 191612	O6 f?p-O8 f?cp	35	3.5	2.5	538	0	$62.0 \pm {}^{0.5}_{0.5}$
HD 57682	O9 V	34	4.0	1.7	64	0	$19.2 \pm {}^{0.3}_{0.3}$
CPD -28 2561	O6.5 f?p	35	4.0	1.7	70	0	$24.3 \pm {}^{1.0}_{0.9}$
HD 37022	O7 Vp	39	4.1	1.1	15	24	$42.9 \pm {}^{0.5}_{0.6}$
HD 148937	O6 f?p	41	4.0	1.0	7	45	$54.0 \pm {}^{0.9}_{0.9}$
HD 108	O8 f?p	35	3.5	0.5	1.8×10^4	0	$64.4 \pm {}^{0.4}_{0.4}$
HD 36861	O8 III((f))	35	3.7	0	–	45	$50.0 \pm {}^{0.3}_{0.3}$

2. Rotation and macroturbulence in massive, hot stars

For most stars, it is not possible to directly measure the rotation rate. Instead one typically infers the *projected* stellar rotation, $v \sin i$ (with inclination angle i), from observed, broadened line-spectra. However, it is since long known that rotation is not the only macroscopic broadening agent operating in hot star atmospheres. The additional broadening is of very large width, typically on order ~ 50 km/s (well in excess of the photospheric speed of sound, ~ 20 km s^{-1}), and the occurrence of this "macroturbulence" seriously complicates deriving accurate $v \sin i$ rates for massive stars that are not too rapidly rotating (e.g., Howarth *et al.* 1997; Simón-Díaz & Herrero 2014). Moreover, since early-type stars lack surface convection associated with hydrogen recombination – which is responsible for such non-thermal broadening in late-type stellar atmospheres (Asplund *et al.* 2000) – the physical origin of macroturbulence in hot stars remains unclear (though see, e.g., Aerts *et al.* 2009). At the present, it is normally treated by simply introducing ad-hoc photospheric velocity fields with Gaussian distributions of speeds, assumed to be either *isotropic* or directed only *radially and/or tangentially* to the stellar surface.

Properties of the magnetic O-stars. Table. 1 summarizes relevant parameters for the sample of magnetic O-stars considered here, including a non-magnetic comparison star (HD 36861). The table includes derived values of characteristic (isotropic) macroturbulent velocities θ_G from Sundqvist *et al.* (2013a), obtained by using information about $v \sin i$ from the measured rotation periods. Note in particular two things from this table: i) The long rotation periods of the magnetic stars indeed imply $v \sin i \approx 0$ km s^{-1} for several of them, and ii) strong macroturbulent line-broadening is present in all but one of the magnetic O-stars, namely NGC 1624-2.

3. Does standard methods overestimate $v \sin i$?

I here follow Sundqvist *et al.* (2013b) and use HD 191612 and HD 108 as test-beds, the two stars in Table 1 with $v \sin i < 1$ km s^{-1} and characteristic macroturbulent velocities $\theta_G > 50$ km s^{-1}. Fig. 1 shows the results from deriving $v \sin i$ and macroturbulent velocities for HD 191612 and HD 108, using the standard Fourier Transform (FT) and Goodness-of-fit (GOF) techniques. As illustrated by the figure, the FT method derives $v \sin i$ from the position of the first minimum in Fourier space, whereas the GOF method convolves synthetic line-profiles for a range of $v \sin i$ and macroturbulent velocities, creating a standard χ^2-landscape from which a best-combination of the two parameters is

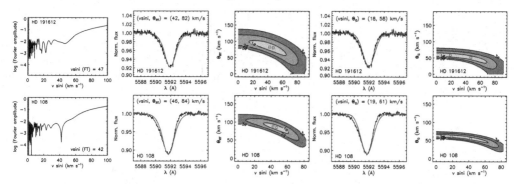

Figure 1. Projected rotation speeds $v \sin i$ and macroturbulent velocities θ for HD 191612 and HD 108, derived using standard FT (left panel) and GOF (middle/right panels) techniques. The contour-maps show $1,2,3\sigma$ confidence intervals for the fits in the $v \sin i$-θ plane. The blue and red squares on the contour-maps indicate the FT derived value and the best GOF model, respectively. The middle panel assumes a radial-tangential macroturbulence with equal contributions from both directions, θ_{RT}, and the right panel assumes isotropic velocity fields, θ_{G}. The true values for $v \sin i$ are < 1 km/s for both stars, see text. Adapted from Sundqvist *et al.* (2013b).

determined (see Simón-Díaz & Herrero 2014 for details). Fig. 1 illustrates how the FT method yields $v \sin i \approx 40 - 50\,\mathrm{km\,s^{-1}}$ for both stars, a severe overestimate compared to the true value $v \sin i < 1\,\mathrm{km\,s^{-1}}$. The best GOF model assuming radial-tangential macroturbulence also gives $v \sin i \approx 40 - 50\,\mathrm{km\,s^{-1}}$, whereas assuming isotropic macroturbulence actually results in lower $v \sin i \approx 20\,\mathrm{km\,s^{-1}}$, although then of course the results from the FT and GOF methods do not agree†. Agreement in the derived $v \sin i$ between these two methods has indeed been used as an argument in favor of the radial-tangential macroturbulence model (e.g., Simón-Díaz & Herrero 2014), but the analysis here shows clearly that such agreement does not necessarily mean the derived $v \sin i$ is correct. The GOF contour-maps in Fig. 1 further display quite wide ranges of allowed values of $v \sin i$. Particularly for isotropic macroturbulence the results are degenerate all the way down to zero rotation, rendering the "best" model from this GOF quite useless (in contrast to the well-constrained values of θ_{G} in Table 1, derived using independent knowledge of $v \sin i$).

Overall, these results demonstrate a big problem regarding deriving $v \sin i$ in the presence of a macroturbulent broadening that significantly influences the appearance of the line profile. In the case here of slow rotators, blindly applying standard methods leads to drastic overestimates of $v \sin i$, where the results also depend on the assumptions made about the unknown velocity fields causing the additional broadening. The next section now shows how we may indeed use the magnetic O-stars to also shed some light on the physical origin of this enigmatic macroturbulence.

4. Constraining the origin of macroturbulence by exploring magnetic inhibition of hot-star sub-surface convection

Using the method described in Sundqvist *et al.* (2013a), the left panel of Fig. 2 shows fitted C IV photospheric line-profiles for three stars in the sample given in Table 1, namely HD 191612, NGC 1624-2, and the non-magnetic comparison star HD 36861. Since in the optical the magnetic broadening due to the Zeeman effect is only $\sim 1 - 2\,\mathrm{km\,s^{-1}}$ per

---

† Note also that the derived characteristic velocities are quite different depending on which form of macroturbulence is assumed, due to the markedly different shapes of a disc-integrated radial-tangential velocity field model and an isotropic one.

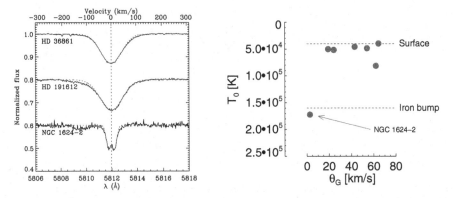

Figure 2. Left panel: Observed (black solid) and fitted (red dashed) C IV λλ5812 line profiles for three stars in our sample, as labeled in the figure. The vertical dashed line marks line center, and the continua in the two lower curves have been shifted downwards by 0.2 and 0.4 normalized flux units. **Right panel:** Atmospheric temperature T_0 (see text) vs. macroturbulent velocity θ_G for the magnetic stars in Table 1, with NGC 1624-2 explicitly labeled. The dashed blue lines mark approximate locations of the stellar surface and the iron-opacity bump. Figure adapted from Sundqvist *et al.* (2013a).

kG, both the magnetic (Table 1) and rotational contribution to the line-broadening is negligible for HD 191612, i.e., the total line broadening may be quite unambiguously associated solely with macroturbulence. Since the comparison star HD 36861 reveals very similar broad lines, this then suggests a common origin of the observed macroturbulence in magnetic and non-magnetic O-stars. By contrast, the observed line in NGC 1624-2 is qualitatively very different, much narrower and with magnetic Zeeman splitting directly visible (due to the very strong surface field, see Table 1). This indicates that the mechanism responsible for the large macroturbulent velocities in the other stars is not effective in NGC 1624-2. The quantitative analysis by Sundqvist *et al.* (2013a) results in $\theta_G = 2.2 \pm^{0.9}_{2.2}$ km/s for NGC 1624-2. Such a very low (consistent with zero) macroturbulent velocity is in stark contrast with the rest of the sample, which displays $\theta_G \approx 20 - 65\,\mathrm{km\,s^{-1}}$ (Table 1). Thus, macroturbulence seems to behave similar in non-magnetic and magnetic O-stars, except for in NGC 1624-2 where it is anomalously low or even completely absent.

A simple model for magnetic inhibition of macroturbulence. In intermediate-mass Ap-stars, it is believed that the strong magnetic field prohibits atmospheric motions between field lines and so suppresses surface convection (e.g. Balmforth *et al.* 2001; J. Landstreet, priv. comm.). The critical parameter controlling the competition between plasma and field in the atmosphere is the so-called "plasma β", the ratio between gas pressure and magnetic pressure, $\beta \equiv \frac{P_G}{P_B} = \frac{P_G}{B^2/(8\pi)}$ for magnetic field strength B. Let us now thus assume the magnetic field stabilizes the atmosphere against motions approximately down to the stellar layer at which $\beta = 1$. By adopting a very simple, classical gray model atmosphere, and assuming a fossil field with no significant horizontal variations in pressure and density between field lines, we obtain an analytic expression for the temperature T_0 in the atmosphere at which $\beta = 1$ (see Sundqvist *et al.* 2013a for details),

$$T_0 = T_{\mathrm{eff}} \left(\frac{3}{32\pi} \frac{B^2 \kappa}{g} + \frac{1}{2} \right)^{1/4} \approx 0.42\, T_{\mathrm{eff}} B^{1/2} (\kappa/g)^{1/4}, \qquad (4.1)$$

where B has units of Gauss and κ (cm^2/g) is a mean mass absorption coefficient. The second expression here neglects the 1/2 within the parenthesis, and so implicitly assumes a field strength significantly stronger than the $B \approx 400 \, (10^{-4} g/\kappa)^{1/2}$ that yields $\beta = 1$ at $T_0 = T_{\mathrm{eff}}$.

To estimate T_0 for the magnetic O-stars, the stellar parameters in Table 1 are used together with the averaged surface field for B. For simplicity, $\kappa = 1$ is further assumed for all stars; inspections of Rosseland opacities in detailed FASTWIND non-LTE model atmospheres (Puls *et al.* 2005) show that for atmospheric layers with $\tau_{\mathrm{Ross}} \geqslant 0.1$, such constant $\kappa \approx 1$ actually is a quite good opacity-estimate for Galactic O-stars that are not too evolved. The right panel of Fig. 2 shows T_0 vs. the θ_{G} values given in Table 1 for the magnetic O stars. The figure illustrates the influence of the magnetic field reaches down to much deeper layers in NGC 1624-2 than in any other star. This suggests that the physical mechanism causing the large macroturbulence in O-stars likely originates in stellar layers between 100 000 K and 200 000 K, consistent with a physical origin in the iron-peak opacity zone located roughly at $T \approx 160\,000$ K. Since the increased opacity in this sub-surface zone is believed to trigger extensive convective motions (e.g., Cantiello *et al.* 2009), this makes the analogy with suppression of surface convection in magnetic Ap-stars quite appealing.

5. Summary and conclusions

Sect. 3 in this paper shows that the presence of significant macroturbulence can result in severe overestimates of $v \sin i$ (at least for slow rotators) when applying standard spectroscopic methods (see also Simón-Díaz & Herrero 2014; Aerts *et al.* 2014). This may have important consequences, e.g., for determining the statistical distribution of rotation rates for populations of massive stars (e.g., Ramírez-Agudelo, this Volume).

The key for obtaining better constrained values of $v \sin i$ is a more robust description of the so-called macroturbulent line-broadening. Following Sundqvist *et al.* (2013a), Sect. 4 places first empirical constraints on the formation depth of such macroturbulence, locating it to the region around the iron opacity-bump at $T \approx 160\,000$ K. An attractive scenario then is that the responsible physical mechanism is related to the convection believed to occur in this region (e.g., Cantiello *et al.* 2009), perhaps via stellar pulsations excited by the convective motions (Aerts *et al.* 2009; Shiode *et al.* 2013).

Acknowledgements

JOS gratefully acknowledges support by the German DFG, under grant PU117/8-1.

References

Aerts, C., Puls, J., Godart, M., & Dupret, M.-A. 2009, *A&A* 508, 409

Aerts, C., Simon-Diaz, S., Groot, P. J., & Degroote, P. 2014, *ArXiv e-prints*

Alecian, E., Wade, G. A., Catala, C., *et al.* 2013, *MNRAS* 429, 1001

Asplund, M., Nordlund, Å., Trampedach, R., Allende Prieto, C., & Stein, R. F. 2000, *A&A* 359, 729

Balmforth, N. J., Cunha, M. S., Dolez, N., Gough, D. O., & Vauclair, S. 2001, *MNRAS* 323, 362

Borra, E. F. & Landstreet, J. D. 1980, *ApJS* 42, 421

Cantiello, M., Langer, N., Brott, I., *et al.* 2009, *A&A* 499, 279

Howarth, I. D., Siebert, K. W., Hussain, G. A. J., & Prinja, R. K. 1997, *MNRAS* 284, 265

Howarth, I. D., Walborn, N. R., Lennon, D. J., *et al.* 2007, *MNRAS* 381, 433

Petit, V., Owocki, S. P., Wade, G. A., *et al.* 2013, *MNRAS* 429, 398

Puls, J., Urbaneja, M. A., Venero, R., *et al.* 2005, *A&A* 435, 669

Shiode, J. H., Quataert, E., Cantiello, M., & Bildsten, L. 2013, *MNRAS* 430, 1736

Simón-Díaz, S. & Herrero, A. 2014, *A&A* 562, A135

Sundqvist, J. O., Petit, V., Owocki, S. P., *et al.* 2013a, *MNRAS* 433, 2497

Sundqvist, J. O., Simón-Díaz, S., Puls, J., & Markova, N. 2013b, *A&A* 559, L10

Wade, G. A. & Grunhut, J. H., MiMeS Collaboration 2012, in A. C. Carciofi & T. Rivinius (eds.), *Circumstellar Dynamics at High Resolution*, Vol. 464 of *Astronomical Society of the Pacific Conference Series*, p. 405

Discussion

HERRERO: I agree that we badly need a better description of the broadening we observe in O-stars. Conserning the accuracy of classical methods (FT, GOF), in the recent paper by Simon-Diaz and myself we show that if there are other broadening mechanisms than $v \sin i$ and macroturbulence, we may obtain too large $v \sin i$ values. On the other hand, in θ^1 Ori C we get $v \sin i$ values that agree with the rotation period derived from spectroscopic variations and B-field inclinations.

SUNDQVIST: Yes, I am aware that in Simón-Díaz & Herrero (2014) you show that $v \sin i$ may be overestimated in the presence of large *micro*-turbulence. I was not aware, however, that you obtained good agreement for θ^1 Ori C. Note that I did not include this star here, since its rotation period 15 days actually implies a "non-zero" rotation speed. As such, the exact value of $v \sin i$ then depends on the uncertain stellar radius. But we should definitely investigate this further.

IBADOV: What can you say about generation of spots on massive stars surfaces, like sun-spots?

SUNDQVIST: Note first that the magnetic fields I have been discussing here are large-scale, organized fields, with a dominant dipolar component and presumably of fossil origin. These fields are quite different from the complex, dynamo-generated fields in the Sun and other cool stars. That said, there have been some investigations regarding how a hypothetical magnetic field generated in the near-surface convection zone of massive stars could give rise to spots on the surface (e.g. Cantiello & Braithwaite 2011). Such spots, however, would be *hot and bright*, since the energy near the surface of massive stars is transported by radiation.

LOBEL: An important spectroscopic characteristic of yellow hypergiants ($T_{\mathrm{eff}} < 10\,kK$) are very broad photospheric absorption lines. They are slow rotators with large supersonic macroturbulence. Would you attribute its physical origin to g-modes in these cool massive stars as well?

SUNDQVIST: That is difficult to say. The situation definitely seems reminiscent of that in blue supergiants, in which g-modes may indeed be the physical origin (e.g., Aerts *et al.* 2009). But without looking further into the situation, I unfortunately cannot say much more than that at the moment.

AERTS: Remark: We are including velocity fields due to 2-D (ideally in the future in 3-D) hydro simulations, and we do get broadened wings as suggested observationally. This is probably best explained with pulsations in gravity modes, as a "natural" explanation for un-evolved B stars near the main sequence.

New windows on massive stars: asteroseismology, interferometry, and spectropolarimetry
Proceedings IAU Symposium No. 307, 2014
G. Meynet, C. Georgy, J. H. Groh & Ph. Stee, eds.

© International Astronomical Union 2015
doi:10.1017/S174392131400708X

Linear line spectropolarimetry as a new window to measure 2D and 3D wind geometries

Jorick S. Vink

Armagh Observatory, College Hill, BT61 9DG Armagh, Northern Ireland
email: jsv@arm.ac.uk

Abstract. Various theories have been proposed to predict how mass loss depends on the stellar rotation rate, both in terms of its strength, as well as its latitudinal dependence, crucial for our understanding of angular momentum evolution. Here we discuss the tool of linear spectropolarimetry that can probe the difference between mass loss from the pole versus the equator. Our results involve several groups of O stars and Wolf-Rayet stars, involving Oe stars, Of?p stars, Onfp stars, as well as the best candidate gamma-ray burst progenitors identified to date.

Keywords. techniques: polarimetric, circumstellar matter, stars: early-type, stars: emission-line, Be, stars: mass loss, stars: rotation, stars: winds, outflows, stars: Wolf-Rayet

1. Introduction

Ultimately, we would like to understand massive stars and their progeny both locally as well as in the distant Universe. What is clear is that rotation, mass loss, and the link between them, play a pivotal role in the fate of massive stars. However, in order to test mass-loss predictions for rotating stars, we need to probe the density contrast between the stellar pole and equator. In the local Universe, this may potentially be achievable through the technique of long-baseline interferometry, as discussed during this meeting. However, in order to determine wind asymmetry in the more distant Universe we necessarily rely on the technique of *linear* spectropolarimetry. The only limiting factor is then the collecting power of the mirror of the largest telescopes.

2. 2D Wind Predictions

Until 3D radiation transfer models with 3D hydrodynamics become available, theorists have necessarily been forced to make assumptions with respect to either the radiative transfer (e.g. by assuming a power law approximation for the line force due to Castor *et al.* 1975) or the hydrodynamics, e.g. by assuming an empirically motivated wind terminal velocity in Monte Carlo predictions (Abbott & Lucy 1985; Vink *et al.* 2000). Albeit recent 1D and 2D models of Müller & Vink (2008, 2014) no longer require the assumption of an empirical terminal wind velocity.

There are 2D wind models on the market that predict the wind mass loss predominately emanating from the equator (Friend & Abbott 1986; Bjorkman & Cassinelli 1993; Lamers & Pauldrach 1991; Pelupessy *et al.* 2000), whilst other models predict higher mass-loss rates from the pole, in particular as a result of the von Zeipel (1924) theorem, resulting in a larger polar Eddington factor than the equatorial Eddington factor (Owocki *et al.* 1996; Petrenz & Puls 2000; Maeder & Meynet 2000; Müller & Vink 2014).

The key point is that mass loss from the equator results in more angular momentum loss than would 1D spherical or 2D polar mass loss, so we need 2D data to test this.

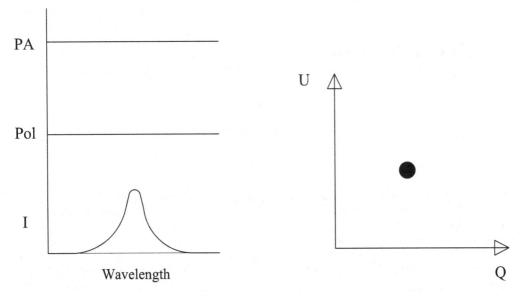

Figure 1. Cartoon indicating "no line effect'. On the left, polarization spectrum "triplot" and a Stokes QU diagram on the right. A typical Stokes I emission is shown in the lower panel of the triplot, the %Pol in the middle panel, while the Position Angle (PA) is sketched in the upper panel of the triplot. See Vink *et al.* (2002) for further details.

3. Line polarization versus depolarization

Whilst *circular* Stokes V spectropolarimetry is oftentimes employed to measure stellar magnetic fields, *linear* Stokes QU polarimetry can be utilized to measure large-scale 2D asymmetry in a stellar wind or any other type of circumstellar medium, such as a disk. In this sense, the Stokes QU plane plays an analogous role to the interferometric UV plane, with the additional advantage that it can measure the smallest spatial scales, such as the inner disk holes of order just a few stellar radii in pre-main sequence (PMS) stars (Vink *et al.* 2005), which would otherwise remain "hidden", or the driving region of stellar winds in massive stars, that we explore in the following.

In principle, linear continuum polarimetry would already be able to inform us about the presence of an asymmetric (e.g. a disk or flattened wind) structure on the sky, but in practice, this issue is complicated by the roles of intervening circumstellar and/or interstellar dust, as well as instrumental polarization. The is one of the reasons linear *spectro*polarimetry, measuring the change in the degree of linear polarization across emission lines is such a powerful tool, as "clean" or "intrinsic" information can be directly obtained from the QU plane. The second reason is the additional bonus that it may provide kinematic information of the flows around PMS as well as massive stars.

Figures 1–3 show linear line polarization cartoons (both in terms of polarization "triplot" spectra and Stokes QU planes) for the case that the spatially unresolved object under consideration is (i) spherically symmetric on the sky showing "no line effect", (ii) asymmetric showing line "depolarization" where the emission line simply acts to "dilute" the polarized continuum, or (iii) cases where the line effects are more subtle, involving position angle (PA) flips across intrinsically polarized lines.

Whilst the third situation of intrinsic line polarization in a rotating disk has been encountered in PMS (see Vink *et al.* 2005), it is the second case of "depolarization" that is most familiar to the massive-star community through its application to classical Be stars, starting as early as the 1970s (see the various works by Poeckert, Marlborough,

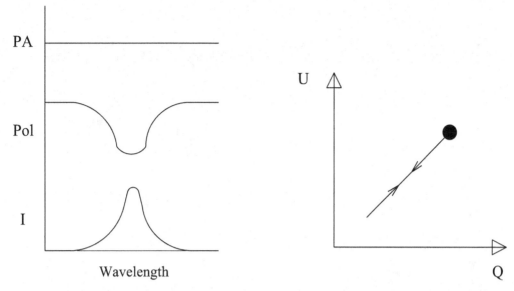

Figure 2. Cartoon indication "depolarization" or "dilution". Note that the depolarisation across the line is as broad as the Stokes *I* emission. Depolarisation translates into Stokes *QU* space as a linear excursion. See Vink *et al.* (2002) for further details.

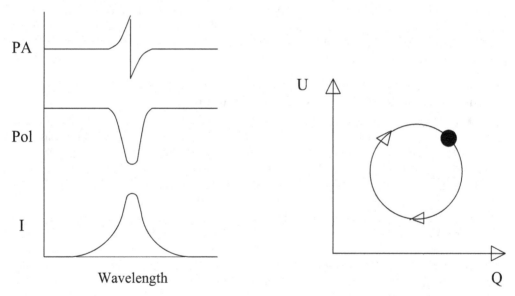

Figure 3. Cartoon indicating a compact source of line photons scattered off a *rotating* disk. Note that the polarisation signatures are relatively narrow compared to the Stokes *I* emission. The PA flip is associated with a loop in Stokes *QU* space. See Vink *et al.* (2002, 2005) for further details.

Brown, Clarke, and McLean). Interestingly, the same method has in more recent years also been applied to Oe stars, the alleged more massive counterparts of Be stars, see Fig. 4. Note that although the Oe star HD 120678 (on the right hand side of Fig. 4) has a significant observed level of linear polarization, the lack of a line effect implies that the object is not intrinsically polarized (Vink *et al.* 2009). This could either mean the object is spherically symmetric or that it has a disk that is too "pole on" to provide intrinsic

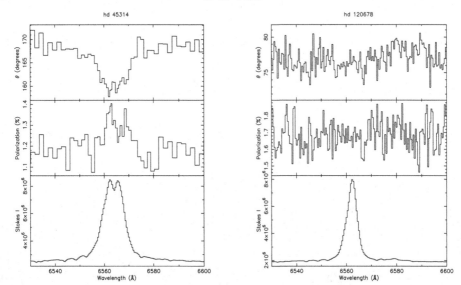

Figure 4. Hα line polarization "triplots" of the Oe stars HD 45314 and HD 120678. HD 45314 shows a line effect indicating that it is intrinsically polarized, but HD 120678 is not intrinsically polarized. See Vink *et al.* (2009) for further details.

polarization. It is for these reasons vital to consider a *sample* of objects. For Oe stars Vink *et al.* (2009) found that the incidence of line effects (1/6) was much lower than for Be stars. This implies that the chance the Oe and the Be stars are drawn from the same parent distribution is small, providing relevant constraints on the formation of Be stars.

4. Survey results of O and Wolf-Rayet winds

We now turn to more massive stars with stronger winds than Oe/Be stars. Linear spectropolarimetry results have been performed on relatively large samples (of order 40-100) for both O (Harries *et al.* 2002; Vink *et al.* 2009) and Wolf-Rayet (WR) stars (Harries *et al.* 1998; Vink 2007), and the key result from these surveys is that the vast majority of 80% of them is to first order spherically symmetric. This is of key importance for the accuracy of mass-loss predictions from 1D models for rotating stars.

However, the above studies also found a number of interesting *exceptions*. With respect to O stars, Vink *et al.* (2009) found that certain O-type subgroups involving Of?p and Onfp class are more likely polarized than the garden-variety of spherical O-stars. For instance, Vink *et al.* (2009) highlighted that HD 108 is linearly polarized, which may be related to its probably magnetic properties. Indeed, it was later found that HD 108 and several other Of?p stars form a magnetic sub-class. The line effects in the Onfp stars (involving famous objects like λ Cep and ζ Pup) may may involve intrinsic line polarization effects due to the rapid rotation of this O-type subgroup in addition to (or instead of) depolarization.

Turning to WR stars, Vink *et al.* (2011) and Gräfener *et al.* (2012) uncovered that the small 20% minority of WR stars that display a depolarization line effect indicating stellar rotation are highly significantly correlated with the subset of WR stars that have ejecta nebulae. These objects have most likely only recently transitioned from a red sugergiant (RSG) or luminous blue variable (LBV) phase. As these presumably youthful WR stars have yet to spin-down, they are the best candidate gamma-ray burst (GRB) progenitors identified to date. However, in our own Milky Way these WR stars are still expected to spin down before explosion (due to WR winds). However, in lower metallicity (Z) environments WR stars are thought to be weaker and WR stars in low Z environments, such as those studied in the Magellanic Clouds may offer the best way to directly pinpoint GRB progenitors (Vink 2007).

5. Future

In addition to the quest for WR GRB progenitors, there is a whole range of interesting wind physics to be constrained from linear spectropolarimetry. The main limitation at this point is still sensitivity. We are currently living in an exciting time as we are at a point where the possibility of extremely large telescopes (ELTs) may become reality. If these telescopes materialize with the required polarization optics, we might – for the first time in history – be able to obtain spectropolarimetric data at a level of precision that has been feasible with 1D Stokes I data for more than a century. It is really important to note that current 3D Monte Carlo radiative transfer is well able to do the required modelling, but the main limitation is the necessary 3D data!

Another interesting future application will involve polarimetric monitoring. Whilst we now know that on large scales the 1D approximation is appropriate for stellar winds, we have also become aware of the intrinsic 3D clumpy nature of stellar winds on smaller scales (but with macroscopic implications!). In particular the existence of wind clumps on small spatial scales near the stellar photosphere (Cantiello *et al.* 2009) has been confirmed by linear polarization variability studies (Davies *et al.* 2007), but to probe further – mapping wind clumps in detail – we need good monitoring data.

References

Abbott, D. C. & Lucy, L. B. 1985, *ApJ* 288, 679
Bjorkman, J. E. & Cassinelli, J. P. 1993, *ApJ* 409, 429
Cantiello, M., Langer, N., Brott, I., *et al.* 2009, *A&A* 499, 279
Castor, J. I., Abbott, D. C., & Klein, R. I. 1975, *ApJ* 195, 157
Davies, B., Vink, J. S., & Oudmaijer, R. D. 2007, *A&A* 469, 1045
Friend, D. B. & Abbott, D. C. 1986, *ApJ* 311, 701
Gräfener, G., Vink, J. S., Harries, T. J., & Langer, N. 2012, *A&A* 547, A83
Harries, T. J., Hillier, D. J., & Howarth, I. D. 1998, *MNRAS* 296, 1072
Harries, T. J., Howarth, I. D., & Evans, C. J. 2002, *MNRAS* 337, 341
Lamers, H. J. G. & Pauldrach, A. W. A. 1991, *A&A* 244, L5
Maeder, A. & Meynet, G. 2000, *A&A* 361, 159
Müller, P. E. & Vink, J. S. 2008, *A&A* 492, 493
Müller, P. E. & Vink, J. S. 2014, *A&A* 564, A57
Owocki, S. P., Cranmer, S. R., & Gayley, K. G. 1996, *ApJ (Letters)* 472, L115
Pelupessy, I., Lamers, H. J. G. L. M., & Vink, J. S. 2000, *A&A* 359, 695
Petrenz, P. & Puls, J. 2000, *A&A* 358, 956
Vink, J. S. 2007, *A&A* 469, 707
Vink, J. S., Davies, B., Harries, T. J., Oudmaijer, R. D., & Walborn, N. R. 2009, *A&A* 505, 743
Vink, J. S., de Koter, A., & Lamers, H. J. G. L. M. 2000, *A&A* 362, 295
Vink, J. S., Drew, J. E., Harries, T. J., & Oudmaijer, R. D. 2002, *MNRAS* 337, 356
Vink, J. S., Gräfener, G., & Harries, T. J. 2011, *A&A* 536, L10
Vink, J. S., Harries, T. J., & Drew, J. E. 2005, *A&A* 430, 213
von Zeipel, H. 1924, *MNRAS* 84, 665

Discussion

HENRICHS: In your very nice talk, you showed polarisation data of the HeII 4686 line of
λ Cep. Since this profile is known to change in less than one hour (see our poster), it
would be very interesting to see how the polarisation would change, in the hope to model
the geometry. What would be a typical exposure time for this bright star? Would it be
feasible?

VINK: It is a bright object, so exposure time will be short. So yes, it is possible!

NAZÉ: A comment: Linear polarisation is important, but must be detected with high
sensitivity. In your 2009 paper, you detected no signal for θ' Ori C while ud Doula
models and all other observations (UV, X-rays, spectropolarimetry) agree on a very
axisymmetric structure of the wind down to photosphere.

VINK: Current linear polarization sensitivity is sufficient to probe density contrasts between the stellar equator and pole of as little as 1.25. Why the θ' Ori C model you refer
to seems capable to reproduce 1D data, but does not seem to reproduce 3D data, is an
issue that needs to be clarified.

MEYNET: You said that O/WR winds are spherical and that this is compatible with
your theory. This would also be compatible with the formula we obtained with Maeder
since most, if not all, these stars are rotating very far from the critical velocity. Would
you agree?

VINK: Yes, I said the 80% majority is spherical because of $V_{\mathrm{rot}} < 300\,\mathrm{km\,s^{-1}}$. We need
to test (by comparison with the data) if your formula also agrees with this. It might well
be possible.

Jorick Vink

New windows on massive stars: asteroseismology, interferometry, and spectropolarimetry
Proceedings IAU Symposium No. 307, 2014
G. Meynet, C. Georgy, J. H. Groh & Ph. Stee, eds.

© International Astronomical Union 2015
doi:10.1017/S1743921314007091

Discovery of Secular Evolution of the Atmospheric Abundances of Ap Stars

J. D. Bailey[1], J. D. Landstreet[2,3] and S. Bagnulo[3]

[1] Max Planck Institut für extraterrestrische Physik, Giessenbachstrasse 1, 85748 Garching, Germany
email: jeffbailey@mpe.mpg.de

[2] Department of Physics and Astronomy, The University of Western Ontario, London, Ontario, N6A 3K7, Canada

[3] Armagh Observatory, College Hill, Armagh, BT61 9DG, Northern Ireland, UK

Abstract. The stars of the middle main-sequence have relatively quiescent outer layers, and unusual chemical abundance patterns may develop in their atmospheres, revealing the action of such subsurface phenomena as gravitational settling and radiatively driven levitation of trace elements, and their competition with mixing processes such as turbulent diffusion. We report the discovery of the time evolution of such chemical tracers through the main-sequence lifetime of magnetic chemically peculiar stars.

Keywords. stars: abundances, stars: chemically peculiar, stars: magnetic fields

1. Introduction

The A– and late B–type main sequence stars do not have any single powerful process acting to force the surface layers to retain essentially their initial chemical composition. As a result, many A and B stars show chemical abundances remarkably different from those measured in the Sun. Among A and B stars there are several families of distinctive compositional patterns that reflect different broad sets of physical conditions. The compositional peculiarities vary rather strongly with effective temperature (and thus with the momentum carried by the outflowing radiation that levitates some trace elements) and are very different, depending on whether the star has a strong magnetic field or not. These chemical peculiarities provide powerful probes of invisible processes occurring beneath the visible layers of all kinds of stars, particularly of upward and downward diffusion (Landstreet 2004). It is therefore of obvious interest to observe and characterise the importance and time-scale of these phenomena.

A major difficulty of interpreting observed chemical signatures has been that the ages of isolated (field) A and B stars can be estimated only approximately. We have dealt with this issue by performing accurate abundance analysis on a sample of high-resolution spectra of magnetic A and B stars that are open cluster members. The ages of cluster members can be determined with much better accuracy than that of individual stars, and the cluster age applies to all its members (which are presumed to have formed essentially contemporaneously) (Bagnulo *et al.* 2006; Landstreet *et al.* 2007). We have determined the chemical compositions of 15 stars with masses of $3.5 \pm 0.5\,M_\odot$ and ages between $5 \cdot 10^6$ yr and $2.5 \cdot 10^8$ yr. In effect, by studying the changes in average atmospheric chemical composition with cluster age, we are observing the time evolution of the chemistry of a $3.5\,M_\odot$ star through its main sequence life.

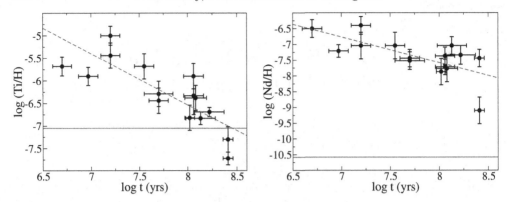

Figure 1. Observed evolution of the abundances of Ti and Nd with main sequence age for a
3.5 $M_\odot$ star. The solar abundance ratio is plotted as a solid red line.

2. Methods

To model the atmospheric abundances of the magnetic Ap stars, the spectrum synthesis program ZEEMAN was used (Landstreet 1988; Wade *et al.* 2001). When performing
the abundance analysis, a uniform atmospheric abundance distribution over the stellar
surface was assumed. For each star, a simple dipolar magnetic field with the polar field
value B_d approximately three time B_{rms} was assumed, with the line-of-sight parallel to
the magnetic field axis. The microturbulence parameter was set to 0 (because the magnetic field suppresses any convective motions). When multiple spectra were available, the
two that exhibit the greatest differences were analysed, and average abundance of the
two modelled spectra was adopted.

3. Results

During the main sequence life of a 3.5 $M_\odot$, magnetic chemically peculiar A–type (Ap)
star, the abundances:
- of the light elements (O, Mg, Si) remain roughly constant. Compared to the Sun, O
and Mg are in general underabundant, whereas Si is always overabundant.
- of the Fe-peak (Ti, Cr, Fe) and rare-earth (Pr, Nd) elements decrease. The former evolve from large overabundances to nearly solar values. The latter exhibit drastic
overabundances in young Ap stars (of the order of 10^4 times greater than in the Sun)
decreasing to values still about 100 times larger than the solar ratios.
- of He unexpectedly increases from about 1% to 10% of the solar He abundance. The
nature of this increase is not yet understood.

Fig. 1 shows two examples of these results. We refer the reader to Bailey *et al.*
(2014),where a complete discussion of this work is provided.

References

Bagnulo, S., Landstreet, J. D., Mason, E., *et al.* 2006, *A&A* 450, 777
Bailey, J. D., Landstreet, J. D., & Bagnulo, S. 2014, *A&A* 561, A147
Landstreet, J. D. 1988, *ApJ* 326, 967
Landstreet, J. D. 2004, in J. Zverko, J. Ziznovsky, S. J. Adelman, & W. W. Weiss (eds.), *The
    A-Star Puzzle*, Vol. 224 of *IAU Symposium*, pp 423–432
Landstreet, J. D., Bagnulo, S., Andretta, V., *et al.* 2007, *A&A* 470, 685
Wade, G. A., Bagnulo, S., Kochukhov, O., *et al.* 2001, *A&A* 374, 265

New windows on massive stars: asteroseismology, interferometry, and spectropolarimetry
Proceedings IAU Symposium No. 307, 2014
G. Meynet, C. Georgy, J. H. Groh & Ph. Stee, eds.

© International Astronomical Union 2015
doi:10.1017/S1743921314007108

The magnetic field of ζ Ori A

A. Blazère[1], C. Neiner[1], J-C. Bouret[2], A. Tkachenko[3]† and the MiMeS collaboration

[1] LESIA, Observatoire de Paris, UMR 8109 du CNRS, UPMC, Université Paris-Diderot, 5 place Jules Janssen, 92195 Meudon, France
email: aurore.blazere@obspm.fr

[2] Aix-Marseille University, CNRS, LAM, UMR 7326, 13388 Marseille, France

[3] Instituut voor Sterrenkunde, KU Leuven, Celestijnenlaan 200D, B-3001 Leuven, Belgium

Abstract. Magnetic fields play a significant role in the evolution of massive stars. About 7% of massive stars are found to be magnetic at a level detectable with current instrumentation (Wade *et al.* 2013) and only a few magnetic O stars are known. Detecting magnetic field in O stars is particularly challenging because they only have few, often broad, lines to measure the field, which leads to a deficit in the knowledge of the basic magnetic properties of O stars. We present new spectropolarimetric Narval observations of ζ Ori A. We also provide a new analysis of both the new and older data taking binarity into account. The aim of this study was to confirm the presence of a magnetic field in ζ Ori A. We identify that it belongs to ζ Ori Aa and characterize it.

Keywords. stars: magnetic fields, stars: early-type, stars: individual (ζ Ori A)

1. Introduction

A magnetic field seems to have been detected in the supergiant O 9.5I star ζ Ori A (Bouret *et al.* 2008). This magnetic field is the weakest ever reported in a massive star. Thanks to this measurement of the magnetic field one can locate ζ Ori A in the magnetic confinement-rotation diagram (Petit *et al.* 2013). ζ Ori A is the only known magnetic massive star with a confinement parameter below 1, i.e. without a magnetosphere. However, Hummel *et al.* (2013) recently found that ζ Ori A is a O9.5I+B1IV binary star with a period of 2687.3 ± 7.0 days, so the magnetic field of ζ Ori A and its confinement parameter might have been wrongly estimated.

2. Spectropolarimetric analysis

Spectropolarimetric data were obtained with Narval at TBL between 2008 and 2012. To improve the signal-to-noise ratio, we have coadded spectra obtained on the same night, leading to 36 average spectra. We applied the well-known and commonly used Least-Squares Deconvolution (LSD) technique (Donati *et al.* 1997) on each average spectrum. The mask is created from a list of lines extracted from VALD (Piskunov *et al.* 1995; Kupka & Ryabchikova 1999) using the effective temperature $T_{\rm eff} = 30000$ K and the $\log g = 3.25$. We then extracted LSD Stokes I and V profiles for each night as well as null (N) polarization profiles to check for spurious signatures. We see clear Zeeman signatures some nights (see left panel of Fig. 1). ζ Ori A is thus confirmed to be magnetic. However, ζ Ori A is a binary and we ignore which component of the binary is magnetic, or if both

† Postdoctoral Fellow of the Fund for Scientific Research (FWO), Flanders, Belgium

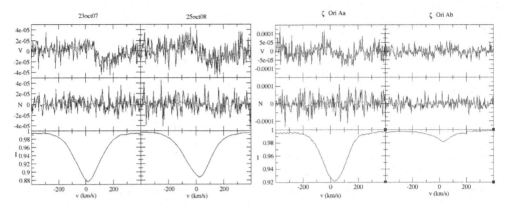

Figure 1. *Left:* Examples of I (bottom), null N (center) and Stokes V (top) profiles for two nights. *Right:* I (bottom), null N (center) and Stokes V (top) profiles for October 25, 2008, for ζ Ori Aa (left) and ζ Ori Ab (right). The red lines represent the averaged signal.

components are magnetic. To provide an answer to this question we must disentangle the composite spectra.

3. Disentangling the spectra

We used the Fourier-based formulation of the spectral disentangling method (Hadrava 1995) as implemented in the FDBINARY code (Ilijic *et al.* 2004) to try to separate the individual spectra of both components of the ζ Ori A binary system. This method failed due to a bad phase coverage of the binary period.

Therefore we used a second method to disentangle the spectra: for each component of ζ Ori A we computed a synthetic spectrum with their respective temperature and log g. Based on these synthetic spectra, we checked which lines in the observations comes from only one of the component. We created two distinct masks with lines from only one of the component and we used LSD with these masks. We find that the magnetic field is present in ζ Ori Aa and not in ζ Ori Ab (see right panel of Fig. 1).

4. Conclusion

We confirm that ζ Ori A is a magnetic star. The magnetic field is present in the supergiant ζ Ori Aa and no magnetic field is detected in its companion ζ Ori Ab.

References

Bouret, J.-C., Donati, J.-F., Martins, F., *et al.* 2008, *MNRAS* 389, 75

Donati, J.-F., Semel, M., Carter, B. D., Rees, D. E., & Collier Cameron, A. 1997, *MNRAS* 291, 658

Hadrava, P. 1995, *A&AS* 114, 393

Hummel, C. A., Rivinius, T., Nieva, M.-F., *et al.* 2013, *A&A* 554, A52

Ilijic, S., Hensberge, H., Pavlovski, K., & Freyhammer, L. M. 2004, Vol. 318 of *ASPCS*, p. 111

Kupka, F. & Ryabchikova, T. A. 1999, *Publications de l'Observatoire Astronomique de Beograd* 65, 223

Petit, V., Owocki, S. P., Wade, G. A., *et al.* 2013, *MNRAS* 429, 398

Piskunov, N. E., Kupka, F., Ryabchikova, T. A., Weiss, W. W., & Jeffery, C. S. 1995, *A&AS* 112, 525

Wade, G. A., Grunhut, J., Alecian, E., *et al.* 2013, *IAUS 302, ArXiv e-prints 1310.3965*

New windows on massive stars: asteroseismology, interferometry, and spectropolarimetry
Proceedings IAU Symposium No. 307, 2014
G. Meynet, C. Georgy, J. H. Groh & Ph. Stee, eds.

© International Astronomical Union 2015
doi:10.1017/S174392131400711X

Spectropolarimetric study of selected cool supergiants

V. Butkovskaya, S. Plachinda and D. Baklanova

Crimean Astrophysical Observatory of Taras Shevchenko National University of Kyiv, 98409,
Nauchny, Crimea, Ukraine
email: varya@crao.crimea.ua

Abstract. Cool supergiants offer a good opportunity to study the interplay of magnetic fields and stellar evolution. We present the results of spectropolarimetric study of the cool supergiants and classical Cepheids η Aql and ζ Gem.

Keywords. Stars: magnetic fields, supergiants, Cepheids, stars: individual (η Aql, ζ Gem)

1. Introduction

Supergiants are the descendants of massive O- and B-type main sequence stars. In recent years, magnetic field in the atmospheres of a small number of such stars has been confirmed by direct measurements (Plachinda 2005, Grunhut 2010, Auriere 2010). Because these fields are usually quite small, the real surface geometry as well as the origin of the magnetic field on cool supergiant stars have not yet been understood. Another interesting but unsolved problem is the magnetic field variability due to pulsation period of Cepheids.

We have analyzed circular polarized spectra of two Cepheids (η Aql and ζ Gem). The spectra were obtained in the wavelength region 6210–6270 Å with the coudé spectrograph mounted on the 2.6-m Shajn telescope of the Crimean Astrophysical Observatory. The technique of longitudinal magnetic field calculation is described in detail by Butkovskaya & Plachinda (2007).

2. Magnetic field and pulsation of η Aql and ζ Gem

Plachinda (2000) and Butkovskaya (2014a,b) obtain that the longitudinal magnetic field of η Aql varies with the radial pulsation period of 7.176726 day in 1991, 2002, and 2004, but find virtually no changes in 2010 and 2012. The amplitude B, mean field B_0, and phases of maximum and minimum field vary from year to year. The reason for this behavior of the magnetic field remains unknown.

A spectropolarimetric study of ζ Gem has been performed during 8 nights in 2003 and 2004. In December 2002–January 2003, the longitudinal magnetic field varies in range from –9 to 26 G, while the difference between maximal and minimal values is 35 G (see left panel of Fig. 1). In December 2003–March 2004, the longitudinal magnetic field varies from 4 to 12 G (mean B per 5 nights is 8.2 ± 1.6 G and $B/\sigma = 5.1$).

In the right panel of Fig. 1, the radial velocity and longitudinal magnetic field are folded in phase with the pulsation period. We use the ephemeris JD $= 2443805.927 + 10.15073E$, where E is the number of pulsation cycle.

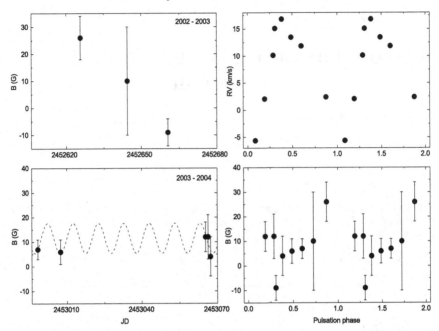

Figure 1. Longitudinal magnetic field of ζ Gem (left panels). Sinusoidal fit is shown by dashed line. Radial velocity (top right panel) and longitudinal magnetic field (bottom right panel) of ζ Gem folded in phase with the 10.15073-day radial pulsation period.

3. Conclusion

Due to long pulsation periods of Cepheids and possibly unstable behavior of the magnetic field from year to year, long sets of observations during every single year are needed.

Acknowledgements

V. V. Butkovskaya acknowledges the support from an IAU travel grant to attend the conference.

References

Auriere, M. e. a. 2010, *A&A* 516, L2
Butkovskaya, V. 2014a, in *Magnetic fields throughout stellar evolution*, Vol. 302 of *IAU Symposium, in press*
Butkovskaya, V. 2014b, in *Precision Asteroseismology*, Vol. 301 of *IAU Symposium*, pp 393–394
Butkovskaya, V. & Plachinda, S. 2007, *A&A* 469, 1069
Grunhut, J. e. a. 2010, *MNRAS* 408, 2290
Plachinda, S. 2000, *A&A* 360, 642
Plachinda, S. 2005, *Astrophysics* 48, 9

New windows on massive stars: asteroseismology, interferometry, and spectropolarimetry
Proceedings IAU Symposium No. 307, 2014
G. Meynet, C. Georgy, J. H. Groh & Ph. Stee, eds.

© International Astronomical Union 2015
doi:10.1017/S1743921314007121

Beam me up, Spotty: Toward a new understanding of the physics of massive star photospheres

Alexandre David-Uraz[1], Gregg Wade[2] and Stan Owocki[3]

[1] Queen's University, Canada
email: adavid-uraz@astro.queensu.ca

[2] Royal Military College, Canada

[3] University of Delaware, USA

Abstract. For 30 years, cyclical wind variability in OB stars has puzzled the astronomical community. Phenomenological models involving co-rotating bright spots provide a potential explanation for the observed variations, but the underlying physics remains unknown. We present recent results from hydrodynamical simulations constraining bright spot properties and compare them to what can be inferred from space-based photometry. We also explore the possibility that these spots are caused by magnetic fields and discuss the detectability of such fields.

Keywords. Hydrodynamics, methods: numerical, stars: winds, outflows

1. Introduction

Massive stars exhibit cyclical spectral variability on timescales of hours to days. One seemingly ubiquitous manifestation of these variations is the presence of so-called "discrete absorption components" (DACs, Howarth & Prinja 1989): additional absorption features within the P Cygni absorption troughs of wind-sensitive UV resonance lines which progress from low to near-terminal velocity. They are believed to be rotationally modulated. The generally-accepted hypothesis which explains their formation is the "corotating interaction regions" model (CIR, Mullan 1986). Bright spots on the surface of the star locally drive an enhanced wind, leading to spiral-like structures (Cranmer & Owocki 1996, henceforth referred to as CO96). The most important feature of this model is the presence of "velocity kinks" in the wind; the driving of extra material by the bright spots cannot be sustained (the wind is overloaded) and that material eventually decelerates, which results in a velocity plateau, thus leading to an accumulation of matter at a given velocity which causes the extra absorption associated with DACs.

2. Modern constraints and revisited model

CO96 provide a phenomenological explanation for the DAC phenomenon. Consequently, the presence of bright spots was not *a priori* physically motivated (nor constrained). However, with the advent of space-based photometry (e.g. the MOST space telescope), the precisely-determined amplitude of light-curve variations places very strong constraints on the size and contrast of putative spots on the surface of the star. Ramiaramanantsoa *et al.* (2014) find a 10 mmag amplitude for ξ Per, constraining any spots present to be much more modest than those used by CO96. It is unclear whether velocity kinks can be formed with such modest spots. One possible way to facilitate kink formation is by including ionization effects, which were not taken into account by CO96.

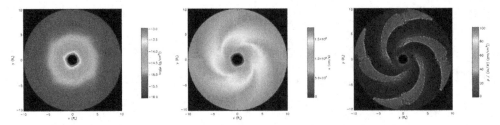

Figure 1. Preliminary 2D hydrodynamic models show that four spots with 10° radius and 33% brightness enhancement allow us to qualitatively reproduce the conditions necessary to obtain DACs. From left to right: density profile, radial velocity profile, density divided by the radial velocity gradient (this proxy variable acts like the optical depth and the bright trailing edges indicate the presence of velocity kinks).

Following up on that work, we aim to study the effects of adding the ionization correction (δ) in hydrodynamic wind simulations, using the VH-1 package. 1D simulations show that for high values of the δ parameter (Abbott 1982), the wind becomes overloaded and we obtain a coasting solution, even without spots. While this work is quite preliminary, 2D hydrodynamical simulations based on realistic stellar parameters (in this case, ξ Per, with 10° spots and a 33% brightness enhancement) and including additional wind physics (ionization effect, finite disk correction) show that we can reproduce CIRs and even obtain the desired velocity kinks (see Fig. 1).

3. The role of magnetic fields

Magnetic fields constitute a popular hypothesis to explain the surface nonuniformities apparently needed to form CIRs. While only a small fraction of OB stars are known to host detectable fields (about 7%, Wade *et al.* 2013), most known magnetic massive stars have a large scale, essentially dipolar field. However, David-Uraz *et al.* (2014) effectively rule out large-scale magnetic fields as the physical cause of DACs. Recent models (e.g. Cantiello *et al.* 2009) suggest the existence of a sub-surface convection layer in massive stars which could produce small-scale surface magnetic fields. Kochukhov & Sudnik (2013) have characterized the detectability of randomly distributed magnetic spots. For $v \sin i < 50\,\mathrm{km\,s^{-1}}$, one would expect to detect 10° magnetic spots with a field strength of over 100 G (with a 0.5 filling factor); to produce a 10 mmag photometric variation, such a spot should have a field strength of about 330 G (using the formula of David-Uraz *et al.* 2014), which should therefore be detectable with current instrumentation. Further work on both the numerical and observational fronts will be required to better understand and constrain the role of small-scale magnetic fields in producing DACs.

References

Abbott, D. C. 1982, *ApJ* 259, 282
Cantiello, M., Langer, N., Brott, I., *et al.* 2009, *A&A* 499, 279
Cranmer, S. R. & Owocki, S. P. 1996, *ApJ* 462, 469
David-Uraz, A., Wade, G. A., Petit, V., *et al.* 2014, *ArXiv e-prints*
Howarth, I. D. & Prinja, R. K. 1989, *ApJS* 69, 527
Kochukhov, O. & Sudnik, N. 2013, *A&A* 554, A93
Mullan, D. J. 1986, *A&A* 165, 157
Ramiaramanantsoa, T., Moffat, A. F. J., Chené, A.-N., *et al.* 2014, *MNRAS* 441, 910
Wade, G. A., Grunhut, J., Alecian, E., *et al.* 2013, *ArXiv e-prints*

New windows on massive stars: asteroseismology, interferometry, and spectropolarimetry
Proceedings IAU Symposium No. 307, 2014
G. Meynet, C. Georgy, J. H. Groh & Ph. Stee, eds.

© International Astronomical Union 2015
doi:10.1017/S1743921314007133

Impact of rotation on the geometrical configurations of fossil magnetic fields

C. Emeriau and S. Mathis

Laboratoire AIM Paris-Saclay, CEA/DSM - CNRS - Université Paris Diderot, IRFU/SAp
Centre de Saclay, F-91191 Gif-sur-Yvette Cedex, France
email: constance.emeriau@cea.fr, stephane.mathis@cea.fr

Abstract. The MiMeS project demonstrated that a small fraction of massive stars (around 7%) presents large-scale, stable, generally dipolar magnetic fields at their surface. These fields that do not present any evident correlations with stellar mass or rotation are supposed to be fossil remnants of the initial phases of stellar evolution. They result from the relaxation to MHD equilibrium states, during the formation of stable radiation zones, of initial fields resulting from a previous convective phase. In this work, we present new theoretical results, where we generalize previous studies by taking rotation into account. The properties of relaxed fossil fields are compared to those obtained when rotation is ignored. Consequences for magnetic fields in the radiative envelope of rotating early-type stars and their stability are finally discussed.

Keywords. magnetic fields, MHD, stars: magnetic fields, stars: rotation, stars: interiors

1. Fossil magnetic fields in early-type stars

Stellar magnetism is one of the key mechanisms that must be studied to understand the evolution of stars. Indeed, magnetic fields impact the structure of massive stars, their activity and winds, and the transport of angular momentum and mixing in their interiors (e.g. Mestel 1999, Maeder, Eggenberger, and Gellert in this volume). In this context, high-resolution spectropolarimetry allows us to study the large variety of stellar magnetic fields as a function of stellar parameters and evolutionary stages (e.g. Donati & Landstreet 2009, Landstreet and Wade in this volume). In upper main-sequence stars with external radiative envelope, no correlations are found between the properties of supposed *fossil* fields, which are often oblique dipoles, and stellar parameters (Wade *et al.* 2011).

In this context, studies led by Tayler (1973) and Markey & Tayler (1973) demonstrated that purely poloidal and toroidal fields are unstable in stably stratified zones. It was then proposed by Tayler (1980) that fossil fields have mixed configurations in order to be stable in such regions. This statement was verified by numerical simulations and theoretical works (Braithwaite & Nordlund 2006; Duez *et al.* 2010), which found that, in non-rotating experiments, the lowest-energy equilibrium states are twisted, non force-free dipolar fields. Moreover, fossil magnetic fields result from the adjustment of the initial field that has been generated in a previous convective phase (for example during the Pre-Main-Sequence) during the transition from the turbulent convective state to a decaying turbulent one when the stable radiation zone is forming (see Fig. 1; Duez & Mathis (2010)). The first numerical simulation of this mechanism during the recession of an external convective envelope has been computed by Arlt (2013)). This process is known as *turbulent relaxation* (e.g. Biskamp 1997).

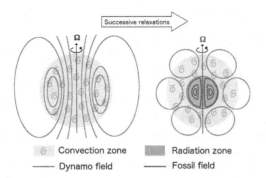

Figure 1. Scheme of the relaxation of fossil magnetic fields during the formation of an internal stably-stratified radiative zone in an initially purely convective star.

2. Modification of relaxed fossil magnetic fields by rotation

In this work, we study the impact of rotation on the properties of fossil magnetic fields by considering rotating magnetohydrodynamic, relaxed equilibrium states using the method introduced by Duez & Mathis (2010). We find that:

• the magnetic field properties are fully determined by the boundary conditions and the properties of their (convective) progenitor, i.e., its magnetic– and cross–helicities and its angular momentum that are almost conserved during the relaxation, while magnetic and kinetic energies decay faster following a selective decay (Biskamp 1997).

• the lowest-energy state for given initial and boundary conditions is the dipolar mode, as in the non-rotating case. In the axisymmetric case, the surface horizontal geometry of the field is dipolar independently of the rotation rate, which modifies only the radial distribution of the magnetic flux. This result is particularly interesting since it matches the observed non-correlation between field geometry and rotation in early-type stars (Wade *et al.* 2011). It also points the strong difference between magnetic fields resulting from turbulent relaxation and of those generated by $\alpha - \Omega$ dynamos, which have a geometry that is directly modified by rotation (e.g. Petit *et al.* 2008).

• the field is non force-free (Reisenegger 2009; Braithwaite & Nordlund 2006; Duez & Mathis 2010). However, contrary to the non-rotating case, the configuration is non-torque free because the azimutal component of the Lorentz force must counterbalance advection in the rotating case.

References

Arlt, R. 2013, *ArXiv:1309.7126*
Biskamp, D. 1997, *Nonlinear Magnetohydrodynamics*, Cambridge University Press
Braithwaite, J. & Nordlund, Å. 2006, *A&A* 450, 1077
Donati, J.-F. & Landstreet, J. D. 2009, *ARA&A* 47, 333
Duez, V., Braithwaite, J., & Mathis, S. 2010, *ApJ (Letters)* 724, L34
Duez, V. & Mathis, S. 2010, *A&A* 517, A58
Markey, P. & Tayler, R. J. 1973, *MNRAS* 163, 77
Mestel, L. 1999, *Stellar magnetism*, Oxford University Press
Petit, P., Dintrans, B., Solanki, S. K., *et al.* 2008, *MNRAS* 388, 80
Reisenegger, A. 2009, *A&A* 499, 557
Tayler, R. J. 1973, *MNRAS* 161, 365
Tayler, R. J. 1980, *MNRAS* 191, 151
Wade, G. A., Alecian, E., Bohlender, D. A., *et al.* 2011, in C. Neiner, G. Wade, G. Meynet, & G. Peters (eds.), *IAU Symposium*, Vol. 272 of *IAU Symposium*, pp 118–123

New windows on massive stars: asteroseismology, interferometry, and spectropolarimetry
Proceedings IAU Symposium No. 307, 2014
G. Meynet, C. Georgy, J. H. Groh & Ph. Stee, eds.

A Simple Mean-Field Diagnostic from Stokes V Spectra

K. G. Gayley[1] and S. P. Owocki[2]

[1] University of Iowa
email: kenneth-gayley@uiowa.edu

[2] University of Delaware

Abstract. It is shown that the diagnostics from an observed circularly polarized line in a rapidly rotating star are *directly* interpretable, not in terms of the observed Stokes V profiles, but in terms of its antiderivative with respect to wavelength (in velocity units if preferred). This also leads to a new mean-field diagnostic that is just as easily obtained as the standard "center of gravity" approach, and is less susceptible to cancellation if the line-of-sight field changes sign over the face of the star.

Keywords. line:profiles, magnetic fields, techniques: polarimetric

1. Introduction

Circular polarization in magnetically sensitive lines is produced by a line-of-sight (LOS) magnetic field via the Zeeman shift, which has the opposite sign for left and right circular polarization. In first order, this leads to a circularly polarized absorption line profile that scales linearly with field strength, and can be used to characterize the LOS field. In a rapid rotator, Doppler shifts connect different locations Ω on the star to different points in the profile x, where x can be in wavelength, velocity, or fractional linewidth units. Rapid rotation also implies something less often appreciated: the polarization signal is now also sensitive to *gradients* in the LOS field at these points. Remarkably, this added complexity is readily navigated by consideration of the antiderivative of the Stokes V profile, as shown below. Since toroidal fields, and magnetic spots, are important examples of LOS field gradients, their presence on rapid rotators is particularly important for the relevance of the results here.

2. The Antiderivative of the Stokes V

The observed Stokes V flux, $F_V(x)$, is given to first order in the field by

$$F_V(x) = \int d\Omega\ h(\Omega)\frac{\partial}{\partial x}I(x,\Omega) = -\frac{\partial}{\partial x}\left\{\int d\Omega\ h(\Omega)\left[I_c(\Omega) - I(x,\Omega)\right]\right\}, \quad (2.1)$$

where h is proportional to the LOS field at locations parametrized by the solid angle Ω, and I_c is the continuum level. Defining an effective mean field as a function of x via

$$\bar{H}(x) = \frac{\int d\Omega\ h(\Omega)\left[I_c(\Omega) - I(x,\Omega)\right]}{\int d\Omega\left[I_c(\Omega) - I(x,\Omega)\right]} = \frac{\int d\Omega\ h(\Omega)\left[I_c(\Omega) - I(x,\Omega)\right]}{\left[F_c - F(x)\right]} \quad (2.2)$$

where F_c and $F(x)$ are the observed continuum and line fluxes, and defining the negative antiderivative

$$A(x) = \int_x^\infty dx'\ F_V(x'), \quad (2.3)$$

we obtain from simple algebra that

$$\bar{H}(x) \;=\; \frac{A(x)}{[F_c \;-\; F(x)]} \;.$$ (2.4)

This is the fundamental expression that connects the observables to what can be known about the mean LOS field, and interestingly, it does not involve $F_V(x)$ directly, but rather its negative antiderivative $A(x)$. In the limit of extreme rotation, $\bar{H}(x)$ becomes the absorption-profile-weighted average of h over the resonant region associated with each x, which can then be inferred from the observables via eq. (2.4).

3. A Simple Alternative Mean-Field Diagnostic

This insight also allows us to use the observed antiderivative to infer a mean-field diagnostic over the whole stellar surface, yielding less opposite-polarity cancellation than the standard "center of gravity" approach when the star is rapidly rotating. The absolute value of $F_V(x)$ cannot be used when signal-to-noise is weak, but the absolute value of $A(x)$ can be used, because noise tends to cancel as signal accumulates, even before the absolute value is taken. This allows a new mean-field diagnostic $< h >$, which suffers less cancellation when the polarity changes across the stellar surface, given by

$$< h > \;=\; \int \mathrm{d}x \; |\bar{H}(x)| \;.$$ (3.1)

In particular, this new diagnostic yields a mean-field measure that gives a nonzero result for a toroidal field, or for a dipole field that is tilted with respect to the rotation axis and is seen from the magnetic equator, both of which would give a zero result in the center-of-gravity approach.

New windows on massive stars: asteroseismology, interferometry, and spectropolarimetry
Proceedings IAU Symposium No. 307, 2014
G. Meynet, C. Georgy, J. H. Groh & Ph. Stee, eds.
© International Astronomical Union 2015
doi:10.1017/S1743921314007157

Linear Polarization and the Dynamics of Circumstellar Disks of Classical Be Stars

Robbie J. Halonen and Carol E. Jones

Department of Physics and Astronomy, The University of Western Ontario, London, ON,
Canada, N6A 3K7
email: `robbie.halonen@gmail.com`

Abstract. The intrinsic linearly polarized light arising from electron scattering of stellar radiation in a non-spherically symmetric distribution of gas is a characterizing feature of classical Be stars. The distinct polarimetric signature provides a mean for directly probing the physical and geometric properties of the gaseous material enveloping these rapidly-rotating massive stars. Using a Monte Carlo radiative transfer computation and a self-consistent radiative equilibrium solution for the circumstellar gas, we explore the role of this observable signature in investigating the dynamical nature of classical Be star disks. In particular, we focus on the potential for using linearly polarized light to develop diagnostics of mass-loss events and to trace the evolution of the gas in a circumstellar disk. An informed context for interpreting the observed linear polarization signature can play an important role in identifying the physical process(es) which govern the formation and dissipation of the gaseous disks surrounding classical Be stars.

Keywords. polarization, circumstellar matter, stars: emission-line, Be

1. Introduction

The non-LTE radiative transfer code BEDISK (Sigut & Jones 2007) computes a self-consistent temperature structure for the circumstellar gas by solving the coupled problems of statistical and radiative equilibrium. BEDISK employs escape probability calculations for an axisymmetric disk of gas with specifiable chemical composition and density parameterized by a base number and a radial power-law. The Monte Carlo code MC-TRACE (Halonen & Jones 2013a) procedure adopts the BEDISK-computed atomic level populations and gas temperatures as the underlying model of a three-dimensional circumstellar envelope and simulates the propagation of photons originating in the radiation field of a star described by a model atmosphere. Combining the thermal solution from the BEDISK code with the Monte Carlo simulation MCTRACE provides an effective computational procedure for producing models of circumstellar environments, such as the disks of classical Be stars.

2. Classical Be Stars

Classical Be stars are rapidly rotating B-type stars surrounded by thin, equatorial, decretion disks of gas. The interaction between the radiation emitted by these hot massive stars with the gaseous circumstellar material that envelops them produces detectable observational properties: a prominent emission line spectrum, an excess in the infrared and radio continuums, and a partial polarization of radiated light. Due mainly to the continually evolving physical state of the disk, fuelled by yet unidentified equatorial mass-loss mechanisms, the observational properties typically vary on timescales of days to decades.

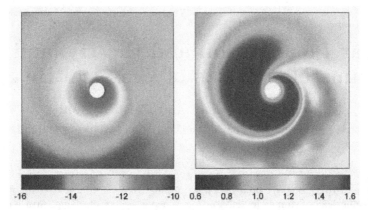

Figure 1. Properties of a modelled one-armed oscillation in the circumstellar disk surrounding a B2V star. The disk density distribution is parametrized by $n = 3.5$ and $\rho_0 = 5.0 \cdot 10^{-11}$ g cm^{-3}. The squares are 20 by 20 stellar radii. The frame on the left shows the equatorial logarithmic density distribution in the disk (g cm^{-3}) while the frame on the right shows the density-weighted gas temperature (10^4 K).

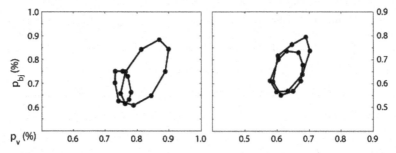

Figure 2. Balmer Jump vs. V-band polarization (BJV) diagrams for a spiral oscillation model produced by MCTRACE. The frame on the left shows a system viewed at inclination 75° while the system in the frame on the right is viewed at inclination 60°. The loops trace out line-of-sight absorptive and scattering opacities in the disk based on the geometry of the structure as seen by the observer. See Halonen & Jones (2013b) for further details of these results.

3. Linear Polarization

The light from classical Be stars is partially polarized due to electron scattering in the disk. The polarized light provides us with a direct probe of the gaseous circumstellar environment in which the scattering occurs. For non-uniform disks, the polarization signature can provide crucial information for tracing the gas in the disk. Figure 1 shows the density and density-weighted gas temperature arising from a global one-armed oscillation in the disk. Polarimetric features originating in different formation regions show phase differences leading to particular behaviour in polarization-color diagrams, such as the Balmer Jump vs. V-band polarization plot illustrated in Fig. 2. Thus, polarimetry represents an important diagnostic of the dynamical nature of the circumstellar gas.

References

Halonen, R. J. & Jones, C. E. 2013a, *ApJ* 765, 17
Halonen, R. J. & Jones, C. E. 2013b, *ApJS* 208, 3
Sigut, T. A. A. & Jones, C. E. 2007, *ApJ* 668, 481

New windows on massive stars: asteroseismology, interferometry, and spectropolarimetry
Proceedings IAU Symposium No. 307, 2014
G. Meynet, C. Georgy, J. H. Groh & Ph. Stee, eds.

© International Astronomical Union 2015
doi:10.1017/S1743921314007169

Multiple, short-lived "stellar prominences" on O stars: the supergiant λ Cephei

H. F. Henrichs[1] and N. Sudnik[2]

[1] University of Amsterdam, Netherlands
email: h.f.henrichs@uva.nl

[2] Saint-Petersburg State University, Russia
email: snata.astro@gmail.com

Abstract. Many OB stars show unexplained cyclical variability in their winds and in many optical lines, which are formed at the base of the wind. For these stars no dipolar magnetic fields have been detected. We propose that these cyclical variations are caused by the presence of multiple, transient, short-lived, corotating magnetic loops, which we call "stellar prominences". We present a simplified model representing these prominences as corotating spherical blobs and fit the rapid variability in the He II $\lambda4686$ line of the O supergiant λ Cep for time-resolved spectra obtained in 1989. Our conclusions are: (1) From model fits we find that the life time of the prominences varies, and is between 2–7 h. (2) The adopted inclination angle is 68° with a rotation period of $\approx$ 4.1 d (but not well constrained). (3) The contribution of non-radial pulsations is negligible (4) Similar behavior is observed in at least 4 other O stars. We propose that prominences are a common phenomenon among O stars.

Keywords. line: profiles, stars: magnetic fields, stars: winds, outflows, stars: individual (λ Cep)

1. Introduction

Wind variability in massive OB stars is a wide spread phenomenon. This variability is not periodic, but cyclic (like sunspots) with often a dominant quasi period which scales with the estimated rotation period (days to weeks), or perhaps an integer fraction thereof, see for instance references in Henrichs & Sudnik (2014), also for the remainder of this contribution. The cause or trigger is unknown. The major time-variable wind features that are observed in the UV (the so-called DACs, discrete absorption components), must start from very near, or at the stellar surface. Possible explanations include non-radial pulsations and/or bright magnetic star spots. Pulsations have been found only for a few O stars, mostly from optical studies but also from space-based photometry. Magnetic dipole fields in such stars (except for the CP stars) have been found since 1998. Both phenomena are expected, however, to cause periodic variations, and are therefore unlikely the cause of the observed cyclical behavior (David-Uraz *et al.* 2014).

New developments in this field are twofold. First, Cantiello *et al.* (2009) showed that in the sub-surface convective layers in massive stars magnetic fields can be generated with a short turnover time (Cantiello *et al.* 2011). Second, space-based photometry of the O giant ξ Per showed rapid variations at the 1 mmag level, incompatible with the known pulsations, but compatible with the presence of a multitude of corotating bright spots, which live only a few days (Ramiaramanantsoa *et al.* 2014). These spots are suggested to be of magnetic origin as described above.

To understand the role of magnetic fields in OB stars is a major challenge. Here we present a simplified model to explain optical wind-line variability in the O star λ Cep.

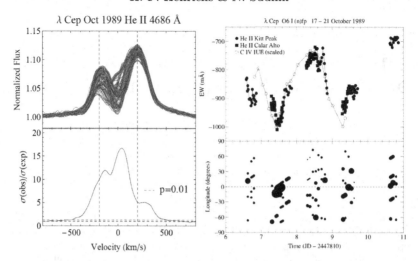

Figure 1. *Left, top:* Overplot of He II 4686 profiles during 5 days. *bottom:* TVS and significance level. *Right:* EW of He II lines and wind C IV blue edge variability (*top*) plotted along with properties of the blobs, with symbol size proportional to optical thickness (*bottom*).

2. Optical wind-line variability in the O6I(n)fp star λ Cep

The bright runaway star O6I(n)fp star λ Cep ($v\sin i \simeq 214$ km s^{-1}, $\log(L/L_\odot) \simeq 5.9$, $T_{\text{eff}} \simeq 36000$ K, $R \simeq 17.5 R_\odot$, $M \simeq 60 M_\odot$), is a nonradial pulsator ($l = 3, P = 12.3$ h; $l = 5, P = 6.6$ h), and shows cyclical DACs in the UV resonance lines. Rapid variability (timescale 10 min) have been observed in the He II emission line in 1989 (and other years as well). The dominant period in the UV and optical lines is $\simeq 2$ d. Only redward moving NRP features have been observed, implying an inclination angle greater than, say, 50°. With the adopted radius, the likely rotation period is then $\simeq 4.1$ d.

The observed optical profile changes extend far beyond $v\sin i$, which requires emitting gas above the surface. We therefore consider the most simplified model to represent a "stellar prominence" by a sphere corotating and touching the surface. The model properties are described by Henrichs & Sudnik (2014). Essential in our procedure is that quotient spectra are fitted, such that the overal shape of the profile cancels out. The procedure is to adopt a fixed inclination angle (matching $v\sin i$ and R_*) which was 68°, and put a number of blobs around the star to fit the first quotient, which is determined by an assumed rotation period. A least-squares fit gives the best parameters. The optical thickness of the fitted blobs are plotted in Fig. 1 (right). The number of blobs needed is between 2 and 6, with typical lifetime in the order of several hours. No clear correlation is apparent. The He I 4713 photospheric line shows only pulsations. Several other data sets of this star, which include other optical lines as well, show very similar behavior. We note that quotient spectra of most lines behave rather similarly. Similar behavior is also observed in other O stars (in progress). This suggests a common phenomenon in O stars.

References

Cantiello, M., Braithwaite, J., Brandenburg, A., *et al.* 2011, in C. Neiner, G. Wade, G. Meynet, & G. Peters (eds.), *IAU Symposium*, Vol. 272 of *IAU Symposium*, pp 32–37

Cantiello, M., Langer, N., Brott, I., *et al.* 2009, *A&A* 499, 279

David-Uraz, A., Wade, G. A., Petit, V., *et al.* 2014, *MNRAS* 444, 429

Henrichs, H. F. & Sudnik, N. P. 2014, in *IAU Symposium*, Vol. 302 of *IAU Symp.*, pp 280–283

Ramiaramanantsoa, T., Moffat, A. F. J., Chené, A.-N., *et al.* 2014, *MNRAS* 441, 910

New windows on massive stars: asteroseismology, interferometry, and
spectropolarimetry
Proceedings IAU Symposium No. 307, 2014
G. Meynet, C. Georgy, J. H. Groh & Ph. Stee, eds.

© International Astronomical Union 2015
doi:10.1017/S1743921314007170

Project VeSElkA : Preliminary results for CP stars recently observed with ESPaDOnS†

Viktor Khalack and Francis LeBlanc

Université de Moncton, Moncton, Canada
email: `khalakv@umoncton.ca`

Abstract. We present the first results for the estimation of gravity and effective temperature of poorly studied chemically peculiar stars recently observed with the spectropolarimeter ESPaDOnS at CFHT in the frame of the VeSElkA (Vertical Stratification of Elements Abundance) project. A grid of theoretical stellar atmosphere models with the corresponding fluxes has been calculated using the PHOENIX code. We have used these fluxes to fit Balmer line profiles employing the code FITSB2 that produces estimates of the effective temperature, surface gravity and radial velocity for each star.

Keywords. stars: atmospheres, stars: chemically peculiar, stars: fundamental parameters

1. Introduction

Detection of vertical abundance stratification of chemical species in atmospheres of chemically peculiar (CP) stars is considered to be an indicator of the effectiveness of the diffusion mechanism in their atmospheres. It is also responsible for the observed peculiarities of chemical abundances. Therefore, we have initiated a new project entitled "Vertical Stratification of Elements Abundance" (VeSElkA – meaning rainbow in Ukrainian) aimed to search for and study the signatures of abundance stratification of chemical species with optical depth in the atmospheres of slowly rotating CP stars. A list of relatively bright and poorly studied CP stars that may be observed with ESPaDOnS has been compiled based on the catalog of CP stars of Renson & Manfroid (2009). In stars without surface convection, their slow rotation supports the hypothesis of a hydrodynamically stable atmosphere. This results in comparatively narrow and unblended line profiles that are suitable for abundance analysis. High resolution (R = 65000) Stokes IV spectra of several CP stars were recently obtained with ESPaDOnS in the spectral domain from 3700Å to 10000Å with maximal S/N $\simeq$ 1000.

2. Grid of models

A new library of high resolution synthetic spectra has been created in order to determine $T_{\rm eff}$ and $\log g$ of stars observed in the frame of the VeSElkA project. A grid of stellar atmosphere models and corresponding fluxes with high spectral resolution (R=60000) have been calculated using version 15 of the PHOENIX code (Hauschildt *et al.* 1997) for: $5000\,{\rm K} < T_{\rm eff} < 9000\,{\rm K}$ with a 250 K step, $9000\,{\rm K} < T_{\rm eff} < 15000\,{\rm K}$ with a 500 K step, $3.5 < \log g < 4.5$ with a 0.5 step, and [Fe/H] = -1.0, -0.5, 0.0, $+0.5$, $+1.0$, $+1.5$.

† Based on observations obtained at the Canada-France-Hawaii Telescope (CFHT) which is operated by the National Research Council of Canada, the Institut National des Sciences de l'Univers of the Centre National de la Recherche Scientifique of France, and the University of Hawaii.

Table 1. List of the observed slowly rotating CP stars.

Star	m_V	Δt (s)	S/N	T_{eff} (K)	$\log g$	V_r (km/s)	χ^2/ν	$V \sin i$ (km/s)
HD 15385	6.2	1600	1100	8230±200	4.00±0.2	22.0±1.0	0.65	29.0
HD 22920	5.5	920	1150	13640±200	3.72±0.2	15.7±2.3	2.08	40.0±2.1
HD 23878	5.2	660	950	8740±200	3.86±0.2	29.5±1.0	0.62	24.0
HD 68351	5.6	920	1100	10050±200	3.22±0.2	18.1±2.0	1.17	33.0
HD 71030	6.1	1140	1100	6780±200	4.04±0.2	38.1±1.0	0.28	9.0±2.0
HD 83373	6.4	1524	1000	9800±200	3.81±0.2	26.5±1.0	0.92	28.0
HD 90277	4.7	520	1000	7250±200	3.62±0.2	14.5±1.0	1.29	34.0
HD 95608	4.4	416	1300	9200±200	4.25±0.2	−10.4±1.0	0.59	17.0±2.0
HD 97633	3.3	152	1300	8750±200	3.45±0.2	8.2±1.0	0.61	23.0
HD 110380	3.6	200	1300	6980±200	4.19±0.2	−17.6±1.0	0.31	23.0
HD 116235	5.9	1040	870	8900±200	4.33±0.2	−10.3±1.0	0.49	20.0±2.0
HD 164584	5.4	880	1300	6800±200	3.54±0.2	−11.2±1.0	1.16	30.0
HD 186568	6.0	1300	1000	11070±200	3.44±0.2	−9.5±1.0	1.81	15.0
HD 209459	5.8	1160	1000	10620±200	3.73±0.2	−0.3±1.0	1.31	4.0
HD 223640	5.2	680	1200	12250±200	3.94±0.2	17.0±2.0	2.04	28.0

To verify the accuracy of the grids of stellar atmosphere models, we have used a spectrum of Vega obtained with ESPaDOnS in the spectropolarimetric mode. The profiles of nine Balmer lines in the non-normalised spectrum of Vega have been fitted with the help of the FITSB2 code (Napiwotzki *et al.* 2004) using the grid of models calculated for metallicity [Fe/H] = −0.5. The best fit is obtained for T_{eff} = 9690 K and $\log g$ = 4.18, resulting in χ^2/ν = 2.01. Exactly the same approach has been applied to fit nine Balmer lines in the same spectrum using the grid of models calculated with version 16 of the PHOENIX code for the metallicity [Fe/H] = −0.5 (see Husser *et al.* (2013) for more detail). The best fit has resulted in T_{eff} = 9630 K and $\log g$ = 4.14 with χ^2/ν = 2.00. These values of effective temperature and surface gravity are close to one another and to the values T_{eff} = 9560 K and $\log g$ = 4.05 obtained for Vega by Hill & Landstreet (1993) using the results of *ubvyH* photometry.

3. Stellar atmosphere parameters for program stars

Nine Balmer line profiles have been fitted in the observed spectra of the selected slowly rotating CP stars to find their effective temperature, surface gravity and radial velocity (see Table 1). The fitting procedure has been performed with the help of FITSB2 code (Napiwotzki *et al.* 2004), employing our grids of stellar atmosphere models (calculated with PHOENIX 15) together with the respective simulated spectra. For each star, the fitting procedure has been performed for the metallicities [Fe/H]= −1.0, −0.5, 0.0, +0.5, +1.0. Among the obtained results, the fundamental parameters corresponding to the fit with the smallest value χ^2/ν are chosen. Our final results are consistent with the previously published results for ten of the stars presented in Tab. 1, and we have determined T_{eff} and $\log g$ for five other CP stars (HD 23878, HD 68351, HD 83373, HD 164584 and HD 186568) for the first time.

References

Hauschildt, P. H., Shore, S. N., Schwarz, G. J., *et al.* 1997, *ApJ* 490, 803
Hill, G. M. & Landstreet, J. D. 1993, *A&A* 276, 142
Husser, T.-O., Wende-von Berg, S., Dreizler, S., *et al.* 2013, *A&A* 553, 6
Napiwotzki, R., Yungelson, L., Nelemans, G., *et al.* 2004, *ASPC* 318, 402
Renson, P. & Manfroid, J. 2009, *A&A* 498, 961

New windows on massive stars: asteroseismology, interferometry, and
spectropolarimetry
Proceedings IAU Symposium No. 307, 2014
G. Meynet, C. Georgy, J. H. Groh & Ph. Stee, eds.

© International Astronomical Union 2015
doi:10.1017/S1743921314007182

Abundance analysis of HD 22920 spectra†

Viktor Khalack and Patrick Poitras

Université de Moncton, Moncton, Canada
email: khalakv@umoncton.ca

Abstract. The new spectropolarimetric observations of HD 22920 with ESPaDOnS at CFHT reveal a strong variability of its spectral line profiles with the phase of stellar rotation. We have obtained $T_{\text{eff}} = 13640\,\text{K}$, $\log g = 3.72$ for this star from the best fit of its nine Balmer line profiles. The respective model of stellar atmosphere was calculated to perform abundance analysis of HD 22920 using the spectra obtained for three different phases of stellar rotation. We have found that silicon and chromium abundances appear to be vertically stratified in the atmosphere of HD 22920. Meanwhile, silicon shows hints for a possible variability of vertical abundance stratification with rotational phase.

Keywords. stars: atmospheres, stars: chemically peculiar, stars: abundances, stars: magnetic fields

1. Introduction

Accumulation or depletion of chemical elements at certain optical depths brought about by atomic diffusion can modify the structure of stellar atmospheres and it is therefore important to gauge the intensity of such stratification. Recently, we have obtained three ESPaDOnS spectra of magnetic CP star HD 22920 with the aim to study vertical stratification of chemical species in its stellar atmosphere. This star shows a weak photometric variability with a period P=3^d.95 (Bartholdy 1988). Its small value of $V \sin i = 30\,\text{km s}^{-1}$ results in comparatively narrow and unblended line profiles which are suitable for abundance analysis. The slow rotation and the presence of a weak magnetic field support the hypotheses of a hydrodynamically stable atmosphere, which is necessary for diffusion to take place.

2. Spectral analysis

The preliminary results of the abundance analysis are presented here for HD 22920. The line profile simulation is performed using the ZEEMAN2 spectrum synthesis code (Landstreet 1982) and LTE stellar atmosphere model calculated with PHOENIX (Hauschildt et al. 1997) for the $T_{\text{eff}} = 13640\,\text{K}$, $\log g = 3.72$. For each element we have selected a sample of unblended lines, which are clearly visible in the analyzed spectra. The element's abundance, radial velocity and $V \sin i$ were fitted using an automatic minimization routine independently for each line profile in each rotational phase, and are presented in Table 1. For each ion presented in the Table 1, the number of analyzed lines is specified in brackets. Our estimates of the radial velocity and $V \sin i$ are consistent with the previously published results for this star.

For each rotational phase we have determined average abundance of oxygen, silicon,

† Based on observations obtained at the Canada-France-Hawaii Telescope (CFHT) which is operated by the National Research Council of Canada, the Institut National des Sciences de l'Univers of the Centre National de la Recherche Scientifique of France, and the University of Hawaii.

V. Khalack & P. Poitras

Table 1. Average abundance of chemical species in the atmosphere of HD 22920 for different rotational phases. Number in parenthesis represents the number of analyzed lines for each ion.

Abundance	$\varphi = 0.0$		$\varphi = 0.497$		$\varphi = 0.763$	
$\log(N(\mathrm{OI})/N_{tot})$	-4.03 ± 0.18	(2)	-4.11 ± 0.20	(3)	-4.07 ± 0.18	(3)
$\log(N(\mathrm{SiII})/N_{tot})$	-3.88 ± 0.38	(15)	-4.34 ± 0.29	(8)	-4.20 ± 0.29	(10)
$\log(N(\mathrm{FeII})/N_{tot})$	-4.26 ± 0.22	(32)	-4.11 ± 0.24	(31)	-4.05 ± 0.27	(35)
$\log(N(\mathrm{CrII})/N_{tot})$	-5.83 ± 0.95	(6)	-5.59 ± 1.31	(5)	-5.27 ± 1.18	(5)
$V\sin i$ (km/s)	36.1 ± 4.3	(55)	37.8 ± 5.4	(47)	37.3 ± 5.0	(53)
V_{r} (km/s)	16.1 ± 4.0	(55)	18.6 ± 4.5	(47)	19.8 ± 4.4	(53)

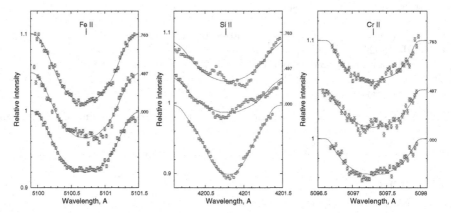

Figure 1. The best fit (continuous line) obtained for the observed FeII 5100Å, SiII 4200Å and CrII 5097Å line profiles (open circles) in HD 22920 at different phases of stellar rotation.

iron and chromium (see Table 1). During simulation of the synthetic line profiles we assume a homogeneous horizontal distribution of elements abundance, which is not correct, but provides an estimate for the average abundance for the given rotational phases. The observed profiles (see Fig. 1) differ significantly from the simulated ones. This and the obvious variability of line profiles with rotational phase argue in favor of a non-homogeneous horizontal abundance distribution for these elements. Among the studied elements, SiII line profiles show the strongest variability with rotational phase.

3. Summary

Here we presented our estimate of $T_{\mathrm{eff}} = 13640\,\mathrm{K}$, $\log g = 3.72$ in HD 22920 and the average abundances of oxygen, silicon, iron and chromium for different rotational phases (see Table 1). All the analyzed elements show variability of their line profiles with rotational phase and seem to have non-uniform horizontal distributions of their abundance. Among the studied elements only silicon and chromium abundances appear to be vertically stratified in the stellar atmosphere of HD 22920.

References

Bartholdy, P. 1988, *Comptes Randues des Journées de Strasbourg* 10, 77
Hauschildt, P. H., Shore, S. N., Schwarz, G. J., et al. 1997, *ApJ* 490, 803
Landstreet, J. D. 1982, *ApJ* 258, 639

New windows on massive stars: asteroseismology, interferometry, and spectropolarimetry
Proceedings IAU Symposium No. 307, 2014
G. Meynet, C. Georgy, J. H. Groh & Ph. Stee, eds.

© International Astronomical Union 2015
doi:10.1017/S1743921314007194

Fundamental properties of single O stars in the MiMeS survey

F. Martins[1], A. Hervé[1], J.-C. Bouret[2], W. L. F. Marcolino[3], G. A. Wade[4], C. Neiner[5], E. Alecian[6] and the MiMeS collaboration

[1]LUPM, Université Montpellier 2, CNRS, Place Eugène Bataillon, F-34095 Montpellier Cedex 05, France
email: `fabrice.martins@univ-montp2.fr`

[2]LAM-UMR 6110, CNRS & Université de Provence, rue Frédéric Joliot-Curie, F-13388, Marseille Cedex 13, France

[3]Observatório do Valongo, Universidade do Rio de Janeiro, Ladeira Pedro António, 43, CEP 20080-090, Brasil

[4]Dept. of Physics, Royal Military College of Canada, PO Box 17000, Stn Forces, Kingston, Ontario K7K 7B4, Canada

[5]LESIA, UMR 8109 du CNRS, Observatoire de Paris, UPMC, Université Paris Diderot, 5 place Jules Janssen, F-92195 Meudon Cedex, France

[6]IPAG, Université Joseph Fourier, CNRS, 414 Rue de la Piscine, Domaine Universitaire, F-38400 St-Martin d'Hères

Abstract. We present preliminary results of the determination of fundamental parameters of single O-type stars in the MiMeS survey. We present the sample and we focus on surface CNO abundances, showing how they change as stars evolve off the zero-age main sequence.

Keywords. stars: fundamental parameters, stars: abundances

1. Introduction

The MiMeS survey of massive stars was designed to establish the magnetic characteristics, including the statistical properties, of the magnetic fields of O and B stars (Wade *et al.* 2012). In total, 100 O stars were observed with at least one of the following instrument: ESPaDOnS (CFHT, Hawaii), NARVAL (Télescope Bernard Lyot, France) and HARPSpol (ESO 3.6m, Chile). A magnetic field was detected in about 7% of the sample (Grunhut *et al.*, in prep.). For the entire sample, the collected high signal-to-noise (S/N>100), high resolution (R > 65000) optical spectra represent a unique database to determine the fundamental properties of O-type stars.

We have undertaken such a task, focussing on single O stars. We have removed all known spectroscopic binaries (SB1 and SB2) from the initial sample. We were left with 65 stars. Special care was taken to derive the surface abundances which provide important constraints on the evolution and internal mixing of massive stars.

2. Spectroscopic analysis and surface abundances

We performed a spectroscopic analysis using atmosphere models and synthetic spectra computed with the code CMFGEN (Hillier & Miller 1998). We determined the following parameters: effective temperature, surface gravity ($\log g$), projected rotational velocity ($v \sin i$), macroturbulent velocity (v_{mac}) and CNO surface abundances. For the latter we

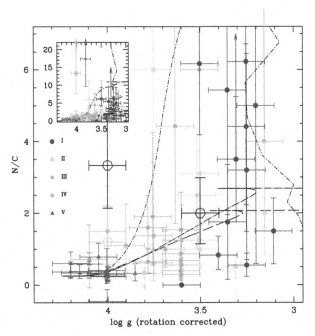

Figure 1. N/C as a function of surface gravity log g (corrected for rotation) for the sample stars. Different symbols/colors correspond to different luminosity classes. Open symbols stand for magnetic stars. The evolutionary tracks of Ekström *et al.* (2012) for initial masses of 20, 25 and 40 $M_\odot$ are overplotted. The upper insert is the same figure on a wider scale.

used between 5 and 15 lines depending on the element and the star's spectral type. The uncertainties on the derived parameters were based on a χ^2 analysis.

Fig. 1 shows the ratio of nitrogen to carbon surface abundances as a function of log g. The error bars on the surface abundances remain large because of uncertainties in the line formation processes. However, a clear trend of higher values of N/C at lower gravities is seen: dwarfs have N/C lower than 1.0, while in supergiants this ratio can reach values up to 6.0. This indicates that the products of hydrogen burning through the CNO cycle are transported to the stellar surface as stars evolve off the zero-age main sequence. The observed N/C values are also broadly consistent with those predicted by the rotating models of Ekström *et al.* (2012).

The open symbols in Fig. 1 correspond to magnetic O stars. They do not stand out as chemically peculiar, except perhaps in one case. Once fundamental parameters are available for the entire sample, we will compare the properties of magnetic and non-magnetic stars to investigate possible effects of magnetism on surface properties.

Acknowledgement. FM thanks the Agence Nationale de la Recherche for financial support under the contract ANR-11-JS56-0007.

References

Ekström, S., Georgy, C., Eggenberger, P., *et al.* 2012, *A&A* 537, A146

Hillier, D. J. & Miller, D. L. 1998, *ApJ* 496, 407

Wade, G. A., Grunhut, J. H., & MiMeS Collaboration 2012, in A. C. Carciofi & T. Rivinius (eds.), *Circumstellar Dynamics at High Resolution*, Vol. 464 of *Astronomical Society of the Pacific Conference Series*, p. 405

New windows on massive stars: asteroseismology, interferometry, and spectropolarimetry
Proceedings IAU Symposium No. 307, 2014
G. Meynet, C. Georgy, J. H. Groh & Ph. Stee, eds.

© International Astronomical Union 2015
doi:10.1017/S1743921314007200

Spectropolarimetry and modeling of WR156

Olga Maryeva

Special Astrophysical Observatory of the Russian Academy of Sciences
email: olga.maryeva@gmail.com

Abstract. For the first time spectropolarimetric observations of Wolf-Rayet star WR156 (WN8h) were conducted. Medium resolution spectropolarimetric data in the range of 3500-7200 ÅÅ were obtained at Russian 6-m telescope of Special Astrophysical Observatory (SAO RAS). These data show that the light from the star is significantly polarized, with the degree of polarization $P = 1.38 \pm 0.06\%$, and the position angle $\Theta = 77.4° \pm 1.2°$. This polarization, most probably, has an interstellar origin, as its magnitude and orientation are similar to the ones of field stars. Also, we present results of numerical modeling of WR156 atmosphere performed using CMFGEN code. According to it, WR156 is the richest hydrogen Wolf-Rayet star of WN8 type in the Galaxy.

Keywords. stars : individual : WR156; stars : Wolf-Rayet; techniques : polarimetric

1. Introduction

Even for spatially unresolvable objects, linear spectropolarimetry with moderate resolution is a powerful tool to detect and investigate the asymmetric structure of the object and to search rapid rotation. The linear polarization (greater than 0.3%) was found in 29 WR stars (i.e. almost in 15% of Galactic WR stars, see Harries *et al.* 1998), but the "line effect" is identified only in six WR stars (Harries *et al.* 1998). Moreover, some WR stars demonstrate polarimetric continuum variability due to small-scale structure of the wind. Combining the results of spectropolarimetric observations with numerical simulations of stellar winds may be the best way for understanding the cause of the appearance of the linear polarization in WR stars, and to reveal the mechanisms of the "line effect".

2. Study of the Galactic Wolf-Rayet star WR 156 (WN8h)

Spectropolarimetry of of WR156 was performed at the Russian 6-m telescope. Measured degree of polarization is $P = 1.38 \pm 0.06\%$, position angle is $\Theta = 77.4 \pm 1.2°$. The spectrum does not show any signs of reduction in polarization at the wavelengths corresponding to emission lines, i.e. "lines effects" are not detected. Estimation of interstellar polarization (ISP) is very important for the study of stellar polarizations. Using the field stars we created a map of the interstellar polarization and we concluded, that, most likely, the polarization measured by us has an interstellar origin.

Using CMFGEN code we have constructed two models of the WR156 atmosphere for different values of the bolometric luminosity of $L_\star$. In the calculation of the first model, we assumed that the $L_\star = 5.3 \cdot 10^5\ L_\odot$ of WR156 is equal to the one of FSZ35, which is located in the galaxy M33 and is also classified as WN8h. The second model was constructed for $L_\star = 3.3 \cdot 10^5\ L_\odot$. Table 1 lists the parameters of both models. In the optical range both models show almost identical spectra. Also the table shows parameters of other WN8h whose are derived using CMFGEN. Luminosity and mass-loss rate of WR156 are similar to the ones of WR40 and WR16 stars, displaying the line effect, but the temperature and the abundance of hydrogen are significantly different from them. Apparently, present

O. Maryeva

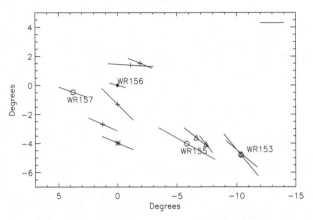

Figure 1. The interstellar polarization map around WR156.

Table 1. Parameters of atmospheres of WR156 and other WN8h stars in the Milky Way and M33 galaxy (FSZ35 and Romano's star, in minimum of brightness). X_H, X_{He}, X_C and X_N are the mass fractions of hydrogen, helium, carbon and nitrogen, respectively.

	$T_\star$ [kK]	$R_\star$ [$R_\odot$]	$\log L_\star$ [$L_\odot$]	$\log \dot{M}_{cl}$ [$M_\odot$ yr^{-1}]	v_∞ [km s^{-1}]	X_H [%]	X_{He} [%]	X_C [%]	X_N [%]	Ref
WR124	32.7	18.0	5.53	−4.7	710	15				[1]
WR40	45.0	10.6	5.61	−4.5	840	15	83	1.2	1.12	[2]
WR16	41.7	12.3	5.68	−4.8	650	23	75			[2]
FSZ35	36.5	19	5.76	−4.58	800	16.5	82	0.75	0.3	[3]
Romano's star	34.0	20.8	5.7	−4.72	360	31.8	67	2	0.7	[4]
WR156	36	18.7	5.72	−4.82	650	33	65.8	6	0.7	[5]
WR156	36	14.8	5.52	−5	650	33	65.8	6	0.7	[5]

Notes: [1]- Crowther *et al.* (1999), [2]- Herald *et al.* (2001), [3]- Maryeva & Abolmasov (2012b), [4]- Maryeva & Abolmasov (2012a), [5]- Maryeva *et al.* (2013)

statistics on WR stars with line effects is so small that it does not allow to define any relation between the line effect and some parameters of the stellar atmosphere.

Results are published in Maryeva *et al.* (2013).

Acknowledgements

Olga Maryeva thanks the International Astronomical Union and Dynasty Foundation for grants. The study was supported by the Russian Foundation for Basic Research (projects no. 14-02-31247,14-02-00291).

References

Crowther, P. A., Pasquali, A., De Marco, O., *et al.* 1999, *A&A* 350, 1007
Harries, T. J., Hillier, D. J., & Howarth, I. D. 1998, *MNRAS* 296, 1072
Herald, J. E., Hillier, D. J., & Schulte-Ladbeck, R. E. 2001, *ApJ* 548, 932
Maryeva, O. & Abolmasov, P. 2012a, *MNRAS* 419, 1455
Maryeva, O. & Abolmasov, P. 2012b, *MNRAS* 421, 1189
Maryeva, O. V., Afanasiev, V. L., & Panchuk, V. E. 2013, *New. Astron.* 25, 27

New windows on massive stars: asteroseismology, interferometry, and
spectropolarimetry
Proceedings IAU Symposium No. 307, 2014
G. Meynet, C. Georgy, J. H. Groh & Ph. Stee, eds.
© International Astronomical Union 2015
doi:10.1017/S1743921314007212

The UVMag space project: UV and visible spectropolarimetry of massive stars

Coralie Neiner and the UVMag consortium

LESIA, Observatoire de Paris, CNRS UMR 8109, UPMC, Université Paris Diderot,
5 place Jules Janssen, 92190 Meudon, France
email: `coralie.neiner@obspm.fr`

Abstract. UVMag is a medium-size space telescope equipped with a high-resolution spectropolarimetrer working in the UV and visible domains. It will be proposed to ESA for a future M mission. It will allow scientists to study all types of stars as well as e.g. exoplanets and the interstellar medium. It will be particularly useful for massive stars, since their spectral energy distribution peaks in the UV. UVMag will allow us to study massive stars and their circumstellar environment (in particular the stellar wind) spectroscopically in great details. Moreover, with UVMag's polarimetric capabilities we will be able, for the first time, to measure the magnetic field of massive stars simultaneously at the stellar surface and in the wind lines, i.e. to completely map their magnetosphere.

Keywords. telescopes, space vehicles, instrumentation: polarimeters, instrumentation: spectrographs, stars: early-type, ultraviolet: stars

1. The UVMag space mission

UVMag is a space mission dedicated to the study of the dynamic 3D environment of stars and planets. It will consist in a 1.3-meter telescope and a spectropolarimeter covering the UV and visible wavelength range from 117 to 870 nm. An option for far-UV (90-117 nm) is also currently explored. The spectral resolution will be at least 25000 in the UV domain and at least 35000 in the visible domain. Full Stokes (IQUV) information, i.e. both circular and linear polarisation, will be obtained. More details on the spectropolarimeter design are available in Pertenaïs *et al.* (2014). The mission will be proposed to ESA this autumn as its M4 mission, for a launch in 2025. It will last 5 years.

2. What can UVMag do for massive stars?

UVMag is particularly well suited for massive stars since it will observe in the UV and visible domains, i.e. where massive stars emit most of their lines. UVMag will mainly target massive stars with magnitudes between V=3 and 10, but massive stars in the Magellanic Clouds will also be reachable with a longer exposure time. This will allow us to probe stars in a different environment. UVMag will allow us to:

• study the stellar wind through UV resonance lines and visible recombination lines, in particular for O stars;

• study wind clumping and the line-driven instabilities;

• study the magnetic field at the stellar surface with improved signal-to-noise. Indeed, thanks to the increased number of photons and of lines available in the UV compared to the visible domain, the signal-to-noise ratio of Zeeman magnetic signatures obtained with the LSD technique in the UV is higher than in the visible domain;

• study the magnetosphere and confinement of material around the star. Emission

from the plasma trapped in the magnetosphere can be observed in visible emission lines, while the confined wind can be studied in the UV resonance lines;
• study linear polarisation and depolarisation effects from circumstellar disks.
With UVMag we will thus be able to study the formation, evolution and environment of massive stars. More details about UVMag's science case for massive stars as well as for other topics can be found in Neiner *et al.* (2014).

3. UVMag's observing program

UVMag will observe three types of targets:
• Mapping targets: 50 to 100 stars (of all types) will be followed over at least one full rotation period with high cadence in order to study them in great details and reconstruct 3D maps of their surface and environment. These targets will be partly secured through the consortium core program and partly chosen following a competitive proposal process. Some targets (in particular solar-type stars) will be reobserve every year to study their variability over activity cycles.
• Survey targets: several thousands stars will be observed once or twice to provide information on their magnetic field, wind and environment. This will include an unbiased magnitude-limited statistical sample and targets selected through a competitive proposal process. These snapshot data will provide statistical results as well as specific inputs (e.g. wind terminal velocity) for stellar modelling.
• A Target of Opportunity (ToO) mode is also planned, in particular for supernovae and outbursting stars such as classical Be stars.
There are $\sim$50000 stars with 3<V<10 observable with UVMag. Among them, there are $\sim$20000 OB stars. Since 7% of OB stars are found to be magnetic (Wade *et al.* 2013), there are statistically $\sim$1400 magnetic OB stars among the $\sim$20000 OB stars. However, only $\sim$100 magnetic OB stars are known as of today. Although this number has been growing significantly since the new generation of spectropolarimeters (Narval at TBL, ESPaDOnS at CFHT and HarpsPol at ESO) is available, it will probably remain rather low in the coming decade. It is therefore probable that all magnetic OB stars known at the time of launch will be observed by UVMag in the survey sample (and several of them will be mapped in details). Of course, non-magnetic OB stars will also be observed.
For massive stars, the signal-to-noise ratio in the intensity spectrum will be above 100 in 20 minutes exposure, both in the UV and visible domains.

4. Conclusions

UVMag is an M-size space mission with a 1.3 meter telescope equipped with a high-resolution spectropolarimeter working in the UV and visible domain simultaneously. This mission will be particularly useful for the study of massive stars which emit most of their light in this wavelength domain. This includes the study of their wind, magnetic field, magnetosphere, disk, clouds as well as of their surface (e.g. spots).

References

Neiner, C., Baade, D., Fullerton, A., *et al.* 2014, *Astrophysics and Space Science*, Volume 354, Issue 1, p 215

Pertenaïs, M., Neiner, C., Pares, L., *et al. Proceedings of the SPIE*, Volume 9144, id. 91443B 9 pp. (2014)

Wade, G. A., Grunhut, J., Alecian, E., *et al.* 2013, in *Magnetic fields throughout stellar evolution*, Vol. 302 of *IAU Symposium*, p 265

New windows on massive stars: asteroseismology, interferometry, and
spectropolarimetry
Proceedings IAU Symposium No. 307, 2014
G. Meynet, C. Georgy, J. H. Groh & Ph. Stee, eds.

© International Astronomical Union 2015
doi:10.1017/S1743921314007224

Magnetic main sequence stars as progenitors of blue supergiants

I. Petermann, N. Castro and N. Langer

Argelander Institute for Astronomy, Bonn, Germany
email: ilka@astro.uni-bonn.de

Abstract. Blue supergiants (BSGs) to the right the main sequence band in the HR diagram can not be reproduced by standard stellar evolution calculations. We investigate whether a reduced convective core mass due to strong internal magnetic fields during the main sequence might be able to recover this population of stars. We perform calculations with a reduced mass of the hydrogen burning convective core of stars in the mass range $3 - 30\,M_\odot$ in a parametric way, which indeed lead to BSGs. It is expected that these BSGs would still show large scale magnetic fields in the order of 10 G.

Keywords. Stars: massive – Stars: supergiants – Stars: magnetic field – Stars: evolution

1. Introduction and setup of calculations

Magnetic massive stars are rare, until recently only three magnetic O-type stars were known (Donati *et al.* 2002, 2006; Bouret *et al.* 2008). However, recently new observations have increased this number significantly (Grunhut *et al.* 2009; Hubrig *et al.* 2011; Fossati *et al.* 2014) and current estimates assume that about ten percent of OB-type stars host a magnetic field. Stable, large-scale magnetic fields are suggested to be of fossil origin, arising from fields in a molecular cloud that are conserved during star formation (Braithwaite & Nordlund 2006). Alternatively, they can be explained as a result of the mergers of two main sequence stars. Binary interaction resets the evolutionary clock of the star. Therefore we assumed that the B-field acts unaltered since the ZAMS of the rejuvenated star (de Mink *et al.* 2014). Many observed magnetic field, mostly of dipole characteristics, are expected to be stabilized by an internal toroidal field (Braithwaite & Spruit 2004). This field is assumed to interact and compress the core convection, resulting in a core with reduced size. Zahn (2009) have shown that even moderate magnetic fields can already induce significant effects. For this work, the stellar evolution code BEC was used, see for example Köhler *et al.* (2012) for details. We simulate a reduced convective core as a result of an interaction with a magnetic field by applying a modified Ledoux criterion

$$\nabla_{\rm rad} < f * \nabla_{\rm Led},$$
(1.1)

with

$$\nabla_{\rm Led} = \nabla_{\rm ad} - \frac{\chi_\mu}{\chi_T}\nabla_\mu,$$
(1.2)

and chose values for a "reduction parameter" f between 1.12 and 1.20. These values are shown to reduce the mass of the convective core sufficiently for the current purpose, although higher values might be necessary for stars with masses lower than $3\,M_\odot$.

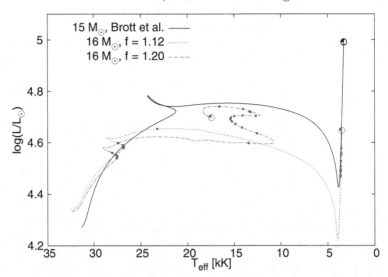

Figure 1. Evolutionary track of a $16\,M_\odot$ star in the HR diagram, computed with two different reduction parameters. For the smaller value, f=1.12, the star continues its evolution as a red supergiant (dashed line), however its final luminosity is significantly smaller than in standard calculations - see for a comparison the track for a $15\,M_\odot$ star (solid line) by Brott *et al.* (2011). For the larger value, $f = 1.20$, the red supergiant stage is avoided (dot-dashed line). The dots indicate timesteps of 10^5 years.

2. Results and discussion

We showed that with a sufficiently reduced convective core, magnetic stars no longer evolve as red supergiants, but remain in the blue part of the HR diagram, thus filling the post-main sequence gap that remains otherwise unpopulated. Those stars that retain a convective core that is still large enough for an evolution as a red supergiant, show a final luminosity that is smaller when compared to standard calculations (see Fig. 1). In general, a trend is seen that for smaller masses, a larger *reduction parameter* is required for avoiding the red supergiant stage. Below $8\,M_\odot$ the red supergiant stage is avoided, even for the largest values of f. With the assumption that stars on the main sequence have surface magnetic fields of the order of 1 kG, those of magnetic blue supergiants are expected to have values around 10 G and are therefore likely observable.

References

Bouret, J.-C., Donati, J.-F., Martins, F., *et al.* 2008, *MNRAS* 389, 75
Braithwaite, J. & Nordlund, Å. 2006, *A&A* 450, 1077
Braithwaite, J. & Spruit, H. C. 2004, *Nature* 431, 819
Brott, I., de Mink, S. E., Cantiello, M., *et al.* 2011, *A&A* 530, A115
de Mink, S. E., Sana, H., Langer, N., Izzard, R. G., & Schneider, F. R. N. 2014, *ApJ* 782, 7
Donati, J.-F., Babel, J., Harries, T. J., *et al.* 2002, *MNRAS* 333, 55
Donati, J.-F., Howarth, I. D., Bouret, J.-C., *et al.* 2006, *MNRAS* 365, L6
Fossati, L., Zwintz, K., Castro, N., *et al.* 2014, *A&A* 562, A143
Grunhut, J. H., Wade, G. A., Marcolino, W. L. F., *et al.* 2009, *MNRAS* 400, L94
Hubrig, S., Schöller, M., Kharchenko, N. V., *et al.* 2011, *A&A* 528, A151
Köhler, K., Borzyszkowski, M., Brott, I., Langer, N., & de Koter, A. 2012, *A&A* 544, A76
Zahn, J.-P. 2009, *Communications in Asteroseismology* 158, 27

New windows on massive stars: asteroseismology, interferometry, and spectropolarimetry
Proceedings IAU Symposium No. 307, 2014
G. Meynet, C. Georgy, J. H. Groh & Ph. Stee, eds.
© International Astronomical Union 2015
doi:10.1017/S1743921314007236

Magnetic CP stars in Orion OB1 association

Iosif I. Romanyuk and Eugene A. Semenko

Special Astrophysical Observatory of the Russian Academy of Sciences, Nizhnii Arkhyz,
Russian Federation, 369167
email: `roman@sao.ru, sea@sao.ru`

Abstract. We present the overview of an observational program carring out on the 6-m telescope of Special Astrophysical Observatory from 2012. This program aims the searches of new Bp stars with surface magnetic field and the detailed study of known magnetic CP star from the stellar association Ori OB1. HD 34736 is the most interesting star that was found as magnetic within the program recently.

Keywords. stars: evolution, open clusters and associations: general, stars: magnetic fields, stars: chemically peculiar

1. General results

A total of 85 CP stars of various types are identified among 814 members of the Orion OB1 association. The fraction of CP stars decreases with age for different subgroups from 15.1% in the youngest subgroup (b) to 7.7% in the oldest one (a).

We selected 59 Bp stars, which account for 13.4% of the total number of B type stars in the association. The fraction of peculiar B-type stars in the Orion OB1 association

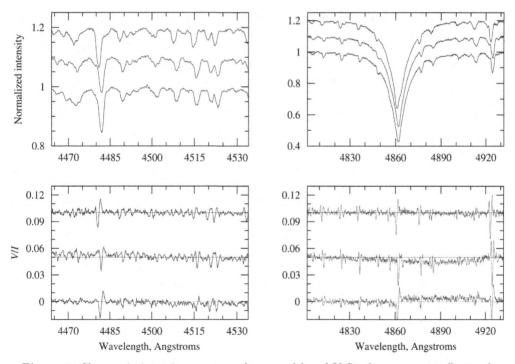

Figure 1. Changes in intensity spectrum (top panels) and *V* Stockes parameter (bottom) during the observations in October and December, 2013.

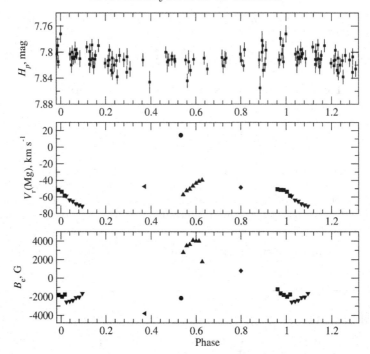

Figure 2. Individual measurements of (from top to bottom) HIPPARCOS photometry, radial velocity as measured from MgII 4481Å, and longitudinal magnetic field, phased with period 1.0808 days. Individual observational sets are marked by different symbols. Errors in V_r and B_e measurement are comparable with corresponding markers size.

is found to be twice higher than of peculiar A-type stars. Magnetic fields are found in 23 stars, 17 of them are objects with anomalous helium lines. Magnetic stars show a well-defined tendency to concentrate in the central region of association. No significant differences are found between the field strength in the Bp type stars of the association and Bp type field stars (Romanyuk *et al.* 2013).

2. HD 34736

Spectropolarimetric observations were carried out with the 6m Russian telescope. We acquired Zeeman spectra for all 59 Bp stars from Orion OB1 association. Strong magnetic field was discovered on a star HD 34736 (Fig. 1). Its longitudinal field B_e exceeds $-4500\,$G (Semenko *et al.* 2014). Hydrogen line profile corresponds to the effective temperature of about 13 700 K. The variable radial velocity of some spectral lines and the presence of lines signatures of at least one additional component testify that HD 34736 is a short periodic binary system. We found the best orbital period has to be about 1.08 days (Fig.2).

Period of longitudinal magnetic field B_e changes is not in coincidence with orbital one. Measured projected rotational velocity of magnetic component is 73 km s^{-1}.

References

Romanyuk, I. I., Semenko, E. A., Yakunin, I. A., & Kudryavtsev, D. O. 2013, *Astrophysical Bulletin* 68, 300
Semenko, E. A., Romanyuk, I. I., Kudryavtsev, D. O., & Yakunin, I. A. 2014, *Astrophysical Bulletin* 69, 191

New windows on massive stars: asteroseismology, interferometry, and spectropolarimetry
Proceedings IAU Symposium No. 307, 2014
G. Meynet, C. Georgy, J. H. Groh & Ph. Stee, eds.

© International Astronomical Union 2015
doi:10.1017/S1743921314007248

Stellar magnetic fields from four Stokes parameter observations

N. Rusomarov, O. Kochukhov and N. Piskunov

Uppsala University
email: naum.rusomarov@physics.uu.se

Abstract. Magnetic Doppler imaging (MDI) from observations of four Stokes parameters can uncover new information that is of interest to the evolution and structure of magnetic fields of intermediate and high-mass stars. Our MDI study of the chemically peculiar star HD 24712 from four Stokes parameter observations, obtained with the HARPSpol instrument at the 3.6-m ESO telescope, revealed a magnetic field with strong dipolar component and weak small-scale contributions. This finding gives evidence for the hypothesis that old Ap stars have predominantly dipolar magnetic fields.

Keywords. stars: magnetic fields, stars: atmospheres, stars: chemically peculiar, stars: imaging

Magnetic Doppler imaging (MDI) from observations of four Stokes parameters contributes to the understanding of the evolution and structure of magnetic fields of intermediate and high-mass main sequence stars. Such studies have recently revealed unprecedented, previously-unknown level of detail of stellar magnetic fields.

Along these lines, we have started a new program aimed at obtaining phase-resolved, spectropolarimetric observations of chemically peculiar stars in all four Stokes parameters, and using these data to perform MDI inversions for these stellar objects. Thus far, we have achieved full rotational phase coverage of the chemically peculiar stars HD 24712, HD 125248, and HD 119419. The observational data for these objects were obtained with the HARPSpol instrument at the 3.6-m ESO telescope at La Silla, Chile. This spectrograph allows full Stokes vector observations with a spectral resolution greater than 10^5. An analysis of the full Stokes vector spectropolarimetric data set of HD 24712 has been published by Rusomarov *et al.* (2013).

Here, we present the results of the MDI analysis of HD 24712. We derived chemical abundance and magnetic distribution maps from a selection of Fe, Nd, and Na lines. The comparison of observed and calculated Stokes profiles of several spectral lines is shown in Fig. 1. The spherical projections of the magnetic maps are shown in Fig. 2.

Our analysis shows that the magnetic field topology of HD 24712 is mostly poloidal and dipolar, with small contributions from higher-order harmonics. In contrast, the only two other stars for which we have MDI studies from four Stokes parameter observations, α^2 CVn (Kochukhov & Wade 2010; Silvester *et al.* 2014) and 53 Cam (Kochukhov *et al.* 2004), exhibit dipole-like fields with considerably more complexity on smaller scales.

Numerical MHD simulations of fossil fields in main sequence stars with radiative envelopes by Braithwaite & Nordlund (2006) showed that stable magnetic field configurations can exist in the form of a poloidal magnetic field wrapped around toroidal flux tubes. These authors also proposed that stars possessing dipole-like fields with little small-scale structures are older than stars with more complex fields.

The finding of dipole, axisymmetric field for HD 24712, which is the oldest object in our three stars sample, in conjunction with the results for α^2 CVn and 53 Cam, supports this hypothesis. In the future, we hope to perform MDI of other Ap/Bp stars with different

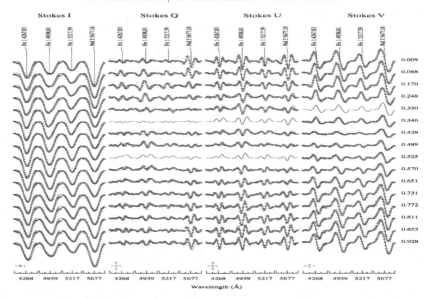

Figure 1. Comparison of the observed (symbols) and calculated (lines) Stokes profiles for HD 24712. The bars at the lower left of each panel show the horizontal and vertical scale (0.2 and 2.5% of the Stokes I continuum intensity).

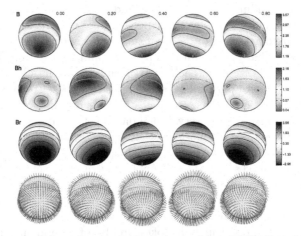

Figure 2. Distribution of the magnetic field on the surface of HD 24712. Magnetic field modulus (first row), horizontal component (second row), radial field component (third row), and field orientation (last row). The color bars indicate field strength in kG. The arrows length is proportional to the field strength. The contours are plotted with 1 kG step.

masses and ages, which will help us further assess this hypothesis and investigate the dependence of the magnetic field geometry on other stellar parameters.

References

Braithwaite, J. & Nordlund, Å. 2006, *A&A* 450, 1077

Kochukhov, O., Bagnulo, S., Wade, G. A., *et al.* 2004, *A&A* 414, 613

Kochukhov, O. & Wade, G. A. 2010, *A&A* 513, A13

Rusomarov, N., Kochukhov, O., Piskunov, N., *et al.* 2013, *A&A* 558, A8

Silvester, J., Kochukhov, O., & Wade, G. A. 2014, *MNRAS* 440, 182

New windows on massive stars: asteroseismology, interferometry, and spectropolarimetry
Proceedings IAU Symposium No. 307, 2014
G. Meynet, C. Georgy, J. H. Groh & Ph. Stee, eds.

© International Astronomical Union 2015
doi:10.1017/S174392131400725X

Plasma Leakage from the Centrifugal Magnetospheres of Magnetic B-Type Stars

Matt Shultz[1,2,3], Gregg Wade[3], Thomas Rivinius[1], Jason Grunhut[1], Véronique Petit[4] and the MiMeS Collaboration

[1] European Southern Observatory

[2] Queen's University, Canada

[3] Royal Military College, Canada

[4] University of Delaware, USA
email: mshultz@eso.org

Abstract. Magnetic B-type stars are often host to *Centrifugal Magnetospheres* (CMs). Here we describe the results of a population study encompassing the full sample of known magnetic early B-type stars, focusing on those with detectable CMs. We present revised rotational and magnetic parameters for some stars, clarifying their positions on the rotation-confinement diagram, and find that plasma densities within their CMs is much lower than those predicted by centrifugal breakout.

Keywords. plasmas, stars: circumstellar matter, stars: magnetic fields, stars: rotation

---

Magnetic confinement of the winds of magnetic OB stars is explained by ud-Doula *et al.*, (this volume) while for a description of CMs and the Rigidly Rotating Magnetosphere (RRM; Townsend *et al.* 2005) model we refer to Oksala *et al.* (this volume).

It is not known how plasma escapes from CMs. The leading proposal is violent ejection in a so-called 'centrifugal breakout' (CB) event (Townsend *et al.* 2005; ud-Doula *et al.* 2008): plasma density increases beyond the ability of the magnetic field to confine the wind, thus rupturing the magnetic field structure. However, no direct evidence of CB has been found (e.g. Townsend *et al.* 2013), motivating a deeper examination of the properties of CMs.

We are conducting a population study of the magnetic B-type stars presented by Petit *et al.* (2013) (see also Fig. 1), aimed at studying the dependence of CM emission on stellar, magnetic, and rotational properties. We are performing systematic follow-up observations of newly discovered or poorly studied magnetic B-type stars, with the intent of determining rotational periods (and hence Kepler radii $R_{\rm K}$) and dipolar magnetic field strengths (and hence Alfvén radii $R_{\rm A}$) for all stars. We also measure the emission strengths and plasma densities of the sub-set of stars with CMs via Hα emission (about 25% of the magnetic B-type population: filled symbols in Fig. 1).

We have measured the emission equivalent width (EW) of CM host stars, in both Hα and UV resonance lines, by subtracting model photospheric spectra from observed spectra. We find clear thresholds dividing stars with and without Hα emission ($R_{\rm A} > 20R_*$, $R_{\rm K} < 3R_*$, $\log R_{\rm A}/R_{\rm K} > 1$), however, beyond these thresholds there are no clear trends. In contrast to Hα, UV line emission is generally stronger, and more widely distributed in the rotation-confinement diagram (outlined symbols in Fig. 1).

Using EWs for Hα, Hβ, and Hγ, we measured the circumstellar plasma density via Balmer decrements (Williams & Shipman 1988) at emission maximum (corresponding to the rotational phase at which the line-of-sight is closest to perpendicular to the

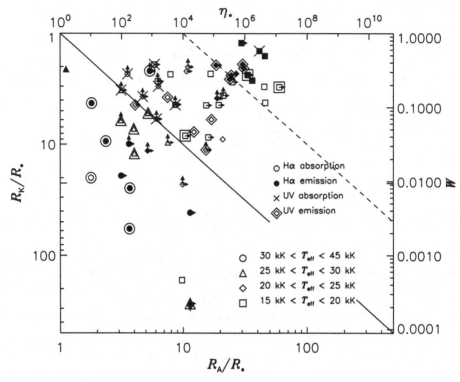

Figure 1. An up-dated rotation-confinement diagram, first presented by Petit *et al.* (2013). Stars are organized by magnetic confinement radius, the Alfvén radius R_A, and the radius at which centrifugal and gravitational forces balance, the Kepler radius R_K, both in terms of the stellar radius R_*. These are respectively proportional to the magnetic wind confinement parameter η_* and the rotation paramter W. Closer analysis of sample stars has moved some stars with Hα emission above $R_A = 10R_K$ (the diagonal dashed line), while some stars with Hα in absorption have been moved below this line. One Hα absorption star, HD 35912, has been removed entirely, due to non-detection of a Zeeman signature in 11 ESPaDOnS observations with $< \sigma_B > \sim 40$ G. A new Hα emission star, ALS 3694, has been discovered with new ESPaDOnS data. Finally, R_K of the early-type β Cep star ξ^1 CMa has been revised sharply upward to >270 R_* (see Shultz *et al.*, this volume).

plane of the magnetosphere). There is no significant variation from the mean density, $\log N \sim 12.5$, with $0.15 < \sigma_{\log N} < 0.7$ dex. CB predicts both substantially higher plasma densities, and clear, strong trends in plasma density, increasing with increasing R_A and decreasing R_K. Furthermore, CB predicts that plasma density should be high enough for Hα emission in numerous stars for which no such emission is detected.

References

Petit, V., Owocki, S. P., Wade, G. A., *et al.* 2013, *MNRAS* 429, 398

Townsend, R., Owocki, S., & Groote, D. 2005, in R. Ignace & K. G. Gayley (eds.), *The Nature and Evolution of Disks Around Hot Stars*, Vol. 337 of *Astronomical Society of the Pacific Conference Series*, p. 314

Townsend, R. H. D., Rivinius, T., Rowe, J. F., *et al.* 2013, *ApJ* 769, 33

ud-Doula, A., Owocki, S. P., & Townsend, R. H. D. 2008, *MNRAS* 385, 97

Williams, G. A. & Shipman, H. L. 1988, *ApJ* 326, 738

New windows on massive stars: asteroseismology, interferometry, and spectropolarimetry
Proceedings IAU Symposium No. 307, 2014
G. Meynet, C. Georgy, J. H. Groh & Ph. Stee, eds.
© International Astronomical Union 2015
doi:10.1017/S1743921314007261

ξ^1 CMa: An Extremely Slowly Rotating Magnetic B0.7 IV Star

Matt Shultz[1,2,3], Gregg Wade[3], Thomas Rivinius[1]
Wagner Marcolino[4], Huib Henrichs[5], Jason Grunhut[1]
and the MiMeS Collaboration

[1] European Southern Observatory

[2] Queen's University, Canada

[3] Royal Military College, Canada

[4] Observatório do Valongo, UFRJ, Brazil

[5] Anton Pannekoek Institute for Astronomy, University of Amsterdam, the Netherlands
email: mshultz@eso.org

Abstract. We present our analysis of 6 years of ESPaDOnS spectropolarimetry of the magnetic β Cep star ξ^1 CMa (B1 III). This high-precision magnetometry is consistent with a rotational period $P_{\rm rot} > 40$ yr. Absorption line profiles can be reproduced with a non-rotating model. We constrain $R_\star$, $L_\star$, and the stellar age via a Baade-Wesselink analysis. Spindown due to angular momentum loss via the magnetosphere predicts an extremely long rotational period if the magnetic dipole $B_{\rm d} > 6$ kG, a strength also inferred by the best-fit sinusoids to the longitudinal magnetic field measurements $B_{\rm Z}$ when phased with a 60-year $P_{\rm rot}$.

Keywords. stars: circumstellar matter, stars: magnetic fields, stars: rotation

1. Magnetometry

The B0.7 IV β Cep star ξ^1 CMa was detected as magnetic by FORS1 (Hubrig *et al.* 2006) and later confirmed by ESPaDOnS (Silvester *et al.* 2009). Further FORS1/2 observations yielded a 2.2 d period (Hubrig *et al.* 2011), however many stars for which periods were reported based on FORS1/2 data were not confirmed to be magnetic by ESPaDOnS data (Shultz *et al.* 2012), casting doubt on the reliability of period analysis using FORS1/2 data. An analysis of an earlier, smaller ESPaDOnS data-set was presented by Fourtune-Ravard *et al.* (2011), who found $P_{\rm rot} \sim 4$ d.

The ESPaDOnS dataset has grown to 34 Stokes V spectra obtained from 2008 to 2014, with annual clusters of up to 16 spectra separated by hours to days, and uniform integration times of 240 s. Sharp spectral lines and high SNR (>440) lead to a mean error bar in $B_{\rm Z}$ of $< \sigma_B > = 6$ G in the single-spectrum least-squares deconvolution (LSD) profiles (Kochukhov *et al.* 2010). $B_{\rm Z}$ declines smoothly from 340 ± 17 G in 2009 to 251 ± 3 G in 2014 (see Fig. 1). There is a systematic error (~15 G) due to a small variation of $B_{\rm Z}$ with the pulsation period. We have also located 2 archival MuSiCoS observations obtained in 2000. In both spectra, the polarity of the Zeeman signature is negative, with $B_{\rm Z} = -137 \pm 32$ G. *The minimum $P_{\rm rot}$ compatible with the combined datasets is 40 years: the longest rotation period yet found for an early B-type star.*

2. Analysis

We used the Si III 455.3 nm line to derive $v \sin i$ and macroturbulence ζ, performing a goodness of fit test using a grid of synthetic line profiles including radial pulsation

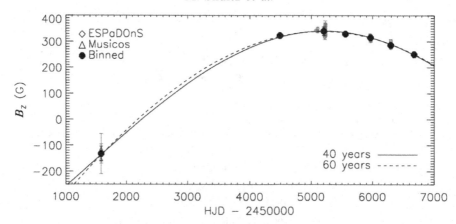

Figure 1. Longitudinal magnetic field $B_{\rm Z}$ measurements as a function of time. Measurements binned by epoch are in solid black circles.

(Saesen *et al.* 2006), finding $v \sin i < 6\ {\rm km\ s^{-1}}$, and $\zeta = 20 \pm 3\ {\rm km\ s^{-1}}$, i.e. the line profile can be be reproduced with a non-rotating model.

To constrain the stellar parameters we performed a Baade-Wesselink analysis, combining the pulsation phase-dependant variations in integrated light (via Hipparcos photometry), radial velocity (via ESPaDonS and CORALIE spectra, Saesen *et al.* 2006), and $T_{\rm eff}$ (via EW ratios for various elemental ionic species). $< T_{\rm eff} >= 25.9 \pm 0.1\ {\rm kK}$, with a $\pm 0.5\ {\rm kK}$ amplitude variation. With the bolometric correction of Nieva (2013), the model light curve reproduces the photometric variability with $R = 8.7 \pm 0.7 R_\odot$, leading to $\log L/L_\odot = 4.46 \pm 0.07$, where the uncertainty is largely a function of the error in the Hipparcos parallax. Isochrones (Ekström *et al.* 2012) indicate an age of 12.6 Myr.

Magnetic stars shed angular momentum via the torque applied to the stellar surface by the corotating magnetosphere (Weber & Davis 1967; ud-Doula *et al.* 2009). Can a magnetic early B-type star spin down to $P_{\rm rot} > 40\,{\rm yrs}$ after only 12.6 Myr? We computed the Alfvén radius (ud-Doula & Owocki 2002) based on the mass-loss rate and wind terminal velocity predicted by the recipe of Vink *et al.* (2001) using the stellar parameters above, for values of $B_{\rm d}$ between 1.2 and 8.0 kG, and from this the spindown time (ud-Doula *et al.* 2009), assuming initially critical rotation. If $B_{\rm d} > 6\,{\rm kG}$, $P_{\rm rot} > 40\,{\rm yrs}$ can be achieved; if $P_{\rm rot} = 60\,{\rm yrs}$, the best-fit sinusoid to $B_{\rm Z}$ implies $B_{\rm d} \sim 6\,{\rm kG}$.

References

Ekström, S., Georgy, C., Eggenberger, P., *et al.* 2012, *A&A* 537, A146
Fourtune-Ravard, C., Wade, G. A., Marcolino, W. L. F., *et al.* 2011, in C. Neiner, G. Wade, G. Meynet, & G. Peters (eds.), *IAU Symposium*, Vol. 272 of *IAU Symposium*, pp 180–181
Hubrig, S., Briquet, M., Schöller, M., *et al.* 2006, *MNRAS* 369, L61
Hubrig, S., Ilyin, I., Schöller, M., *et al.* 2011, *ApJ (Letters)* 726, L5
Kochukhov, O., Makaganiuk, V., & Piskunov, N. 2010, *A&A* 524, A5
Nieva, M.-F. 2013, *A&A* 550, A26
Saesen, S., Briquet, M., & Aerts, C. 2006, *Communications in Asteroseismology* 147, 109
Shultz, M., Wade, G. A., Grunhut, J., *et al.* 2012, *ApJ* 750, 2
Silvester, J., Neiner, C., Henrichs, H. F., *et al.* 2009, *MNRAS* 398, 1505
ud-Doula, A. & Owocki, S. P. 2002, *ApJ* 576, 413
ud-Doula, A., Owocki, S. P., & Townsend, R. H. D. 2009, *MNRAS* 392, 1022
Vink, J. S., de Koter, A., & Lamers, H. J. G. L. M. 2001, *A&A* 369, 574
Weber, E. J. & Davis, Jr., L. 1967, *ApJ* 148, 217

*New windows on massive stars: asteroseismology, interferometry, and
spectropolarimetry*
Proceedings IAU Symposium No. 307, 2014
G. Meynet, C. Georgy, J. H. Groh & Ph. Stee, eds.

© International Astronomical Union 2015
doi:10.1017/S1743921314007273

Magnetic fields and internal mixing
of main sequence B stars

G. A. Wade[1], C. P. Folsom[2], J. Grunhut[3], J. D. Landstreet[4,5] and V. Petit[6]

[1] RMC, Canada

[2] IRAP, France

[3] ESO, Germany

[4] UWO, Canada

[5] Armagh, U.K.

[6] U. Delaware, USA

Abstract. We have obtained high-quality magnetic field measurements of 19 sharp-lined B-type stars with precisely-measured N/C abundance ratios (Nieva & Przybilla 2012). Our primary goal is to test the idea (Meynet *et al.* 2011) that a magnetic field may explain extra drag (through the wind) on the surface rotation, thus producing more internal shear and mixing, and thus could provide an explanation for the appearance of slowly rotating N-rich main sequence B stars.

Keywords. stars: abundances, stars: early-type, stars: evolution, stars: magnetic fields

1. Introduction and motivation

Observations of chemically-enriched main sequence stars (e.g. Gies & Lambert 1992) have led to the idea that rotationally induced mixing may bring fusion products from the cores of massive stars to their surfaces during phases of core hydrogen burning. Measurements of surface chemical abundances - in particular those of light elements - of early-type main sequence stars can therefore be leveraged to place constraints on the associated circulation currents. A surprising result of these studies is the identification of a significant population of stars with enhanced nitrogen abundances with apparently slow rotation (Hunter *et al.* 2008; Morel *et al.* 2006). The origin of the nitrogen enhancement in these stars is not understood, and may be related to various physical processes, e.g. true slow rotation, pulsation (Aerts *et al.* 2014), binarity (Langer 2008) or magnetic fields (Morel *et al.* 2008).

2. The sample

Our sample consists of 19 of the the 20 nearby main sequence B stars for which Nieva & Przybilla (2012) determined precise, homogeneous abundances of CNO elements. All of the sample stars are sharp-lined, and 3 stars (ζ Cas, τ Sco, and β Cep) are known magnetic stars. The HR diagram of the sample is shown in Fig. 1. We identify 6 N-rich stars, those with N/C differing by $> 2\sigma$ from the solar value of -0.60 ± 0.07 (Asplund *et al.* 2009): the 3 magnetic stars, as well as HD 61068 (N/C $= -0.27$, $\sigma_B = 9\,\mathrm{G}$), HD 16582 (N/C $= +0.02$, $\sigma_B = 3\,\mathrm{G}$) and HD 35708 (N/C $= -0.08$, $\sigma_B = 8\,\mathrm{G}$).

3. Magnetic diagnosis

Magnetic fields were diagnosed for 19 stars using high resolution Stokes V spectropolarimetry obtained using the ESPaDOnS, Narval and Harpspol spectropolarimeters.

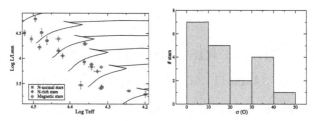

Figure 1. *Left:* HR diagram of the sample stars. Evolutionary tracks (Schaller *et al.* 1992) span $6 - 25\,M_{\odot}$. *Right:* Histogram of longitudinal magnetic field measurement standard errors.

Some observations were obtained as part of the MiMeS Large Programs, while others were obtained from a dedicated PI program. Least-Squares Deconvolution (LSD) using tailored line masks was used to obtain high SNR mean Stokes I, V and diagnostic null line profiles for each star. Typically, 1-3 observations per target were obtained. The presence of a magnetic field was evaluated using the χ^2 detection criterion of Donati *et al.* (1997), as well as via measurement of the longitudinal magnetic field. No new magnetic stars are detected, with a median longitudinal field error bar of 13 G (see histogram in Fig. 1). We have also inferred upper limits on the surface dipole fields of the non-magnetic stars using the method of Petit & Wade (2012). All odds ratios favour the non-magnetic model, and all probability density functions characterizing the dipole field strength are consistent with zero within 68% confidence.

4. Results, conclusions and future work

We find that all magnetic stars in the sample show significant N enrichment. We therefore conclude that the proposed mechanism by which magnetic braking increases mixing from the deep interior (Meynet *et al.* 2011) does seem to be supported by our analysis. However, Morel (2011) reported normal N/C ratios for two magnetic early-type stars (NGC 2244-201, HD 57682), albeit with somewhat worse precision. Taking this at face value, the results appear presently to be ambiguous. Moreover, a small number of confidently non-magnetic stars in our sample are also observed to be significantly N-rich. In particular, the non-magnetic star HD 16582 (B 2IV) shows the largest N/C enrichment of the sample. To explain the non-magnetic N-rich stars, additional mechanisms must be at play. Considering the small size of the current sample, an obvious extension will be to obtain similarly precise N/C ratios and magnetic data for a much larger sample of magnetic and non-magnetic early B/late O stars.

References

Aerts, C., Molenberghs, G., Kenward, M. G., & Neiner, C. 2014, *ApJ* 781, 88
Asplund, M., Grevesse, N., Sauval, A. J., & Scott, P. 2009, *ARA&A* 47, 481
Donati, J.-F., Semel, M., Carter, B. D., Rees, D. E., & Collier Cameron, A. 1997, *MNRAS* 291, 658
Gies, D. R. & Lambert, D. L. 1992, *ApJ* 387, 673
Hunter, I., Brott, I., Lennon, D. J., *et al.* 2008, *ApJ (Letters)* 676, L29
Langer, N. 2008, in L. Deng & K. L. Chan (eds.), *IAU Symposium*, Vol. 252 of *IAU Symposium*, pp 467–473
Meynet, G., Eggenberger, P., & Maeder, A. 2011, *A&A* 525, L11
Morel, T. 2011, in C. Neiner, G. Wade, G. Meynet, & G. Peters (eds.), *IAU Symposium*, Vol. 272 of *IAU Symposium*, pp 97–98
Morel, T., Butler, K., Aerts, C., Neiner, C., & Briquet, M. 2006, *A&A* 457, 651

Morel, T., Hubrig, S., & Briquet, M. 2008, *A&A* 481, 453
Nieva, M.-F. & Przybilla, N. 2012, *A&A* 539, A143
Petit, V. & Wade, G. A. 2012, *MNRAS* 420, 773
Schaller, G., Schaerer, D., Meynet, G., & Maeder, A. 1992, *A&AS* 96, 269

General view of the conference room

Inside the globe at CERN

New windows on massive stars: asteroseismology, interferometry, and spectropolarimetry
Proceedings IAU Symposium No. 307, 2014
G. Meynet, C. Georgy, J. H. Groh & Ph. Stee, eds.

© International Astronomical Union 2015
doi:10.1017/S1743921314007285

Links between surface magnetic fields, abundances, and surface rotation in clusters and in the field

Norbert Przybilla

Institute for Astro- and Particle Physics, University of Innsbruck,
Technikerstr. 25/8, A-6020 Innsbruck, Austria
email: norbert.przybilla@uibk.ac.at

Abstract. Theory predicts that hydrodynamical instabilities transport angular momentum and chemical elements in rotating massive stars. An interplay of rotation and a magnetic field affects these transport processes. The complexity of the problem imposes that a comprehensive description cannot be developed on theoretical grounds alone, progress in the understanding of the evolution of massive stars has to be guided by observations. The challenge lies both in the derivation of accurate and precise observational constraints as well as in the extraction of the relevant information for identifying possible correlations – like between surface magnetic fields, abundances, and surface rotation – from a multivariate function of the many parameters involved. I review the most important steps recently made based on detailed studies of massive stars both in the field and in clusters towards finding such links that ultimately may guide the further development of the models.

Keywords. stars: abundances, stars: early-type, stars: evolution, stars: magnetic fields, stars: rotation

---

1. Introduction

The first grids of evolution models for rotating massive stars (Meynet & Maeder 2000; Heger & Langer 2000) indicated changes in all the outputs when compared to classical models: lifetimes, evolution scenarios, nucleosynthesis, etc. They were very successful in explaining many observed phenomena, however many issues still lack a comprehensive understanding, like the ratio of blue-to-red supergiants in galaxies, supernova channels, long-soft γ-ray burst progenitors and the rotation rates of stellar remnants. Important in this context is that many uncertainties remain in the input physics of the models, e.g. in the mass-loss rates, in the nature of transport mechanisms and the effects of magnetic fields, which furthermore may be generated by some dynamo effect or be of fossil nature (e.g. Langer 2014). In particular challenging are quantitative predictions for the post-main sequence evolution, but one has to keep in mind that the basis for the further developments is already laid on the main-sequence by the details of the core evolution. The topic is of such complexity that progress requires strong guidance by observation.

The most sensitive variables for the comparison with observation are light element abundances, in particular nitrogen, which shows much larger variations like carbon, oxygen or helium (for recent results see e.g. Ekström *et al.* 2012; Chieffi & Limongi 2013). Boron is also a highly sensitive indicator in particular for rotational mixing in the early evolution of massive stars (e.g. Frischknecht *et al.* 2010), however the determination of B abundances requires UV spectroscopy, therefore few results are available.

The development of the surface abundances, e.g. the nitrogen excess, is a multivariate function depending on mass M, age (τ_{evol}), projected rotational velocity ($v \sin i$), metallicity (Z) and multiplicity (Maeder *et al.* 2009): $\Delta(\mathrm{N/H}) = f(M, \tau_{\mathrm{evol}}, v \sin i, Z,$ multiplicity). On the modeling side further 'variables' complicate the problem, in particular in form of the input physics considered and the values adopted for various input parameters like e.g. the overshoot parameter or diffusion coefficients. Therefore, theory needs to be guided by observation if meaningful models shall be developed. Observational constraints should not be only accurate and precise, but they should also be restricted in input parameter range as much as possible in order to facilitate the extraction of relevant information.

Massive stars in clusters should therefore be preferred as targets, as they have the same age and the same initial chemical composition, and they are located at the same distance. However, observationally they are more challenging for high-S/N spectroscopy because any Galactic cluster with a significant massive star population is much more distant than the numerous massive stars in the nearby field and associations, which consequently can be studied in greater detail. Ideally, a statistical approach is used to compare observations with models, employing population synthesis (e.g. Georgy *et al.* 2014; Schneider *et al.* 2014), and to find correlations between observational variables (Aerts *et al.* 2014).

The present work aims at reviewing and evaluating the current status in the literature on observational constraints for establishing a possible correlation between surface magnetic fields, abundances, and rotation. These are required for guiding the inclusion of the interaction of rotation with magnetic fields in models of massive star evolution (e.g. Heger *et al.* 2005; Maeder & Meynet 2005; Meynet *et al.* 2011; Brott *et al.* 2011).

2. Magnetic fields in massive stars

The presence of magnetic fields in massive stars is difficult to be established. Interior magnetic fields may be ubiquitous in differentially rotating stars where the Tayler-Spruit dynamo (Spruit 2002) may be operational. However, they do not reach the stellar surface to become observable (Maeder & Meynet 2005). Their effects on the evolution of rotating massive stars (enhanced transport of angular momentum and chemical elements) can therefore be constrained only indirectly. Small-scale magnetic fields may also be generated in a sub-surface convection zone. They may reach the surface, manifesting as hot spots that give rise to discrete absorption components in the stellar wind (Cantiello *et al.* 2009). The sub-surface convection zone may be important for stellar evolution by inducing wind clumping, which impacts the mass-loss rates (see e.g. Puls, these proceedings).

Large-scale magnetic fields of fossil nature – either a relic of the interstellar magnetic field present during star formation, or generated in a binary merger – are the ones that can be detected by spectropolarimetry. Detections of a surface magnetic field require much higher S/N (>500) than typically achieved for classical quantitative spectroscopy. Therefore, limiting magnitudes for spectropolarimetric studies are significantly lower than for spectroscopy, and detections near the sensitivity limit are often not undisputed.

The knowledge about the occurrence of magnetism in hot, massive stars has benefited tremendously in the past years from large observational surveys like MiMeS (Wade *et al.* 2014) or the complementary BOB survey (Morel *et al.*, these proceedings) in addition to individual PI projects (e.g. Hubrig *et al.* 2011). An extensive (though in the meantime no longer complete) list of magnetic OB stars is given by Petit *et al.* (2013). Overall, an incidence of surface magnetic fields in about 7% of the OB stars is indicated, with dipole field strengths varying from several hundred G to $\sim$20 kG (Wade *et al.* 2014) and a likely extension towards lower field strengths at increased detection sensitivity (Morel *et al.*, these proceedings). The full characterisation of the properties of magnetic fields is much more demanding than the detection, such that results are available only for a limited number of objects (e.g. Donati *et al.* 2006; Neiner *et al.* 2012; Henrichs *et al.* 2013).

3. Towards correlations between surface magnetic fields, abundances and rotational velocity in OB stars: state-of-the-art

Despite the importance of elemental abundances as the most sensitive observational constraints on evolution models, relatively little is known on light element abundances (He, B, CNO) in magnetic massive stars. The topic was addressed first by Morel *et al.* (2008), suggesting a higher incidence of chemical peculiarities in early B-type stars with (weak) magnetic fields, which in several cases could also be confirmed as genuine slow rotators by determining rotational periods either from the occurrence of phase-locked UV wind line-profile variations, or from asteroseismic modelling. N-rich objects were found to be B-poor, in agreement with expectations from theory. Note that this study is the only one considering boron abundances to date. Further studies found additional N-rich magnetic early-B and late-O stars (Morel 2012; Henrichs *et al.* 2012), but also four N-normal magnetic stars (Morel 2011; Fossati *et al.* 2014). A strict correlation between the presence of magnetic fields and N-enrichment appears no longer assumable, additional parameters seem to play a rôle as well. The initial angular momentum is crucial, and possibly the star's feasibility to retain differential rotation in the presence of a large-scale magnetic field, vs. rigid rotation (Meynet *et al.* 2011). In particular, magnetic stars with fossil fields (relics of the interstellar-medium field) that were born with negligible angular momentum are not expected to show mixing signatures.

It is interesting in this context that no signatures of CNO-mixing were found in the components of detached eclipsing OB-type binaries so far (for a summary see Pavlovski & Southworth 2013). Likewise, magnetic fields appear to be absent in massive binaries as indicated by the BinaMIcS project (Alecian *et al.*, these proceedings).

Several O-type stars were analysed by Martins *et al.* (2012a) in order to investigate observational effects of magnetism in more massive objects. For the majority of the stars, slow rotation was found and N-enrichment beyond the values typical for mixed early B-stars. Note that HD 57682 (O9 IV) is in common with the study of Morel (2011), however a nitrogen abundance higher by $\sim$0.6 dex was derived, suggesting it to be N-rich. This discrepancy needs to be resolved by further investigations. The sample of Martins *et al.* (2012a) includes three Of?p stars, the magnetic stars that have the strongest fields among O stars. They show high levels of surface N-enrichment, yet not unusually high compared to other O stars. They are also all slow rotators (Wade *et al.* 2012, and references therein), apparently a consequence of magnetic braking (Ud-Doula *et al.* 2009).

ζ Ori A (O9.5 I) is the only magnetic hot supergiant known to date. It is a N-normal fast rotator (Bouret *et al.* 2008; Martins *et al.* 2012a), which has been identified as a double-lined spectroscopic binary (SB2) by Hummel *et al.* (2013). The supergiant component ζ Ori Aa has been proven to be the magnetic star (Blazère *et al.*, these proceedings).

As mentioned in the previous section, internal dynamo-generated magnetic fields may be ubiquitous among massive stars. It is therefore equally important to investigate possible correlations between rotation and surface abundances for samples of stars without detected surface fields. The by far most extensive study in the past years of $\sim$300 cluster early B-stars was the FLAMES Survey of Massive Stars (Hunter *et al.* 2009). An unexpectedly large number of slowly-rotating N-rich stars was found, as well as a class of non-enriched rapid rotators. Thus, many objects appear to remain unexplained by evolution models for single stars, however see Sect. 5. Nitrogen surface abundances in Galactic O stars indicate a wider range of enrichment compared to B stars and a clear trend toward more enrichment in higher mass stars (Martins *et al.* 2012b). The bulk of the LMC O-stars studied by Rivero González *et al.* (2012) appears to be strongly N-enriched, and a clear correlation of nitrogen and helium enrichment was found.

The so far – in terms of proper statistical treatment – most comprehensive search for correlations among many variables in OB stars has been performed by Aerts *et al.* (2014), finding that $v \sin i$ and the rotational frequency, as well as the magnetic field strength have no predictive power for nitrogen abundances. Furthermore, comprehensive comparisons of observational constraints extracted from numerous literature sources with predictions from various stellar evolution models were presented by Martins & Palacios (2013). Note that both studies draw their conclusions from highly heterogeneous observational datasets; these may require (partial) revision in view of the following discussion.

4. Observational challenges for OB stars

Quantitative analyses of OB stars are not only demanding in terms of model atmospheres and line formation, they also have to face several observational challenges.

Multi-object spectroscopy has improved the efficiency of cluster studies enormously, however it brings its own challenges, as experienced e.g. within the FLAMES Survey of Massive Stars (Evans *et al.* 2005). Crucial issues are the large dynamical range between faint main-sequence stars and bright (super)giants, which can span $\sim$4 to 7 mag in V (this can be solved by observing a few objects with a different spectrograph), and fibre/slit positioning limitations within the field-of-view. The latter is in particular a problem for observations of distant clusters like in the Magellanic Clouds, where the sampling of objects in the denser parts of a cluster may in fact be sparse (Evans *et al.* 2006). Instead, many objects from the field surrounding the cluster may be covered.

The probably most important challenge for observations of OB stars is the identification of contamination of a spectrum by light from another star. It has turned out in the past years that the binary fraction among OB stars may reach 50-70% (Sana *et al.* 2012; Chini *et al.* 2012), many of which show SB2 characteristics. Signatures for the presence of second light may be subtle and can often be recognised only in high-resolution, high-S/N spectra. Second light changes line depths relative to the continuum flux and thus also equivalent widths. An analysis of a binary composite spectrum with standard techniques – therefore assuming it to be single – results in erroneous stellar parameters and elemental abundances. However, a correct analysis can be achieved using either spectrum disentangling (e.g. Pavlovski & Southworth 2013) or spectrum synthesis accounting for both sources (e.g. Irrgang *et al.* 2014).

Chemically peculiar stars are rare among OB stars and comprise He-rich stars and He-weak stars (see e.g. Smith 1996 for a discussion). They are of limited use for obtaining observational constraints on massive star evolution, as their atmospheres show either spotted elemental concentrations, or vertical stratification because of atomic diffusion resulting from the interplay of radiative levitation and gravitational settling.

Finally, second light from the disk complicates the analysis of Be stars. When continuum and line emission from a disc overlaps with the stellar spectrum, this gives rise to a variety of spectral morphologies depending on disc parameters, disc size, and viewing angle on the star+disc system (see e.g. Rivinius *et al.* 2013). Moreover, Be-stars rotate close to the breakup velocity, leading to a distortion from spherical geometry and non-uniform surface temperatures (hot poles and cool equator) and densities. At present, analyses based on a comprehensive model of the stellar atmosphere+disk cannot be performed, therefore any observational constraints suffer from limited accuracy and precision.

Any star sample to be analysed with standard model atmosphere techniques needs to be carefully checked for the presence of binary, chemically peculiar or Be stars. Such objects need to be excluded in order to avoid the introduction of observational bias.

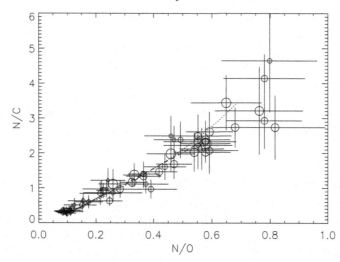

Figure 1. N/C vs. N/O abundance ratios (by mass) for homogeneously analysed stars from the samples of Przybilla *et al.* (2010), Nieva & Przybilla (2012) and Nieva & Simón-Díaz (2011), and Firnstein & Przybilla (2012, including preliminary data). B-type main-sequence stars are displayed as diamonds, BA-type supergiants as circles. The symbol size encodes the stellar mass and error bars give 1σ-uncertainties. The lines represent predictions from evolution calculations, for a rotating $15\,M_\odot$ star, $v_{\rm rot}^{\rm ini} = 300\,{\rm km\,s}^{-1}$, (Meynet & Maeder 2003, solid line: until the end of the main sequence; dashed line: until the end of He burning) and for a star of the same mass and $v_{\rm rot}^{\rm ini}$ that in addition takes the interaction of rotation and a magnetic dynamo into account (Maeder & Meynet 2005, dotted line: until the end of the main sequence). Note the clump of 20 stars at the base of the tracks that results from objects with pristine CNO abundances.

5. New developments in the quantitative spectroscopy of OB stars

The question of the presence of systematical bias in observational constraints is a crucial one for the comparison with stellar evolution models. Such bias can be of observational nature (see above) or it can be produced by flaws in the spectral analysis, and – at first glance – it may appear difficult to establish the presence of bias. However, a simple and objective test exists, which can be used to assess the accuracy and precision of analysis results: the N/C vs. N/O diagram (Przybilla *et al.* 2010; Maeder *et al.* 2014).

The ratios of the participants in the CNO cycles at the stellar surface depend both on the changes produced by nuclear reactions and the dilution effects produced by mixing. Analytical relations can be derived for the predicted nuclear path, which indicate an initially almost linear slope of e.g. $\mathrm{d(N/C)}/\mathrm{d(N/O)} \approx 4$ for typical pristine CNO abundances as found in the solar neighbourhood at present day (Przybilla *et al.* 2008; Nieva & Przybilla 2012), which is matched well by detailed stellar evolution models. In particular, the N/C vs. N/O plot shows little dependence on the initial stellar masses, rotation velocities, and nature of the mixing processes up to relative enrichment of N/O by a factor of about four. Figure 1 shows an example how tight is the match between a homogeneously analysed star sample with high-quality spectra – that have been scrutinised in terms of observational bias – and the predictions by stellar evolution models. The observational sample consists of 29 early B-stars on the main sequence and 35 BA-type supergiants of similar stellar masses, which have been analysed using well-tested hybrid non-LTE models (Nieva & Przybilla 2007; Przybilla *et al.* 2006, 2011), an iterative analysis technique (Nieva & Przybilla 2008) that employs almost the entire observed spectrum and a consistent set of comprehensive model atoms (for CNO: Przybilla *et al.* 2000, 2001; Przybilla & Butler 2001; Nieva & Przybilla 2008; Becker & Butler 1988, the latter updated).

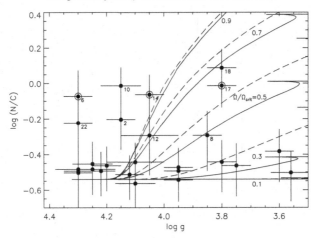

Figure 2. Sample stars of Nieva & Przybilla (2014, dots) in the $\log g$-$\log(N/C)$-plane. Predictions from evolution models for rotating stars by Georgy *et al.* (2013) are shown for a $9\,M_\odot$ (full) and a $15\,M_\odot$ model (dashed line) at $Z = 0.014$, respectively, for initial $\Omega/\Omega_{\rm crit}$ varying from 0.1 to 0.9, as indicated. About 1/3 of the observed objects show signatures of CN-mixing ($\log(\mathrm{N/C}) \gtrsim -0.3$). Objects #6, 14 and 17 are magnetic.

Both, high accuracy and precision can be achieved in this approach, with 1σ-uncertainties reduced to 1–2% in $T_{\rm eff}$, to 0.05–0.10 dex in surface gravity $\log g$ and $\sim$10-20% for absolute elemental abundances. Thus, almost the entire observed visual and near-IR spectra can be reliably reproduced and various additional constraints are also met, like a good match of the observed spectral energy distribution or the good agreement between spectroscopic and Hipparcos parallaxes (Nieva & Przybilla 2012). The most important finding from this study is that young massive stars in the solar neighbourhood show a high degree of chemical homogeneity (except in nitrogen abundance), facilitating a present-day cosmic abundance standard (CAS) to be established. Moreover, evolutionary masses, radii and luminosities can be determined in this approach to better than typically 5%, 10%, and 20% uncertainty, respectively, allowing the mass-radius and mass-luminosity relationships to be recovered for single core H-burning objects with a similar precision (within a factor 2-4) as derived from detached eclipsing binaries (Nieva & Przybilla 2014).

While the N/C vs. N/O diagram provides a simple quality test for observational results, it is not suited itself for assessing the quality of stellar evolution models. This has to be done by other means, e.g. employing the $\log g$–$\log(\mathrm{N/C})$ diagram (see Fig. 2), which represents a proxy for the evolution of the surface abundance as a function of the evolutionary age of a star. The models evolve in Fig. 2 from the zero-age main sequence on the left to the terminal-age main sequence (the hook at the right), with stars of higher initial mass and/or higher angular velocity Ω producing stronger mixing. Noticeable effects of mixing are predicted by the particular models to occur when the stars attain $\log g$-values in the range $\sim$4.0 to 3.9 for ratios of initial to critical angular velocity $\Omega/\Omega_{\rm crit} \gtrsim 0.4$. In this case, most of the apparently slowly-rotating sample stars are consistent with the model predictions: many appear unmixed and some more evolved objects show signatures of mixing within the expectations. However, for a small group of apparently unevolved stars with low $v \sin i$ (#2, 6, 10, 22) other scenarios have to be invoked to explain their degree of mixing, possibly related to binarity and/or magnetic fields. In particular, the unusual characteristics of τ Sco (#6) point to a blue straggler nature, due to a binary merger (Nieva & Przybilla 2014). Obviously, further such high-quality investigations are required, encompassing also faster-rotating stars to obtain reliable conclusions.

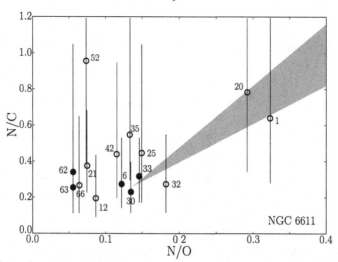

Figure 3. N/C vs. N/O abundance ratios (by number) in the Galactic open cluster NGC 6611 from observations by Hunter *et al.* (2009). The shaded area shows the domain predicted by the stellar models of Ekström *et al.* (2012), between 5 and 60 $M_\odot$ at $Z = 0.014$. Error bars on the N/O ratios are of the same order of magnitude as those shown for the N/C ratios. The numbers attached to the data points correspond to the identifiers employed within the FLAMES Survey. Normal stars are marked by a filled symbol, binaries (or binaries candidates) or Be stars, i.e. stars with biased results, by an open symbol. According to Maeder *et al.* (2014).

At this point we can resume the discussion of the reliability of observational constraints reported in the literature so far (see Sect. 3). Unfortunately, abundances of all three elements carbon, nitrogen and oxygen are rarely provided. However, for the important FLAMES Massive Star Survey, one can perform the test employing the N/C vs. N/O diagram. Overall, there is agreement of the data with model predictions within the (large) observational uncertainties (Maeder *et al.* 2014), see also Fig. 3, but the underlying scatter is much larger than one may expect from considering changes in the initial mass, rotation, and model details on the nuclear path. In order to assess the significance of the scatter, one can aim at confirming the results by Hunter *et al.* (2009). This can be achieved under the premise that no matter the details of the model implementations, similar analysis results will be derived if they are sufficiently realistic. In particular, a good match of the synthetic with the observed spectra should be obtained globally, and in detail, as demonstrated by Nieva & Przybilla (2012). Thus, model spectra computed with the methods introduced earlier in this Section for atmospheric parameters and abundances as derived by Hunter *et al.* (2009) are expected to match the observed spectra from the FLAMES Survey, which are publically available. The test was made for a sub-sample of stars, including all stars in the Galactic cluster NGC 6611 analysed by the FLAMES Survey. Again, an overall good match was found for many spectral features. However, a close inspection of the observed spectra showed the presence of observational bias: many stars were identified as (candidate) SB2 systems, some objects turned out as Be stars and one chemically peculiar He-strong star was identified. For the case of NGC 6611 all the potentially biased stars are identified in Fig. 3, which consequently should be removed from the interpretation, reducing the observed scatter. The main conclusion from the re-assessment by Maeder *et al.* (2014) is that a careful and critical re-investigation of the FLAMES Survey data presented by Hunter *et al.* (2009) is clearly required. This may resolve the unusual characteristics of chemical abundances of the FLAMES Survey objects when compared to other studies of Galactic stars (Martins & Palacios 2013).

In the case of the heterogeneous sample of observational constraints collected by Aerts *et al.* (2014), full CNO abundances are not available. However, there are other indicators for the presence of bias. The nitrogen abundances scatter by over 2 dex; in particular, the presence of a considerable fraction of N-poor OB stars in the solar neighbourhood, some objects with values of ~1/10 of the CAS baseline abundance (i.e. comparable to the metallicity of halo objects), suggest that some of the data needs re-evaluation.

6. Conclusions & Recommendations

In summary, it is found that the currently available observational constraints for testing stellar evolution models suffer from the presence of bias in many cases. The data clearly support that mixing of the surface layers with CNO-processed material has taken place in some stars, however their precision and accuracy is often insufficient to distinguish between different stellar evolution models. There exists a class of slowly-rotating massive stars on the main sequence that shows evidence for mixing, unexplained by standard evolution of single rotating stars. The available data also implies that binary interactions, including mergers, need to be considered for explaining the properties of a sub-set of stars.

Currently, the exploration of the effects of magnetic fields on the transport of angular momentum and chemical elements in massive stars is just beginning. Efforts need to be undertaken to characterise the properties of the ~70 magnetic OB-stars known so far as accurate and precise as possible in a homogeneous way. Moreover, existing surveys need to be continued to search for additional objects in order to improve the still poor number statistics. This is because of the different potential formation channels and many variables involved. Population synthesis (and therefore a large database) is required if one aims at establishing correlations between surface magnetic fields, abundances and surface rotation. It will also be necessary to extend the analysed star samples to both apparently single stars and objects identified as binaries, irrespective of $v \sin i$. The complexity and scope of the topic is such that the collaboration of both large teams currently working on this (BOB, MiMeS) will be required to achieve breakthrough results.

Acknowledgements

I would like to thank M.-F. Nieva, A. Irrgang, M. Firnstein and K. Butler for collaboration and discussions on the quantitative spectroscopy of massive stars over the past years, and A. Maeder, G. Meynet, S. Ekström, P. Eggenberger and C. Georgy for stimulating insights into the interpretation of these findings in the context of stellar evolution.

References

Aerts, C., Molenberghs, G., Kenward, M. G., & Neiner, C. 2014, *ApJ* 781, 88
Becker, S. R. & Butler, K. 1988, *A&A* 201, 232
Bouret, J.-C., Donati, J.-F., Martins, F., *et al.* 2008, *MNRAS* 389, 75
Brott, I., de Mink, S. E., Cantiello, M., *et al.* 2011, *A&A* 530, A115
Cantiello, M., Langer, N., Brott, I., *et al.* 2009, *A&A* 499, 279
Chieffi, A. & Limongi, M. 2013, *ApJ* 764, 21
Chini, R., Hoffmeister, V. H., Nasseri, A., Stahl, O., & Zinnecker, H. 2012, *MNRAS* 424, 1925
Donati, J.-F., Howarth, I. D., Jardine, M. M., *et al.* 2006, *MNRAS* 370, 629
Ekström, S., Georgy, C., Eggenberger, P., *et al.* 2012, *A&A* 537, A146
Evans, C. J., Lennon, D. J., Smartt, S. J., & Trundle, C. 2006, *A&A* 456, 623
Evans, C. J., Smartt, S. J., Lee, J.-K., *et al.* 2005, *A&A* 437, 467
Firnstein, M. & Przybilla, N. 2012, *A&A* 543, A80
Fossati, L., Zwintz, K., Castro, N., *et al.* 2014, *A&A* 562, A143

Frischknecht, U., Hirschi, R., Meynet, G., *et al.* 2010, *A&A* 522, A39

Georgy, C., Ekström, S., Granada, A., *et al.* 2013, *A&A* 553, A24

Georgy, C., Granada, A., Ekström, S., *et al.* 2014, *A&A* 566, A21

Heger, A. & Langer, N. 2000, *ApJ* 544, 1016

Heger, A., Woosley, S. E., & Spruit, H. C. 2005, *ApJ* 626, 350

Henrichs, H. F., de Jong, J. A., Verdugo, E., *et al.* 2013, *A&A* 555, A46

Henrichs, H. F., Kolenberg, K., Plaggenborg, B., *et al.* 2012, *A&A* 545, A119

Hubrig, S., Schöller, M., Kharchenko, N. V., *et al.* 2011, *A&A* 528, A151

Hummel, C. A., Rivinius, T., Nieva, M.-F., *et al.* 2013, *A&A* 554, A52

Hunter, I., Brott, I., Langer, N., *et al.* 2009, *A&A* 496, 841

Irrgang, A., Przybilla, N., Heber, U., *et al.* 2014, *A&A* 565, A63

Langer, N. 2014, in *IAU Symposium Ser.*, Vol. 302, p. 1

Maeder, A. & Meynet, G. 2005, *A&A* 440, 1041

Maeder, A., Meynet, G., Ekström, S., & Georgy, C. 2009, *Comm. Asteroseism.* 158, 72

Maeder, A., Przybilla, N., Nieva, M.-F., *et al.* 2014, *A&A* 565, A39

Martins, F., Escolano, C., Wade, G. A., *et al.* 2012a, *A&A* 538, A29

Martins, F., Mahy, L., Hillier, D. J., & Rauw, G. 2012b, *A&A* 538, A39

Martins, F. & Palacios, A. 2013, *A&A* 560, A16

Meynet, G., Eggenberger, P., & Maeder, A. 2011, *A&A* 525, L11

Meynet, G. & Maeder, A. 2000, *A&A* 361, 101

Meynet, G. & Maeder, A. 2003, *A&A* 404, 975

Morel, T. 2011, *Bulletin de la Societe Royale des Sciences de Liege* 80, 405

Morel, T. 2012, *ASP Conf. Ser.* 465, p. 54

Morel, T., Hubrig, S., & Briquet, M. 2008, *A&A* 481, 453

Neiner, C., Alecian, E., Briquet, M., *et al.* 2012, *A&A* 537, A148

Nieva, M. F. & Przybilla, N. 2007, *A&A* 467, 295

Nieva, M. F. & Przybilla, N. 2008, *A&A* 481, 199

Nieva, M.-F. & Przybilla, N. 2012, *A&A* 539, A143

Nieva, M.-F. & Przybilla, N. 2014, *A&A* 566, A7

Nieva, M.-F. & Simón-Díaz, S. 2011, *A&A* 532, A2

Pavlovski, K. & Southworth, J. 2013, in *EAS Publ. Ser. 64*, p. 29

Petit, V., Owocki, S. P., Wade, G. A., *et al.* 2013, *MNRAS* 429, 398

Przybilla, N. & Butler, K. 2001, *A&A* 379, 955

Przybilla, N., Butler, K., Becker, S. R., & Kudritzki, R. P. 2006, *A&A* 445, 1099

Przybilla, N., Butler, K., Becker, S. R., Kudritzki, R. P., & Venn, K. A. 2000, *A&A* 359, 1085

Przybilla, N., Butler, K., & Kudritzki, R. P. 2001, *A&A* 379, 936

Przybilla, N., Firnstein, M., Nieva, M. F., Meynet, G., & Maeder, A. 2010, *A&A* 517, A38

Przybilla, N., Nieva, M.-F., & Butler, K. 2008, *ApJ (Letters)* 688, L103

Przybilla, N., Nieva, M.-F., & Butler, K. 2011, *J. Phys. Conf. Ser.* 328, 012015

Rivero González, J. G., Puls, J., Najarro, F., & Brott, I. 2012, *A&A* 537, A79

Rivinius, T., Carciofi, A. C., & Martayan, C. 2013, *A&A Rev.* 21, 69

Sana, H., de Mink, S. E., de Koter, A., *et al.* 2012, *Science* 337, 444

Schneider, F. R. N., Langer, N., de Koter, A., *et al.* 2014, *A&A*, 570, A66

Smith, K. C. 1996, *Ap&SS* 237, 77

Spruit, H. C. 2002, *A&A* 381, 923

Ud-Doula, A., Owocki, S. P., & Townsend, R. H. D. 2009, *MNRAS* 392, 1022

Wade, G. A., Grunhut, J., Alecian, E., *et al.* 2014, in *IAU Symposium Ser.*, Vol. 302, p. 265

Wade, G. A., Maíz Apellániz, J., Martins, F., *et al.* 2012, *MNRAS* 425, 1278

Discussion

ANDERSON: Do you think using the $\frac{N/C}{N/O}$ ratio could be a viable constraint to improve the accuracy of statistical analyses of CNO abundances such as that by the FLAMES Survey?

PRZYBILLA: The $\frac{N/C}{N/O}$ ratios for a star sample should follow tightly the predicted nuclear path. This is an objective quality indicator for stellar analyses, and it can be used to detect potential systematic error in the analyses. Efforts should therefore be made not only to derive nitrogen abundances – undoubtedly the most sensitive indicator for chemical mixing – but also carbon and oxygen abundances.

AERTS: Completely agreed that we would like to re-apply the methodology of Aerts *et al.* (2014, ApJ, 781, 88) on a precise and accurate sample, but we did the best we could to assemble a large enough database at the present day...

PULS: A comment: In order to apply the correlation analysis (Aerts *et al.*) successfully, we cannot only use B-type objects which can be analysed with a very high precision, but also the more massive/hotter stars. For these objects the errors become necessarily larger (mass-loss, complete NLTE description, etc.).

SUNDQUIST: I admit I do not fully understand the high accuracy (as opposed to precision) you seem to claim in your abundances. To me, systematic uncertainties in atomic data and model atmospheres should enter here. Could you perhaps comment on this?

PRZYBILLA: We have made extensive comparisons of different model atmospheres and found overall good agreement in the stratifications and emerging fluxes for the kind of stars under study. We also investigated the effects of varying input atomic data for the analyses. The resulting systematic uncertainties from uncertainties in the atmospheric parameters and atomic data typically amount to ∼0.07-0.12 dex in abundance (depending on element and stellar parameters), which is comparable to the 1σ statistical uncertainties (∼0.05-0.10 dex).

NAJARRO: 1) Could you comment on systematics for the $\log g$, T_{eff} determinations for the B0 and O9 V's? 2) Are your high $\log g$, highly N-rich data early-type (one was τ Sco)?

PRZYBILLA: 1) Admittedly, the analysis becomes more complicated for the B0 Vs and in particular for (weak-wind) O9 Vs. Overall, the scatter in the abundances increases, but the ability to establish still multiple ionization equilibria simultaneously raises hope that the stellar parameter determination is not affected seriously by systematic error. We start to see systematic deviations in abundances from some lines of some elements, indicating that the employed models are just about sufficiently realistic. And, we are still missing several of the higher ionization stages that would be useful for the analysis. Work on this is in progress. 2) The fraction of high $\log g$, N-rich objects among the stars with spectral type earlier than B0.5 is about the same as elsewhere in the sample, ∼1/3.

MASSEY: Just one comment. These fits you showed are beautiful. It forces me to remember the question raised by Georges in the general discussion Tuesday: are the analysis pipelines used in large survey sufficiently accurate? I know of recent large studies of finer data that have used automatic fitting techniques to find the "best" fits – but these best fits are not necessarily *good* fits. The quality of the fit must also be considered.

HERRERO: I wanted to follow Phil's comment before and say that we need pipelines for data analysis if we want to have large amounts of data and good statistics. We then have to be open and provide all the information, check carefully all aspects and use the pipelines not as black boxes. Then we will learn from our mistakes and improve our badly needed pipelines.

New windows on massive stars: asteroseismology, interferometry, and spectropolarimetry
Proceedings IAU Symposium No. 307, 2014
G. Meynet, C. Georgy, J. H. Groh & Ph. Stee, eds.

© International Astronomical Union 2015
doi:10.1017/S1743921314007297

Massive Star Astrophysics with the new Magellanic Cloud photometric survey MCSF

Dominik J. Bomans[1,2], Alexander Becker[1] and Kerstin Weis[1]

[1]Astronomical Institute of the Ruhr-University Bochum
email: bomans@astro.rub.de, abeck@astro.rub.de, kweis@astro.rub.de

[2]RUB Research Department "Plasmas with complex interactions"

Abstract. Surveys of the resolved stellar content of entire galaxies are the natural tool to study fast evolutionary phases of massive stars. Therefore we launched the Magellanic Clouds Massive Stars and Feedback Survey (MCSF) and periodically imaged for 3 years the entire Small and Large Magellanic Cloud in u, B, V, R, I and Hα, [OIII], [SII] using the twin telescope RoBoTT at the University Observatory of the Ruhr-University Bochum at Cerro Armazones, Chile. Observations with short exposure times are included to ensure brightest stars not to be saturated, yielding a full coverage in luminosity. With this unique dataset we can study the massive stellar populations up to $M_B \sim -10$ mag and their feedback. Upon completion a high quality photometric and spatially complete catalog of the Magellanic Clouds will be established which is be comparable (or even beyond) the quality of HST based photometry of nearby galaxies.

Keywords. (stars:) supergiants, stars: Wolf-Rayet, (stars:) Hertzsprung-Russell diagram, stars: emission-line, ISM: bubbles, surveys, (galaxies:) Magellanic Clouds, galaxies: stellar content, methods: statistical

1. Introduction

Our closest actively star forming neighbor galaxies, the LMC and the SMC are surprime laboratories of stellar evolution and feedback of stars on the interstellar medium. They are nearby enough to be easily resolved into individual stars, are at relatively high galactic latitude (low foreground absorption), have somewhat (LMC) and significantly (SMC) lower metallicity of compared to the Milky Way, and very well determined distances. Still, due to their proximity, they extend on a quite large area on the sky, posing special problems to derive global parameters and compile complete samples in the area of CCD-based optical observing.

Several CCD based survey have already be conducted on the Magellanic Clouds (MCs), but there are some drawbacks to all these data sets. There are saturation problems at the brightest magnitudes (typically above ~ 12 mag) e.g. in MCPS and OGLE (Zaritsky & Harris 2004; Zaritsky *et al.* 2002; Udalski *et al.* 2008a,b), limitations in terms of filters used, e.g. MCELS (Smith *et al.* 2000), or the surveys do not cover the whole extend of the Magellanic Clouds (Massey 2002).

This motivated us to undertake a new survey specifically designed to analyse the massive stars and their feedback on the interstellar medium. The survey goal is a coherent data set in both broad band filters (for stellar parameters) and narrow band filters (structure and ionization of the ISM), deep enough to be comparable to HST data set of more distant Local Group galaxies. The idea of **MCSF**, the "**M**agellanic **C**louds Massive **S**tars and their **F**eedback" survey was born.

During this conference several examples of the power of complete datasets were presented, e.g. by P. Massey, E. Levesque, and M. Kourniotis. Another example is the recent analysis of Wolf-Rayet stars in the Magellanic Clouds (Hainich *et al.* 2014). When combining the optical data pf MCSF with similar resolution data in the NIR (2MASS (Skrutskie *et al.* 2006)), MIR (SAGE (Meixner *et al.* 2006; Gordon *et al.* 2011)), UV (GALEX and SWIFT) (Simons *et al.* 2014; Hagen *et al.* 2014), or X-ray (XMM-Newton) (Sturm *et al.* 2013; Bomans 2013), synergy effects allow an even broader set of science projects, as shown below.

2. Observation and reduction

The data where taken mostly between 2011 and 2013 with the RoBoTT twin telescopes at the observatory of the Ruhr-University Bochum (Cerro Armazones, Chile). The system is a robotic telescope with predefined queue operation. RoBoTT consists of two Takahashi 15cm refractors on a common mount. The Apogee Alta 4096^2 CCDs give a field of view of $2.42° \times 2.42°$ each, approximately co-spatial on the sky. The resulting pixel scale of the raw images is 2.34"/pix. We observed two overlapping fields for the SMC covering approximately $4.5° \times 2.5°$ and 9 fields in the LMC, covering about $8° \times 8°$ with many slightly dithered exposures spread over the three Magellanic Clouds observing seasons. Images where taken in u, B, V, R, I, as well as Hα, [OIII], and [SII] filters with complete area coverage. Some limited time resolution is possible due to 9 individual visits to the two SMC fields. Due to the large number of fields only few return visits could be scheduled for the LMC. We also took short exposure data in all broad band filters and Hα to ensure non-saturated images of even the brightest Magellanic Cloud member stars.

Basic data reduction was performed with the latest version of the THELI (Erben *et al.* 2005) pipeline. Making use of the many individual images, we could drizzle the images to a pixel scale of $1''$, significantly improving the sampling of our PSF. Pointspread function fitting photometry was performed using a modifed version of DOPHOT (Alonso-Garcia *et al.* 2012), which allowed a spatially variable PSF. We also performed extensive artificial star experiments to test our completeness. Spatial variation of the crowding in both the LMC and SMC fields is large, but we still reached our goal of better than 90% completeness in both B and R for stars brighter than $\sim$ 19.3. With these parameters MCSF data of the massive stars and the ionized gas of the Magellanic Clouds are comparable to the HST data of other Local Group galaxies at factor >10 larger distances.

3. First science results

Currently we are in the optimisation phase of the data reduction, therefore only a few glimpses of what we plan to do and already started to do with MCSF are presented here.

3.1. *The Magellanic Clouds as external galaxies: Color magnitude/Hess diagrams*

In Fig.1 we plot the B-R,B color magnitude diagrams of the LMC (left) and SMC (right). By color coding the density of source at each position we extend the CMDs to Hess diagrams (Hess 1924). These diagrams allow not only to study the location of structures in the CMD, but also the number density of stars in the corresponding evolutionary phase. This enables us studies of the spatial distribution of stars in the same evolutionary phase and (using stellar evolution grids) their spatial change with evolution time. The MCSF CMDs show a split of the apparent main sequence and the blue supergiants. The tip of the red supergiants is well defined, but some stars are apparently located above the predictions of stellar evolution (Levesque *et al.* 2006) and the Humphreys-Davidson limit (Humphreys & Davidson 1979). The tip of the RGB and the extension towards very red stars (AGB, carbon stars) are also clearly visible.

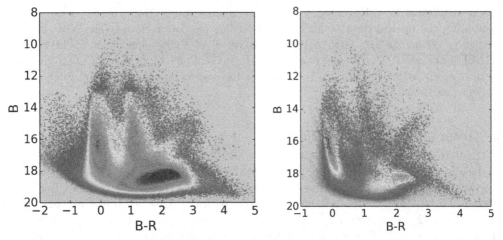

Figure 1. MCSF based B-R,B color magnitude diagrams of the whole LMC and SMC, with added color / grayscale coding for the number density of the stars at the position in the diagram (Hess 1924).

The region between the main sequence and red supergiant branches is dominated by the galactic foreground. Correction of this unwanted feature is the most critical point to be improved, yet. The CMDs plotted in Fig. 1 are not strongly cleaned for photometric error or shape parameters in order to give a full impression of the data. Clearly some areas (e.g. the very blue objects in the LMC CMD) are artifacts.

3.2. Be stars

With not only broad band filter photometry in hand, but also deep Hα and [OIII], we can analyse the emission line stars content of both Magellanic Clouds. We use our photometric catalog in B and R to merge it it with the source catalog in Hα. We then rejected all objects which are bright in [OIII] as probable compact HII regions. Afterwards we use the color - magnitude diagram to define the locus main sequence stars and supergiants.

The resulting histogram of the ratio of emission line objects normalized to total objects as function of brightness is plotted in the left panel of in Fig.2. As observed several times before (e.g. Iqbal & Keller 2013), the main sequence Be/B star ratio is higher in the SMC than in the LMC. The difference in our data seems to be larger for the Oe stars than the Be stars and the ratio is about constant from B0 to the later B stars, or even slightly rising going to lower masses. The absolute number of the the the Be/(Be+B) ratio we derive from our data is smaller than the ratio derived by Martayan *et al.* (2010), but consistent with the value derived by Iqbal & Keller (2013). The reason for this difference may be that slitless spectroscopy used in Martayan *et al.* (2010) is more efficient to detect faint Balmer emission than the narrow band filter method employed by us. Another explanation of the difference may be related to the definition of the main sequence.

Beside checking our selection criteria used, and performing a star by star comparison with the samples of Sabogal *et al.* (2005); Martayan *et al.* (2010), we will also look into differences between clusters and field (Keller *et al.* 1999), color effects, and variability (Beaulieu *et al.* 2001; Keller *et al.* 2002) of our sample in comparison to earlier ones. With the full coverage and good photometry at hand we will also look at the Oe stars and supergiant emission line stars and investigate the apparent downturn of the ratio in the LMC in the O emission line star regime. This downturn appears to be much weaker in the SMC (see Fig.2a).

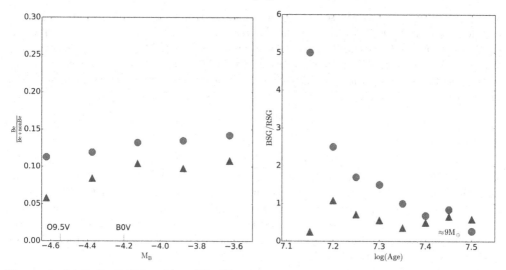

Figure 2. (a) Ratio Oe/(Oe+O) and Be/(Be+B) star ratio for LMC and SMC based on MCSF Hα and broad band photometry. LMC data are plotted as triangles, SMC data are plotted as circles. (b) Blue to red supergiant ratio of of the LMC and SMC. Markers are the same as the left panel.

3.3. *Blue to red supergiant ratio*

The stars going in their evolution trough both red and blue supergiant phase are exactly the stars which are progenitors to the majority of the type II supernovae. Understanding their evolution is critical, since they provide the largest part of the kinetic energy feedback into the interstellar medium and a significant part of the heavy metals. This ratio depends on a large number of aspects of stellar structure and evolution (e.g. Maeder & Meynet 2001). MCSF is a well suided database to explore the the physics of this ratio. For a first attempt, we counted stars along lines of same age using Geneva isochrones (Lejeune & Schaerer 2001; Georgy *et al.* 2013) and alternatively used the cuts along the same luminosity to derive the ratios. The results of the first method are plotted in Fig.2b. The SMC shows higher BSG/RSG ratio than the LMC in all mass bins, exept the very lowest. Interstingly, the difference between the SMC and LMC ratio appears to decrease with decreasing mass. While the models with rotation are a much better fit to the ratios than models without rotation, the agreement between observation and models is still not satisfactory. Either the models miss some relevant physics, or our selection criteria do not yet catch stars at well defined ages, or both.

3.4. *Wolf-Rayet and other hot evolved stars*

Supplementing the MCSF multiband photometry with UV (GALEX, SWIFT) und IR (2MASS, Spitzer) data creates a large set of very wide stellar spectral energy distribution (SED). This data set we crossmatched with the "Catalogue of Stellar Spectral Classifications" (Skiff, B. A. 2013) generating a sample of Magellanic Cloud stars with known spectral classification. This sample was used to train a machine learning algorithm that is able to estimate an "semi-spectral" classification of a star based on its SED. The algorithm is able to reproduce 98% of the training sample. We applied it to ∼ 36000 stars in the SMC and ∼ 45000 stars in the LMC to assess their "spectral" type. First tests of this method showed a generally good agreement within a few spectral subtypes. We used the derived classifications to compile a list of 150 new Wolf-Rayet stars and very hot supergiant candidates in the LMC and SMC. A comparison with the Wolf-Rayet stars found in a new search program by Massey *et al.* (2014) showed that the astro-informatics method is able to find new Wolf-Rayet stars: in the very small overlap between our sample and the Massey *et al.* (2014) sample, we correctly predicted one (LMC143-1) of two new spectroscopically confirmed Wolf-Rayet stars. Both our MCSF plus astro-informatics based result and the search project by Massey *et al.* (2014) point at the same direction: The inventory of the Magellanic Clouds Wolf-Rayet stars is far from complete, yet.

4. The archive

We set up a dedicated web site for MCSF to serve as first stop for status, results, and the actual data: http://www.astro.rub.de/Astronomie/MCSF. At the time of writing (summer 2014) we are in the middle of quality tests for our photometry. We already started to upload technical details, some plots, and pretty pictures to the archive page. Starting from fall 2014, we plan to upload the first photometric catalogs and ancilliary data. Full catalogs and finally the images will follow, as we write science papers. In case you need some data fast, please do not hesitate to contact us.

References

Alonso-Garcia, J., Mateo, M., Sen, B., *et al.* 2012, *AJ* 143, 70

Beaulieu, J.-P., de Wit, W. J., Lamers, H. J. G. L. M., *et al.* 2001, *A&A* 380, 168

Bomans, D. J. 2013, in G. Pugliese, A. de Koter, & M. Wijburg (eds.), *370 Years of Astronomy in Utrecht*, Vol. 470 of *ASP Conference Series*, p. 245

Erben, T., Schirmer, M., Dietrich, J. P., *et al.* 2005, *Astron. Nachr.* 326, 432

Georgy, C., Ekström, S., Eggenberger, P., *et al.* 2013, *A&A* 558, A103

Gordon, K. D., Meixner, M., Meade, M. R., *et al.* 2011, *AJ* 142, 102

Hagen, L., Siegel, M., Gronwall, C., Hoversten, E. A., & Immler, S. 2014, in *American Astronomical Society Meeting Abstracts #223*, Vol. 223, p. 442.30

Hainich, R., Rühling, U., Todt, H., *et al.* 2014, *A&A* 565, A27

Hess, R. 1924, in *Probleme der Astronomie. Festschrift fur Hugo v. Seeliger*, p. 265, Springer, Berlin

Humphreys, R. M. & Davidson, K. 1979, *ApJ* 232, 409

Iqbal, S. & Keller, S. C. 2013, *MNRAS* 435, 3103

Keller, S. C., Bessell, M. S., Cook, K. H., Geha, M., & Syphers, D. 2002, *AJ* 124, 2039

Keller, S. C., Wood, P. R., & Bessell, M. S. 1999, *A&AS* 134, 489

Lejeune, T. & Schaerer, D. 2001, *A&A* 366, 538

Levesque, E. M., Massey, P., Olsen, K. A. G., *et al.* 2006, *ApJ* 645, 1102

Maeder, A. & Meynet, G. 2001, *A&A* 373, 555

Martayan, C., Baade, D., & Fabregat, J. 2010, *A&A* 509, A11

Massey, P. 2002, *ApJS* 141, 81

Massey, P., Neugent, K. F., Morrell, N., & Hillier, D. J. 2014, *ApJ* 788, 83

Meixner, M., Gordon, K. D., Indebetouw, R., *et al.* 2006, *AJ* 132, 2268

Sabogal, B. E., Mennickent, R. E., Pietrzyński, G., & Gieren, W. 2005, *MNRAS* 361, 1055

Simons, R., Thilker, D., Bianchi, L., & Wyder, T. 2014, *Advances in Space Research* 53, 939

Skiff, B. A. 2013, *VizieR Online Data Catalog* 1, 2023

Skrutskie, M. F., Cutri, R. M., Stiening, R., *et al.* 2006, *AJ* 131, 1163

Smith, C., Leiton, R., & Pizarro, S. 2000, in D. Alloin, K. Olsen, & G. Galaz (eds.), *Stars, Gas and Dust in Galaxies: Exploring the Links*, Vol. 221 of *ASP Conference Series*, p. 83

Sturm, R., Haberl, F., Pietsch, W., *et al.* 2013, *A&A* 558, A3

Udalski, A., Soszynski, I., Szymanski, M. K., *et al.* 2008a, *AcA* 58, 89

Udalski, A., Soszyński, I., Szymański, M. K., *et al.* 2008b, *AcA* 58, 329

Zaritsky, D. & Harris, J. 2004, *ApJ* 604, 167

Zaritsky, D., Harris, J., Thompson, I. B., Grebel, E. K., & Massey, P. 2002, *AJ* 123, 855

Discussion

MASSEY: Absolutely beautiful work - I'm glad to see someone doing this right. One question: I'm surprised by the number of very luminous red supergiant candidates you have. In Neugent *et al.* (2012) we don't see any of these. I'm wondering of you check the proper motions from the UCAC4 survey? This might help weed out the foreground stars.

BOMANS: Up to now, we tried to correct for the foregound by using statistical subtraction from Magellanic Cloud star free (or nearly so) parts of fields. Both this and using predictions from galactic models did not work fully satisfactory. So, I guess a lot of the RSG candidate may well be foreground stars. Thanks for the suggestion, we will try with the proper motions data to clear this CMD area.

MAEDER: The comparison of the numbers of blue to red supergiants are very important. As said long ago by Kippenhahn: this is a magnifying glass of all the mistakes we have made in the models. As a side remark, I point out that AGB at the top luminosity may influence the distribution of luminosities.

BOMANS: We are working to improve our selection of especially the blue supergiants to make the blue to red supergiant ratios even cleaner from interlopers. The effect of the brightest AGB star contributon to the RSG numbers may explain the small bump in the BSG/RSG ratio we see near the luminosity of the faintest RSGs. Thanks for the comment.

Dominik Bomans

New windows on massive stars: asteroseismology, interferometry, and spectropolarimetry
Proceedings IAU Symposium No. 307, 2014
G. Meynet, C. Georgy, J. H. Groh & Ph. Stee, eds.

© International Astronomical Union 2015
doi:10.1017/S1743921314007303

Asteroseismology and spectropolarimetry: opening new windows on the internal dynamics of massive stars

S. Mathis[1,2] and C. Neiner[2]

[1]Laboratoire AIM Paris-Saclay, CEA/DSM - CNRS - Université Paris Diderot, IRFU/SAp
Centre de Saclay, F-91191 Gif-sur-Yvette Cedex, France
email: stephane.mathis@cea.fr

[2]LESIA, Observatoire de Paris, CNRS UMR 8109, UPMC, Univ. Paris-Diderot,
5 place Jules Janssen, 92195 Meudon, France

Abstract. In this article, we show how asteroseismology and spectropolarimetry allow to probe dynamical processes in massive star interiors. First, we give a summary of the state-of-the-art. Second, we recall the MHD mechanisms that take place in massive stars. Next, we show how asteroseismology gives strong constraints on the internal mixing and transport of angular momentum while spectropolarimetry allows to unravel the role played by magnetic fields.

Keywords. hydrodynamics, turbulence, waves, MHD, stars: oscillations (including pulsations), stars: interiors, stars: rotation, stars: magnetic field, stars: evolution

1. New probes of the dynamics of massive stars

Massive stars are dynamical objects: they are rotating, turbulent, pulsating, magnetic for 7% of them, and they strongly impact their environment, because of their winds (ud-Doula, this volume) and tides (Auclair-Desrotour *et al.*, this volume; Langer, this volume). These dynamical processes drastically affect their evolution and those of their galactic environment (Maeder 2009). Therefore, they must be modeled with a high degree of realism (Mathis 2013) taking into account the best available observational constraints. In this context, our knowledge of stellar dynamics has undergone a revolution thanks to the seismology of the Sun and of stars (e.g. SOHO, MOST, CoRoT, and *Kepler*; Aerts *et al.* 2010) and ground-based high-resolution spectropolarimetry that characterizes the magnetic field of stars at their surface (e.g. ESPaDOnS/CFHT, Narval/TBL, HARP-Spol/ESO; Donati & Landstreet 2009). On one hand, *Kepler* demonstrated that very powerful mechanisms are acting to extract angular momentum from the radiative cores of sub-giant and red giant stars during their evolution (Beck *et al.* 2012; Deheuvels *et al.* 2012; Mosser *et al.* 2012; Deheuvels *et al.* 2014). This result creates a bridge with the case of main-sequence stars where such mechanisms are also needed to understand the flat rotation profile of the solar radiative core until $0.2 R_\odot$ discovered by helioseismology (García *et al.* 2007) and the weak differential rotation recently revealed by asteroseismology in some intermediate-mass and massive stars (Kurtz *et al.* 2014; Pápics *et al.* 2014, for KIC 11145123 and KIC 10526294 respectively). On the other hand, the MiMeS spectropolarimetric survey (Wade *et al.* 2014) has demonstrated that only 7% of massive stars host a large-scale magnetic field. Because of their stability along time, their simple geometry, which is often an oblique dipole, and their presence on the PMS with the same proportion of occurrence (Alecian *et al.* 2013), we concluded that they must be of fossil origin. These groundbreaking discoveries thus open a golden age for the study of dynamical processes driving the evolution of stars. In this context, the selection of new space and ground-based facilities, among which K2 (NASA), TESS (2017/NASA), SPIRou (2017/CFHT), and PLATO (2024/ESA), calls for an urgent progress in these fields of investigation and the corresponding theoretical modeling effort along this roadmap.

2. Dynamical processes in massive stars

On one hand, the core of massive stars is convective because of the strong exo-thermic behavior of the CNO cycle. Because of their inertia, turbulent coherent convective structures, the plumes, penetrate into the surrounding radiative envelope. This is the so-called overshoot. Then, mixing occurs at the convection-radiation boundary and internal gravity waves (hereafter IGWs) are stochastically excited (e.g. Browning *et al.* 2004, Meakin, this volume; Arnett, this volume; Mathis & Neiner, this volume). Moreover, magnetic fields are generated in the core by a dynamo action because of sustained helical flows and differential rotation (Brun *et al.* 2005). If the dynamo field interacts with a surrounding fossil field, a strong-dynamo regime is obtained and its amplitude is increased (Featherstone *et al.* 2009). Finally, it tends to inhibit the overshoot.

On the other hand, four major transport mechanisms in stellar radiation zones must be studied. First, large-scale meridional flows, which are driven by applied torques and internal stresses, must be properly modelled (e.g. Zahn 1992; Mathis & Zahn 2004). Then, various hydrodynamical instabilities that generate turbulence in stably stratified layers must be identified and characterized (e.g. Zahn 1983; Mathis *et al.* 2004). Next, the effects of fossil magnetic fields and of their instabilities should be examined (Mathis & Zahn 2005; Zahn *et al.* 2007, Gellert, this volume). Finally, IGWs must be taken into account. They are excited at the interfaces between convective and radiative regions by penetrative convection and by the κ-mechanism if stars are in the instability strip. They propagate in the radiation zone until they deposit/extract angular momentum where they are damped at their co-rotation layers (e.g. Mathis & de Brye 2012; Alvan *et al.* 2013). In this context, dynamical stellar evolution codes that include differential rotation, meridional flows, shear-induced turbulence, and IGWs in a coherent way have been developed for fifteen years (Meynet & Maeder 2000; Talon & Charbonnel 2005; Decressin *et al.* 2009; Mathis *et al.* 2013).

3. Asteroseismology: new windows on internal mixing and transport of angular momentum in massive stars

Asteroseismology is currently revolutionizing our vision of stars. It is now able, thanks to high-precision space photometry and ground-based spectroscopy, to provide stratified information on the structure, chemical composition, and dynamics of stellar interiors, which would be unavailable otherwise.

3.1. *Internal mixing in massive stars: the case of HD 181231 and HD 175869*

HD 181231 and HD 175869 are two late rapidly rotating Be stars, which have been observed using high-precision photometry with the CoRoT satellite during about five consecutive months and 27 consecutive days, respectively. An analysis of their light curves, by Neiner *et al.* (2009) and Gutiérrez-Soto *et al.* (2009) respectively, showed that several independent pulsation g-modes are present in these stars. Fundamental parameters have also been determined by these authors using spectroscopy. In Neiner *et al.* (2012), we succeeded to model these results to infer seismic properties of HD 181231

and HD 175869, and constrain internal transport and mixing processes of rapidly rotating massive stars. We used the non-adiabatic Tohoku oscillation code that accounts
for the combined action of Coriolis and centrifugal accelerations on stellar pulsations as
needed for rapid rotator modelling (Lee & Baraffe 1995). The action of "non-standard"
mixing processes was parametrized with the mixing parameter α_{ov}, which represents the
"non-standard" extension of the convective core, and is determined by matching observed
pulsation frequencies assuming a single star evolution scenario. We find that extra mixing
of $\alpha_{ov} = 0.3 - 0.35\,H_P$, where H_P is the pressure scale-height, is needed in HD 81231 and
HD 175869 to match the observed frequencies with those of prograde sectoral g-modes.
We also detect the possible presence of r-modes. We investigated the respective contributions of several transport processes to this mixing. First, we used Geneva evolution
models (Meynet & Maeder 2000) to evaluate the contribution of the secular rotational
transport and mixing processes in the radiative envelope, which is due to the combined
action of differential rotation and of the shear-induced turbulence and meridional circulation it induces. Next, a Monte Carlo analysis of spectropolarimetric data was performed
to examine the role of a potential fossil magnetic field. Finally, based on state-of-the-art
modeling of penetrative convection at the top of the convective core and IGWs (Browning
et al. 2004), we unravelled their respective contribution to the needed "non-standard"
mixing. We showed that the extension of the convective core needed to match observations and models may be explained by mixing induced by the penetrative movements at
the bottom of the radiative envelope ($\alpha_{ov} = 0.2\,H_P$) and by the secular hydrodynamical
transport processes induced by the rotation in the envelope ($\alpha_{ov} = 0.15\,H_P$) (see fig. 1).
This work showed how asteroseismology now opens a new door to probe transport and
mixing processes in massive stars.

3.2. *Angular momentum in early-type stars as revealed by Kepler: a strong transport mechanism is needed*

The seismology of the Sun has allowed to probe its rotation profile until $0.2\,R_\odot$ and
a quasiuniform rotation in the radiative core has been found, which strongly contrasts
with the surrounding differentially rotating convective envelope. More recently, asteroseismology has allowed us to put constraints on rotation profiles in stars in the whole
Hertzsprung-Russel diagram. In particular, Deheuvels et al. (2012, 2014) and Beck et al.
(2012) have probed surface-to-core differential rotation in subgiant and red giant stars.
Simultaneously, Mosser et al. (2012) has demonstrated the spin-down of the core of red
giants along their evolution. For upper-main sequence stars, Aerts et al. (2003), Pamyatnykh et al. (2004), Briquet et al. (2007), Dziembowski & Pamyatnykh (2008), Kurtz
et al. (2014), and Pápics et al. (2014) have provided constraints on the internal differential rotation. Finally, both rigid rotation (Charpinet et al. 2009) and differential
rotation (Córsico et al. 2011) have been discovered in white dwarfs. Therefore, as written
by Kurtz et al. (2014), these results *"have made angular momentum transport in stars
throughout their life time an observational science"*. This new transformational results
all demonstrate that:

• i) Strong mechanism(s) for angular momentum transport must be acting along stellar
evolution. Indeed, the simple conservation of angular momentum or purely hydrodynamical mechanisms related to differential rotation would lead to stronger angular velocity
contrasts than those observed in stars (e.g. Ceillier et al. 2013, Eggenberger, this volume). Moreover their remnants (white dwarfs and neutron stars) would rotate faster than
observed (Heger et al. 2005; Suijs et al. 2008).

• ii) As demonstrated for example by the cases of the main-sequence A and SPB
stars studied by Kurtz et al. (2014) and Pápics et al. (2014) respectively, it is absolutely

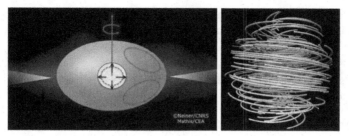

Figure 1. Left: non-standard mixing processes in rapidly rotating massive stars: convective overshoot (black arrows) and rotational mixing (the red loops represent the large-scale meridional circulation driven by the differential rotation). The green dashed line corresponds to the "standard" limit of the convective core. **Right:** Mixed stable fossil field configuration in a stellar radiation zone (Taken from Duez *et al.* 2010).

necessary to go beyond models that treat the transport of angular momentum as purely diffusive processes as often done in the literature. Results have already been obtained theoretically by Meynet & Maeder (2000), Talon & Charbonnel (2005), and Mathis & Zahn (2005) for the transport of angular momentum by large-scale meridional flows, IGWs, and magnetic fields in stellar radiation zones, respectively. In the cases of the two stars studied by Kurtz *et al.* (2014) and Pápics *et al.* (2014), the action of IGWs stochastically excited by the convective core (see Mathis & Neiner, this volume) is a good candidate to explain the properties of observed rotation profiles (Rogers *et al.* 2013; Lee *et al.* 2014).

4. Spectropolarimetry in upper main-sequence stars: a guide to understand fossil magnetism and its impact in stellar interiors

Magnetic fields are now detected at the surface of $\sim 7\%$ of main-sequence and Pre-Main-Sequence intermediate-mass and massive stars (Wade *et al.* 2014; Alecian *et al.* 2013, Grunhut, this volume). Indeed, fields (300 G to 30 kG) are observed in some fraction of Herbig stars, A stars (the Ap stars), as well as in B stars and in a handful of O stars. Furthermore, non-convective neutron stars display fields strength of $10^8 - 10^{15}$ G. These magnetic fields in stably stratified non-convective stellar regions deeply modify the evolution of massive stars since their formation to their late stages. Their large-scale, stable, and ordered nature (often approximately dipolar) and the non-correlation of their properties with those of their host stars favour a fossil hypothesis (even if a dynamo is present in the convective core), whose origin has to be investigated. The first important point is to understand the topology of these large-scale magnetic fields. To survive since the star's formation or the PMS stage, they must be stable. It was suggested by Tayler (1980) that a stellar magnetic field in stable axisymmetric equilibrium must contain both meridional and azimuthal components, since both are unstable on their own (Tayler 1973; Markey & Tayler 1973). This was confirmed by numerical simulations and theoretical works (see fig. 1 and Braithwaite & Nordlund 2006; Duez & Mathis 2010, Emeriau & Mathis, this volume) who showed that initial stochastic helical fields evolve on an Alfvén timescale into stable mixed configurations. This phenomenon well known in plasma physics is a MHD turbulent relaxation (*i.e.* a self-organization process involving magnetic reconnections in resistive MHD). After their formation, these fields deeply modify the transport of angular momentum and the mixing of chemicals. Indeed, they enforced a uniform rotation along field lines and the mixing is inhibited. This phenomena have been observed in the case of the star V2052 Oph studied using the combination of asteroseismology and spectropolarimetry (Briquet *et al.* 2012, Neiner *et al.* this volume).

References

Aerts, C., Christensen-Dalsgaard, J., & Kurtz, D. W. 2010, *Asteroseismology*
Aerts, C., Thoul, A., Daszyńska, J., *et al.* 2003, *Science* 300, 1926
Alecian, E., Wade, G. A., Catala, C., *et al.* 2013, *MNRAS* 429, 1001
Alvan, L., Mathis, S., & Decressin, T. 2013, *A&A* 553, A86
Beck, P. G., Montalban, J., Kallinger, T., *et al.* 2012, *Nature* 481, 55
Braithwaite, J. & Nordlund, Å. 2006, *A&A* 450, 1077
Briquet, M., Morel, T., Thoul, A., *et al.* 2007, *MNRAS* 381, 1482
Briquet, M., Neiner, C., Aerts, C., *et al.* 2012, *MNRAS* 427, 483
Browning, M. K., Brun, A. S., & Toomre, J. 2004, *ApJ* 601, 512
Brun, A. S., Browning, M. K., & Toomre, J. 2005, *ApJ* 629, 461
Ceillier, T., Eggenberger, P., García, R. A., & Mathis, S. 2013, *A&A* 555, A54
Charpinet, S., Fontaine, G., & Brassard, P. 2009, *Nature* 461, 501
Córsico, A. H., Althaus, L. G., Kawaler, S. D., *et al.* 2011, *MNRAS* 418, 2519
Decressin, T., Mathis, S., Palacios, A., *et al.* 2009, *A&A* 495, 271
Deheuvels, S., Doğan, G., Goupil, M. J., *et al.* 2014, *A&A* 564, A27
Deheuvels, S., García, R. A., Chaplin, W. J., *et al.* 2012, *ApJ* 756, 19
Donati, J.-F. & Landstreet, J. D. 2009, *ARA&A* 47, 333
Duez, V., Braithwaite, J., & Mathis, S. 2010, *ApJ (Letters)* 724, L34
Duez, V. & Mathis, S. 2010, *A&A* 517, A58
Dziembowski, W. A. & Pamyatnykh, A. A. 2008, *MNRAS* 385, 2061
Featherstone, N. A., Browning, M. K., Brun, A. S., & Toomre, J. 2009, *ApJ* 705, 1000
García, R. A., Turck-Chièze, S., Jiménez-Reyes, S. J., *et al.* 2007, *Science* 316, 1591
Gutiérrez-Soto, J., Floquet, M., Samadi, R., *et al.* 2009, *A&A* 506, 133
Heger, A., Woosley, S. E., & Spruit, H. C. 2005, *ApJ* 626, 350
Kurtz, D. W., Saio, H., Takata, M., *et al.* 2014, *ArXiv e-prints*
Lee, U. & Baraffe, I. 1995, *A&A* 301, 419
Lee, U., Neiner, C., & Mathis, S. 2014, *ArXiv e-prints*
Maeder, A. 2009, *Physics, Formation and Evolution of Rotating Stars*
Markey, P. & Tayler, R. J. 1973, *MNRAS* 163, 77
Mathis, S. 2013, in M. Goupil, K. Belkacem, C. Neiner, F. Lignières, & J. J. Green (eds.),
    Lecture Notes in Physics, Berlin Springer Verlag, Vol. 865 of *Lecture Notes in Physics,
    Berlin Springer Verlag*, p. 23
Mathis, S. & de Brye, N. 2012, *A&A* 540, A37
Mathis, S., Decressin, T., Eggenberger, P., & Charbonnel, C. 2013, *A&A* 558, A11
Mathis, S., Palacios, A., & Zahn, J.-P. 2004, *A&A* 425, 243
Mathis, S. & Zahn, J.-P. 2004, *A&A* 425, 229
Mathis, S. & Zahn, J.-P. 2005, *A&A* 440, 653
Meynet, G. & Maeder, A. 2000, *A&A* 361, 101
Mosser, B., Goupil, M. J., Belkacem, K., *et al.* 2012, *A&A* 548, A10
Neiner, C., Gutiérrez-Soto, J., Baudin, F., *et al.* 2009, *A&A* 506, 143
Neiner, C., Mathis, S., Saio, H., *et al.* 2012, *A&A* 539, A90
Pamyatnykh, A. A., Handler, G., & Dziembowski, W. A. 2004, *MNRAS* 350, 1022
Pápics, P. I., Moravveji, E., Aerts, C., *et al.* 2014, *ArXiv e-prints*
Rogers, T. M., Lin, D. N. C., McElwaine, J. N., & Lau, H. H. B. 2013, *ApJ* 772, 21
Suijs, M. P. L., Langer, N., Poelarends, A.-J., *et al.* 2008, *A&A* 481, L87
Talon, S. & Charbonnel, C. 2005, *A&A* 440, 981
Tayler, R. J. 1973, *MNRAS* 161, 365
Tayler, R. J. 1980, *MNRAS* 191, 151
Wade, G. A., Grunhut, J., Alecian, E., *et al.* 2014, Magnetic Fields throughout Stellar Evolution,
    Proceedings of the International Astronomical Union, IAU Symposium, Volume 302, p 265

Zahn, J.-P. 1983, in A. N. Cox, S. Vauclair, & J. P. Zahn (eds.), *Saas-Fee Advanced Course 13: Astrophysical Processes in Upper Main Sequence Stars*, p. 253
Zahn, J.-P. 1992, *A&A* 265, 115
Zahn, J.-P., Brun, A. S., & Mathis, S. 2007, *A&A* 474, 145

Discussion

PULS: Could you please comment on the magnetic field originating from the core? How important for stellar structure is it to account for this field or to neglect it?

MATHIS: First, we can think to include the Lorentz force in the hydrostatic balance. But after evaluation we can see it is negligible. The much interesting point is to use 3D numerical simulations of convective core with magnetic fields inside and around and to look at the evolution of the length of convective penetration (overshoot) as a function of their amplitude.

UD-DOULA: Is that reasonable to assume that star formation in a region of uniform magnetic field will lead to "unstable" field in the star?

MATHIS: The phases of formation of fossil field is much complex. First, the protostar is completely convective. Next, we have the birth of the radiative zone, and at this point we get the relaxation on a mixed quasi-static stable magnetic configuration.

Stéphane Mathis

Francisco Najarro

New windows on massive stars: asteroseismology, interferometry, and spectropolarimetry
Proceedings IAU Symposium No. 307, 2014
G. Meynet, C. Georgy, J. H. Groh & Ph. Stee, eds.

© International Astronomical Union 2015
doi:10.1017/S1743921314007315

The Massive Star Population at the Center of the Milky Way

Francisco Najarro[1], Diego de la Fuente[1], Tom R. Geballe[2], Don F. Figer[3] and D. John Hillier[4]

[1] Centro de Astrobiología (CSIC/INTA), ctra. de Ajalvir km. 4, 28850 Torrejón de Ardoz, Madrid, Spain

[2] Gemini Observatory, 670 N. A'ohoku Place, Hilo, HI 96720, USA

[3] Center for Detectors, Rochester Institute of Technology, 54 Lomb Memorial Drive, Rochester, NY 14623, USA

[4] Department of Physics and Astronomy, University of Pittsburgh, 3941 O'Hara Street, Pittsburgh, PA 15260

Abstract. Recent detection of a large number of apparently isolated massive stars within the inner 80 pc of the Galactic Center has raised fundamental questions regarding massive star formation in a such a dense and harsh environment. Are these isolated stars the results of tidal interactions between clusters, are they escapees from a disrupted cluster, or do they represent a new mode of massive star formation in isolation? Noting that most of the isolated massive stars have spectral analogs in the Quintuplet Cluster, we have undertaken a combined analysis of the infrared spectra of both selected Quintuplet stars and the isolated objects using Gemini North spectroscopy. We present preliminary results, aiming at α-elements vs iron abundances, stellar properties, ages and radial velocities which will differentiate the top-heavy and star-formation scenarios.

Keywords. stars: early-type –stars: mass loss – stars: winds – stars: abundances – Galaxy: center – infrared: stars

1. Massive stars in clusters and isolation at the Galactic Center. Star formation scenario

The Galactic Center (GC), hosting three of the most massive young clusters in the Local Group (Central, Arches and Quintuplet), is a unique laboratory to investigate massive stars and massive star formation. Given the intense radiation fields and extreme stellar densities present in the GC, one may question if star formation occurs in the same manner as seen in the giant molecular clouds elsewhere in the Galaxy. Recent results have revealed the presence of a large number of isolated massive stars (Mauerhan *et al.* 2010a,b) at the GC which is comparable to the massive star population of each of the clusters (Figer *et al.* 1999, 2002). Such detection of apparently isolated massive stars in this region has raised a further fundamental issue - whether these massive field stars are results of tidal interactions among clusters, are escapees from a disrupted cluster, or represent a new mode of massive star formation in isolation (Dong *et al.* 2014). The first option has been investigated (Habibi *et al.* 2014) based on the spectral analogy of these "field" stars with those in the Quintuplet and Arches clusters suggesting that they could have been physically associated with the clusters. Indeed, following the numerical dynamical simulations from Harfst *et al.* (2010), and including the effects of stellar evolution and the orbit of the Arches cluster in the Galactic Center potential, Habibi *et al.* (2014) models were able to account for $\sim 60\%$ of the isolated sources within the

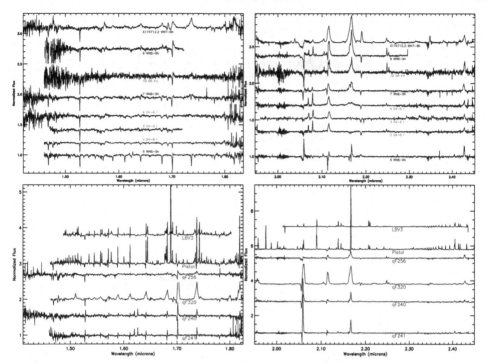

Figure 1. A representative sample of H- and K-band spectra isolated (**top**) and Quintuplet (**bottom**) massive stars, showing the diversity of Wolf-Rayet (magenta), OIf (green) and LBVs (blue) stellar population.

central 100 pc as sources drifted away from the center of the clusters. On the other hand, radial velocity measurements of eight objects in the vicinity of the Arches cluster (Dong *et al.* 2014) and comparisons with those of the cluster, nearby ionized gas and molecular clouds suggest that two of them could have been associated with the cluster while other two likely formed in isolation. Also, from radial velocity studies of WR102ka and a deep integral-field spectroscopy survey of its surroundings Oskinova *et al.* (2013) concluded that the star likely formed in isolation. However, we note that radial velocity estimates of these objects (mainly OIf+ and WNh) may be subject to high uncertainties, as the spectral lines utilized in these studies are severely contaminated by the stellar winds (see below). Thus, further detailed evidence for or against these scenarios is still lacking and awaits precise proper motion measurements (currently underway) providing 3-D velocities of the sources relative to the clusters.

A major step to differentiate among the above scenarios can be achieved through spectroscopic studies of the isolated sources, yielding stellar properties, ages and abundances. Comparison of the results of the quantitative model-atmosphere analysis to theoretical isochrones will allow us to determine if these stars were born in single co-eval cold molecular cloud event or formed over an extended (e.g., 1-10 Myr) period. Obtaining metal abundances from these "field" objects is crucial, not only to understand the metal enrichment history of the GC, but also to test whether these isolated stars have followed a metal-enrichment scenario different than those in the GC clusters. The Quintuplet and Arches clusters provide the stellar reference sources to perform such studies. At its current evolutionary phase (age ∼4 Myr, Figer *et al.* 1999) the massive members of the Quintuplet Cluster are currently WN9-10h (plus a WN6), weak lined WC9, OIf+ stars and LBVs (Figer *et al.* 1999; Liermann *et al.* 2009), while the massive population of the

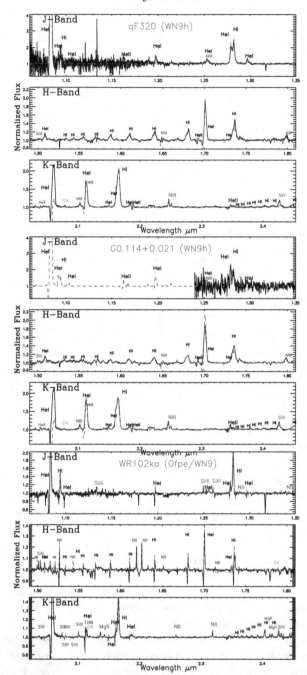

Figure 2. Preliminary fits to the NIR spectra of the Quintuplet star qF320 (WN9h) and the isolated massive stars G0114+0.021 WN8-9h and WR102ka (Ofpe/WN9).

younger Arches cluster is dominated by WN8-9h and OIf+ stars (Figer *et al.* 2002; Najarro *et al.* 2004; Martins *et al.* 2008). These spectral types basically encompass all the isolated evolved massive stars identified from recent follow-up spectroscopic observations of Paα emission objects in the GC (Mauerhan *et al.* 2010a,b) which show quite similar spectral morphology to those present in the Quintuplet Cluster and some of the brightest Arches members.

Another current hot-topic is whether the IMFs of massive star clusters are top-heavy. In such occurrences the larger number of type II supernovae produce enhanced yields of α-elements, resulting in an increase of α-element vs Fe (the main suppliers of iron are type Ia SNe). Najarro *et al.* (2004, 2009) have shown that quantitative NIR spectroscopy of high-mass stars may provide estimates of both absolute abundances and abundance ratios, telling us about the global integrated enrichment history up to the present, placing constraints on models of galactic chemical evolution, and acting as clocks by which chemical evolution can be measured. Abundance analyses may thus help to distinguish between top-heavy and standard star formation in the region.

Finally, the presence of three LBVs in the Quintuplet cluster, well in excess of the empirical Humphreys-Davidson limit implying ZAMS masses $> 100\,M_\odot$, requires further investigation. In theory the existence of such stars is problematic, and thus it is important to test for multiplicity in order to address topics such as the high mass cutoff to the formation of massive stars.

2. Observations and ongoing analysis

Bearing the above scientific challenges in mind, we started in 2010 an observing program at GEMINI North (currently ongoing) which has been obtaining high S/N, medium-resolution spectra of the most massive stars in the Quintuplet as well as isolated massive stars in the inner GC. So far, around 20 massive stars in the Quintuplet cluster and GC inner region have been observed since 2011 with GEMINI NIFS and GNIRS near infrared spectrographs in the H and K bands at medium-resolution (R 5000). The brightest targets were also secured at the shorter X and J Bands. A representative sample, consisting in LBVs, OIfs, and WNL stars is presented in Fig. 1. We expect to complete our sample by Aug-Sep 2015. We are currently in the process of modeling the early-type spectra with the CMFGEN code (Hillier & Miller 1998), to obtain physical and chemical properties. Figure 2 displays preliminary fits to three stars of our sample (qF320, a WN9h star in the Quintuplet cluster, G0.114+0.021, a heavily reddened WN8-9h star close to the Arches and WR102ka, a Ofpe/WN9 object relatively far from the three clusters). At this stage, we may anticipate below some results from our ongoing analysis:

• We obtain a clear α-element enrichment from the analysis of the Quintuplet WNh stars, consistent with the results derived for the LBVs (Najarro *et al.* 2009) which denoted a clearly enhanced α/Fe = 2 ratio with respect to solar.

• Stellar abundances of the isolated objects seem to show a similar trend on average, with the presence of even higher α-element enrichment in some cases.

• Our new high S/N spectroscopic J, H and K data provide important diagnostic lines (NII-III, SiII-IV, CII-IV, etc.), which are crucial not only for abundance determinations but also to constrain stellar properties. As an example, previously found uncertainties in $T_{\rm eff}$ (±6000 K) for the Ofpe/WN9 objects at the GC (Najarro *et al.* 1997; Martins *et al.* 2007) due to the lack of HeII lines in the spectra, are drastically reduced (±1000 K)by making use of the NII/NIII and SiIII/SiIV ionization equilibria (e.g. WR102ka).

• When available the HI and He lines at Pβ provide an excellent He/H ratio diagnostic, allowing much more accurate He abundance determinations than those performed by means of K-Band spectra. Simple blue (HeI) to red (HI) peak ratios may be used (see qF320 and G0.114+0.021 Pβ complex in Fig. 2)

• Radial velocity estimates, if obtained for OIf+ and WNh stars, require detailed modeling of the observed spectra (Figer *et al.* 2004). For the OIf+ stars, the HeII absorption lines, which are decent diagnostics for O V and for some O I stars with weak-to-moderate stellar winds start to be filled by the stellar wind, producing an effective blue-shift as high as $80 - 90\,\mathrm{km\,s^{-1}}$. This may have important consequences when associating the radial velocities of these objects with the nearby gas and clusters. Further, even quantitative modeling may suffer from high uncertainties. Our preliminary fits to WR 102ka making use of the full J, H and K band spectra reveal a radial velocity of $\sim 100\,\mathrm{km\,s^{-1}}$. This value differs significantly from the $60\,\mathrm{km\,s^{-1}}$ obtained by Oskinova *et al.* (2013) by means of only K-band spectra and with a slightly (25%) lower spectral resolution.

Acknowledgements

F. N. and D.dF. acknowledge grants AYA2010-21697-C05-01 and FIS2012-39162-C06-01 and ESP2013-47809-C3-1-R.

References

Dong, H., Mauerhan, J., Morris, M. R., Wang, Q. D., & Cotera, A. 2014, in L. O. Sjouwerman, C. C. Lang, & J. Ott (eds.), *IAU Symposium*, Vol. 303 of *IAU Symposium*, pp 230–234
Figer, D. F., McLean, I. S., & Morris, M. 1999, *ApJ* 514, 202
Figer, D. F., Najarro, F., Gilmore, D., *et al.* 2002, *ApJ* 581, 258
Figer, D. F., Najarro, F., & Kudritzki, R. P. 2004, *ApJ (Letters)* 610, L109
Habibi, M., Stolte, A., & Harfst, S. 2014, *A&A* 566, A6
Harfst, S., Portegies Zwart, S., & Stolte, A. 2010, *MNRAS* 409, 628
Hillier, D. J. & Miller, D. L. 1998, *ApJ* 496, 407
Liermann, A., Hamann, W.-R., & Oskinova, L. M. 2009, *A&A* 494, 1137
Martins, F., Genzel, R., Hillier, D. J., *et al.* 2007, *A&A* 468, 233
Martins, F., Hillier, D. J., Paumard, T., *et al.* 2008, *A&A* 478, 219
Mauerhan, J. C., Cotera, A., Dong, H., *et al.* 2010a, *ApJ* 725, 188
Mauerhan, J. C., Muno, M. P., Morris, M. R., Stolovy, S. R., & Cotera, A. 2010b, *ApJ* 710, 706
Najarro, F., Figer, D. F., Hillier, D. J., Geballe, T. R., & Kudritzki, R. P. 2009, *ApJ* 691, 1816
Najarro, F., Figer, D. F., Hillier, D. J., & Kudritzki, R. P. 2004, *ApJ (Letters)* 611, L105
Najarro, F., Krabbe, A., Genzel, R., *et al.* 1997, *A&A* 325, 700
Oskinova, L. M., Steinke, M., Hamann, W.-R., *et al.* 2013, *MNRAS* 436, 3357

Discussion

NIEVA: Comment. I don't think that the Sun is an appropriate reference for chemical abundances in massive stars. Since a few years we have a reference derived from massive stars which is able to reproduce better the CNO nuclear path than the Asplund's values. Ref: Nieva & Przybilla (2012), confirmed by Irrgang *et al.* (in prep) for a larger sample. The stellar evolution models could also take this new reference into account.

NAJARRO: I agree, but so far we have to compare to evolutionary models and their assumed abundances. It would be nice to have models for massive stars with twice α and solar Fe group.

LOBEL: You indicate that the Pistol star could be binary (Martayan, in prep.). This is a very important result since the number of LBVs detected in binary systems is growing (four are known). It could hint at a link between the LBV phenomenon and binarity.

NAJARRO: There could be, but there are LBVs like P Cyg or AG Car which have proven not to be binaries. On the other hand I recall the LBV close to SGR1806-20 to be a spectroscopic binary.

New windows on massive stars: asteroseismology, interferometry, and spectropolarimetry
Proceedings IAU Symposium No. 307, 2014
G. Meynet, C. Georgy, J. H. Groh & Ph. Stee, eds.

© International Astronomical Union 2015
doi:10.1017/S1743921314007327

Accretion Signatures on Massive Young Stellar Objects

F. Navarete[1], A. Damineli[1], C. L. Barbosa[2] and R. D. Blum[3]

[1]IAG-USP, Rua do Matão, 1226, 05508-900, São Paulo, SP Brazil
email: navarete@usp.br

[2]UNIVAP, São José dos Campos, SP, Brazil
[3]NOAO, 950 N Cherry Ave., Tuczon, AZ 85719 USA

Abstract. We present preliminary results from a survey of molecular H_2 (2.12 μm) emission in massive young stellar objects (MYSO) candidates selected from the Red *MSX* Source survey. We observed 354 MYSO candidates through the H_2 S(1) 1-0 transition (2.12 μm) and an adjacent continuum narrow-band filters using the Spartan/SOAR and WIRCam/CFHT cameras. The continuum-subtracted H_2 maps were analyzed and extended H_2 emission was found in 50% of the sample (178 sources), and 38% of them (66) have polar morphology, suggesting collimated outflows. The polar-like structures are more likely to be driven on radio-quiet sources, indicating that these structures occur during the pre-ultra compact H II phase. We analyzed the continuum images and found that 54% (191) of the sample displayed extended continuum emission and only ∼23% (80) were associated to stellar clusters. The extended continuum emission is correlated to the H_2 emission and those sources within stellar clusters does display diffuse H_2 emission, which may be due to fluorescent H_2 emission. These results support the accretion scenario for massive star formation, since the merging of low-mass stars would not produce jet-like structures. Also, the correlation between jet-like structures and radio-quiet sources indicates that higher inflow rates are required to form massive stars in a typical timescale less than 10^5 years.

Keywords. Stars: formation, Stars: early-type, Stars: pre-main sequence, ISM: jets and outflows.

1. Introduction

The formation of massive stars ($M > 8\,M_\odot$) is one of the most important problems in stellar astrophysics and is still poorly understood. Two scenarios currently dominate the discussions, assuming that high mass stars are *a*) formed by accretion through a disk (Krumholz *et al.* 2005); or *b*) via coalescence of low mass stars (Bonnell *et al.* 2001). Low and intermediate mass stars are formed by the gravitational collapse of the parental giant molecular cloud (GMC), followed by the accretion process (Palla 1996). However, when a young stellar object (YSO) reaches $8\,M_\odot$, the radiative flux is intenser than in the previous case and may interrupt the accretion flow. A process that collimates the radiation field is required to overcome this effect, such as the bipolar outflows observed in massive YSOs (MYSO). In the second scenario, massive stars are formed by coalescence of low-mass stars in dense clusters (Bonnell *et al.* 2001). Low-mass stars are formed under the accretion scenario, interact with each other, and collide to form stars of larger masses (Stahler *et al.* 2000; Bally *et al.* 2002).

While there is no evidence for stellar mergers in clusters, a growing number of both observational evidences (Bik & Thi 2004; Blum *et al.* 2004) and simulations (Krumholz *et al.* 2009) support the accretion scenario. Recently, MYSO candidates were observed in the H_2 narrow filter and collimated jets were identified, suggesting accretion discs around these objects (Varricatt *et al.* 2010). Although this work presents observational evidences for the accretion scenario, only a few MYSOs were confirmed on this sample. Although the scenario of an accretion disk may apply for all massive stars, the details are lacking. Instead of doing detailed study of a small number of potential candidates that might harbor a disk, we are moving toward a large statistical study which will point to accretion signatures (or not) of a well selected sample of 354 MYSO candidates, selected by the Red *MSX* Source (RMS) survey (Lumsden *et al.* 2002; Mottram *et al.* 2011, and references therein).

2. Observations and Data Reduction

Our sample of 354 MYSO candidates comprises 135 sources located in the Northern hemisphere ($\delta > 0°$) and 219 from the Southern one ($\delta < 0°$). Most of the Northern sources were observed using the WIRCam camera deployed at the Canada-France-Hawaii Telescope (CFHT, Hawaii) while the Southern objects were observed using the Spartan and OSIRIS cameras at the Southern Astrophysical Research Telescope (SOAR, Chile). Each source was observed through the H_2 ($\lambda \approx 2.12\,\mu m$, $\Delta\lambda \approx 0.03\,\mu m$) and K-band continuum ($\lambda \approx 2.2\,\mu m$, $\Delta\lambda \approx 0.03\,\mu m$) narrow-band filters, using a total exposure time of ~ 600 seconds per filter. The data was processed using THELI, an instrument-independent pipeline for automated reduction of astronomical images (Erben *et al.* 2005; Schirmer 2013). Each H_2 processed image was continuum-subtracted, scaling its narrow-band K-band continuum image. This procedure was adopted in order to subtract the continuum emission at the same wavelength range and the usage of narrow filters avoided the contamination of the H_2 emission by nebular features such as Brackett-γ.

3. Results

The continuum-subtracted H_2 maps were analyzed and each source was classified according to the following properties:

a) Classification of Extended H_2 emission: four classes were defined to classify the H_2 emission (each type is illustrated in Figure 1): 1) Polar emission (BPn): those extended H_2 emission are likely to be jet-like structures (the "n" corresponds to the number of polar structures: BP1 is monopolar, BP2 is bipolar and BP5 is multipolar emission); 2) Nodal emission (K): K-type morphology corresponds to those sources that display non-aligned knots of H_2 emission; 3) Diffuse emission (D): D-type sources are associated to diffuse H_2 emission, possibly originated by fluorescent excitation of the H_2 molecules; 4) No emission (N): corresponds to the non-detections.

b) NIR Classification of the environment: The environment surrounding each RMS source was classified according to the following properties: *i*) presence of extended continuum emission; and *ii*) association with a stellar cluster.

3.1. *Global analysis of the sample*

The general statistics of the sample is shown in Figure 2. We found extended H_2 emission towards $\approx 50\%$ of the targets (148 sources) while $\sim 54\%$ are associated to extended continuum emission (191). Only $\approx 23\%$ of the sample (80) are apparent members of stellar clusters. The relatively low fraction of sources associated to stellar clusters is an evidence that confronts the merging scenario proposed by Bonnell *et al.* (1998), which cannot explain the formation of roughly $\sim 77\%$ of the observed sources. This result indicates that massive stars require a mechanism that allow their formation even isolated from other stars. Figure 2b displays the morphological classification of the extended H_2 emission as defined on subsection 3. It shows that $\approx 50\%$ of the sample were classified

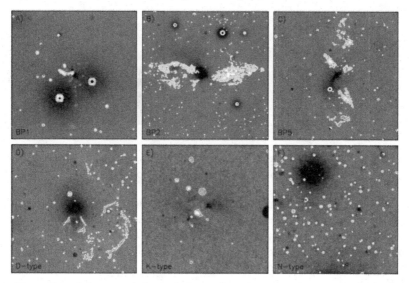

Figure 1. Examples of the morphological classes based on the H_2 emission. The contours are placed at $\sigma = 1.0$ and 3.0. *(a)*, *(b)*, and *(c)*: BP1-, BP2-, and BP5-type sources, respectively. *(d)*, *(e)*, and *(f)*: D-, K-, and N-type sources, respectively.

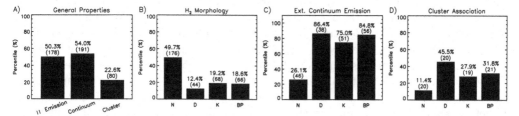

Figure 2. Global properties of the sample. *(a)*: detection of extended H_2 and continuum emission and association with stellar clusters; *(b)*: morphology of the H_2 emission. *(c)*: detection of extended continuum emission as a function of the H_2 morphology. *(d)*: association with stellar clusters as a function of the H_2 morphology. The numbers shown at the top of each bar corresponds to the percentage of the sample.

as N-type objects (176 sources), about $\approx 12\%$ displays diffuse H_2 emission (44), $\approx 19\%$ were classified as K-type sources (68) and $\approx 19\%$ sources (66) are associated with polar structures (monopolar, bipolar or multipolar) and were classified as BP-type. Figure 2c displays a correlation between the sources associated to extended H_2 emission (BP-, K- and D-type) with those which display extended continuum emission (roughly $\sim 80\%$ compared to $\sim 25\%$ found for N-type sources). The positive detection of extended continuum emission could be due to two processes: *i)* nebular emission or *ii)* dust-scattered radiation. Our strategy does not allow us to resolve the nature of the continuum emission. Additional observations in atomic transition lines (such as Brackett-γ at $2.16\,\mu$m) could trace the gaseous emission associated to such regions. Figure 2d indicates that most stellar clusters with MYSOs also exhibit extended H_2 emission.

3.2. *Radio quiet/loud phases and H_2 morphology*

The RMS sources were classified as radio quiet (YSO) or radio loud (HII) sources based on 6 cm observations made by Urquhart *et al.* (2007, 2009). Our sample consists of 282 radio quiet sources and 72 radio loud ones. Figure 3 displays the global results of the survey but separated between radio quiet/loud sources. Panel (a) shows the fraction of

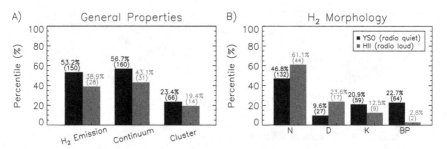

Figure 3. Same as *(a)* and *(b)* of Fig. 2, but separated between radio quiet (YSO, in black) and radio loud (HII, in grey) sources.

radio quiet sources associated to both extended H_2 and continuum emission is $\sim 10\%$ larger than the values found for radio loud ones. The fraction of radio quiet or radio loud sources associated to stellar clusters is almost the same ($\sim 20\%$). Panel (b) indicates the fraction of radio loud sources increases in the opposite direction of the radio quiet ones: YSOs are more often associated to BP-type emission while D-type emission is found towards HII regions.

3.3. *Polar structures and comparison with other samples*

Varricatt *et al.* (2010) present a similar survey based on a smaller sample of 50 MYSO candidates. They found extended H_2 emission towards 38 of them (76%) and 25 sources were associated to polar H_2 structures, suggesting collimated jets. The fraction of the sources associated with extended H_2 emission and those that present polar structures are 1.5 and 2.7 times the values found on the present work (50.3 and 18.6%, respectively).

Figure 4 displays the distribution of the projected length (ℓ_{proj}) and the aspect ratio (R) of the polar structures as a function of the bolometric luminosity of the RMS source. Panel (a) shows a slight tendency that $\ell_{\mathrm{proj}} \propto (L/L_\odot)$ on di-log scales. The plot also indicates that the data from Varricatt *et al.* (2010) display $\ell_{\mathrm{proj}} \lesssim 2\,\mathrm{pc}$, while those from our sample are up to ~ 10 times bigger. The overall conclusion from the plot is that high-luminosity sources can drive massive and powerful structures that extends up to $\sim 10\,\mathrm{pc}$ while low-luminosity ones cannot produce very extended structures. This result also indicates the size of the jets are related to intrinsic characteristics of their driven source.

Figure 4b indicates that high-collimated ($R \gtrsim 10$) structures are found on the entire range of luminosities on the present work while the sample from Varricatt *et al.* (2010) displays only three high-collimated structures associated to sources with $3 \lesssim \log(L/L_\odot) \lesssim 5$. Our sample presents several polar structures driven by high-luminosity sources ($\log(L/L_\odot) \gtrsim 4.0$) that are less collimated ($R \lesssim 2.0$) than those found by Varricatt *et al.* (2010). On the other hand, there is no correlation between R values as a function of the luminosity of the central source, indicating that outflows driven by MYSOs have a similar collimation mechanism than those found toward YSOs.

4. Summary and Conclusions

We observed a sample of 354 MYSOs in the H_2 narrow-filter at $2.12\,\mu\mathrm{m}$ and an adjacent continuum narrow-band filter. The analysis of each H_2-map revealed that:

(a) 50% of our sources display extended H_2 emission: 19% exhibit polar structures, 19% are associated to nodal or amorphous emission and $\approx 12\%$ corresponds to diffuse H_2 emission;

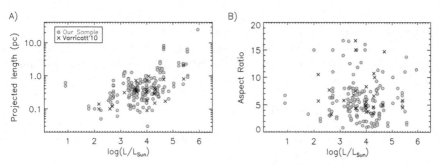

Figure 4. Distribution of the projected length of the structures classified as BP-type *a* and their aspect ratio (*b*) as a function of the logarithm of the luminosity of the driven RMS source. Data from this work is shown as filled grey circles while those from Varricatt *et al.* (2010) are shown as × symbols.

(*b*) We found a considerable number of objects (66) associated to collimated structures and bipolar outflows and most of them are radio quiet sources. This is an observational evidence for the accretion scenario and suggests that MYSOs are formed through discs and most of the accretion occurs on timescales less than 10^5 years;

(*c*) We found a tendency that larger structures are found towards high-luminosity sources. The low-luminosity counterpart cannot drive very extended jets or even be associated to diffuse emission, which requires higher radiative fluxes;

(*d*) We could not establish a correlation between the collimation factor of the polar structures and the bolometric luminosity of their driven sources;

(*e*) The number of sources associated to stellar clusters (high density environments) are much less than the expected for the coalescence scenario, indicating that the merging of low mass stars cannot be the main scenario for massive star formation.

References

Bally, J., Reipurth, B., Walawender, J., & Armond, T. 2002, *AJ* 124, 2152

Bik, A. & Thi, W. F. 2004, *A&A* 427, L13

Blum, R. D., Barbosa, C. L., Damineli, A., Conti, P. S., & Ridgway, S. 2004, *ApJ* 617, 1167

Bonnell, I. A., Bate, M. R., Clarke, C. J., & Pringle, J. E. 2001, *MNRAS* 323, 785

Bonnell, I. A., Bate, M. R., & Zinnecker, H. 1998, *MNRAS* 298, 93

Erben, T., Schirmer, M., Dietrich, J. P., *et al.* 2005, *Astronomische Nachrichten* 326, 432

Krumholz, M. R., Klein, R. I., McKee, C. F., Offner, S. S. R., & Cunningham, A. J. 2009, *Science* 323, 754

Krumholz, M. R., McKee, C. F., & Klein, R. I. 2005, *ApJ* 618, L33

Lumsden, S. L., Hoare, M. G., Oudmaijer, R. D., & Richards, D. 2002, *MNRAS* 336, 621

Mottram, J. C. *et al.* 2011, *A&A* 525, A149

Palla, F. 1996, in S. Beckwith, J. Staude, A. Quetz, & A. Natta (eds.), *Disks and Outflows Around Young Stars*, Vol. 465 of *Lecture Notes in Physics, Berlin Springer Verlag*, p. 143

Schirmer, M. 2013, *ApJS* 209, 21

Stahler, S. W., Palla, F., & Ho, P. T. P. 2000, *Protostars and Planets IV* p. 327

Urquhart, J. S., Busfield, A. L., Hoare, M. G., *et al.* 2007, *A&A* 461, 11

Urquhart, J. S. *et al.* 2009, *A&A* 501, 539

Varricatt, W. P., Davis, C. J., Ramsay, S., & Todd, S. P. 2010, *MNRAS* 404, 661

Discussion

IBADOV: Have you any manifestations of the fall of large comet-like objects onto young massive stars?

NAVARETE: Since we are working with large-scale structures (a few parsecs), I don't think it would be possible to observe these comet-like objects in our maps. It would be better to work with higher angular resolution observations in order to resolve the innermost region around the circumstellar medium of MYSOs and –perhaps– observe these kind of phenomena.

IBADOV: Will you study them in the future?

NAVARETE: We are planning to work with high angular resolution data from NIFS (Gemini north) to study the accretion process "in situ" and then move to longer wavelength (APEX, ALMA) to conclude the work. If there are any evidences of flares or even the comet-like objects associated to the accretion process, these data could reveal them.

GROH: Could you comment on how accurate the luminosity and distance determinations are?

NAVARETE: The luminosity of the sources are based in the bolometric fluxes derived by Mottram *et al.* (2011) and the kinematic distances derived from ^{13}CO observations by Urquhart *et al.* (2007, 2008). The major source of uncertainty of the bolometric fluxes determination comes from the fittings that do not have far-infrared fluxes. The worst cases could have an error up to 50% of the bolometric flux value. Regarding the distance determination, the ^{13}CO line is more likely to be associated to the source itself than the ^{12}CO one, which could display a complex structure due to the presence of larger structures on the line-of-sight.

Felipe Navarete

New windows on massive stars: asteroseismology, interferometry, and
spectropolarimetry
Proceedings IAU Symposium No. 307, 2014
G. Meynet, C. Georgy, J. H. Groh & Ph. Stee, eds.

© International Astronomical Union 2015
doi:10.1017/S1743921314007339

The X-ray properties of magnetic massive stars

Yaël Nazé[1], Véronique Petit[2], Melanie Rinbrand[2], David Cohen[3], Stan Owocki[2], Asif ud-Doula[4] and Gregg Wade[5]

[1] FNRS/ULg, Dept AGO, Allée du 6 Août 17, B5C, 4000-Liège, Belgium
email: `naze@astro.ulg.ac.be`

[2] Dept of Physics & Astronomy, Univ. of Delaware, Bartol Res. Inst., Newark, DE 19716, USA

[3] Dept of Physics & Astronomy, Swarthmore College, Swarthmore, PA 19081, USA

[4] Penn State Worthington Scranton, Dunmore, PA 18512, USA

[5] Dept of Physics, RMC, PO Box 17000, Station Forces, Kingston, ON K7K 4B4, Canada

Abstract. Early-type stars are well-known to be sources of soft X-rays. However, this high-energy emission can be supplemented by bright and hard X-rays when magnetically confined winds are present. In an attempt to clarify the systematics of the observed X-ray properties of this phenomenon, a large series of Chandra and XMM observations was analyzed, over 100 exposures of 60% of the known magnetic massive stars listed recently by Petit *et al.* (2013). It is found that the X-ray luminosity is strongly correlated with mass-loss rate, in agreement with predictions of magnetically confined wind models, though the predictions of higher temperature are not always verified. We also investigated the behaviour of other X-ray properties (absorption, variability), yielding additional constraints on models. This work not only advances our knowledge of the X-ray emission of massive stars, but also suggests new observational and theoretical avenues to further explore magnetically confined winds.

Keywords. stars: early-type, stars: magnetic fields, X-rays: stars

1. Introduction

In the last decade, tens of massive stars were found to be strongly magnetic (e.g. Donati *et al.* 2002; Hubrig *et al.* 2011; Wade *et al.* 2012). Their overall properties appear quite similar to those of magnetic AB stars: the fields appear strong, stable, and organized on large scales (e.g. dipoles) but quite rare. Such magnetic fields are able to channel the stellar winds of massive stars towards the magnetic equator, giving rise to magnetically confined winds (MCWs). The shocks between the wind flows should give rise to an intense X-ray emission (Babel & Montmerle 1997). Two MCWs prototypes, σ Ori E and θ^1 Ori C, were readily found. Their high-energy emission appears in line with expectations (Gagné *et al.* 2005; Townsend *et al.* 2007), but the situation appears more confused for other objects (Oskinova *et al.* 2011), like for example the Of?p stars (Nazé *et al.* 2010): overluminosity and hard X-rays are not always the rule. As previous studies have focused on a limited number of objects, we have undertaken a systematic study of the full sample of magnetic massive stars, with the hope to derive the overall high-energy properties of MCWs (Nazé *et al.*, submitted).

2. X-ray data

To ensure a high homogeneity as well as to maximize the number of detections, we focused on CCD spectra in the 0.5–10. keV range coming from XMM-EPIC and

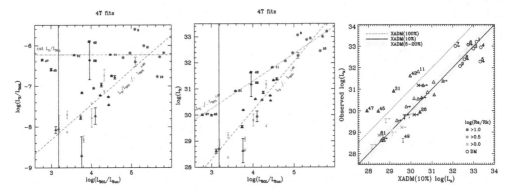

Figure 1. *Left and middle:* $\log(L_X/L_{BOL})$ ratio and X-ray luminosity as a function of bolometric luminosity, along with best-fit relations. *Right:* Comparison of observed and predicted (using ud-Doula *et al.* models) luminosities.

Chandra-ACIS. High-resolution data are indeed few in number, ROSAT data do not sample the hard X-ray emission deemed to exist in such objects, ASCA has a poor resolution and cannot isolate the X-ray emission from our targets (which often lie in clusters), and Swift or Suzaku archives did not yield additional detections. Thus concentrating on Chandra and XMM data, we found that over a hundred exposures are available for 39 out of the 64 magnetic stars in the Petit *et al.* (2013) catalog. In the magnetic confinement-rotation diagram (Petit *et al.* 2013), these targets are well distributed, so that our sample is not a particular subpopulation of the full catalog.

Amongst the objects covered by X-ray observations, there are 6 non-detections, 5 faint detections (only count rate available), and 28 detections bright enough for spectra to be extracted. These spectra were fitted by absorbed optically-thin plasma (with free temperatures or a set of fixed temperatures, both methods yielding similar results).

3. X-ray luminosity

If MCWs are responsible for producing X-rays, then some correlations between X-ray observables and stellar/wind/magnetic properties are expected. We first investigated the link between the X-ray emission levels and the mass-loss rates since shocks extract some of the wind kinetic energy to produce X-rays. Our sources can be separated in two groups (Fig. 1).

The first group comprises all O-stars and six B-stars which display $\log(L_X/L_{BOL}) = -6.23 \pm 0.07$. This is equivalent to the relation $L_X \propto \dot{M}^{0.6}$, which is typical of the X-ray emission from O-stars and their radiative stellar winds. However, the magnetic objects display a significantly brighter X-ray emission than "normal" O-stars which have $\log(L_X/L_{BOL}) \sim -7$ (e.g. Nazé *et al.* 2011, and references therein). Amongst O-stars, note that there are two deviant cases, indicating a probable non-magnetic origin for the X-rays: ζ Ori, which has a very weak field hence a "normal", fainter X-ray emission, and Plaskett's star, whose brighter X-ray emission is probably linked to binary interactions rather than MCWs.

The second group, comprising most B-stars, displays $L_X \propto \dot{M}^{1.4}$ or equivalently, since the winds are radiation-driven, $L_X \propto \dot{L}_{BOL}^{1.9}$ (though it may be noted that relations with L_{BOL} display a larger scatter than relations with $\dot{M}$). The trend is here much steeper, but it is still shallower than the $L_X \propto \dot{M}^2$ relation for adiabatic shocks, adequate for embedded-wind shocks in low-density winds (Owocki *et al.* 2013). Indeed, wind confinement is likely to reduce significantly the amount of adiabatic cooling, since material in

MCW shocks is trapped in closed loops and so does not undergo the adiabatic expansion cooling of an outflowing stellar wind.

The best-fit relations for O- and B-stars are shown in the right panel of Fig. 1. Note that faint/non-detections, though not included in the derivation of these best-fit relations, agree well with them.

We have examined alternative relations (with wind density or Hα emission strength) as well as more complicated relations (e.g. considering dependences on both $\dot{M}$ and magnetic field strength) but none was found significantly better.

Are these relations theoretically understood? Babel & Montmerle (1997) proposed $L_\mathrm{X} \propto \dot{M} \times B^{0.4} \times v_\infty$ but this scaling seems to be inadequate for the lowest and highest mass-loss rates. Besides, the X-ray emission level predicted by Babel & Montmerle formula is too high by 1.8 dex compared to observations (which probably comes from the nearly perfect conversion efficiency of that model). New MHD simulations by ud-Doula *et al.* (2014) re-investigated the problem of the X-ray emission of MCWs, demonstrating the impact of shock-retreat effects especially in low-density winds. Their predictions appear much better in line with the observational results, both for the emission levels and the observed trends with $\dot{M}$ (Fig. 1). A few discrepancies remain, though, which could not yet be explained. They concern the few B-stars with bright L_X but low $\dot{M}$ belonging to group 1. Their predicted luminosities are much lower than their observed ones. Unfortunately, we could not clearly identify a common feature amongst these objects. For example some of them are fast rotators but not all, and some supposed "twins" of these targets do not show a similar elevated X-ray emission.

4. X-ray spectral shape

Hardness ratios and average temperatures were also derived from the spectral fits. Strongly magnetic O-stars clearly are harder/hotter than their non-magnetic siblings, while most of the B-stars show low hardness ratios. However, no correlation was found with stellar/wind/magnetic parameters despite the fact that hotter plasma is theoretically expected for higher confinement values or higher mass-loss rates (ud-Doula *et al.* 2014). Also, the overluminous (compared to MHD predictions) B-stars do not necessarily stand up with particularly high or low temperatures. The drivers for X-ray luminosity and plasma temperature certainly appear different.

5. X-ray absorption

Spectral fits also allowed for the possibility of absorption in addition to the interstellar one, e.g. arising in the dense magnetospheres. In general, magnetic stars appear to behave as non-magnetic objects: an additional absorption is not needed for magnetic B-stars (as is usual for the "normal" ones), while O-stars require an amount comparable to what is needed for "normal" O-stars without MCWs. No correlation between that absorption and stellar/wind/magnetic parameters is found. The only special case is NGC1624-2, the most magnetic O-star, whose extremely dense magnetosphere leads to the presence of very high absorption (Petit *et al.*, submitted). Again, as for temperatures, the overluminous B-stars do not show, as a group, particular values of the absorption.

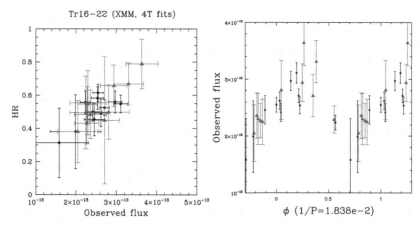

Figure 2. Variations of Tr16-22 in XMM data. *Left:* Correlation between flux and hardness changes. *Right:* Flux changes phased with the newly discovered 54d period. The 2003 data are shown in red.

6. Variation of the X-ray emission

When several exposures are available, we have examined the X-ray variability. Three behaviours are detected:

- Constant emission, as expected from the properties of the targets (pole-on geometry for HD148937, large magnetosphere for σ Ori E, non-magnetic origin for the X-rays of ζ Ori).
- Flux changes without spectral changes (HD47777 and β Cep), which cannot be linked to MCWs with current data. Note that these changes include the detection of flares in HD47777.
- Flux and spectral changes (Tr16-22, HD191612, NU Ori, θ^1 Ori C), often found to be in harmony with the stellar rotation (HD191612, θ^1 Ori C). In this context, the large number of X-ray observations of Tr16-22 enabled us to identify its period, about 54d (Fig. 2, Nazé *et al.*, submitted). The observed variations can be explained by occultation effects in stratified MCWs. Note however that two trends are seen: the X-ray emissions are either harder (Tr16-22, HD191612, NU Ori) or softer (θ^1 Ori C) when brighter. This suggests that the structure of MCWs may vary amongst magnetic objects, even if their stellar/wind/magnetic properties are apparently quite similar. Detailed models are now needed to explain this intriguing behaviour.

7. Conclusion

Using archival and dedicated XMM and Chandra observations, we have examined the X-ray behaviour of magnetic massive stars. More than half of the stars in Petit *et al.* catalog have been surveyed, and this sample is representative of the whole population.

We found that the X-ray luminosity is strongly correlated to the bolometric luminosity or, equivalently, to the mass-loss rate. However, two groups of objects can be defined: one with constant $\log(L_X/L_{BOL}) \sim -6.2$, and one with $L_X \propto \dot{M}^{1.4}$. No alternative or more complex correlation was found significantly better. MHD models are able to reproduce the levels of X-ray emissions and the above relations for most objects - the exceptions are a few B-stars with an observed luminosity much brighter than expected on the basis of their mass-loss rate. No specific, common property could be identified for these outliers, whose origin remains unknown.

The X-ray spectral shapes appear varied amongst targets, but no correlation with stellar, wind, or magnetic properties could be derived. In particular, the prediction of harder X-rays for higher mass-loss or wind confinement is not verified, and the overluminous outliers do not stand up as peculiarly soft or hard X-ray sources. Similarly, the local absorptions do not appear to be correlated with stellar, wind, or magnetic properties.

Finally, we have examined the variability of the targets when several exposures were available. Constancy of the X-ray emission is found for some cases, and can be explained by the properties of the targets. At the other extreme, correlated flux and spectral changes are derived for some objects. These variations appear to be coherent with the stellar rotation period, hence can be explained by regular occultations of the X-ray emission regions by the stellar body. In that case, the presence of simultaneous spectral changes requires the MCWs to be stratified in temperature.

References
Babel, J. & Montmerle, T. 1997, *A&A* 323, 121
Donati, J.-F., Babel, J., Harries, T. J., *et al.* 2002, *MNRAS* 333, 55
Gagné, M., Oksala, M. E., Cohen, D. H., *et al.* 2005, *ApJ* 628, 986
Hubrig, S., Schöller, M., Kharchenko, N. V., *et al.* 2011, *A&A* 528, A151
Nazé, Y., Broos, P. S., Oskinova, L., *et al.* 2011, *ApJS* 194, 7
Nazé, Y., Ud-Doula, A., Spano, M., *et al.* 2010, *A&A* 520, A59
Oskinova, L. M., Todt, H., Ignace, R., *et al.* 2011, *MNRAS* 416, 1456
Owocki, S. P., Sundqvist, J. O., Cohen, D. H., & Gayley, K. G. 2013, *MNRAS* 429, 3379
Petit, V., Owocki, S. P., Wade, G. A., *et al.* 2013, *MNRAS* 429, 398
Townsend, R. H. D., Owocki, S. P., & Ud-Doula, A. 2007, *MNRAS* 382, 139
ud-Doula, A., Owocki, S., Townsend, R., Petit, V., & Cohen, D. 2014, *MNRAS* 441, 3600
Wade, G. A., Grunhut, J., Gräfener, G., *et al.* 2012, *MNRAS* 419, 2459

Discussion

ALECIAN: Why do you think a Herbig Be star would flare?

NAZÉ: It may not be from the Herbig Be star itself - it is not uncommon for the X-ray emission of "normal" B stars to be considered as coming from (or contaminated by) that of a companion. If the star is an Herbig, then a companion would also be young, explaining the flaring activity. With the current data, it is not possible to firmly link the flares and the confined winds, as I said.

AERTS: Did you try to compare your X-ray data with EUVE data for the few pulsating B stars that now turn out to be magnetic, e.g., β CMa (low amplitude non-radial pulsator), ξ^1 CMa (large amplitude radial pulsator)?

NAZÉ: We didn't do this comparison yet because EUVE data were not available for most of our targets, but I agree such a study would be valuable.

PULS: I just want to suggest (in future publications) that you call your "mass-loss rate" the "theoretical feeding rate", since the actual mass-loss rate is lower due to infall.

NAZÉ: Fully agree.

WEIS: You expect the X-ray emission to origin in the compression zone. If however this axis of rotation and the magnetic field are not aligned, this region is precessing and might be variable.

NAZÉ: This is a possibility, yes, but which has not be explored (and cannot yet with the current data).

Yaël Nazé

Norbert Przybilla

New windows on massive stars: asteroseismology, interferometry, and spectropolarimetry
Proceedings IAU Symposium No. 307, 2014
G. Meynet, C. Georgy, J. H. Groh & Ph. Stee, eds.

© International Astronomical Union 2015
doi:10.1017/S1743921314007340

Combining seismology and spectropolarimetry of hot stars

Coralie Neiner[1], Maryline Briquet[2,1], Stéphane Mathis[3,1] and Pieter Degroote[4,1]

[1]LESIA, Observatoire de Paris, CNRS UMR 8109, UPMC, Université Paris Diderot,
5 place Jules Janssen, 92190 Meudon, France
email: coralie.neiner@obspm.fr

[2]Institut d'Astrophysique et de Géophysique, Université de Liège,
Allée du 6 Août 17, Bât B5c, 4000 Liège, Belgium

[3]Laboratoire AIM Paris-Saclay, CEA/DSM - CNRS - Université Paris Diderot,
IRFU/SAp Centre de Saclay, 91191 Gif-sur-Yvette, France

[4]Instituut voor Sterrenkunde, Celestijnenlaan 200D, 3001 Heverlee, Belgium

Abstract. Asteroseismology and spectropolarimetry have allowed us to progress significantly in our understanding of the physics of hot stars over the last decade. It is now possible to combine these two techniques to learn even more information about hot stars and constrain their models. While only a few magnetic pulsating hot stars are known as of today and have been studied with both seismology and spectropolarimetry, new opportunities - in particular *Kepler*2 and BRITE - are emerging and will allow us to rapidly obtain new combined results.

Keywords. stars: early-type, stars: magnetic fields, stars: oscillations (including pulsations)

1. Introduction

Over the last decade, two major steps have been done in parallel leading to substantial progress in the field of hot stars:

1) asteroseismology, in particular with space-based facilities such as MOST, CoRoT and *Kepler* but also with multi-site spectroscopic campaigns, has allowed us to study the pulsations of hot stars with high precision and thus their internal structure (see Aerts, this volume),

2) spectropolarimetry, with the new generation of high-resolution instruments Narval at TBL, ESPaDOnS at CFHT and HarpsPol at ESO, has allowed us to study the magnetic fields of hot stars and their circumstellar magnetospheres (see Grunhut, this volume).

We have now reached a point where these two techniques can be and should be combined to increase the physics we can probe in hot stars.

About 10% of hot stars are found to be magnetic with oblique dipolar fields above $\sim 100\,\mathrm{G}$ at the pole (Wade *et al.* 2013). A similar occurence of magnetic fields is observed in pulsating hot stars and in non-pulsating ones (Neiner *et al.* 2011). The presence of this magnetic field impacts the power spectrum of pulsations that we observe in β Cep, Slowly Pulsating B (SPB) and roAp stars. In particular magnetic splitting of the pulsation modes occurs. The width of this splitting is directly related to the strength of the field, but the amplitude of the components of the splitted multiplet depends on the obliquity of the dipolar magnetic field compared to the rotation/pulsation axis (see e.g. Shibahashi & Aerts 2000). Therefore the number of peaks that can be detected in the power spectrum with current instrumentation depends on the magnetic geometry.

Knowing this information is very important to perform a proper pulsation mode identification and thus compute a correct seismic model.

Conversely, knowing the pulsation properties of a hot star is important when studying its magnetic field. Indeed, the Zeeman Stokes signature of the magnetic field observed in spectropolarimetry, which allows us to derive the longitudinal field value and magnetic configuration, depends on the intensity of the line profiles and thus on how the lines are deformed by pulsations. In particular, variations of the line profiles due to pulsations should be taken into account when measuring the longitudinal magnetic field. To derive proper magnetic field strength and geometry, one thus needs to model the intensity line profiles taking pulsations into account.

2. Magnetic modelling of pulsating stars

The Phoebe2.0 code (Degroote *et al.* 2013) allows users to model various surface observables such as the light curve or line profiles of a star with pulsations, spots, one or more companions, etc. The code has recently been modified to also provide the possibility to include an oblique magnetic dipole field at the surface of the modeled star.

Thanks to this code, we have modeled the line profile variations due to pulsations of the star β Cep and the corresponding Stokes V profiles due to the magnetic field (see Neiner *et al.* 2013).

β Cep is the first pulsating B star discovered to host a magnetic field (Henrichs *et al.* 2000) and its field configuration has thus been well studied since then. Donati *et al.* (2001) proposed that $B_{pol} = 360$ G, the inclination angle i is 60° and the obliquity angle β is 85°. Henrichs *et al.* (2013) showed that $B_{pol} > 306$ G (since $B_{max} = 97$ G), with the same inclination angle $i = 60°$ but $\beta = 96°$.

We have used 56 new Narval observations of β Cep and applied the LSD technique (Donati *et al.* 1997) to produce LSD Stokes I and V profiles. The I profiles vary mostly with the pulsation periods, while the Stokes V profiles clearly vary with both the pulsation periods and rotation period (because of the rotational modulation due to the field obliquity).

We fitted these LSD I and V profiles with Phoebe2.0. Our model results in $B_{pol} = 370$ G and $\beta = 92°$, compatible with the values published in the literature when ignoring pulsations. However, the most striking result is that the inclination angle we find is different from the one published so far. We find that $i = 46°$ rather than $i = 60°$ (Neiner *et al.*, in prep.).

This shows that ignoring pulsations in magnetic studies of pulsating stars introduces large errors in the geometrical configuration that is deduced from the Stokes profiles. Therefore, it is necessary to take pulsations into account, especially when their amplitude is large and in the case of radial modes, to obtain the proper geometry.

3. Seismic modelling of magnetic stars

The second pulsating B star discovered to host a magnetic field is V2052 Oph (Neiner *et al.* 2003). Its magnetic field was recently analysed in more details thanks to new Narval observations (Neiner *et al.* 2012). It was found that the magnetic field of V2052 Oph has a polar strength of $B_{pol} \sim 400$ G with an obliquity of $\beta \sim 35°$ and an inclination of $i \sim 70°$.

A multi-site spectroscopic (Briquet *et al.* 2012) and photometric (Handler *et al.* 2012) campaign also allowed us to reanalyse its pulsations in more details. The stellar pulsations are dominated by a radial mode but two non-radial low-amplitude prograde modes are also detected. Rotational modulation is also clearly present due to the oblique magnetic field. Moreover, we found that the seismic models that reproduce best the observations are those in which no or very small overshooting is added ($\alpha = 0.07 \pm 0.08\,\mathrm{H_p}$). Such a low overshooting parameter is not common in β Cep stars especially when they rotate as fast as V2052 Oph ($v \sin i = 80\,\mathrm{km\,s^{-1}}$) and when rotational mixing is thus expected.

The reason for this lack of macroscopic internal mixing is the presence of the magnetic field. Indeed, using the criteria proposed by Zahn (2011) based on Mathis & Zahn (2005) or by Spruit (1999), we derive that mixing is inhibited in V2052 Oph when $\mathrm{B_{pol}} \sim 70\,\mathrm{G}$ and 40 G, respectively (Briquet *et al.* 2012). This critical field value is 6 to 10 times weaker than the observed polar field strength of V2052 Oph. Therefore, the seismic model of V2052 Oph is the first observational proof that internal mixing is inhibited by magnetic field.

The criteria proposed by Zahn (2011) and Spruit (1999), however, use several crude approximations. We have thus started to devise a new more precise critical field criterion, computing the magnetic torque with realistic non-axisymmetric geometry, i.e. taking obliquity into account (Mathis *et al.*, in prep.). This criterion is of the form:

$$\mathrm{B_{crit}^2} = 4\pi\rho R^2\Omega/t_\mathrm{AM} * F(\beta), \qquad (3.1)$$

where ρ is the density of the star, R is its radius, Ω is its angular velocity, t_AM is the characteristic time for the angular momentum evolution (due to the wind or to structural adjustments), and F is a function of the obliquity angle β.

We will apply this criterion to all known magnetic hot stars. This will allow us to provide a new constraint for seismic models from spectropolarimetry, i.e. to tell if overshoot should be added in the seismic models or fixed to 0. Spectropolarimetry also already allows us to constraint these models by providing the geometry and in particular the inclination angle.

4. New opportunities for seismic studies of magnetic stars

MOST observed several bright pulsating OB stars and a few magnetic hot stars, but had not observed any pulsating magnetic hot star so far. While MOST will ceased its observations at the end of August 2014, one of its last target from May 29 to June 27, 2014, was V2052 Oph. Therefore a MOST light curve of this well-studied magnetic pulsating star has just become available (see Fig. 1). We hope to be able to identify magnetic splittings in the power spectrum of the pulsations of this star.

CoRoT observed many OB stars. None of them were known to be magnetic at the time of observations. However, targets observed in the "seismo" fields of CoRoT are sufficiently bright to be observed with high-resolution spectropolarimetry. This was done with Narval for all hot stars. Among these targets, one of them, HD 43317, was found to be magnetic (Briquet *et al.* 2013). HD 43317 is a hybrid (β Cep + SPB) pulsator and thus exhibits both p and g pulsation modes. In addition, regular frequency spacing was detected in its power spectrum (Pápics *et al.* 2012). A first seismic model was computed (Savonije 2013). However, the subsequent discovery of the magnetic field in HD 43317 calls for a reinterpretation of the observations, in particular in terms of magnetic splittings, and a new seismic model taking the magnetic field into account. The fact that HD 43317 shows p modes, g modes, rotational modulation and a magnetic field provides a wealth of constraints for such a model.

Finally, *Kepler* observed only one field of view optimised for the detection of exoplanets around solar-type stars. This field unfortunately did not contain O stars nor early-B stars.

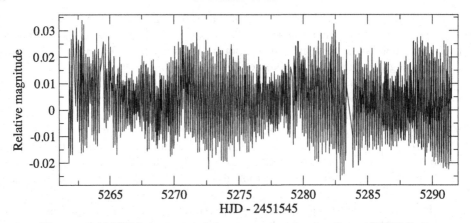

Figure 1. MOST light curve of the magnetic pulsating B star V2052 Oph.

It did contain some late-B stars and thus SPB stars, but none of them are known to be magnetic and these targets are too faint for current spectropolarimeters to detect a field unless it would be many kG in strength.

As a consequence, only very few magnetic pulsating stars could be observed so far with space-based photometry. New opportunities are however opening now: *Kepler*2 and BRITE.

4.1. *Kepler2*

Following the failure of several reaction wheels in *Kepler*, the spacecraft's objectives have been redefined into a new mission called *Kepler*2 (or K2). K2 is limited to pointing fields near the ecliptic plane and has to change field about every 3 months. As a consequence, the new fields of K2 include all kinds of targets.

In particular, Field 0 included 35 magnetic hot stars, among which 7 were observed between March 8 and May 30, 2014. Field 2 also contains 5 magnetic hot stars, which have been requested for observation in the second semester of 2014. Several magnetic hot stars will also be observable in the forthcoming Fields 4, 5, 7 and 9.

These K2 data will multiply tenfold the number of magnetic hot stars observed with high-precision space photometry. Many of these targets are expected to pulsate and will thus provide a unique opportunity for combined seismic and spectropolarimetric studies.

4.2. *BRITE*

The BRIght Target Explorer (BRITE) is an Austrian, Polish and Canadian constellation of nano-satellites (Weiss *et al.* 2014). Each participating country builds two nano-satellites: one providing photometry in the red band and the other one in the blue band. Using two colours allows one to more easily identify the pulsation modes of hot stars thanks to the difference in mode amplitude in the two colours.

The constellation is already composed of two Austrian, one Polish and one Canadian nano-satellites. The second Polish nano-satellite should be launched soon, while the second Canadian nano-satellite unfortunately failed to separate from the upper-stage of the launch vehicle during its launch in June 2014.

BRITE concentrates on targets brighter than V=4. As a consequence it mostly observes hot stars and evolved stars. There are ~ 600 stars with $V \leqslant 4$, among which ~ 300 are hotter than F5, i.e. are stars in which a fossil magnetic field can be detected.

As of now, we know 8 magnetic pulsating OB stars with $V \leqslant 4$: β Cen, ζ Ori A, τ Sco, β Cep, ϵ Lup, ζ Cas, ϕ Cen and β CMa.

Since BRITE targets are very bright, they can very easily be observed in spectropolarimetry with a very good magnetic field detection threshold. We have thus started a large survey of all stars with V$\leqslant 4$ with Narval (for stars with a declination $\delta > -20°$; PI Neiner), ESPaDOnS (for stars with $-45 < \delta < -20°$; PI Wade) and HarpsPol (for stars with $\delta < -45°$; PI Neiner). We aim at a detection threshold of $B_{pol} = 50\,G$ for all fossil field stars i.e. hotter than F5, and $B_{pol} = 5\,G$ for cooler stars with a magnetic field of dynamo origin. These very high quality data will also provide an ideal spectroscopic database, e.g. for the determination of the stellar parameters of all BRITE targets.

The spectropolarimetric observations of BRITE targets started in February 2014. We have already discovered 9 new magnetic stars. However, these are mainly cool stars. The hottest one, ι Peg, is a F5V star and a SB2 spectroscopic binary.

5. Conclusion

Combined magnetic and seismic observational studies can now be performed on bright hot stars. Such studies provide better and stronger constraints for models, e.g. on internal mixing or on the geometrical configuration of the star (inclination and field obliquity). They thus allow us to better understand the physics at work inside hot stars.

Conversely, ignoring one or the other aspect (pulsations or magnetism) can lead to substantially wrong results, therefore combined studies should be performed whenever possible. To this aim, *Kepler*2 and BRITE will provide new opportunities.

References

Briquet, M., Neiner, C., Aerts, C., *et al.* 2012, *MNRAS* 427, 483
Briquet, M., Neiner, C., Leroy, B., & Pápics, P. I. 2013, *A&A* 557, L16
Degroote, P., Conroy, K., Hambleton, K., *et al.* 2013, in *EAS Publications Series*, Vol. 64 of *EAS Publications Series*, p. 277
Donati, J.-F., Semel, M., Carter, B. D., Rees, D. E., & Collier Cameron, A. 1997, *MNRAS* 291, 658
Donati, J.-F., Wade, G. A., Babel, J., *et al.* 2001, *MNRAS* 326, 1265
Handler, G., Shobbrook, R. R., Uytterhoeven, K., *et al.* 2012, *MNRAS* 424, 2380
Henrichs, H. F., de Jong, J. A., Donati, J.-F., *et al.* 2000, in M. A. Smith, H. F. Henrichs, & J. Fabregat (eds.), *IAU Colloq. 175: The Be Phenomenon in Early-Type Stars*, Vol. 214 of *Astronomical Society of the Pacific Conference Series*, p. 324
Henrichs, H. F., de Jong, J. A., Verdugo, E., *et al.* 2013, *A&A* 555, A46
Mathis, S. & Zahn, J.-P. 2005, *A&A* 440, 653
Neiner, C., Alecian, E., Briquet, M., *et al.* 2012, *A&A* 537, A148
Neiner, C., Alecian, E., & Mathis, S. 2011, in G. Alecian, K. Belkacem, R. Samadi, & D. Valls-Gabaud (eds.), *SF2A-2011: Proceedings of the Annual meeting of the French Society of Astronomy and Astrophysics*, p. 509
Neiner, C., Degroote, P., Coste, B., Briquet, M., & Mathis, S. 2013, in *Magnetic fields throughout stellar evolution*, Vol. 302 of *IAU Symposium*, *ArXiv 1311.2262*
Neiner, C., Henrichs, H. F., Floquet, M., *et al.* 2003, *A&A* 411, 565
Pápics, P. I., Briquet, M., Baglin, A., *et al.* 2012, *A&A* 542, A55
Savonije, G. J. 2013, *A&A* 559, A25
Shibahashi, H. & Aerts, C. 2000, *ApJ (Letters)* 531, L143
Spruit, H. C. 1999, *A&A* 349, 189
Wade, G. A., Grunhut, J., Alecian, E., *et al.* 2013, in *Magnetic fields throughout stellar evolution*, Vol. 302 of *IAU Symposium*, p 265
Weiss, W. W., Rucinski, S. M., Moffat, A. F. J., *et al.* 2014, *PASP* 126, 573
Zahn, J.-P. 2011, in C. Neiner, G. Wade, G. Meynet, & G. Peters (eds.), *IAU Symposium*, Vol. 272 of *IAU Symposium*, p. 14

Discussion

HENRICHS: Congratulations with this beautiful work. Now that there is a new determination of the inclination angle of β Cep, I wonder if the UV wind behavior with its asymmetry could be modelled as well. In particular the asymmetry of the equivalent width at opposite orientation pointed to an off-centered wind configuration. Would this be reproduced? And what about the wind profiles themselves?

NEINER: In the Phoebe2.0 model of β Cep, we introduced as many observational constraints as possible. This includes not only the variable intensity line profiles and the magnetic Stokes V profiles, but also, e.g., constraints on the stellar parameters such as the distance or mass obtained thanks to speckle interferometry and radial velocities. As far as the UV is concerned, we have used the rotational modulation that you derived from the UV wind variations, but we have not modeled the wind line profiles themselves. Our model points to a mainly dipolar (but oblique) magnetic field.

MAEDER: What may be the typical value of this new factor F that accounts for the obliquity in the expression of the critical field?

NEINER: This factor is composed of trigonometric functions of the obliquity β corresponding to the projection of the Lorentz force in the reference frame associated to the rotation axis. The exact details of this factor are still under investigation by Stéphane Mathis.

MEYNET: When you say that in the magnetic star the mixing is inhibited, do you mean that the core does not seem to present any extension due to an overshoot or do you mean that there is no extension of the convective core and no mixing at all in the radiative zone, or no mixing in the radiative zone?

NEINER: In the CLES code that we used to model V2052 Oph, α only corresponds to core overshooting, so this model shows that there is no extension of the convective core. However, the critical field criteria concern any kind of mixing in the whole radiative zone, not only the extension of the core, and thus these criteria show that any mixing is inhibited in this star by the magnetic field. (See Mathis *et al.*, this volume, for more details.)

Coralie Neiner

New windows on massive stars: asteroseismology, interferometry, and
spectropolarimetry
Proceedings IAU Symposium No. 307, 2014
G. Meynet, C. Georgy, J. H. Groh & Ph. Stee, eds.
© International Astronomical Union 2015
doi:10.1017/S1743921314007352

X-rays From Centrifugal Magnetospheres in Massive Stars

Christopher Bard and Richard Townsend

University of Wisconsin, Madison
corresponding email: bard@astro.wisc.edu

Abstract. In the subset of massive OB stars with strong global magnetic fields, X-rays arise
from magnetically confined wind shocks (Babel & Montmerle 1997). However, it is not yet clear
what the effect of stellar rotation and mass-loss rate is on these wind shocks and resulting X-
rays. Here, we present results from a grid of Arbitrary Rigid-Field Hydrodynamic simulations
(ARFHD) of a B-star centrifugal magnetosphere with an eye towards quantifying the effect of
stellar rotation and mass-loss rate on the level of X-ray emission. The results are also compared
to a generalized XADM model for X-rays in dynamical magnetospheres (ud-Doula et al. 2014).

1. Arbitrary Rigid-Field Hydrodynamic Simulations

The strong magnetic fields of centrifugal magnetospheres produce Alfvén speeds so
fast that MHD simulations are impractical. To address this issue, Townsend et al. (2007)
developed a Rigid-Field Hydrodynamics (RFHD) approach to simulate magnetospheres
in the limit of strong magnetic confinement. The three-dimensional stellar outflow is
approximated as many quasi-one-dimensional flows along individual field lines subject
to radiative and gravitocentrifugal forces. We simulate each line separately and stitch
together the results to form a picture of the overall magnetosphere.

We have since extended the RFHD technique to incorporate completely arbitrary magnetic configurations (*arbitrary* RFHD, or ARFHD; Bard & Townsend, in prep.), though
all of the simulations presented here focus on the simplest case of a rotation-aligned
dipole field. Our goal is to understand the effect of stellar rotation and mass-loss rate on
the level of X-ray emission in a B-star centrifugal magnetosphere.

Towards this end, we simulate a star with an aligned dipole magnetic field ($\beta = 0°$)
and other fundamental parameters based on the archetype σ Ori E (Townsend et al.
2013): $M = 8.3\ M_\odot$, $R = 3.8\ R_\odot$, $B = 11$ kG, and $T_{\rm eff} = 22500$ K. To understand
the effect of rotation and mass-loss rate ($\dot{M}$) on the X-ray luminosity, we simulate every
possible combination of six critical rotation fractions W $(0.0, 0.2, 0.35, 0.5, 0.65, 0.8)$ and
seven $\dot{M}$ values $(1 \cdot 10^{-5}, 1 \cdot 10^{-6}, 1 \cdot 10^{-7}, 1 \cdot 10^{-8}, 1 \cdot 10^{-9}, 1 \cdot 10^{-10}, 1 \cdot 10^{-11}\ M_\odot/{\rm yr})$. We
vary $\dot{M}$ through the Q opacity parameter; see Gayley (1995) for a discussion.

2. Results and Discussion

In general, X-ray luminosity increases with both rotation and mass-loss rate (Fig 1).
This is expected, since increasing the rotation rate provides a higher acceleration of the
plasma along the field lines, resulting in a higher shock velocity and more X-ray emitting
gas. Increasing the mass-loss rate increases the amount of density in the magnetosphere,
which in turn increases the X-ray emission. Additionally, the results match up well with a
semi-analytic XADM model (ud-Doula et al. 2014) calculated for a non-rotating dynamical magnetosphere (dashed line in Fig. 1). Although the XADM model was developed for
dynamical magnetospheres, it applies well to our centrifugal magnetosphere simulations.

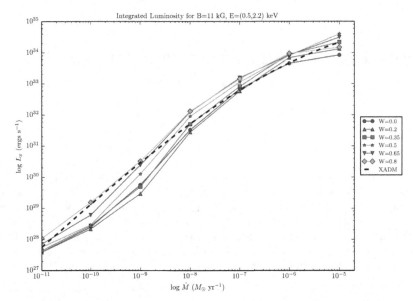

Figure 1. The relationship between integrated luminosity and mass-loss rate for varying critical rotation fraction. The dashed line is the predicted X-ray luminosity from a non-rotating XADM scaling law for B = 11 kG. The results generally agree with the scaling law.

The XADM model does not include rotation, however. To better understand the dependence of X-ray luminosity on rotation rate, we fit our centrifugal magnetosphere simulation data as a power law $L_x \propto W^\alpha$ for each mass-loss rate. The average α is 0.88, though there is a wide range (min=0.17, max=1.84) among mass-loss rates. We believe this wide variation to be a result of siphon flows (e.g. Cargill & Priest 1980) within our simulations. It is unclear whether these siphon flows are physical or simply a unphysical numerical solution for our quasi-1D field lines.

We derive a back-of-envelope $L_x \propto W^\alpha$ for comparison with our simulation power-law fit (with subscript 1 indicating pre-shock and 2 post-shock):

(a) In a strong shock, $T_2 \propto v_1^2$.

(b) In a situation where only the rotation rate varies, we have $g_{\text{cen}} \propto W^2$, where g_{cen} is the centrifugal acceleration along the field line.

(c) Making the simplifying assumption that g_{cen} is independent of field line position, we get from kinematics that $v_1^2 \propto g_{\text{cen}}$.

(d) X-ray emission in the magnetosphere is dominated by line emission: $L_x \propto T_2$.

From these relations, we naively expect $L_x \propto W^2$, but our simulation yields power-law coefficients smaller than 2 for every mass-loss rate. Future research will be needed to resolve this apparent discrepancy.

References

Babel, J. & Montmerle, T. 1997, *A&A* 323, 121

Cargill, P. J. & Priest, E. R. 1980, *Sol. Phys.* 65, 251

Gayley, K. G. 1995, *ApJ* 454, 410

Townsend, R. H. D., Owocki, S. P., & Ud-Doula, A. 2007, *MNRAS* 382, 139

Townsend, R. H. D., Rivinius, T., Rowe, J. F., *et al.* 2013, *ApJ* 769, 33

ud-Doula, A., Owocki, S., Townsend, R., Petit, V., & Cohen, D. 2014, *MNRAS* 441, 3600

New windows on massive stars: asteroseismology, interferometry, and spectropolarimetry
Proceedings IAU Symposium No. 307, 2014
G. Meynet, C. Georgy, J. H. Groh & Ph. Stee, eds.

© International Astronomical Union 2015
doi:10.1017/S1743921314007364

Abundance study of two magnetic B-type stars in the Orion Nebula Cluster

T. Morel

Institut d'Astrophysique et de Géophysique, Allée du 6 Août, 4000 Liège, Belgium
email: morel@astro.ulg.ac.be

Abstract. We present the results of an abundance analysis of two magnetic B-type stars in the Orion Nebula Cluster that support the lack of a direct relationship between the existence of a magnetic field and a nitrogen excess in the photosphere.

1. Motivation

The effects of magnetic fields on internal mixing in massive stars remain largely unknown. Collecting He and CNO abundances for magnetic OB stars is needed to assess the efficiency of rotational mixing and to constrain evolutionary and interior models (e.g. Meynet *et al.* 2011). We present the determination of the atmospheric parameters and chemical abundances of two magnetic B-type stars in the very young Orion Nebula Cluster (ONC) based on high-resolution optical spectra acquired with FIES at NOT.

2. The targets

The two ONC stars analysed are LP Ori (HD 36982; B1.5 Vp) and NU Ori (HD 37061; B0.5 V). They have been shown to host a strong magnetic field with a polar strength in the range 600-900 G assuming a dipolar geometry (Petit *et al.* 2008; Petit & Wade 2012). LP Ori and NU Ori are moderate to fast rotators with $v \sin i \sim 80$ and 200 km s^{-1}, respectively (Petit & Wade 2012; Simón-Díaz *et al.* 2011). NU Ori is of particular interest because it is one of the fastest-rotating magnetic B stars known.

3. Determination of parameters and chemical abundances

The analysis is based on the non-LTE line-formation code DETAIL/SURFACE and Kurucz model atmospheres. For LP Ori, $T_{\rm eff}$ is derived from SiII/III ionisation balance and $\log g$ from fitting the Balmer line wings. The abundances are inferred from a classical curve-of-growth analysis (Morel 2012, and references therein). The set of unblended lines is selected from a comparison with a spectrum of a slowly-rotating analogue (δ Ceti). A specific method is used for NU Ori because of the high $v \sin i$ (see details in Rauw *et al.* 2012, and Cazorla *et al.*, these proceedings). The atmospheric parameters and abundances are estimated by finding the best match between a set of observed H, He, and CNO line profiles and a grid of rotationally-broadened, synthetic profiles (see Fig.1 for some illustrative fits). The star is part of an SB1 binary system with $\mathcal{P} \sim 19$ d and a mass ratio above 0.19 (Abt *et al.* 1991). To explore the impact on our results, we have repeated the analysis using composite, synthetic spectra. We assumed a luminosity ratio of 30%, i.e., that the secondary is a main-sequence B1 star of $\sim 10\,M_\odot$ with $T_{\rm eff} = 26\,000$ K, $\log g = 4.3$, and abundances typical of B stars. This may be considered as a rather extreme case (the lines of the secondary are not visible in our spectrum; see Fig.1).

Table 1. Results of the analysis.

	LP Ori	NU Ori	
		Companion ignored	Companion taken into account
T_{eff} [K]	21 500±1000	30 800±1000	31 600±1000
$\log g$	4.30±0.15	4.30±0.15	4.30±0.15
ξ [km s^{-1}]	[5] (fixed)	[10] (fixed)	[10] (fixed)
y	0.071±0.018	0.097±0.030	0.101±0.030
$\log \epsilon(\mathrm{C})$	8.27±0.13	7.96±0.19	8.04±0.19
$\log \epsilon(\mathrm{N})$	7.91±0.36	7.68±0.26	7.55±0.26
$\log \epsilon(\mathrm{O})$	8.70±0.42	8.48±0.27	8.53±0.27
[N/C]	−0.36±0.29	−0.28±0.32	−0.49±0.32
[N/O]	−0.79±0.33	−0.80±0.17	−0.98±0.17

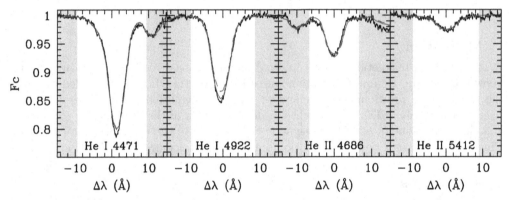

Figure 1. Examples of fit to the He lines in NU Ori (black and red lines: observed and best-fitting synthetic profiles, respectively). The dashed, red lines show the line profiles computed for the final, mean parameters. The quality of the fit was not evaluated within the grey-shaded areas.

4. Results and discussion

The analysis proved difficult because of the substantial line broadening due to rotation. The abundances are weakly constrained, but neither star shows evidence for a He or a N excess (Table 1). The data for NU Ori are compatible with a star that just arrived on the ZAMS and is too young to have dredged up CNO-cycled material to the surface or dramatically spun down because of magnetic braking. These results confirm the diversity of surface nitrogen abundances found in magnetic, early B-type stars (Morel 2012).

Acknowledgements

I acknowledge financial support from Belspo for contract PRODEX GAIA-DPAC.

References

Abt, H. A., Wang, R., & Cardona, O. 1991, *ApJ* 367, 155
Meynet, G., Eggenberger, P., & Maeder, A. 2011, *A&A* 525, L11
Morel, T. 2012, Vol. 465 of *Astronomical Society of the Pacific Conference Series*, p. 54
Petit, V. & Wade, G. A. 2012, *MNRAS* 420, 773
Petit, V., Wade, G. A., Drissen, L., Montmerle, T., & Alecian, E. 2008, *MNRAS* 387, L23
Rauw, G., Morel, T., & Palate, M. 2012, *A&A* 546, A77
Simón-Díaz, S., García-Rojas, J., Esteban, C., *et al.* 2011, *A&A* 530, A57

New windows on massive stars: asteroseismology, interferometry, and spectropolarimetry
Proceedings IAU Symposium No. 307, 2014
G. Meynet, C. Georgy, J. H. Groh & Ph. Stee, eds.

© International Astronomical Union 2015
doi:10.1017/S1743921314007376

Circumstellar Environments of MYSOs Revealed by IFU Spectroscopy

F. Navarete[1], A. Damineli[1], C. L. Barbosa[2] and R. D. Blum[3]

[1]IAG-USP, Rua do Matão, 1226, 05508-900, São Paulo, SP Brazil
email: navarete@usp.br

[2]UNIVAP, São José dos Campos, SP, Brazil
[3]NOAO, 950 N Cherry Ave., Tuczon, AZ 85719 USA

Abstract. Formation of massive stars ($M > 8\,M_\odot$) is still not well understood and lacks of observational constraints. We observed 7 MYSO candidates using the NIFS spectrometer at Gemini North Telescope to study the accretion process at high angular resolution (~ 50 mas) and very closer to the central star. Preliminary results for 2 sources have revealed circumstellar structures traced by Brackett-Gamma, CO lines and extended H_2 emission. Both sources present kinematics in the CO absorption lines, suggesting rotating structures. The next step will derive the central mass of each source by applying a keplerian model for these CO features.

Keywords. Stars: formation, Stars: early-type, Stars: pre-main sequence, Stars: circumstellar matter, Techniques: spectroscopic, Techniques: high angular resolution

1. Introduction

The formation mechanisms of massive stars is one of the most important problems in stellar astrophysics and still poorly understood. While low mass young stellar objects (YSOs) reach the zero age main sequence (ZAMS) after the accretion period is finished, the massive ones (MYSOs) reach the main sequence while the accretion is still ongoing.

Recently, Navarete *et al.* (2014) carried a survey of extended H_2 emission towards a well defined sample of MYSO candidates. They found that $\sim 20\%$ of the sample is associated to jet-like structures, suggesting bipolar outflows. These observational evidences supports the accretion scenario (Krumholz *et al.* 2005) and also indicate that massive stars cannot be merely formed by merging of low mass stars (Bonnell *et al.* 2001). Although the scenario of an accretion disk may apply for all massive stars, the details are lacking. There are a few examples of well documented accretion disks around massive (10-20 $M_\odot$) forming stars (Davies *et al.* 2010; Murakawa *et al.* 2013). Disks have been also identified through profile fitting of emission features seen in high spectral resolution data (Bik & Thi 2004; Blum *et al.* 2004). In order to study the details of the circumstellar environment around MYSOs, we present K-band integral field spectroscopy of six MYSOs associated to large scale H_2 outflows, identified by Navarete *et al.* (2014); and one source selected from Varricatt *et al.* (2010).

2. Observations

K-band ($\lambda \approx 2.2$ μm, R≈ 5200) spectra of 7 MYSO candidates were obtained with the Near-infrared Integral Field Spectrograph (NIFS) deployed at the 8-m Gemini North Telescope (Hawaii).

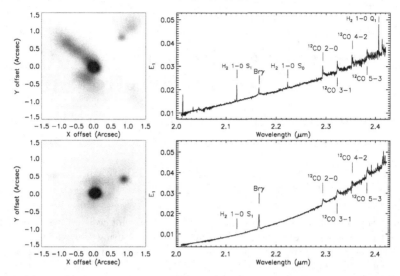

Figure 1. PCA results for two sources. Each component consists in a tomogram (left image) and an eigenspectrum (right plot). Upper panel: First component of the PCA analysis for source #58. Lower Panel: First component for source #A49.

Data reduction was performed using the same methodology described in Menezes *et al.* (2014) and the procedure for applying the PCA analysis on the datacubes are described in Steiner *et al.* (2009).

2.1. *Results*

Figure 1 presents the tomogram (left) and eigenspectrum (right) for the first component of #58 source (upper panel), and components 1 and 4 of #A49 source (middle and bottom panels). The first component of the PCA results basically displays the stellar spectrum, which corresponds to $\approx 90 - 99\%$ of the datacube variance. The source #58 exhibits a series of H_2 transitions, a relatively broad Brγ feature and CO bandhead, all identified in emission. The eigenvector associated to the first component of the #A49 source (middle panel) exhibits emission of both Brγ and CO bandhead. Several absorption features were identified at $\sim 2.4\,\mu$m.

2.2. *Summary and Conclusions*

The K-band IFU spectroscopy of #58 and #A49 sources revealed that: *i*) The circumstellar environment of MYSOs are mainly traced by Br-γ, H_2 and CO bandhead emission (and CO absorption) features; *ii*) there is no clear signs of kinematics in molecular lines except for CO absorption lines at $\sim 2.4\mu$m.

References

Bik, A. & Thi, W. F. 2004, *AAp* 427, L13
Blum, R. D., Barbosa, C. L., Damineli, A., Conti, P. S., & Ridgway, S. 2004, *ApJ* 617, 1167
Bonnell, I. A., Bate, M. R., Clarke, C. J., & Pringle, J. E. 2001, *MNRAS* 323, 785
Davies, B., Lumsden, S. L., Hoare, M. G., *et al.* 2010, *MNRAS* 402, 1504
Krumholz, M. R., McKee, C. F., & Klein, R. I. 2005, *ApJ* 618, L33
Menezes, R. B., Steiner, J. E., & Ricci, T. V. 2014, *MNRAS* 438, 2597
Murakawa, K., Lumsden, S. L., Oudmaijer, R. D., *et al.* 2013, *MNRAS* 436, 511
Navarete, F., Damineli, A., Barbosa, C. L., & Blum, R. D. 2014, *MNRAS*, In preparation
Steiner, J. E., Menezes, R. B., Ricci, T. V., & Oliveira, A. S. 2009, *MNRAS* 395, 64
Varricatt, W. P., Davis, C. J., Ramsay, S., & Todd, S. P. 2010, *MNRAS* 404, 661

New windows on massive stars: asteroseismology, interferometry, and
spectropolarimetry
Proceedings IAU Symposium No. 307, 2014
G. Meynet, C. Georgy, J. H. Groh & Ph. Stee, eds.
© International Astronomical Union 2015
doi:10.1017/S1743921314007388

An X-ray surprise in a magnetic pulsator

Yaël Nazé

FNRS/ULg, Dept AGO, Allée du 6 Août 17, B5C, 4000-Liège, Belgium
email: naze@astro.ulg.ac.be

Abstract. ξ^1 CMa is a rare β Cep star with a strong magnetic field. To gain new insight on this
object, a dedicated campaign using XMM-Newton was performed. These data reveal a new type
of variations, X-ray pulsations, posing a new challenge to our understanding of stellar winds.

Keywords. stars: early-type, stars: oscillations, X-rays: stars, stars: individual (ξ^1 CMa)

1. Introduction

ξ^1 CMa is a monoperiodic radial pulsator with a stable period and a strong magnetic
field ($\sim$5kG) - see Oskinova et al. (2014) and ref. therein. It was detected in the X-ray
range by Einstein, and then observed by ROSAT and XMM (8ks snapshot in 2009): it
is a moderately bright source ($F_X \sim 10^{-12}$ erg cm^{-2} s^{-1}). It was reobserved with XMM
during a long (110ks) run in Oct. 2012 (ObsID=0691900101). The thick filter was used,
to avoid contamination by opt/UV light. The XMM data were reduced in a standard
way with SAS. Their full analysis is presented in Oskinova et al. (2014), here is discussed
the timing analysis (my contribution to the latter paper, with some further details).

2. The X-ray lightcurve

Equivalent on-axis, full PSF lightcurves were extracted for ξ^1 CMa using SAS tasks
(evselect, epiclccorr) for 3 time bins (100, 500, 1000s) and 6 energy bands (0.2–10.,
0.2–0.5, 0.5–0.7, 0.7–1., 0.2–1., 1.–10. keV). First, χ^2 tests were performed for several
hypotheses (constancy, linear variation, quadratic variation, Nazé et al. 2013): all pn
lightcurves are significantly variable (SL <0.01), as are the MOS1 lightcurves with 100s
bins ; remaining MOS1 and MOS2 lightcurves show no significant variations. This difference comes from the many more counts in the pn data which allow more sensitive
detections (see next section). Second, as the pn lightcurves in 0.2–10. keV band seem
to display a periodic signal (Fig. 1), a period search was performed. The calculation of
the autocorrelation function led to the detection of P=17.1±1.5ks, whereas a Fourier
algorithm found a clear peak at P=17.6±0.3ks (Fig. 1). The spectral window is clean,
a prewhitening by this period leaves only noise in the periodogram, and the false error probability associated to the peak is below 2×10^{-8}: the period is therefore real.
Finally, the best-fit sinusoid to the pn lightcurve in 0.2–10. keV band was determined
using χ^2 (Fig. 1). It yields an amplitude of 0.034±0.004 ct/s (the same value as in the
periodogram), corresponding to a pulsed fraction (=amplitude/mean) of 5.0±0.6%, and
a time HJD_0 for the maximum X-ray flux of 2456216.682±0.004.

3. Pulsations at different energies

The Hipparcos catalog lists the following ephemeris for ξ^1 CMa: P=0.209577±0.000001d
and JD_0=2448500.028±0.001. This period corresponds to 18.10745±0.00009ks: the optical period is therefore fully compatible with the X-ray period (Oskinova et al. 2014).

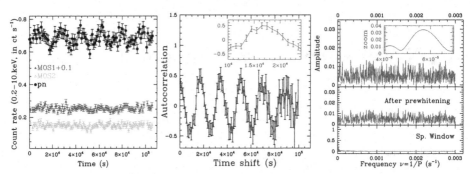

Figure 1. EPIC lightcurves of ξ^1 CMa, with best-fit sinusoid, autocorrelation, and periodogram for pn data.

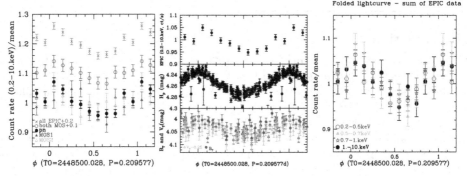

Figure 2. Folded pn, MOS, EPIC, and Hipparcos lightcurves.

The X-ray lightcurves were folded using these ephemeris (Fig. 2). The folded lightcurves of the individual and combined MOS are fully compatible with the presence of the periodic signal, both in amplitude and shape. There is thus no incompatibility between pn and MOS data. The folded lightcurves for the combined EPIC data yields a pulsed fraction of 4.1±0.5% in 0.2–10. keV energy band. Besides, the lightcurves clearly are energy-independent (Fig. 2). The Hipparcos lightcurve is very similar in shape to the combined EPIC lighcurve (Oskinova *et al.* 2014), but with a 3× lower amplitude (min–max=0.037mag, or a pulsed fraction of 1.73±0.07% in flux). A small shift (∼0.08 in phase) is detected, but is not significant in view of the ephemeris uncertainties and the large number of cycles (∼37000) between the observations.

4. Conclusion

X-ray variability has been detected in several massive stars: phase-locked changes in colliding-wind binaries, occultation effects in magnetically confined winds, and variations due to large-scale structures (e.g. CIRs) in single stars. A new type of variations, X-ray pulsations, is now discovered; their origin remains to be explained. Such simultaneous optical/X-ray variations were not seen in other pulsators (e.g. β Cep, β Cen) nor are predicted by theory (Oskinova *et al.* 2014).

References

Nazé, Y., Oskinova, L. M., & Gosset, E. 2013, *ApJ* 763, 143
Oskinova, L. M., Nazé, Y., Todt, H., *et al.* 2014, *Nature Communications* 5, 4024

New windows on massive stars: asteroseismology, interferometry, and spectropolarimetry
Proceedings IAU Symposium No. 307, 2014
G. Meynet, C. Georgy, J. H. Groh & Ph. Stee, eds.

© International Astronomical Union 2015
doi:10.1017/S174392131400739X

New insights on Be shell stars from modelling their Hα emission profiles

J. Silaj[1], C. E. Jones[1], T. A. A. Sigut[1] and C. Tycner[2]

[1]Dept. of Physics and Astronomy, *The* University of Western Ontario
London, ON, Canada N6A 3K7
email: `jsilaj@uwo.ca`

[2]Dept. of Physics, Central Michigan University
Mt. Pleasant, MI 48859, USA

Abstract.
Be shell stars are believed to be ordinary Be stars seen edge-on, which makes them particularly desirable objects for study since the uncertainty in the inclination of the rotation axis is largely eliminated. We have recently modelled high resolution Hα spectroscopic observations for eight Be shell stars, using the non-local thermodynamical equilibrium radiative transfer code BEDISK (Sigut & Jones 2007) and the new spectral synthesis package BERAY (Sigut 2011). Generally, we confirm that these systems are oriented at high inclination angles, although we find that they are not necessarily as close to edge-on as initially expected.

Keywords. stars: emission-line, Be, line: profiles, radiative transfer, circumstellar matter

1. Introduction

The emission that characterizes Be stars arises in a circumstellar disk of gas which receives photoionizing radiation from the central B-type star. Observed Hα profiles take a variety of shapes, but are generally grouped into one of three broad categories: singly-peaked, doubly-peaked, or shell spectra. The different profile shapes are thought to be an effect of the system's orientation to the observer's line of sight, with singly-peaked profiles corresponding to (near) pole-on orientations, doubly-peaked profiles to mid inclination angles, and shell spectra to (near) edge-on orientations (i.e. $i \sim 90°$).

Our disk simulations assume that the density in the equatorial plane takes the form $\rho(r) = \rho_0(r/R_*)^{-n}$, where ρ_0 is the initial disk density at the stellar surface, and n is the power-law index that governs its decrease with increasing distance (r) from the central star. The disk is assumed to be axisymmetric about the star's rotation axis, as well as symmetric about the midplane of the disk. BEDISK computes the thermal structure and atomic level populations of the disk, and BERAY then solves the transfer equation along a series of rays ($\approx 10^5$) through the star+disk system to compute synthetic Hα profiles.

2. Results

Some preliminary results of our modelling are shown in Fig. 1 (full results are in Silaj *et al.* 2014, submitted). Table 1 lists the parameters of the best fit models for these stars. We generally confirm that these stars are oriented at high inclination angles, as the adopted value of i directly affects the central absorption depth of the model, and no suitable models for these observations could be found at low i values. However, in some cases, the best fit value of i is not as close to 90° as initially expected, and in the case of 4 Aql, the observation could only be matched with models created at mid inclination angles.

Table 1. Best fit parameters for the observations shown in Fig. 1.

HR	HD	Name	Sp.Type	n	ρ_0 (g cm^{-3})	i ($^\circ$)
8053	200310	60 Cyg	B1V	3.5	1.8×10^{-11}	65
7040	173370	4 Aql	B8V	2.5	3.3×10^{-12}	46
7836	195325	1 Del	B9V	3.5	3.0×10^{-11}	84
8260	205637	ϵ Cap	B3V	3.0	1.7×10^{-11}	80

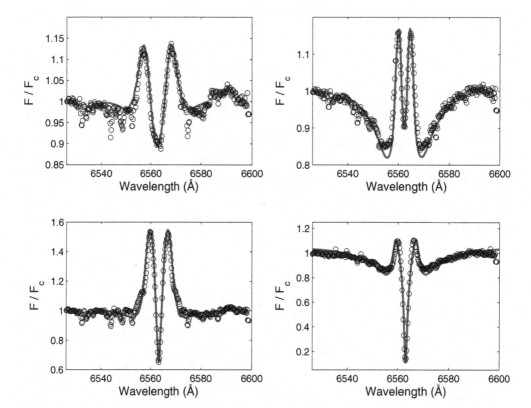

Figure 1. Observed Hα emission lines (open circles) of 60 Cyg, 4 Aql, 1 Del, and ϵ Cap (clockwise from top left). Model spectra, computed with BERAY, are shown in solid lines.

3. Future Work

We are currently modifying MCTRACE (Halonen & Jones 2013) – a Monte Carlo code that uses the output of BEDISK to compute continuum linear polarization – to include spectral lines, thereby allowing us to examine the polarization across the line. We also plan on using this new tool to model asymmetric emission profiles (which are believed to arise from non-axisymmetric density distributions) since non-axisymmetric geometries are relatively easy to adopt in Monte Carlo based simulations.

References

Halonen, R. J. & Jones, C. E. 2013, *ApJ* 765, 17
Sigut, T. A. A. 2011, in C. Neiner, G. Wade, G. Meynet, & G. Peters (eds.), *IAU Symposium*, Vol. 272 of *IAU Symposium*, pp 426–427
Sigut, T. A. A. & Jones, C. E. 2007, *ApJ* 668, 481

New windows on massive stars: asteroseismology, interferometry, and spectropolarimetry
Proceedings IAU Symposium No. 307, 2014
G. Meynet, C. Georgy, J. H. Groh & Ph. Stee, eds.

© International Astronomical Union 2015
doi:10.1017/S1743921314007406

3D and Some Other Things Missing from the Theory of Massive Star Evolution

W. David Arnett

Steward Observatory
University of Arizona
Tucson AZ 85721, USA
email: darnett@as.arizona.edu

Abstract. This is a sketch of a 321D approximation which is nonlocal, and thus has nonzero fluxes of KE (to be published in more detail elsewhere). We plan to add this as an option to MESA. Inclusion of KE fluxes seems to help resolve the solar abundance problem (Asplund *et al.* 2009). Smaller cores may ease the explosion problems with core collapse supernova simulations.

Keywords. convection, turbulence, Sun: abundances, stars: evolution, supernovae

1. Introduction

Stars are three dimensional (3D), turbulent plasma, and much more complex than the simplified one dimensional (1D) models we use for stellar evolution. Computer power is inadequate† at present for adequately resolved (i.e., turbulent) 3D simulations of whole stars. In his review talk, Meakin (2015) has illustrated this complexity and shown how it may be tamed by use of 3D simulations and Reynolds-Averaged Navier-Stokes (RANS) equations.

A minimalist, more approximate step may be easier to implement in stellar evolutionary codes, and instructive for the closure problem. Formally, the RANS equations are incomplete unless taken to infinite order; they must be *closed* by truncation at low order to be useful. This may be due to the nature of the Reynolds averaging, which allows *all* fluctuations rather than restriction only to *dynamically consistent* ones. Closure requires additional information to remove these new, extraneous solutions. As a complement to the RANS approach, approximations which focus on dynamics are emphasized here. In the cascade, turbulent kinetic energy and momentum are concentrated in the largest eddies. We introduce a simple dynamically self-consistent model which contains the largest eddies and the Kolmogorov cascade, and examine the consequences.

Here we will concentrate on the following issues:

(*a*) the turbulent cascade (Kolmogorov 1941, 1962),

(*b*) chaos (Lorenz 1963) and time dependence (Arnett & Meakin 2011b),

(*c*) boundary layers (Prandtl & Tietjens 1934) and surfaces of separation (Landau & Lifshitz 1959),

(*d*) combined mixing and burning, and

(*e*) sensitivity of core collapse to progenitor structure

The important and related issues of coherent treatment of pulsations, eruptions and explosions, rotation and magnetic fields, and variable mass and angular momentum must be deferred.

† But perhaps close; see Herwig *et al.* (2013).

459

Table 1. A Few Examples of Three-dimensional Simulations of Convection in Stars.

Attribute	3D Atmospheres	Solar Convection Zone	Stellar Interiors
representative	Stein & Nordlund[1]	J. Toomre[2]	Meakin & Arnett[3]
photosphere	Y(yes)	N(no)	N(no)
composition gradient	N	N	Y
nuclear burning	N	N	Y
magnetic field	Y	Y	N
rotation	N	Y	N
driving	top	top	top or bottom
geometry	box in star	CZ in box	box in star
boundary inside grid	top only	No	Yes
hydro	compressible	anelastic	compressible

Notes:
[1] See Stein & Nordlund (1998), and Magic *et al.* (2013, 2014) which also refers to other recent work.
[2] See Brun *et al.* (2011, 2004), and many papers in the *IAU Symp.* 271 proceedings (Brun 2011).
[3] See Meakin & Arnett (2007); Viallet *et al.* (2013), and Arnett *et al.* (2014) for an overview.

Erika Böhm-Vitense developed mixing-length theory in the 1950's (Vitense 1953; Böhm-Vitense 1958), prior to the publication in the west of Andrey Kolmogorov's theory of the turbulent cascade (Kolmogorov 1941, 1962). MLT might have been different had she been aware of the original work Kolmogorov had done in 1941. Edward Lorenz showed that a simple convective roll had chaotic behavior (a strange attractor, Lorenz 1963). Ludwig Prandtl developed the theory of boundary layers (Prandtl & Tietjens 1934). All these ideas will be relevant to our discussion, which is based upon theory and 3D simulations.

1.1. *3D Simulations*

Numerical approaches for fluid dynamics simulations include implicit large eddy simulations (ILES) and direct numerical simulation (DNS). DNS resolves the smallest relevant scales, and proceeds to larger scales until limited by computer power available. At present this limit is far smaller than stellar scales, so DNS is important in stellar problems for points of principle (e.g. Wood *et al.* 2013). ILES includes the largest scales in the system, and proceeds to smaller scales until limited by computer power available. Implicit is the assumption that sub-grid scale phenomena are correctly treated. This seems to be valid for 3D turbulence using state-of-the-art methods (Grinstein, F. F. *et al.* 2007).

In order to be useful in a 1D stellar evolution code, 3D information must be projected onto a 1D coordinate system, hence "321D"†. Table 1 gives a brief comparison of features of some 3D simulations which are relevant to designing 1D stellar algorithms. The simulations have different strengths and weaknesses, and tend to complement each other.

3D atmospheres. Simulations of stellar atmospheres in 3D were pioneered by Stein & Nordlund (1998), see also Magic *et al.* (2013) and references therein. These simulations are one of the great successes of radiation hydrodynamics, removing the necessity for the micro-turbulence and macro-turbulent fudge factors previously used to calculate stellar spectra. It is tempting to use MLT to fit such 3D simulations as a "stellar engineering" exercise to connect atmospheres to interiors, because stellar evolution codes are almost always formulated in the language of MLT. Magic *et al.* (2014) give a clear discussion of this process, and show that MLT must be modified in at least one respect to make the identification: a ram pressure term must be added to MLT. It appears that such fits necessarily ignore a quantity important for stellar interiors which comes from the same

† The acronym "321D" for "projection of three dimensions to one dimension" is due to John Lattanzio.

term but in the energy equation. This is the flux of turbulent kinetic energy, which is incorrectly defined as zero in MLT.

These 3D stellar atmospheres generally do not contain the whole convective zone, but use a lower boundary condition which has been shown to have little effect on the predicted spectra. By the same token, this means that these simulations are not a sensitive probe of the convection at the bottom of the convection zone.

Convection, rotation and MHD. Juri Toomre and his students and collaborators have pioneered anelastic simulations of the solar convection zone, focusing on global MHD and differential rotation. For 2D see (Hurlburt *et al.* 1984), and 3D (Brun *et al.* 2004). Unlike the "box in a star" grids used in 3D atmospheres and stellar interiors, such global solutions make more serious demands on computer resources because of their larger extent, so that, other things being equal, their zoning tends to be coarser. Recent results suggest that the numerical viscosity is sufficiently low so that the flows are becoming realistically turbulent (Brun *et al.* 2011). Merging of the insights from these simulations with stellar evolution with rotation is a present challenge.

Stellar Interiors. Another option for use of limited computational ressources is to examine deep interiors in which composition gradients and nuclear burning occur. By using a "box in a star" approach it is possible to include convective zone boundaries within the computational grid. The treatment of radiation flow is simpler (radiative diffusion). Due to neutrino cooling the thermal time scales become shorter, and make thermal relaxation easier to deal with; also higher luminosity as found in deep layers of red giants reduces the thermal time scale. Deep zones (large stratification) and pressure dilatation (Viallet *et al.* 2013) have been examined. The "box in a star" approach truncates the lowest order modes, so that a natural complement is the pioneering work of Paul Woodward (e.g. Herwig *et al.* 2013) to put a "whole star in a box".

2. A 321D Algorithm

This proceeds in several steps:
(*a*) add turbulent cascade,
(*b*) add dynamic (acceleration) equation for integral scale,
(*c*) balance between driving, damping, and the role of turbulent KE flux,
(*d*) make quantitative connections to MLT and the Lorenz model,
(*e*) use the steady-state Lorenz model to approximate average behavior,
(*f*) use acceleration equation to define boundary behavior, and
(*g*) add composition effects.
Because an acceleration equation is used, it is straight-forward (in principle at least) to add inertial forces (centrifugal and Coriolis), Lorenz forces and differential rotation.

2.1. *The Turbulent Cascade*

Arnett *et al.* (2014) estimate the Reynolds number to be $Re \sim 10^{18}$ at the base of the solar convection zone. Numerical simulations and laboratory experiments become turbulent for $Re \sim 10^3$, so fluid flows in stars are strongly turbulent if, as we assume for the moment, rotational and magnetic field effects may be neglected. This special, simpler case is thought to be widely but not universally appropriate to stellar interiors.

For homogeneous, isotropic, and steady-state turbulence, the Kolmogorov relation between the dissipation rate of turbulent kinetic energy, velocity, and length scale is,

$$\epsilon \sim v^3/\ell. \qquad (2.1)$$

This is a global constraint, and applies to each length scale λ in the turbulent cascade, so

$$\epsilon \sim (\Delta v_\lambda)^3/\lambda, \tag{2.2}$$

for all scales λ, or,

$$\Delta v_\lambda \sim (\epsilon\lambda)^{\frac{1}{3}}. \tag{2.3}$$

so that the velocity variation across a scale λ is Δv_λ, and increases as $\lambda^{\frac{1}{3}}$. A description of the cascade needs both large and small scales; Eq. 2.3 implies that the largest (integral) scales have most of the KE and momentum while the smallest have the fastest relaxation times. Simulations confirm this (Arnett *et al.* 2009).

2.2. *Dynamics*

In MLT the buoyant force is approximately integrated over a mixing length to obtain an average velocity u (e.g., Smith & Arnett (2014)),

$$u^2 = g\beta_T \Delta\nabla\left(\frac{\ell_{MLT}^2}{8H_P}\right). \tag{2.4}$$

Working backward, this may be expressed as

$$du/dt = g\beta_T\Delta\nabla - u/\tau, \tag{2.5}$$

where $\ell_d \equiv \ell_{\mathrm{MLT}}^2/8H_P$ and $\tau = \ell_d/|u|$, and for $\Delta\nabla > 0$. Multiplying by u gives a kinetic energy equation,

$$d(u^2/2)/dt = u \cdot g\beta_T\Delta\nabla - u^2/\tau, \tag{2.6}$$

for which the steady-state solution† is Eq. 2.4 (only positive $\Delta\nabla$ are allowed).

Had it been available, Böhm-Vitense might have identified the damping term with the Kolmogorov value (Eq. 2.1). However, Kolmogorov found the damping length ℓ_d to be the depth of the turbulent region, so that it is not a free parameter, unlike MLT. There is a further issue; ϵ is the *average* dissipation rate, not the instantaneous value (u^3/ℓ_d) which fluctuates over time; that is, $u \neq v$ except on average over τ (see Fig. 4 in Meakin & Arnett 2007); we return to this below.

2.3. *Kinetic Energy Flux*

The flow is relative to the grid of the background stellar evolution code, so

$$d(u^2/2)/dt = \partial_t(\mathbf{u}\cdot\mathbf{u})/2 + \nabla\cdot\mathbf{F_{KE}}, \tag{2.7}$$

where $\mathbf{F_{KE}} = \rho\mathbf{u}(\mathbf{u}\cdot\mathbf{u})/2$ is a flux of kinetic energy. The generation of the divergence of a kinetic energy flux in this way is robust for dynamic models; it occurs in the more precise RANS approach as well (Meakin & Arnett 2007). It does not occur in MLT, which assumes symmetry in velocity between upflows and downflows, a condition that is violated at even weak levels of stratification (Meakin & Arnett 2010; Viallet *et al.* 2013). Alternatively, MLT is equivalent to Eq. 2.5 with $du/dt = 0$, the *local* approximation.

Viallet *et al.* (2013); Mocák *et al.* (2014) find a *global* balance between buoyant driving (at the largest scales) and turbulent damping (at the smallest). Because of the large separation of these length scales, they are weakly coupled, giving rise to fluctuating behavior typical of turbulence. This balance (on average only) is a new condition beyond MLT, and allows the elimination of the free parameter in MLT, the mixing length, or $\alpha = \ell/H_P$. The actual flow patterns that correspond to this global balance depend upon the details of the turbulent energy input; nuclear heating and photospheric cooling give different flows (Meakin & Arnett 2010), i.e., fluxes of kinetic energy.

† Care must be taken with the sign of the transit time τ and the deceleration for negative u.

We choose a vector equation for the integral scale velocity **u**,

$$\partial\mathbf{u}/\partial t + (\mathbf{u}\cdot\nabla)\mathbf{u} = \mathbf{g}\beta_T\Delta\nabla - \mathbf{u}/\tau. \tag{2.8}$$

The damping term is chosen to give the Kolmogorov expression for ϵ, the turbulent dissipation, if Eq. 2.8 is dotted by **u** and averaged over a turnover time τ.

2.4. *Connections to MLT and Lorenz*

In the *local, steady-state*, limiting case, the left-hand side of Eq. 2.8 vanishes, and an equation similar to Eq. 2.4 results, but with the mixing length replaced by the turbulent damping length (essentially the lesser of the depth of the convective zone or 4 pressure scale heights, Arnett & Meakin 2011b). With this change, *the cubic equation of Böhm-Vitense may be derived* (Smith & Arnett 2014), and we recover a form of MLT.

If it is assumed that the integral scale motion is that of a convective roll, *Eq. 2.8 may be reduced to the form of the classic Lorenz equations*, but with a nonlinear damping term provided by the Kolmogorov cascade (Arnett & Meakin 2011b). Because of the time lag, the modified equations become even more unstable than the original ones.

For 321D the stellar evolution code must be supplied a smooth time-averaged value for the convective variables; we approximate that by the steady state. The weak coupling between large scale driving and dissipation at the small scale causes time dependent fluctuations of significant amplitude in luminosity and turbulent velocity; see Meakin & Arnett (2007), Fig. 4, for fluctuations in KE in an oxygen burning shell. The term $\partial\mathbf{u}/\partial t$ in Eq. 2.8 is needed for chaotic fluctuations, wave generation and large scale dynamic behavior. A stellar evolution code must step over the shorter turnover time scales (weather) to solve for the evolutionary times (climate); this requires an average over active cells, so that there is a cellular structure in both space and time. The steady-state limit of the Lorenz equation gives a reasonable approximation to its average behavior, filtering out the fluctuations (Arnett & Meakin 2011b); we apply the same approximation to Eq. 2.8 for slow stages of stellar evolution.

2.5. *Boundary Layers*

Fluid motion in a star may be separated into two fundamentally different flows: solenoidal flow (divergence free) and potential flow (curl free) (Landau & Lifshitz 1959). Fig. 1 shows the striking separation in flow velocities at such boundary layers (see Prandtl & Tietjens 1934). We do not claim that we have resolved these boundary layers yet†, but their structure and nature are important to the rate at which turbulent flow moves into or from non-turbulent regions—the entrainment rate.

Peter Eggleton took an early step in dealing with steep gradients in composition with the introduction of an admittedly ad-hoc diffusion operator (Eggleton 1973); this numerically advantageous procedure has been widely adopted for stellar evolution.

A more physical picture results from consideration of the dynamics of the motion. At its most elemental level, the velocity vector must turn at boundaries to maintain solenoidal flow. Most of the momentum is at the largest scales, where dissipation is least and the flow almost adiabatic, so this velocity must be the integral scale velocity. The magnitude of the acceleration required is just the centrifugal value v^2/w where w is the width of the turning region and v the relevant velocity. Using Eq. 2.8 in the steady state limit, and taking $w \sim \Delta r << \ell$, the radial component of the acceleration equation becomes

$$u_r\,\partial u_r/\partial r \sim \Delta(\frac{1}{2}u_r^2)/\Delta r \sim g\beta_T\Delta\nabla. \tag{2.9}$$

† We have fewer than 10 zones across the lower boundary layer; numerical viscosity may affect the computed entrainment rate.

Nonconvective →

← Boundary layer

Body of
Convective
Region

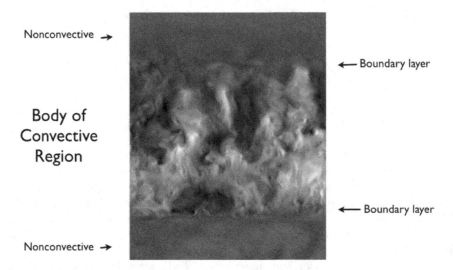

← Boundary layer

Nonconvective →

A snapshot of turbulent convection
Viallet et al. 2013

Figure 1. Boundary Layers enclosing Convection (after Viallet *et al.* (2013).

Sorting out the signs we see that the buoyancy must be negative for the radial kinetic energy to decrease, so that negative buoyancy decelerates the motion, and gives the required turning. This defines a layer at which the velocity goes to zero on average. The "overshoot" region has a width w; this material is mixed because it moves back into the convective region after it turns.

There are fluctuations, so that the layer undulates, generating waves in the neighboring stable region. In a coordinate system moving with the undulating layer, there is a boundary between the mixed (overshoot) region and the overlying layer which is stably stratified. The overshoot region has solenoidal flow which flow in the stably stratified region is potential (wave) flow. By Kelvin's theorem (conservation of circulation), entropy change (e.g., dissipation, radiative diffusion) is required to move matter from one region to the other. Most of the turbulent momentum is in the largest scale convective motions, the integral scale. In simulations we see that these couple well with gravity waves†, whose speed increases with wavelength (Landau & Lifshitz 1959, §12), so that the longer wavelengths carry most of the energy. The shorter wavelengths are more dissipative; they are generated by nonlinear interaction of the long wavelengths (wave breaking; think water waves). This is an example of a mechanism for changing entropy at the boundary, allowing the turbulent region to grow. If this example is representative, such entrainment rates are not universal but depend upon local conditions at the boundary as well as turbulent velocities. See also Viallet *et al.* (2013) discussion of radiative heating at bottom of a deep (strongly stratified) convection zone.

Grolsch (2015) approaches this issue from a quite different point of view, but comes to some similar conclusions.

---

† Sound waves couple well only for higher Mach number flow (Landau & Lifshitz 1959, §64).

2.6. *Mixing and Burning*

Change in composition due to nuclear burning is a fundamental feature of stellar evolution. In general stars do not have uniform composition. In stellar evolution, mixing is determined according to the criterion of Schwarzschild, or of Ledoux. The Schwarzschild criterion is defined as $S = \nabla - \nabla_e$, and the Ledoux criterion is $\mathcal{L} = S - \nabla_Y$, so $\Delta\nabla = \mathcal{L}$ which reverts to S if there is no composition gradient ($\nabla_Y = 0$).

Mixing motions (solenoidal flow) result from buoyancy, pressure perturbations, or differential rotation. We focus on the buoyant acceleration, $-\mathbf{g}\rho'/\rho$. Composition gradients enter the buoyancy on an equal basis with entropy gradients; for the simplest case of an ideal gas, $\rho'/\rho = T'/T + Y'/Y - P'/P$, where $Y = 1/\mu$ is the number of free particles per baryon, or inverse mean molecular weight. For a more general equation of state, there are multiplicative factors of order unity on the r.h.s. Traditionally P'/P is taken to be zero even though this is incorrect for strongly stratified convection (Viallet *et al.* 2013).

Brunt frequency, which is important for asteroseismology, is $N^2 = -\mathcal{L}(g\beta_T/H_P)$, indicating a fundamental connection with boundary conditions (Aerts, C. *et al.* 2010).

3. Solar Abundances

In the steady-state but non-local case, the $(\mathbf{u}\cdot\nabla)\mathbf{u}$ term gives a coupling between driving regions and damping regions in the form of a flux of turbulent kinetic energy (Meakin & Arnett 2007, 2010); this is assumed to be zero by symmetry in MLT (a major flaw). *Even moderate stratification breaks the symmetry, and gives a finite flux of turbulent kinetic energy.*

Any model of the Sun which uses MLT will neglect the effect of turbulent kinetic energy fluxes. To estimate the sign and size of the necessary changes, we scale from the simulations in Viallet *et al.* (2013) for a convection zone which like the Sun is highly stratified. The luminosity due to kinetic energy is $L_{KE} \sim -0.35L$ near the base of the convection zone, which is significant. This negative luminosity must be balanced by increased positive radiative luminosity, to maintain a solar luminosity, so $L^{new}/L_\odot = 1.35$. This may obtained by a reduction in opacity. The opacity is well known, and depends upon the composition, primarily the metal abundance. If we simply assume that the opacity scales with metalicity, $\kappa^{new}/\kappa^{old} \sim 3/4$, and the actual metalicity should shift from the stellar evolution value (from the "standard solar model", SSM) to $z^{new} \sim 0.02(3/4) \sim 0.015$, which is the value determined from 3D atmospheres. Asplund *et al.* (2009) conclude that the discrepancy between abundances determined from 3D atmospheres, and from stellar evolution is unidentified. The argument just given suggests that there is a significant flaw in the physics of SSM, correction for which tends to resolve the discrepancy. Until stellar models including kinetic energy flux are constructed, it seems reasonable to take the 3D atmospheric abundances as our best estimate, and attribute the disagreement to use of a local convection theory.

4. Sensitivity to Progenitor Structure

Simulations of core collapse may be sensitive to convection algorithms used to construct precollapse models. Couch & Ott (2013) get increased tendency toward explosion simply by adding a *nonradial* velocity component to the progenitor model (Arnett & Meakin 2011a), as turbulence requires. This also suggests changes to the initial models used by Kochanek (2014).

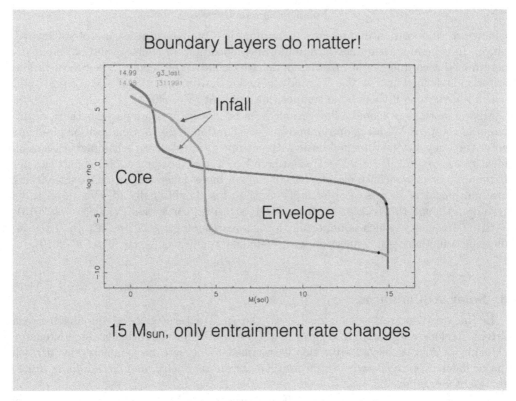

Figure 2. Changes in Density structure of $15\,M_\odot$ with Entrainment.

Fig. 2 shows two TYCHO models at the end of oxygen burning. The density structure is sensitive to the assumed entrainment physics. The two curves represent stars of $15\,M_\odot$, with only the entrainment physics changed. Not only are the sizes of the cores changed, but the carbon/oxygen ratio is affected by entrainment as well.

Core-collapse simulations often fail because of the high rate of infall of mantle matter onto the newly formed, nuclear density core. This matter must be photo-dissociated by the explosion shock if an explosion is to occur. Lower rates of infall aid the explosion. The rate of infall is

$$\dot{M}_{in} = u_{in}4\pi r_{in}^2 \rho_{in}. \tag{4.1}$$

The infall velocity is essentially the local sound speed in the progenitor mantle; matter falls in as a rarefaction wave moves out. To make the rate of mass infall $\dot{M}_{in}$ small, the initial density ρ_{in} should be small. The critical time occurs during the infall of the mantle (labeled "Infall"); the envelope falls in too slowly to affect the explosion shock. Clearly the curve in Fig. 2 with the larger core will have larger ρ_{in}, and be harder to explode. This large core resulted from taking the maximum entrainment rate that was energetically allowed, throughout the evolution from the main sequence (for simplicity no mass loss or binary stripping were considered).

The small core case resulted from an entrainment rate of 0.01 of the maximum, so that the boundary dynamics was almost elastic, and consistent with analytic estimates. Smaller rates and smaller cores are possible. The highest resolution 3D simulations we have done were the most nearly elastic in the boundary layers, so we regard this as the more reasonable case. It has smaller cores that conventional progenitor models; ad-hoc diffusion smoothes gradients, leading to larger cores.

5. Conclusions

This is a sketch of a 321D approximation which is nonlocal, and thus has nonzero fluxes of KE (to be published in more detail elsewhere). We plan to add this as an option to MESA. Inclusion of KE fluxes seems to help resolve the solar abundance problem (Asplund *et al.* 2009). Smaller cores may ease the explosion problems with core collapse supernova simulations.

References

Aerts, C., Christensen-Dalsgaard, J., & Kurtz, D. W. 2010, *Asteroseismology*, Springer
Arnett, D., Meakin, C., & Young, P. A. 2009, *ApJ* 690, 1715
Arnett, W. D. & Meakin, C. 2011a, *ApJ* 733, 78
Arnett, W. D. & Meakin, C. 2011b, *ApJ* 741, 33
Arnett, W. D., Meakin, C., & Viallet, M. 2014, *AIP Advances* 4(4), 041010
Asplund, M., Grevesse, N., Sauval, A. J., & Scott, P. 2009, *ARA&A* 47, 481
Böhm-Vitense, E. 1958, *ZfA* 46, 108
Brun, A. S. (ed.) 2011, *Astrophysical Dynamics: From Stars to Galaxies*, Vol. S 271, International Astronomical Union, CUP
Brun, A. S., Miesch, M. S., & Toomre, J. 2004, *ApJ* 614, 1073
Brun, A. S., Miesch, M. S., & Toomre, J. 2011, *ApJ* 742, 79
Couch, S. M. & Ott, C. D. 2013, *ApJ (Letters)* 778, L7
Eggleton, P. P. 1973, *MNRAS* 163, 279
Grinstein, F. F., Margolin, L. G., & Rider, W. J. (eds.) 2007, *Implicit Large Eddy Simulations*, Cambridge University Press
Grolsch, A. 2015, in *Proceedings of this conference*
Herwig, F., Woodward, P. R., Lin, P.-H., Knox, M., & Fryer, C. 2013, *ArXiv e-prints*
Hurlburt, N. E., Toomre, J., & Massaguer, J. M. 1984, *ApJ* 282, 557
Kochanek, C. S. 2014, *MNRAS*
Kolmogorov, A. 1941, *Akademiia Nauk SSSR Doklady* 30, 301
Kolmogorov, A. N. 1962, *Journal of Fluid Mechanics* 13, 82
Landau, L. D. & Lifshitz, E. M. 1959, *Fluid mechanics*, Pergamon
Lorenz, E. N. 1963, *Journal of Atmospheric Sciences* 20, 130
Maeder, André 1999, *Physics, Formation and Evolution of Rotating Stars*, Springer
Magic, Z., Collet, R., Asplund, M., *et al.* 2013, *A&A* 557, A26
Magic, Z., Weiss, A., & Asplund, M. 2014, *ArXiv e-prints*
Meakin, C. 2015, in *Proceedings of this conference*
Meakin, C. A. & Arnett, D. 2007, *ApJ* 667, 448
Meakin, C. A. & Arnett, W. D. 2010, *Ap&SS* 328, 221
Mocák, M., Meakin, C., Viallet, M., & Arnett, D. 2014, *ArXiv e-prints*
Prandtl, L. & Tietjens, O. G. 1934, *Applied Hydro-& Aeromechanics*, Dover Publications, Inc.
Smith, N. & Arnett, W. D. 2014, *ApJ* 785, 82
Stein, R. F. & Nordlund, Å. 1998, *ApJ* 499, 914
Viallet, M., Meakin, C., Arnett, D., & Mocák, M. 2013, *ApJ* 769, 1
Vitense, E. 1953, *ZfA* 32, 135
Wood, T. S., Garaud, P., & Stellmach, S. 2013, *ApJ* 768, 157

Discussion

MAEDER: Analytical studies suggest that transport processes by meridional circulation and by shear diffusion are strongly affected by the horizontal turbulence in differentially rotating stars. What can the 3D simulations tell us about the possible value of the horizontal turbulence?

ARNETT: Great question! We do not seem to see Jean-Paul Zahn's strongly asymmetric diffusive mixing velocities. However the overturn is effective at mixing the whole convective region, except for Si burning, which has "cellular" burning regions. The shellular approximation may be fine for earlier stages, up to and including oxygen burning, if we can get the boundary motion—the entrainment rates—right. The shellular approximation will break down during explosive or eruptive events due to 3D instabilities, of course.

Now suppose we spin up a non-rotating star. Turbulence is turbulence, whether driven by buoyancy or by differential rotation. At first we get slight distortions of spherical equipotentials, leading to meridional circulation, and the solenoidal flow also reacts to the inertial accelerations. Eq. 2.8 is a 3D vector equation, and contains these effects; the velocity includes the velocity of meridional circulation and the horizontal turbulence, at least in principle. An interesting issue is the relative importance of dissipation in the boundary layers versus in the bulk of the turbulent flow.

As the spin increases (Rossby number decreases) it will have an increasingly important influence on flow, convection, and angular momentum transport, and the importance of MHD becomes an issue (our code does not have MHD active at present). One can think of a continuous sequence from non-rotating star to accretion disk, parameterized by angular momentum, and the literature contains examples of 3D simulations all along this sequence. That said, the results are complex, and it seems that we still have much to learn.

KHALAK: How may stellar rotation be introduced into the equation for turbulence?

ARNETT: It is already in Eq. 2.8 if we do the usual transformation to a rotating frame. This is a deceptively simple *vector* equation (the Navier-Stokes equation in the turbulent limit), which has deep connections to a lot of theoretical work on rotating stars. André Maeder has written a beautiful development of the physics of rotating stars (Maeder, André 1999) which starts from the Navier-Stokes equation (his Eq. 1.2).

MORAVVEJI: In the very vicinity of the fully mixed core and the radiative layers (braking and/or entrainment layer?), what is the behavior of $\Delta\nabla$ term in your energy balance equation?

ARNETT: Thanks, this is an important point that I went through too quickly. The answer may be implicit in §2.3 and §2.4. The $\Delta\nabla$ is a factor in the buoyant acceleration, and can change sign. It depends upon both the temperature gradient and the composition gradient. However mixing makes the composition gradient tend to zero, which changes $\Delta\nabla$ itself; this problem must be solved implicitly. In the simulations there is a transition layer in which the composition does change, but it may not yet be resolved numerically (the "boundary layer" in Fig. 1). This is the layer where the velocity field changes from solenoidal to potential flow, i.e., from convection to waves. See also the contribution by Arlette Noels (Grolsch 2015).

The simulations show additional complexity: the boundary is dynamic and has vigorous and fluctuating wave motion, features not in MLT. If we assume that the boundary dynamics has only a slow secular variation on average (an "entrainment" model, Meakin & Arnett 2007), we might use Eq. 2.9 and asteroseismology to try to make progress.

Dave Arnett

New windows on massive stars: asteroseismology, interferometry, and spectropolarimetry
Proceedings IAU Symposium No. 307, 2014
G. Meynet, C. Georgy, J. H. Groh & Ph. Stee, eds.

© International Astronomical Union 2015
doi:10.1017/S1743921314007418

Asteroseismology of Massive Stars: Some Words of Caution

A. Noels[1], M. Godart[2], S. J. A. J. Salmon[1], M. Gabriel[1], J. Montalbán[1] and A. Miglio[3]

[1] Institut d'Astrophysique et de Géophysique, Liège University, Allée du 6 Août, 17, B-4000 Liège, Belgium
email: Arlette.Noels@ulg.ac.be

[2] Dept. of Astronomy, The University of Tokyo, 7-3-1 Hongo, Bunkyo-ku Tokyo, 113-0033, Japan

[3] School of Physics and Astronomy, University of Birmingham, Birmingham, B15 2TT, UK

Abstract. Although playing a key role in the understanding of the supernova phenomenon, the evolution of massive stars still suffers from uncertainties in their structure, even during their "quiet" main sequence phase and later on during their subgiant and helium burning phases. What is the extent of the mixed central region? In the local mixing length theory (LMLT) frame, are there structural differences using Schwarzschild or Ledoux convection criterion? Where are located the convective zone boundaries? Are there intermediate convection zones during MS and post-MS phase, and what is their extent and location? We discuss these points and show how asteroseismology could bring some light on these questions.

Keywords. stars: evolution, stars: interiors, stars: oscillations (including pulsations), stars: variables: massive stars, stars: mass loss,

1. Introduction

The key role played by asteroseismology in probing and understanding the stellar structure of massive stars is undeniable. Thanks to excited modes, in particular those propagating in the deep interior, the extent of the mixed central region can be constrained. This region consists of the fully mixed convective core surrounded by an "extra-mixed" region, either fully or partially mixed. The physical origin of this region can be rotation, overshooting... What asteroseismology will test is of course not the physics but the resulting chemical composition profile. In Sect. 2, we show how the extent of the extra-mixed region can only be reliably determined by asteroseismology provided some rather strong requirements are fulfilled not only on the number of well identified modes but also on the detailed elemental abundances of the observed star (Salmon 2014).

One of the big issues extensively discussed in this symposium is the treatment of convection in massive stars. However, most stellar evolution codes still compute convective regions in the frame of the local mixing length theory (LMLT) with either Ledoux or Schwarzschild criterion. We first show that whatever the adopted criterion, the extent of convective cores must be identical (see Sect. 3.1).

Asteroseismic analyses require fully consistent models, in particular models with correctly located convective boundaries, *i.e.* satisfying the convection criterion on its convective side. As shown in Gabriel *et al.* (2014), a departure from this requirement leads to too small convective cores and erroneous Brunt-Väisälä frequency distributions, with a possible misleading of the asteroseismic interpretation. This is discussed in Sect. 3.2.

In Sect. 3.3 we address the important point of the exact location and extent of convective shells in the vicinity of the hydrogen burning shell in supergiant B stars and we discuss the different aspects of these shells resulting from the choice of Ledoux or Schwarzschild convection criterion. We raise some problems (semi-convection, overshooting, non-existence of a static solution) that can appear when a convective shell develops (Gabriel *et al.* 2014).

We then show in Sect. 4 how recent interesting asteroseismic analyses can help bring new constraints on these shells (Saio *et al.* 2006; Gautschy 2009; Godart *et al.* 2009, 2014; Saio *et al.* 2013; Georgy *et al.* 2014).

Sect. 5 addresses the problem of exciting β Cephei- and SPB-type modes in the Small Magellanic Cloud, with a metallicity so low that no such modes (only very few SPB-type) are indeed theoretically expected. This could be due to an underestimation of the Ni opacity in the "metal opacity bump" (Salmon *et al.* 2012).

2. Amount of extra-mixing in MS and core helium burning stars

In what follows we shall define the extra-mixing as an additional full mixing taking place on top of the convective core, resulting from various physical causes such as convective overshooting or penetration, rotation, semi-convection, diffusion... The extent of this extra-mixing region is referred to as a fraction α_{em} of the pressure scale height.

As was shown and discussed in session II of this symposium (Aerts 2014), several asteroseismic analyses of β Cephei stars have revealed quite a large range of values for this parameter α_{em}, from 0 to ~ 0.5. The method used is a χ^2 minimisation of the differences between the observed frequencies and those found in a large grid of models with various stellar parameters, in particular various values of α_{em}. No direct relation between this latter parameter and another stellar property, such as the rotation velocity, seems to exist.

In an attempt to better understand the link between α_{em} and the fitting method, Salmon (2014) has realized a series of *hare-and-hound* exercises. Targets were computed with various choices of stellar parameters $(M, R, X, Z, \alpha_{em})$ and different physical assumptions, such as the solar mixtures AGSS05 (Asplund *et al.* 2005) or GN93 (Grevesse & Noels 1993) and the opacities OPAL (Iglesias & Rogers 1996) or OP (Badnell *et al.* 2005). A set of non-adiabatic frequencies were computed for each target, to serve as "observed" frequencies. The grid of models used to minimize χ^2 for each target had fixed physical assumptions, *i.e.* AGSS05 for the solar mixture and OP for the opacities. Adiabatic frequencies were available for each model of the grid. We cite below some of the conclusions drawn from these exercises:

• If the physics is the same in the target and in the grid, five *well identified* (well known) observed frequencies are required to recover the stellar parameters, in particular α_{em}. With only three frequencies, the error on α_{em} can reach a factor two.

• If among those five frequencies, one of them is not correctly identified, the errors are as large as in the case where only three frequencies are observed. The minimum of χ^2 is however much larger than when a good solution is obtained.

• If the target is built with the solar mixture GN93 instead of that of the grid (AGSS05), a 50 % error is found on α_{em}. Here also the minimum of χ^2 presents a rather large value that could be used as an alarm criterion to detect an incorrect solution.

It should however be noted that, without asteroseismic data, the amount of extra-mixing in massive stars is impossible to obtain since there is a well-known degeneracy in M and α_{em} for a given set of T_{eff} and L. But we think useful to stress the importance of combining asteroseismic data with spectroscopic and photometric data in order to reach a higher level of constraints on the model. Moreover, "best" solutions obtained with a too large χ^2 should be regarded with caution.

3. Convection in LMLT frame

In this section, we would like to emphasize, and call attention on some problems related to convective zones and, in particular to convective boundaries, which seem to be met in some stellar evolution codes. Since some codes, like the MESA code for instance (Paxton *et al.* 2013), are now being used on a larger and larger scale by the astrophysical community, extra care should be given to the outputs in order to help the developers to still increase the performances of the code. Another reason to be extremely careful is that an asteroseismic analysis requires a fully consistent model, at the risk of misinterpreting ateroseismic data if this exigence is not fully met.

3.1. *Convection criterions*

In spherically symmetric stars, the condition to define the boundary of a convective zone is (see Gabriel *et al.* 2014, and references therein)

$$V_r = 0 \qquad (3.1)$$

where V_r is the radial component of the convection velocity. In the frame of the LMLT, this implies

$$L_{\rm rad} = L(m) \text{ and } \nabla_{\rm rad} = \nabla_{\rm ad} \qquad (3.2)$$

where $L_{\rm rad}$ is the radiative luminosity and $L(m)$, the total luminosity at mass fraction m; ∇ stands for usual temperature gradient and the indices *rad* and *ad* refer to radiative and adiabatic. This is the so-called Schwarzschild criterion. Let us visualize a convective boundary as a two-sided spherical surface, with a "convective" and a "radiative" side. The condition 3.2 is obviously only meaningful in a convective region and thus it must be applied *on the convective side* of the boundary.

On the radiative side of the convective boundary, the situation is different since condition 3.2 becomes

$$L_{\rm rad} = L(m) \text{ and } \nabla_{\rm rad} = \text{ or } \neq \nabla_{\rm ad} \qquad (3.3)$$

depending on the presence, or not, of a discontinuity in chemical composition.

For a radiative layer to become convective, the Ledoux criterion† should be applied. For an equation of state $P = \mathcal{R}\rho T/\beta\mu$ with $\beta = P_{\rm g}/P$ and μ, the mean molecular weight, it writes

$$\nabla_{\rm rad} > \nabla_{\rm ad} + \left(\frac{\beta}{4-3\beta}\right)\frac{{\rm d}\ln\mu}{{\rm d}\ln P} = \nabla_{\rm Ldx} \qquad (3.4)$$

If the physical conditions within a layer are such that $\nabla_{\rm rad}$ is in between $\nabla_{\rm ad}$ and $\nabla_{\rm Ldx}$, the layer is generally assumed to be semi-convective‡. In the absence of a special treatment of semi-convection in the code, such a layer will be assumed to be convective (radiative) if the Schwarzschild (Ledoux) criterion is adopted throughout the code. In MS massive stars for instance, the μ-gradient region is characterized by a quasi-equality of $\nabla_{\rm rad}$ and $\nabla_{\rm ad}$. Small convective zones, leading to a step-like hydrogen profile, will generally appear with the Schwarzschild criterion while a smooth hydrogen profile is maintained throughout MS with the Ledoux criterion.

---

† Convection is the result of the non-linear development of the linear instability of dynamically unstable gravity modes, which is only the case if condition 3.4 is satisfied.

‡ Semi-convection is the result of the non-linear development of the linear vibrational instability of dynamically stable gravity modes.

3.2. *Location of convective boundaries*

Whatever the convection criterion adopted in the code, the convective boundary is located by finding the zero of the function y, *i.e.*

$$\begin{aligned} y &= \nabla_{\text{rad}} - \nabla_{\text{ad}} = 0 \quad \text{or} \\ y &= \nabla_{\text{rad}} - \nabla_{\text{Ldx}} = 0 \end{aligned} \tag{3.5}$$

on the convective side of the boundary. However, a discontinuity in y may be present at the boundary. With the Schwarzschild criterion, this happens when the chemical composition is discontinuous, *i.e.* during the early MS phase of low mass stars ($\sim 1.1 - 1.6 M_\odot$) when the convective core is growing in mass, and during core helium-burning phase. With the Ledoux criterion, this happens all the time since, except in homogeneous models, the μ-gradient is discontinuous at the boundary.

Finding the zero of a discontinuous function by interpolation or by locating a change of sign (or checking the sign layer by layer) leads to a infinity of solutions. Whatever the location of the discontinuity, *i.e.* the location of the boundary, the condition on the change of sign will be satisfied. However, the only acceptable solution in the frame of LMLT is the one for which $\nabla_{\text{rad}} = \nabla_{\text{ad}}$ *on the convective side* of the boundary. This indeed means that the radial convective velocity is zero and that the convective flux is accordingly zero at the boundary. This condition leads to $L_{\text{rad}} = L(m)$, which must be satisfied on both sides of the boundary. The way to proceed is to *extrapolate* from points located exclusively in the convective region (see the more complete discussion in Gabriel *et al.* 2014).

The extent of convective cores of MS massive stars should be exactly identical, whatever the adopted convection criterion, if an extrapolation procedure is applied. With an interpolation (or change of sign) scheme however, the mass extent can be very different. If, at a given iteration, the estimated convective boundary is smaller than the real boundary, such a scheme will not be able to move the estimated boundary outward since the condition 3.5 will appear to be satisfied. This (too small) location will be accepted by the code as the correct one since from one iteration to the next, it will not move anymore.

Fig. 1 illustrates this behavior in the case of a $16\,M_\odot$ MS star computed with the code CLÉS (Scuflaire *et al.* 2008) and the Ledoux criterion. The hydrogen profile is drawn in long dashed line, ∇_{rad} in full line, ∇_{ad} in dashed line and ∇_{Ldx} in dotted line. In the left (right) panel, the location of the convective core boundary has been obtained through an extrapolation (interpolation) scheme. One can easily see that the extent of the convective core (fully mixed region) is smaller when an interpolation scheme is used. At the end of MS, the difference can reach $\sim 25\%$ depending on the distribution of mesh points for instance. It is also very clear that the model displayed in the right panel is not coherent since at the convective boundary, condition 3.5 is in fact not satisfied. Moreover, on top of the too small convective core, a misidentified semi-convective region covering the whole μ-gradient region is found. This of course will lead to significantly different Brunt-Väisälä frequency distributions (see the behavior of ∇_{Ldx} in both panels).

Fig. 2 shows the helium profile (dotted line), ∇_{rad} (full line) and ∇_{ad} (dashed line) in a core helium-burning star of $8\,M_\odot$ computed with the code CLÉS and the Schwarzschild criterion. In the left (right) panel, the location of the convective core boundary results from an extrapolation (interpolation) scheme. The boundary of the convective core displayed in the left panel is fully consistent since it satisfies condition 3.5. A convective boundary similar to that displayed in the right panel of Fig. 2 was met and discussed by Castellani *et al.* (1971). With ∇_{rad} larger than ∇_{ad} at the boundary, V_r is not equal to 0 and L_{rad} is still smaller than $L(m)$ on the convective side of the boundary, while it must

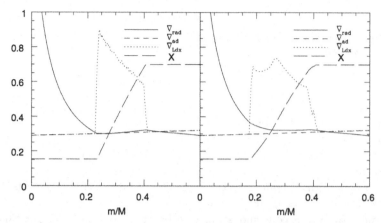

Figure 1. Hydrogen profile (long dashed line), radiative (full line), adiabatic (dashed line) and Ledoux (dotted line) temperature gradients, as a function of the fractional mass m/M, for an MS model of $16\,M_\odot$, computed with an extrapolation scheme (left panel) and computed with an interpolation scheme (right panel), using in both cases the Ledoux criterion

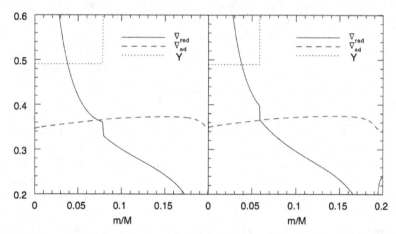

Figure 2. Helium profile (dotted line), radiative (full line), adiabatic (dashed line) temperature gradients, as a function of the fractional mass m/M, for a core helium-burning model of $8\,M_\odot$, computed with an extrapolation scheme (left panel) and computed with an interpolation scheme (right panel), using in both cases the Schwarzschild criterion

obviously be equal to $L(m)$ on the radiative side. Indeed with the transformation of helium into carbon and oxygen, the discontinuity becomes larger and larger and Castellani *et al.* (1971) showed that an increase of the convective core mass was the only way out of such an unstable situation. This is indeed obtained when an extrapolation procedure from convective points only is implemented in the code.

It is clear that, with an underestimation of the extent of the convective core, the Brunt-Väisälä frequency distribution will be affected and as a result, the asteroseismic properties expected from core helium-burning stars will also be impacted.

3.3. *Convective shells*

If the Ledoux criterion predicts a convective shell in a μ-gradient region, the consistency of the model can be very difficult to achieve (see Sect. 7 in Gabriel *et al.* 2014). Fig. 3 illustrates the change in the hydrogen profile arising from the occurrence of such a convective shell located in between mesh points j_1 and j_2. The opacity is noted κ and

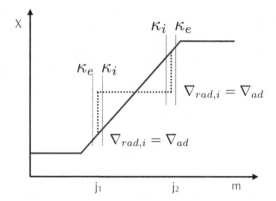

Figure 3. Schematic illustration of a convective shell developing in a μ-gradient region described by the hydrogen profile X as a function of m. κ stands for the opacity coefficient and the indices i and e refer to the inner and external sides of the convective boundaries

the indices i and e refer to the inner and external sides of the convective boundaries. The main problems are listed below:

• Two discontinuities arise except if the opacity, κ, is independent of the chemical composition. One of them will necessarily be such that $\kappa_e > \kappa_i$ and since $\nabla_{rad,i} = \nabla_{ad}$ on the inner side of both boundaries, $\nabla_{rad,e}$ will be greater than ∇_{ad} at the external side of either j_1 or j_2. This means that a semi-convective region may develop either below j_1 or above j_2.

• Since consistent boundaries imply $L_{rad} = L(m)$ on both sides, L_{rad} must necessarily decrease above j_1 and increase when reaching j_2. This might not be the case except maybe in nuclear burning shells or in the vicinity of an opacity peak.

• When consistency is not met, one of the boundaries is such that $\nabla_{rad} > \nabla_{ad}$ and $V_r \neq 0$. An overshooting or undershooting will take place and the chemical composition will change inside the shell, which will move along the μ-gradient region.

Once more, the consistency should be checked at each convective boundary in order to correctly interpret the asteroseismic analyses.

4. Intermediate convective zones (ICZ) in B supergiants

As can be seen in the left panel of Fig. 1, MS models of massive stars are characterized by a near equality of ∇_{rad} and ∇_{ad} in the μ-gradient region. Once the H-burning shell develops, the rapid increase of $L(m)/m$ creates a region where $\nabla_{Ldx} > \nabla_{rad} > \nabla_{ad}$, *i.e.* a semi-convective zone, which is treated as convective if the Schwarzschild criterion is used. This intermediate convective zone (ICZ) overlaps the region of nuclear energy production.

Fig. 4 shows such an ICZ in a post-MS star of $16\,M_\odot$ computed with CLÉS and the Schwarzschild criterion. The dashed line is the hydrogen profile, ∇_{rad} and ∇_{ad} are respectively drawn in black and gray full lines and the Brunt-Väisälä frequency (in log) is shown in dotted line.

The left panel illustrates a model computed without any extra-mixing nor any mass loss. If an extra-mixing is added on top of the convective core during MS, the near equality of both temperature gradients is replaced by $\nabla_{rad} < \nabla_{ad}$ in the μ-gradient region and for large enough extra-mixing, the ICZ can become very small and be disconnected from the nuclear burning region. The middle panel of Fig.4 is an illustration of this influence of an extra-mixing ($\alpha_{em} = 0.2$). If the models are computed with mass loss, the near

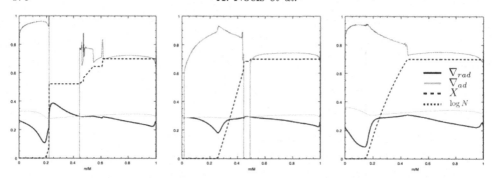

Figure 4. Hydrogen profile (dashed line), radiative (full line) and adiabatic (gray line) temperature gradients, and log N (dotted line) (N is the Brunt-Väisälä frequency) for a post-MS model of 16 $M_\odot$ still quite close to the MS turn-off, , computed with the Schwarzschild criterion. Left panel: no extra mixing, no mass loss. Middle panel: extra-mixing with $\alpha_{em} = 0.2$. Right panel: mass loss with $\dot{M} = 2 \cdot 10^{-7} M_\odot/\mathrm{yr}$. The regions where N is equal to zero are convective

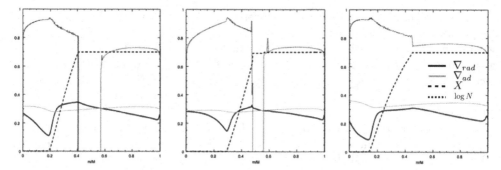

Figure 5. Hydrogen profile (dashed line), radiative (full line) and adiabatic (gray line) temperature gradients, and log N (dotted line) (N is the Brunt-Väisälä frequency) for a post-MS model of 16 $M_\odot$ still quite close to the MS turn-off, computed with the Ledoux criterion. Left panel: no extra mixing, no mass loss. Middle panel: extra-mixing with $\alpha_{em} = 0.3$. Right panel: mass loss with $\dot{M} = 2 \cdot 10^{-7} M_\odot/\mathrm{yr}$. The regions where N is equal to zero are convective

equality is also affected since the μ-profile is less steep and this lessens the increase of the opacity, leading to smaller values of ∇_{rad}. With a high enough mass loss rate, the ICZ does not appear, as can be seen in the right panel of Fig. 4 ($\dot{M} = 2 \cdot 10^{-7} M_\odot/\mathrm{yr}$).

When the Ledoux criterion is used, the semi-convective zone is treated as radiative and the ICZ is limited to the base of the homogeneous hydrogen-rich envelope. This is illustrated in the left panel of Fig. 5. For the reasons discussed here above, the ICZ becomes smaller when an extra-mixing is added during MS (middle panel, $\alpha_{em} = 0.2$), and it disappears when a rather high mass loss rate is applied (right panel, $\dot{M} = 2 \cdot 10^{-7} M_\odot/\mathrm{yr}$). With the Ledoux criterion, whatever the extent of the ICZ, its location is disconnected from the nuclear burning region.

The importance of an ICZ in post-MS stars has been emphasized by the discovery of excited g-modes in a B supergiant observed by the MOST satellite (Saio *et al.* 2006). The role played by the ICZ is that of a barrier preventing g-modes from entering a very stabilizing helium core. The huge density in the central layers of such post-MS stars is indeed responsible for a strong radiative damping and in a star devoid of an ICZ, no g-modes can be excited (see also Gautschy 2009; Godart *et al.* 2009). The effects of extra-mixing and mass loss have been investigated and thoroughly discussed by Godart *et al.* (2009) (see also Godart *et al.* 2014, and references therein).

Radial pulsations observed in α Cygni variables are also good indicators of the presence of an ICZ. As discussed in Saio *et al.* (2013), a high value of L/M is required in order to obtain excited radial modes with periods compatible with those observed in α Cygni stars. Such high L/M can only be met in massive stars coming back from the red supergiant (RSG) stage, after a heavy mass loss undergone as a RSG. Moreover for two stars, Rigel and Deneb, the superficial N/C and N/O ratios show nitrogen and oxygen enrichments typical of CNO burning. With models computed with the Schwarzschild criterion, Saio *et al.* (2013) obtained far too strong such enrichments, resulting from the overlap of the ICZ and the H-burning shell. More recently, Georgy *et al.* (2014) computed similar models with the Ledoux criterion and showed that, with the disconnection of the ICZ and the H-burning shell, the enrichments were much closer to the observations.

The *power of asteroseismology* is here clearly seen in those two examples:

• One single g-mode observed in a B supergiant star is a signature of an ICZ and therefore brings constraints on the amount of extra-mixing and mass loss.

• The detection of radial modes, coupled with a detailed spectroscopic analysis, in an α Cygni variable star, not only lifts the degeneracy between supergiant models crossing the Hertzsprung-Russell diagram from blue to red and vice-versa but also imposes constraints on the location and mass extent of the ICZ.

5. Opacity in the metal opacity bump

Since the excitation mechanism in β Cephei and SPB stars is the κ-mechanism acting in the metal opacity bump at $T \simeq 2 \cdot 10^5$ K, it is no surprise that metallicity plays a key role in the extension of their instability strips in the Hertzsprung-Russell diagram. It was indeed shown by Miglio *et al.* (2007) that at $Z = 0.005$ the instability strip for β Cephei stars does no longer exist while a very narrow strip still remains for SPB stars. Those results actually predict that no β Cephei and very few SPB stars should be expected in the Magellanic Clouds.

Challenging observations of such stars in the Small Magellanic Cloud (SMC) have however been presented by Karoff *et al.* (2008); Diago *et al.* (2008); Kourniotis *et al.* (2014). The possible explanations for this apparent contradiction have been investigated by Salmon *et al.* (2012) who concluded their analysis by suggesting that *the current opacity data are underestimating the stellar opacity due to nickel by a factor* ~ 2 *(in the metal opacity bump)*. This has indeed been an additional incentive element to proceed to new revisions of the stellar opacities. Recent comparisons between different opacity codes presented by Turck-Chièze & Gilles (2013) are very encouraging since they evoke a possible increase of the nickel opacity in the metal opacity bump by a factor similar to that proposed by Salmon *et al.* (2012).

This is not the first time that variable stars have demanded a revision in opacity data. Simon (1982)'s plea was followed by a successful update, which turned out to be the key factor in explaining the excitation mechanism in β Cephei stars. As for now, the OP, OPAC and OPAL teams are undertaking new opacity computations, especially in physical conditions typical of B-type and solar-like stars.

6. Conclusions

Asteroseismology is definitely a powerful tool to bring some light on some unsolved problems affecting the structure and evolution of massive stars. However, we have tried with a few selected examples to call attention on some aspects of stellar modeling as well as asteroseismic analyses, which could lead to misinterpretations of seismic data:

- Our *first word of caution* relates to the asteroseismic estimation of the extent of extra-mixing on top of the convective core in β Cephei stars. Although not sufficient the following requirements must be met in order to have a really reliable determination:
  - at least 4-5 *well identified* frequencies must be observed;
  - the metallicity, and more precisely the detailed elemental abundances, should be known.
- Our *second word of caution* sets in the LMLT frame of convection. For a theoretically asteroseismic analysis to be reliable, fully consistent models are absolutely required. This implies that:
  - the condition on convective neutrality defining the boundary of a convective region must be applied on the *convective side* of the boundary;
  - when this boundary is affected by a discontinuity in the chemical composition (Schwarzschild criterion) or in the μ-gradient (Ledoux criterion), the location of the boundary must result from an *extrapolation* from convective points only;
  - the boundaries of each intermediate convective zones should systematically be checked for consistency.
- Our *third word of caution* addresses the physical data entering stellar model computations. It certainly is another success of asteroseismology to have been at the origin of a new, and still in progress, opacity revision. This stresses the necessity to be open-minded to all the tools available in order to unveil the physical processes inside massive stars, *i.e.* not only asteroseismology but also spectroscopy and photometry for detailed elemental abundances and global stellar properties, as well as the most updated physical data related to nuclear reactions and opacities.

References

Aerts, C. 2014, in G. Meynet, C. Georgy, J. H. Groh, & P. Stee (eds.), *IAU Symposium*, Vol. 307 of *IAU Symposium*

Asplund, M., Grevesse, N., & Sauval, A. J. 2005, in T. G. Barnes, III & F. N. Bash (eds.), *Cosmic Abundances as Records of Stellar Evolution and Nucleosynthesis*, Vol. 336 of *Astronomical Society of the Pacific Conference Series*, p. 25

Badnell, N. R., Bautista, M. A., Butler, K., *et al.* 2005, *MNRAS* 360, 458

Castellani, V., Giannone, P., & Renzini, A. 1971, *Ap&SS* 10, 340

Diago, P. D., Gutiérrez-Soto, J., Fabregat, J., & Martayan, C. 2008, *A&A* 480, 179

Gabriel, M., Noels, A., Montalban, J., & Miglio, A. 2014, *ArXiv e-prints*

Gautschy, A. 2009, *A&A* 498, 273

Georgy, C., Saio, H., & Meynet, G. 2014, *MNRAS* 439, L6

Godart, M., Grotsch-Noels, A., & Dupret, M.-A. 2014, in J. A. Guzik, W. J. Chaplin, G. Handler, & A. Pigulski (eds.), *IAU Symposium*, Vol. 301 of *IAU Symposium*, pp 313–320

Godart, M., Noels, A., Dupret, M.-A., & Lebreton, Y. 2009, *MNRAS* 396, 1833

Grevesse, N. & Noels, A. 1993, in B. Hauck, S. Paltani, & D. Raboud (eds.), *Perfectionnement de l'Association Vaudoise des Chercheurs en Physique*, pp 205–257

Iglesias, C. A. & Rogers, F. J. 1996, *ApJ* 464, 943

Karoff, C., Arentoft, T., Glowienka, L., *et al.* 2008, *MNRAS* 386, 1085

Kourniotis, M., Bonanos, A. Z., Soszyński, I., *et al.* 2014, *A&A* 562, A125

Miglio, A., Montalbán, J., & Dupret, M.-A. 2007, *MNRAS* 375, L21

Paxton, B., Cantiello, M., Arras, P., *et al.* 2013, *ApJS* 208, 4

Saio, H., Georgy, C., & Meynet, G. 2013, *MNRAS* 433, 1246

Saio, H., Kuschnig, R., Gautschy, A., *et al.* 2006, *ApJ* 650, 1111

Salmon, S. 2014, *Ph.D. thesis*, University of Liège, Belgium

Salmon, S., Montalbán, J., Morel, T., *et al.* 2012, *MNRAS* 422, 3460

Scuflaire, R., Théado, S., Montalbán, J., *et al.* 2008, *Ap&SS* 316, 83

Simon, N. R. 1982, *ApJ (Letters)* 260, L87

Turck-Chièze, S. & Gilles, D. 2013, in *European Physical Journal Web of Conferences*, Vol. 43 of *European Physical Journal Web of Conferences*, p. 1003

Discussion

AERTS: Just a general remark for the majority of the audience about the hare-and-hound exercises: you should also compare the uncertainty on the model parameters (M, X, Z, α_{em}) between the case where you have only T_{eff} and $\log g$, with errors of 1000 K and 0.2 dex, and the case where you also have a few identified modes, including, or not, a misidentification and/or wrong input physics.

NOELS: I fully agree that most of the time, the addition of 4 or 5 well identified modes will highly increase the level of constraints on the theoretical model best fitting the observed star. This is indeed the power of asteroseismology to probe deep inside the star, down to the centrally mixed region. This lifts the well-known degeneracy of possible M and α_{em} for a given set of T_{eff} and L. However, I want to call attention on a possible misinterpretation of asteroseismic data if the physics is not the same in the grid and in the observed target. It is of course impossible to check all the possible differences but an extra care should be given to the detailed chemical composition. This is indeed, in my opinion, an *important word of caution*: to increase the relevance and the power of asteroseismic analyses, all the tools should enter the game, and among them, detailed spectroscopic and photometric analyses definitely play key roles.

Arlette Noels

New windows on massive stars: asteroseismology, interferometry, and spectropolarimetry
Proceedings IAU Symposium No. 307, 2014
G. Meynet, C. Georgy, J. H. Groh & Ph. Stee, eds.

© International Astronomical Union 2015
doi:10.1017/S174392131400742X

Interferometry of massive stars: the next step

Ph. Stee[1] A. Meilland[1] and O. L. Creevey[2]

[1] Observatoire de la Côte d'Azur - Boulevard de l'Observatoire - CS 34229 - F 06304 Nice
Cedex 4 - France
email: Philippe.Stee@oca.eu

[2] Institut d'Astrophysique Spatiale, Université Paris XI, UMR 8617, CNRS, Batiment 121,
91405 Orsay Cedex, France

Abstract. We present some new and interesting results on the complementarity between asteroseismology and interferometry, the detection of non-radial pulsations in massive stars and the possibility for evidencing differential rotation on the surface of Bn stars. We also discuss the curretn interferometric facilities, namely the Very Large Telescope Interferometer (VLTI)/AMBER, VLTI/MIDI, VLTI/PIONIER within the European Southern Observatory (ESO) context and the Center for High Angular Resolution Astronomy (CHARA) array with their current limitations. The forthcoming second-generation VLTI instruments GRAVITY and MATISSE are presented as well as the FRIEND prototype in the visible spectral domain and an update of the Navy Precision Optical Interferometer (NPOI). A conclusion is presented with a special emphasis on the foreseen difficulties for a third generation of interferometric instruments within the (budget limited) Extremely Large Telescope framework and the need for strong science cases to push a future visible beam combiner.

Keywords. instrumentation: interferometers, instrumentation: high angular resolution, techniques: interferometric, stars: fundamental parameters, stars: kinematics, stars: emission-line, Be, stars: winds, outflows, stars: rotation, stars: spots, stars: imaging

---

1. Introduction

I will not talk about the future of interferometry in space (Fridlund 2000b) since this is a very long term prospective and most interesting missions in space have been rejected up to now, e.g. DARWIN (Fridlund 2000a), PEGASE (Ollivier *et al.* 2009), SIM (Marr 2006), SIM-Lite (Goullioud *et al.* 2008). I will rather present the forthcoming future instruments which will be operated on the Very Large Telescope Interferometer (VLTI) and on the Center for High Angular Resolution Astronomy (CHARA) array as well as a brief overview of the current improvement of the Navy Precision Optical Interferometer (NPOI). But we first start with an example on how important interferometric radius measurements are to improve the determination of stellar masses. We will demonstrate that interferometry is really complementary to asteroseismic analysis and can help to better constrain stellar fundamental parameters.

2. Asteroseismology & Interferometry

2.1. *Improving constraints on fundamental parameters.*

As already outlined by Creevey *et al.* (2007), interferometry and asteroseismology are providing complementary observations to determine the mass of stars. This is certainly true for solar type main sequence stars for which we have many asteroiseismic measurements. For massive stars the situation is more difficult since we don't have many asteroseismic constraints, which reinforce the importance of an accurate radius determination by interferometric measurements. Creevey *et al.* (2007) demonstrate that coupling a radius measurement of solar type stars with oscillation data improves the determination of the star's fundamental parameters while also reducing their uncertainties. While solar-like oscillations have been detected in hundreds to thousands of stars, these have been observed mostly from space and range in magnitudes from $7 < V < 14$. Interferometry in the visible or infrared (IR) bands, however, has a limiting magnitude of roughly $V = 7$. With new instruments under development, e.g. FRIEND (Berio *et al.* in prep.), and adaptive optics currently being installed on the CHARA Array (ten Brummelaar *et al.* 2005), which hosts some of the longest baselines to observe the smallest angular diameters, the magnitude limit could be extended to $V < 11$. On the other end, the recently selected ESA M3 mission *PLATO* (Rauer *et al.* 2013), with a planned launch at the end of 2024, will search for planetary transits around bright stars.

The complementarity of interferometry and asteroseismology is based on the fact that the measured p-mode oscillations and their global seismic quantities that characterise the spectrum provides a direct relation with the mean density of the star, since the measured frequency spacing $\langle \Delta \nu \rangle$ from asteroseismology is proportional to the square root of the mean density leading to Eq. 2.1.

$$\langle \Delta \nu \rangle \simeq \sqrt{\frac{M}{R^3}} \langle \Delta \nu \rangle_\odot \qquad (2.1)$$

If the radius R is measured independently and we apply Eq. 2.1 directly to determine the mass M of the star, we find that for a star similar to the Sun the mass can be determined with a theoretical precision as good as 2.5% if we consider the best possible scenario $[\sigma(R) = 0.5\%$ and $\sigma(\Delta \nu) = 1\%]$, or to 5–6% for more conservative uncertainties $[\sigma(R) \sim 1 - 1.5\%$ and $\sigma(\Delta \nu) = 2\%]$. This is already an improvement of at least a factor of two from using the scaling relations alone. Needless to say, the use of individual frequency separations and the frequencies themselves along with the measured radius yields uncertainties in the mass as good as 1–2%, but again model-dependent as looked in detail below.

Creevey *et al.* (2007) conducted a sensitivity analysis to study the impact of an interferometric radius measurement on the determination of stellar model parameters by invoking structure and evolution models and using different types of seismic information. Asteroseismic data provide so much more information characterizing the structure of solar type stars that we are now in a position to determine these properties independent of any external assumptions. However, asteroseismic data is sensitive to the density structure of the star and thus can constrain well the determination of the relative M-to-R relation. With R measured independently, the determination of M and hence the other stellar parameters can improve.

In Fig. 1 we show the predictions of the mass uncertainty (y-axis) as the precision in the radius improves (x-axis) for a star like the Sun. The different lines represent the results when different observational constraints are at hand. In all cases, spectroscopic parameters were included (effective temperature T_{eff}, metallicity [M/H], surface gravity $\log g$) with errors of 100 K, 0.05 dex and 0.1 dex, respectively. These values are typical of what can be found in the recent literature. The circles show the results when the asteroseismic data consist of just $\langle \Delta \nu \rangle$ and $\langle \delta \nu \rangle$, the squares using the full range of individual $l = 0$ frequency separations $\Delta \nu_{l=0}$, and the diamonds when the individual $l = 0$ frequencies are used. The predictions illustrate that the inclusion of the radius as a constraint is always very useful. Even including several individual oscillation modes,

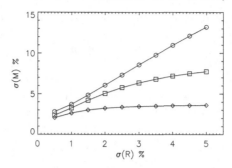

Figure 1. Predictions of the mass uncertainty (y-axis) as the precision in the measured radius improves (x-axis), when spectroscopic and asteroseismic data are also at hand. The three curves include spectroscopic constraints and the asteroseismic constraints differ from curve to curve. Circles correspond to using just the global seismic quantities, squares to including the $l = 0$ large frequency separations, and diamonds to including the $l = 0$ individual modes.

$< 2\%$ precision in the radius leads directly to an improvement in the mass uncertainty and in fact for a 0.5% radius error, the mass can be determined to 2%.

As this sensitivity study is a linear approach that depends just on the local description of the χ^2 surface, such results were then tested with simulations in Creevey et al. (2007). In the example that follows a slightly more massive star of $1.03\,M_\odot$ is considered with an age of 1 billion years (one quarter that of the Sun). Using a theoretical set of observables (frequencies, $T_{\rm eff}$, .. etc.), observational data were simulated by adding typical errors that were scaled by a random number drawn from a Gaussian distribution. These simulated data were then fitted to model data by employing a Levenberg-Marquardt minimization. The results are a set of optimal model parameters (mass M, age, metallicity Z, initial H mass fraction X, mixing-length parameter α). This was repeated 50 times for each set of observations and associated errors. The observations always included the radius and spectroscopic constraints but also 1) the large frequency separations $\Delta\nu_{l,n}$ with typical errors on the individual frequencies of 0.5 μHz and and 2) the small frequency separations $\delta\nu_{l,n}$ with errors on the frequencies typical of ground-based data (1.3 μHz).

The results from the simulations are shown in Fig. 2. The continuous lines illustrate case 1 and the dashed line for the α parameter denotes case 2. Each point in the figure corresponds to the standard deviation of the fitted parameter for a given radius error. Four of the five parameters are shown in this figure and they are marked on the right side of the figure. Inspecting the results for Z it can be seen that the radius has no impact on improving its uncertainty, it remains at a constant $\sim 10\%$, with a similar result for α ($\sim 8\%$). However, it is clear to see that the uncertainty in the mass does improve from an average of 3% when the seismic information dominates to 1.5% when the error in the radius reaches 0.5%. This much higher precision in the mass leads directly to an improvement in the determination of X ($< 2\%$) or, similarly, the initial He surface abundance (remember $X+Y+Z = 1$). Including the small frequency separations, we find that the determination of α is also influenced by an improved radius measurement and reaches a mere 4% when this parameter is otherwise entirely unconstrained in models.

While this study highlights how important a radius measurement is to improve the determination of the stellar mass, it is of equal importance to make a direct comparison of interferometrically-determined radii with those predicted from detailed asteroseismic analysis alone.

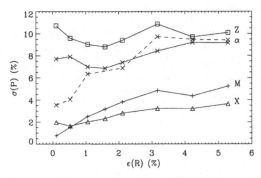

Figure 2. The uncertainties in stellar parameters (mass M, initial metallicity Z, initial hydrogen mass fraction X, and mixing-length parameter α) as the precision in the star's measured radius improves, while considering seismic data comprising the large frequency separations (continuous) and including the small frequency separations (dashed).

2.2. *Direct detection of Non Radial Pulsations of massive stars.*

The study of the surface of rapidly rotating stars with strong non-radial oscillations using Differential Speckle Interferometry (DSI) technique was already proposed by Vakili & Percheron (1991). The application of this technique to the study of surface structure is particularly well justified for the stars at high inclinations (Petrov 1988). Jankov *et al.* (2001) treated explicitly the case of non-radial stellar pulsations, for which the cancellation of opposite sign temperature or velocity fields introduces difficulties, and showed that interferometric constraint introduces the crucial improvement. In fact, the photocenter shift provides the first order moment of the spatial brightness distribution and (comparing to the zero order moment spectroscopic information) the corresponding stellar regions are reinforced by weighting with the coordinate parallel or orthogonal to rotation. Consequently, the modes that are cancelled in flux spectrum should appear in the spectrally resolved photocenter shift data. Of course, the correct detection and identification of modes present in the star is crucial for a credible asteroseismological analysis.

Following this technique, Jankov *et al.* (2001) propose to detect non-radial pulsations by Differential Interferometry using the dynamic spectra of photocenter shift variability characterized by bumps traveling from blue to red within the spectral lines. The theoretical estimation of expected signal-to-noise ratios in differential speckle interferometry (Chelli 1989) demonstrated the practical applicability of the technique to a wide number of sources. For instance, the high precision CHARA/VEGA or VLTI/AMBER measurements of differential fringe phase corresponding to $\sim 10\ \mu$as photocenter shift (standard limit), makes measurements feasible for the low order pulsation of $20\ \mathrm{km\,s^{-1}}$ and considering a star at an intermediate inclination ($i = 45°$), which should yield a signal of about $\sim 25\ \mu$as for the typical angular diameter of η Cen ($\sim 0.5\,$mas). Moreover, for η Cen $i \sim 70°$ and, since the star is tilted at high inclinations, the expected signals, calculated by Jankov *et al.* (2001), should be even stronger, particularly for low order pulsations, improving the feasibility.

3. Measuring differential rotation on the surface of Bn stars

Be stars are non-supergiant stars that have exhibited at least once Balmer lines in emission. These emission lines are produced in a gaseous circumstellar environment. Be stars are very rapid rotators. However, the characteristics of their rotational rate (rigid or differential) and their role in the formation of circumstellar discs are still highly debated

(Frémat *et al.* 2005; Cranmer 2005). Consequently, the role of rotation in the ejection of stellar matter and the formation of the circumstellar environment from it is still an open issue (Stee & Meilland 2009). In the past few years, significant progresses in the understanding of the structure of the circumstellar environment of Be stars were made using the new generation of spectro-interferometric instruments, mainly AMBER installed on the VLTI and VEGA installed on CHARA. For the first time, these instruments allowed to probe quantitatively the kinematics of the circumstellar gas. All the studies that were conducted (Meilland *et al.* 2007a; Meilland *et al.* 2007b; Carciofi *et al.* 2009; Delaa *et al.* 2011; Meilland *et al.* 2011) concluded that the disk kinematics was dominated by rotation and that expansion velocities were too small to be detected (i.e. of the order of $1\,\mathrm{km\,s^{-1}}$). Finally, the first statistical study of the disks kinematics by Meilland *et al.* (2012) showed that the rotation law within the disk was nearly Keplerian and that the inner boundary of the disk (i.e. near the stellar surface) was rotating very close to the stellar critical velocity, with an angular rotation velocity relative to critical of $\frac{\Omega}{\Omega_c} \sim 0.95$.

However, even though $\frac{\Omega}{\Omega_c}$ can be very high it still underestimates the actual stellar rotation, which is more clearly evidenced by the η parameter. η is defined as the ratio of the centrifugal and the gravitational forces at the stellar equator. Thus, the measurement of η is crucial if we would like to know how close are Be stars to critical rotation. Unfortunately, Be stars can have photospheric spectral lines marred by emission/absorptions due to their circumstellar disc. Bn stars, which are nearly as fast rotators as Be stars, don't have spectra perturbed by circumstellar matter, so that the study of their apparent geometry can be carried out more properly and reliably.

Due to the rapid rotation, the surface geometry of Bn and Be stars is highly deformed. The surface angular velocity can be dependent on the stellar latitude. Then, not only the centrifugal force acting on the stellar surface, but also the effective temperature distribution (Von Zeipel effect) should depend on the surface rotation law. Up to now, these effects have been measured only for few stars (Domiciano de Souza *et al.* 2003; Monnier *et al.* 2007; Zhao *et al.* 2009; Che *et al.* 2011) and different values of the β exponent of the gravity darkening law were deduced using image reconstruction techniques(Monnier *et al.* 2007; Zhao *et al.* 2009; Che *et al.* 2011). Delaa *et al.* (2013) has demonstrated that a differential rotation at the surface of fast rotating star may affect the brightness distribution of the stellar disc and, so, also modify the value of the β exponent. Consequently, a value of β different from 0.25 (for a fully a radiative envelope; von Zeipel 1924) can also be due to a non-conservative rotational law in the outermost stellar layers (Zorec *et al.* 2011) and can be constrained by spectrally resolved interferometric measurements. The latitude dependent angular velocity also introduces peculiar characteristics to the spectral line differential phases. These last are sensitive to the stellar inclination angle and to the surface angular velocity law (Delaa *et al.* 2013).

4. The Very Large Telescope Interferometer (VLTI)

The VLTI, located on Paranal (Chile), is equipped with two general-user focal instruments: the Astronomical Multi-BEam combineR (AMBER; Petrov *et al.* 2007; Robbe-Dubois *et al.* 2007) and the MID-infrared Interferometric instrument (MIDI; Leinert *et al.* 2003) beam combiners. MIDI can combine only 2 telescopes in the $8-13\,\mu\mathrm{m}$ spectral bandwidth with resolving power R=30 and 230, and can measure visibilities and differential phases. The spatial resolution is about 10 mas. AMBER operates in the near-IR bands (J, H,and K) and can combine 3 telescopes simultaneously with 3 spectral resolution modes: R= 30, 1500 and 12000. It can measure visibilities, differential phases and phase closure. Its spatial resolution is about 2 mas.

4.1. *Current strengths and limitations*

The VLTI is currently the best angular resolution facility accessible in Europe. Equipped with 8m class telescopes (UT), the magnitude limit in the K band (2.2 μm) is now ∼ 11 (Petrov *et al.* 2012). After a huge investment, it is now working well and the 1.8 m Auxiliary Telescopes (AT) array is very powerful. It is operated within the ESO framework and thus benefits from various facilities such as service mode observations, archives, maintenance and good stability. This instrument is very powerful for the study of multiplicity (e. g., massive stars, brown dwarfs, YSOs), temporal variability of various objects (e. g., YSOs, Novae, evolved stars, MIRAS, Be stars) and benefits from a very active and well organized community (e. g., Schools, European Interferometric Initiative, Jean-Marie Mariotti Center – JMMC). Two new instruments are coming up (MATISSE and GRAVITY) and the VLTI offers a potential synergy with ALMA.

On the other hand, the community is small and partly "interferocentric". The accessibility to reduced data is still a problem 10 years after the start of operations. Long-baselines are missing, which is a strong limitation for resolving most of the stellar surfaces (a typical diameter for a B-type star is about 0.5 mas within our Galaxy). The funding context within the Extremely Large Telescope (ELT) era is hostile and the ESO current organization context (no more Paranal VLTI group, probably no more Garching VLTI group) weakens the possibility to develop the VLTI in a holistic way. Regarding the present focal instruments, MIDI can only estimates sizes of objects (no real information on the geometry with only two telescopes). It has no imaging capabilities nor (or few) gas study possibilities. AMBER has a limited accuracy and limited imaging capabilities and dust analysis are limited to (very) hot dust. In addition, the $K-$ and $N-$bands have been used a lot and we need to open new wavelength windows. We also have to notice that with 2 telescopes we can measure only two quantities (1 visibility + 1 differential phase), with 3 telescopes we have access to 7 measurements (3 visibilities, 3 differential phases, 1 closure phase), with 4 telescopes we have 16 measurements accessible (6 visibilities, 6 differential phases, 3 closure phases and 1 closure amplitude) and thus going from 3 to 4 telescopes more than doubles the number of available measurements. This is currently the success of the new visiting Precision Integrated-Optics Near-Infrared Imaging ExpeRiment (PIONIER) instrument which can combine the light from 4 telescopes simultaneously in the near-infrared (1.5 − 2.4 μm) domain (Le Bouquin *et al.* 2011).

4.2. *The forthcoming VLTI instruments GRAVITY and MATISSE*

GRAVITY is a four-way beam combination, second generation instrument for the VLTI. Its main operation mode makes use of all four 8 m Unit Telescopes to measure astrometric distances between objects located within the 2" field-of-view of the VLTI (Eisenhauer *et al.* 2011). With the sensitivity of the UTs (8 m) and the ∼ 10 μas astrometric precision (for a 5 mn integration), it will allow to measure orbital motions within the galactic center with unprecedented precision. GRAVITY is dedicated to the study of the back hole at the center of our Galaxy and will carry out the ultimate empirical test to show whether or not the Galactic Centre harbors a black hole of four million solar masses. It will also provide a high-precision narrow-angle astrometry and phase-referenced interferometric imaging in the astronomical K-band. The resolving power will be $R = 22$, 500, and 4000, which will be very interesting for the study of circumstellar disks around massive stars and interacting binaries or Be-X ray sources. Other modes of the instrument will allow imaging and the use of the ATs. It will allow fringe tracking up to

K $\sim$ 10 with the UTs and K $\sim$ 7 with the ATs. It will be able to produce images up to K $\sim$ 16 (UTs) and K $\sim$ 13 (ATs) for a 100 s integration. It will start operations in mid-2015.

The Multi AperTure mid-Infrared SpectroScopic Experiment (MATISSE) is a mid-infrared ($L-$, $M-$ and $N-$bands; i.e. $3.7 - 5\,\mu$m and $8 - 13\,\mu$m) spectro-interferometer project combining up to four UTs/ATs beams of the VLTI. It will measure closure phase relations, thus offering an efficient capability for image reconstruction. This mid-infrared instrument will include several efficient spectroscopic modes ranging from $R = 20$ to 1500. The typical spatial resolution will be 5–40 mas for baselines ranging from 8 to 150 m. MATISSE is supposed to be operational just after GRAVITY in 2016.

5. Update of other interferometric arrays

5.1. *The CHARA array*

CHARA is located at Mount Wilson Observatory, California and is an optical interferometric array of six 1m telescopes. The longest available baseline of 330m makes the CHARA array the highest spatial resolution instrument in the world. Actually, 7 beam combiners are operational:

- CHARA Classic: 2 telescopes open air beam combiner in J, H and K band, limited magnitude of 6.5 in K.
- CHARA CLIMB: 3 telescopes open air in J,H,K
- JouFLU (upgrade of FLUOR: 2 telescopes, fiber based in K band, limited magnitude of 4-5 in K
- MIRC: 6 telescopes fiber based imager in H and K, spectral resolution R: 35, 150, 450, limited magnitude of 4-5 in H
- VEGA: 4 telescopes, open air in V, R and I with a spectral resolution up to 30000, limited magnitude of 7 in V
- PAVO: 3 telescopes, aperture plane in V, R and I bands, limited magnitude of 8 in V
- CHAMP: 6 telescopes pair wise fringe tracker in H and K

At the Observatoire de la Côte d'Azur we are trying to improve the VEGA beam combiner limitations, such as mainly no closure phase possible, difficulties to measure low visibilities and limitations of the measurement's accuracy due to the multimode regime, the saturation effect of the photon counting detector and the photon centroiding "hole" (Mourard et al. 2009; Mourard et al. 2011). We are developing a prototype named "FRIEND" for Fibered and spectrally Resolved Interferometric Experiment - New Design (Berio et al. 2014, in preparation) that can benefit from the forthcoming adaptive optics on the CHARA array. The FRIEND prototype will combine 3 telescopes with spectrally dispersed fringes, i.e. R=1500 (as VEGA and AMBER) in the R band ($0.6 - 0.8\,\mu$m). It will use spatial filtering thanks to fibers (as MIRC and AMBER), a V-groove (as MIRC) and simultaneous photometry (as AMBER). Finally the fringes will be sampled thanks to a low noise ($<$ rev) visible detector OCAM2. The predicted performances will be as good as $m_v \sim 10$ with the full CHARA adaptive optics system. First tests of FRIEND at CHARA are foreseen mid december 2014 with tip-tilt only. The long term perspective of FRIEND is to combine up to 6 telescopes (from the VLTI or CHARA) in the visible with 300 m baselines which will enable stellar surface imaging of massive stars.

5.2. *The NPOI interferometer*

NPOI is a six 12.5 cm apertures array on a Y shape with baselines up to 79m located near Flagstaff, Arizona (Armstrong *et al.* 1998). It has been recently upgraded with the completion of 6 stations "imaging" (portable) siderostat array: new enclosures for star acquisition and tip-tilt optics have been installed for 5 of 6 stations. The news domes are installed for 5 of 6 imaging siderostats and 2 more imaging stations will be commissioned in 2014. A long baseline up to 432 m and a "compact" configuration will be available. The "classic" combiner operates at low spectral resolution (16 spectral channels from 550 nm to 850 nm) with a limiting magnitude of $m_v \sim 6$. The second beam combiner VISION uses single-mode polarization-preserving fibers to spatially filter the incoming beams and a EMCCD (Electron Multiplying Charge Coupled Device). It allows a full 6-way combination with flexible spectral resolution ($R = 500, 2000$) and a visibility measurement precision 10 times better than the "classic" beam combiner. Last but not least, they will replace the 12.5 cm siderostats by 1.8m telescopes that have been built by NASA for the Keck interferometer. The site install will start in October 2014.

6. Conclusion

With the forthcoming second generation of VLTI instruments, the CHARA array equipped with adaptive optics and the promising NPOI interferometer, the massive star research will benefit from new imaging facilities, good spectral coverage and very high spatial resolution that will enable the imaging of massive star surfaces in our Galaxy. Nevertheless, within the ELT framework and the limited budget allocated to future projects, the third generation of interferometers will be very competitive. For instance, due to the limitation of the Paranal mountain, no baselines larger than ~ 200 m are foreseen. Thus a possible way to increase the spatial resolution without increasing the baseline is to observe at smaller wavelengths. I.e. going from the near-IR to visible will increase the spatial resolution by a factor 4! On the other side, third generation instruments will only be possible if we are able to extend the community to non-interferometry people and convince ESO or other institutions that we have very strong science cases for a visible instrument. Thus, if you are interested, don't hesitate to join us and contact our group for future collaborations.

References

Armstrong, J. T., Mozurkewich, D., Rickard, L. J., *et al.* 1998, *ApJ* 496, 550
Carciofi, A. C., Okazaki, A. T., Le Bouquin, J.-B., *et al.* 2009, *A&A* 504, 915
Che, X., Monnier, J. D., Zhao, M., *et al.* 2011, *ApJ* 732, 68
Chelli, A. 1989, *A&A* 225, 277
Cranmer, S. R. 2005, *ApJ* 634, 585
Creevey, O. L., Monteiro, M. J. P. F. G., Metcalfe, T. S., *et al.* 2007, *ApJ* 659, 616
Delaa, O., Stee, P., Meilland, A., *et al.* 2011, *A&A* 529, A87
Delaa, O., Zorec, J., Domiciano de Souza, A., *et al.* 2013, *A&A* 555, A100
Domiciano de Souza, A., Kervella, P., Jankov, S., *et al.* 2003, *A&A* 407, L47
Eisenhauer, F., Perrin, G., Brandner, W., *et al.* 2011, *The Messenger* 143, 16
Frémat, Y., Zorec, J., Hubert, A.-M., & Floquet, M. 2005, *A&A* 440, 305
Fridlund, C. V. M. 2000a, in B. Schürmann (ed.), *Darwin and Astronomy : the Infrared Space Interferometer*, Vol. 451 of *ESA Special Publication*, p. 11
Fridlund, C. V. M. 2000b, in P. Léna & A. Quirrenbach (eds.), *Interferometry in Optical Astronomy*, Vol. 4006 of *Society of Photo-Optical Instrumentation Engineers (SPIE) Conference Series*, pp 762–771

Goullioud, R., Catanzarite, J. H., Dekens, F. G., Shao, M., & Marr, IV, J. C. 2008, in *Society of Photo-Optical Instrumentation Engineers (SPIE) Conference Series*, Vol. 7013 of *Society of Photo-Optical Instrumentation Engineers (SPIE) Conference Series*

Jankov, S., Vakili, F., Domiciano de Souza, Jr., A., & Janot-Pacheco, E. 2001, *A&A* 377, 721

Le Bouquin, J.-B., Berger, J.-P., Lazareff, B., *et al.* 2011, *A&A* 535, A67

Leinert, C., Graser, U., Przygodda, F., *et al.* 2003, *Ap&SS* 286, 73

Ligi, R., Mourard, D., Lagrange, A. M., *et al.* 2012, *A&A* 545, A5

Marr, IV, J. C. 2006, in *Society of Photo-Optical Instrumentation Engineers (SPIE) Conference Series*, Vol. 6268 of *Society of Photo-Optical Instrumentation Engineers (SPIE) Conference Series*

Meilland, A., Delaa, O., Stee, P., *et al.* 2011, *A&A* 532, A80

Meilland, A., Millour, F., Kanaan, S., *et al.* 2012, *A&A* 538, A110

Meilland, A., Millour, F., Stee, P., *et al.* 2007a, *A&A* 464, 73

Meilland, A., Stee, P., Vannier, M., *et al.* 2007b, *A&A* 464, 59

Monnier, J. D., Zhao, M., Pedretti, E., *et al.* 2007, *Science* 317, 342

Mourard, D., Bério, P., Perraut, K., *et al.* 2011, *A&A* 531, A110

Mourard, D., Clausse, J. M., Marcotto, A., *et al.* 2009, *A&A* 508, 1073

Ollivier, M., Absil, O., Allard, F., *et al.* 2009, *Experimental Astronomy* 23, 403

Petrov, R. G. 1988, in D. M. Alloin & J.-M. Mariotti (eds.), *Diffraction-Limit.Imaging/ Very Large Telescopes*, p. 249

Petrov, R. G., Malbet, F., Weigelt, G., *et al.* 2007, *A&A* 464, 1

Petrov, R. G., Millour, F., Lagarde, S., *et al.* 2012, in *Society of Photo-Optical Instrumentation Engineers (SPIE) Conference Series*, Vol. 8445 of *Society of Photo-Optical Instrumentation Engineers (SPIE) Conference Series*

Rauer, H., Catala, C., Aerts, C., *et al.* 2013, *ArXiv e-prints*

Robbe-Dubois, S., Lagarde, S., Petrov, R. G., *et al.* 2007, *A&A* 464, 13

Stee, P., Delaa, O., Monnier, J. D., *et al.* 2012, *A&A* 545, A59

Stee, P. & Meilland, A. 2009, in J.-P. Rozelot & C. Neiner (eds.), *The Rotation of Sun and Stars*, Vol. 765 of *Lecture Notes in Physics, Berlin Springer Verlag*, pp 195–205

ten Brummelaar, T. A., McAlister, H. A., Ridgway, S. T., *et al.* 2005, *ApJ* 628, 453

Vakili, F. & Percheron, I. 1991, in D. Baade (ed.), *European Southern Observatory Conference and Workshop Proceedings*, Vol. 36 of *European Southern Observatory Conference and Workshop Proceedings*, p. 77

von Zeipel, H. 1924, *MNRAS* 84, 665

Zhao, M., Monnier, J. D., Pedretti, E., *et al.* 2009, *ApJ* 701, 209

Zorec, J., Frémat, Y., Domiciano de Souza, A., *et al.* 2011, *A&A* 526, A87

Discussion

WADE: When will we get circularly polarized spectro-photometric interferometry?

STEE: It is not planned for the second generation VLTI instruments and the foreseen projects. But you certainly know that we have already implemented a polarimeter within the VEGA focal instrument on the CHARA array composed of a Wollaston prism to separate two orthogonal polarization states and a movable quarter wave plate. A fixed quarter wave plate is placed after the Wollaston prism to transform the two linearly polarized output beams into two circularly polarized beams. See Mourard *et al.* (2009) and Stee *et al.* (2012) for more details.

AERTS: Could you elaborate on the dependence of the data and the calibrators on the assumptions on the limb darkening models and how that propagates into an uncertainty on the stellar radius?

STEE: Calibrating interferometric measurements is a science in itself and finding a good calibrator is not an easy task. Fortunately the JMMC provide a tool named "SearchCal" to assist the astronomers in this calibrator selection process for long baseline interferometric observations. To constrain the limb darkening you need first to access to second and third visibility "lobes", i.e. access to very faint and small visibility amplitudes ($\sim 2-3\%$). It means a good photometric calibration is needed in order to reach a precision of a few percent in the limb-darkened diameter. This is clearly something that is better suited for fiber-based interferometers with well-controlled photometry as the FLUOR or MIRC beam combiners. In order to constrain the limb-darkened diameter, people have often used the Claret & Bloemen tables but they have to take care because the limb darkening is also wavelength dependent and thus there is not only "one" limb-darkened radius. Of course, even if you have very accurate interferometric measurements, you also need a very accurate calibrator diameter, otherwise your final uncertainty will be due to the error on the calibrator diameter. This is clearly the case when you are calibrating data in the visible with very large baselines (larger than 200 m) where it is very difficult to find unresolved calibrators and very often your calibrator is more resolved than your science star! Fortunately, the forthcoming *GAIA* measurements will help to better constrain the stellar distances and thus the precisions on the calibrator's diameters. A nice work and a discussion about interferometric limb-darkened diameters of exoplanet host stars can be found in Ligi *et al.* (2012).

Philippe Stee

New windows on massive stars: asteroseismology, interferometry, and spectropolarimetry
Proceedings IAU Symposium No. 307, 2014
G. Meynet, C. Georgy, J. H. Groh & Ph. Stee, eds.

© International Astronomical Union 2015
doi:10.1017/S1743921314007431

Spectropolarimetry of massive stars: Requirements and potential from today to 2030

G. A. Wade

Royal Military College of Canada, Kingston, Ontario, Canada

Abstract. We develop the requirements and potential of spectropolarimetry as applied to understanding the physics of massive stars during the immediate, intermediate-term and long-term future.

Keywords. Stars: magnetic fields, Stars: massive, Techniques: polarimetry

1. Introduction and scope

Spectropolarimetry is a key technique for investigating the atmospheres, winds and envelopes of massive stars. Through the polarization of the continuum, and polarization and depolarization of spectral lines through scattering, Zeeman effect and Hanlé effects, polarimetry yields unique information about the geometry, density and magnetic fields of astrophysical plasma structures that are often impossible to obtain using other approaches.

In this paper we review the principal problematic themes and key questions related to massive stars that are currently being addressed using spectropolarimetric instrumentation. We then discuss forthcoming instrumentation and technical developments that will influence the field during the next 5 years, and ultimately for the period 2020–2030.

2. Today's tools and methods

Today, the astronomical spectropolarimetric instrumental landscape is dominated by a small number of efficient, broadband optical spectropolarimeters on 2-8m class telescopes. The principal players, summarized in Table 1, are the FORS2 multi-mode instrument on the 8m Very Large Telescope (VLT) UT1, the HARPSpol instrument at the ESO La Silla 3.6m telescope, ESPaDOnS at the 3.6m Canada-France-Hawaii Telescope (CFHT), and the Narval instrument at the 2m Bernard Lyot telescope (TBL). The basic characteristics of these instruments are summarized in Table 1.

Today's general-purpose spectropolarimeters can be roughly classed into two categories: low resolving power instruments, such as FORS2, designed for the measurement of both continuum and line polarization with moderate precision, and high resolving power instruments, such as ESPaDOnS, designed principally to measure line polarization with high precision.

The fundamental design difference in these two categories of instruments is the situation of the spectrograph. In the low resolving power instruments, the entire polarimetric and spectroscopic unit is mounted at the Cassegrain focus of the telescope, leaving it subject to variable instrumental flexures (Bagnulo *et al.* 2012). In contrast, the high resolution instruments employ bench-mounted, isolated spectrographs that are fed from the

	High resolution optical	Low resolution optical
Examples	Espadons, Narval, Harpspol	FORS2, ISIS
Telescopes	2-4m class	4-8m class
Bandpass	Complete optical	Partial optical
Resolving power	65K-100K	< 2.6K
Stokes parameters	IQUV, $\sigma_B \sim 1$ G	Stokes IQUV , $\sigma_B \sim 100$ G
Throughput	$5 - 20\%$	$\sim 20\%$
Observational strategies	PI, LP	PI, LP
Lines/continuum	Lines only	Lines & continuum
Analysis approach	Ind. lines, LSD	Multiline

Table 1. Optical spectropolarimeters in the current period.

Cassegrain-mounted polarimetric unit using optical fibres. As a consequence, the low resolution instruments are likely limited to the detection and measurement of longitudinal magnetic fields stronger that ~ 100 G (Bagnulo *et al.* 2012), whereas the high resolution instruments are capable of measuring fields of order 1 G (e.g. Konstantinova-Antova *et al.* 2014; Aurière *et al.* 2010).

On the other hand, due to their fibre design the high resolution instruments struggle to measure continuum polarization accurately. As a consequence, exploitation of these instruments for studies of continuum polarization and line depolarization (relative to a polarized continuum) have been limited.

3. Today's main themes and questions

Much of the research performed with today's spectropolarimetric instrumentation has been focused on broadening our understanding of stellar magnetism. Magnetometry relies principally on the use of circular (Stokes V) spectropolarimetry, exploiting the longitudinal Zeeman effect to detect the line-of-sight component of magnetic fields in stellar photospheres. Through both PI programs and Large Programs, circular spectropolarimetry has been employed to investigate the statistics of strong, organised fields of massive stars at the late pre-main sequence, main sequence, and post-main sequence phases (Alecian *et al.* 2008; Petit *et al.* 2011b; Grunhut *et al.* 2010). The surface geometries of magnetic fields of individual stars have been mapped in impressive detail using time series of Stokes V (and sometime Stokes Q and U) measurements interpreted using the Zeeman Doppler Imaging approach (e.g. Kochukhov *et al.* 2011, and Fig. 1). In some cases, the surface field distributions have been extrapolated based on simplifying assumptions in order to model the 3 dimensional structure of the magnetic field and magnetosphere (e.g. Fig. 1).

The relationships between magnetic field characteristics and other stellar properties have been explored. A particularly energetic field of investigation has been the wind-field interaction, which has been explored from both observational and theoretical perspectives, for individual stars and populations (e.g. Petit *et al.* 2013). Extensive theoretical investigations using magneto-hydrodynamic (MHD) simulations and semi-analytic magneto-hydrostatic models (ud-Doula & Owocki 2002; Townsend & Owocki 2005) have provided a rich theoretical context for interpretation of empirical results.

The origin of the magnetic fields of OB stars has been a major driving question during the last decade. Analytic (Duez *et al.* 2010) and numerical (Braithwaite & Spruit 2004) investigations have demonstrated that stable fossil magnetic field configurations can

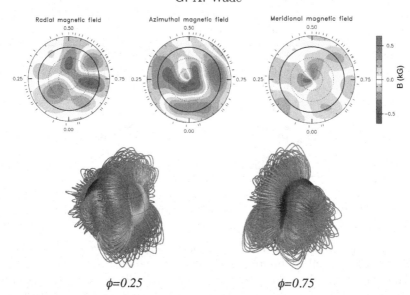

Figure 1. *Top* - Zeeman Doppler maps of the surface magnetic field components of the magnetic B-type star τ Sco. *Bottom* - Potential extrapolation of the radial component of the surface field above the stellar surface, illustrating an idealized structure of the closed loops of the stellar magnetosphere. (Kochukhov presentation at this meeting, and priv. comm.)

result naturally in stellar radiative zones from the relaxation of an initially random seed field.

Linear polarimetry, primarily in the continuum, has been employed to investigated scattering geometries of circumstellar discs and envelopes, to probe wind structures and oblateness, and to provide independent constraints on the structure of stellar magnetospheres (e.g. Carciofi *et al.* 2013). Linear polarimetry of spectral lines has been exploited to a limited extent (e.g. Vink 2012; Harrington & Kuhn 2007), although useful interpretation of those results has often remained a unsolved problem.

4. Today's limitations

While today's spectropolarimetric technologies represent a major step forward compared to the previous generation of instruments, they are still limited in many ways that constrain our ability to address important questions. For example, while today's instruments enjoy broad bandpasses, they are all still confined to the optical window. Most instruments are designed for characterization of continuum or line polarization, but not both. The high resolution instruments are limited principally to bright stars ($V = 12$ is a stretch), restricting them typically to nearby, unextincted, mainly field stars, at least at high signal-to-noise ratio. In particular, these instruments are unable to make the leap beyond the Galaxy, even to the Magellanic clouds. On the other hand, very deep spectropolarimetry of bright stars is limited by readout overheads, which are typically of the order of 1 minute per sub exposure (whereas exposure times to saturation may only be a few seconds).

The vectorial nature of polarimetry makes is susceptible to cancellation effects. Hence most current observations are sensitive to (relatively) organised magnetic fields. Due to the thin zone of formation of the large majority of optical spectral diagnostics (i.e. in the photosphere), magnetometry is essentially confined to two dimensions (the stellar surface), as are associated models.

The large majority of investigations today rely on the interpretation of averaged peudo line profiles (e.g. Least-Squares Deconvolved, or LSD profiles). Line averaging improves precision, but the interpretation of pseudo line profiles is uncertain, leading to reduced accuracy in their interpretation.

The largest LP allocations dedicated to stellar magnetism are of order 100 nights.

Finally, the most productive workhorse instruments (ESPaDOnS and Narval) are both located in the northern hemisphere; capabilities in the southern hemisphere are somewhat more limited.

5. Tomorrow's questions: 2015–2020

It seems likely that the questions to be seriously addressed during the next years will be many of those those viewed as the most challenging and exotic of today. For example, Lignières *et al.* (2009) reported the detection of a very weak (0.3 G) longitudinal magnetic field at the surface of the bright A0V star Vega that Petit *et al.* (2010) interpreted using Zeeman-Doppler Imaging as the result of a ~ 10 G magnetic spot located near the star's rotational pole (Fig. 2). Additional evidence for similarly weak fields (e.g. Petit *et al.* 2011a, Morel and the BOB collaboration, this conference) has been reported in a small number of stars. Such fields may represent a much more common weak component of the magnetic field distribution of intermediate-mass and massive stars, and may help to understand the so-called "magnetic desert" and the processes leading to the observed distribution of fields in higher-mass stars. However, the detection of such fields is at the very limit of the capabilities of todays instruments.

Investigations of the evolution of magnetic fields through stellar evolution, both before, on and after the main sequence, have been initiated during the last 10 years. However, the relative faintness of stars at the earliest pre-main sequence stages of evolution, and those located in environments very different (in terms of age, or metallicity) from the local Galactic field, has limited those studies. It is therefore likely that there will be significant pressure to extend these studies to explore characteristics of diverse stellar populations, including distant open clusters, the Galactic halo, and nearby galaxies such as the LMC and SMC (see Najarro presentation at this meeting.)

While today's science has focused on the magnetic fields at the surfaces of stars, the next years may witness the extension of direct field constraints into stellar interiors (with the help of asteroseismology), and into winds, discs and envelopes by taking advantage of spectral diagnostics formed in these environments and present in other regions of the electromagnetic spectrum.

The demonstration that broadband linear polarization measurements with precision of the order of 10^{-4} or even 10^{-5} of the continuum flux (e.g. Carciofi *et al.* 2013, Faes this conference), combined with increasingly sophisticated models that allow for interpretation of these data, will lead to important advances in studying of stellar rotation and oblateness (Faes presentation), wind/envelope asymmetries (Lomax and Halonen presentations), and the distribution of plasma in stellar magnetospheres (Oksala and Shultz presentations).

6. Tomorrow's tools: 2015–2020

A major limitation of the high resolution instruments today is their association with 4m class telescopes. Because these instruments are capable of achieving much greater

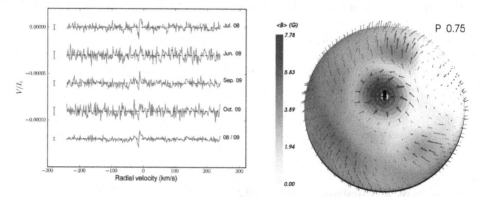

Figure 2. *Left* - Detection of a very weak Stokes V signature in the mean spectral line of the rapidly-rotating A1V star Vega (Lignières *et al.* 2009; Petit *et al.* 2010), corresponding to a surface magnetic field intensity of about 8 G (Petit *et al.*, submitted; Wade *et al.* 2014). The top 4 curves represent observations obtained at different epochs with different instruments. The bottom curve is the mean. *Right* - Illustration of the complex topology of Vega's magnetic field (Adapted from Wade *et al.* 2014).

magnetic precision per collected photon than the current low resolution instruments, a future priority must be to equip 8m-class telescopes in both the northern and southern hemispheres with instruments with capabilities (at least) similar to those of e.g. ESPaDOnS. For comparison, ESPaDOnS on an 8m telescope would yield a gain of 1.7 mag, corresponding to a decrease of 5x in exposure time, or increase of 2.2x in signal-to-noise ratio. It is reasonable to imagine that existing spectroscopic instrumentation, such as UVES or X-shooter on the VLT, could be equipped with polarimeters (Snik *et al.* 2013). Alternatively, existing projects such as GRACES (Gemini Remote Access to CFHT ESPaDOnS Spectrograph; Tollestrup *et al.* 2012) might naturally lead to the transfer of existing spectropolarimetric instruments from 4m to 8m class telescopes as new instrumentation comes online. While efforts such as these will represent major steps forward, they will still not be sufficient to permit e.g. the extension of high precision spectropolarimetry to stars outside of the Galaxy.

A major instrumental theme of the 2015–2020 era will be the introduction of high performance infrared (IR) spectropolarimeters. The two instruments that are currently in development are the "CRIRES+" upgrade to the existing CRIRES spectrograph at the ESO VLT, and the new "SPIRou" instrument under development for the Canada-France-Hawaii Telescope. The basic characteristics of these instruments are summarized in Table 2.

6.1. *CRIRES+*

CRIRES+ represents a major upgrade to the current CRIRES (Cryogenic high-resolution InfraRed Echelle Spectrograph) installed at the VLT (Oliva *et al.* 2014). This transformation will see many improvements to the existing AO-assisted instrument, while retaining its industry-leading ability to capture long-slit, high-resolution ($R \sim 100$K) spectra over a wide spectral range (0.95-5.2 μm). The most significant upgrade is the introduction of a cross-disperser unit and larger detectors that will increase the single-exposure wavelength coverage by about a factor of 10 compared to the existing setup. Furthermore, CRIRES+ will see the introduction of polarimetric optics that will allow for the high-precision measurement of circular (Stokes V) and linear (Stokes Q and U) polarization within spectral lines in the range of 1-2.7 μm. In addition to these significant changes,

	CRIRES+	Spirou
Telescopes	8m VLT	3.6m CFHT
Bandpass	1-2.7 μm (single exp.)	0.98-2.35 μm (multiple exp.)
Resolving power	100K	75K
Stokes parameters	IQUV	IQUV
Throughput	$\sim 5\%$	$\sim 15\%$
Commissioning	2017	2017

Table 2. IR spectropolarimeters in the period 2015-2020.

CRIRES+ will undergo many other smaller upgrades that will enable it do new and exciting science: the search for super-Earths in the habitable zone of low-mass stars; the atmospheric characterization of transiting planets; the origin and evolution of stellar magnetic fields. CRIRES+ will remain a general-purpose user facility and will offer comprehensive calibration and data-reduction support. Commissioning of CRIRES+ is projected for 2017.

6.2. *SPIRou*

SPIRou is a near-infrared spectropolarimeter / velocimeter currently under construction for CFHT. SPIRou aims in particular at becoming the world-leader on two forefront science topics, (i) the quest for habitable Earth-like planets around very- low-mass stars, and (ii) the study of low-mass star and planet formation in the presence of magnetic fields. In addition to these two main goals, SPIRou will be able to tackle many key programs, from weather patterns on brown dwarf to solar-system planet atmospheres, to dynamo processes in fully-convective bodies and planet habitability. The science programs that SPIRou proposes to tackle are forefront (identified as first priorities by most research agencies worldwide), ambitious (competitive and complementary with science programs carried out on much larger facilities, such as ALMA and JWST) and timely (ideally phased with complementary space missions like TESS and CHEOPS). SPIRou is designed to carry out its science mission with maximum efficiency and optimum precision. More specifically, SPIRou will be able to cover a very wide single-shot nIR spectral domain (0.98-2.35 μm) at a resolving power of about 75K, providing unpolarized and polarized spectra of low-mass stars with a $\sim 15\%$ average throughput and a radial velocity (RV) precision of 1 m/s. Commissioning of SPIRou is projected for 2017.

7. Around the bend: 2020–2030

7.1. *Extremely large telescopes*

The defining tools of ground-based astronomy in the 2020-2030 timeframe will be the so-called "giant" telescopes - the 39m European Extremely Large Telescope (E-ELT), the 30m Thirty Metre Telescope (TMT), and the 24.5m Giant Magellan Telescope (GMT). The former two facilities will be located in the southern hemisphere (northern Chile), while the latter will be in the North (Hawaii).

Equipping any of these telescopes with a moderate resolution ($R \gtrsim 10000$) fibre fed optical spectropolarimeter would be an utter game-changer, yielding an improvement of 5 magnitudes relative to existing high resolution instrumentation, with minimal reduction in polarimetric or magnetic precision. This translates into a gain of 10x in signal-to-noise ratio, or 100x in exposure time. Such an instrument could potentially achieve a magnetic

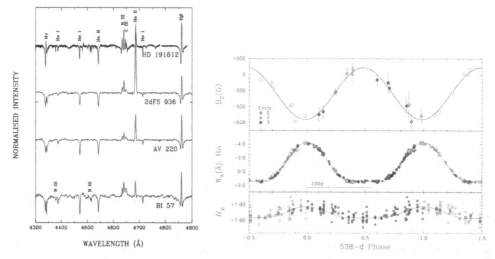

Figure 3. *Left* - Optical spectra of Of?p stars identified in the LMC and SMC (Walborn & Fitzpatrick 2000; Massey & Duffy 2001; Evans *et al.* 2004, Howarth *et al.*, in prep), as compared to the Galactic Of?p star HD 191612. *Right* - Illustration of the variations of the magnetic field, Hα equivalent width and Hipparcos magnitude of the Galactic Of?p star HD 191612 (Howarth *et al.* 2007; Wade *et al.* 2011).

detection precision of a few hundredths of a G, routinely measure Zeeman Stokes QUV polarization in individual spectral lines of typical magnetic O stars similar to HD 191612 (see Fig. 3) as faint as 10th magnitude, and detect fields comparable to that of HD 191612 in early O stars in the Magellanic clouds. As candidate extra-Galactic magnetic O stars have already been identified spectroscopically (see Fig. 3), this capability is a natural next step in understanding the environmental factors influencing massive-star magnetism.

Unfortunately, there is currently no high resolution spectropolarimeter planned for any of the giant optical telescopes.

7.2. *UVMag*

UVMag (Neiner *et al.* 2014, and Coralie Neiner's presentation at this meeting) is a planned intermediate-sized space telescope (1.3 m) supporting optical and UV spectropolarimetry with spectral resolution sufficient for Zeeman Doppler surface imaging. UVMag would enable continuous, high cadence line and continuum spectropolarimetry from 117-900 nm at high ($\sim 25,000$) resolving power. The project is currently supported by France for an ESA M-class mission (under the mission name "Arago") with planned launch in the late 2020s. The unique observations that UVmag will provide will allow a simultaneous and continuous view of stellar environments from the deep photosphere to the peripheries of their winds, discs and magnetospheres, potentially exploiting signal from scattering, longitudinal and transverse Zeeman effects, and Hanlé effect.

Recent spectral synthesis calculations by Colin Folsom (Fig. 4) demonstrate the power and richness of the UV domain for detection and characterization of fields in hot, massive stars.

Simultaneous with these new ground and space technologies, new techniques, strategies and modeling approaches must be developed [e.g. Spectro Polarimetric Interferometry (SPIn; Chesneau *et al.* 2003), large-scale, holistic models combining asteroseismology and

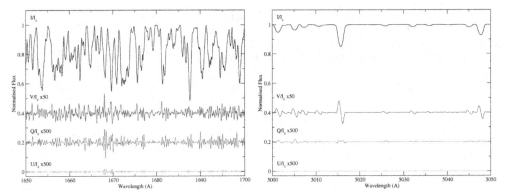

Figure 4. Spectral regions of width 50 Å computed in all 4 Stokes $IQUV$ parameters for a B1V star with a 1 kG dipolar magnetic field in its photosphere (Folsom *et al.*, priv. comm.). *Left* - Ultraviolet. *Right* - Optical. Note both the richness and the strength of the UV polarimetric signal.

spectropolarimetry, leading to sophisticated, realistic predictions]. Perhaps new observational strategies will be developed, with different scales of large programs.

8. Conclusion

In conclusion, today's suite of front-line spectropolarimetric instrumentation retains significant potential for scientific productivity and discovery. During the next 5 years, we will witness the introduction of powerful infrared spectropolarimeters as the leading instruments. At the same time, commissioning of high resolution optical instruments on 8m class telescopes should be a priority. In the era 2020-2030 we may encounter the first space spectropolarimeter working at both optical and ultraviolet wavelengths. We will witness ground-based optical astronomy dominated by a new era of giant telescopes. However, there is currently no plan to equip any of these observatories with high resolution spectropolarimetric equipment.

References

Alecian, E., Wade, G. A., Catala, C., *et al.* 2008, *A&A* 481, L99

Aurière, M., Donati, J.-F., Konstantinova-Antova, R., *et al.* 2010, *A&A* 516, L2

Bagnulo, S., Landstreet, J. D., Fossati, L., & Kochukhov, O. 2012, *A&A* 538, A129

Braithwaite, J. & Spruit, H. C. 2004, *Nature* 431, 819

Carciofi, A. C., Faes, D. M., Townsend, R. H. D., & Bjorkman, J. E. 2013, *ApJ (Letters)* 766, L9

Chesneau, O., Wolf, S., Rousselet-Perraut, K., *et al.* 2003, in S. Fineschi (ed.), *Polarimetry in Astronomy*, Vol. 4843 of *Society of Photo-Optical Instrumentation Engineers (SPIE) Conference Series*, pp 484–491

Duez, V., Braithwaite, J., & Mathis, S. 2010, *ApJ (Letters)* 724, L34

Evans, C. J., Howarth, I. D., Irwin, M. J., Burnley, A. W., & Harries, T. J. 2004, *MNRAS* 353, 601

Grunhut, J. H., Wade, G. A., Hanes, D. A., & Alecian, E. 2010, *MNRAS* 408, 2290

Harrington, D. M. & Kuhn, J. R. 2007, *ApJ (Letters)* 667, L89

Howarth, I. D., Walborn, N. R., Lennon, D. J., *et al.* 2007, *MNRAS* 381, 433

Kochukhov, O., Lundin, A., Romanyuk, I., & Kudryavtsev, D. 2011, *ApJ* 726, 24

Konstantinova-Antova, R., Aurière, M., Charbonnel, C., *et al.* 2014, in *IAU Symposium*, Vol. 302 of *IAU Symposium*, pp 373–376

498   G. A. Wade

Lignières, F., Petit, P., Böhm, T., & Aurière, M. 2009, *A&A* 500, L41
Massey, P. & Duffy, A. S. 2001, *ApJ* 550, 713
Neiner, C., Baade, D., Fullerton, A., *et al.* 2014, *Ap&SS*
Oliva, E., Tozzi, A., Ferruzzi, D., *et al.* 2014, *ArXiv e-prints*
Petit, P., Lignières, F., Aurière, M., *et al.* 2011a, *A&A* 532, L13
Petit, P., Lignières, F., Wade, G. A., *et al.* 2010, *A&A* 523, A41
Petit, V., Massa, D. L., Marcolino, W. L. F., *et al.* 2011b, *MNRAS* 412, L45
Petit, V., Owocki, S. P., Wade, G. A., *et al.* 2013, *MNRAS* 429, 398
Snik, F. & Harpspol Team, X-Shooter-Pol Team 2013, in G. Pugliese, A. de Koter, & M. Wijburg
    (eds.), *370 Years of Astronomy in Utrecht*, Vol. 470 of *Astronomical Society of the Pacific
    Conference Series*, p. 401
Tollestrup, E. V., Pazder, J., Barrick, G., *et al.* 2012, in *Society of Photo-Optical Instrumentation
    Engineers (SPIE) Conference Series*, Vol. 8446 of *Society of Photo-Optical Instrumentation
    Engineers (SPIE) Conference Series*
Townsend, R. H. D. & Owocki, S. P. 2005, *MNRAS* 357, 251
ud-Doula, A. & Owocki, S. P. 2002, *ApJ* 576, 413
Vink, J. S. 2012, in J. L. Hoffman, J. Bjorkman, & B. Whitney (eds.), *American Institute of
    Physics Conference Series*, Vol. 1429 of *American Institute of Physics Conference Series*,
    pp 147–158
Wade, G. A., Folsom, C. P., Petit, P., *et al.* 2014, *ArXiv e-prints*
Wade, G. A., Howarth, I. D., Townsend, R. H. D., *et al.* 2011, *MNRAS* 416, 3160
Walborn, N. R. & Fitzpatrick, E. L. 2000, *PASP* 112, 50

Discussion

KHALACK: Could you, please, comment on the use of the LSD method for UV space
spectropolarimetry in the region 117-400 nm where a lot of lines is strongly blended?

WADE: Colin Folsom's calculations using synthetic spectra and realistic atomic data
suggest the outlook is favourable!

Gregg Wade

New windows on massive stars: asteroseismology, interferometry, and spectropolarimetry
Proceedings IAU Symposium No. 307, 2014
G. Meynet, C. Georgy, J. H. Groh & Ph. Stee, eds.

© International Astronomical Union 2015
doi:10.1017/S1743921314007443

Observing programs, what are the priorities?

Georges Meynet[1] and Huib Henrichs[2]

[1] University of Geneva, Switzerland
email: Georges.Meynet@unige.ch

[2] University of Amsterdam, Netherlands
email: h.f.henrichs@uva.nl

Abstract. This brief note reflects the exchanges that were made between the participants during the general discussion on the observing programs. Because of time constraints only two questions have been addressed: 1) What is the priority between large programs and detailed observations of one or only a few objects? 2) Large surveys implies some automatic pipelines: what is their reliability? Elements of answers are given below. One interesting unexpected outcome of this discussion was the realization that a recent and up-to-date textbook on spectral synthesis is missing, while it would be an extremely useful tool for the community.

Keywords. stars: observations

1. Large programs vs. dedicate programs; applying for telescope time

The discussion first addressed the question of large programs (LP) versus dedicated programs. Large programs seem to be favored by Time Allocation Committees (TAC), likely because such programs are of interest for a large community, a point which is reflected by the fact that their results attract in general a lot of citations. Moreover, results obtained by LP can trigger further focused works on individual objects if the results are promising enough. For this to be possible however, the data should be reduced and made available quickly to all.

The pressure on focused programs is less high than on LP, but, sometimes, TAC may tend to reject such programs because of the existence of previous LP that may give the impression that the subject/object has already received all attention it deserves.

Although LP can provide extremely useful general trends, they are often not sufficient to fully answer a scientific question. The fact that, for LP, important amounts of data have to be analyzed, often implies shorter exposures and also some automatic pipeline for the reduction and analysis. In some cases, although some general trends can be correct, strong and useful conclusions may still need additional observations and/or reanalysis of the data. It happens that a small number of very carefully observed and analyzed objects brings new views that would have been out of reach by very large programs, due simply, as underlined above, to the limitations on the quality of the observables and of their analysis imposed by the too large numbers of objects to be treated in LPs. It happens also that some targeted observations can provide clues for physical processes having a very general application. In this respect the observation of the internal rotation of red giants is a very good example of an observation dealing with a small number of objects but potentially providing clues on the general process of angular momentum transport in stars.

From what precedes, one sees that large programs and dedicated ones have actually different purposes: while LP can provide a first global view of the question, targeted proposal can provide some refined clues. Thus it appeared important that observing time is obtained for both large and focused programs. In this respect, it was underlined that the project MiMeS (Magnetism in Massive Stars) was a good example, since it had both approaches, allowing to obtain general trends for a whole population as well as detailed and accurate results for a selected cases.

2. Reliability of automatic pipeline reduction

The second question was then addressed: large surveys imply automatic pipelines. Is their reliability sufficiently tested? It was said that people in charge of LP have to keep testing the reliability all the time. But of course they have to stop at some point because they need to publish and communicate their results. In these publications, they need to explain what has been done and the limits of the technique.

The discussion then focused on the case of abundance determinations, since they represent key data for testing theories and models. Indeed, the surface abundances are crucial since different models predict different abundances. Without the surface abundances, asteroseismic models would not have much meaning. Also it is important for mass loss determinations.

The question of reliability of automatic pipelines and how to test its quality control becomes a concern for the ESO-GAIA survey in view of the number of planned targets that has to be analyzed. While for cool stars some automatic procedure is working fine, for hot stars all the work needs to be done by hand. Therefore the progress is very slow about which GAIA people are complaining, but it is necessary to do it carefully for hot stars. To get proper error bars on abundances, it is necessary to invest large amounts of time and it does appear very dangerous to have an automatic pipeline. Such automatic pipeline may work for a few stars and not for others, hence the need to do it case by case, and to do proper error analysis.

3. The lack of a textbook on spectral synthesis

It emerged from the discussion that there is no recent and up-to-date textbook on spectral synthesis. A call was made to anyone who has good notes on this topic to distribute them for the benefit of all of us. One participant mentioned that Y. Hubeny (with Mihalas) has finished their book on "Theory of Stellar Atmospheres". Among the important points that should be discussed in such a textbook are the big differences that are produced on the lines when NLTE effects are accounted for. The field of spectral synthesis is evolving rapidly. As an example, nitrogen-line formation has been explained only four years ago. Such progress is true of all active research fields and likely to have a textbook making the synthesis of the present status of the research in that area available would be an extremely useful tool for students and researchers. It was strongly hoped that people in this field could find a way to get organized to put such a, possibly multi-author, book together.

We thank Coralie Neiner and Philippe Stee for having taken written notes during the discussions as well as all the intervening participants: Conny Aerts, Gérard van Belle, Artemio Herrero, Viktor Khalak, Alex de Koter, Arlette Noels, Thierry Morel, John Landstreet, Joachim Puls, Jon Sundqvist, Andrew Tkachenko.

New windows on massive stars: asteroseismology, interferometry, and spectropolarimetry
Proceedings IAU Symposium No. 307, 2014
G. Meynet, C. Georgy, J. H. Groh & Ph. Stee, eds.

© International Astronomical Union 2015
doi:10.1017/S1743921314007455

Stellar Models: What is the future direction?

Asif ud-Doula

Penn State Worthington Scranton, Dunmore, PA 18512, USA
email: asif@psu.edu

1. Introduction

This was one of two general discussions at the conference and it focussed primarily on stellar modelling. In particular, we were interested in the quality of the models, how we can check the models and what the direction of future modelling will be given the rise in popularity of MESA.

Since there are no transcripts of the discussions available, the actual words of the speakers are paraphrased here briefly. In what follows I summarize the discussion in several separate themes which might or might not have occurred in the same order during the lively discussion session.

Discussion
Quality of Stellar Models

UD-DOULA: Over the past few days we have been talking a lot about stellar evolution and modelling in general. Let us begin our discussion with how good our models are. Are they self-consistent?

CHIEFFI: A stellar evolutionary code is a very complex tool that should be used only by those who actively contribute to write it. It is very easy and extremely dangerous to simply push a button and get a result because there are so many choices and things that must always be kept in mind when computing even the simplest model that only an expert can really (try to) understand the *meaning*. It should be evident to everyone that different codes very often give different results, and since it is not easy at all to understand where such differences come from, only a comparison among similar computations obtained with totally different codes can really give us an idea of the robustness of a prediction, i.e that at least that there are no hidden numerical mistakes. For this reason I consider extremely dangerous the growing use of an open source stellar evolutionary code. Everyone can modify and compute without any control and this means that sometimes models may come out computed after having made nonsense changes to the code. No one can control this and since it is much easier to push a button instead of spending hours, weeks, years coding in front of a terminal, in the future I think that most of the people will just use such codes with the final total loss of knowledge (understanding) and control in this field. It is already evident with tools like MESA which are now being used like black boxes.

NOELS: I fully agree with Alessandro. According to Bill Paxton's own words MESA is *just a tool*, not a theory, a very convenient and efficient tool indeed but it should be not used as a black box. Consistency has to be checked, especially when models are used to interpret asteroseismic observations. It should be recalled that asteroseismology help bring some constraints on the chemical profile, not directly on the physical processes that have led to such a profile (see Andreas talk). An essential prerequisite is thus to have a fully consistent model in the frame of the physics adopted in the computation.

During previous sessions in this meeting, we already discussed some problems related to the boundary of convective cores in massive stars (see also my talk for more details). An easy check is to verify that the radiative temperature gradient is equal to the adiabatic temperature gradient on the convective side of the boundary. If this is not the case, the position of the boundary is not correct and the inferences that will be brought from an asteroseismic analyses will be unreliable. This problem is mainly encountered in core helium burning models computed with the Schwarzschild criterion. As a result of helium burning, a discontinuity in helium abundance forms at the convective core boundary. If the equality of the temperature gradients is imposed on the radiative side of the boundary, convective neutrality is not met on the convective side. The convective core is then kept from growing, as it should normally do, and since the discontinuity in helium abundance increases with time, the incoherency becomes larger and larger. In red clump stars, this leads to a clear disagreement between seismic observations and theoretical models (see, e.g., Montalbán *et al.* 2013). With the Ledoux criterion the problem is already encountered in main-sequence models since a discontinuity in μ-gradient exists at the core boundary. When the boundary is searched for through an interpolation process, notwithstanding the presence of such a discontinuity, convective cores are too small, have inconsistent boundaries, and are surrounded by apparently semi-convective layers, which should indeed be part of the convective cores if the location of the core boundary were properly determined.

CHIEFFI: The difference between the size of the 'theoretical' convective cores and the ones derived by the analysis of the seismic data is usually interpreted in terms of "overshooting", like if a good or a bad match between the theoretical and 'observed' values would deny/prove the presence of the *physical* phenomenon 'overshooting'. What is worst is that in all (most) papers that address such a comparison, the size of the convective core is not mentioned at all, the comparison is directly expressed in in terms of α (the free parameter that multiply the pressure scale height in the standard Mixing Length Theory). This approach assumes that each stellar code provides the same standard size of the convective core and that such a size is a firm theoretical prediction. This is false. The size of the convective core depends on both numerical and physical choices that may vary with time and from author to author. For example in the 80's the size of the 'overshooted region' was much larger than presently adopted just because the old opacity tables lead to smaller standard convective cores. Moreover a 'possible' extra mixing could be due to different physical mechanisms (e.g. rotation). Hence in my opinion the word overshooting should be totally dropped in this context and the differences should be discussed simply in terms of 'discrepancies/differences' between stellar models and real stars. The papers that address the comparison between theoretical and 'observed' convective cores should explicitly mention and publish the sizes of their standard convective cores and their chosen 'extended' ones in *solar masses* (not in terms of α). Only in this way these data could be really fruitful for the community of the stellar modelers.

MORAVVEJI: We fortunately live in an era of ultra high-precision space photometry, but the question is whether or not stellar models are accurate enough to explain such observations. Currently, there are several flavours of stellar structure and evolution codes available, MESA being just one of them. I believe it is very timely for code developers to set up benchmarking exercises, to gauge the theoretical uncertainties on our best models, yet keeping the flavour of different codes. Such an exercise was already performed in 2008 in preparation for the analysis of CoRoT asteroseismic data (Lebreton *et al.* 2008). With a much improved frequency resolution from the Kepler 4-year observations, we still need to organize comparison and benchmarking exercises to improve our theoretical models.

PULS: I agree that the cross-checks are important and I am willing to take on this task. However, having enough time in hand is the problem. But I too caution against using black boxes. You need to talk to experts before using these tools.

AERTS: But talking to experts takes time, students need to get out the work. Can they afford it?

HERRERO: I think we should continue to use codes like MESA for estimates and to get a feeling of the physics involved, or even take the results when we are using well stablished conditions. But when we are using new physics or results or deriving important consequences we should ask the experts to understand their limitations. But I certainly believe we should not be using any code, whether for evolutionary, atmospheric or any other kind of models, as black boxes.

TKACHENKO: I agree with Alessandro's earlier comment that junior scientists should rely on senior researchers, as they are the ones who can understand the meaning better when computing even the simplest model. But even then one should *always* learn from experts to gain the necessary expertise otherwise there will be no experts in stellar modelling left anymore in 20-30 years from now, when all current experts retire!

CHIEFFI: As a final remark about self-consistency, let me say that this is something that cannot be fixed once for ever but it must be checked continuously every time the code or some physical ingredient (EOS, opacity, etc.) is changed. Some skill and experience is necessary to do that.

Mass Loss Rates

SUNDQVIST: Remember that high-mass stars can also have cool surfaces. The RSG case presumably needs 3-D, and especially with respect to their stellar winds RSGs have not been much investigated so far. Considering the importance of mass loss in this stage for the late evolution and fate of massive stars, this is an area where some big efforts seem to be quite urgently needed.

KHALACK: Analysis of stellar spectra obtained in the UV, visual and IR spectral regions provides information about physical conditions in stellar atmosphere and stellar wind (if it is strong enough). Each star is unique and its spectrum has some particular features, but one can always trace general tendencies (e.g., chemical peculiarity, presence of magnetic field, strong emission lines, P-Cyg profiles, etc.) that together with estimates of effective temperature and surface gravity might help identify the type of star and its stage of evolution. However, it would be useful if models provided more information (predictions) about phenomena observed during certain stages of stellar evolution in outer parts of stars that can be studied by the methods of spectral analysis, photometry and polarimetry. In this way one could verify a validity of model predictions and treat properly the observed particular features in stellar spectra (e.g., presence of weak emission lines, line profile asymmetry, abundance stratification, wind clumping, etc.) employing more realistic physical conditions in stellar atmosphere and/or wind of studied star.

Dimensionality

VINK: The majority (>80%) of massive stars have large-scale winds that are more or less spherical, so here 1D is not too bad an approximation (though 2D needs to be considered for the exceptions). However, because winds are intrinsically clumpy, 3D effects are still needed in the end. And these small-scale 3D effects may have crucial large-scale consequences!

SUNDQVIST: I think it is important to consider also the outermost layers in cases where radiation pressure dominates the support against gravity. Several radiation-hydrodynamic instabilities are expected when $\Gamma \to 1$, presumably leading to a highly structured atmosphere. So multi-dimensionality and corresponding simulations are probably key here.

MATHIS: Most of the processes that drive transport of angular momentum and chemicals mixing in stellar interiors are 3D (see e.g. hydrodynamical and MHD instabilities and turbulence, internal gravity waves, etc., e.g. Brun (2011) and the review by Meakin (2008)). Today, we are able to simulate them on dynamical time-scales using LES for global geometry and DNS in local boxes. However, these 3D non-linear simulations that use super-calculators have a large computation time for one star and cannot yet be computed for evolution time-scales. Therefore, to tackle the effects of 3D MHD mechanisms on stellar evolution, it is necessary to extract prescriptions and scaling laws that can be implemented in 1- or 2-D stellar evolution codes that treat the large-scale transport of angular momentum on secular time-scales along the whole evolution of stars (e.g. Meynet & Maeder 2000; Mathis et al. 2013; Rieutord & Espinosa 2013). Moreover, these "secular" dynamical stellar evolution codes are those that allow today to compute the grids of stellar models necessary to interpret the large amount of data coming for example from asteroseismology, spectroscopy, and spectropolarimetry.

VINK: There is sometimes still a tendency in the community to be happier with detections of magnetic field rather than with non-detections (i.e. detections get published, and non-detections do not) whether or not this is about wind asphericity or magnetic field incidence. But the point is that non-detections can scientifically be as interesting as detections and we need to discuss ramifications of such non-detections more broadly.

References
Brun, A. S. 2011, in H. Wozniak & G. Hensler (eds.), *EAS Publications Series*, Vol. 44 of *EAS Publications Series*, pp 81–95
Lebreton, Y., Montalbán, J., Christensen-Dalsgaard, J., Roxburgh, I. W., & Weiss, A. 2008, *Ap&SS* 316, 187
Mathis, S., Decressin, T., Eggenberger, P., & Charbonnel, C. 2013, *A&A* 558, A11
Meakin, C. A. 2008, in L. Deng & K. L. Chan (eds.), *IAU Symposium*, Vol. 252 of *IAU Symposium*, pp 439–449
Meynet, G. & Maeder, A. 2000, *A&A* 361, 101
Montalbán, J., Miglio, A., Noels, A., et al. 2013, *ApJ* 766, 118
Rieutord, M. & Espinosa, L. F. 2013, in L. Cambresy, F. Martins, E. Nuss, & A. Palacios (eds.), *SF2A-2013: Proceedings of the Annual meeting of the French Society of Astronomy and Astrophysics*, pp 101–104

Author Index

Abellan, F. J. – 280
Aerts, C. – 154
Aggrawal, R. – 218
Alecian, E. – 330, 385
Anderson, R. I. – 206, 286
Araya, I. – 104, 125
Arcos, C. – 104
Arlt, R. – 342
Arnett, D. – 98, 459
Aroui, H. – 288
Arroyo-Torres, B. – 280
Auclair-Desrotour, P. – 208

Bagnulo, S. – 365
Bailey, J. D. – 365
Baklanova, D. – 369
Barbosa, C. L. – 431, 453
Barbá, R. – 342
Bard, C. – 449
Becker, A. – 414
Bendjoya, P. – 261
Bensby, T. – 90
Berio, P. – 111
Bersten, M. C. – 37
Bischoff-Kim, A. – 211
Bisol, A. C. – 224
Blazère, A. 367
Blomme, R. – 88, 140
Blum, R. D. – 431, 453
Boffin, H. M. J. – 295
Bohlender, D. – 330
Bomans, D. J. – 148, 414
Bonanos, A. Z. – 92, 171
Borges Fernandes, M. – 90, 291
Bouret, J.-C. – 367, 385
Bragança, G. A. – 90
Briquet, M. – 342, 443
Britavskiy, N. – 92
Burggraf, B. – 148
Butkovskaya, V. – 369

Cabezas, M. – 125
Calzoletti, L. – 137
Camacho, I. – 106
Carciofi, A. C. – 133, 261, 291
Carroll, T. – 342
Castro, N. – 106, 342, 391
Cazorla, C. – 94
Challouf, M. – 288
Chantereau, W. – 96
Charbonnel, C. – 96
Chené, A. – 100

Chesneau, O. – 291
Chiavassa, A. – 273, 280
Chieffi, A. – 1
Cidale, L. – 104
Cohen, D. – 437
Corcoran, M. – 115
Creevey, O. L. – 480
Cristini, A. – 98
Crowther, P. A. – 135
Cséki, A. – 125
Cunha, K. – 90
Curé, M. – 100, 104, 125
Cébron, D. – 330

D'souza, D. – 213
D. de la Fuente 426
Daflon, S. – 90
Damineli, A. – 431, 453
Daszyńska-Daszkiewicz, J. – 226, 232
David-Uraz, A. – 371
de Koter, A. – 76, 144, 342
de Mink, S. E. – 76
de Wit, W. J. – 121, 295, 297
Degroote, P. – 443
Demers, Z. – 297
Djurašević, G. – 125
Domiciano de Souza, A. – 261, 288, 291
Dufton, P. – 76, 342
Dupret, M.-A. – 188, 215
Dushin, V. – 113

Eenens, P. – 237
Eggenberger, P. – 165, 188
Ekström, S. – 102, 142, 206
Emeriau, C. – 373
Engle, S. – 224
Evans, C. J. – 76
Eyer, L. – 206

Fabrika, S. – 146
Faes, D. M. – 261
Figer, D. – 137, 426
Folsom, C. P. – 330, 401
Fossati, L. – 342
Freytag, B. – 280
Frémat, Y. – 88, 115, 140

Gabriel, M. – 470
García, M. – 41, 106
Garmany, C. D. – 90
Gayley, K. G. – 375
Geballe, T. R. – 426

Georgy, C. – 47, 98, 102, 142, 206, 230
Gies, D. – 131, 293
Glaspey, J. W. – 90
Godart, M. – 215, 470
González, J. F. – 342
Gordon, K. – 293
Gormaz-Matamala, A. C. – 100
Gosset, E. – 88, 140
Granada, A. – 102, 142
Groh, J. H. – 115, 267
Grundstrom, E. D. – 131
Grunhut, J. – 301, 330, 397, 399, 401
Gräfener, G. – 52, 76, 144
Guieu, S. – 295
Guinan, E. – 224
Guzik, J. A. – 176

H. Spruit, 342
Haemmerlé, L. – 102
Halonen, R. J. – 377
Hamann, W.-R. – 342
Handler, G. – 239
Haubois, X. – 133
Haucke, M. – 104
Hauschildt, P. H. – 280
Henrichs, H. – 379, 399, 499
Herrero, A. – 41, 76, 88, 106, 137, 342
Hervé, A. – 100, 385
Hillier, D. J. – 64, 426
Hirschi, R. – 98
Hubeny, I. – 90
Hubrig, S. – 342
Humphreys, R. M. – 148

Ibadov, S. – 108
Ibodov, F. S. – 108
Ilyin, I. – 342
Irrgang, A. – 342

J. S. Vink, 76
Jamialahmadi, N. – 111
Jones, C. E. – 377, 457
Joshi, G. C. – 218
Joshi, S. – 218
Joshi, Y. C. – 218

Kanaan, S. – 104
Kaper, L. – 144
Kervella, P. – 273
Khalack, V. – 381, 383
Kharchenko, N. – 342
Kholtygin, A. – 113, 342
Kochukhov, O. – 348, 395
Kourniotis, M. – 171
Kołaczkowski, Z. – 222
Kraus, M. – 104, 235
Krikelis, G. – 171

Krtička, J. – 348

Landstreet, J. D. – 311, 365, 401
Langer, N. – 76, 144, 342, 391
Lanz, T. – 90
Le Bouquin, J.-B. – 273, 295, 330
LeBlanc, F. – 381
Lennon, D. J. – 41, 76
Lesur, G. – 70
Levesque, E. M. – 57
Lignières, F. – 70
Limongi, M. – 1
Lobel, A. – 88, 115, 140
Lomax, J. R. – 336
Lopez, B. – 111
Lovekin, C. C. – 176

Maeder, A. – 9
Maíz Apellániz, J. – 76, 88, 342
Marcaide, J. M. – 280
Marcolino, T. R. W. – 399
Marcolino, W. L. F. – 385
Markova, N. – 25, 76, 117
Martayan, C. – 115, 295
Martin-Pintado, J. – 137
Martins, F. – 385
Maryeva, O. – 119, 387
Massey, P. – 57, 64, 127
Mathis, S. – 208, 220, 330, 373, 420, 443
Mathys, G. – 342
Meakin, C. A. – 20, 98
Mehner, A. – 92, 121, 295
Meilland, A. – 111, 241, 288, 480
Melnick, J. – 123
Mennickent, R. E. – 100, 125
Meynet, G. – 9, 47, 96, 102, 142, 206, 230, 499
Miglio, A. – 470
Mikulášek, Z. – 348
Montalbán, J. – 188, 470
Montargès, M. – 273
Moravveji, E. – 182
Morel, T. – 88, 94, 140, 342, 451
Morrell, N. – 57, 64
Mourard, D. – 288
Mowlavi, N. – 206
Moździerski, D. – 222

Najarro, F. – 41, 76, 137, 426
Nardetto, N. – 288
Navarete, F. – 431, 453
Nazé, Y. – 94, 437, 455
Negueruela, I. – 88
Neilson, H. R. – 224
Neiner, C. - 220, 330, 367, 385, 389, 420, 443

Neugent, K. F. – 64, 127
Niemczura, E. – 125
Nieva, M.-F. – 129, 342
Noels, A. – 470

Oey, M. S. – 90
Oksala, M. E. – 235, 348
Oskinova, L. – 342
Ostrowski, J. – 226
Owocki, S. – 371, 375, 437

Parfenov, S. – 119
Perrin, G. – 273
Petermann, I. – 391
Peters, G. J. – 131
Petit, V. – 330, 397, 401, 437
Piskunov, A. – 342
Piskunov, N. – 395
Plachinda, S. – 369
Poitras, P. – 383
Poleski, R. – 171
Poncin-Lafitte, C. L. – 208
Prat, V. – 70
Prvák, M. – 348
Przybilla, N. – 342, 404
Puls, J. – 25, 76, 117, 137

Quirrenbach, A. – 297

Ramírez-Agudelo, O. H. – 76
Rauw, G. – 94, 237
Reese, D. R. – 188
Reisenegger, A. – 342
Rinbrand, M. – 437
Rivinius, T. - 121, 133, 228, 295, 297,
    397, 399
Romanyuk, I. I. – 393
Rosales, J. – 125
Rosslowe, C. K. – 135
Rubio-Díez, M. M. – 137
Rusomarov, N. – 395
Rímulo, L. R. – 133

Sabín-Sanjulián, C. – 106
Saio, H. – 47, 230
Salmon, S. J. A. J. – 188, 470
Sana, H. – 76, 144, 330, 342
Schaefer, G. – 293
Schneider, F. R. N. – 342
Scholz, M. – 280
Scholz, R.-D. – 342
Schöller, M. – 342
Semaan, T. – 88, 140
Semenko, E. A. – 393

Shibahashi, H. – 215
Sholukhova, O. – 146
Shultz, M. – 228, 397, 399
Sigut, T. A. A. – 457
Silaj, J. – 457
Simon Díaz, S. – 342
Simoniello, R. – 142
Simón-Díaz, S. – 76, 88, 106, 194
Soszyński, I. – 171
Stee, P. – 111, 241, 288, 480
Straal, S. M. – 144
Sudnik, N. – 113, 379
Sundqvist, J. O. – 25, 137, 353
Szewczuk, W. – 232

Takahashi, K. – 82
Taylor, W. D. – 76
Tkachenko, A. – 200, 330, 367
Tomić, S. – 235
Townsend, R. – 449
Traficante, A. – 137
Tramper, F. – 76, 144
Tycner, C. – 457

ud-Doula, A. – 321, 330, 437, 504
Umeda, H. – 82
Urbaneja, M. A. – 41
Uuh-Sonda, J. M. – 237

Valeev, A. F. – 146
van Belle, G. T. – 252
Venero, R. – 104
Viallet, M. – 98
Vink, J. S. – 144, 359
Volpi, D. – 88

Wade, G. - 228, 330, 371, 385, 397, 399,
    401, 437, 490
Walczak, P. – 239
Wang, L. – 131
Weis, K. – 148, 414
Weiss, A. – 213
Wittkowski, M. – 280
Wood, P. R. – 280

Yamada, S. – 150
Yasutake, N. – 150
Yoon, S.-C. – 342
Yoshida, T. – 82
Yusof, N. – 152

Zahajkiewicz, E. – 222
Zorec, J. – 140
Żytkow, A. N. 57

Printed in the United States
by Baker & Taylor Publisher Services